Lecture Notes in Computer Science 8573

Commenced Publication in 1973
Founding and Former Series Editors:
Gerhard Goos, Juris Hartmanis, and Jan van Leeuwen

Advanced Research in Computing and Software Science

Subline of Lectures Notes in Computer Science

T0215769

Javier Esparza   Pierre Fraigniaud
Thore Husfeldt   Elias Koutsoupias (Eds.)

Automata, Languages, and Programming

41st International Colloquium, ICALP 2014
Copenhagen, Denmark, July 8-11, 2014
Proceedings, Part II

Springer

Volume Editors

Javier Esparza
Technische Universität München, Germany
E-mail: esparza@in.tum.de

Pierre Fraigniaud
LIAFA, Université Paris Diderot-Paris 7, France
E-mail: pierre.fraigniaud@liafa.univ-paris-diderot.fr

Thore Husfeldt
IT University of Copenhagen, Denmark
E-mail: thore@itu.dk

Elias Koutsoupias
University of Oxford, UK
E-mail: elias@cs.ox.ac.uk

ISSN 0302-9743   e-ISSN 1611-3349
ISBN 978-3-662-43950-0   e-ISBN 978-3-662-43951-7
DOI 10.1007/978-3-662-43951-7
Springer Heidelberg New York Dordrecht London

Library of Congress Control Number: 2014941781

LNCS Sublibrary: SL 1 – Theoretical Computer Science and General Issues

Typesetting: Camera-ready by author, data conversion by Scientific Publishing Services, Chennai, India

Printed on acid-free paper

Springer is part of Springer Science+Business Media (www.springer.com)

Preface

This volume contains the papers presented at ICALP 2014: the 41st International Colloquium on Automata, Languages and Programming, held during July 8–11, 2014, at IT University of Copenhagen. ICALP is the main conference and annual meeting of the European Association for Theoretical Computer Science (EATCS) and first took place in 1972. This year the ICALP program consisted of three tracks:

- Track A: Algorithms, Complexity, and Games
- Track B: Logic, Semantics, Automata, and Theory of Programming
- Track C: Foundations of Networked Computation

In response to the call for papers, the three Program Committees received 484 submissions, a record number for ICALP. Track A received 319 submissions (another record), track B received 106 submissions, and track C received 59 submissions. Each submission was reviewed by at least three Program Committee members, aided by many subreviewers. The committee decided to accept 136 papers, which are collected in these proceedings. The selection was made by the Program Committees based on originality, quality, and relevance to theoretical computer science. The quality of the submissions was very high indeed, and many deserving papers could not be selected.

The EATCS sponsored awards for both a best paper and a best student paper for each of the three tracks, selected by the Program Committees.

The best paper awards were given to the following papers:

- Track A: Andreas Björklund and Thore Husfeldt, "Shortest Two Disjoint Paths in Polynomial Time"
- Track B: Joel Ouaknine and James Worrell. "Ultimate Positivity Is Decidable for Simple Linear Recurrence Sequences"
- Track C: Oliver Göbel, Martin Hoefer, Thomas Kesselheim, Thomas Schleiden, and Berthold Vöcking, "Online Independent Set Beyond the Worst-Case: Secretaries, Prophets, and Periods"

The best student paper awards, for papers that are solely authored by students, were given to the following papers:

- Track A: Sune K. Jakobsen, "Information Theoretical Cryptogenography"
- Track B: Michael Wehar, "Hardness Results for Intersection Non-Emptiness"
- Track C: Mohsen Ghaffari, "Near-Optimal Distributed Approximation of Minimum-Weight Connected Dominating Set"

Apart from the contributed talks, the conference included invited presentations by Sanjeev Arora, Maurice Herlihy, Viktor Kuncak, and Claire Mathieu. Abstracts of their talks are included in these proceedings as well.

The program of ICALP 2014 also included presentation of the Presburger Award 2014 to David Woodruff, the EATCS Award 2014 to Gordon Plotkin, and the Gödel Prize to Ronald Fagin, Amnon Lotem, and Moni Naor.

Two satellite events of ICALP were held on 7 July, 2014:

- Trends in Online Algorithms (TOLA 2014)
- Young Researcher Workshop on Automata, Languages and Programming (YR-ICALP 2014)

We wish to thank all the authors who submitted extended abstracts for consideration, the members of the three Program Committees for their scholarly efforts, and all additional reviewers who assisted the Program Committees in the evaluation process. We thank the sponsors Springer-Verlag, EATCS, CWI Amsterdam, and Statens Kunstfond for their support, and the IT University of Copenhagen for hosting ICALP 2014.

We are also grateful to all members of the Organizing Committee and to their support staff.

The conference-management system EasyChair was used to handle the submissions, to conduct the electronic Program Committee meetings, and to assist with the assembly of the proceedings.

May 2014

Javier Esparza
Pierre Fraigniaud
Thore Husfeldt
Elias Koutsoupias

Organization

Program Committee

Dimitris Achlioptas	UC, Santa Cruz, USA
Pankaj Agrawal	Duke University, USA
Paolo Baldan	Università di Padova, Italy
Nikhil Bansal	Eindhoven University of Technology, The Netherlands
Michele Boreale	Università di Firenze, Italy
Tomas Brazdil	Masaryk University, Czech Republic
Gerth Stølting Brodal	Aarhus University, Denmark
Véronique Bruyère	University of Mons, Belgium
Jean Cardinal	Université libre de Bruxelles, Belgium
Ning Chen	Nanyang Technological University, Singapore
Giorgos Christodoulou	University of Liverpool, UK
Andrea Clementi	University of Rome Tor Vergata, Italy
Veronique Cortier	CNRS, Loria, France
Anuj Dawar	University of Cambridge, UK
Xiaotie Deng	Shanghai Jiaotong University, China
Ilias Diakonikolas	University of Edinburgh, UK
Benjamin Doerr	MPI Saarbrücken, Germany
Chaled Elbassioni	Masdar Institute, Abu Dhabi
Javier Esparza	TU München, Germany
Kousha Etessami	University of Edinburgh, UK
Panagiota Fatourou	University of Crete, Greece
Michal Feldman	Hebrew University, Israel
Maribel Fernandez	Kings College London, UK
Antonio Fernández Anta	Universidad Rey Juan Carlos, Spain
Amos Fiat	Tel Aviv University, Israel
Pierre Fraigniaud	CNRS and University of Paris Diderot, France
David Frutos Escrig	Complutense University of Madrid, Spain
Pierre Ganty	IMDEA Software Institute, Spain
Leszek Gasieniec	University of Liverpool, UK
Phillip Gibbons	Intel Labs, USA
Leslie Goldberg	University of Oxford, UK
Vipul Goyal	Microsoft, India
Peter Habermehl	LIAFA, University of Paris 7, France
Magnus Halldorsson	Reykjavik University, Iceland
Giuseppe Italiano	University of Rome Tor Vergata, Italy
Marcin Kaminski	University of Warsaw, Poland

Haim Kaplan	Tel Aviv University, Israel
Anna Karlin	University of Washington, USA
Ioardanis Kerenidis	University of Paris Diderot, France
Anne-Marie Kermarrec	Inria Rennes, France
Robert Kleinberg	Cornell University, USA
Michal Koucky	Czech Academy of Sciences, Czech Republic
Elias Koutsoupias	University of Oxford, UK
Robert Krauthgamer	Weizmann Institute, Israel
Manfred Kufleitner	University of Stuttgart, Germany
Sławomir Lasota	Warsaw University, Poland
James Lee	University of Washington, USA
Oded Maler	CNRS-VERIMAG, France
Sebastian Maneth	NICTA and UNSW, Australia
Madhavan Mukund	Chennai Mathematical Institute, India
Ashwin Nayak	University of Waterloo, Canada
Jens Palsberg	UCLA, USA
Gopal Pandurangan	Nanyang Technological University, Singapore
Boaz Patt-Shamir	Tel Aviv University, Israel
Andrea Pietracaprina	Università di Padova, Italy
Andrea Richa	Arizona State University, USA
Luís Rodrigues	Universidade Técnica de Lisboa, Portugal
Jared Saia	University of New Mexico, USA
Piotr Sankowski	University of Warsaw, Poland
Christian Scheideler	Universität Paderborn, Germany
Thomas Schwentick	TU Dortmund, Germany
Maria Serna	UP Catalunya, Spain
Sonja Smets	University of Amsterdam, The Netherlands
Christian Sohler	TU Dortmund, Germany
Jiri Srba	Aalborg University, Denmark
Jukka Suomela	Aalto University, Finland
Ryan Williams	Stanford University, USA
Philipp Woelfel	University of Calgary, Canada
Steve Zdancewic	University of Pennsylvania, USA

Additional Reviewers

Aaronson, Scott	Agarwal, Rachit
Abe, Masayuki	Aghazadeh, Zahra
Abraham, Ittai	Agrawal, Shweta
Aceto, Luca	Ajwani, Deepak
Adler, Isolde	Akutsu, Tatsuya
Adsul, Bharat	Al-Humaimeedy, Abeer
Afshani, Peyman	Alamdari, Soroush
Agarwal, Alekh	Alglave, Jade

Allender, Eric
Alon, Noga
Althaus, Ernst
Alves, Sandra
An, Hyung-Chan
Anagnostopoulos, Aris
Ananth, Prabhanjan
Andoni, Alex
Andoni, Alexandr
Ardenboim, Alon
Arkhipov, Alex
Asarin, Eugene
Aspnes, James
Atig, Mohamed Faouzi
Atserias, Albert
Augustine, John
Avron, Haim
Babichenko, Yakov
Bacci, Giorgio
Bacci, Giovanni
Bach, Eric
Balabonski, Thibaut
Banerjee, Abhishek
Barrington, David
Bartoletti, Massimo
Basset, Nicolas
Bavarian, Mohammad
Beame, Paul
Becchetti, Luca
Bei, Xiaohui
Belmonte, Rémy
Ben Avraham, Rinat
Ben-Amram, Amir
Berger, Eli
Berry, Jonathan
Bertrand, Nathalie
Berwanger, Dietmar
Bhaskar, Umang
Bitansky, Nir
Blazy, Olivier
Blesa, Maria J.
Blömer, Johannes
Bodirsky, Manuel
Bodlaender, Hans L.
Bodlaender, Marijke

Bogdanov, Andrej
Bojanczyk, Mikolaj
Boker, Udi
Bollig, Beate
Bollig, Benedikt
Bonamy, Marthe
Bonchi, Filippo
Boneh, Dan
Bonifaci, Vincenzo
Bonnet, Edouard
Bonsangue, Marcello
Bonsma, Paul
Borgström, Johannes
Boutsidis, Christos
Boyar, Joan
Boyle, Elette
Brakerski, Zvika
Brandstadt, Andreas
Braverman, Mark
Bremner, Michael
Brettell, Nick
Briet, Jop
Brihaye, Thomas
Broadbent, Anne
Brody, Joshua
Bruni, Roberto
Brzuska, Christina
Buchbinder, Niv
Buchin, Kevin
Buhrman, Harry
Byrka, Jaroslaw
Böhl, Florian
Cai, Yang
Caltais, Georgiana
Canetti, Ran
Canonne, Clément
Cao, Yixin
Carraro, Alberto
Cash, David
Ceccarello, Matteo
Chakrabarti, Amit
Chakraborty, Supratik
Chalermsook, Parinya
Chan, Hubert
Chan, Siu On

Chan, Timothy
Chandran, Nishanth
Charatonik, Witold
Chase, Melissa
Chatterjee, Krishnendu
Chechik, Shiri
Chekuri, Chandra
Chen, Jing
Chen, Xujin
Chen, Zhou
Cheval, Vincent
Choudhury, Ashish
Chow, Sherman S.M.
Chrobak, Marek
Chung, Kai-Min
Ciancia, Vincenzo
Cicalese, Ferdinando
Clavier, Christophe
Clemente, Lorenzo
Codenotti, Paolo
Cohen, Edith
Cohen, Sarel
Cohn, Henry
Colcombet, Thomas
Colini Baldeschi, Riccardo
Costello, Craig
Crescenzi, Pierluigi
Cryan, Mary
Cygan, Marek
Czerwiński, Wojciech
Dalmau, Victor
Damaschke, Peter
Damgård, Ivan
Dang, Thao
Dani, Varsha
Dasgupta, Bhaskar
Datta, Samir
David, Alexandre
De Bonis, Annalisa
de Caro, Angelo
De Caro, Angelo
De Liguoro, Ugo
de Wolf, Ronald
Decker, Normann
Degorre, Aldric

Delahaye, Benoit
Delling, Daniel
Delvenne, Jean-Charles
Delzanno, Giorgio
Denysyuk, Oksana
Dereniowski, Dariusz
Devanur, Nikhil
Devroye, Luc
Diaz, Josep
Dietzfelbinger, Martin
Diks, Krzysztof
Dima, Catalin
Diochnos, Dimitris
Dobrev, Stefan
Doerr, Carola
Doyen, Laurent
Driemel, Anne
Duflot, Marie
Dumitrescu, Adrian
Dupuis, Frédéric
Durand, Arnaud
Durand-Gasselin, Antoine
Durnoga, Konrad
Dvir, Zeev
Dyer, Martin
Edmonds, Jeff
Efremenko, Klim
Efthymiou, Charilaos
Ehrgott, Matthias
Ehsanfar, Ebrahim
Elbassioni, Khaled
Elberfeld, Michael
Elmasry, Amr
Elsässer, Robert
Emmi, Michael
Ene, Alina
Enea, Constantin
Enqvist, Sebastian
Eppstein, David
Epstein, Leah
Erlebach, Thomas
Escoffier, Bruno
Even, Guy
Fahrenberg, Uli
Fanelli, Angelo

Farshim, Pooya
Fefferman, Bill
Feige, Uriel
Fekete, Sándor
Fernau, Henning
Fijalkow, Nathanaël
Filiot, Emmanuel
Filmus, Yuval
Fiorini, Samuel
Firmani, Donatella
Fisman, Dana
Flammini, Michele
Forbes, Michael A.
Forejt, Vojtech
Fortnow, Lance
Fotakis, Dimitris
Fountoulakis, Nikolaos
Franciosa, Paolo
Frati, Fabrizio
Frieze, Alan
Fu, Hu
Fu, Zhiguo
Fábregas, Ignacio
Gaboardi, Marco
Gadducci, Fabio
Gaertner, Bernd
Galanis, Andreas
Galesi, Nicola
Gambs, Sebastien
Garg, Ankit
Gaspers, Serge
Gastin, Paul
Gavinsky, Dmitry
Gawrychowski, Pawel
Geck, Gaetano
Geeraerts, Gilles
Gelles, Ran
Genest, Blaise
Ghaffari, Mohsen
Giakkoupis, George
Giannakopoulos, Yiannis
Giannopoulou, Archontia
Giaquinta, Emanuele
Gierasimczuk, Nina
Gilbert, Seth

Gille, Marc
Giunti, Marco
Gkatzelis, Vasilis
Glacet, Christian
Glen, Amy
Gmyr, Robert
Gogacz, Tomasz
Goldberg, Paul
Gonzalez Vasco, Maria Isabel
Gopalan, Parikshit
Gorbunov, Sergey
Gorecki, Pawel
Gorgunov, Sergey
Gorla, Daniele
Grandoni, Fabrizio
Greco, Gianluigi
Green, Oded
Grenet, Bruno
Grigorescu, Elena
Grigoryev, Dmitry
Grossi, Roberto
GualÃ, Luciano
Guo, Heng
Guo, Jiong
Guo, Siyao
Guruswami, Venkatesan
Gutwenger, Carsten
Göbel, Andreas
Haeupler, Bernhard
Haghpanah, Nima
Haitner, Iftach
Hajiaghayi, Mohammadtaghi
Hansen, Kristoffer Arnsfelt
Hansen, Thomas Dueholm
Hardt, Moritz
Harju, Tero
Harrow, Aram
Harsha, Prahladh
Hatami, Hamed
Haviv, Ishay
Hayes, Thomas
Hazay, Carmit
He, Meng
Heam, Pierre-Cyrille
Heggernes, Pinar

Helmi, Maryam
Hirschkoff, Daniel
Hlout, Loc
Hoefer, Martin
Hoffmann, Hella-Franziska
Hofheinz, Dennis
Hofman, Piotr
Huang, Chien-Chung
Huang, Sangxia
Huang, Xiangru
Huang, Zhiyi
Hunter, Paul
Husfeldt, Thore
Im, Hyeonseung
Indyk, Piotr
Iovino, Vincenzo
Irani, Sandy
Isopi, Marco
Ito, Takehiro
Jacob, Riko
Jain, Rahul
Jansen, Bart M.P.
Jao, David
Jerrum, Mark
Jeż, Artur
Jeż, Łukasz
Jiang, Minghui
Jiang, Zhansheng
Joret, Gwenaël
Joux, Antoine
Jurdzinski, Tomasz
Jørgensen, Allan Grønlund
Kakimura, Naonori
Kantor, Erez
Kao, Ming-Yang
Kapralov, Michael
Kapur, Deepak
Kara, Ahmet
Karakostas, George
Karhumäki, Juhani
Kausch, Jonathan
Kavitha, Telikepalli
Kawamura, Akitoshi
Kayal, Neeraj
Keller, Orgad

Kerber, Michael
Kesselheim, Thomas
Khandekar, Rohit
Kiefer, Stefan
King, Valerie
Kiraly, Tamas
Klauck, Hartmut
Klein, Philip
Klima, Ondrej
Klin, Bartek
Klivans, Adam
Kniesburges, Sebastian
Kobayashi, Yusuke
Kobourov, Stephen
Koebler, Johannes
Koiran, Pascal
Kolay, Sudeshna
Kolliopoulos, Stavros
Komjathy, Julia
Kontchakov, Roman
Kopczyński, Eryk
Kopelowitz, Tsvi
Kopparty, Swastik
Kortsarz, Guy
Kosowski, Adrian
Kosub, Sven
Kothari, Nishad
Kothari, Pravesh
Koutis, Ioannis
Koutsopoulos, Andreas
Kovacs, Annamaria
Kratsch, Stefan
Krcal, Jan
Kretinsky, Jan
Krishnaswamy, Ravishankar
Krivosija, Amer
Krug, Robert
Krysta, Piotr
Kucera, Antonin
Kulikov, Alexander
Kulkarni, Janardhan
Kulkarni, Raghav
Kumar, Akash
Kumar, Amit
Kumar, K. Narayan

Kuperberg, Denis
Kurz, Denis
Kyropoulou, Maria
Labourel, Arnauld
Lachish, Oded
Laekhanukit, Bundit
Lagniez, Jean Marie
Lanik, Jan
Laura, Luigi
Lauria, Massimo
Lauriere, Mathieu
Laursen, Simon
Lauser, Alexander
Le Gall, Francois
Le Scouarnec, Nicolas
Lee, James
Lee, Troy
Leonardos, Nikos
Lerays, Virginie
Leroux, Jerome
Levavi, Ariel
Levin, Asaf
Levy, Jean-Jacques
Lewenstein, Moshe
Li, Jian
Li, Minming
Li, Shi
Li, Yi
Li, Yingkai
Libert, Benoit
Libkin, Leonid
Lime, Didier
Lin, Anthony Widjaja
Lin, Chengyu
Liu, Feng-Hao
Llana, Luis
Lodaya, Kamal
Lohrey, Markus
Lopez-Ortiz, Alejandro
Loreti, Michele
Lotker, Zvi
Lovett, Shachar
Lozin, Vadim
Lu, Pinyan
Lu, Steve

Lucier, Brendan
Löding, Christof
M.S., Ramanujan
Ma, Minghui
Magniez, Frederic
Mahdian, Mohammad
Mahmoody, Mohammad
Makarychev, Konstantin
Makarychev, Yury
Maletti, Andreas
Malizia, Enrico
Mallmann-Trenn, Frederik
Manea, Florin
Maneva, Elitza
Mansour, Yishay
Mardare, Radu
Markey, Nicolas
Markou, Euripides
Martens, Wim
Martin, Barnaby
Martin, Russell
Marx, Dániel
Marx, Maarten
Masopust, Tomas
Mathieson, Luke
Matulef, Kevin
May, Alexander
Mayr, Richard
McColl, Robert
McGregor, Andrew
McSherry, Frank
Megow, Nicole
Meier, Arne
Meiklejohn, Sarah
Meir, O.
Mendel, Manor
Meng, Xianmeng
Mens, Irini-Eleftheria
Mertzios, George
Meunier, Pierre-Etienne
Miao, Peihan
Michail, Dimitrios
Michalak, Tomasz
Mignot, Ludovic
Milanic, Martin

Milchtaich, Igal
Miltersen, Peter Bro
Misra, Pranabendu
Molinero, Xavier
Monemizadeh, Morteza
Monmege, Benjamin
Montanaro, Ashley
Montecchiani, Fabrizio
Montenegro, Ravi
Moore, Cristopher
Moran, Tal
Morere, Philippe
Morris, Ben
Morsy, Ehab
Moseley, Benjamin
Movahedi, Mahnush
Mucha, Marcin
Munagala, Kamesh
Munteanu, Alexander
Murawski, Andrzej
Murlak, Filip
Muscholl, Anca
Mvprao
Nagaj, Daniel
Nanongkai, Danupon
Narayan Kumar, K.
Narodytska, Nina
Natale, Emanuele
Nathan, Lemons
Navara, Mirko
Navarra, Alfredo
Nederlof, Jesper
Neiman, Ofer
Nekrich, Yakov
Newman, Alantha
Nguyen, Hung Son
Nguyen, Huy
Nguyen, Trung Thanh
Niehren, Joachim
Nielsen, Jesper Sindahl
Niewerth, Matthias
Nikishkin, Vladimir
Nikoletseas, Sotiris
Nikolov, Aleksandar
Nissim, Kobbi

Niwinski, Damian
Nordstrom, Jakob
Novotný, Petr
Nowotka, Dirk
Nutov, Zeev
Nuñez Chiroque, Luis
O'Donnell, Ryan
O'Neill, Adam
Obdrzalek, Jan
Ogierman, Adrian
Olesen, Mads C.
Oliveira, Igor
Onak, Krzysztof
Ong, Luke
Ortega-Mallén, Yolanda
Ortmann, Mark
Ossona De Mendez, Patrice
Oualhadj, Youssouf
Paes Leme, Renato
Pagh, Rasmus
Palomino, Miguel
Paluch, Katarzyna
Pan, Jiangwei
Pan, Jiaxin
Panagiotou, Konstantinos
Panangaden, Prakash
Pandey, Omkant
Panigrahi, Debmalya
Papadopoulos, Dimitrios
Papakonstantinou, Periklis
Paparas, Dimitris
Parys, Pawel
Pasquale, Francesco
Pastro, Valerio
Patt-Shamir, Boaz
Paulusma, Daniel
Pauly, Arno
Pavan, A.
Peikert, Christopher
Peng, Pan
Peressotti, Marco
Peretz, Ron
Perez, Guillermo
Perifel, Sylvain
Perrin, Dominique

Peserico, Enoch
Pettie, Seth
Peña, Ricardo
Picaronny, Claudine
Pieris, Andreas
Pighizzini, Giovanni
Pilipczuk, Marcin
Pilipczuk, Michal
Pin, Jean-Eric
Plandowski, Wojciech
Polychroniadou, Antigoni
Pottier, Franois
Pottonen, Olli
Pous, Damien
Pozzato, Gian Luca
Prabhakar, Pavithra
Praveen, M.
Price, Eric
Pruhs, Kirk
Pucci, Geppino
Pulina, Luca
Pérez, Jorge A.
Qiang, Ruixin
Qiao, Youming
Quyen, Vuong Anh
Rabani, Yuval
Rabie, Mikael
Raecke, Harald
Raghavendra, Prasad
Raghunathan, Ananth
Raghvendra, Sharathkumar
Rahaman, Anisur
Rampersad, Narad
Raskin, Jean-François
Raz, Ran
Regev, Oded
Rehak, Vojtech
Reynier, Pierre-Alain
Riba, Colin
Richerby, David
Riondato, Matteo
Robinson, Peter
Roditty, Liam
Rodriguez, Ismael
Roetteler, Martin

Roland, Jérémie
Romano, Paolo
Ron, Dana
Rosa-Velardo, Fernando
Rosołek, Robert
Rossi, Gianluca
Rossmanith, Peter
Rosulek, Michael
Rothvoss, Thomas
Rubin, Natan
Rubio, Fernando
Ruppert, Eric
Saad, George
Sablik, Mathieu
Sack, Joshua
Sadrzadeh, Mehrnoosh
Saha, Chandan
Salvati, Sylvain
Sammartino, Matteo
Sangnier, Arnaud
Sankur, Ocan
Santaroni, Federico
Santhanam, Rahul
Santocanale, Luigi
Santos, Nuno
Saptharishi, Ramprasad
Sarkar, Susmit
Satti, Srinivasa Rao
Sau, Ignasi
Sauerwald, Thomas
Saurabh, Saket
Sawada, Joe
Saxena, Nitin
Scarpa, Giannicola
Scheder, Dominik
Schmidt, Melanie
Schmidt-Schauss, Manfred
Schmitz, Sylvain
Schneider, Stefan
Schroder, Dominique
Schröder, Lutz
Schuster, Martin
Schwartz, Roy
Schweikardt, Nicole
Schwiegelshohn, Chris

Schwoon, Stefan
Servais, Frédéric
Servedio, Rocco
Seshadhri, C.
Setzer, Alexander
Shah, Rahul
Shah, Simoni
Shamir, Ohad
Sharma, Vikram
Shen, Alexandre
Shenoy R., Gautham
Shpilka, Amir
Shraibman, Adi
Sidiropoulos, Anastasios
Siebertz, Sebastian
Sikdar, Somnath
Silva, Alexandra
Silvestri, Riccardo
Singh, Mohit
Sitchinava, Nodari
Sitters, Rene
Skowron, Piotr
Sokolova, Ana
Solomon, Shay
Sommer, Christian
Sousi, Perla
Spoerhase, Joachim
Sramek, Rastislav
Srinivasan, Srikanth
Srivastava, Piyush
Srivathsan, B.
Stachowiak, Grzegorz
Staiger, Ludwig
Stainer, Julien
Starikovskaya, Tatiana
Stefankovic, Daniel
Stehle, Damien
Stephan, Frank
Stergiou, Christos
Stoddard, Greg
Strassburger, Lutz
Straubing, Howard
Strefler, Mario
Strejcek, Jan
Strothmann, Thim

Struth, Georg
Su, Le
Suchy, Ondrej
Sun, Xiaoming
Sun, Xiaorui
Suomela, Jukka
Suresh, S.P.
Syrgkanis, Vasilis
Sénizergues, Géraud
Ta-Shma, Amnon
Tamaki, Suguru
Tamir, Tami
Tan, Li-Yang
Tang, Bo
Tao, Yufei
Tarjan, Robert
Tavenas, Sébastien
Telle, Jan Arne
Terhal, Barbara
Terui, Kazushige
Terzi, Evimaria
Thaler, Justin
Thanh, Nguyen
Thapper, Johan
Thiagarajan, P.S.
Thilikos, Dimitrios
Thorup, Mikkel
Thraves, Christopher
Toledo, Sivan
Toledoii, Sivan
Tompits, Hans
Torres Vieira, Hugo
Torunczyk, Szymon
Toruńczyk, Szymon
Trevisan, Luca
Trivedi, Ashutosh
Tschudi, Daniel
Tulsiani, Madhur
Uehara, Ryuhei
Ulus, Dogan
Umans, Chris
Umboh, Seeun
Uno, Yushi
Upadhyay, Jalaj
Valiant, Gregory

Valiente, Gabriel
Valiron, Benoît
van Breugel, Franck
van Melkebeek, Dieter
Van Melkebeek, Dieter
van Stee, Rob
Varacca, Daniele
Vassilevska Williams, Virginia
Vegh, Laszlo
Velickovic, Boban
Venkitasubramaniam,
    Muthuramakrishnan
Ventre, Carmine
Verschae, Jose
Vidick, Thomas
Viet Tung, Hoang
Viglietta, Giovanni
Vijayaraghavan, Aravindan
Vilaça, Xavier
Visconti, Ivan
Viswanathan, Mahesh
Vogler, Walter
Volkovich, Ilya
Vrgoc, Domagoj
Wachter-Zeh, Antonia
Wahlström, Magnus
Walter, Tobias
Walukiewicz, Igor
Wang, Juntao
Wang, Kainan
Wanka, Rolf
Watson, Thomas
Wee, Hoeteck
Weinstein, Omri
Weiss, Armin
Westermann, Matthias
Whistler, William
Wieder, Udi
Wiese, Andreas
Wilkinson, Bryan T.
Wilson, David
Winslow, Andrew
Witek, Maximilian
Witkowski, Piotr
Wollan, Paul

Wong, Prudence W.H.
Woodruff, David
Wootters, Mary
Wright, John
Wrochna, Marcin
Wu, Xiaodi
Wulff-Nilsen, Christian
Wullschleger, Juerg
Xia, Ge
Xiao, Tao
Xie, Ning
Xing, Chaoping
Xu, Xiaoming
Xue, Guoliang
Yamada, Shota
Yamakami, Tomoyuki
Yamauchi, Yukiko
Yang, Kaiyu
Yao, Penghui
Yaroslavtsev, Grigory
Ye, Tao
Yekhanin, Sergey
Yi, Ke
Yiannakopoulos, Yiannis
Yin, Yitong
Yoshida, Yuichi
Young, Max
Yu, Huacheng
Yuen, Tsz Hon
Zacharias, Thomas
Zamani, Mahdi
Zang, Wenan
Zeh, Norbert
Zhang, Bingsheng
Zhang, Chihao
Zhang, Hongyang
Zhang, Jialin
Zhang, Jie
Zhang, Jin
Zhang, Shengyu
Zhang, Wuzhou
Zhang, Yong
Zhao, Zhiguang
Zhou, Hong-Sheng
Zhou, Yuan

Table of Contents – Part II

Track B: Logic, Semantics, Automata, and Theory of Programming

Track C: Foundations of Networked Computing

Table of Contents – Part I

Symmetric Groups and Quotient Complexity of Boolean Operations*

Jason Bell[1], Janusz Brzozowski[2], Nelma Moreira[3], and Rogério Reis[3]

[1] Department of Pure Mathematics, University of Waterloo,
Waterloo, ON, Canada N2L 3G1
jpbell@uwaterloo.ca
[2] David R. Cheriton School of Computer Science, University of Waterloo,
Waterloo, ON, Canada N2L 3G1
brzozo@uwaterloo.ca
[3] CMUP & DCC, Faculdade de Ciências da Universidade do Porto,
Rua do Campo Alegre, 4169–007 Porto, Portugal
{nam,rvr}@dcc.fc.up.pt

Abstract. The quotient complexity of a regular language L is the number of left quotients of L, which is the same as the state complexity of L. Suppose that L and L' are binary regular languages with quotient complexities m and n, and that the subgroups of permutations in the transition semigroups of the minimal deterministic automata accepting L and L' are the symmetric groups S_m and S_n of degrees m and n, respectively. Denote by $\circ$ any binary boolean operation that is not a constant and not a function of one argument only. For $m, n \geq 2$ with $(m, n) \notin \{(2, 2), (3, 4), (4, 3), (4, 4)\}$ we prove that the quotient complexity of $L \circ L'$ is mn if and only either (a) $m \neq n$ or (b) $m = n$ and the bases (ordered pairs of generators) of S_m and S_n are not conjugate. For $(m, n) \in \{(2, 2), (3, 4), (4, 3), (4, 4)\}$ we give examples to show that this need not hold. In proving these results we generalize the notion of uniform minimality to direct products of automata. We also establish a non-trivial connection between complexity of boolean operations and group theory.

Keywords: Boolean operation, quotient complexity, regular language, state complexity, symmetric group, transition semigroup.

1 Motivation

The *left quotient,* or simply *quotient,* of a regular language L over an alphabet Σ by a word $w \in \Sigma^*$ is the regular language $w^{-1}L = \{x \in \Sigma^* : wx \in L\}$. It is well known that a language is regular if and only if it has a finite number of quotients. Consequently, the number of quotients of a regular language, its *quotient complexity* [1], is a natural measure of complexity of the language. Quotient complexity is also known as *state complexity* [15], which is the number of

---

* For a complete version of this work see http://arxiv.org/abs/1310.1841.

J. Esparza et al. (Eds.): ICALP 2014, Part II, LNCS 8573, pp. 1–12, 2014.
© Springer-Verlag Berlin Heidelberg 2014

states in the complete minimal *deterministic finite automaton* (*DFA*) recognizing the language. We prefer quotient complexity because it is a language-theoretic concept, and we refer to it simply as *complexity*.

The problem of determining the complexity of an operation [1,8,15,16] on regular languages has received much attention. It is defined as the maximal complexity of the language resulting from the operation, taken as a function of the complexities of the operands. When operations are performed on large automata it is important to have some information about the size of the result and the time it will take to compute it. The quotient complexity of an operation gives an upper bound on its time and space complexity [15].

Languages that meet the upper bound on the complexity of an operation are *witnesses* for this operation. Although witnesses for common operations on regular languages are well known, there are occasions when one has to look for new witnesses:

1. One may be interested in a *class* of languages that have the same complexity with respect to a given operation. For example, let $\Sigma = \{a, b\}$ and let $|w|_a$ be the number of times the letter a appears in the word $w \in \Sigma^*$. Then the intersection of the languages $L = \{w \in \Sigma^*: |w|_a \equiv m - 1 \bmod m\}$ and $L' = \{w \in \Sigma^*: |w|_b \equiv n - 1 \bmod n\}$ has complexity mn. The languages $K = (b^*a)^{m-1}\Sigma^*$ and $K' = (a^*b)^{n-1}\Sigma^*$ also meet this bound; hence (L, L') and (K, K') are in the same complexity class with respect to intersection.

2. Whenever one studies complexity within a *proper subclass* of regular languages, one usually needs to find new witnesses. For example, in the class of regular right ideals – languages $L \subseteq \Sigma^*$ satisfying $L = L\Sigma^*$ – languages K and K' are appropriate, but L and L' are not. The main result of the present paper has been applied to right ideals in [4], where the proof that the witnesses used there meet the bounds for boolean operations was greatly simplified with the aid of our theorem.

3. When one studies *combined operations* – operations that involve more than one basic operation, for example, the intersection of reversed languages – one again need new witnesses [7].

Before stating our result, we provide some additional background information. The *syntactic congruence* $\leftrightarrow_L$ of L is defined as follows: For all $x, y \in \Sigma^*$, $x \leftrightarrow_L y$ if and only if $uxv \in L \Leftrightarrow uyv \in L$ for all $u, v \in \Sigma^*$. The set $\Sigma^+ / \leftrightarrow_L$ of equivalence classes of the relation $\leftrightarrow_L$ is a semigroup with concatenation as the operation; it is called the *syntactic semigroup* of L, which we denote by S_L. It is well known that the syntactic semigroup is isomorphic to the semigroup S_D of transformations performed by non-empty words on the set of states in the minimal DFA $\mathcal{D}$ recognizing L; this semigroup is known as the *transition semigroup* of $\mathcal{D}$. If $\mathcal{D}$ has n states, the cardinality of the transition semigroup is bounded from above by n^n, and this bound is reachable.

The *atoms* [5,6] of a regular language are non-empty intersections of all left quotients of the language, some or all of which may be complemented. A regular language has at most 2^n atoms, and their quotient complexities are known [5].

The *reverse* of a word is defined inductively: the reverse of the empty word ε is $\varepsilon^R = \varepsilon$, and the reverse of wa with $w \in \Sigma^*$ and $a \in \Sigma$ is $(wa)^R = aw^R$. The reverse of a language L is $L^R = \{w^R \colon w \in L\}$. For L with complexity n the maximal complexity of L^R is 2^n, and this bound is reachable.

Whenever new witnesses are used, it is necessary to prove that these witnesses meet the required bound. It would be very useful to have results stating that *if the languages in question have some property P, then they meet the upper bound for a given operation*. Some results of this type are now briefly discussed.

Let **MSC** denote the class of languages with *maximal syntactic complexity* (languages with largest syntactic semigroups), let **STT** denote the class of languages whose minimal DFAs have *set-transitive transition semigroups* (for any two sets of states of the same cardinality there is a transformation that maps one set to the other), let **MAL** denote the class of *maximally atomic languages* (languages that have all 2^n atoms, all of which have maximal possible quotient complexity), let **MNA** denote the class of languages with the *maximal number* (2^n) of *atoms*, and let **MCR** denote the class of languages with a *maximally complex reverse* (reverse of complexity 2^n). The following relations hold [3]:

$$\mathbf{MSC} \subset \mathbf{STT} = \mathbf{MAL} \subset \mathbf{MNA} = \mathbf{MCR}.$$

The fact that **MSC** $\subset$ **MCR** is a result of A. Salomaa, Wood, and Yu [12], and the observation that **MNA** = **MCR** was made by Brzozowski and Tamm [6].

Our main theorem relates the complexity of proper binary boolean operations on regular languages to the nature of the syntactic semigroups of the languages. A boolean operation is *proper* if it is not a constant and not a function of one variable only.

Let S_n denote the symmetric group of degree n. A *basis* [9] of S_n is an ordered pair (s, t) of distinct transformations of $Q_n = \{0, \ldots, n - 1\}$ that generate S_n. Two bases (s, t) and (s', t') of S_n are *conjugate* if there exists a transformation $r \in S_n$ such that $rsr^{-1} = s'$, and $rtr^{-1} = t'$.

Assume that a DFA $\mathcal{D}$ (respectively, $\mathcal{D}'$) has state set Q_m (Q_n), and transition semigroup S_m (S_n). Let L (L') be the language accepted by $\mathcal{D}$ ($\mathcal{D}'$). Our main theorem is a generalization of a result of Brzozowski and Liu [2]:

Theorem 1. *Let $\mathcal{D}$ and $\mathcal{D}'$ be binary DFAs with m and n states respectively, where $m, n \geq 2$ and $(m, n) \notin \{(2, 2), (3, 4), (4, 3), (4, 4)\}$. If the subgroups of permutations in the transition semigroups of $\mathcal{D}$ and $\mathcal{D}'$ are S_m and S_n respectively, and $\circ$ is a proper binary boolean operation, then the complexity of $L \circ L'$ is mn, unless $m = n$ and the bases of the transition semigroups of $\mathcal{D}$ and $\mathcal{D}'$ are conjugate, in which case the quotient complexity of $L \circ L'$ is at most $m = n$.*

The proof that the complexity of a binary boolean operation is maximal involves two steps. First, one proves that the direct product of the minimal DFAs of the languages is connected, meaning that all of its states are reachable from the initial state. Second, one verifies that any two states in the direct product are distinguishable by some word, that is, that they are not equivalent. Since both reachability and distinguishability will be proved using only permutations, it is convenient to ignore other transformations and assume that the transition semigroups of the DFAs we deal with are symmetric groups.

The remainder of the paper is structured as follows: Section 2 defines our terminology and notation. Section 3 deals with the conditions under which the direct product of two automata is connected. Section 4 studies uniformly minimal semiautomata (automata without final states), that is, semiautomata which become minimal DFAs if one adds an arbitrary set of final states, other than the empty set and the set of all states. Section 5 contains our main result relating symmetric groups to the complexity of boolean operations for all except a few cases. Section 6 concludes the paper.

2   Preliminaries

Groups. Our results rely heavily on the theory of finite groups. We refer the reader to [11,13], for example, for basic facts about groups.

Transformations. A *transformation* of a set Q is a mapping of Q into itself. We deal only with finite non-empty sets and, without loss of generality, assume that $Q = Q_n = \{0, 1, \ldots, n-1\}$. If t is a transformation of Q_n and $i \in Q_n$, then $t(i)$ is the image of i under t. An arbitrary transformation is written in the form

$$t = \begin{pmatrix} 0 & 1 & \ldots & n-2 & n-1 \\ i_0 & i_1 & \ldots & i_{n-2} & i_{n-1} \end{pmatrix},$$

where $i_k = t(k)$, $0 \le k \le n-1$, and $i_k \in Q_n$. The *composition* of two transformations t_1 and t_2 of Q_n is a transformation $t_1 \circ t_2$ such that $(t_1 \circ t_2)(i) = t_1(t_2(i))$ for all $i \in Q_n$. We usually omit the composition operator and write $t_1 t_2$. The set of all transformations of Q_n is a monoid under composititon with the identity transformation acting as the unit element $\mathbf{1}$.

A *permutation* is a mapping of Q_n onto itself. A permutation t is a *cycle of of length k* or a *k-cycle* , where $k \ge 2$, if there exist pairwise different elements $i_1, \ldots, i_k$ such that $t(i_1) = i_2$, $t(i_2) = i_3$, $\ldots$, $t(i_{k-1}) = i_k$, and $t(i_k) = i_1$, and t does not affect any other elements. A cycle is denoted by $(i_1, i_2, \ldots, i_k)$. A *transposition* is a 2-cycle. Every permutation is a product (composition) of transpositions, and the parity of the number of transpositions in the factorization is an invariant. A permutation is *odd* (*even*) if its factorization has an odd (even) number of factors. The *symmetric group* S_n of *degree* n is the set of all permutations of Q_n, with composition as the group operation, and the identity as $\mathbf{1}$. The *alternating group* A_n is the set of all even permutations of S_n.

Given a subgroup H of S_n, we say that H *acts transitively* on Q_n if for each $i, j \in Q_n$ there is some $t \in H$ such that $t(i) = j$. We say that H *acts doubly transitively* on Q_n if whenever $i, j, k, \ell \in Q_n$ with $i \ne j$ and $k \ne \ell$ there is some $t \in H$ such that $t(i) = k$, $t(j) = \ell$.

Semiautomata and Automata. A *deterministic finite semiautomaton (DFS)* is a quadruple $\mathcal{A} = (Q, \Sigma, \delta, q_0)$, where Q is a finite set of *states*, Σ is a finite non-empty *alphabet*, $\delta \colon Q \times \Sigma \to Q$ is the *transition function*, and q_0 is the *initial state*. We extend δ to $Q \times \Sigma^*$ in the usual way. A state q is *reachable* from the initial state if there is a word w such that $q = \delta(q_0, w)$. A DFS is *connected* if every state $q \in Q$ is reachable.

For a DFS $\mathcal{A} = (Q, \Sigma, \delta, q_0)$ and a word $w \in \Sigma^*$, the transition function $\delta(\cdot, w)$ is a transformation of Q, the transformation *induced by w*. The set of all transformations induced by non-empty words is the *transition semigroup* $S_\mathcal{A}$ of $\mathcal{A}$. For $w \in \Sigma^+$, we denote by $w\colon t$ the transformation t of Q_n induced by w.

Given semiautomata $\mathcal{A} = (Q, \Sigma, \delta, q_0)$ and $\mathcal{A}' = (Q', \Sigma, \delta', q_0')$, we define their direct product to be the DFS $\mathcal{A} \times \mathcal{A}' = (Q \times Q', \Sigma, (\delta, \delta'), (q_0, q_0'))$.

A *deterministic finite automaton (DFA)* is a quintuple $\mathcal{D} = (Q, \Sigma, \delta, q_0, F)$, where (Q, Σ, δ, q_0) is a DFS and $F \subseteq Q$ is the set of *final states*. The DFA $\mathcal{D}$ *accepts* a word $w \in \Sigma^*$ if $\delta(q_0, w) \in F$. The set of all words accepted by $\mathcal{D}$ is the *language* $L(\mathcal{D})$ of $\mathcal{D}$. The *language accepted from a state q* of a DFA is the language $L_q(\mathcal{D})$ accepted by the DFA $(Q, \Sigma, \delta, q, F)$. Two states of a DFA are *distinguishable* if there exists a word w which is accepted from one of the states and rejected from the other. Otherwise, the two states are *equivalent*. A DFA is *minimal* if all of its states are reachable from the initial state and no two states are equivalent. Note that if $|Q| \geq 2$ and $\mathcal{D}$ is minimal, then $\emptyset \subsetneq F \subsetneq Q$.

3 Connectedness

From now on we are interested in semiautomata $\mathcal{A}$ and $\mathcal{A}'$ whose transition semigroups are symmetric groups generated by two-element bases. We assume that permutations s and s' are induced by a in $\mathcal{A}$ and $\mathcal{A}'$, and permutations t and t' by b, that is, $a\colon s$, $b\colon t$ in $\mathcal{A}$ and $a\colon s'$, $b\colon t'$ in $\mathcal{A}'$.

Example 1. Let $\Sigma = \{a, b\}$, $\mathcal{A} = (Q_3, \Sigma, \delta, 0)$, and $\mathcal{A}' = (Q_3, \Sigma, \delta', 0)$, where $a\colon s = (0, 1, 2)$, $b\colon t = (0, 1)$ in $\mathcal{A}$, and $a\colon s' = (0, 1, 2)$, $b\colon t' = (1, 2)$ in $\mathcal{A}'$. Then (s, t) and (s', t') are conjugate, since $rsr^{-1} = s'$ and $rtr^{-1} = t'$ for $r = (0, 1, 2)$. If $\mathcal{A}''$ has $s'' = (0, 1)$ and $t'' = (0, 1, 2)$, then (s, t) and (s'', t'') are not conjugate.

The transition semigroups of $\mathcal{A}$, $\mathcal{A}'$ and $\mathcal{A}''$ all have 6 elements. Those of $\mathcal{A}$ and $\mathcal{A}'$, when viewed as semigroups generated by a and b, are identical, but those of $\mathcal{A}$ and $\mathcal{A}''$ are not: for example, $a^3 = 1$ in $S_\mathcal{A}$ but $a^2 = 1$ in $S_{\mathcal{A}''}$. ∎

Theorem 2. *Let $\Sigma = \{a, b\}$, let $\mathcal{A} = (Q_m, \Sigma, \delta, 0)$ and $\mathcal{A}' = (Q_n, \Sigma, \delta', 0)$ be semiautomata with transition semigroups that are symmetric groups of degrees m and n respectively, and let the corresponding bases be B and B'. For $m, n \geq 1$, the direct product $\mathcal{A} \times \mathcal{A}'$ is connected if and only if either (1) $m \neq n$ or (2) $m = n$ and B and B' are not conjugate.*

Proof. Without loss of generality, assume that $m \leq n$. Let H denote the transition semigroup of $\mathcal{A} \times \mathcal{A}'$; then H is a subgroup of $S_m \times S_n$. Define homomorphisms $\pi_1\colon H \to S_m$ and $\pi_2\colon H \to S_n$ by $\pi_1((s, t)) = s$ and $\pi_2((s, t)) = t$. Observe that π_1 and π_2 are surjective, since the transition semigroups of $\mathcal{A}$ and $\mathcal{A}'$ are S_m and S_n respectively. We let H_0 denote the subgroup of H consisting of all elements that map the set $\{0\} \times Q_n$ to itself. Then H_0 has index m in H and thus $\pi_2(H_0)$ has index at most m in $\pi_2(H) = S_n$. Thus the order of $\pi_2(H_0)$ is at least $n!/m \geq (n-1)!$.

Since a subgroup of S_n that does not act transitively on Q_n is necessarily iso-morphic to a subgroup of $S_i \times S_{n-i}$ for some $i \in \{1, \ldots, n-1\}$ [14, Section 2.5.1], a subgroup of S_n whose order is strictly greater than $(n-1)!$ acts transitively on Q_n. Moreover, a subgroup of order $(n-1)!$ that does not act transitively on Q_n is isomorphic to $S_1 \times S_{n-1}$; that is, it is the stabilizer of a point. Thus $\pi_2(H_0)$ fails to act transitively on Q_n if and only if $m = n$ and $\pi_2(H_0)$ is the stabilizer of a point.

Suppose that $m < n$ or $m = n$ and $\pi_2(H_0)$ is not the stabilizer of a point, which is equivalent to assuming that $\pi_2(H_0)$ acts transitively on Q_n. We claim that the direct product $\mathcal{A} \times \mathcal{A}'$ is connected. To see this, notice that given (i, j) and (i', j') in $Q_m \times Q_n$, we can find t (respectively t') in H that sends (i, j) to $(0, k)$ (respectively (i', j') to $(0, k')$) for some k (respectively k') in Q_n, since $\pi_1(H) = S_m$ acts transitively on Q_m. Since we have assumed that $\pi_2(H_0)$ acts transitively on Q_n, we can find $t'' \in H$ such that $\pi_2(t'') \in \pi_2(H_0)$ sends $(0, k)$ to $(0, k')$. Hence $(t')^{-1} t'' t$ sends (i, j) to (i', j'), and so $\mathcal{A} \times \mathcal{A}'$ is connected.

Suppose next that $m = n$ and $\pi_2(H_0)$ is the stabilizer of a point. By relabelling if necessary, we may assume that $\pi_2(H_0)$ stabilizes 0. Then H cannot send $(0, 0)$ to $(0, i)$ for $i \neq 0$ and so $\mathcal{A} \times \mathcal{A}'$ is not connected. We claim that the bases B and B' are conjugate.

To prove this claim, note that H has the property that if $(s, t) \in H \subseteq S_n \times S_n$ and $s(0) = 0$, then $t(0) = 0$. We claim there is a permutation $u \in S_n$ with $u(0) = 0$ such that if $(s, t) \in H$ sends $(0, 0)$ to (j, k), then $k = u(j)$. First suppose that $k_1, k_2 \in Q_n$ have the property that there is some $j \in Q_n$ such that (j, k_1) and (j, k_2) are in the orbit of $(0, 0)$ under the action of H. Then we can pick h in H such that $\pi_1(h)(j) = 0$. Then $(0, \pi_2(h)(k_1))$ and $(0, \pi_2(h)(k_2))$ are both in the orbit of $(0, 0)$, which means that $\pi_2(h)(k_1) = \pi_2(h)(k_2) = 0$, giving $k_1 = k_2$. It follows that there is a map $u \colon Q_n \to Q_n$ with $u(0) = 0$ such that, if $(s, t) \in H$ sends $(0, 0)$ to (j, k), then $k = u(j)$. Since $\pi_2(H)$ acts transitively on Q_n, the map u must be surjective and hence is a permutation, as claimed.

Let $s_1, s_2 \in S_n$ denote the elements in the transition semigroup corresponding to $a \in \Sigma$, and let $t_1, t_2 \in S_n$ correspond to $b \in \Sigma$. Let H' be the group generated by $(s_1, u^{-1} t_1 u), (s_2, u^{-1} t_2 u)$. Then H' is conjugate to H (we conjugate H by $(1, u)$ to obtain H'); furthermore, H' has the property that if $(s, t) \in H'$ sends $(0, 0)$ to (i, j), then $i = j$. Thus H' acts transitively on the diagonal of $Q_n \times Q_n$; if $(s, t) \in H'$ then $s(i) = t(i)$ for all $i \in Q_n$, which gives that $s = t$. Hence, if $(s, t') \in H$, then $u^{-1} t' u = s$ and so the bases B and B' are conjugate. Thus if $\mathcal{A} \times \mathcal{A}'$ is not connected, then $m = n$ and the bases B and B' are conjugate.

Now we show the converse: If $m = n$ and the bases $B = (s, t)$ and $B' = (s', t')$ are conjugate, then $\mathcal{A} \times \mathcal{A}'$ is not connected. If $rsr^{-1} = s'$, and $rtr^{-1} = t'$, let $\psi_r \colon \{s, t\}^+ \to \{s', t'\}^+$ be the mapping that assigns to $x \in \{s, t\}^+$ the element $rxr^{-1} \in \{s', t'\}^+$. For any $x, y \in \{s, t\}^+$, if $xy = z$, then $\psi_r(x) \psi_r(y) = (rxr^{-1})(ryr^{-1}) = r(xy)r^{-1} = \psi_r(z)$. Hence the transition semigroups of $\mathcal{A}$ and $\mathcal{A}'$ are isomorphic.

The direct product $\mathcal{A} \times \mathcal{A}'$ is defined by $(Q_n \times Q_n, \{a, b\}, (\delta, \delta'), (0, 0))$, where $(\delta, \delta')((i, j), a) = (s(i), rsr^{-1}(j))$ and $(\delta, \delta')((i, j), b) = (t(i), rtr^{-1}(j))$ for any $i, j \in Q_n$.

If $\mathcal{A} \times \mathcal{A}'$ is connected, then for all $(i, j) \in Q_n \times Q_n$ there must exist a word $w \in \Sigma^+$ such that $(\delta, \delta')((0, 0), w) = (i, j)$ or, equivalently, there exists a permutation p such that $p(0) = i$ and $rpr^{-1}(0) = j$. There are now two cases:

1. If $r^{-1}(0) \neq 0$, we prove that state $(i, r(i))$ is unreachable for all $i \in Q_n$. If $(i, r(i))$ is reachable, then there exists a permutation p such that $p(0) = i$ and $rpr^{-1}(0) = r(i)$. But then $r^{-1}rpr^{-1}(0) = pr^{-1}(0) = i = p(0)$, and so $p^{-1}pr^{-1}(0) = r^{-1}(0) = 0$, which is a contradiction.

2. If $r^{-1}(0) = 0$, we prove that state (i, i) is unreachable for some $i \in Q_n$. Since r cannot be the identity, there must exist an i such that $r(i) \neq i$. Suppose (i, i) is reachable for that i. Then there exists a permutation p such that $p(0) = i$ and $rpr^{-1}(0) = i$. Thus $i = rpr^{-1}(0) = rp(0) = p(0)$ and $r(i) = i$, which is a contradiction.

Hence $\mathcal{A} \times \mathcal{A}'$ cannot be connected. □

Remark 1. If $\mathcal{A} \times \mathcal{A}'$ is connected, then it is strongly connected, since the transition semigroup of $\mathcal{A} \times \mathcal{A}'$ is a group.

4   Uniformly Minimal Semiautomata

Semiautomata that result in minimal DFAs under any non-trivial assignment of final states were studied by Restivo and Vaglica [10]. We modify their definitions slightly to suit our purposes. A strongly connected DFS $\mathcal{A} = (Q, \Sigma, \delta, q_0)$ with $|Q| \geq 2$ is *uniformly minimal* if the DFA $\mathcal{D} = (Q, \Sigma, \delta, q_0, F)$ is minimal for each set F of final states, where $\emptyset \subsetneq F \subsetneq Q$.

Given a DFS $\mathcal{A} = (Q, \Sigma, \delta, q_0)$, we define the *pair graph* of $\mathcal{A}$ to be the directed graph $G_{\mathcal{A}} = (V_{\mathcal{A}}, E_{\mathcal{A}})$, where the set $V_{\mathcal{A}}$ of vertices is the set of all two-element subsets $\{p, q\}$ of Q, and the set $E_{\mathcal{A}}$ of edges consists of unordered pairs $(\{p, q\}, \{p', q'\})$ such that $\{\delta(p, a), \delta(q, a)\} = \{p', q'\}$. The following is from [10]:

Proposition 1 (Restivo and Vaglica). *Let $\mathcal{A} = (Q, \Sigma, \delta, q_0)$ be a strongly connected DFS with at least two states. If the pair graph $(V_{\mathcal{D}}, E_{\mathcal{D}})$ is strongly connected, then $\mathcal{A}$ is uniformly minimal.*

We prove a similar result for semiautomata with symmetric groups.

Proposition 2. *Suppose that $\mathcal{A} = (Q_n, \Sigma, \delta, q_0)$ is a DFS and the transition semigroup $S_{\mathcal{A}}$ of $\mathcal{A}$ is the symmetric group S_n. Then $\mathcal{A}$ is strongly connected and uniformly minimal.*

Proof. If $S_{\mathcal{A}} = S_n$, then $S_{\mathcal{A}}$ contains all permutations of Q_n, in particular, the cycle $(0, \ldots, n-1)$; hence $\mathcal{A}$ is strongly connected. For any $(i, j), (k, \ell) \in Q_n \times Q_n$ with $i \neq j$, $k \neq \ell$, and $\{i, j\} \neq \{k, \ell\}$, any permutation that maps i to k and j to ℓ connects $\{i, j\}$ to $\{k, \ell\}$ in the pair graph of $\mathcal{A}$. Hence the pair graph is strongly connected, and $\mathcal{A}$ is uniformly minimal by Proposition 1. □

Let the truth values of propositions be 1 (true) and 0 (false). Let $\circ\colon \{0,1\} \times \{0,1\} \to \{0,1\}$ be a binary boolean function. Extend $\circ$ to a function $\circ\colon 2^{\Sigma^*} \times 2^{\Sigma^*} \to 2^{\Sigma^*}$: If $w \in \Sigma^*$ and $L, L' \subseteq \Sigma^*$, then $w \in (L \circ L') \Leftrightarrow (w \in L) \circ (w \in L')$. Also, extend $\circ$ to a function $\circ\colon 2^{Q_m} \times 2^{Q_n} \to 2^{Q_m \times Q_n}$: If $q \in Q_m$, $q' \in Q_n$, $F \subseteq Q_m$, and $F' \subseteq Q_n$, then $(q, q') \in (F \circ F') \Leftrightarrow (q \in F) \circ (q' \in F')$.

Suppose that $\mathcal{A} = (Q, \Sigma, \delta, 0)$ and $\mathcal{A}' = (Q', \Sigma, \delta', 0)$ with $|Q| = m$ and $|Q'| = n$ are uniformly minimal DFSs, and $\circ$ is any proper boolean function. The pair $(\mathcal{A}, \mathcal{A}')$ is *uniformly minimal for* $\circ$ if the direct product $\mathcal{P} = (Q \times Q', \Sigma, (\delta, \delta'), (0, 0), F \circ F')$ is minimal for all *valid assignments* of sets F and F' of final states to $\mathcal{A}$ and $\mathcal{A}'$, that is, sets such that $\emptyset \subsetneq F \subsetneq Q$ and $\emptyset \subsetneq F' \subsetneq Q'$.

If $n = 1$, then $\mathcal{A} \times \mathcal{A}'$ is isomorphic to $\mathcal{A}$ and no boolean function $\circ$ is proper. Hence this case, and also the case with $m = 1$, is of no interest. Henceforth we assume that $m, n \geq 2$.

We now consider pair graphs of DFSs with symmetric groups as their transition semigroups.

Example 2. Suppose that $m = n = 2$, and $\mathcal{A}$ and $\mathcal{A}'$ both have S_2 as their transition semigroup. There are two permutations in S_2: $(0, 1)$ and $\mathbf{1}$, and there are three bases: $B_1 = (a\colon (0,1), b\colon (0,1))$, $B_2 = (a\colon (0,1), b\colon \mathbf{1})$, and $B_3 = (a\colon \mathbf{1}, b\colon (0,1))$. Note that no two of these bases are conjugate.

For each basis, there are two possible final states, 0 or 1, and hence two DFAs; thus there are six different DFAs. There are then twelve direct products $\mathcal{D}_j^i \times \mathcal{D}_\ell^k$ with non-conjugate bases, where $\mathcal{D}_j^i$ ($\mathcal{D}_\ell^k$) uses basis B_i (B_k) and has j (ℓ) as final state, for $i, k = 1, 2, 3$ and $j, \ell = 1, 2$.

For each pair of DFAs accepting languages L and L' respectively, we tested the complexity of five boolean functions: $L \cup L'$, $L \cap L'$, $L \oplus L'$, $L \setminus L'$ and $L' \setminus L$. Note that the complexity of each remaining proper boolean function is the same as that of one of these five functions. For all twelve direct products of DFAs with non-conjugate bases, all proper boolean functions reach the maximal complexity 4, except for the functions $L \oplus L'$ and $\overline{L \oplus L'}$, which fail in all twelve cases. Thus any two DFAs $\mathcal{D} = (Q_2, \Sigma, \delta_i, 0, F)$ and $\mathcal{D}' = (Q_2, \Sigma, \delta_k, 0, F')$, where $Q_2 = \{0, 1\}$, $\Sigma = \{a, b\}$, δ_i (δ_k) is defined by basis B_i (B_k), $F = \{j\}$ and $F' = \{\ell\}$, are uniformly minimal for all proper boolean functions, except $\oplus$ and its complement. So our main result applies only in some cases if $m = n = 2$. ∎

Proposition 3. *Let* $\mathcal{A} = (Q_m, \Sigma, \delta, 0)$ *and* $\mathcal{A}' = (Q_n, \Sigma, \delta', 0)$, *with* $m, n \geq 2$ *and* $\max(m, n) \geq 3$, *be DFSs with transition semigroups that are symmetric groups, and let* $\mathcal{P}$ *be their direct product. Then the following hold:*

1. The pair graph of $\mathcal{P}$ *consists of strongly connected components – which we will call simply components – of one of the following three types:*

(a) $T_1 \subseteq C_1 = \{\{(i, j), (k, \ell)\}\colon i \neq k, j \neq \ell\}$,

(b) $T_2 \subseteq C_2 = \{\{(i, j), (i, \ell)\}\colon j \neq \ell\}$,

(c) $T_3 \subseteq C_3 = \{\{(i, j), (k, j)\}\colon i \neq k\}$.

2. Every state (i, j) *of the direct product* $\mathcal{P}$ *appears in at least one pair in each component.*

3. Each component has at least $mn/2 \geq 3$ *pairs.*

Proof. The first claim follows since the transition semigroup of $\mathcal{P}$ is a group. The second claim holds because the direct product is strongly connected, by Remark 1. For the third claim, note that there are mn states in $\mathcal{P}$, but they can appear in pairs; hence the bound $mn/2$. Since we are assuming that $mn \geq 6$, the last claim follows. □

Now consider DFAs $\mathcal{D} = (Q_m, \Sigma, \delta, 0, F)$ and $\mathcal{D}' = (Q_n, \Sigma, \delta', 0, F')$, where $\emptyset \subsetneq F \subsetneq Q_m$ and $\emptyset \subsetneq F' \subsetneq Q_n$. A state $\{(i,j), (k,\ell)\}$ of the pair graph of the direct product $\mathcal{P}$ of $\mathcal{D}$ and $\mathcal{D}'$ is *distinguishing* if and only if (i,j) is final and (k,ℓ) is not, or *vice versa*.

Example 3. Suppose $m = 3$, $n = 4$, δ is defined by the basis $(a\colon (0,1), b\colon (0,1,2))$ of S_3, and δ' by the basis $(a\colon (0,1), b\colon (1,3,2))$ of S_4. One verifies that these bases are not conjugate. The direct product $\mathcal{P}$ is connected and has twelve states.

If $F = \{2\}$, $F' = \{0,1\}$ and intersection is the boolean function, then there are no distinguishing pairs in the component of the pair graph T containing $\{(0,0), (0,3)\}$. Hence any two states appearing in the same pair of T are equivalent. Indeed, the minimal version of $\mathcal{P}$ has only six states. ∎

Example 4. Suppose $m = n = 4$, δ is defined by the basis $(a\colon (0,1,2), b\colon (2,3))$, and δ' by the basis $(a\colon (1,3,2), b\colon (0,2,1,3))$. If $F = \{0,1\}$ and $F' = \{0,1\}$, then the complexity of $L \oplus L'$ is 4, but all the other complexities are 12. ∎

Lemma 1. *Let $\mathcal{D} = (Q, \Sigma, \delta, 0, F)$ and $\mathcal{D}' = (Q', \Sigma, \delta', 0, F')$, with $|Q|, |Q'| \geq 2$, be DFAs with transition semigroups that are groups, and let $\mathcal{P} = (Q \times Q', \Sigma, (\delta, \delta'), (0,0), F \circ F')$ be their direct product. Then $\mathcal{P}$ is minimal if and only if every component of the pair graph $G_{\mathcal{P}}$ of $\mathcal{P}$ has a distinguishing pair.*

5 Symmetric Groups and Boolean Operations

We begin with a well-known but apparently unpublished result.

Lemma 2. *Let n be a positive integer, let G be either S_n or A_n, and let H be a subgroup of G of index $m \leq n$. Then the following hold:*

(i) if $n \neq 4$ and $m < n$, then H is either A_n or S_n;

(ii) if $m = n$ and $n \neq 6$, then there is some $i \in Q_n$ such that H is the set of permutations in G that fix i.

(iii) if $m = n = 6$, then there is an automorphism ϕ of S_6 such that $\phi(H)$ is the set of elements that fix 0.

The following lemma, like Theorem 2, deals with reachability. The conditions in the lemma, however, are useful for determining reachability in the pair graph of $\mathcal{A} \times \mathcal{A}'$, rather than in $\mathcal{A} \times \mathcal{A}'$ itself.

Lemma 3. *Let $\Sigma = \{a, b\}$, let $\mathcal{A} = (Q_m, \Sigma, \delta, 0)$ and $\mathcal{A}' = (Q_n, \Sigma, \delta', 0)$ be semiautomata with transition semigroups that are symmetric groups of degrees m and n respectively with $m \leq n$, $n \neq 4$ and $(m,n) \neq (6,6)$. Let H be the transition semigroup of $\mathcal{A} \times \mathcal{A}'$, and let π_1 and π_2 be the natural projections from H onto S_m and S_n respectively. If $H_0 = \{h \in H\colon \pi_1(h)(0) = 0\}$, then*

1. $\pi_2(H_0)$ is either S_n or A_n, or is the stabilizer of a point in Q_n.

2. $\pi_2(H_0)$ is the stabilizer of a point if and only if $m = n$, and in this case the direct product $\mathcal{A} \times \mathcal{A}'$ is not connected.

Proof. For Part 1, since $\pi_1(H) = S_m$, for each $i \in \{0, \ldots, m - 1\}$ there is some $h_i \in H$ such that $\pi_1(h_i)$ takes 0 to i. For a given $h \in H$, $\pi_1(h)$ takes 0 to j for some $j \in \{0, 1, \ldots, m - 1\}$, and thus $h_j^{-1}h \in H_0$ and so $h \in h_j H_0$. However, since $\pi_1(h)$ takes 0 to j, we have $h_i^{-1}h \notin H_0$ and thus $h \notin h_i H_0$ for $i \neq j$. Thus the cosets $h_0 H, \ldots, h_{m-1}H$ are distinct, and H_0 has index m in H. Since $\pi_2(H) \subseteq \bigcup_{i=0}^{m-1} \pi_2(h_i)\pi_2(H_0)$, $\pi_2(H_0)$ has index at most m in $\pi_2(H) = S_n$. If $n \neq 4$ and $m < n$, then $\pi_2(H_0)$ is either A_n or S_n by Lemma 2. If $m = n$ and $n \neq 6$, then $\pi_2(H_0)$ has index n in S_n and hence must be the stabilizer of a some $i \in Q_n$ by Lemma 2.

For Part 2, suppose that $m = n$ and $\pi_2(H_0)$ is the stabilizer of a point in Q_n. By relabelling if necessary, we may assume that $\pi_2(H_0)$ stabilizes 0. Hence, if $h \in H$ sends $(0,0)$ to $(0,j)$ then $j = 0$. In particular, there is no $h \in H$ that sends $(0,0)$ to $(0,1)$ or that sends $(0,1)$ to $(0,0)$, and so $\mathcal{A} \times \mathcal{A}'$ is necessarily not connected. □

Lemma 4. *Let $\mathcal{A} = (Q_m, \Sigma, \delta, 0)$ and $\mathcal{A}' = (Q_n, \Sigma, \delta', 0)$ be semiautomata with transition semigroups that are the symmetric groups of degrees m and n, respectively with $m \leq n$, $m \geq 2$, $n \geq 5$, and $(m, n) \neq (6, 6)$. If $\mathcal{A} \times \mathcal{A}'$ is connected, then the pair graph of $\mathcal{A} \times \mathcal{A}'$ has exactly three connected components: $C_1 = \{\{(i,j), (k,\ell)\} : i \neq k, j \neq \ell\}$, $C_2 = \{\{(i,j), (i,\ell)\} : j \neq \ell\}$, and $C_3 = \{\{(i,j), (k,j)\} : i \neq k\}$.*

Proof. We let H denote the transition semigroup of $\mathcal{A} \times \mathcal{A}'$. We show that each of C_1, C_2, C_3 is strongly connected. Note that each of C_1, C_2, C_3 is necessarily a union of connected components.

We show that C_1 is strongly connected. Suppose we have pairs $\{(i,j), (k,\ell)\}$ and $\{(i',j'), (k',\ell')\}$ with i, k distinct, i', k' distinct, j, ℓ distinct, and j', ℓ' distinct. Since S_m acts doubly transitively on Q_m when $m \geq 2$, there is some $s \in H$ that sends (i,j) to (i', j'') and (k, ℓ) to (k', ℓ'') for some $j'', \ell'' \in Q_n$.

Thus we may assume without loss of generality that $i' = i$ and $k' = k$. Let H_0 be the subgroup of $S_m \times S_n$ consisting of all $x \in H$ such that $\pi_1(x)$ fixes i. By Lemma 3, since we assume that $\mathcal{A} \times \mathcal{A}'$ is connected, $\pi_2(H_0)$ is not a stabilizer of a point in Q_n. Hence $\pi_2(H_0)$ is either S_n or A_n. Let H_1 denote the subgroup of $S_m \times S_n$ consisting of all $x \in H$ such that $\pi_1(x)$ fixes i and k. By the argument used in Lemma 3 to show that $\{h \in H : \pi_1(h)(0) = 0\}$ has index m in H, we see that $\pi_2(H_1)$ has index at most $m - 1$ in $\pi_2(H_0)$. Thus $\pi_2(H_1)$ is a subgroup of A_n or S_n of index at most $n - 1$, and hence must again be A_n or S_n by Lemma 2. Since A_n and S_n both act doubly transitively on Q_n, there is some $h \in H$ that sends (i,j) to (i,j') and (k,ℓ) to (k,ℓ') whenever ℓ and ℓ' are distinct. This proves that C_1 is indeed a strongly connected component.

Next, consider pairs $\{(i,j), (i,k)\}$ with j, k distinct. For given $\{(i',j'), (i',k')\}$ with j', k' distinct, there is some element $s \in H$ such that $\pi_1(s)(i) = i'$ and thus s sends (i,j) to (i',j'') and (i,k) to (i',k'') for some $j'', k'' \in Q_n$ with $j'' \neq k''$.

Now note that $\pi_2(\{x \in H \colon \pi_1(x)(i') = i'\})$ is either S_n or A_n by Lemma 3, and thus acts doubly transitively on Q_n. It follows that there is some $s' \in H$ such that s' sends (i', j'') to (i', j') and (i', k'') to (i', k'). Then $s's$ sends $\{(i, j), (i, k)\}$ to $\{(i', j'), (i', k')\}$ and thus C_2 is strongly connected.

Finally, consider pairs $\{(i, j), (k, j)\}$ and $\{(i', j'), (k', j')\}$ with i, k distinct and i', k' distinct. From the argument used in proving C_1 is strongly connected, we see that we can find $s \in H$ that sends $\{(i, j), (k, j)\}$ to $\{(i', j''), (k', j'')\}$ for some j''. As in the proof that C_1 is strongly connected, we see that the image of the set of $h \in H$ for which $\pi_1(h)$ stabilizes both i' and k' under π_2 acts transitively on Q_n; hence we can find $s' \in H$ that sends $\{(i', j''), (k', j'')\}$ to $\{(i', j'), (k', j')\}$. Thus C_3 is strongly connected. □

Corollary 1. *Let m and n be positive integers with $n \geq m \geq 2$, $n \geq 5$, and $(m, n) \neq (6, 6)$, and let $\mathcal{A} = (Q_m, \Sigma, \delta, 0)$ and $\mathcal{A}' = (Q_n, \Sigma, \delta', 0)$ be semiautomata with transition semigroups that are the symmetric groups of degrees m and n. Suppose that the direct product $\mathcal{A} \times \mathcal{A}'$ is connected and assume further that sets of final states are added to $\mathcal{A}$ and $\mathcal{A}'$ and that $\circ$ is a proper binary boolean function that defines the set of final states of the direct product $\mathcal{P}$. Then $\mathcal{P}$ is minimal for any such $\circ$.*

Proof. By Lemma 4, the pair graph of $\mathcal{A} \times \mathcal{A}'$ has three strongly connected components: $C_1 = \{\{(i, j), (k, \ell)\} \colon i \neq k, j \neq \ell\}$, $C_2 = \{\{(i, j), (i, \ell)\} \colon j \neq \ell\}$, and $C_3 = \{\{(i, j), (k, j)\} \colon i \neq k\}$.

For $(i, j) \in Q_m \times Q_n$, define $f((i, j))$ to be 1 if (i, j) is a final state, and 0, otherwise. We first claim that C_1 has a distinguishing pair, that is, there are pairs (i, j) and (k, ℓ) in $Q_m \times Q_n$ with $i \neq k$ and $j \neq \ell$ such that $f((i, j)) \neq f((k, \ell))$.

Suppose no distinguishing pair exists in C_1. Assume without loss of generality that $f((0, 0)) = 0$. then $f((i, j)) = 0$ whenever $i \neq 0$ and $j \neq 0$. Given $k \in Q_n$, we pick $\ell \in Q_n \setminus \{0, k\}$; this is always possible since $n \geq 3$. Since $\{(0, k), (1, \ell)\}$ is in C_1 and we have assumed that C_1 has no distinguishing pairs, we must have $f((0, k)) = f((1, \ell))$. But $f(1, \ell)$ must be 0, for otherwise we would have the distinguishing pair $\{(0, 0), (1, \ell)\}$. Hence $f((0, k)) = f((1, \ell)) = 0$. Thus we have $f((i, j)) = 0$ for every $i \in Q_m$ and every $j \in Q_n \setminus \{0\}$. Similarly, we must have $f((i, 0)) = f((0, 1)) = 0$ for $i \in Q_m \setminus \{0\}$, and hence f is the zero function, a contradiction.

The fact that C_2 and C_3 both have distinguishing pairs follows from the fact that $\circ$ is a proper boolean function. By Lemma 1, we conclude that $\mathcal{A} \times \mathcal{A}'$ is uniformly minimal. □

We have proved our main result in the case that $m \leq n$ and $n \geq 5$ if $(m, n) \neq (6, 6)$. By symmetry we may always assume that $m \leq n$. The case $(m, n) = (2, 2)$ was handled in Example 2, that of $(m, n) = (3, 4)$, in Example 3, and that of $(m, n) = (4, 4)$, in Example 4. So the only cases to consider are those with $(m, n) \in \{(2, 3), (2, 4), (3, 3), (6, 6)\}$; these cases are covered at http://arxiv.org/abs/1310.1841.

6   Conclusions

We have shown that if the inputs of two DFAs induce transformations that constitute non-conjugate bases of symmetric groups, then the quotient complexity of all non-trivial boolean operations on the languages accepted by the DFAs is maximal, except for a few special cases when the sizes of the DFAs are small. We believe that other similar results are possible and deserve further study.

Acknowledgments. This work was supported by the Natural Sciences and Engineering Research Council of Canada under grants No. 611456 and OGP0000871, by the European Regional Development Fund through the programme COMPETE, and by the Portuguese Government through the FCT under projects PEst-C/MAT/UI0144/2011 and CANTE-PTDC/EIA-CCO/101904/2008. We thank Gareth Davies for his careful proofreading and constructive comments.

References

1. Brzozowski, J.: Quotient complexity of regular languages. J. Autom. Lang. Comb. 15(1/2), 71–89 (2010)
2. Brzozowski, J.: In search of the most complex regular languages. Int. J. Found. Comput. Sc. 24(6), 691–708 (2013)
3. Brzozowski, J., Davies, G.: Maximally atomic languages. In: Ésik, Z., Fülop, Z. (eds.) 14th International Conference Automata and Formal Languages, AFL 2014, Szeged, Hungary, May 27-29. EPTCS, vol. 151, pp. 151–161 (2014)
4. Brzozowski, J., Davies, G.: Most complex regular right-ideal languages. In: 16th International Workshop on Descriptional Complexity of Formal Systems, DCFS 2014, Turku, Finland, August 5-8. LNCS 8614 (to appear, 2014)
5. Brzozowski, J., Tamm, H.: Complexity of atoms of regular languages. Int. J. Found. Comput. Sc. 24(7), 1009–1027 (2013)
6. Brzozowski, J., Tamm, H.: Theory of átomata. Theoret. Comput. Sci. (article in press, 2014)
7. Liu, G., Martin-Vide, C., Salomaa, A., Yu, S.: State complexity of basic language operations combined with reversal. Inform. and Comput. 206, 1178–1186 (2008)
8. Maslov, A.N.: Estimates of the number of states of finite automata. Dokl. Akad. Nauk SSSR 194, 1266–1268 (1970) (Russian); English Translation: Soviet Math. Dokl. 11, 1373–1375 (1970)
9. Piccard, S.: Sur les bases du groupe symétrique. Časopis Pro Pěstování Matematiky a Fysiky 68(1), 15–30 (1939)
10. Restivo, A., Vaglica, R.: A graph theoretic approach to automata minimality. Theoret. Comput. Sc. 429, 282–291 (2012)
11. Rotman, J.: The Theory of Groups: An Introduction. Allyn and Bacon, Inc., Boston (1965)
12. Salomaa, A., Wood, D., Yu, S.: On the state complexity of reversals of regular languages. Theoret. Comput. Sci. 320, 315–329 (2004)
13. Suzuki, M.: Group Theory, vol. 1. Springer, Berlin (1982)
14. Wilson, R.: The Finite Simple Groups. Springer, Berlin (2009)
15. Yu, S.: State complexity of regular languages. J. Autom. Lang. Comb. 6, 221–234 (2001)
16. Yu, S., Zhuang, Q., Salomaa, K.: The state complexities of some basic operations on regular languages. Theoret. Comput. Sci. 125(2), 315–328 (1994)

Handling Infinitely Branching WSTS*

Michael Blondin[1], Alain Finkel[2], and Pierre McKenzie[1]

[1] Université de Montréal and ENS Cachan
{blondimi,mckenzie}@iro.umontreal.ca
[2] ENS Cachan
alain.finkel@lsv.ens-cachan.fr

Abstract. Most decidability results concerning well-structured transition systems apply to the *finitely branching* variant. Yet some models (inserting automata, ω-Petri nets, ...) are naturally infinitely branching. Here we develop tools to handle infinitely branching WSTS by exploiting the crucial property that in the (ideal) completion of a well-quasi-ordered set, downward-closed sets are *finite* unions of ideals. Then, using these tools, we derive decidability results and we delineate the undecidability frontier in the case of the termination, the control-state maintainability and the coverability problems. Coverability and boundedness under new effectivity conditions are shown decidable.

1 Introduction

Well-structured transition systems (WSTS) [12,11,2] as a general class of infinite-state systems have spawned decidability results for important problems such as termination, boundedness, control-state maintainability and coverability. WSTS consist of a (usually infinite) well ordered set of states, together with a monotone transition relation. WSTS have found multiple *uses*: in settling the decidability status of reachability and coverability for graph transformation systems [4,22], in the forward analysis of depth-bounded processes [26,27], in the verification of parameterized protocols [10] and the verification of multi-threaded asynchronous software [21]. WSTS remain under development and are actively being investigated [13,14,18,25,5,24].

Most existing decidability results for WSTS apply to the *finitely branching* variant. However, WSTS such as inserting FIFO automata [7], inserting automata [6] and ω-Petri nets [17], that can arbitrarily increase some values, are intrinsically *infinitely branching*, and any finitely branching WSTS parameterized with an infinite set of initial states (such as broadcast protocols [10]) also inherits an infinitely branching state. For instance, Geeraerts, Heußner, Praveen and Raskin argue in [17] that parametric concurrent systems with dynamic

---

* Supported by the French Agence Nationale de la Recherche, REACHARD (grant ANR-11-BS02-001), by the Fonds québécois de la recherche sur la nature et les technologies, by the Natural Sciences and Engineering Research Council of Canada and by the "Chaire DIGITEO, ENS Cachan - École Polytechnique".

J. Esparza et al. (Eds.): ICALP 2014, Part II, LNCS 8573, pp. 13–25, 2014.

thread creation can naturally be modelled by some classes of infinitely branching systems, like ω-Petri nets, i.e. Petri net with arcs that can consume/create arbitrarily many tokens.

An outcome of our work is that the finite tree construction technique can be recovered, even in the infinitely branching case, for the purpose of deciding the boundedness problem for example.

The primary motivation for this paper is to explore the decidability status of the termination, boundedness, control-state maintainability and coverability problems for infinitely branching (general) WSTS. For the coverability problem, known to be decidable for WSTS fulfilling the so-called prebasis computability hypothesis [2], we wish to draw from the recent algebra-theoretic characterizations of downward-closed sets [13] and conceive of a post-oriented computability hypothesis suitable for the design of a forward algorithm. (Indeed, forward algorithms are arguably more intuitive than backward algorithms and post-oriented computability more easily verified than prebasis computability, where prebasis computability means computability of a finite basis of the upward closure of the set of immediate predecessors, the testing of which is provably undecidable in some WSTS.) Our contributions are the following:

1. As technical tools, we simplify and extend the analysis of the completion of a general WSTS and we relate the behavior of a WSTS to that of its completion. In particular, we provide a general presentation of the completion that is much less daunting than the presentations currently available in the literature. This sets the stage for exploiting the main property of the completion of a WSTS, namely, the expressibility of any downward-closed set as a (unique, as shown here) finite union of ideals, in the design of algorithms.

2. We uncover a new termination property (called *strong* termination) that is computationally equivalent to the usual termination property for finitely branching WSTS but that subtly differs from it in the presence of infinite branching. Indeed, we exhibit WSTS for which strong termination is decidable yet the usual termination is undecidable. A similar subtle issue arises as well in our generalization of the maintainability problem to infinitely branching.

3. We generalize most decidability results mentioned for finitely branching WSTS earlier to the infinitely branching case. This requires carefully tracking the effectiveness and the monotonicity conditions which support decidability. When possible, we delineate the frontier between decidability for a problem and the undecidabilty that results from dropping one of these conditions. The new decidability results for (strong) termination and (strong) maintainability exploit the completion. The new algorithm for coverability uses a forward strategy coupled with a post-oriented computability hypothesis.

Our work further highlights the naturalness of the class of ω^2-WSTS. Indeed our decidability results apply in one blow to known classes of infinitely branching WSTS like inserting FIFO automata [7], inserting automata [6], ω-Petri nets [17] and broadcast protocols [10].

Section 2 below introduces notation and preliminaries. Section 3 surveys known decidabilities and exhibits some undecidabilities. Section 4 develops our tools to handle infinite branching. Section 5 contains the bulk of our decidability results for infinitely branching WSTS. Section 6 summarizes our contribution and suggests future work.

2   WSTS

Let X be a set and $\le$ a quasi-ordering on X ($\le$ reflexive and transitive), then $\le$ is a *well-quasi-ordering* (wqo) if for every infinite sequence $x_0, x_1, \ldots$ of elements $x_n \in X$, there exist $i < j$ such that $x_i \le x_j$. It is well-known that $\mathbb{N}^d$ is well-quasi-ordered under $(x_1, \ldots, x_d) \le_{\mathbb{N}^d} (x'_1, \ldots, x'_d)$ where the latter means that $\forall i\ x_i \le x'_i$ (Dickson's Lemma). We extend $\mathbb{N}$ to $\mathbb{N}_\omega$ by adding an element ω verifying $\omega \ge_{\mathbb{N}_\omega} x$ for all $x \in \mathbb{N}_\omega$. The set $\mathbb{N}_\omega^d$ is also well-quasi-ordered. We simply write $\le$ for $\le_{\mathbb{N}}$ and $\le_{\mathbb{N}_\omega}$ when there is no ambiguity.

Recall that a *WSTS* is an ordered transition system $S = (X, \to_S, \le)$ such that $\le$ is a well-quasi-ordering on X, and the relation $\to_S \subseteq X \times X$ is monotone (or compatible) with $\le$ meaning that for all x, y, x' such that $x \to_S y$ and $x \le x'$, there exists a state y' such that $x' \xrightarrow{*}_S y'$ and $y \le y'$. WSTS thus satisfy a general monotony by definition. There exist other variations of monotony:

strong: $x \to_S y \wedge x' \ge x \implies x' \to_S y' \ge y,$

stuttering: $x \to_S y \wedge x' \ge x \implies x' = x'_0 \to_S \ldots \to_S x'_k \to_S y' \ge y, \forall i\ x'_i \ge x,$

transitive: $x \to_S y \wedge x' \ge x \implies x' \xrightarrow{+}_S y' \ge y,$

strict: $x \to_S y \wedge x' > x \implies x' \xrightarrow{*}_S y' > y.$

Strong monotony implies stuttering monotony which implies transitive monotony.

We denote, as usual, $\mathrm{Pre}_S(x) = \{y : y \to_S x\}$, $\mathrm{Post}_S(x) = \{y : x \to_S y\}$, $\mathrm{Pre}_S(T) = \bigcup_{x \in T} \mathrm{Pre}_S(x)$ and $\mathrm{Post}_S(T) = \bigcup_{x \in T} \mathrm{Post}_S(x)$.

Throughout this paper, WSTS will be assumed *effective* in the following sense: (1) the set of states X is r.e. (which suffices to compute $\mathrm{Post}_S(x)$ when $|\mathrm{Post}_S(x)|$ is known and finite); (2) the transition relation is decidable, i.e., the WSTS comes equipped with an algorithm that can decide, given $x, y \in X$, whether $x \to_S y$ or equivalently whether $y \in \mathrm{Post}_S(x)$; (3) the quasi-ordering $\le$ is decidable, i.e., the WSTS also comes equipped with an algorithm that can decide, given $x, y \in X$, whether $x \le y$. Forward analysis techniques for (finitely branching) WSTS typically compute the finite set $\mathrm{Post}_S(x)$, which is made possible by assuming Post_S computable. Because our new setting allows $\mathrm{Post}_S(x)$ to be infinite, we need to adapt this assumption. Our "post-effectivity" notion mildly weakens the usual hypothesis of "being able to compute Post_S":

Definition 2.1. *A transition system* $S = (X, \to_S)$ *is* post-effective *if S is effective and $f : X \to \mathbb{N} \cup \{$ "infinite"$\}$ given by $f(x) = |\mathrm{Post}_S(x)|$ is computable.*

Transition systems defined by a finite set of recursive functions are typical examples of finitely branching systems and they will be called *functional*. Let $\mathcal{F}_d$

denote the set of WSTS whose transitions relation is prescribed by finitely many increasing functions f from $\mathbb{N}^d$ to $\mathbb{N}^d$ (i.e. $x \leq y \implies f(x) \leq f(y)$) which are also recursive (i.e., given by halting Turing machines); these WSTS are finitely branching and post-effective. Inserting FIFO automata [7], inserting automata [6] and ω-Petri nets [17] are post-effective infinitely branching WSTS.

Recall that an effective ordered transition system is said *essentially finite branching* [2] if the subset $maxpost(x)$ of maximal elements of $\text{Post}_S(x)$ is non empty, finite and computable. Some WSTS, e.g. ω-Petri nets, are post-effective but are not essentially finite branching and conversely, we can exhibit essentially finite branching WSTS that are not post-effective.

Post-effectivity (Definition 2.1) is a weaker notion than "having a finite and computable Post_S". The weaker notion does imply "computable Post_S" for effective WSTS that are finitely branching. Hence it is natural to ask whether the finitely branching property is decidable for post-effective WSTS. It is not:

Proposition 2.2. *Testing, given a post-effective WSTS S and $x_0 \in X$, whether there exists an execution $x_0 \xrightarrow{*}_S x$ such that $\text{Post}_S(x)$ is infinite is undecidable.*

Let $\uparrow T$ and $\downarrow T$ stand respectively for the set of states that are $\geq$ and $\leq$ some state in T. A set T is *upward closed* if $T = \uparrow T$ and *downward closed* if $T = \downarrow T$. An *upward basis* of a set T is a set B such that $T = \uparrow B$. An *ideal* I is a downward closed set that is also *directed*, i.e., $\forall a, b \in I, \exists c \in I$ such that $a \leq c$ and $b \leq c$. We note $\text{Ideals}(X)$ the set of ideals of an ordered set X. A *directed complete partial ordering* (dcpo) is an ordered set $(X, \leq)$ such that every directed set $D \subseteq X$ has a least upper bound (lub) in X: for instance, $(\mathbb{N}, \leq)$, with the usual notations, is not a dcpo since the directed set $\mathbb{N}$ has no lub in $\mathbb{N}$; if we add the lub ω to $\mathbb{N}$, then $(\mathbb{N}_\omega, \leq)$ is a dcpo. There is a way to add all lubs to any ordered set $(X, \leq)$, that is called the *ideal completion*, since each element $x \in X$ can be identified with $\downarrow x \in \text{Ideals}(X)$ and since it is well-known that $(\text{Ideals}(X), \subseteq)$ is a dcpo [3,13]. We will consider the following problems for WSTS, where the input to each problem is an effective WSTS $S = (X, \to_S, \leq)$ and a state $x_0 \in X$, together with an $x \in X$ in the case of coverability, and a set $t_1, \ldots, t_n \in X$ in the case of the maintainability problem:

- Coverability: $\exists$ execution $x_0 \to_S x_1 \to_S \ldots \to_S x_k \geq x$?
- Boundedness: $\text{Post}_S^*(x_0)$ is infinite?
- Termination: $\nexists$ infinite execution $x_0 \to_S x_1 \to_S \ldots$?
- Strong termination: $\exists k \in \mathbb{N}$ s.t. $x_0 \to_S x_1 \to_S \ldots \to_S x_m \implies m \leq k$?
- Control-state maintainability: $\exists$ computation (i.e. an infinite execution $x_0 \to_S x_1 \to_S \ldots$ or a finite execution $x_0 \to_S x_1 \to_S \ldots \to_S x_k$ that cannot be further extended) such that $\forall i \ x_i \in \uparrow \{t_1, \ldots, t_n\}$?
- Strong control-state maintainability: $\forall k \in \mathbb{N}, \exists$ execution $x_0 \to_S x_1 \to_S \ldots \to_S x_m$ such that $m \geq k$ and $\forall i \ x_i \in \uparrow \{t_1, \ldots, t_n\}$?

3   Decidability for WSTS

Recall that a WSTS $S = (X, \to_S, \leq)$ has a *computable prebasis* [11,2] if the WSTS comes equipped with a computable function that maps each $x \in X$ to some finite basis of the upward closed set $\uparrow \mathrm{Pre}_S(\uparrow x)$. We summarize the four main decidability results known about (essentially) finite branching WSTS:

Theorem 3.1 ([12,11,2]).

- *Termination is decidable for post-effective finitely branching WSTS with transitive monotony [12], and for essentially finite branching effective WSTS with strong monotony [2].*
- *Boundedness is decidable for post-effective finitely branching WSTS with strict transitive monotony and well partial ordering [11].*
- *Control-state maintainability is decidable for post-effective finitely branching WSTS with stuttering monotony [11], and for essentially finite branching effective WSTS with strong monotony [2].*
- *Coverability is decidable for effective WSTS with prebasis computability [11,2].*

Theorem 3.1 states results exactly as they appear in the literature, but it would not be difficult to unify some of the hypotheses made here. For instance, termination can be shown decidable for essentially finite branching effective WSTS with transitive monotony. We defer a systematic treatment of this unification to a future version of the present paper.

Our goals in this paper are to extend the decidability of termination, boundedness and maintainability given by Theorem 3.1 to the more general case of *infinitely branching* WSTS. Our goal for the coverability problem is to investigate alternative effectivity hypotheses. We first note:

Theorem 3.2. *Termination is undecidable for post-effective WSTS with transitive (and even strong and strict) monotony.*

In Sect. 5, we prove boundedness decidable for *post*-effective infinitely branching WSTS with strict monotony and well partial ordering. By contrast, as exemplified by Petri nets with Reset [8], boundedness is well known to be undecidable for post-effective finitely branching WSTS with *non-strict* yet transitive (even strong) monotony and with well partial ordering. Concerning maintainability,

Theorem 3.3. *Control-state maintainability is undecidable for post-effective WSTS with stuttering (and even strong and strict) monotony.*

We now turn to coverability. Existing proofs that coverability is decidable need the prebasis hypothesis: Abdulla et al. use a backward algorithm [11,1] that computes a finite basis of $\uparrow \mathrm{Pre}^*(\uparrow x)$ and Geeraerts et al. use a forward algorithm [18] that requires further hypotheses (i.e. restriction to an adequate domain of limits, a *mathematical* hypothesis subsequently shown superfluous [16,13]). Note that coverability for post-effective (even finitely branching) WSTS becomes undecidable without the prebasis hypothesis, as is the case for instance for WSTS in $\mathcal{F}_2$ (recall definition from Sect 2, i.e., WSTS composed of recursive increasing functions from $\mathbb{N}^2$ to $\mathbb{N}^2$) [15].

Prebasis computability is *sufficient* to ensure decidability of coverability. However, as we show in Prop. 3.4 below, prebasis computability is not *necessary*: there is a class of WSTS, namely $\mathcal{F}_1$, for which coverability is decidable yet *no* prebasis function is computable.

Proposition 3.4. *Coverability for $\mathcal{F}_1$ is decidable, but no algorithm that takes as input $S \in \mathcal{F}_1$ and $x \in \mathbb{N}$ can systematically output a finite basis of $\uparrow Pre_S(\uparrow x)$.*

4 Handling Infinite Branching Finitely

In this section we prepare the ground for developping decision procedures capable of handling, under natural hypotheses, infinitely branching systems. First we would like the ability to compute finite representations of each term in the sequence $\downarrow x, \downarrow \mathrm{Post}_S(\downarrow x), \downarrow \mathrm{Post}_S(\downarrow \mathrm{Post}_S(\downarrow x)), \ldots$. This requires finitely representing downward closed sets, which is possible for wqo. This section describes how this is done and presents effective tools for doing it.

4.1 Downward Closed Sets and Ideals

It has long been known that in a wqo, any upward closed set has a finite basis; this is Dickson's lemma in $(\mathbb{N}^k, \leq)$ and it is Higman's lemma in $(\Sigma^*, \leq)$ when $\leq$ is the subword relation. It has recently been discovered that a similar situation occurs for downward closed sets in wqo.

Theorem 4.1. *[13] Any downward closed subset in a wqo X is a finite union of ideals.*

The original proof of Theorem 4.1 needs a technical bridge between topological completions and ordering completions of a set. A short and self-contained proof of Theorem 4.1 was given by Goubault-Larrecq [19].

Theorem 4.3 below slightly refines Theorem 4.1. It shows that any downward closed set uniquely decomposes as a certain finite union of ideals. This requires:

Proposition 4.2. *Any ideal contained in a finite union of ideals is contained in one of these ideals. In particular, testing the inclusion of an ideal I in a union $J_1 \cup J_2 \cup \ldots \cup J_k$ of ideals is equivalent to testing whether $I \subseteq J_j$ for some j such that $1 \leq j \leq k$.*

A finite union $D = \bigcup_{i=1}^{m} I_i$ of ideals will be said to *canonically* decompose D if the I_i's are pairwise incomparable under inclusion. This terminology is justified:

Theorem 4.3. *Any downward closed subset in a wqo X admits a unique decomposition as a finite union of pairwise incomparable ideals. Therefore, a downward closed subset decomposes canonically as the union of its maximal ideals.*

Ideals in a wqo cannot necessarily be manipulated effectively. For instance, there exist some ordered countable sets X such that $\mathrm{Ideals}(X)$ is not countable. Consider $X = \Sigma^*$, with the prefix ordering. Then $\mathrm{Ideals}(X)$ is isomorphic to $\Sigma^* \cup \Sigma^\omega$ and is not countable when Σ contains at least two letters. However:

Proposition 4.4. *A wqo X is countable iff $\mathrm{Ideals}(X)$ is countable.*

Fortunately, inclusion between ideals is decidable for well-quasi-ordered sets obtained by closing finite sets and closing naturals numbers under finite products, disjoint sums, multiset operator and Kleene star (respectively with their natural associated orderings) [13]. Therefore inclusion of ideals of $\mathbb{N}^d$ and inclusion of ideals of Σ^* are decidable.

4.2   Completion of WSTS

Recall that for a functional WSTS $S = (X, \overset{F}{\rightarrow}, \leq)$ where F is a finite set of increasing recursive functions $f : X \rightarrow X$, the *functional completion* [14] is defined by $\overline{S} = (\overline{X}, \overset{\overline{F}}{\rightarrow}, \subseteq)$ where $\overline{X} = \text{Ideals}(X)$ and $\overline{F}$ is the set of functions $\overline{f} : \text{Ideals}(X) \rightarrow \text{Ideals}(X)$ defined by $\overline{f}(I) \overset{def}{=} \downarrow f(I)$ for every $f \in F$. We note that $\overline{f}(I)$ is an ideal if I is an ideal. Here we extend the completion process to any (infinitely) branching WSTS:

Definition 4.5. *The* completion $\widehat{S}$ *of a WSTS* $S = (X, \rightarrow_S, \leq)$ *is the ordered transition system* $\widehat{S} = (\widehat{X}, \rightarrow_{\widehat{S}}, \subseteq)$ *where* $\widehat{X} = \text{Ideals}(X)$, *and* $I \rightarrow_{\widehat{S}} J$ *if* J *appears in the canonical decomposition of* $\downarrow \text{Post}(I)$.

Let $S = (X, \overset{F}{\rightarrow}, \leq)$ be a functional WSTS, then the following relation holds between $\overline{S}, \widehat{S}$ and S for every ideal $I \in \text{Ideals}(X)$:

$$\text{Post}_{\overline{S}}(I) = \bigcup_{\overline{f} \in \overline{F}} \overline{f}(I) = \bigcup_{f \in F} \downarrow f(I) = \bigcup_{J \in \text{Post}_{\widehat{S}}(I)} J = \downarrow \text{Post}_S(I).$$

Another good news is that:

Proposition 4.6. *The completion* $\widehat{S}$ *of any WSTS* S *is finitely branching.*

Moreover the completion computes exactly the downward closure of the reachability set of its original system.

Proposition 4.7. *Let* $S = (X, \rightarrow_S, \leq)$ *be a WSTS and* $\text{Post}_{\widehat{S}}^*(\downarrow x) = \{J_1, \ldots, J_n\}$. *We have* $\downarrow \text{Post}_S^*(x) = J_1 \cup \ldots \cup J_n$.

A natural question that arises is whether the completion of a WSTS is also a WSTS. It does indeed have monotony:

Proposition 4.8. *Let* $S = (X, \rightarrow_S, \leq)$ *then* $\widehat{S}$ *has strong monotony.*

However, $(\text{Ideals}(X), \subseteq)$ is not always a wqo and therefore the completion is not always a WSTS. In fact, it is known to be a wqo iff $(X, \leq)$ is a so-called ω^2-wqo, a notion we will not define here. In general, a wqo is not necessarily a ω^2-wqo and the typical counter-example is the Rado ordering [20]. Now, a result from Jancar [20] simplifies the characterization of ω^2-wqos as follows: a wqo $\leq$ is a ω^2-wqo iff $\leq^\#$ is a wqo, where $\leq^\#$ is the Hoare ordering defined by $A \leq^\# B$ iff $\uparrow B \subseteq \uparrow A$.

Extending the terminology to WSTS, we obtain the following result generalizing the known result for functional WSTS [14]:

Theorem 4.9. *Let* S *be a WSTS, then* $\widehat{S}$ *is a WSTS iff* S *is a* ω^2-*WSTS.*

We end this section with the observations that a WSTS inherits the strict monotony of its completion but not conversely, and that post-effectivity of a WSTS is independent from the post-effectivity of its completion.

Proposition 4.10. *Let* $(X, \to_S, \leq)$ *be a WSTS. If* $\widehat{S}$ *has strict monotony, then so does* S. *However, if* S *has strict monotony then* $\widehat{S}$ *doesn't necessarily have it.*

Proposition 4.11. *There exists a post-effective WSTS whose completion is not post-effective. Conversely, there exists a non post-effective WSTS whose completion is post-effective.*

4.3   Post-effectiveness of Completions in Concrete Examples

An *affine net* S is a WSTS in $\mathcal{F}_d$ in which the recursive functions are affine and a Petri net can be seen as an affine net where all matrices are the identity. An ω-*Petri net* [17] is an (extended) Petri net in which arcs can be labelled by positive integers *or by* ω. The completions of affine nets, ω-Petri nets and Lossy Channel Systems can be shown post-effective.

5   Decidability in Infinitely Branching Post-effective WSTS

5.1   (Strong) Termination

We are able to strengthen the hypotheses of Theorem 3.2 and to obtain: *termination is undecidable, even for post-effective* ω^2-*WSTS with strong and strict monotony, and with post-effective completion* by reducing from structural termination for Transfer Petri nets [9].

When a WSTS is infinitely branching, its termination problem differs in a subtle way from its strong termination problem. We show the latter decidable under suitable hypotheses:

Theorem 5.1. *Strong termination is decidable for* ω^2-*WSTS with transitive monotony and post-effective completion.*

Proving Theorem 5.1 requires comparing executions in a system with executions in its completion:

Proposition 5.2. *Let* $S = (X, \to_S, \leq)$ *be a WSTS, and* $I, J \in \widehat{X}$. *If* $I \xrightarrow{k}_{\widehat{S}} J$, *then for every* $x_J \in J$ *there exists* $x_I \in I$, $y \in \uparrow x_J$ *and* $k' \in \mathbb{N}$ *such that* $x_I \xrightarrow{k'}_S y$. *Moreover, if* S *has transitive monotony then* $k' \geq k$; *if* S *has strong monotony then* $k' = k$.

Proposition 5.3. *Let* $S = (X, \to_S, \leq)$ *be a WSTS and* $x, y \in X$. *If* $x \xrightarrow{k}_S y$, *then for every ideal* $I \supseteq \downarrow x$ *there exists an ideal* $J \supseteq \downarrow y$ *such that* $I \xrightarrow{k}_{\widehat{S}} J$.

Proof sketch of Theorem 5.1. Consider a ω^2-WSTS $S = (X, \to_S, \leq)$ such that $\widehat{S}$ is post-effective. Finkel and Schnoebelen [11, Theorem 4.6] show that termination, and thus strong termination, is decidable for post-effective WSTS having transitive monotony. By hypothesis, $\widehat{S}$ is a WSTS and $\widehat{S}$ has strong (and transitive) monotony by Prop. 4.8. Therefore, strong termination for $\widehat{S}$ is decidable. From Prop. 5.2 and Prop. 5.3, no bound on the length of executions from x_0 exists in S iff no bound on the length of executions from $\downarrow x_0$ exists in $\widehat{S}$. Hence decidability of strong termination from x_0 in S follows from being able to decide strong termination from $\downarrow x_0$ in $\widehat{S}$. Note that we have implicitly assumed that a representation of $\downarrow x_0$ can be effectively computed.   □

5.2   Boundedness

Drawing from [8], we know that *boundedness is undecidable, even for finitely branching post-effective ω^2-WSTS with strong (but not strict) monotony and post-effective completion.* Petri net with reset arcs are such a class.

It is known that for finitely branching post-effective WSTS with strict transitive monotony and a well partial ordering (wpo), the boundedness problem is decidable [11]. We generalize this result to (possibly) infinitely branching WSTS and we note that the hypothesis of transitive monotony was not necessary in the proof of [11]. The proof follows [11] by building a finite reachability tree, with the extra step of testing whether $\mathrm{Post}_S(x)$ is infinite for each new node.

Theorem 5.4. *Boundedness is decidable for post-effective WSTS with strict monotony and with well partial ordering.*

5.3   (Strong) Control-State Maintainability

By a reduction from the termination problem, the hypotheses of Theorem 3.3 can be strengthened: *control-state maintainability is undecidable, even for post-effective ω^2-WSTS with strong and strict monotony, and with post-effective completion.* By contrast, the *strong* variant of the problem introduced in this paper is decidable, under suitable hypotheses, for infinitely branching WSTS:

Theorem 5.5. *Strong control-state maintainability is decidable for ω^2-WSTS with strong monotony and a post-effective completion.*

Before proving Theorem 5.5, we need Prop. 5.6 and Prop. 5.7 to relate covering executions in a WSTS to covering executions in its completion.

Proposition 5.6. *Let $S = (X, \rightarrow_S, \leq)$ be a WSTS with strong monotony and $\{t_1, \ldots, t_n\} \subseteq X$. Let $I_0 \rightarrow_{\widehat{S}} I_1 \rightarrow_{\widehat{S}} \ldots \rightarrow_{\widehat{S}} I_k$ be an execution such that for all $0 \leq j \leq k$ we have $I_j \in \uparrow_{\widehat{X}} \{\downarrow t_1, \ldots, \downarrow t_n\}$. Then for every $y \in I_k$ there exists an execution $x_0 \rightarrow_S x_1 \rightarrow_S \ldots \rightarrow_S x_k$ such that $x_0 \in I_0$, $x_k \in \uparrow y$ and for all $0 \leq j \leq k$ we have $x_j \in \uparrow \{t_1, \ldots, t_n\}$.*

Proof. Let I_0 be an execution of length 0 in $\widehat{S}$ as described in the proposition, and let $y \in I_0$. By hypothesis, there exists t_i such that $\downarrow t_i \subseteq I_0$ and thus $t_i \in I_0$. Since I_0 is an ideal, there exists $x_0 \in I_0$ such that $x_0 \geq y$ and $x_0 \geq t_i$. Therefore the execution x_0 of length 0 in S meets all requirements.

Let $I_0 \to_{\widehat{S}} I_1 \to_{\widehat{S}} \dots \to_{\widehat{S}} I_k$ be an execution of length $k > 0$ in $\widehat{S}$ as described in the proposition. By induction, for every $y \in I_k$ there exists an execution $x_1 \to_S x_2 \to_S \dots \to_S x_k$ such that $x_1 \in I_1$, $x_k \in \uparrow y$ and for all $1 \le j \le k$ we have $x_j \in \uparrow \{t_1, \dots, t_n\}$.

Since $x_1 \in I_1 \subseteq \downarrow \mathrm{Post}_S(I_0)$, there exists $x_0 \in I_0$ and $y' \in \uparrow x_1$ such that $x_0 \to_S y'$. By hypothesis, there exists t_i such that $\downarrow t_i \subseteq I_0$ and thus $t_i \in I_0$. Since I_0 is an ideal, there exists $x_0' \in I_0$ such that $x_0' \ge x_0$ and $x_0 \ge t_i$. By strong monotony, there exists $x_1' \ge y'$ such that $x_0' \to_S x_1'$.

Moreover, applying strong monotony to $x_1 \to_S x_2 \to_S \dots \to_S x_k$ with $x_1' \ge x_1$, we obtain an execution $x_1' \to_S x_2' \to_S \dots \to_S x_k'$ such that for all $1 \le j \le k$ we have $x_j' \ge x_j$. Therefore, $x_0' \to_S x_1' \to_S \dots \to_S x_k'$, $x_0' \in I_0$, $x_k' \in \uparrow y$ and for all $0 \le j \le k$ we have $x_j' \in \uparrow \{t_1, \dots, t_n\}$. $\qquad\square$

Proposition 5.7. *Let $S = (X, \to_S, \le)$ be a WSTS and $\{t_1, \dots, t_n\} \subseteq X$. Let $x_0 \to_S x_1 \to_S \dots \to_S x_k$ be an execution such that for all $0 \le j \le k$ we have $x_i \in \uparrow \{t_1, \dots, t_n\}$. Then for every ideal $I_0 \supseteq \downarrow x_0$ there exists an execution $I_0 \to_{\widehat{S}} I_1 \to_{\widehat{S}} \dots \to_{\widehat{S}} I_k$ such that $I_k \supseteq \downarrow x_k$ and for all $0 \le j \le k$ we have $I_j \in \uparrow_{\widehat{X}} \{\downarrow t_1, \dots, \downarrow t_n\}$.*

Proof of Theorem 5.5. By Prop. 5.6 and Prop. 5.7 there exists an execution $x_0 \to_S x_1 \to_S \dots \to_S x_k$ such that for all $0 \le j \le k$ we have $x_j \in \uparrow \{t_1, \dots, t_n\}$ iff there exists an execution $I_0 \to_{\widehat{S}} I_1 \to_{\widehat{S}} \dots \to_{\widehat{S}} I_k$ such that for all $0 \le i \le k$ we have $I_j \in \uparrow_{\widehat{X}} \{\downarrow t_1, \dots, \downarrow t_n\}$. Therefore, it suffices to solve the problem in $\widehat{S}$ with $\downarrow x_0$ and $\{\downarrow t_1, \dots, \downarrow t_n\}$.

The algorithm from [11] solving the control-state maintainability problem for finitely branching post-effective WSTS with stuttering monotony can easily be adapted to solve strong control-state maintainability for finitely branching WSTS. Since $\widehat{S}$ is a post-effective WSTS by hypothesis and has strong (and stuttering) monotony by Prop. 4.8, we obtain an algorithm.

More specifically, it suffices to build the finite reachability tree of $\widehat{S}$ and verify that it contains a maximal path labelled $I_0, I_1, \dots, I_k$ with $I_j \in \uparrow_{\widehat{X}} \{\downarrow t_1, \dots, \downarrow t_n\}$ for every $0 \le j \le k$ and $I_j \subseteq I_k$ for some $0 \le j < k$. $\qquad\square$

5.4   Coverability

Some classes of WSTS admit both post-effective completions and prebasis computability, e.g., WSTS from $\mathcal{F}_d$ where the recursive increasing functions have computable limits (called ω-well-structured nets in [15]). Therefore, coverability was already known to be decidable for these classes. However, the following Theorem 5.8 yields an algorithm that relies on evaluating Post_S on ideals rather than Pre_S on upward closed sets. Often this is more efficient, e.g., it is easier to evaluate affine functions in $\mathbb{N}_\omega^d$ than inverting them.

Theorem 5.8. *Coverability is decidable for WSTS having a post-effective completion.*

Proof. Let $S = (X, \to_S, \leq)$ be a post-effective WSTS and $x_0 \in X$.

Coverability is semi-decidable by iteratively building larger portions of the reachability tree looking for a path with some state $x' \geq x$.

We note that x is coverable from x_0 in S iff there exists an ideal $I \supseteq \downarrow x$ reachable from$\downarrow x_0$ in $\widehat{S}$. To prove that non-coverability is semi-decidable, one enumerates all the downward closed sets D_i (as finite unions of ideals) that are inductive invariants, i.e., such that $x_0 \in D_i$ and $\downarrow \mathrm{Post}_S(D_i) \subseteq D_i$. If x is not coverable, a downward closed set D_i such that $x \notin D_i$ will inevitably be found.

The inclusion $\downarrow \mathrm{Post}_S(D_i) \subseteq D_i$ is decidable for WSTS whose completion is post-effective since there is an algorithm, which runs $\mathrm{Post}_{\widehat{S}}$ on D_i (expressed as the union $J_1 \cup \ldots \cup J_m$ of ideals) to obtain ideals $I_1, \ldots, I_n$ such that $\downarrow \mathrm{Post}_S(D_i) = \cup_{1 \leq i \leq m} \cup_{I \in \mathrm{Post}_{\widehat{S}}(J_i)} \downarrow I = I_1 \cup \ldots \cup I_n$. Now Prop. 4.2 says that this inclusion $I_1 \cup I_2 \cup \ldots \cup I_n \subseteq J_1 \cup J_2 \cup \ldots \cup J_m$ is decidable. $\square$

The technique of enumerating inductive invariants, used in our coverability algorithm, was already used by Pachl in 1982 to provide a witness of non-reachability for finite automata communicating through fifo channels, having recognizable reachability sets (Corollary 9.6 in [23]). More recently, Raskin et al. [18,16] also used enumeration of inductive invariants to provide forward algorithms for deciding coverability of WSTS. Note that their forward algorithms use the prebasis hypothesis while we appeal to post-effective completion.

6   Conclusion and Further Work

Here we have continued the development of tools to manipulate completions of wqos and we have applied these tools together with new ideas to deduce the following decidabilities: strong termination for ω^2-WSTS with transitive monotony and post-effective completion, boundedness for post-effective WSTS with strict transitive monotony and with well partial ordering, strong control-state maintainability for ω^2-WSTS with strong monotony and a post-effective completion and finally, coverability for WSTS having a post-effective completion.

Future work should apply these decidabilities to parameterized WSTS and should investigate algorithmic aspects of these decidabilities, including a comparison of the relative efficiencies of backward and forward strategies.

Acknowledgements. We thank the referees for helpful comments and pointers.

References

1. Abdulla, P.A., Cerans, K., Jonsson, B., Tsay, Y.K.: General decidability theorems for infinite-state systems. In: LICS, pp. 313–321 (1996)
2. Abdulla, P.A., Cerans, K., Jonsson, B., Tsay, Y.K.: Algorithmic analysis of programs with well quasi-ordered domains. Inf. Comput. 160(1-2), 109–127 (2000)
3. Abramsky, S., Jung, A.: Domain theory. In: Handbook of Logic in Comp. Sci., vol. 3, pp. 1–168. Oxford University Press (1994)

4. Bertrand, N., Delzanno, G., König, B., Sangnier, A., Stückrath, J.: On the decidability status of reachability and coverability in graph transformation systems. In: RTA, pp. 101–116 (2012)
5. Bertrand, N., Schnoebelen, P.: Computable fixpoints in well-structured symbolic model checking. Formal Methods in System Design 43(2), 233–267 (2013)
6. Bouyer, P., Markey, N., Ouaknine, J., Schnoebelen, P., Worrell, J.: On termination and invariance for faulty channel systems. FAC 24(4-6), 595–607 (2012)
7. Cécé, G., Finkel, A., Iyer, S.P.: Unreliable channels are easier to verify than perfect channels. Inf. Comput. 124(1), 20–31 (1996)
8. Dufourd, C., Finkel, A., Schnoebelen, P.: Reset nets between decidability and undecidability. In: Larsen, K.G., Skyum, S., Winskel, G. (eds.) ICALP 1998. LNCS, vol. 1443, pp. 103–115. Springer, Heidelberg (1998)
9. Dufourd, C., Jančar, P., Schnoebelen, P.: Boundedness of reset P/T nets. In: Wiedermann, J., Van Emde Boas, P., Nielsen, M. (eds.) ICALP 1999. LNCS, vol. 1644, pp. 301–310. Springer, Heidelberg (1999)
10. Esparza, J., Finkel, A., Mayr, R.: On the verification of broadcast protocols. In: LICS, pp. 352–359 (1999)
11. Finkel, A., Schnoebelen, P.: Well-structured transition systems everywhere! Theoret. Comput. Sci. 256(1–2), 63–92 (2001)
12. Finkel, A.: Reduction and covering of infinite reachability trees. Information and Computation 89(2), 144–179 (1990)
13. Finkel, A., Goubault-Larrecq, J.: Forward analysis for WSTS, part I: Completions. In: STACS, pp. 433–444 (2009)
14. Finkel, A., Goubault-Larrecq, J.: Forward analysis for WSTS, Part II: Complete WSTS. In: Albers, S., Marchetti-Spaccamela, A., Matias, Y., Nikoletseas, S., Thomas, W. (eds.) ICALP 2009, Part II. LNCS, vol. 5556, pp. 188–199. Springer, Heidelberg (2009)
15. Finkel, A., McKenzie, P., Picaronny, C.: A well-structured framework for analysing Petri net extensions. Information and Computation 195(1-2), 1–29 (2004)
16. Ganty, P., Raskin, J.-F., Van Begin, L.: A complete abstract interpretation framework for coverability properties of WSTS. In: Emerson, E.A., Namjoshi, K.S. (eds.) VMCAI 2006. LNCS, vol. 3855, pp. 49–64. Springer, Heidelberg (2006)
17. Geeraerts, G., Heußner, A., Praveen, M., Raskin, J.F.: ω-Petri nets. In: Petri Nets, pp. 49–69 (2013)
18. Geeraerts, G., Raskin, J.F., Begin, L.V.: Expand, enlarge and check: New algorithms for the coverability problem of WSTS. JCSS 72(1), 180–203 (2006)
19. Goubault-Larrecq, J., Schnoebelen, P.: Personal communication (October 2013)
20. Jancar, P.: A note on well quasi-orderings for powersets. Inf. Process. Lett. 72(5-6), 155–160 (1999)
21. Kaiser, A., Kroening, D., Wahl, T.: Efficient coverability analysis by proof minimization. In: Koutny, M., Ulidowski, I. (eds.) CONCUR 2012. LNCS, vol. 7454, pp. 500–515. Springer, Heidelberg (2012)
22. König, B., Stückrath, J.: Well-structured graph transformation systems with negative application conditions. In: Ehrig, H., Engels, G., Kreowski, H.-J., Rozenberg, G. (eds.) ICGT 2012. LNCS, vol. 7562, pp. 81–95. Springer, Heidelberg (2012)
23. Pachl, J.K.: Reachability problems for communicating finite state machines. Technical Report CS-82-12, University of Waterloo (1982)
24. Schmitz, S., Schnoebelen, P.: Multiply-recursive upper bounds with Higman's lemma. In: Aceto, L., Henzinger, M., Sgall, J. (eds.) ICALP 2011, Part II. LNCS, vol. 6756, pp. 441–452. Springer, Heidelberg (2011)

25. Schmitz, S., Schnoebelen, P.: The power of well-structured systems. In: D'Argenio, P.R., Melgratti, H. (eds.) CONCUR 2013 – Concurrency Theory. LNCS, vol. 8052, pp. 5–24. Springer, Heidelberg (2013)
26. Wies, T., Zufferey, D., Henzinger, T.A.: Forward analysis of depth-bounded processes. In: Ong, L. (ed.) FOSSACS 2010. LNCS, vol. 6014, pp. 94–108. Springer, Heidelberg (2010)
27. Zufferey, D., Wies, T., Henzinger, T.A.: Ideal abstractions for well-structured transition systems. In: Kuncak, V., Rybalchenko, A. (eds.) VMCAI 2012. LNCS, vol. 7148, pp. 445–460. Springer, Heidelberg (2012)

Transducers with Origin Information

Mikołaj Bojańczyk*

University of Warsaw

Abstract. Call a string-to-string function *regular* if it can be realised by one of the following equivalent models: MSO transductions, two-way deterministic automata with output, and streaming transducers with registers. This paper proposes to treat origin information as part of the semantics of a regular string-to-string function. With such semantics, the model admits a machine-independent characterisation, Angluin-style learning in polynomial time, as well as effective characterisations of natural subclasses such as one-way transducers or first-order definable transducers.

This paper is about string-to-string functions which can be described by deterministic two-way automata with output [AU70]. As shown in [EH01], this model is equivalent to MSO definable string transductions. Another equivalent model, used in [AC10], is a deterministic one-way automaton with registers that store parts of the output[1]. Examples of such functions include: duplication $w \mapsto ww$; reversing $w \mapsto w^R$; a function $w \mapsto ww^R$ which maps an input to a palindrome whose first half is w; and a function which duplicates inputs of even length and reverses inputs of odd length. As witnessed by the multiple equivalent definitions, this class of string-to-string function is robust, and therefore, following [AC10], we call it the class of *regular string-to-string functions*. Regular string-to-string functions have good closure properties. For instance, if f and g are regular, then the composition $w \mapsto f(g(w))$ is also regular, which is straightforward if the MSO definition is used, but nontrivial if the two-way automata definition is used [CJ77]. Also the concatenation $w \mapsto f(w) \cdot g(w)$ is regular, which is apparent in any of the three definitions. Equivalence of regular string-to-string functions is decidable, as was shown in [Gur82] using the two-way automata definition.

Origins. The motivation of this paper is the simple observation that the models discussed above, namely deterministic two-way automata with output, MSO definable string transductions, and automata with registers, provide more than just a function from strings to strings. In each case, one can also reconstruct *origin information*, which says how positions of the output string originate from positions in the input string. How do we reconstruct the origin of a position x in

---

* Supported by ERC Starting Grant "Sosna".
[1] Registers are similar to attributes in attribute grammars. The equivalence of MSO definable transductions with a form of attribute grammars, in the tree-to-tree case, was shown in [BE00]. In the special case of string-to-string functions, the attribute grammars from [BE00] correspond to left-to-right deterministic automata with registers and regular lookahead.

an output string? In the case of a deterministic two-way automaton, this is the position of the head when x was output. In the case of an MSO definable transducer, this is the position in which x is interpreted. In the case of an automaton with registers, this is the position in the input when the letter x was first loaded into a register. In other words, for a transducer we can consider two semantics: the *standard semantics*, where the output is a string, and the *origin semantics*, where the output is a string with origin information. The second semantics is finer in the sense that some transducers might be equivalent under the standard semantics, but not under the origin semantics.

Tracking origin information for transducers has been studied before, for instance in the programming language community, see e.g. [vDKT93]. Origin information has also been used as a technical tool in the study of tree-to-tree transducers. Examples include [EM03], where origin information is used to characterise those macro tree transducers which are MSO definable, and [LMN10], where origin information is used to get a Myhill-Nerode characterisation of deterministic top-down tree transducers. The novelty of this paper is that origin information is built into the semantics of a transducer.

Origin semantics. To illustrate the difference between the two semantics (standard and origin) of a string-to-string transducer, consider a transducer which is the identity on the string ab, and which maps other strings to the empty string. If we care about origins, then this description is incomplete, and can be instantiated in four different ways depicted below.

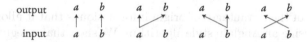

For example, the second diagram above describes a two-way automaton that first reads it input to determine if it is ab, and then moves its head to the first position, where it outputs both a and b.

Another example is the identity function on strings over a one letter alphabet, which can be realised by copying the input left-to-right or right-to-left. Actually the function can be realised in infinitely many different ways once origins are taken into account: consider an automaton that outputs n letters in input positions divisible by n, and then outputs the remainder under division by n in the last input position.

This paper is a study of the more refined semantics. Almost any "natural" construction for transducers will respect origin information. For instance, the translation from [EH01] which converts an MSO interpretation into a deterministic two-way automaton remains correct when the origin information is taken into account. The same holds for the other translations between the three models. In other words, one can also talk about *regular string-to-string functions with origin information*. Various closure properties, such as composition and concatenation, are retained when origins are taken into account. Some results become easier to prove, e.g. decidability of equivalence of string-to-string transducers.

A machine independent characterisation. The main contribution of this paper is a machine independent characterisation of regular string-to-string functions with origin information, which is given in Theorem 1. The characterisation is similar to the Myhill-Nerode theorem, which says that a language L is regular if and only if it has finitely many left derivatives of the form

$$w^{-1}L \stackrel{\text{def}}{=} \{v : wv \in L\}.$$

From the usual Myhill-Nerode theorem for regular languages one obtains a canonical device, which is the minimal deterministic automaton. The situation is similar here. We define a notion of left and right derivatives for string-to-string functions with origin information, and show that a function is regular if and only if it has finitely many left and right derivatives (finitely many left derivatives is not enough, same for right derivatives). The proof of the theorem yields a canonical device, which is obtained from the function itself and not its representation as a two-way automaton, MSO transduction, or machine with registers. One use for the canonical device is testing equivalence: two devices are equivalent if and only if they yield the same canonical machine.

Another use of the canonical device is that it is easy to see when the underlying function actually belongs to a restricted class, e.g. if it can be defined by a deterministic one-way automaton with output (see Theorem 4), or by functional nondeterministic one-way automaton with output (see Theorem 3). A more advanced application is given in Theorem 5, which characterises the first-order fragment of MSO definable transducers with origin information.

Learning. One of the advantages of origin information is that it allows functions to be learned, using an Angluin style algorithm. We show that a regular string-to-string function with origin information can be learned with a number of queries that is polynomial in the size of the canonical device. The queries are of two types: the learner can ask for the output on a given input string; or the learner can propose a transducer with origin information, and in case this is not the correct one, then the teacher gives a counterexample string where the proposed transducer produces a wrong output.

In the algorithm, the learner uses the origin information. However, it seems that the learner's advantage from the origin information does not come at any significant cost to the teacher. Suppose that we want to learn a transducer inside a text editor, e.g. the user wants to teach the text editor that she is thinking of the transducer which replaces every = by :=. If a user is trying to show an example of this transducer on some input, then she will probably place the cursor on occurrences of = in the input, delete them, and retype :=, thus giving origin information to the algorithm. A user who backspaces the whole input and retypes a new version will possibly be thinking of some different transformation. It would be wasteful to ignore this additional information supplied by the user.

Thank you. I would like to thank Sebastian Maneth and the anonymous referees for their valuable feedback; Anca Muscholl, Szymon Toruńczyk and Igor

Walukiewicz for discussions about the model; and Rajeev Alur for asking the question about a machine-independent characterisation of transducers.

1   Regular String to String Transducers

A *string-to-string function* is any function from strings over some fixed input alphabet to strings over some fixed output alphabet, such that the empty string is mapped to the empty string. A *string-to-string function with origin information* is defined in the same way, but for every input string w it provides not only an output string $f(w)$, but also origin information, which is a function from positions in $f(w)$ to positions in w. We consider total functions, although the results can easily be adapted to partial functions. In this section we recall three equivalent models recognising string-to-string functions.

Streaming transducer. Following [AC10], a *streaming transducer* is defined as follows. It has finite *input* and *output alphabets*. There is a finite set of *control states* with a distinguished *initial state*, and a finite set of *registers*, with a distinguished *output register*. The transition function inputs a control state and an input letter, and outputs a new control state and a *register update*, which is a sequence of register operations of two possible types:

- *Concatenate.* Replace the contents of register r with rs, and replace the contents of register s by the empty string;
- *Create.* Replace the contents of register r with output letter b.

Finally, there is an *end of input function*, which maps each state to a sequence of register operations of the first type[2].

When given an input string, the transducer works as follows. It begins in the initial state with all registers containing the empty string. Then it processes each input letter from left to right, updating the control state and the registers according to the transition function. Once the whole input has been processed, the end of input function is applied to the last state, yielding another sequence of register operations, and finally the value of the transducer is extracted from the output register. For the origin semantics, we observe that every letter in a register is created once using an operation of type *create*, and then moved around using operations of type *concatenate*. The origin of an output letter is defined to be the input position which triggered the transition whose register update contained the appropriate *create* operation.

Observe that the register operations do not allow copying registers. This is an important restriction which guarantees, among other things, that the size of the output is linear in the size of the input.

Example 1. By composing the atomic register operations and using additional registers, we can recover additional register operations such as "add letter b to

---

[2] The end of input function is prohibited to produce new output letters so that the origin information can be assigned. Alternatively, one could assume that the positions produced by the end of input function have a special origin, "created out of nothing".

the end of register r", "add letter b to the beginning of register r", "move register r to register s, leaving r empty". The examples use the additional operations.

Consider the function $w \mapsto ww^R$, where w^R is the reverse of w. The transducer has one control state and two registers, used to store w and w^R. When it reads an input letter a, the transducer adds a to the end of the register storing w and adds a to the beginning of the register storing w^R. The end of input update concatenates both registers, and puts the result in the first register, which is the output register.

A transducer for the duplication function is obtained in a similar way. Observe that since the register operations do not allow copying, it is still necessary to have two registers, both storing w.

Deterministic two-way automaton with output. A deterministic two-way automaton with output is like a deterministic two-way finite automaton, except that every transition is additionally labelled by a string (possibly empty) over the output alphabet. A run over an input w can be seen as a sequence of pairs $(\delta_1, x_1), \ldots, (\delta_n, x_n)$ where δ_i is a transition and x_i a position in the string $\vdash w \dashv$. The transition δ_i reads the label of position x_i and the state generated by the previous transition, and chooses the new position x_{i+1}, a new state, and what will be appended to the output. The output of the automaton is the concatenation of the strings labelling the transitions $\delta_1, \ldots, \delta_n$. The origin of a position in the output string that is generated by the transition δ_i is defined to be the position x_i. To make the origin well-defined, we require that every output letter is produced for transitions that have their source in input letters, and not over the markers $\vdash$ and $\dashv$.

MSO *transduction.* Following [Tho97], a string over an alphabet A can be treated as a relational structure, whose universe is the positions of the string, and which has a binary position order predicate $x < y$ and label predicates $a(x)$ for the letters of the alphabet. To transform strings into strings, we can use MSO interpretations. An MSO *interpretation* is a function from structures over some fixed input vocabulary (set of relation names with their arities) to structures over some fixed output vocabulary, which is specified by a system of MSO formulas, as follows. There is a universe formula with one free variable over the input vocabulary, which selects the elements of the universe of the input structure that will appear in the universe of the output structure. Furthermore, for every predicate of the output vocabulary there is a formula over the input vocabulary of the same arity, which says how the predicates are defined in the output structure.

Another function from structures to structures is called k-*copying*; which maps a structure to k disjoint copies of itself, together with binary relations $1(x, y), \ldots, k(x, y)$ such that $i(x, y)$ holds if y is the i-th copy of x. A *copying* MSO *transduction* consists of first a copying function, followed by an MSO interpretation. A string-to-string function f is called MSO-*definable* if there is some copying MSO transduction such that for every input string w, the transduction transforms the relational structure corresponding to w into a relational structure corresponding to $f(w)$. The origin information in such a transducer is defined in the natural way: a position in the output string is interpreted in some copy of a position in the input string, the latter is defined to be the origin.

Equivalence of the models. Deterministic two-way automata with output define the same translations as copying MSO transductions in [EH01]. The same proof works if the semantics with origin information is used. Streaming transducers are shown to be equivalent to the previous two models in [AC10]; the same proof also works with the origin semantics. A string-to-string function with origin information is called a *regular string-to-string function with origin information* if it can be defined by any one of the three models mentioned above.

2   A Machine Independent Characterisation

In this section we present a Myhill-Nerode style characterisation of regular transducers with origin information.

Factorised output. Suppose that f is a string-to-string function with origin information and output alphabet B. A *factorised input* is a tuple of strings $w_1, \ldots, w_n$ over the input alphabet, which is meant to describe an input string factorised into n blocks. Given such a factorised input, define an *output block of type i* to be a maximal connected subset of positions in the output $f(w_1 \cdots w_n)$ that originates in w_i. Define the *factorised output* corresponding to a factorised input $w_1, \ldots, w_n$, denoted by

$$f(w_1| \ldots |w_n) \in \big(\{1, \ldots, n\} \times B^+\big)^*.$$

to be the sequence of output blocks read from left to right, with each block described by its type and corresponding part of the output. In particular, if we concatenate all of the strings coming from B^+, we obtain the output string $f(w_1 \cdots w_n)$. When $n = 3$, instead of numbers 1, 2, 3 we use "left", "middle" and "right" to indicate types of blocks. We use fraction-style notation for output blocks, with the lower part indicating the type, and the upper part describing the output. For instance, if f is the duplicating function, then

$$f(ab|cd|e) = \text{left}^{ab}\ \text{middle}^{cd}\ \text{right}^{e}\ \text{left}^{ab}\ \text{middle}^{cd}\ \text{right}^{e}.$$

Some input blocks might be empty, as in the following example:

$$f(ab||e) = \text{left}^{ab}\ \text{right}^{e}\ \text{left}^{ab}\ \text{right}^{e}.$$

If some of the input blocks are underlined, then in the output we just keep the information that there is a nonempty output block, but we do not store the actual output strings which originate in the underlined blocks. For example,

$$f(\underline{ab}|cd|\underline{e}) = \text{left}\ \text{middle}^{cd}\ \text{right}\ \text{left}\ \text{middle}^{cd}\ \text{right}.$$

Note that we will never have two consecutive blocks of the same type, e.g. left left, in the factorised output, since blocks are maximal. In particular, for underlined input blocks we lose track of how long their corresponding output blocks are.

Derivatives. Define a *two-sided derivative* of string-to-string function with origin information f to be any function of the form

$$f_{u\_w} \overset{\text{def}}{=} v \mapsto f(\underline{u}|v|\underline{w}),$$

for some choice of strings u and w over the input alphabet. *Left derivatives* and *right derivatives* are the special cases of the two-sided derivative when either u or w is empty, i.e. they are functions of the forms:

$$f_{u\_} \overset{\text{def}}{=} v \mapsto f(\underline{u}|v) \qquad f_{\_w} \overset{\text{def}}{=} v \mapsto f(v|\underline{w}).$$

Example 2. Let f be the function $w \mapsto w^R w$. Then

$$f_{u\_w}(v) = \text{right middle } \overset{v^R}{\text{left}} \text{ middle } \overset{v}{\text{right}}$$

for every nonempty strings u or w. When the string u is empty, then the left block disappears, likewise when w is empty then the right blocks disappear. In particular, this function has four possible values for the two-sided derivative. There are two possible values for the left derivative $f_{u\_}$, namely the functions

$$v \mapsto \overset{v^R v}{\text{right}} \qquad\qquad v \mapsto \text{right } \overset{v^R}{\text{left}} \overset{v}{\text{right}}.$$

Example 3. Let f be the function which is the identity on strings of even length, and which maps strings of odd length to the empty string. This function has three possible left derivatives $f_{v\_}$, depending on whether v is empty, nonempty and even length, or odd length. Below is the derivative for the last case.

$$w \mapsto \begin{cases} \overset{w}{\text{left right}} & \text{if } w \text{ has odd length} \\ \epsilon & \text{otherwise} \end{cases}$$

Example 4. Here is a function with finitely many right derivatives, but infinitely many left derivatives. Consider first the function which scans its input from left to right, and outputs only those letters whose position is a prime number

$$f(a_1 \cdots a_n) = w_1 \cdots w_n \qquad \text{where } w_i = \begin{cases} a_i & \text{if } i \text{ is a prime number} \\ \epsilon & \text{otherwise.} \end{cases}$$

This particular function has infinitely many right derivatives, since

$$f_{\_w}(v) = \begin{cases} \overset{f(v)}{\text{left right}} & \text{if there is a prime number in } \{|v|+1, \ldots, |vw|\} \\ \overset{f(v)}{\text{left}} & \text{otherwise.} \end{cases}$$

However finitely many right derivatives can be obtained by making the last position to be output unconditionally, i.e. in the string-to-string function

$$g(a_1 \cdots a_n) = f(a_1 \cdots a_{n-1})a_n.$$

In this case, g has only two right derivatives, namely

$$v \mapsto \overset{g(v)}{\text{left}} \qquad v \mapsto \overset{f(v)}{\text{left right.}}$$

The function has infinitely many left derivatives g_{v_-} because the criterion "i is a prime number" needs to be replaced by "$i + |v|$ is a prime number".

To present our machine independent characterisation, we need a notion of regularity for functions from tuples of strings to a finite set. Define the *language encoding* of a function $f : (A^*)^n \to C$, with C finite, to be

$$\{w_1 \# w_2 \# \cdots \# w_n \# f(w_1, \ldots, w_n) : w_1, \ldots, w_n \in A^*\} \subseteq (A \cup C \cup \{\#\})^*.$$

assuming $\#$ is a symbol outside $A \cup C$. The function f is called a *regular colouring* if its language encoding is regular. Among several models of automata reading tuples of strings, regular colourings correspond to the weakest model, called recognisable. For instance, the equality function, seen as a colouring of string pairs by "equal" or "not equal", is not a regular colouring.

Theorem 1 (Machine Independent Characterisation). *For a string-to-string function f with origin information, the following conditions are equivalent*

1. *f is regular;*
2. *f has finitely many left derivatives and finitely many right derivatives;*
3. *for every letter a in the input alphabet, the following is a regular colouring*

$$(v, w) \mapsto f(\underline{v}|a|\underline{w}).$$

The function $(v, a, w) \mapsto f(\underline{v}|a|\underline{w})$, where v, w are words and a is a letter over the input alphabet, is called the *characteristic function* of f.

Proof (rough sketch). The implication from 1 to 2 is shown by using deterministic two-way automata with output. For the implication from 2 to 3, one observes that the functions $v \mapsto f_{v_-}$ and $w \mapsto f_{-w}$ are regular colorings, and that $f(\underline{v}|a|\underline{w})$ is uniquely determined by f_{v_-}, a and f_{-w}. For the implication from 3 to 1, one shows that an arbitrary string-to-string function with origin information can be uniquely reconstructed based on its characteristic function, and when the characteristic function happens to be a regular coloring then this reconstruction can be done by a finite state device.

Since a string-to-string function is uniquely determined by its characteristic function, instead of studying string-to-string functions, one can study their characteristic functions. This is the case in the learning algorithm from Section 3, and the studies of subclasses of transducers in Sections 4 and 5. The characteristic function can be computed based on a representation as a transducer model, e.g. from a copying MSO transduction. In particular, Theorem 1 gives a conceptually simple equivalence check for origin semantics: compute the characteristic functions and test if they are equal. The complexity of this algorithm, especially in the case when the function is given by streaming transducers, is left open.

As shown in Example 4, it is not enough to require finitely many derivatives of one kind, say right derivatives, since a function might have finitely many derivatives of one kind, but infinitely many derivatives of the other kind[3].

3   Learning

This section shows that transducers with origin information can be learned. We first recall the Angluin algorithm for regular languages, which will be used as a black box in our learning algorithm for learning transducers. The setup for the Angluin algorithm is as follows. A teacher knows a regular language. A learner wants to learn this language, by asking two kinds of queries. In a *membership query*, the learner gives a string and the teacher responds whether this string is in the language. In an *equivalence query*, the learner proposes a candidate for the teacher's language, and the teacher either says that this candidate is correct, in which case the protocol is finished by learner's success, or otherwise the teacher returns a *counterexample*, which is a string in the symmetric difference between the candidate and teacher's languages.

Angluin proposed an algorithm [Ang87], in which the learner learns the language by asking a number of queries which is polynomial in the minimal deterministic automaton for the teacher's language, and the size of the counterexamples given during the interaction. Theorem 2 shows that a variant of this algorithm works for regular string-to-string transducers with origin information. In the case of transducers, the membership query becomes a *value query*, where the learner gives a string and the teacher responds with the output of the transducer on that string. In the equivalence query, the counterexample becomes a string where the transducer proposed by the learner gives a different value than the transducer of the teacher. In both the value query and in the counterexample, the teacher also provides the origin information.

Theorem 2. *A regular string-to-string function with origin information can be learned using value and equivalence queries in polynomial time (both number of queries and computation time) in terms of the number of left and right derivatives, and the size of the counterexamples given by the teacher.*

4   Order-Preserving Transducers

In this section, we present two characterisations of subclasses of transducers. For semantics without origins, [FGRS13] shows how to decide if a deterministic two-way transducer is equivalent to a nondeterministic one-way transducer,

---

[3]   It does follow from the theorem that a function with finitely many left and right derivatives has finitely many two-sided derivatives. This is because every regular string-to-string function has finitely many two-sided derivatives.

while [WK94] shows how to decide (in polynomial time) if a nondeterministic one-way transducer is equivalent to a deterministic one-way transducer. This section shows analogous results for the origin semantics. Unlike [FGRS13, WK94], the characterisations for the origin semantics are self-evident, which shows how changing the semantics (and therefore changing the problem) makes some technical problems go away. A more difficult characterisation, about first-order definable transducers, is presented in the next section.

In the following theorem, a string-to-string function with origin information is called *order preserving* if for every input positions $x < y$, every output position corresponding to x is before every output position corresponding to y.

Theorem 3. *For a regular string-to-string function with origin information f, the following conditions are equivalent.*

1. *f is order-preserving.*
2. *$f(\underline{v}|\underline{w})$ is one of ϵ, left, right or left right for all input strings v, w.*
3. *f is recognised by a streaming transducer with lookahead which has only one register, and which only appends output letters to that register.*
4. *f is recognised by a nondeterministic one-way automaton with output, which has exactly one run over every input string.*

Proof. The implication from item 1 to item 2 follows straight from the definition. For the implication from item 2 to 3, we observe that if condition 1 is satisfied, then the transducer constructed in the proof of Theorem 1 will only have one register, and it will only append letters to that register during the run. For the implication from item 3 to item 4, we observe that a nondeterministic one-way automaton with output can guess, for each position of the input, what the lookahead will say. Since the lookahead is computed by a deterministic right-to-left automaton, this leads to a unique run on every input string. The implication from item 4 to item 1 also follows straight from the definition.

Observe that the condition in item 2 can be decided, even in polynomial time, when the characteristic function of the transducer is known.

We can further restrict the model by requiring that the transducer in item 3 does not use any lookahead, or equivalently, by requiring that the automaton in item 4 be deterministic. This restricted model is characterised in the following theorem.

Theorem 4. *Let f be a regular string-to-string function which satisfies any of the equivalent conditions in Theorem 3. Then f is defined by a left-to-right deterministic automaton with output if and only if all input strings u, v, w satisfy*

$$f(u|\underline{v}) = f(u|\underline{w}).$$

Proof. The left-to-right implication is immediate. For the right-to-left implication, we observe that the assumption implies that

$$f(\underline{u}|a|\underline{v})$$

does not depend on v, but only on f_{u_-} and the letter a. Furthermore, since f satisfies the assumptions from Theorem 3, the above value is of the form

$$\text{left } \overset{x}{\text{right}},$$

where each block is possibly missing. After reading input u, the automaton stores in its control state the derivative f_{u_-}. When it reads a letter a, it updates its control state, and outputs the string w, which depends only on the control state and input letter a.

5   First-Order Definable Transducers

In this section we consider first-order definable transducers. Recall that when coding a string as a relational structure, we have a predicate for the order. We underline this because, unlike for MSO, for first-order logic order is more powerful than successor. The notion of first-order definability makes sense for:

– *languages:* there is a first-order formula that is true in the strings from the language and false in strings from outside the language.
– *regular colourings:* the language encoding is first-order definable.
– *string-to-string functions with origin information:* the same definition as for MSO-definable ones, except that set quantification is disallowed.

Theorem 5. *The following conditions are equivalent for a regular string-to-string function f with origin information.*

1. *it is definable by a first-order string-to-string transduction.*
2. *the colourings $w \mapsto f_{w_-}$ and $w \mapsto f_{_-w}$ are first-order definable.*
3. *for every letters a, b, the following is a first-order definable colouring*

$$(u, v, w) \mapsto f(\underline{u}|a|\underline{v}|b|\underline{w})$$

Before proving the theorem, we observe that condition in item 2 is effective. Using a straightforward extension of the the Schützenberger-McNaughton-Papert Theorem, one can decide if a regular colouring is first-order definable. By applying the decision procedure to the functions $w \mapsto f_{w_-}$ and $w \mapsto f_{_-w}$, we can decide if a regular string-to-string function with origin semantics is first-order definable. It is unclear if this sheds any light for the analogous question for semantics without origins.

Without origin information, a variant of first-order definable transducers was considered in [MSTV06], namely the transducers which are first-order definable in the sense of Theorem 5 and simultaneously order preserving in the sense of Theorem 3. For instance, the doubling transducer $w \mapsto ww$ is first-order definable in the sense of Theorem 5, but not in the sense of [MSTV06], because it is not order preserving. By testing for both Condition 2 from Theorem 5 and Condition 2 of Theorem 3, we get an effective characterisation of the origin version of the transducers from [MSTV06].

6   Further Work

Preliminary work indicates that the ideas in this paper extend to MSO-definable tree-to-tree transducers; this should be followed up. Another direction for further study is the computational complexity of equivalence with respect to origin semantics; in particular finding models for which equivalence is polynomial time.

References

[AC10]   Alur, R., Cerný, P.: Expressiveness of streaming string transducers. In: FSTTCS 2010, pp. 1–12 (2010)

[Ang87]   Angluin, D.: Learning regular sets from queries and counterexamples. Inf. Comput. 75(2), 87–106 (1987)

[AU70]   Aho, A.V., Ullman, J.D.: A characterization of two-way deterministic classes of languages. J. Comput. Syst. Sci. 4(6), 523–538 (1970)

[BE00]   Bloem, R., Engelfriet, J.: A comparison of tree transductions defined by monadic second order logic and by attribute grammars. J. Comput. Syst. Sci. 61(1), 1–50 (2000)

[CJ77]   Chytil, M., Jákl, V.: Serial composition of 2-way finite-state transducers and simple programs on strings. In: Salomaa, A., Steinby, M. (eds.) ICALP 1977. LNCS, vol. 52, pp. 135–147. Springer, Heidelberg (1977)

[EH01]   Engelfriet, J., Hoogeboom, H.J.: MSO definable string transductions and two-way finite-state transducers. ACM Trans. Comput. Log. 2(2), 216–254 (2001)

[EM03]   Engelfriet, J., Maneth, S.: Macro tree translations of linear size increase are MSO definable. SIAM J. Comput. 32(4), 950–1006 (2003)

[FGRS13]   Filiot, E., Gauwin, O., Reynier, P.-A., Servais, F.: From two-way to one-way finite state transducers. In: LICS, pp. 468–477. IEEE Computer Society (2013)

[Gur82]   Gurari, E.M.: The equivalence problem for deterministic two-way sequential transducers is decidable. SIAM J. Comput. 11(3), 448–452 (1982)

[LMN10]   Lemay, A., Maneth, S., Niehren, J.: A learning algorithm for top-down XML transformations. In: Paredaens, J., Van Gucht, D. (eds.) PODS, pp. 285–296. ACM (2010)

[MSTV06]   McKenzie, P., Schwentick, T., Thérien, D., Vollmer, H.: The many faces of a translation. J. Comput. Syst. Sci. 72(1), 163–179 (2006)

[Tho97]   Thomas, W.: Languages, automata, and logic. In: Handbook of Formal Language Theory, vol. III, pp. 389–455. Springer (1997)

[vDKT93]   van Deursen, A., Klint, P., Tip, F.: Origin tracking. J. Symb. Comput. 15(5/6), 523–545 (1993)

[WK94]   Weber, A., Klemm, R.: Economy of description for single-valued transducers. In: Enjalbert, P., Mayr, E.W., Wagner, K.W. (eds.) STACS 1994. LNCS, vol. 775, pp. 607–618. Springer, Heidelberg (1994)

Weak MSO+U with Path Quantifiers over Infinite Trees*

Mikołaj Bojańczyk**

University of Warsaw

Abstract. This paper shows that over infinite trees, satisfiability is decidable for weak monadic second-order logic extended by the unbounding quantifier U and quantification over infinite paths. The proof is by reduction to emptiness for a certain automaton model, while emptiness for the automaton model is decided using profinite trees.

This paper presents a logic over infinite trees with decidable satisfiability. The logic is *weak monadic second-order logic with* U *and path quantifiers* (WMSO+UP). A formula of the logic is evaluated in an infinite binary labelled tree. The logic can quantify over: nodes, finite sets of nodes, and paths (a path is a possibly infinite set of nodes totally ordered by the descendant relation and connected with respect to the child relation). The predicates are as usual in MSO for trees: a unary predicate for every letter of the input alphabet, binary left and right child predicates, and membership of a node in a set (which is either a path or a finite set). Finally, formulas can use the *unbounding quantifier*, denoted by

$$\mathsf{U}X \ \varphi(X),$$

which says that $\varphi(X)$ holds for arbitrarily large finite sets X. As usual with quantifiers, the formula $\varphi(X)$ might have other free variables except for X. The main contribution of the paper is the following theorem.

Theorem 1. *Satisfiability is decidable for* WMSO+UP *over infinite trees.*

Background. This paper is part of a program researching the logic MSO+U, i.e. monadic second-order logic extended with the U quantifier. The logic was introduced in [1], where it was shown that satisfiability is decidable over infinite trees as long as the U quantifier is used once and not under the scope of set quantification. A significantly more powerful fragment of the logic, albeit for infinite words, was shown decidable in [3] using automata with counters. These automata where further developed into the theory of cost functions initiated by Colcombet in [8]. Cost functions can be seen as a special case of MSO+U in the sense that decision problems regarding cost functions, such as limitedness or domination, can be easily encoded into satisfiability of MSO+U formulas. This

---

* Full version of this paper with proofs is at `arxiv.org/abs/1404.7278`.
** Supported by ERC Starting Grant "Sosna".

J. Esparza et al. (Eds.): ICALP 2014, Part II, LNCS 8573, pp. 38–49, 2014.
© Springer-Verlag Berlin Heidelberg 2014

encoding need not be helpful, since the unsolved problems for cost functions get encoded into unsolved problems from MSO+U.

The logic MSO+U can be used to solve problems that do not have a simple solution in MSO alone. One example (discussed later in Example 1) is the finite model problem for the two-way μ-calculus [1]. A more famous problem is the star height problem, which can be solved by a reduction to the satisfiability of MSO+U on infinite words; the particular fragment of MSO+U used in this reduction is decidable by [3]. In Section 1 we give more examples of problems which can be reduced to satisfiability for MSO+U, examples which use the fragment that is solved in this paper. An example of an unsolved problem that reduces to MSO+U is the decidability of the nondeterministic parity index problem, see [9].

The first strong evidence that MSO+U can be too expressive was given in [11], where it was shown that MSO+U can define languages of infinite words that are arbitrarily high in the projective hierarchy. In [4], the result from [11] is used to show that there is no algorithm which decides satisfiability of MSO+U on infinite trees and has a correctness proof using the axioms of ZFC. A challenging open question is whether satisfiability of MSO+U is decidable on infinite words.

The principal reason for the undecidability result above is that MSO+U can define languages of high topological complexity. Such problems go away in the weak variant, where only quantification over finite sets is allowed, because weak quantification can only define Borel languages. Indeed, satisfiability is decidable for WMSO+U over infinite words [2] and infinite trees [6]. This paper continues the research on weak fragments from [2,6]. Note that WMSO+UP can, unlike WMSO+U, define non Borel-languages, e.g. "finitely many a's on every path", which is complete for level $\mathbf{\Pi}^1_1$ of the projective hierarchy. The automaton characterization of WMSO+UP in this paper implies that WMSO+UP definable languages are contained in level $\mathbf{\Delta}^1_2$.

What is the added value of path quantifiers? One answer is given in the following section, where we show how WMSO+UP can be used to solve games winning conditions definable in WMSO+U; here the use of path quantifiers is crucial. Another answer is that solving a logic with path quantifiers is a step in the direction of tackling one of the most notorious difficulties when dealing with the unbounding quantifier, namely the interaction between quantitative properties (e.g. some counters have small values) with qualitative limit properties (e.g. the parity condition). The difficulty of this interaction is one of the reasons why the boundedness problem for cost-parity automata on infinite trees remains open [9]. Such interaction is also a source of difficulty in the present paper, arguably more so than in the previous paper on WMSO+U for infinite trees [6]. One of the main contributions of the paper is a set of tools that can be used to tackle this interaction. The tools use profinite trees.

1   Notation and Some Applications

Let us begin by fixing notation for trees and parity automata. Notions of root, leaf, sibling, descendant, ancestor, parent are used in the usual sense. A tree in this paper is labelled, binary, possibly infinite and not necessarily complete. In other words, a tree is a partial function from $\{0, 1\}^*$ to the input alphabet, whose domain is closed under parents and siblings. The logic WMSO+UP, as defined in the introduction, is used to define languages of such trees. To recognise properties of trees, we use the following variant of parity automata. A parity automaton is given by an input alphabet A, a set of states Q, an initial state, a total order on the states, a set of accepting states, and finite sets of transitions

$$\delta_0 \subseteq Q \times A \quad \text{and} \quad \delta_2 \subseteq Q \times A \times Q^2.$$

A run of the automaton is a labeling of the input tree by states such that for every node with $i \in \{0, 2\}$ children, the set δ_i contains the tuple consisting of the node's state, label and the sequence of states in its children. A run is accepting if it has the initial state in the root, and on every infinite path, the maximal state appearing infinitely often is accepting. Parity automata defined this way have the same expressive power as MSO.

Before continuing, we underline the distinction between paths, which are connected sets of nodes totally ordered by the ancestor relation, and chains which can be possibly disconnected. Having chain quantification and the U quantifier would be sufficient to express all properties of the leftmost path definable in MSO+U, and therefore its decidability would imply decidability of MSO+U on infinite words, which is open.

The rest of this section is devoted to describing some consequences of Theorem 1, which says that satisfiability is decidable for WMSO+UP on infinite trees.

Stronger than MSO. When deciding satisfiability of WMSO+UP in Theorem 1, we ask for the existence of a tree labelled by the input alphabet. Since the labelling is quantified existentially in the satisfiability problem, the decidability result immediately extends to formulas of *existential* WMSO+UP , which are obtained from formulas of WMSO+UP by adding a prefix of existential quantifiers over arbitrary, possibly infinite, sets. A result equivalent to Theorem 1 is that the existential WMSO+UP theory of the unlabeled complete binary tree is decidable.

Existential WMSO+UP contains all of MSO, because it can express that a parity tree automaton has an accepting run. The existential prefix is used to guess the accepting run, while the path quantifiers are used to say that it is accepting. One can prove a stronger result. Define WMSO+UP *with* MSO *subformulas*, to be the extension of WMSO+UP where quantification over arbitrary sets is allowed under the following condition: if a subformula $\exists X \; \varphi(X)$ quantifies over an arbitrary set X, then $\varphi(X)$ does not use the unbounding quantifier.

Fact 1. WMSO+UP *with* MSO *subformulas is contained in existential* WMSO+UP.

The idea behind the fact is to use the existential prefix to label each node with the MSO-theory of its subtree.

Example 1. Consider the modal μ-calculus with backward modalities, as introduced in [16]. As shown in [1], for every formula φ of the modal μ-calculus with backward modalities, one can compute a formula $\psi(X)$ of MSO such that φ is true in some finite Kripke structure if and only if

$$\mathsf{U}X\varphi(X) \tag{1}$$

is true in some infinite tree. The paper [1] gives a direct algorithm for testing satisfiability of formulas of the form as in (1). Since this formula is in WMSO+UP with MSO subformulas, Theorem 1 can be used instead.

By inspecting the proofs of [6], one can show that also [6] would be enough for the above example. This is no longer the case for the following example.

Example 2. Consider a two-player game over an arena with a finite set of vertices V, where the winning condition is a subset of V^ω defined in WMSO+U over infinite words. For instance, the winning condition could say that a node $v \in V$ is visited infinitely often, but the time between visits is unbounded. A winning strategy for player 1 in such a game is a subset $\sigma \subseteq V^*$, which can be visualized as a tree of branching at most V. The properties required of a strategy can be formalised in WMSO+UP over infinite trees, using path quantifiers to range over strategies of the opposing player. Therefore, one can write a formula of WMSO+UP over infinite trees, which is true in some tree if and only if player 1 has a winning strategy in the game. Therefore Theorem 1 implies that one can decide the winner in games over finite arenas with WMSO+U winning conditions.

The games described in Example 2 generalize cost-parity games from [10] or energy consumption games from [7], so Theorem 1 implies the decidability results from those papers (but not the optimal complexities).

Example 3. Consider a game as in the previous example, but where the winning condition is defined by a formula φ of WMSO+U which can also use a binary predicate "x and y are close". For $n \in \mathbb{N}$, consider the winning condition φ_n to be the formula φ with "x and y are close" replaced by "the distance between x and y is at most n". Consider the following problem: is there some $n \in \mathbb{N}$, such that player 1 has a winning strategy according to the winning condition φ_n? This problem can also be reduced to satisfiability of WMSO+UP on infinite trees. The idea is to guess a strategy $\sigma \subseteq V^*$, and a set of nodes $X \subseteq \sigma$, such that 1) there is a common upper bound on the length of finite paths that do not contain nodes from X; 2) every infinite path consistent with σ satisfies the formula φ with "x and y are close" replaced by "between x and y there is at most one node from X". Using the same idea, one can solve the realizability problem for Prompt LTL [12].

2 Automata

In this section, we define an automaton model with the same expressive power as existential WMSO+UP, which is called a WMSO+UP *automaton*. The automaton uses a labellings of trees by counter operations called *counter trees*, so we begin by describing these.

Counter trees. Let C be a finite set of counters. A *counter tree* over a set of counters C is defined to be a tree where every node is labelled by a subset of

$$C \times \{\text{parent}, \text{self}\} \times \{\text{increment}, \text{transfer}\} \times C \times \{\text{parent}, \text{self}\}, \qquad (2)$$

where every tuple contains "self" at least once. The counter tree induces a graph with edges labelled by "increment" or "transfer", called its associated *counter configuration graph*. The vertices of this graph, called *counter configurations*, are pairs (x, c) where x is a node of the counter tree and c is a counter. The counter configuration graph contains an edge from (x_0, c_0) to (x_1, c_1) labelled by o if and only if there exists a node x in the counter tree whose label contains a tuple

$$(c_0, \tau_0, o, c_1, \tau_1) \qquad \text{with} \qquad \tau_0, \tau_1 \in \{\text{parent}, \text{self}\}$$

such that x_i is x or its parent depending on whether τ_i is "self" or "parent".

A path in the counter configuration graph, using possibly both kinds of edges, is called a *counter path*. Its value is defined to be the number of "increment" edges. The value of a counter configuration is defined to be the supremum of values of counter paths that end in it. When t is a counter tree, then we write $[\![t]\!]$ for the tree with the same nodes but with alphabet $\bar{\mathbb{N}}^C$, where the label of a node x maps $c \in C$ to the value of (x, c) in the associated counter graph.

WMSO+UP *automata.* We now present the automaton model used to decide WMSO+UP. The syntax of a WMSO+UP *consists* of:

1. A parity automaton;
2. A set of counters C, partitioned into *bounded* and *unbounded* counters;
3. For every state q of the parity automaton:
   (a) a set $cut(q)$ of bounded counters, called the counters cut by q;
   (b) a set $check(q)$ of unbounded counters, called the counters checked by q;
   (c) a subset $counterops(q)$ of the set in (2).

The automaton inputs a tree over the input alphabet of the parity automaton in the first item. A *run* of the automaton is a labelling of the input tree by states, consistent with the transition relation of the parity automaton. Using the sets $counterops(q)$, we get a counter tree with counters C, call it $counterops(\rho)$. By abuse of notation, we write $[\![\rho]\!]$ for the tree $[\![counterops(\rho)]\!]$, which is a tree over $\bar{\mathbb{N}}^C$. Using the sets $cut(q)$ and $check(q)$, we can talk about the nodes in a run where a bounded counter gets cut, or an unbounded counter gets checked. A run is accepting if it has the initial state in the root, and it satisfies all three acceptance conditions defined below. In the conditions, we define the limsup of a function ranging over a countable set to be

$$\limsup_{x \in X} f(x) \stackrel{\text{def}}{=} \limsup_{n \in \mathbb{N}} f(x_n) \qquad \text{for some enumeration of } X = \{x_1, x_2, \ldots\},$$

which is well-defined because it does not depend on the enumeration.

- *Parity.* On every path the maximal state seen infinitely often is accepting.
- *Boundedness.* If a bounded counter c is never cut in a connected[1] set of nodes X, then

$$\limsup_{x \in X} [\![\rho]\!](x, c) < \infty$$

- *Unboundedness.* If an unbounded counter c is checked infinitely often on a path π, then

$$\limsup_{x} [\![\rho]\!](x, c) = \infty$$

with x ranging over those nodes in π where c is checked.

The automaton accepts an input tree if it admits an accepting run.

Equivalence to logic and emptiness. Below are the two main technical results about WMSO+UP automata. The two results immediately imply that satisfiability is decidable for WMSO+UP logic.

Theorem 2. *For every formula of existential* WMSO+UP *one can compute a* WMSO+UP *automaton that accepts the same trees, and vice versa.*

Theorem 3. *Emptiness is decidable for* WMSO+UP *automata.*

The proof of Theorem 2 is in the appendix. The rest of this paper is devoted to describing the proof of Theorem 3. The proof itself is described in Section 4, while the next section is about profinite trees, which are used in the proof.

Remark 1. If in the definition of the unboundedness acceptance condition, we replace lim sup by lim inf, we get a more powerful model. The same proof as for Theorem 3 also shows that this more powerful model has decidable emptiness.

3   Profinite Trees and Automata on them

In the emptiness algorithm for WMSO+UP automata, we use profinite trees. The connection between boundedness problems and profiniteness was already explored in [14], in the case of words. Profinite trees are similar to profinite words, because the recognizers are MSO formulas, the difference is that the objects are (infinite) trees. Consider an input alphabet A. Fix an enumeration of all MSO formulas over the alphabet A. We define the distance between two trees to be $1/n$ where n is the smallest number such that the n-th formula is true in one of the trees but not the other. The distance itself depends on the enumeration, but the notion of an open set or Cauchy sequence does not. Cauchy sequences are considered equivalent if some (equivalently, every) shuffle of them is also a Cauchy sequence. A *profinite tree* is defined to be an equivalence class of Cauchy sequences. To avoid confusion with profinite trees, we use from now on the term *real tree* instead of tree. Therefore, a profinite tree is a limit of a sequence of real trees. Every real tree is also a profinite tree, as a limit of a constant sequence.

---

[1] It suffices to restrict attention to maximal connected sets of nodes where c is not cut, such sets are called c-cut factors.

Evaluating MSO *formulas on profinite trees.* A Cauchy sequence is said to satisfy an MSO formula if almost all trees in the sequence satisfy it. A Cauchy sequence satisfies either an MSO formula, or its negation. Equivalent Cauchy sequences satisfy the same MSO formulas, and therefore satisfaction of MSO formulas is meaningful for profinite trees: a profinite tree is said to satisfy an MSO formula if this is true for some (equivalently, every) Cauchy sequence that tends to it. Formulas of MSO are the only ones that can be extended to profinite trees in this way; one can show that if L is a set of real trees that is not MSO-definable (for instance, L is defined by a formula of WMSO+UP that is not in MSO), then there is a Cauchy sequence which has infinitely many elements in L and infinitely many elements outside L. Summing up, it makes sense to ask if a profinite tree satisfies a formula of MSO, but it does not make sense to ask if it satisfies a formula of WMSO+UP.

Profinite subtrees. The *topological closure* of a binary relation on real trees is defined to be the pairs of profinite trees that are limits of pairs of real trees in the binary relation; with the metric in the product being the maximum of distances over coordinates. Define the *profinite subtree* relation to be the topological closure of the subtree relation. A real tree might have profinite subtrees that are not real. For example, consider a real tree t such that for every n, some subtree s_n of t has exactly one a, which occurs at depth n on the leftmost branch. By compactness, the sequence $s_1, s_2, \ldots$ has a convergent subsequence, whose limit is not a real tree, but is a profinite subtree of t.

Partially colored trees. Let A and Q be finite sets. A *partially Q-colored tree over A* is a tree, possibly profinite, over the alphabet $A \times (Q \cup \{\bot\})$. Suppose that ρ is a real partially Q-colored tree over A. If a node has second coordinate $q \in Q$, then we say that it is colored by q. When the second coordinate is $\bot$, then the node is called uncolored. A *color zone* of ρ is a connected set of nodes X in ρ such that:

- the unique minimal element of X is either the root of ρ or is colored;
- maximal elements of X are either leaves of ρ or are colored;
- all other elements of X are uncolored.

A real tree is called *real factor* of ρ if it is obtained from ρ by only keeping the nodes in some color zone. These notions are illustrated in Figure 1. The notions of defined color, color zone and real factor are only meaningful when ρ is a real tree. When ρ is not a real tree, then we can still use MSO-definable properties, such as "the root has undefined color" or "only the leaves and root have defined color". Define the *profinite factor* relation to be the topological closure of the real factor relation.

Generalized parity automata. A transition in a parity automaton can be visualized as a little tree, with one or three nodes, all of them colored by states. We introduce a generalized model, where transitions can be arbitrary trees, possibly infinite, and possibly profinite. A *generalized parity automaton* consists of: a totally ordered set of states Q, a subset of accepting states, an input alphabet, and a set of transitions, which is an arbitrary set of Q-colored profinite trees over the input alphabet. An

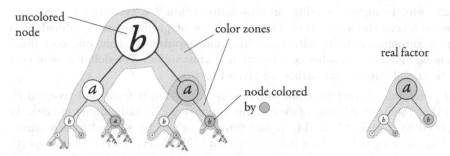

Fig. 1. A real $\{\bullet, \circ\}$-colored tree over $\{a, b\}$, together with a real factor. Uncolored nodes are white. Note how color zones overlap on colored nodes.

input to the automaton is a profinite tree over the input alphabet. A run over such an input is a partially Q-colored profinite tree over the input alphabet, call it ρ, which projects to the input on the coordinate corresponding to the input alphabet. By projection we mean the topological closure of the projection relation on real trees. A run ρ is accepting if all of its profinite factors are transitions, and it satisfies the MSO properties "the root is uncolored" and "on every infinite path where colored nodes appear infinitely often, the maximal color seen infinitely often is accepting". (The transitions where the root is uncolored play the role of the initial state.) There might be some infinite paths which have colors finitely often, because some transitions might have infinite paths. Every profinite factor of a run will necessarily satisfy the MSO property "every node that is not the root or a leaf is uncolored", therefore it only makes sense to have transitions that satisfy this property. It is not difficult to show that if a run satisfies the property "the root is uncolored", which is the case for every accepting run, then the run has a unique profinite factor that satisfies this property.

A run is called *regular* if it has finitely many profinite subtrees rooted in colored nodes. For a generalized parity automaton $\mathcal{A}$, define $L(\mathcal{A})$ to be the set of profinite trees accepted by $\mathcal{A}$, and let $L_{\mathrm{reg}}(\mathcal{A})$ be the subset of those profinite trees which are accepted via a regular run. The following theorem shows that two sets have the same topological closure (denoted by a bar on top), i.e. the smaller set is dense in the bigger one.

Theorem 4. $\overline{L_{\mathrm{reg}}(\mathcal{A})} = \overline{L(\mathcal{A})}$ *holds for every generalized parity automaton* $\mathcal{A}$.

3.1   Automaton Chains

Generalised parity automata are too general to be useful. For instance, every set of profinite trees is recognised by a generalised parity automaton, which has no states, and uses the recognised set as its transitions. Also, these automata do not allow a finite representation, and therefore cannot be used in algorithms. The emptiness algorithm for WMSO+UP automata uses a special case of generalised parity automata, called *automaton chains*, which can be represented in a finite way. Roughly speaking, an automaton chain is a generalised parity automaton where the set of transitions is the set of profinite trees defined by a simpler automaton chain, with the additional requirement that one cost function is bounded and another cost function is unbounded. The definitions of cost functions and automaton chains are given below.

Cost functions. A *cost function* on trees is a function α from real trees to $\bar{\mathbb{N}}$, such that the inverse image of every finite number $n \in \mathbb{N}$ is definable in MSO. As proposed by Toruńczyk in [14], a cost function α can be applied to a profinite tree t by defining $\alpha(t)$ to be a finite number $n \in \mathbb{N}$ if t satisfies the MSO property "has value n under α", and to be ∞ otherwise. Cost functions on finite words were introduced by Colcombet in [8] and then extended to finite trees, infinite words and infinite trees. The specific variant of cost functions that we use is the logic *cost* WMSO that was proposed by Vanden Boom in [15]. A sentence of this logic is built the same way as a sentence of WMSO over infinite trees, except that it can use an additional predicate "X is small", which takes a set X as a parameter, and can only be used under an even number of negations. The predicate can be used for different sets, like in the following example, call it α:

$$\exists X \ \exists Y \ X \text{ is small} \land Y \text{ is small} \land (\forall x \ a(x) \Rightarrow x \in X) \land (\forall y \ b(y) \Rightarrow y \in Y)$$

The cost function defined by a sentence of cost WMSO maps a tree to the smallest number n such that the sentence becomes true after "X is small" is replaced by $|X| < n$. If such a number does not exist, the result is ∞. In the case of the example α above, the function maps a tree to the number of a's or to the number of b's, whichever is bigger.

Automaton chains. We now define automaton chains, by induction on a parameter called *depth*. A automaton chain of depth 0 is any parity automaton. For $n > 0$, an automaton chain of depth n is a generalised parity automaton $\mathcal{A}$ whose set of transitions is

$$\{t : t \text{ is accepted by } \mathcal{B} \text{ and } \alpha(t) < \infty \text{ and } \beta(t) = \infty\}$$

for some automaton chain $\mathcal{B}$ of smaller depth and some cost functions α, β that are definable in cost WMSO. An automaton chain can be represented in a finite way and therefore used as an input for an algorithm, such as in the following lemmas.

Lemma 1. *Nonemptiness is decidable for automaton chains.*

Lemma 2. *Automaton chains are effectively closed under intersection with* MSO.

4   Emptiness of WMSO+UP Automata

In this section, we describe the proof of Theorem 3, which says that emptiness is decidable for WMSO+UP automata. We reduce emptiness for WMSO+UP automata to emptiness of automaton chains, which is decidable by Lemma 1.

A normal form. We begin by normalising the automaton. A counter c is called *separated* in a counter tree if the counter tree does not contain edges that involve c and and some other counter. A counter c is called *root-directed* if every counter edge involving c is directed toward the root. A WMSO+UP automaton is said to be in *normal form* if:

(a) for every run, in the counter graph generated by the automaton, every bounded counter is separated and root-directed.

(b) there is a total order on the states which is consistent with the order from the parity condition, and a mapping which maps every state q to sets of counters $larcut(q)$ and $larcheck(q)$ with the following property. For every run and every finite path in the run that starts and ends in state q and does not visit bigger states in the meantime,
   − the counters checked in the path are exactly $larcheck(q)$;
   − the counters cut in the path are exactly $larcut(q)$.

Lemma 3. *For every* WMSO+UP *automaton one can compute an equivalent one in normal form.*

In the proof, to achieve property (b), we use the latest appearance record data structure introduced by McNaughton in [13].

Partial runs. Let $\mathcal{A}$ be a WMSO+UP automaton that we want to test for emptiness. Thanks to Lemma 3, we assume without loss of generality that it is in normal form. In the emptiness algorithm, we describe properties of pieces of runs of $\mathcal{A}$, called partial runs, and defined below. Recall that in a parity automaton, there are two types of transitions δ_0 and δ_2, for leaves and non-leaves, respectively. A *partial run* of a parity automaton is a labelling of the input tree by states which respect δ_2 in nodes with two children, but need not respect δ_0 in leaves. A *partial run* of a WMSO+UP automaton is a partial run of the underlying parity automaton. A partial run is called *accepting* if it satisfies the parity, boundedness and unboundedness acceptance conditions. An accepting run of $\mathcal{A}$ is a partial accepting run where the root has the initial state and for every leaf, its (state, label) pair is in δ_0. Note that every finite partial run is an accepting partial run.

Chain automata recognising accepting runs. For a state q of $\mathcal{A}$, consider the following sets of real trees over the alphabet $A \times Q$, where A is the input alphabet of $\mathcal{A}$ and Q is its state space:

R_q accepting partial runs where states strictly bigger than q appear only in nodes with finitely many descendants;

R_{q*} the subset of R_q where state q is allowed only finitely often on every path.

Note that if q is a parity-rejecting state of the automaton $\mathcal{A}$, then $R_q = R_{q*}$. By induction on q in the order on states from the assumption on $\mathcal{A}$ being in normal form we define automaton chains $\mathcal{R}_q$ and $\mathcal{R}_{q*}$ such that

$$\overline{R_{q*}} = \overline{L(\mathcal{R}_{q*})} \qquad \text{and} \qquad \overline{R_q} = \overline{L(\mathcal{R}_q)}. \tag{3}$$

The definition of $\mathcal{R}_q$ and $\mathcal{R}_{q*}$ is given below. The proof of (3) is in the appendix.

The automaton $\mathcal{R}_{q*}$. The automaton $\mathcal{R}_{q*}$ has a unique state, call it "state", which is rejecting, meaning that it must appear finitely often on every path. A transition of this automaton is any profinite partially {"state"}-colored tree σ over $A \times Q$ such that:

1. the projection of σ onto the $A \times Q$ coordinate belongs to $\overline{R_p}$, where p is the predecessor of q in the order on states; and
2. for every root-to-leaf path in σ which ends in a leaf with defined color "state", the maximal value of the Q coordinate is q.

Property 1 is recognised by an automaton chain by the induction assumption. Property 2 is MSO-definable, and therefore the conjunction of properties 1 and 2 is recognised by an automaton chain thanks to Lemma 2. It follows that R_{q*} is a degenerate form of an automaton chain where the cost functions α and β are not used. This degenerate form is a special case of an automaton chain, by taking α to be the constant 0 and β to be the constant ∞.

The automaton R_q. If q is a parity-rejecting state of A, then R_q is equal to R_{q*}. Otherwise, it is defined as follows. The automaton R_q has a unique state, call it "state", which is accepting, meaning that it can appear infinitely often on a path. A transition of this automaton is any profinite partially {"state"}-colored tree σ over $A \times Q$ such that:

1. the projection of σ onto the $A \times Q$ coordinate belongs to $\overline{R_{q*}}$; and
2. for every root-to-leaf path in σ which ends in a leaf with defined color "state", the maximal value of the Q coordinate is q.
3. $\alpha(\sigma) < \infty$ holds for the cost function defined by

$$\alpha(\sigma) = \max_c \max_x \ [\![\sigma]\!](x,c)$$

with c ranging over bounded counters not in $larcut(q)$ and x ranging over nodes which do not have an ancestor where c is cut.
4. $\beta(\sigma) = \infty$ holds for the cost function defined by

$$\beta(\sigma) = \begin{cases} \min_c \min_x \max_y \ [\![\sigma]\!](y,c) & \text{if the root of } \sigma \text{ has defined color "state"} \\ \infty & \text{otherwise} \end{cases}$$

with c ranging over unbounded counters in $larcheck(q)$, x ranging over leaves with defined color "state", and y ranging over ancestors of x where c is checked.

As for the automaton R_{q*}, the conjunction of properties 1 and 2 is recognised by an automaton chain, and therefore R_q is an automaton chain.

Proof (of Theorem 3). If q is the maximal state of A, then R_q is the set of all partial accepting runs. Therefore, the automaton A is nonempty if and only if R_q accepts some tree which is an accepting run of the underlying parity automaton in A. This is decidable by Lemmas 1 and 2   □

5   Conclusions

This paper shows that satisfiability is decidable for WMSO+UP on infinite trees. We conjecture the logic remains decidable after adding the R quantifier from [5]. We also conjecture that the methods developed here, maybe the automaton mentioned in Remark 1, can be used to decide satisfiability of tree languages of the form "every path is in L", with L being ωB- or ωS-regular languages of infinite words, as defined in [3].

Acknowledgment. I would like to thank Szymon Toruńczyk and Martin Zimmermann for months of discussions about this paper; in particular Szymon Toruńczyk suggested the use of profinite trees. Also, I would like to thank the anonymous referees for their comments.

References

1. Bojańczyk, M.: A bounding quantifier. In: Marcinkowski, J., Tarlecki, A. (eds.) CSL 2004. LNCS, vol. 3210, pp. 41–55. Springer, Heidelberg (2004)
2. Bojańczyk, M.: Weak MSO with the unbounding quantifier. Theory Comput. Syst. 48(3), 554–576 (2011)
3. Bojańczyk, M., Colcombet, T.: Bounds in ω-regularity. In: LICS, pp. 285–296 (2006)
4. Bojańczyk, M., Gogacz, T., Michalewski, H., Skrzypczak, M.: On the decidability of MSO+U on infinite trees. In: Esparza, J., Fraigniaud, P., Husfeldt, T., Koutsoupias, E. (eds.) ICALP 2014, Part II. LNCS, vol. 8573, pp. 50–61. Springer, Heidelberg (2014)
5. Bojańczyk, M., Toruńczyk, S.: Deterministic automata and extensions of weak MSO. In: FSTTCS, pp. 73–84 (2009)
6. Bojańczyk, M., Toruńczyk, S.: WMSO+U over infinite trees. In: STACS, pp. 648–660 (2012)
7. Brázdil, T., Chatterjee, K., Kučera, A., Novotný, P.: Efficient controller synthesis for consumption games with multiple resource types. In: Madhusudan, P., Seshia, S.A. (eds.) CAV 2012. LNCS, vol. 7358, pp. 23–38. Springer, Heidelberg (2012)
8. Colcombet, T.: The theory of stabilisation monoids and regular cost functions. In: Albers, S., Marchetti-Spaccamela, A., Matias, Y., Nikoletseas, S., Thomas, W. (eds.) ICALP 2009, Part II. LNCS, vol. 5556, pp. 139–150. Springer, Heidelberg (2009)
9. Colcombet, T., Löding, C.: The non-deterministic Mostowski hierarchy and distance-parity automata. In: Aceto, L., Damgård, I., Goldberg, L.A., Halldórsson, M.M., Ingólfsdóttir, A., Walukiewicz, I. (eds.) ICALP 2008, Part II. LNCS, vol. 5126, pp. 398–409. Springer, Heidelberg (2008)
10. Fijalkow, N., Zimmermann, M.: Cost-parity and cost-Streett games. In: FSTTCS, pp. 124–135 (2012)
11. Hummel, S., Skrzypczak, M.: The topological complexity of MSO+U and related automata models. Fundam. Inform. 119(1), 87–111 (2012)
12. Kupferman, O., Piterman, N., Vardi, M.Y.: From liveness to promptness. Formal Methods in System Design 34(2), 83–103 (2009)
13. McNaughton, R.: Finite state infinite games. Project MAC Report, MIT (1965)
14. Toruńczyk, S.: Languages of profinite words and the limitedness problem. In: Czumaj, A., Mehlhorn, K., Pitts, A., Wattenhofer, R. (eds.) ICALP 2012, Part II. LNCS, vol. 7392, pp. 377–389. Springer, Heidelberg (2012)
15. Vanden Boom, M.: Weak cost monadic logic over infinite trees. In: Murlak, F., Sankowski, P. (eds.) MFCS 2011. LNCS, vol. 6907, pp. 580–591. Springer, Heidelberg (2011)
16. Vardi, M.Y.: Reasoning about the past with two-way automata. In: Larsen, K.G., Skyum, S., Winskel, G. (eds.) ICALP 1998. LNCS, vol. 1443, pp. 628–641. Springer, Heidelberg (1998)

On the Decidability of MSO+U on Infinite Trees

Mikołaj Bojańczyk[1], Tomasz Gogacz[2],
Henryk Michalewski[1], and Michał Skrzypczak[1,*]

[1] University of Warsaw
[2] University of Wrocław

Abstract. This paper is about MSO+U, an extension of monadic second-order logic, which has a quantifier that can express that a property of sets is true for arbitrarily large sets. We conjecture that the MSO+U theory of the complete binary tree is undecidable. We prove a weaker statement: there is no algorithm which decides this theory and has a correctness proof in ZFC. This is because the theory is undecidable, under a set-theoretic assumption consistent with ZFC, namely that there exists of projective well-ordering of 2^ω of type ω_1. We use Shelah's undecidability proof of the MSO theory of the real numbers.

1 Introduction

This paper is about MSO+U, which is the extension of MSO by the *unbounding quantifier*. The unbounding quantifier, denoted by

$$\mathsf{U}X. \ \varphi(X),$$

says that $\varphi(X)$ holds for arbitrarily large finite sets X. As usual with quantifiers, the formula $\varphi(X)$ might have other free variables except for X. The main contribution of the paper is the following theorem, which talks about the complete binary tree 2^*.

Theorem 1.1. *Assuming that there exists a projective well-ordering of 2^ω of type ω_1, it is undecidable if a given sentence of MSO+U is true in the complete binary tree.*

The assumption on the projective ordering can be seen as a set theory axiom. The assumption follows from the axiom V=L, which is relatively consistent with ZFC. Therefore, if ZFC has a model, then it has one where the assumption of Theorem 1.1 is true, and therefore it has a model where the MSO+U theory of the complete binary tree is undecidable. In particular, there is no algorithm which decides the MSO+U theory of the complete binary tree, and has a correctness proof in ZFC. Although the theorem stops short of full undecidability, which we conjecture to be the case, it seems to settle the decidability question for all practical purposes.

* The first and fourth author are supported by ERC Starting Grant "Sosna", the third author is supported by the Polish NCN grant DEC-2012/07/D/ST6/02443.

J. Esparza et al. (Eds.): ICALP 2014, Part II, LNCS 8573, pp. 50–61, 2014.

Background. This paper is part of a programme researching the logic MSO+U, i.e. monadic second-order logic extended with the U quantifier. The logic was introduced in [Boj04], where it was shown that satisfiability is decidable for formulae on infinite trees where the U quantifier is used once and not under the scope of set quantification. A significantly more powerful fragment of the logic, albeit for infinite words, was shown decidable in [BC06] using automata with counters. These automata where further developed into the theory of cost functions initiated by Colcombet in [Col09]. Cost functions can be seen as a special case of MSO+U in the sense that decision problems regarding cost functions, such as limitedness or domination, can be easily encoded into satisfiability of MSO+U formulae. This encoding need not be helpful, since the unsolved problems for cost functions get encoded into unsolved problems from MSO+U.

The added expressive power of MSO+U can be used to solve problems that do not have a simple solution in MSO alone. An example is the star height problem, one of the most difficult problems in the theory of automata, which can be straightforwardly reduced to the satisfiability of MSO+U on infinite words; the particular fragment of MSO+U used in this reduction is decidable by [BC06]. An example of an important unsolved problem that reduces to MSO+U is the decidability of the nondeterministic parity index problem [CL08].

So far, most research on MSO+U has focussed on the weak variant, call it WMSO+U, where only quantification over finite sets is allowed. Satisfiability is decidable for WMSO+U over infinite words [Boj11] and infinite trees [BT12]. In a parallel submission to this conference, it is shown that WMSO+U remains decidable over infinite trees even after adding quantification over infinite paths. The decidability proofs use automata with counters.

Undecidability. The first strong evidence that MSO+U can be too expressive was given in [HS12], where it was shown that MSO+U can define languages of infinite words that are arbitrarily high in the projective hierarchy from descriptive set theory. The present paper builds on that observation. We show that, using the languages from [HS12], one can use MSO+U on the complete binary tree 2^* to simulate a variant of MSO on the Cantor set 2^ω, which we call *projective* MSO. Projective MSO is like MSO, except that set quantification is restricted to projective sets. As shown by Shelah in [She75], the MSO theory of 2^ω is undecidable. From the proof of Shelah it follows that, under the assumption that there exists a projective well-ordering of 2^ω, already projective MSO is undecidable on 2^ω. Therefore, thanks to our reduction, MSO+U is undecidable on 2^*.

2 MSO+U on 2^*

We consider the following logical structures: the complete binary tree 2^*, the Cantor set 2^ω, and the union of the two $2^{\leq\omega}$. In the complete binary tree 2^*, the universe consists of finite strings over $\{0,1\}$, called *nodes*, and there are predicates for the lexicographic and prefix orders. The prefix order corresponds to the ancestor relation. In the Cantor set 2^ω, the universe consists of infinite strings over $\{0,1\}$, called *branches*, and there is a predicate for the lexicographic order. Finally, in $2^{\leq\omega}$, the universe consists of both nodes and branches, and there are predicates for the prefix and lexicographic order. In $2^{\leq\omega}$, the prefix relation can hold between two nodes, or between a node and a branch. The lexicographic order is a total order on both nodes and branches, e.g. $0 < 0^\omega < 01$.

Two fundamental theorems about MSO are that the MSO theory is decidable for 2^*, but undecidable for 2^ω, and therefore also undecidable for $2^{\leq\omega}$. The decidability was shown by Rabin in [Rab69], while the undecidability was shown by Shelah in [She75] conditionally on the Continuum Hypothesis, and by Shelah and Gurevich in [GS82] without any conditions.

The projective hierarchy. Consider a topological space X. The family of Borel sets is the least family of subsets of X that contains open sets, and is closed under complements and countable unions. Define the family of *projective sets* to be the least family of subsets of X which contains the Borel sets, and is closed under complements and images under continuous functions. The projective sets can be organised into a hierarchy, called the *projective hierarchy*, where $\mathbf{\Sigma}_0^1 = \mathbf{\Pi}_0^1$ is the class of Borel sets, $\mathbf{\Pi}_n^1$ is the class of complements of sets from $\mathbf{\Sigma}_n^1$, and $\mathbf{\Sigma}_{n+1}^1$ is the class of images of sets from $\mathbf{\Pi}_n^1$ under continuous functions. Additionally, $\mathbf{\Delta}_n^1$ is the intersection of $\mathbf{\Sigma}_n^1$ and $\mathbf{\Pi}_n^1$. When the space X is not clear from the context, we add it in parentheses, e.g. $\mathbf{\Sigma}_n^1(X)$

We are mostly interested in the projective hierarchy for the space 2^ω with the topology of the Cantor set. This topology is induced by a metric, where the distance between two infinite bit strings is the inverse of the first position where they differ. We write $\mathbf{\Sigma}_n^1(2^\omega)$ for the subsets of 2^ω that are in level $\mathbf{\Sigma}_n^1$ of the projective hierarchy under this topology.

The main result. The main result of this paper is Theorem 1.1 from the introduction, which says that the MSO+U theory of 2^* is undecidable. The proof of Theorem 1.1 is by a reduction from the undecidability of MSO on 2^ω. Our proof uses a stronger undecidability version of MSO on 2^ω, where instead of full MSO we have a logic called *projective* MSO , where quantification is restricted to projective sets, as defined later in Section 2.1. We are unable to prove the projective MSO theory of 2^ω to be undecidable without any conditions, or even conditionally on the Continuum Hypothesis, but only assuming the stronger assumption that there exists a projective well-ordering of 2^ω of type ω_1.

This assumption can be seen as a conjunction of two assumptions: the Continuum Hypothesis (the type ω_1 part) and that the well-ordering is "definable" in some sense (the projective part). As shown in [GS82] the MSO theory of 2^ω remains undecidable even without the Continuum Hypothesis. This does not help us, because our reduction to MSO+U crucially depends on the definability.

Before proving the theorem, we observe the following corollary.

Corollary 2.1. *If* ZFC *is consistent, then there is no algorithm which decides the* MSO+U *theory of* 2^* *and has a proof of correctness in* ZFC.

Proof. [The following proof is in ZFC] If ZFC is consistent, then Gödel's constructible universe L is a model of ZFC, as shown by Gödel (for a modern treatment of this topic see Chapter 13 and specifically Theorem 13.6 in [Jec02]). In Gödel's constructible universe, there exists a well-ordering of 2^ω of type ω_1 that is in level Δ_2^1 of the projective hierarchy on $2^\omega \times 2^\omega$ ([Jec02, Theorem 25.26]). Therefore, if ZFC is consistent, then by Theorem 1.1 it has a model where the MSO+U theory of 2^* is undecidable.   □

2.1   Projective MSO on $2^{\leq \omega}$, and its Reduction to MSO+U on 2^*

For $n \leq \omega$, define the syntax of MSO_n to be the same as the syntax of MSO, except that instead of one pair of set quantifiers $\exists X$ and $\forall X$, there is a pair of quantifiers $\exists_i X$ and $\forall_i X$ for every $i \leq n$. To evaluate a sentence of MSO_n over a structure, we need a sequence $\{\mathcal{X}_j\}_{j \leq i}$ of families of sets, called the *monadic domains*. The semantics are then the same as for MSO, except that the quantifiers $\exists_j$ and $\forall_j$ are interpreted to range over subsets of the universe that belong to $\mathcal{X}_j$. First-order quantification is as usual, it can quantify over arbitrary elements of the universe. We write $MSO[\mathcal{X}_1, \mathcal{X}_2, \ldots]$ for the above logic with the monadic domains being fixed to $\mathcal{X}_1, \mathcal{X}_2, \ldots$. Standard MSO for structures with a universe Ω is the same as $MSO[P(\Omega)]$, i.e. there is one monadic domain for the powerset of the universe. If Ω is equipped with a topology, we define *projective* MSO over Ω to be

$$MSO[\Sigma_1^1(\Omega), \Sigma_2^1(\Omega), \ldots]$$

The expressive power of projective MSO is incomparable with the expressive power of MSO. Although projective MSO cannot quantify over arbitrary subsets, it can express that a set is in, say, Σ_1^1.

Example 2.2. In the structure $2^{\leq \omega}$, being a node is first-order definable: a node is an element of the universe that is a proper prefix of some other element. Since there are countably many nodes, every set of nodes is Borel, and therefore in $\Sigma_1^1(2^{\leq \omega})$. Therefore, in projective MSO on $2^{\leq \omega}$ one can quantify over arbitrary sets of nodes. It is easy to see that a subset of $2^{\leq \omega}$ is in $\Sigma_n^1(2^{\leq \omega})$ if and only if it is a union of a set of nodes and a set from $\Sigma_n^1(2^\omega)$. It follows that projective MSO on $2^{\leq \omega}$ has the same expressive power as the logic

$$MSO[P(2^*), \Sigma_1^1(2^\omega), \Sigma_2^1(2^\omega), \ldots].$$

Example 2.3. In projective MSO on $2^{\leq \omega}$, one can say that a set of branches is countable. This is by using notions of interval, closed set, and perfect. A set of branches is open if and only if for every element, it contains some open interval around that element. A *perfect* is a set of branches which is closed (i.e. its complement is open) and contains no isolated points. The notions of open interval, closed set, and perfect are first-order definable. By [Kec95, Theorem 29.1], a set of branches is countable if and only if it is in $\Sigma_1^1(2^\omega)$ and does not contain any perfect subset, which is a property definable in projective MSO.

The following lemma shows that the projective MSO theory of $2^{\leq\omega}$ can be reduced to the MSO+U theory of 2^*.

Lemma 2.4. *For every sentence of projective* MSO *on* $2^{\leq\omega}$, *one can compute an equivalently satisfiable sentence of* MSO+U *on* 2^*.

The proof uses Theorem 5.1 from [HS12] and the following lemma.

Lemma 2.5. *Suppose that* $L_1, L_2, \ldots \subseteq A^\omega$ *are definable in* MSO+U, *and let*

$$\mathcal{X}_i \stackrel{def}{=} \{f^{-1}(L_i) \,|\, f : 2^\omega \to A^\omega \text{ is a continuous function}\}.$$

Then for every sentence of MSO$[\mathsf{P}(2^*), \mathcal{X}_1, \mathcal{X}_2, \ldots]$ *on* $2^{\leq\omega}$, *one can compute an equivalently satisfiable sentence of* MSO+U *on* 2^*.

Proof. The proof of this lemma is based on the observation that, using quantification over sets of nodes, one can quantify over continuous functions $2^\omega \to A^\omega$.

Call a mapping $f : 2^* \to A \cup \{\epsilon\}$ *proper* if on every infinite path in 2^*, the labelling f contains infinitely many letters different than ϵ. If f is proper then define $\hat{f} : 2^\omega \to A^\omega$ to be the function that maps a branch to the concatenation of values under f of nodes on the branch. It is not difficult to see that a function $g : 2^\omega \to A^\omega$ is continuous if and only if there exists a proper f such that $g = \hat{f}$, see e.g. Proposition 2.6 in [Kec95]. Since a mapping $f : 2^* \to A \cup \{\epsilon\}$ can be encoded as a family of disjoint sets $\{X_a \subseteq 2^*\}_{a \in A}$, one can use quantification over sets of nodes to simulate quantification over continuous functions $g : 2^\omega \to A^\omega$.

The reduction in the statement of the lemma works as follows. First-order quantification over branches is replaced by (monadic second-order) quantification over paths, i.e. subsets of 2^* that are totally ordered and maximal for that property. For a formula $\exists X \in \mathcal{X}_i.\ \varphi$, we replace the quantifier by existential quantification over a family of disjoint subsets $\{X_a\}_{a \in A}$ which encode a continuous function. In the formula φ, we replace a subformula $x \in X$, where x is now encoded as a path, by a formula which says that the image of x, under the function encoded by $\{X_a\}_{a \in A}$, belongs to the language L_i. In order to verify if a given element belongs to the language L_i definable in MSO+U on infinite words, we can use a formula of MSO+U on infinite trees.   □

Proof (of Lemma 2.4). Theorem 5.1 of [HS12] shows that there is an alphabet A such that for every $i \geq 1$, there is a language $L_i \subseteq A^\omega$ which is definable in MSO+U on infinite words and hard for $\Sigma^1_i(2^\omega)$. It is easy to check (see the full version) that L_i is in fact complete for $\Sigma^1_i(2^\omega)$. Apply Lemma 2.5 to these languages. By their completeness, the classes $\mathcal{X}_1, \mathcal{X}_2, \ldots$ in Lemma 2.5 are exactly the projective hierarchy on 2^ω, and therefore Lemma 2.4 follows thanks to the observation at the end of Example 2.2.   □

Before we move on, we present an example of a nontrivial property that can be expressed in the projective MSO on $2^{\leq\omega}$.

Example: projective determinacy. A Gale-Stewart game with winning condition $W \subseteq 2^\omega$ is the following two-player game. For ω rounds, the players propose bits in an alternating fashion, with the first player proposing a bit in even-numbered rounds, and the second player proposing a bit in odd-numbered rounds. At the end of such a play, an infinite sequence of bits is produced, and the first player wins if this sequence belongs to W, otherwise the second player wins. Such a game is called *determined* if either the first or the second player has a winning strategy, see [Kec95, Chapter 20] or [Jec02, Chapter 33] for a broader reference. Martin [Mar75] proved that the games are determined if W is a Borel set.

It is not difficult to see that for every $i > 0$, the statement

"every Gale-Stewart game with a winning condition in Σ_i^1 is determined"   (1)

can be formalised as a sentence φ_{det}^i of projective MSO on $2^{\leq\omega}$ (see the full version). As we show below, the ability to formalise determinacy of Gale-Stewart games with winning conditions in Σ_1^1 already indicates that it is unlikely that projective MSO on $2^{\leq\omega}$ is decidable.

Indeed, suppose that there is an algorithm P deciding the projective MSO theory of $2^{\leq\omega}$ with a correctness proof in ZFC. Note that by Lemma 2.4, this would be the case if there was an algorithm deciding the MSO+U theory of 2^* with a correctness proof in ZFC. Run the algorithm on φ_{det}^1 obtaining an answer, either "yes" or "no". The algorithm together with its proof of correctness and the run on φ_{det}^1 form a proof in ZFC resolving Statement (1) for $i = 1$. The determinacy of all Σ_1^1 games cannot be proved in ZFC, because it does not hold if V=L, see [Jec02, Corollary 25.37 and Section 33.9], and therefore P must answer "no" given input φ_{det}^1.

This means that a proof of correctness for P would imply a ZFC proof that Statement (1) is false for $i = 1$. Such a possibility is considered very unlikely by set theorists, see [FFMS00] for a discussion of plausible axioms extending the standard set of ZFC axioms. A similar example regarding MSO($\mathbb{R}$) and the Continuum Hypothesis was provided in [She75].

3   Undecidability of Projective MSO on 2^ω

In this section we show that projective MSO is undecidable already on 2^ω with the lexicographic order. From the discussion in Example 2.2 it follows that the projective MSO theory of 2^ω reduces to the projective MSO theory of $2^{\leq\omega}$. Therefore, the undecidability result for 2^ω is stronger than for $2^{\leq\omega}$, in particular it implies the undecidability result for MSO+U from Theorem 1.1.

Theorem 3.1. *Assume that there is a projective well-ordering of 2^ω of type ω_1. Then the projective MSO theory of 2^ω is undecidable.*

The proof of Theorem 3.1 is a minor adaptation of Shelah's proof [She75] that, assuming the Continuum Hypothesis, the MSO theory of 2^ω is undecidable. In fact, Shelah already observed that such an adaptation is possible, in the following remark on p. 410: "Aside from countable sets, we can use only a set constructible from any well-ordering of the reals." To make the paper self-contained, we include a proof of Theorem 3.1.

Proof strategy. We use the name $\forall^*\exists^*$ *sentence* for a sentence of first-order logic in the prenex normal form that has a $\forall^*\exists^*$ quantification pattern. The vocabulary of graphs is defined to be the vocabulary with one binary predicate $E(x,y)$. Finally, an equality-free formula is one that does not use equality. The proof is by a reduction from the following satisfiability problem:

- **Input.** An equality-free $\forall^*\exists^*$ sentence over the vocabulary of graphs.
- **Question.** Is the sentence true in some undirected simple graph?

The above problem is undecidable by Theorem 1 in Section 9 of [Gur80].

Reducing from the above problem is one of the main differences between our proof and Shelah's proof, which uses a reduction from the first-order theory of arithmetic $(\mathbb{N}, +, *)$. The other main difference is that we introduce two definitions, which we call modal graphs and Shelah graphs, which are only implicit in Shelah's proof. Our intention behind these definitions is to give the reader a better intuition of what exactly is being coded into the MSO theory of 2^ω.

3.1 Modal Graphs

Instead of encoding undirected simple graphs in projective MSO, it will be more convenient to encode a less rigid structure, which we call a *modal graph*[1]. A modal graph consists of

- a partially ordered set of *worlds* with a least element;
- for every world I a set of *local vertices*[2] V_I;
- for every world I a set of *local edges* $E_I \subseteq V_I \times V_I$

subject to the monotonicity property that $V_I \subseteq V_J$ and $E_I \subseteq E_J$ holds for every worlds $I \leq J$. Furthermore, for every I the local edges E_I are a symmetric irreflexive relation, i.e. modal graphs are simple and undirected.

We use first-order logic to describe properties of modal graphs, with the semantics relation denoted by

$$\mathcal{G}, I, val \models \varphi, \tag{2}$$

where φ is a formula of first-order logic, $\mathcal{G}$ is a modal graph, I is a world in the modal graph, and val is a valuation that maps the free variables of φ to the local vertices V_I of the world I. The definition is by induction on the formula:

---

[1] Another take on modality is presented in [GS82] using the language of forcing.
[2] We will only construct graphs where every world has the same local vertices, but we give the more general definition to match Kripke models for intuitionistic logic.

$$\mathcal{G}, I, val \models E(x, y) \qquad \text{iff} \qquad (val(x), val(y)) \in E_I$$
$$\mathcal{G}, I, val \models \varphi \wedge \psi \qquad \text{iff} \qquad \mathcal{G}, I, val \models \varphi \text{ and } \mathcal{G}, I, val \models \psi$$
$$\mathcal{G}, I, val \models \varphi \vee \psi \qquad \text{iff} \qquad \mathcal{G}, I, val \models \varphi \text{ or } \mathcal{G}, I, val \models \psi$$
$$\mathcal{G}, I, val \models \neg \varphi \qquad \text{iff} \qquad \mathcal{G}, J, val \not\models \varphi \text{ for every } J \geq I$$
$$\mathcal{G}, I, val \models \exists x \, \varphi \qquad \text{iff} \qquad \mathcal{G}, J, val[x \to v] \models \varphi \text{ for some } J \geq I \text{ and } v \in V_J$$
$$\mathcal{G}, I, val \models \forall x \, \varphi \qquad \text{iff} \qquad \mathcal{G}, J, val[x \to v] \models \varphi \text{ for every } J \geq I \text{ and } v \in V_J$$

The definition above is almost the same as Kripke's semantics for intuitionistic logic [Kri65]. The only difference is in the $\exists$ quantifier: Kripke requires the world J to be equal to I. We say that a sentence (i.e. a formula without free variables) is satisfied in a modal graph if (2) holds with I being the least world and val being the empty valuation.

Example 3.2. A modal graph with one world is the same thing as an undirected simple graph. In this case, the standard semantics of first-order logic coincide with the semantics on modal graphs.

Example 3.3. Modal graphs satisfy more sentences of first-order logic than undirected simple graphs. In particular, if two existentially quantified sentences are satisfied in (possibly different) modal graphs, then their conjunction is also satisfied in the modal graph obtained by joining the two modal graphs by a common least world where the are no local edges.

The following lemma shows that for $\forall^* \exists^*$-sentences, the answers are the same for the satisfiability problem in modal graphs and the satisfiability problem in simple undirected graphs. The same lemma would hold for directed graphs, and also for vocabularies with more predicates.

Lemma 3.4. *For every $\forall^* \exists^*$ sentence η over the vocabulary of graphs, η is satisfied in some undirected simple graph if and only if it is satisfied in some modal graph.*

Proof. The left-to-right implication is true for all sentences, not just $\forall^* \exists^*$ sentences, and follows from Example 3.2.

For the right-to-left implication, consider a $\forall^* \exists^*$ sentence

$$\eta = \forall x_1, \ldots, x_k. \, \exists x_{k+1}, \ldots, x_n. \, \alpha$$

where α is quantifier-free. For directed graphs G and H, we say that H is an η-extension of G if G is an induced subgraph of H, and for every valuation of the universally quantified variables of η that uses only vertices of G, there is a valuation of the existentially quantified variables of η which makes the formula α true, but possibly uses vertices from H.

Suppose that $\mathcal{G}$ is a modal graph. For a world I and a subset V of the local vertices V_I, define $G_{I,V}$ to be the undirected simple graph where the vertices are V and the edges are local edges E_I restricted to $V \times V$. By monotonicity of local edges, the set of edges in $G_{I,V}$ grows or stays equal as I grows. We say that $G_{I,V}$ is *stable* if $G_{I,V} = G_{J,V}$ holds for every $J \geq I$. The key properties of being stable are:

1. If $G_{I,V}$ is stable then for every valuation $val : \{x_1, \ldots, x_n\} \to V$,

$$\mathcal{G}, I, val \models \alpha \qquad \text{iff} \qquad G_{I,V}, val \models \alpha.$$

In the equivalence above, the left side talks about semantics in modal graphs and the right side talks about semantics in simple undirected graphs.
2. For every world I and finite $V \subseteq V_I$, there exists a world $J \geq I$ such that $G_{I,V}$ is stable;
3. If $I \leq J$ are worlds and $V \subseteq W$ are such that $G_{I,V}$ and $G_{J,W}$ are stable, then $G_{I,V}$ is an induced subgraph of $G_{J,W}$.

Suppose that η is satisfied in $\mathcal{G}$.

Claim. There exists a sequence of worlds $I_1 \leq I_2 \leq \ldots$ and a sequence $V_1 \subseteq V_2 \subseteq \cdots$ of finite sets of vertices such that G_{I_i, V_i} is stable and η-extended by $G_{I_{i+1}, V_{i+1}}$ for every i.

This claim proves the lemma, since the limit, i.e. union, of the graphs G_{I_i, V_i} is a simple undirected graph that satisfies η.

Proof (of the claim). The sequence is constructed by induction; we only show the induction step. Suppose that I_i and V_i have already been defined. Let Γ_i be the finite set of valuations from the universally quantified variables $x_1, \ldots, x_k$ to the vertices V_i. Repeatedly using the assumption that $\mathcal{G}$ satisfies η for every valuation in Γ_i, one shows that there exists a world $J \geq I_i$ such that every valuation $val \in \Gamma_i$ extends to a valuation

$$val' : \{x_1, \ldots, x_n\} \to V_J \qquad \text{such that} \qquad \mathcal{G}, J, val' \models \alpha.$$

Define $V_{i+1} \subseteq V_J$ to be the finite set of vertices that are used by valuations of the form val' with val ranging over elements of Γ_i. Define $I_{i+1} \geq I_i$ to be the world, which exists by property 2 of stability, such that $G_{I_{i+1}, V_{i+1}}$ is stable. For quantifier-free formulas, the semantics in modal graphs are preserved when going into bigger worlds, and therefore

$$\mathcal{G}, I_{i+1}, val' \models \alpha$$

holds for every $val \in \Gamma_i$. By property 1 of stability, it follows that

$$G_{I_{i+1}, V_{i+1}}, val' \models \alpha.$$

Together with property 3 of stability, this implies that G_{I_i, V_i} is η-extended by $G_{I_{i+1}, I_{i+1}}$. $\qquad\square$

$\square$

3.2   Coding a Modal Graph in 2^ω

In this section, we describe how a modal graph can be coded in 2^ω. We use the name *interval* for a subset of 2^ω which consists of all branches that are lexicographically between some two distinct branches. Intervals defined this way are homeomorphic with 2^ω. Intervals are denoted I, J, K.

Define a *Shelah graph* to be two families $\mathcal{V}, \mathcal{E}$ of subsets of 2^ω such that every set in $\mathcal{V}$ is dense. For a Shelah graph, define its *associated modal graph* as follows. The worlds are the intervals in 2^ω, ordered by the opposite of inclusion, in particular the least world is the whole space 2^ω. The local vertices do not depend on the worlds: for every interval I, the local vertices V_I are are $\mathcal{V}$ (in particular a vertex is a subset of 2^ω). For an interval I and $V, W \in \mathcal{V}$, the local edge set E_I contains (V, W) if and only if

$$I \cap V \cap W = \emptyset \tag{3}$$
$$I \cap (V \cup W) = I \cap E \qquad \text{for some } E \in \mathcal{E}. \tag{4}$$

It is easy to see that $E_I \subseteq E_J$ when interval J is included in interval I. Since worlds are ordered by the opposite of inclusion, this means that $I \leq J$ implies $E_I \subseteq E_J$. Every local edge set is symmetric because it is defined in terms of union and intersection. Every local edge is irreflexive because (3) implies $V \neq W$ (here we use density, since the dense sets V, W must have nonempty intersections with I). In other words the associated modal graph is a modal graph.

For a sentence φ of MSO$_2$, and families $\mathcal{V}, \mathcal{E}$ of subsets in 2^ω, we write

$$2^\omega, \mathcal{V}, \mathcal{E} \models \varphi$$

if φ holds, with the quantifiers $\exists_1 X$ and $\forall_1 X$ interpreted to range over sets in $\mathcal{V}$, and the quantifiers $\exists_2 X$ and $\forall_2 X$ interpreted to range over sets in $\mathcal{E}$. By using logic to formalise the definition of a Shelah graph, its associated modal graph, and the semantics of first-order logic on modal graphs, we get the following lemma.

Lemma 3.5. *For every sentence η of first-order logic over the vocabulary of graphs, one can compute a sentence $\hat{\eta}$ of MSO$_2$ such that*

$$2^\omega, \mathcal{V}, \mathcal{E} \models \hat{\eta}$$

if and only if $(\mathcal{V}, \mathcal{E})$ is a Shelah graph whose associated modal graph satisfies η.

The general idea in the undecidability result is to use $\hat{\eta}$ from the above lemma. The main problem is that a projective MSO sentence cannot begin saying "there exists a Shelah graph", because a Shelah graph is described by an infinite (even uncountable) family of subsets of 2^ω. The solution to this problem, and the technical heart of the undecidability proof, is Proposition 3.6 below, which shows how to describe the infinite families $(\mathcal{V}, \mathcal{E})$ by using just four sets. The corresponding part in Shelah's paper [She75] consists of Lemmas 7.6–7.9.

Proposition 3.6. *Assume that there exists a well-ordering of 2^ω of type ω_1 which belongs to $\mathbf{\Delta}_k^1(2^\omega \times 2^\omega)$ for some k.*

Then there is a formula $\varphi_{\mathrm{elem}}(V, Q, S)$ *of projective* MSO *on* 2^ω *with the following property. If G is a countable undirected simple graph, then there are sets*

$$Q_V, Q_E, S_V, S_E \subseteq 2^\omega, \tag{5}$$

such that the families

$$\mathcal{V} = \{V \subseteq 2^\omega : \varphi_{\mathrm{elem}}(V, Q_V, S_V)\}, \quad \mathcal{E} = \{E \subseteq 2^\omega : \varphi_{\mathrm{elem}}(E, Q_E, S_E)\} \tag{6}$$

are a Shelah graph whose associated modal graph satisfies the same equality-free $\forall^*\exists^*$ *sentences as G.*

Furthermore, the formula φ_{elem} *quantifies only over* $\mathbf{\Sigma}_1^1$ *sets; the sets from (5) are in* $\mathbf{\Sigma}_{k+4}^1$, *and the families from (6) contain only countable sets.*

We now use the proposition and the previous results to show the undecidability of projective MSO from Theorem 3.1.

Corollary 3.7. *Assume that there exists a projective well-ordering of 2^ω of type ω_1. Let η be an equality-free $\forall^*\exists^*$ sentence over the vocabulary of graphs. Then the following conditions are equivalent:*

1. *η is true in some undirected simple graph, with standard semantics of logic.*
2. *There are sets as in (5) such that the families $\mathcal{V}, \mathcal{E}$ from (6) satisfy*

$$2^\omega, \mathcal{V}, \mathcal{E} \models \hat\eta$$

   where $\hat\eta$ is the sentence defined in Lemma 3.5.
3. *η is true in some modal graph, with semantics of logic on modal graphs.*

Proof. By the Löwenheim-Skolem theorem, if η is true in some undirected simple graph, then it is true in some countable undirected simple graph. Therefore, the implication $1 \Rightarrow 2$ follows from Proposition 3.6 and Lemma 3.5.

The implication $2 \Rightarrow 3$ follows from Lemma 3.5, which implies that η is true in some modal graph, namely the modal graph associated to the Shelah graph given by formula (6). The implication $3 \Rightarrow 1$ is the right-to-left implication in Lemma 3.4. □

Proof (of Theorem 3.1). Condition 2 in the above corollary can be formalised by the formula of projective MSO on 2^ω

$$\exists S_V, Q_V, S_E, Q_E \in \mathbf{\Sigma}_{k+4}^1. \; \tilde\eta$$

where k is the natural number from Proposition 3.6 and $\tilde\eta$ is the same as $\hat\eta$, except that instead of quantifying over a set $V \in \mathcal{V}$, it quantifies over a countable set V satisfying $\varphi_{\mathrm{elem}}(V, Q_V, S_V)$; likewise for quantifying over $E \in \mathcal{E}$.

We have thus shown a reduction from the undecidable satisfiability problem for equality-free $\forall^*\exists^*$ sentences over undirected simple graphs to the theory of projective MSO on 2^ω. Therefore, the latter is undecidable. □

4 Conclusions

We have shown that the MSO+U theory of 2^* is undecidable, conditional on the existence of a projective well-ordering of 2^ω of type ω_1. Apart from the obvious question about unconditional undecidability, a natural question is about the decidability of MSO+U on infinite words: is the MSO+U theory of the natural numbers with successor decidable? The methods used in this paper are strongly reliant on trees, so an undecidability proof would need new ideas to be adapted to the word case. Evidence for undecidability is that the topological hardness of MSO+U on words is shown in [HS12] by encoding trees in words.

An interesting related problem [She75, Conjecture 7a] is the decidability of MSO[Borel] on $2^{\leq\omega}$, i.e. the logic defined analogously to projective MSO except, that set quantification is over Borel sets only.

References

[BC06]   Bojańczyk, M., Colcombet, T.: Bounds in ω-regularity. In: LICS, pp. 285–296 (2006)
[Boj04]   Bojańczyk, M.: A bounding quantifier. In: Marcinkowski, J., Tarlecki, A. (eds.) CSL 2004. LNCS, vol. 3210, pp. 41–55. Springer, Heidelberg (2004)
[Boj11]   Bojańczyk, M.: Weak MSO with the unbounding quantifier. Theory Comput. Syst. 48(3), 554–576 (2011)
[BT12]   Bojańczyk, M., Toruńczyk, S.: Weak MSO+U over infinite trees. In: STACS, pp. 648–660 (2012)
[CL08]   Colcombet, T., Löding, C.: The non-deterministic Mostowski hierarchy and distance-parity automata. In: Aceto, L., Damgård, I., Goldberg, L.A., Halldórsson, M.M., Ingólfsdóttir, A., Walukiewicz, I. (eds.) ICALP 2008, Part II. LNCS, vol. 5126, pp. 398–409. Springer, Heidelberg (2008)
[Col09]   Colcombet, T.: The theory of stabilisation monoids and regular cost functions. In: Albers, S., Marchetti-Spaccamela, A., Matias, Y., Nikoletseas, S., Thomas, W. (eds.) ICALP 2009, Part II. LNCS, vol. 5556, pp. 139–150. Springer, Heidelberg (2009)
[FFMS00] Feferman, S., Friedman, H.M., Maddy, P., Steel, J.R.: Does mathematics need new axioms? The Bulletin of Symbolic Logic 6(4), 401–446 (2000)
[GS82]   Gurevich, Y., Shelah, S.: Monadic theory of order and topology in ZFC. Annals of Mathematical Logic 23(2-3), 179–198 (1982)
[Gur80]   Gurevich, Y.: Existential interpretation II. Archiv für Mathematische Logik und Grundlagenforschung 22(3-4), 103–120 (1980)
[HS12]   Hummel, S., Skrzypczak, M.: The topological complexity of MSO+U and related automata models. Fundamenta Informaticae 119(1), 87–111 (2012)
[Jec02]   Jech, T.: Set Theory. Springer (2002)
[Kec95]   Kechris, A.: Classical descriptive set theory. Springer, New York (1995)
[Kri65]   Kripke, S.A.: Semantical analysis of intuitionistic logic I. Studies in Logic and the Foundations of Mathematics (1965)
[Mar75]   Martin, D.A.: Borel determinacy. Annals of Mathematics 102(2), 363–371 (1975)
[Rab69]   Rabin, M.O.: Decidability of second-order theories and automata on infinite trees. Trans. of the American Math. Soc. 141, 1–35 (1969)
[She75]   Shelah, S.: The monadic theory of order. The Annals of Mathematics 102(3), 379–419 (1975)

A Coalgebraic Foundation
for Coinductive Union Types

Marcello Bonsangue[1,2], Jurriaan Rot[1,2,*], Davide Ancona[3],
Frank de Boer[2,1], and Jan Rutten[2,4]

[1] LIACS — Leiden University, The Netherlands
[2] Formal Methods — Centrum Wiskunde en Informatica, The Netherlands
[3] DIBRIS — Universita di Genova, Italy
[4] ICIS — Radboud University Nijmegen, The Netherlands

Abstract. This paper introduces a coalgebraic foundation for coinductive types, interpreted as sets of values and extended with set theoretic union. We give a sound and complete characterization of semantic subtyping in terms of inclusion of maximal traces. Further, we provide a technique for reducing subtyping to inclusion between sets of finite traces, based on approximation. We obtain inclusion of tree languages as a sound and complete method to show semantic subtyping of recursive types with basic types, product and union, interpreted coinductively.

1 Introduction

Basically all programming languages today support recursion to manipulate inductively defined data structures such as linked lists and trees. Whereas induction deals with finite but unbounded data, its dual, coinduction, deals with possibly infinite data. The relevant distinction here concerns traditional algebraic data structures which can be fully unfolded by a recursive program, and coalgebraic data structures which can be manipulated while they unfold, even if this process may never terminate. The interest in theoretical foundations for coinductive types and reasoning techniques is rapidly growing. Practical applications of coinductive types are found in the world of functional languages with lazy evaluation. Moreover a coinductive interpretation of structural recursively defined types with record, product and union type constructors allows one to assign types to coinductive data, such as infinite and circular lists of objects in object-oriented languages [3]. Union types allow a more precise analysis than disjoint sum [6], for example to type constructs like if-then-else. Consider, for instance, the recursive type definition below.

$$x_1 \mapsto \mathsf{null} \vee < \mathsf{elm:\ int,\ nxt:}\ x_1 > . \tag{1}$$

Here null and int are primitive type constants for representing the empty list and the integer values, respectively, and $< \mathsf{elm:}\ x,\ \mathsf{nxt:}\ y >$ represents the (tagged) product of the type variables x and y.

---

* The research of this author has been funded by the Netherlands Organisation for Scientific Research (NWO), CoRE project, dossier number: 612.063.920.

J. Esparza et al. (Eds.): ICALP 2014, Part II, LNCS 8573, pp. 62–73, 2014.

Intuitively, the type defined by (1) is a recursive type representing all finite and infinite linked lists of integer values. More formally, the type definition in (1) can be interpreted both syntactically and semantically. Syntactically, (1) can be interpreted as the set of finite and infinite closed terms over the alphabet consisting of the constants null and int, obtained by unfolding. Semantically, the set theoretic interpretation of the type definition (1) is based on a given semantic interpretation of type constructors. The usual interpretation of null and int is the set containing the empty list and the set of all integers, respectively. The product type constructor then corresponds to the Cartesian product, and the union type to set theoretic union. Recursion is interpreted by fixed points. Since the interpretations of the union and product type constructors are monotonic functions, by the Knaster-Tarski theorem we have that (1) admits both the least and the greatest fixed point, that is, the equation can be interpreted either inductively, or coinductively. The inductive interpretation yields the set of integer linked lists of finite length. Notably, cyclic and other infinite lists are not captured. In contrast, the coinductive interpretation consists of finite and infinite lists.

Moreover, the inductive interpretation of a type definition

$$x_2 \mapsto < \text{elm: int, nxt: } x_2 > \tag{2}$$

is the empty set. In a setting where cyclic lists can be built (e.g., in an object-oriented program) it is unsound to give an inductive type as above to cyclic lists. In fact, in the semantic subtyping approach an empty type cannot be inhabited by any value, otherwise the system becomes unsound: any such value can have an arbitrary type, by subsumption. In order to guarantee soundness either cyclic values are banned, or cyclic values are allowed but have less precise types. For instance, an acceptable inductive type for a cyclic list would be x_1 from (1). This, however, is not very precise, since accessing the n-th element of the list in a type safe way would require n non-emptiness checks which are useless in the case of a cyclic list.

The above argument shows that we have to consider a coinductive interpretation of recursive types (yielding, for example, for x_2, the set of infinite lists), and define subtyping semantically as set inclusion of coinductive interpretations. The main challenge is to provide an equivalent syntactic interpretation of recursive type declarations, and a corresponding sound and complete method for proving subtyping. Note that such a syntactic representation cannot be inductive either, because we are dealing with infinite terms. Existing coinductive proof methods such as [3–5] are incomplete and involve complex soundness proofs.

The theory of *coalgebras* has emerged as a general framework for a transparent and uniform study of coinduction (the basics are recalled in Section 2). Our aim therefore is to develop a coalgebraic approach to coinductive types, providing a single framework for the formalization of both canonical syntactic interpretations and equivalent semantic interpretations.

To achieve this goal we first focus on the basic notion of coinductive types without union (Section 3). This allows us to derive a natural syntactic interpretation of coinductive types by final coalgebras with bisimulation as a sound and complete proof method for equivalence of coinductive types. Further, this basic class of coinductive types allows us to focus on the general development of a final coalgebra of values from which we derive, in our framework, a semantic interpretation equivalent to the syntactic one.

The main challenge for a coalgebraic formalization of *union* types is to capture the distributivity of the union constructor over the product type constructor. In the setting of coalgebras this problem is reflected in the difference between bisimilarity and trace semantics. Our solution uses a coalgebraic approach to trace semantics based on [10, 18, 21], to extend the case of types without union to a precise characterization of semantic subtyping as inclusion between *subsets* of the final coalgebra, thus incorporating union types (Section 4).

Finally, we show how to reduce subtyping to inclusion between sets of finite traces, based on *approximation* of maximal traces by finite ones (Section 5). Such a reduction does not hold for arbitrary types of systems, but we devise a general coinductive proof technique for showing that it does apply in mildly restricted settings. This technique is instantiated to Moore automata and tree automata, yielding sound and complete methods for proving subtyping.

The contributions of this paper are as follows. We provide a structural and natural coalgebraic semantics for semantic subtyping of coinductive union types, which is parametric in the type constructors and abstracts away from a specific choice of syntax. We extend the theory of coalgebraic trace semantics with a novel coinductive method for finitely approximating maximal traces. We apply this technique to give the first sound and complete method for deciding semantic subtyping of coinductively interpreted recursive types with product and union.

2   Coalgebras

For an extensive introduction to the theory of universal coalgebra see [23]. We denote by Set the category of sets and functions and by Id the identity functor. Given a functor $F \colon \text{Set} \to \text{Set}$, an F-*coalgebra* is a pair (X, c) of a set X and a function $c \colon X \to FX$. A *homomorphism* between two coalgebras (X, c) and (Y, d) is a function $h \colon X \to Y$ such that $d \circ h = Fh \circ c$. An F-*bisimulation* between two F-coalgebras (X, c) and (Y, d) is a relation $R \subseteq X \times Y$ that can be equipped with an F-coalgebra structure γ turning both projections $\pi_l \colon R \to X$ and $\pi_r \colon R \to Y$ into coalgebra homomorphisms. Two elements $x \in X$ and $y \in Y$ are F-*bisimilar*, denoted by $x \sim_F y$, if there exists a bisimulation R containing the pair (x, y). If F is clear from the context we write $\sim$ instead of $\sim_F$.

Example 2.1. Let A be a set. For the functor $A \times \text{Id}$, a coalgebra consists of a set X and a function $\langle o, \delta \rangle \colon X \to A \times X$. Here $\langle o, \delta \rangle$ denotes the pairing of the *output* function $o \colon X \to A$ and the *next state* function $\delta \colon X \to X$. Given sets A and B, coalgebras for the functor $LX = B + (A \times X)$ are representations of infinite lists over A and finite lists over A with termination in B.

A (single-sorted) *signature* $\Sigma = (\Sigma_n)_{n \in \mathbb{N}}$ can be represented by a polynomial Set functor defined by $H_\Sigma(X) = \coprod_{n \in \mathbb{N}} \Sigma_n \times X^n$. A Σ-coalgebra over the set of variables X is given by a function assigning each $x \in X$ to a term $\sigma(x_1, \ldots, x_n)$, where $\sigma \in \Sigma_n$ is an operator of arity $n \geq 0$, and $x_i \in X$ for all $1 \leq i \leq n$.

For a given functor F, the *final coalgebra* (Ω, ξ_F) (if it exists) is a canonical domain of behaviour of F-coalgebras, with the property that for any F-coalgebra (X, c) there exists a unique homomorphism $h \colon X \to \Omega$ into it [23]. Final coalgebras exist under mild conditions on the functor.

Example 2.2. The carrier of the final coalgebra for the functor $A \times \mathsf{Id}$ consists of the set of all infinite lists over A. For $LX = B + (A \times X)$, the final coalgebra consists of all finite lists in A^*B and infinite lists in A^ω. It is thus given by the set $A^*B \cup A^\omega$ with coalgebra map $\zeta \colon A^*B \cup A^\omega \to B + (A \times (A^*B \cup A^\omega))$ defined by $\zeta(b) = b$ and $\zeta(aw) = \langle a, w \rangle$ for all $a \in A$, $b \in B$ and $w \in A^*B \cup A^\omega$. The final coalgebra of a signature functor H_Σ is given by $\omega \colon T_\Sigma^\infty \to \Sigma(T_\Sigma^\infty)$ where T_Σ^∞ is the set of all finite and infinite Σ-trees (see, for instance, [1]).

One of the central elements of the theory of coalgebras is the (proof) principle of *coinduction*, which says that bisimilar states are mapped to the same element of the final coalgebra: if $x \sim y$ then $h(x) = h(y)$. Establishing bisimulations is a concrete proof method for bisimilarity, and thus, by the above principle, for equality in the final coalgebra. If the functor preserves weak pullbacks, a rather mild condition satisfied by all of the above examples, the converse holds as well [23], i.e., $h(x) = h(y)$ implies $x \sim y$. In the following sections we implicitly assume all functors to preserve weak pullbacks.

3 A Semantic Approach to Coinductive Types

In this section we propose a framework for coinductive types without union. We use two functors F and G as follows: F-coalgebras are interpreted as (recursive) *type* definitions, whereas G-coalgebras are (recursive) *value* definitions. We assume that the final coalgebras of F and G exist. The carrier $\mathbb{T}$ of the final F-coalgebra $(\mathbb{T}, \xi_F)$ consists of all coinductive types. The carrier $\mathbb{V}$ of the final G-coalgebra $(\mathbb{V}, \xi_G)$ is the set of all coinductive values.

Example 3.1. A type definition such as $x \mapsto <$ elm: int, nxt: $y >$ together with $y \mapsto <$ elm: bool, nxt: $x >$, can be given as a coalgebra for $\{\mathsf{int}, \mathsf{bool}\} \times \mathsf{Id}$. The homomorphism into the final coalgebra maps x to $\mathsf{int}, \mathsf{bool}, \mathsf{int}, \mathsf{bool}, \ldots \in \mathbb{T}$.

An infinite recursive definition $p_i \mapsto (i, q_i)$ and $q_i \mapsto (true, p_{i+1})$ for $i \in \mathbb{N}$ can be represented as a coalgebra for the functor $(\mathbb{N} + \mathbb{B}) \times \mathsf{Id}$, where $\mathbb{B} = \{true, false\}$ is the set of Boolean values, and $\mathbb{N}$ is the set of non-negative integers. Then p_0 is mapped to the infinite list $0, true, 1, true, 2, \ldots \in \mathbb{V}$ in the final coalgebra.

The functors F and G will be connected by a *natural transformation*. A natural transformation $\alpha \colon G \Rightarrow F$ associates to every set X a function $\alpha_X \colon GX \to FX$ such that for any function $f \colon X \to Y$ we have $Ff \circ \alpha_X = \alpha_Y \circ Gf$. In order to assign types to values we assume given a natural transformation $\alpha \colon G \Rightarrow F$, which represents an assignment of types to basic values. We will exhibit an example below, but first we set up the general framework, which coinductively assigns types to values. More precisely, by applying the natural transformation α to the final G-coalgebra, we turn it into an F-coalgebra and thus obtain a unique F-coalgebra homomorphism from coinductive values to coinductive types. This is depicted in the middle of the diagram below.

The map τ defined by finality gives the assignment of types to values. The left and the right side of the diagram are representations:

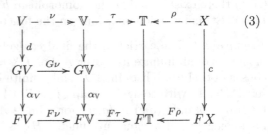

$$(3)$$

– Given a *value representation* $d\colon V \to GV$ we let $\nu\colon V \to \mathbb{V}$ be the unique coalgebra homomorphism and extend d, again using α, to a F-coalgebra. This is depicted in the two commuting squares on the left side of the diagram.

– Given a *type representation* $c\colon X \to FX$ we let $\rho\colon X \to \mathbb{T}$ be the unique homomorphism into $\mathbb{T}$, as depicted on the right side of the diagram. A *typing* relation between V and X is then defined in the obvious way: given $p \in V$ and $x \in X$ we let $p : x$ iff $\tau(\nu(p)) = \rho(x)$.

Example 3.2. Continuing the above Example 3.1, we can define $\alpha\colon ((\mathbb{N} + \mathbb{B}) \times \mathsf{Id}) \Rightarrow (\{\mathsf{int}, \mathsf{bool}\} \times \mathsf{Id})$ for every set S simply by putting $\alpha_S((n, s)) = (\mathsf{int}, s)$ and $\alpha_S((b, s)) = (\mathsf{bool}, s)$ for all $n \in \mathbb{N}$, $b \in \mathbb{B}$, and $s \in S$. For the concrete type and value definitions x and p_0 respectively, of Example 3.1, it is easy to check that $\tau(\nu(p_0)) = \rho(x)$, so $p_0 : x$ as expected. In fact, as we will see below, this can be checked by establishing a bisimulation.

In the above approach the meaning of a type declaration $c\colon X \to FX$ is given by finality, in terms of the unique homomorphism $\rho\colon X \to \mathbb{T}$. It is thus independent of the language of values. Next we interpret types semantically, as sets of values, and subsequently we relate the two interpretations.

Definition 3.1. *Types are interpreted as sets of values by* $[\![-]\!]\colon \mathbb{T} \to \mathcal{P}(\mathbb{V})$, *defined as the inverse of* τ, *i.e.,* $[\![t]\!] = \{v \in \mathbb{V} \mid \tau(v) = t\}$ *for any* $t \in \mathbb{T}$.

It follows from the above definition that if $[\![t_1]\!] = [\![t_2]\!]$ and both $[\![t_1]\!]$ and $[\![t_2]\!]$ are non-empty, then $t_1 = t_2$. Types are inhabited by values (thus non-empty) if the natural transformation $\alpha\colon G \Rightarrow F$ mapping values to types is surjective in all of its components, i.e., α_X is surjective for any set X.

Lemma 3.1. *If* α *is a surjective natural transformation then* τ *is surjective.*

Corollary 3.1. *If* α *is a surjective natural transformation then* $t_1 = t_2$ *if and only if* $[\![t_1]\!] = [\![t_2]\!]$ *for all* $t_1, t_2 \in \mathbb{T}$.

Note that if $[\![t_1]\!] \subseteq [\![t_2]\!]$ then $[\![t_1]\!] = [\![t_2]\!]$. Subtyping will become relevant in the next section, where we consider subsets of $\mathbb{T}$.

To see why surjectivity is a natural condition, consider the type definitions x and y from Example 3.1, and $\alpha\colon (\mathbb{N} \times \mathsf{Id}) \Rightarrow (\{\mathsf{int}, \mathsf{bool}\} \times \mathsf{Id})$ given by $\alpha_S(n, s) = (\mathsf{int}, s)$. In this case clearly $\rho(x) \neq \rho(y)$, whereas $[\![\rho(x)]\!] = \emptyset = [\![\rho(y)]\!]$.

Equality of types coincides with *bisimilarity*, by coinduction: $\mathbb{T}$ is a final coalgebra. Thus we obtain the following soundness and completeness result.

Theorem 3.1. *Using the setting of* (3), *let* $c\colon X \to FX$ *be a coalgebra, and* α *surjective. For all* $x, y \in X$ *we have* $[\![\rho(x)]\!] = [\![\rho(y)]\!]$ *iff* $\rho(x) = \rho(y)$ *iff* $x \sim_F y$.

If F is a polynomial functor (constructed by finite sum and product) and we restrict to type declarations using only finitely many variables (so that types essentially represent rational trees over a signature) then bisimulation is not only a sound and complete proof method for type equality, but it is also decidable [7]. We note that in the above framework, computing the typing relation can be seen as a special case of type equality (by turning G-coalgebras into F-coalgebras), and therefore it can also be computed using bisimulations.

4   Coinductive Union Types

In the previous section we have introduced a coalgebraic semantics, where types, i.e., elements of the final coalgebra $\mathbb{T}$, are equal if and only if they represent the same sets of values. Types and values can be represented by coalgebras, and bisimulation provides a concrete proof principle for type equivalence. In the current section we are interested in extending these results to *union* types, that is, *subsets* of $\mathbb{T}$. By $\mathcal{P}(X)$ we denote the power set functor applied to a set X, i.e., the set of subsets of X; for a function $f\colon X \to \mathcal{P}(Y)$ we write $f^\sharp\colon \mathcal{P}(X) \to \mathcal{P}(Y)$ for its direct image. In the previous section we have coinductively constructed a map $\tau\colon \mathbb{V} \to \mathbb{T}$ from values to types, from which the semantics $[\![t]\!]$ of a type $t \in \mathbb{T}$ as a set of values can be defined simply by using the inverse. In order to have a natural counterpart of Theorem 3.1 in the setting of *subtyping* we extend the semantics to sets of types using direct image $[\![-]\!]^\sharp\colon \mathcal{P}(\mathbb{T}) \to \mathcal{P}(\mathbb{V})$, i.e., $[\![S]\!]^\sharp = \{v \in \mathbb{V} \mid \tau(v) \in S\}$.

Theorem 4.1. *If* α *is a surjective natural transformation then* $T_1 \subseteq T_2$ *if and only if* $[\![T_1]\!]^\sharp \subseteq [\![T_2]\!]^\sharp$, *for all* $T_1, T_2 \subseteq \mathbb{T}$.

One of the main problems is to *represent* elements of $\mathcal{P}(\mathbb{T})$ as coalgebras. In the previous section we have seen how an F-coalgebra represents a type definition; it is natural to consider a $\mathcal{P}F$-coalgebra instead, in the case of union types, adding a top-level union constructor. The problem here is that the *branching* of $\mathcal{P}F$-coalgebras should not be considered. Indeed, $\mathcal{P}(\mathbb{T})$ is not the final coalgebra of $\mathcal{P}F$—in fact, $\mathcal{P}F$ does not even have a final coalgebra for cardinality reasons. But even if we restrict ourselves to $\mathcal{P}_f F$ (where $\mathcal{P}_f(X)$ is the set of finite subsets of X), then the final coalgebra consists of finitely branching synchronization trees labelled in a and quotiented by strong bisimilarity. Instead, we need the *trace semantics* of $\mathcal{P}F$-coalgebras. To this end we base ourselves on the coalgebraic trace semantics of [21].

Definition 4.1. *Let* $c\colon X \to \mathcal{P}(FX)$ *be a coalgebra, and* $(\mathbb{T}, \xi_{\mathbb{T}})$ *the final* F-*coalgebra. A trace map* tr *is a map that makes the following diagram commute:*

$$
\begin{array}{ccc}
X & \xrightarrow{\;\;tr\;\;} & \mathcal{P}(\mathbb{T}) \\
{\scriptstyle c}\big\downarrow & & \big\downarrow{\scriptstyle \mathcal{P}(\xi_{\mathbb{T}})} \\
\mathcal{P}(FX) & \xrightarrow{\;(\bar{F}tr)^{\sharp}\;} & \mathcal{P}(F\mathbb{T})
\end{array}
$$

where $\bar{F}(tr)$ *is defined by relation lifting [21]. If the diagram does not commute but* $\mathcal{P}(\xi_{\mathbb{T}}) \circ tr \subseteq (\bar{F}tr)^{\sharp} \circ c$, *then we say* tr *is a* quasi *trace map.*

Instead of recalling the definition of relation lifting, we introduce it by examples.

Example 4.1. Consider the functor $FX = B + (A \times X)$. Then $\mathbb{T} = A^*B \cup A^\omega$ (see Example 2.2). A coalgebra $c\colon X \to \mathcal{P}FX$ is a *nondeterministic Moore automaton*. A *trace map* is a map $tr\colon X \to \mathcal{P}(A^*B \cup A^\omega)$ such that for all $b \in B$: $b \in tr(x)$ iff $b \in c(x)$, and for all $aw \in A(A^*B \cup A^\omega)$: $aw \in tr(x)$ iff $(a, y) \in c(x)$ and $w \in tr(y)$ for some $y \in X$. For a quasi trace map, these equivalences are relaxed to implications from left to right.

Given any signature functor H_Σ (Example 2.1), a $\mathcal{P}H_\Sigma$-coalgebra is a *nondeterministic top-down tree automaton*. The trace map associated with a coalgebra $c\colon X \to \mathcal{P}(H_\Sigma X)$ satisfies the following: $\sigma \in tr(x)$ iff $\sigma \in \Sigma_0 \cap c(x)$, and $\sigma(k_1, \ldots, k_n) \in tr(x)$ iff $\langle \sigma, x_1, \ldots, x_n \rangle \in \Sigma_n \times X^n \cap c(x)$ and $k_i \in tr(x_i)$ for $1 \leq i \leq n$. Again, for a quasi trace map these are implications from left to right.

The set of maps of type $X \to \mathcal{P}(\mathbb{T})$ forms a complete lattice, by pointwise extension of the subset inclusion order on $\mathcal{P}(\mathbb{T})$. A trace map can be viewed as a fixpoint of a map on this complete lattice; since relation lifting is monotone, this is a monotone map, and therefore, by the Knaster-Tarski theorem, for a fixed $\mathcal{P}F$-coalgebra the greatest trace map as well as the least trace map exist (a similar approach is taken in [10]). To model *coinductive* types we are interested in this greatest trace map (in the sequel typically denoted by T and called *maximal traces*). Moreover, we get the following proof principle: if tr is a quasi trace map, then it is a post-fixed point of the above monotone map, so it is (pointwise) included in the greatest one: $tr \subseteq T$. This proof technique is applied in Section 5.

Example 4.2. Continuing Example 4.1, the *least* trace map t for a non-deterministic Moore automaton assigns to a state the standard definition of its finite traces in A^*B. The *greatest* trace map T assigns to a state the finite traces as well as the infinite traces in A^ω. For example, recall the type definition x_1 from equation (1) of the introduction, representing finite and infinite lists of integers, and x_2 from equation (2) representing infinite lists of integers. They clearly define Moore automata. For the least trace map t we have $t(x_1) = \mathsf{int}^*\mathsf{null}$ and $t(x_2) = \emptyset$. For the greatest trace map T we have $T(x_1) = t(x_1) \cup \mathsf{int}^\omega$ and $T(x_2) = \mathsf{int}^\omega$ (i.e. the desired coinductive types of definitions x_1 and x_2).

For a non-deterministic (top-down) tree automaton, the least trace map is simply the standard semantics of tree automata, assigning a tree language (of finite trees) to each state. The greatest trace map contains this language as well as all *infinite* trees such that, when parsed, the automaton does not block. These tree automata can be used to represent type definitions, similarly to Moore automata, but generalizing this to arbitrary (finite) use of the product constructor.

Corollary 4.1. *For any coalgebra $c\colon X \to \mathcal{P}(FX)$ and any $x, y \in X$ we have $T(x) \subseteq T(y)$ iff $[\![T(x)]\!]^\sharp \subseteq [\![T(y)]\!]^\sharp$ (given that α is surjective).*

Thus, subset inclusion between syntactic unfoldings of sets of types is sound and complete with respect to *semantic subtyping*, i.e., inclusion between types interpreted as sets of values. Unfortunately, since $\mathcal{P}(\mathbb{T})$ is not a final coalgebra, we do not obtain bisimilarity (or similarity) as a proof principle, as was the case in the framework of Section 3. We address the problem of proving subtyping in the following section.

5 Approximating Coinductive Union Types

By the main results of the previous section, semantic subtyping coincides with subtyping between sets of maximal traces, that is, syntactic unfoldings of type definitions. In this section we provide a generally applicable technique to reduce subtyping to inclusion between *finite* traces. This is based on finite *approximation* of maximal traces, which we introduce below.

We fix a functor $F\colon \mathsf{Set} \to \mathsf{Set}$ (preserving weak pullbacks) and a coalgebra $c\colon X \to \mathcal{P}FX$. In order to define approximation, consider the functor $F_\bot = F + \{\bot\}$, and the natural transformation $\gamma\colon \mathcal{P}F \Rightarrow \mathcal{P}F_\bot$ given by $\gamma_X(S) = S \cup \{\bot\}$. We can now turn c into the $F_\bot$-coalgebra $\gamma_X \circ c$. It is our aim to use the finite traces of $\gamma_X \circ c$ to approximate the maximal traces of c. We use the approach of [18] to finite trace semantics via finality in the category Rel, where objects are sets and morphisms are relations (represented as functions $X \to \mathcal{P}(Y)$).

Central to this approach is the *initial algebra* of $F_\bot$, which we denote by $\iota\colon F_\bot \mathbb{I} \to \mathbb{I}$. By Lambek's lemma ι is an isomorphism. Now, by [18, Theorem 3.8], $\mathbb{I}$ is the *final coalgebra* in Rel, for the functor $\bar{F}_\bot$ defined by relation lifting. Thus, for any $F_\bot$-coalgebra in Rel, that is, a $\mathcal{P}F_\bot$-coalgebra in Set, we obtain a unique map into $\mathcal{P}(\mathbb{I})$. Applying this to a coalgebra $\gamma_X \circ c\colon X \to \mathcal{P}F_\bot X$ as constructed above, we get a unique map $t_\bot\colon X \to \mathcal{P}(\mathbb{I})$ as in (4).

$$
\begin{array}{ccc}
X & \xrightarrow{\;\;t_\bot\;\;} & \mathcal{P}(\mathbb{I}) \quad (4)\\
{\scriptstyle c}\downarrow & & \downarrow\\
\mathcal{P}(FX) & & \\
{\scriptstyle \gamma_X}\downarrow & & \downarrow\\
\mathcal{P}(F_\bot X) & \xrightarrow{(\bar{F}_\bot t_\bot)^\sharp} & \mathcal{P}(F_\bot \mathbb{I})
\end{array}
$$

Example 5.1. For a non-deterministic Moore automaton $c\colon X \to \mathcal{P}(B + (A \times X))$, the above construction yields the finite trace semantics for $\gamma_X \circ c$, which is the Moore automaton obtained by adding the output $\bot$ to each state. We regard a word $w\bot$ as a *prefix* of a word $v \in A^*B \cup A^\omega$ if $wv' = v$ for some v'; in this sense, $t_\bot(x)$ is prefix-closed.

Applying the above construction to $\mathbb{T}$, we get a map $approx\colon \mathbb{T} \to \mathcal{P}(\mathbb{I})$. This map, informally, computes the approximations of maximal traces. Consider now the following map defined from it: $maxtr\colon \mathcal{P}(\mathbb{I}) \to \mathcal{P}(\mathbb{T})$, given by $maxtr(S) = \{w \in \mathbb{T} \mid approx(w) \subseteq S\}$. The map $maxtr$ computes the set of maximal traces represented by a set of approximations. The following lemma states that the function $t_\perp$ can be represented as the approximation of maximal traces.

Lemma 5.1. $t_\perp = approx^\sharp \circ T$.

This follows from the fact that $\mathbb{I}$ is final in Rel. As a simple consequence of this result and the fact that $maxtr$ is defined as the (upper) inverse of $approx$, we now obtain the following:

Corollary 5.1. $T \subseteq maxtr \circ t_\perp$.

The converse of the above corollary does not hold in general. There is a standard counterexample (e.g., [17]): take a non-deterministic Moore automaton containing a state x that accepts all *finite* traces of the form $a^n b$ (for some b and all $n \geq 0$), but not the infinite trace $a^\omega = aaa\ldots$ (such an automaton can be realized using infinite branching). Then $maxtr \circ t_\perp(x)$ contains a^ω, whereas $T(x)$ does not.

To prove the converse for restricted classes of coalgebras, we use that T is a greatest fixpoint. Under the condition that $maxtr \circ t_\perp$ is a quasi trace map, we obtain the soundness and completeness of finite traces w.r.t. (semantic) subtyping.

Theorem 5.1. *Let* $c\colon X \to \mathcal{P}(FX)$ *be a coalgebra such that* $maxtr \circ t_\perp$ *is a quasi trace map. Then for any* $x, y \in X\colon t_\perp(x) \subseteq t_\perp(y)$ *iff* $T(x) \subseteq T(y)$.

Proof. Suppose $t_\perp(x) \subseteq t_\perp(y)$. If $maxtr \circ t_\perp$ is a quasi trace map then $maxtr \circ t_\perp \subseteq T$; combined with Corollary 5.1, this yields $maxtr \circ t_\perp = T$. Conversely, if $T(x) \subseteq T(y)$ then $approx^\sharp \circ T(x) \subseteq approx^\sharp \circ T(y)$, so $t_\perp(x) \subseteq t_\perp(y)$ by Lemma 5.1.

Moore automata. As shown in Example 4.1, non-deterministic Moore automata can be used to represent types for finite and infinite lists. However, in general they do not satisfy the condition of Theorem 5.1; we need to make an appropriate restriction on the branching behaviour. We say $c\colon X \to \mathcal{P}(B + (A \times X))$ is *image-finite* when for any $x \in X$ and any $a \in A\colon c(x)$ may contain finitely many elements of the form (a, x) (but infinitely many of B, and A may itself be infinite).

Proposition 5.1. *For any image-finite Moore automaton:* $t_\perp(x) \subseteq t_\perp(y)$ *iff* $T(x) \subseteq T(y)$.

Proof. Let c be image-finite. Using Example 4.1, we see that to prove that $maxtr \circ t_\perp$ is a trace map, is to prove that 1) $b \in maxtr \circ t_\perp(x)$ implies $b \in c(x)$, and 2) for all $aw \in A(A^*B \cup A^\omega)$: if $aw \in maxtr \circ t_\perp(x)$ then $(a, y) \in c(x)$ and $w \in maxtr \circ t_\perp(y)$ for some $y \in X$. The first part 1) is easy: $b \in maxtr \circ t_\perp(x)$ implies $b \in t_\perp(x)$, which in turn implies $b \in c(x)$. For 2), suppose $aw \in maxtr \circ t_\perp(x)$. Then $w \in \bigcup_{(a,y) \in c(x)} maxtr \circ t_\perp(y)$; by image-finiteness, this is a *finite* union. The case that w is finite is straightforward; suppose w is infinite. Then $approx(w)$ is infinite; and thus there is some y for which infinitely many prefixes of w are contained in $t_\perp(y)$. But $t_\perp(y)$ is prefix-closed; so $w \in maxtr \circ t_\perp(y)$.

Example 5.2. Consider the following type definition.

$$x_3 \mapsto < \text{elm: int, nxt: } x_4 >$$
$$x_4 \mapsto \text{null} \vee < \text{elm: int, nxt: } x_4 > \vee < \text{elm: bool, nxt: } x_4 > . \tag{5}$$

In the coinductive interpretation this represents all finite and infinite lists of integers and booleans that start with an integer. Consider the types below:

$$x_5 \mapsto < \text{elm: int, nxt: } x_6 > \vee < \text{elm: int, nxt: } x_7 > \vee < \text{elm: int, nxt: } x_8 >$$
$$x_6 \mapsto < \text{elm: bool, nxt: } x_8 > \vee < \text{elm: bool, nxt: } x_6 > \vee < \text{elm: int, nxt: } x_6 >$$
$$x_7 \mapsto < \text{elm: int, nxt: } x_8 > \vee < \text{elm: bool, nxt: } x_7 > \vee < \text{elm: int, nxt: } x_7 >$$
$$x_8 \mapsto \text{null} .$$
$$\tag{6}$$

Here x_6 and x_7 represent infinite lists, as well as finite lists ending with bool and int, respectively. We can now prove that $T(x_3) \subseteq T(x_5)$ by reducing it to $t_\perp(x_3) \subseteq t_\perp(x_5)$, which is a simple case of language inclusion.

Tree automata. A tree automaton $c\colon X \to \mathcal{P}^+(H_\Sigma X)$ is said to be *image finite* if for all $x \in X$ and $\sigma \in \Sigma^\perp$ there are only finitely many tuples $\langle \sigma, x_1, \ldots x_n \rangle \in c(x)$, where n is the arity of σ.

Proposition 5.2. *For any image-finite tree automaton:* $t_\perp(x) \subseteq t_\perp(y)$ *iff* $T(x) \subseteq T(y)$.

The proof is a straightforward extension of the case of Moore automata. Thus, we obtain inclusion of *tree languages* (of finite trees) as a sound and complete method to show semantic subtyping of recursive types with product and union, interpreted coinductively. For regular tree languages, i.e., languages accepted by a top-down non-deterministic tree automaton with finitely many states, language inclusion (and thus subtyping) is decidable, although it is EXPTIME-complete [11].

6 Related Work

Axiomatizations and algorithms for subtyping on recursive types interpreted coinductively have been proposed by Amadio and Cardelli [2] in the context of functional programming; subsequently, a more concise sound and complete axiomatization has been proposed by Brandt and Henglein [9], with a novel rule for a finitary coinduction principle. In these papers types are interpreted as ideals in a universal domain, hence they do not follow the semantic subtyping approach where subtyping corresponds to the subset relation. Furthermore, types have no Boolean operators; as we will see, introducing union types makes sound and complete axiomatization of subtyping more challenging.

Damm [12] proves decidability of subtyping between recursive types with intersection, union, and function types, by reduction to the problem of inclusion between regular tree expressions. However, the paper does not consider record types, and, more importantly, types are interpreted inductively, rather than coinductively, over a rather complex metric space of ideals. As a consequence, the corresponding subtyping relation is not comparable with ours. Di Cosmo et al. [13] study subtyping of recursive types up to associativity and commutativity of products; their definition of subtyping is fully axiomatic, and only products and arrow types are considered, no Boolean operators. A nice introduction to the fundamental theory of recursive types and subtyping can be found in the work by Gapeyev et al. [16]; the survey does not consider Boolean operators, and subtyping is defined axiomatically, hence a type interpretation is not introduced.

Semantic subtyping in the presence of Boolean operators and product or record type constructors has been intensively studied in the context of the XDuce [20] and CDuce [6] programming languages. As in our case, the subtyping relation corresponds to a natural semantic notion: types denote sets of documents (that is, sets of finite trees), and subtyping coincides with inclusion between the sets denoted by two types. The main difference with coinductive types is their interpretation: types in both XDuce and CDuce are interpreted inductively, therefore a type definition as (2) corresponds to the empty set of values; as a matter of fact, types in XDuce and CDuce fail to capture cyclic values. Even though CDuce supports references, and, hence, it is possible to create cycles, the types that can be correctly assigned to cyclic values are "inductive".

Semantic subtyping with union and coinductive types has been studied in the context of precise static type analysis for object-oriented programming [3]. Sound but not complete axiomatizations of subtyping have been defined in [4, 5].

7   Future Work

The coalgebraic framework presented in this paper provides the basis for an extensive, structured investigation of subtyping for coinductive union types.

The subtyping relation could be refined by allowing subtyping between primitive types (e.g., nat is a subtype of int) as well as depth and width subtyping between records. Technically, this could be achieved by moving our framework from the category Set to the category of partially ordered sets.

The methods in [8, 14] allow to canonically derive sound and complete axiomatizations for the rational subset of the final coalgebra of a polynomial functor. For example, one can easily obtain a calculus for subtyping, by combining the axiomatisation of tree regular expressions of [14] with the approximation results of Section 5 of the present paper.

In our framework we abstracted from concrete calculi of expressions evaluating to values. It would be interesting to integrate the bialgebraic approach [22] (defining syntax and semantics of expressions) within our framework by allowing the specification of typing rules for each operator.

References

1. Aczel, P., Adámek, J., Milius, S., Velebil, J.: Infinite trees and completely iterative theories: A coalgebraic view. Theoretical Computer Science 300(1-3), 1–45 (2003)
2. Amadio, R., Cardelli, L.: Subtyping recursive types. ACM Transactions on Programming Languages and Systems 15(4) (1993)

3. Ancona, D., Lagorio, G.: Coinductive type systems for object-oriented languages. In: Drossopoulou, S. (ed.) ECOOP 2009. LNCS, vol. 5653, pp. 2–26. Springer, Heidelberg (2009)
4. Ancona, D., Lagorio, G.: Coinductive subtyping for abstract compilation of object-oriented languages into Horn formulas. In: GandALF 2010. EPTCS, vol. 25 (2010)
5. Ancona, D., Lagorio, G.: Complete coinductive subtyping for abstract compilation of object-oriented languages. In: FTfJP 2010. ACM Digital Library (2010)
6. Benzaken, V., Castagna, G., Frisch, A.: CDuce: An XML-Centric General-Purpose Language. In: ICFP (2003)
7. Bonsangue, M., Caltais, G., Goriac, E.-I., Lucanu, D., Rutten, J., Silva, A.: Automatic equivalence proofs for non-deterministic coalgebras. Science of Computer Programming 798(9), 1324–1345 (2013)
8. Bonsangue, M., Milius, S., Silva, A.: Sound and Complete Axiomatizations of Coalgebraic Language Equivalence. ACM Trans. on Comp. Logic 14(1), 7 (2013)
9. Brandt, M., Henglein, F.: Coinductive axiomatization of recursive type equality and subtyping. Fundamentae Informatica 33(4) (1998)
10. Cîrstea, C.: From Branching to Linear Time, Coalgebraically. In: FICS 2013. EPTCS, vol. 126, pp. 11–27 (2013)
11. Comon, H., Dauchet, M., Gilleron, R., Löding, C., Jacquemard, F., Lugiez, D., Tison, S., Tommasi, M.: Tree Automata Techniques and Applications, http://www.grappa.univ-lille3.fr/tata
12. Damm, F.: Subtyping with Union Types, Intersection Types and Recursive Types. In: Hagiya, M., Mitchell, J.C. (eds.) TACS 1994. LNCS, vol. 789, pp. 687–706. Springer, Heidelberg (1994)
13. Di Cosmo, R., Pottier, F., Rémy, D.: Subtyping Recursive Types Modulo Associative Commutative Products. In: Urzyczyn, P. (ed.) TLCA 2005. LNCS, vol. 3461, pp. 179–193. Springer, Heidelberg (2005)
14. Ésik, Z.: Axiomatizing the equational theory of regular tree languages. Journal of Logic and Algebraic Programming 79(2), 189–213 (2010)
15. Frisch, A., Castagna, G., Benzaken, V.: Semantic Subtyping: dealing set-theoretically with function, union, intersection, and negation types. The Journal of the ACM (2008)
16. Gapeyev, V., Levin, M.Y., Pierce, B.C.: Recursive subtyping revealed. The Journal of Functional Programming 12(6), 511–548 (2002)
17. van Glabbeek, R.: The linear time - branching time spectrum I. The semantics of concrete, sequential processes. In: Handbook of Process Algebra, pp. 3–99 (2001)
18. Hasuo, I., Jacobs, B., Sokolova, A.: Generic trace semantics via coinduction. Logical Methods in Computer Science 3, 1–36 (2007)
19. Hosoya, H., Vouillon, J., Pierce, B.C.: Regular expression types for XML. ACM Trans. Program. Lang. Syst. 27(1), 46–90 (2005)
20. Hosoya, H., Pierce, B.C.: XDuce. A statically typed XML processing language. ACM Trans. Internet Techn. 3(2), 117–148 (2003)
21. Jacobs, B.: Trace Semantics for Coalgebras. In: CMCS 2004. ENTCS, vol. 106 (2004)
22. Klin, B.: Bialgebras for structural operational semantics: An introduction. Theoretical Computer Science 412(38), 5043–5069 (2011)
23. Rutten, J.: Universal coalgebra: a theory of systems. Theoretical Computer Science 249 (2000)

Turing Degrees of Limit Sets of Cellular Automata[*]

Alex Borello, Julien Cervelle, and Pascal Vanier

Laboratoire d'algorithmique, complexité et logique
Université de Paris-Est, LACL, UPEC, France

Abstract. Cellular automata are discrete dynamical systems and a model
of computation. The limit set of a cellular automaton consists of the configurations having an infinite sequence of preimages. It is well known that
these always contain a computable point and that any non-trivial property
on them is undecidable. We go one step further in this article by giving a
full characterization of the sets of Turing degrees of limit sets of cellular automata: they are the same as the sets of Turing degrees of effectively closed
sets containing a computable point.

1 Introduction

A d-dimensional cellular automaton (CA for short) consists of cells aligned on $\mathbb{Z}^d$
that may be in a finite number of states, and are updated synchronously with a
local rule, i.e. depending only on a finite neighborhood. All cells operate under
the same local rule. The state of all cells at some time step is called a configuration. CAs are very well known for being simple systems that may exhibit
complicated behavior.

A d-dimensional subshift of finite type (SFT for short) is a set of colorings
of $\mathbb{Z}^d$ by a finite number of colors containing no pattern from a finite family of
forbidden patterns. Most proofs of undecidability concerning CAs involve the
use of SFTs, so both topics are very intertwined [Kar90; Kar92; Kar94a; Mey08].
A recent trend in the study of SFTs has been to give computational characterizations of dynamical properties, which has been followed by the study of their
computational structure and in particular the comparison with the computational structure of effectively closed sets, which are the subsets of $\{0,1\}^{\mathbb{N}}$ on
which some Turing machine does not halt. It is quite easy to see that SFTs are
such sets.

In this paper, we follow this trend and study limit sets of CAs, which consist
of all the configurations of a given CA that can occur after arbitrarily long
computations; they were introduced by [CPY89] in order to classify CAs. It
has been proved that non-trivial properties on these sets are undecidable by
[Kar94b] for CAs of all dimensions. Limit sets of CAs are subshifts, and the
question of which subshifts may be limit sets of CA has been a thriving topic,

[*] This work was sponsored by grants EQINOCS ANR 11 BS02 004 03 and TARMAC
ANR 12 BS02 007 01.

see [Hur87; Hur90a; Hur90b; Maa95; FK07; LM09; BGK11]. However, most of these results are on the language of the limit set or on simple limit sets. Our aim here is to study the configurations themselves.

In dimension 1, limit sets are effectively closed sets, so it is quite natural to compare them from a computational point of view. The natural measure of complexity for effectively closed sets is the Medvedev degree [Sim11], which, informally, is a measure of the complexity of the simplest points of the set. As limit sets always contain a uniform configuration (wherein all cells are in the same state), they always contain a computable point and have Medvedev degree $\mathbf{0}$. Thus, if we want to study their computational structure, we need a finer measure; in this sense, the set of Turing degrees is appropriate.

It turns out that for SFTs, there is a characterization of their sets of Turing degrees found by [JV13b], which states that one may construct SFTs with the same Turing degrees as any effectively closed set containing a computable point. In the case of limit sets, such a characterization would be perfect, as limit sets always contain a computable point[1]. This is exactly what we achieve in this article:

Theorem 1. *For any effectively closed set S, there exists a one-dimensional cellular automaton $\mathcal{A}$ such that*

$$\deg_T \Omega\left(\mathcal{A}\right) = \deg_T S \cup \{\mathbf{0}\}.$$

In the way to achieve this theorem, we introduce a new construction that allows us some control over the limit set. We hope that this construction will lead to other unrelated results on limit sets of CAs, as was the case for the construction in [JV13b], see [JV13a].

The paper is organized as follows: in section 2 we recall the usual definitions concerning CAs and Turing degrees, then in section 3 we give the reasons for each trait of the construction that allows us to prove theorem 1, and section 4 gives the actual construction. The choice has been made to have colored figures, which are best viewed onscreen.

2   Preliminary Definitions

A (1-dimensional) *cellular automaton* is a triple $\mathcal{A} = (Q, r, \delta)$, where Q is the finite set of *states*, $r > 0$ is the *radius* and $\delta : Q^{2r+1} \to Q$ the *local transition function*.

An element of $i \in \mathbb{Z}$ is called a *cell*, and the set $[\![i-r, i+r]\!]$ is the *neighborhood* of i (the elements of which are the *neighbors* of i). A *configuration* is a function $\mathfrak{c} : \mathbb{Z} \to Q$. The local transition function induces a *global transition function* (that can be regarded as the automaton itself, hence the notation), which associates to any configuration $\mathfrak{c}$ its *successor*:

$$\mathcal{A}(\mathfrak{c}) : \begin{cases} \mathbb{Z} \to Q \\ i \mapsto \delta(\mathfrak{c}(i-r), \dots, \mathfrak{c}(i-1), \mathfrak{c}(i), \mathfrak{c}(i+1), \dots, \mathfrak{c}(i+r)). \end{cases}$$

---

[1] Note that this is not the case for subshifts: there exist non-empty effective subshifts containing only non-computable points.

In other words, all cells are finite automata that update their states in parallel, according to the same local transition rule, transforming a configuration into its successor.

If we draw some configuration as a horizontal bi-infinite line of cells, then add its successor above it, then the successor of the latter and so on, we obtain a *space-time diagram*, which is a two-dimensional representation of some computation performed by $\mathcal{A}$.

A *site* $(i, t) \in \mathbb{Z}^2$ is a cell i at a certain time step t of the computation we consider (hereinafter there will never be any ambiguity on the automaton nor on the computation considered).

The *limit set* of $\mathcal{A}$, denoted by $\Omega(\mathcal{A})$, is the set of all the configurations that can appear after arbitrarily many computation steps:

$$\Omega(\mathcal{A}) = \bigcap_{k \in \mathbb{N}} \mathcal{A}^k(Q^{\mathbb{Z}}).$$

For surjective CAs, the limit set is the set of all possible configurations $Q^{\mathbb{Z}}$, while for non-surjective CAs, it is the set of all configurations containing no orphan of any order, see [Hur90b]. An *orphan of order n* is a finite word w which has no preimage by $\mathcal{A}^n_{Q^{|w|}}$.

An *effectively closed set*, or Π^0_1 *class*, is a subset S of $\{0,1\}^{\mathbb{N}}$ for which there exists a Turing machine that, given any $x \in \{0,1\}^{\mathbb{N}}$, halts if and only if $x \notin S$. Equivalently, a class $S \subseteq \{0,1\}^{\mathbb{N}}$ is Π^0_1 if there exists a computable set L such that $x \in S$ if and only if no prefix of x is in L. It is then quite easy to see that limit sets of CAs are Π^0_1 classes: for any limit set, the set of forbidden patterns is the set of all orphans of all orders, which form a recursively enumerable set, since it is computable to check whether a finite word is an orphan.

For $x, y \in \{0,1\}^{\mathbb{N}}$, we say that $x \leq_T y$ if x is computable by a Turing machine using y as an oracle. If $x \leq_T y$ and $x \geq_T y$, x and y are said to be Turing-equivalent, which is noted $x \equiv_T y$. The *Turing degree* of x, noted $\deg_T x$, is its equivalence class under relation $\equiv_T$. The Turing degrees form a join semi-lattice whose bottom is $\mathbf{0}$, the Turing degree of computable sequences. For a set $S \subseteq \{0,1\}^{\mathbb{N}}$, we note $\deg_T S$ the set of Turing degrees of all points of S.

Effectively closed sets are quite well understood from a computational point of view, and there has been numerous contributions concerning their Turing degrees, see the book of [CR98] for a survey. One of the most interesting results may be that there exist Π^0_1 classes whose members are two-by-two Turing incomparable [JS72].

3   Requirements of the Construction

The idea to prove Theorem 1 is to make a construction that embeds computations of a Turing machine that will check a read-only oracle tape containing a member of the Π^0_1 class S that will have to appear completely in a configuration of the limit set. The following constraints have to be addressed.

- Since CAs are intrinsically deterministic, the oracle will have to appear in the "past", i.e. from the "limit" of the preimages.
- The oracle tape, the element of $\{0,1\}^{\mathbb{N}}$ that is to be checked, needs to appear entirely on at least one configuration of the limit set.
- Each configuration of the limit set containing the oracle tape needs to have exactly one head of the Turing machine, in order to ensure that there really is a computation going on in the associated space-time diagram.
- The construction, without any computation, needs to have a very simple limit set, i.e. it needs to be computable, and in particular countable; this to ensure that no complexity overhead will be added to any configuration containing the oracle tape, and that "unuseful" configurations of the limit set—the configurations that do not appear in a space-time diagram corresponding to a computation—will be computable.
- The computation of the embedded Turing machine needs to go backwards, this to ensure that we can have the non-determinism. And an error in the computation must ensure that there is no infinite sequence of preimages.
- The computation needs to have a beginning (also to ensure the presence of a head), hence it requires to mark it, and the representation of the oracle and work tapes in the construction need to disappear at this point, otherwise by compactness the part without any computation could be extended bi-infinitely to contain any member of $\{0,1\}^{\mathbb{N}}$, thus leading to the full set of Turing degrees.

There are other constraints that we will discuss during the construction, as they arise.

In order to make a construction complying to all these constraints, we reuse, with heavy modifications, an idea of [JV13b], which is to construct a sparse grid. However, their construction, being meant for subshifts, requires to be completely rethought in order to work for CAs. In particular, there was no determinism in this construction, and the oracle tape did not need to appear on a single column/row, since their result was on two-dimensional subshifts.

4 The Construction

4.1 A Self-Vanishing Sparse Grid

In order to have space-time diagrams that constitute sparse grids, the idea is to have columns of squares, each of these columns containing less and less squares as we move to the left, see fig. 1. The CA has three categories of states:

- a *killer state*, which is a spreading state that erases anything on its path; it appears whenever some neighborhood not on fig. 1 appears.
- a *quiescent state*, represented in white in the figures; its sole purpose is to mark the spaces that are "outside" the construction;
- some *construction states*, which will be constituted of signals and background colors.

In order to ensure that just with the signals themselves it is not possible to encode anything non-computable in the limit set, all signals will need to have, at all points, at any time, different colors on their left and right, otherwise the local rule will have a killer state arise. Here are the main signals.

- Vertical lines: serve as boundaries between columns of squares and form the left/right sides of the squares.
- SW-NE and SE-NW diagonals: used to mark the corners of the squares, they are signals of respective speeds 1 and −1. Each time they collide with a vertical line (except for the last square of the row), they bounce and start the converse diagonal of the next square.
- Counting signal: counts the number of squares inside a column; every time it crosses the SW-NE diagonal of a square it will shift to the left. When it is superimposed to a vertical line, it means that the square is the last of its column, so when it crosses the next SE-NW diagonal, it vanishes and with it the vertical line.
- Starting signals: used to start the next column to the left, at the bottom of one column. Here is how they work.
    - The bottommost signal, of speed $-\frac{1}{4}$, is at the boundary between the empty part of the space-time diagram and the construction. It is started 4 time steps after the collision with the signal of speed $-\frac{1}{3}$.
    - The signal of speed $-\frac{1}{3}$ is started just after the vertical line sees the incoming SE-NW diagonal of the first square of the row on the right, at distance 3 (the diagonal will collide with the vertical line 2 time steps after the start of that signal)[2].
    - At the same time as the signal of speed $-\frac{1}{3}$ is created, a signal of speed $-\frac{1}{2}$ is generated. When this signal collides with the bottommost signal, it bounces into a signal of speed $\frac{1}{4}$ that will create the first SE-NW diagonal of the first square of the row of squares of the left, 4 time steps after its collision with the vertical line.

On top of the construction states, except on the vertical lines, we add a parity layer $\{0, 1\}$: on a configuration, two neighboring cells of the construction must have different parity bits, otherwise a killer state appears. On the left of a vertical line there has to be parity 1 and on the right parity 0, otherwise the killer state pops up again. This is to ensure that the columns will always contain an even number of squares.

The following lemmas address which types of configurations may occur in the limit set of this CA. First note that any configuration in which the construction states do not appear in the right order do not have a preimage.

Lemma 1. *The sequence of preimages of a segment ended by vertical lines (and containing none) appearing in the limit set is a slice of a column of squares of even side.*

---

[2] That can be done, provided the radius of the CA is large enough.

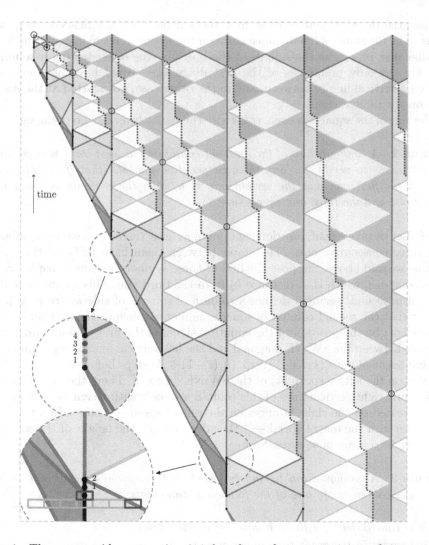

Fig. 1. The sparse grid construction: it is based on columns containing a finite number of squares, whose number decreases when we go left. Note that the figure is squeezed vertically.

Proof. Suppose a configuration contains two vertical-line symbols, then to be in the limit set, in between these two symbols there needs to be two diagonal symbols, one for the SE-NW one and one for SW-NE one, a symbol for the counting signal, and in between these signals there needs to be the appropriate colors: there is only one possibility for each of them. If this is not the case, then the configuration has no preimage since the rule enforces the appearance of a killer state when they are not correctly ordered.

Also, the distance between the first vertical line and the SE-NW diagonal needs to be the same as the distance between the second vertical line and the SW-NE diagonal, otherwise the signals at the bottom—the ones starting a column, that are the only preimages of the first diagonals—would have, in one case, created a vertical line in between, and in the other case, not started at the same time on the right vertical.

The side of the squares is even, otherwise the parity layer has no preimage. $\square$

Lemma 2. *A configuration of the limit set containing at least three vertical-line symbols needs to verify, for any three consecutive symbols, that if the distance between the first one and the second one is k, then the distance between the second one and the third one equals $(k + 2)$.*

Proof. Let us take a configuration containing at least three vertical-line symbols, take three consecutive ones. The states between them have to be of the right form as we said above. Suppose the first of these symbols is at distance k_1 from the second one, which is at distance k_2 from the third one. This means that the first (resp. second) segment defines a column of squares of side k_1 (resp. k_2). It is clear that the second column of squares cannot end before the first one.

Now let i be the position of the counting signal of the first column and j the distance between the SW-NE diagonal and the left vertical line. The preimage of the first segment ends $(k_1 i + j)$ (resp. $(k_1(i-1) + j)$) steps before if the counting signal is on the left (resp. right) of the SW-NE diagonal. Then, the preimages of the left and right vertical lines of this column are the creating signals. Before the signal created on the right bounces on the one of speed $-\frac{1}{4}$ created on the left, it collides with the one of speed $-\frac{1}{3}$, thus determining the height of the squares on the right column of squares. So $k_1 = k_2 - 2$. $\square$

Lemma 3. *A configuration having two vertical-line symbols pertaining to the limit set needs to verify one of the following statements.*

- *It is constituted of a finite number of vertical lines.*
- *It appears in the space-time diagram of fig. 1.*
- *It is constituted of an infinite number of vertical lines, then starting from some position it is equal on the right to some (shifted) line of fig. 1.*

Proof. We place ourselves in the case of a configuration of the limit set. Because of lemma 1, two consecutive vertical lines at distance k from each other define a column of squares. In a space-time diagram they belong to, on their left there is necessarily another column of squares, because of the starting signal generated at the beginning of the left vertical line, except when $k = 3$, in which case there is nothing on the left. In this column, the vertical lines are at distance $(k - 2)$, see lemma 2. So, if there is an infinite number of vertical lines, either it is of the form of fig. 1, or there is some killer state coming from infinity on the left and "eating" the construction. $\square$

4.2 Backward Computation Inside the Grid

We now wish to embed the computation of a reversible Turing machine inside the aforementioned sparse grid, which for this purpose is better seen as a lattice. The fact that the TM is reversible allows us to embed it backwards in the CA. Below we will denote by *TM time* (resp. *CA time*) the time going forward for the Turing machine (resp. the CA); on a space-time diagram, TM time goes from top to bottom, while CA time goes from bottom to top (cf. arrows in fig. 2a). That way, the beginning of the computation of the TM will occur in the first (topmost) square of the first (leftmost) column of squares.

We have to ensure that any computation of the TM is possible, and in particular ensure that such a computation is consistent over time; the idea is that at the first TM time step, i.e. the moment the sparse grid disappears, the tape is on each of the vertical line symbols, but since these all appeared a finite number of CA steps before (the height of any column of squares being finite), we have to compel all tape cells to shift to the right regularly as TM time increases.

Moreover, we want to force the presence of exactly one head (there could be none if it were, for instance, infinitely far right). To do that, the grid is divided into three parts that must appear in this order (from left to right): the left of the head, the right of the head (together referred to as the computation zone), and the unreachable zone (where no computation can ever be performed, because of the absence of a work tape), resp. in blue, yellow and green in fig. 2a.

The vertices of our lattice are the top left corners of the squares, each one marked by the rebound of a SE-NW diagonal on a vertical line, while the top right corners will just serve as intermediate points for signals. More precisely, for any $i, j \in \mathbb{N}$, the respective sites for the top left and top right corners of $s_{i,j}$, the $(j+1)$-th square of the $(i+1)$-th column, are the following (cf. fig. 2a):

$$\begin{cases} s_{i,j}^\ell = s_{0,0}^\ell + (i(i+1), -2(i+1)j) \\ s_{i,j}^r = s_{i,j}^\ell + (2(i+1), 0). \end{cases}$$

Fig. 2b illustrates a computation by the TM, with the three aforementioned zones, as it would be embedded the usual way (but with reverse time) into a CA, with site $(i, -t)$ corresponding to the content of the tape at $i \in \mathbb{N}$ and TM time $t \in \mathbb{N}$.

Fig. 2c represents another, still simple, embedding, which is a distortion of the previous one: the head moves every even time step within a tape that is shifted every odd time steps, so that instead of site $(i, -t)$, we have two sites, $(i+t, -2t)$ and $(i+t, -2t-1)$, resp. the *computation site* (big circle on fig. 2c) and the *shifting site* (small circle on fig. 2c). The head only reads the content of the oracle when it lies on a computation site. This type of embedding can easily be realized forwards or backwards (provided the TM is reversible).

Our embedding, derived from the latter, is drawn on fig. 2a. The "only" difference is the replacement of sites $(i+t, -2t)$ and $(i+t, -2t-1)$ by sites $s_{i,t}^\ell$ and $s_{i,t+1}^\ell$. Notice that as the number of squares in a column is always finite, each square can "know" whether its top left corner is a computation or a shifting site with a parity bit. More precisely, the j-th square (from bottom to top) of a column has a computation site on its top left if and only if j is even.

Let $s_{i,j}$ be a square of our construction. Its top left $s^\ell_{i,j}$ is either a computation site or a shifting site. In the latter case, it is supposed to receive the content of a cell of the TM tape with an incoming signal of speed -1. All it has to do is to send it to $s^\ell_{i,j-1}$ (at speed 0), which is a computation site. In the former case, however, things are slightly more complicated. The content of the tape has to be transmitted to $s^\ell_{i-1,j-1}$ (which is a shifting site). To do that, a signal of speed 0 is sent and waits for site $s^r_{i-1,j}$, which sends the content to $s^\ell_{i-1,j-1}$ with a signal of speed -1 along the SE-NW diagonal. The problem is to recognize which s^r site is the correct one. Fortunately, there are only two possibilities: it is either the first or the second s^r site to appear after (in CA time, of course) $s^\ell_{i,j}$ on the vertical line. The first case corresponds exactly to the unreachable zone (where $j \leq i$), hence the result if the three zones are marked. The lack of other cases is due to the number of s_i squares, which is only $2(i+1)$.

Another issue is the superposition of such signals. Here again, there are only two cases: in the unreachable zone there is none, whereas in the computation zone a signal of speed 0 from a computation site can be superimposed to the signal of speed 0 sent by the shifting site just above it. As aforesaid, there is no other case because of the limited number of s_i squares. Thus, there is no problem to keep the number of states of the CA finite, since the number of signals going through a same cell is limited to two at the same time.

The two parts of the computation zone are separated by the presence of a head, while the unreachable zone is easily hardcoded as the right of the path corresponding to a TM head that would always move rightwards : this is done simply by seeing whether the counting signal is on the left or right of the crossing of the SE-NW and SW-NE diagonals.

Now only the movements of the head remain to be described (in black on fig. 2a). Let $s^\ell_{i,j}$ be a computation site containing the head.

- If the previous move of the head (previous because we are in CA time, that is, in reverse TM time) was to the left, the next computation site is the one just above, that is, $s^\ell_{i,j-2}$. The head is thus transferred by a simple signal of speed 0.
- If the previous move was to stand still, the next computation site is $s^\ell_{i-1,j-2}$. It can be reached by a signal of speed 0 until the second next s^r site, from which a signal of speed -1 (along a SE-NW diagonal) is launched, to be replaced by another signal of speed 0 from $s^\ell_{i-1,j-1}$ on.
- If the previous move was to the right, the next computation site is $s^\ell_{i-2,j-2}$. It can be reached by a signal of speed 0 until the second next s^r site, from which a signal of speed -1 (along a SE-NW diagonal) is launched, to be replaced by another signal of speed 0 from $s^\ell_{i-1,j-1}$ on, which itself waits for the next s^r site (which is $s^r_{i-2,j}$) to start another signal of speed 1 (along a SW-NE diagonal) that is finally succeeded to by a last signal of speed 0 from $s^\ell_{i-2,j-1}$ on.

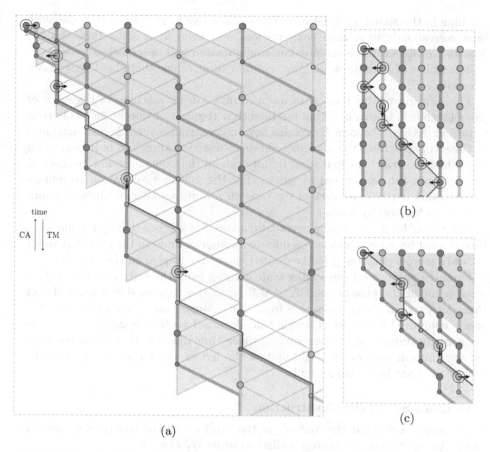

(a)

(b)

(c)

Fig. 2. The embedding of a Turing machine computation in the sparse grid (2a), compared to the usual embedding (2b) and one with a simple shift of the oracle tape (2c). The paths followed by the content of each cell of the oracle tape are in red and orange (two colors just to keep track of the signals when they are superimposed), while the one of the head is in black. The arrows indicate the next move of the head (for TM time, going towards the bottom). The green background denotes the zone the head cannot reach, while the computation zone is in blue on the left of the head and in yellow on its right.

4.3   The Computation Itself

As we said before, the computation will take place on the computation sites, which will contain two tape cells: one for the oracle and one for the work. In the unreachable zone there are only oracle cells, which do not change over time except for the shifting. Now we want to eliminate all space-time diagrams corresponding to rejecting computations of some Turing machine M. [Ben73] proved that for any Turing machine, we can construct a reversible one computing the same function. So, a first idea would just be to encode this reversible Turing machine in the sparse grid; however there is no way to guarantee that the work tape corresponds to a valid computation, and even if at some time the CA detects a problem, the corresponding configuration will nevertheless have an infinite number of preimages, and may contain some oracle that should have been rejected.

The solution to this issue is to use a robust Turing machine in the sense of [Hoo66], that is to say, a Turing machine that regularly rechecks its whole computation. [KO08, Theorem 7] explains how to construct reversible such machines. In these constructions the machines obtained were working on a bi-infinite tape, which had the drawback that some infinite side of the tape might not be checked; here it is not the case, hence we can modify the machine so that on an infinite computation it checks all cells of the tape an infinite number of times (we omit the details for brevity's sake).

In terms of limit sets, this means that if some oracle is rejected by the machine, then it must have been rejected an infinite number of times in the past (CA time). So, only oracles pertaining to the desired class may appear in the limit set.

Furthermore, even if some killer state coming from the right eats the grid, at some point in the past of the CA, it will be in the unreachable zone, and stay there for ever, so the computation from that moment on even ensures that the oracle computed is correct. Though, that does not matter, because in this case the configurations of the corresponding space-time diagram that are in the limit set are uniform both on the right and on the left except for a finite part in the middle, and are hence computable.

4.4 Limit Set of the Construction

Let us now check what the contents of the limit set of the construction are for some Turing machine M correspondind to some Π_1^0 class S :

- Given some $s \in S$, it is easy to see that there are several configurations in the limit set of the CA with the same Turing degree. The configuration corresponding to a "perfect" space-time diagram, the configurations corresponding to the same space-time diagram, but "eaten" on the left or right by a killer state, and the configurations corresponding to the same space-time diagram, but where the beginning of the computation is not at the right height in the construction.
- In section 4.1 we made sure that the only configurations of the limit set containing an infinite alternation of vertical-line symbols were of the last form described in lemma 3. This means that the only way to encode something non-computable in the limit set is through the oracle of the construction. The backward computation ensures that only oracles allowed by M may appear.

References

[Ben73] Bennett, C.H.: Logical Reversibility of Computation. IBM J. Res. Dev. 17(6), 525–532 (1973)

[BGK11] Ballier, A., Guillon, P., Kari, J.: Limit Sets of Stable and Unstable Cellular Automata. Fundam. Inform. 110(1-4), 45–57 (2011)

[CR98] Cenzer, D., Remmel, J.: Π_1^0 classes in mathematics. In: Handbook of Recursive Mathematics - Volume 2: Recursive Algebra, Analysis and Combinatorics. Studies in Logic and the Foundations of Mathematics, ch. 13, vol. 139, pp. 623–821 (1998)

[FK07] Formenti, E., Kůrka, P.: Subshift attractors of cellular automata. Nonlinearity 20, 105–117 (2007)

[Hoo66] Hooper, P.K.: The Undecidability of the Turing Machine Immortality Problem. Journal of Symbolic Logic 31(2), 219–234 (1966)

[Hur87] Hurd, L.P.: Formal Language Characterization of Cellular Automaton Limit Sets. Complex Systems 1(1), 69–80 (1987)

[Hur90a] Hurd, L.P.: Nonrecursive Cellular Automata Invariant Sets. Complex Systems 4(2), 131–138 (1990)

[Hur90b] Hurd, L.P.: Recursive Cellular Automata Invariant Sets. Complex Systems 4(2), 131–138 (1990)

[JS72] Jockusch, C.G., Soare, R.I.: Degrees of members of classes Π_1^0. Pacific J. Math. 40(3), 605–616 (1972)

[JV13a] Jeandel, E., Vanier, P.: Hardness of Conjugacy, Embedding and Factorization of multidimensional Subshifts of Finite Type. In: STACS. LIPIcs, vol. 20, pp. 490–501 (2013)

[JV13b] Jeandel, E., Vanier, P.: Turing degrees of multidimensional SFTs. In: Theoretical Computer Science 505.0. Theory and Applications of Models of Computation, pp. 81–92 (2011)

[Kar90] Kari, J.: Reversibility of 2D cellular automata is undecidable. Physica D: Nonlinear Phenomena 45(1-3), 379–385 (1990)

[Kar92] Kari, J.: The Nilpotency Problem of One-Dimensional Cellular Automata. SIAM Journal on Computing 21(3), 571–586 (1992)

[Kar94a] Kari, J.: Reversibility and surjectivity problems of cellular automata. Journal of Computer and System Sciences 48(1), 149–182 (1994)

[Kar94b] Kari, J.: Rice's theorem for the limit sets of cellular automata. Theoretical Computer Science 127(2), 229–254 (1994)

[KO08] Kari, J., Ollinger, N.: Periodicity and Immortality in Reversible Computing. In: Ochmański, E., Tyszkiewicz, J. (eds.) MFCS 2008. LNCS, vol. 5162, pp. 419–430. Springer, Heidelberg (2008)

[LM09] Lena, P.D., Margara, L.: Undecidable Properties of Limit Set Dynamics of Cellular Automata. In: 26th International Symposium on Theoretical Aspects of Computer Science. Leibniz International Proceedings in Informatics (LIPIcs), vol. 3, pp. 337–348 (2009)

[Maa95] Maass, A.: On the sofic limit sets of cellular automata. Ergodic Theory and Dynamical Systems 15(04), 663–684 (1995)

[Mey08] Meyerovitch, T.: Finite entropy for multidimensional cellular automata. Ergodic Theory and Dynamical Systems 28(04), 1243–1260 (2008)

[Sim11] Simpson, S.G.: Mass problems associated with effectively closed sets. Tohoku Mathematical Journal 63(4), 489–517 (2011)

[CPY89] Čulik, K., Pachl, J., Yu, S.: On the limit sets of cellular automata. SIAM Journal on Computing 18(4), 831–842 (1989)

On the Complexity of Temporal-Logic Path Checking*

Daniel Bundala and Joël Ouaknine

Department of Computer Science, University of Oxford
Wolfson Building, Parks Road, Oxford, OX1 3QD, UK

Abstract. Given a formula in a temporal logic such as LTL or MTL, a fundamental problem is the complexity of evaluating the formula on a given finite word. For LTL, the complexity of this task was recently shown to be in NC [9]. In this paper, we present an NC algorithm for MTL, a quantitative (or metric) extension of LTL, and give an AC^1 algorithm for UTL, the unary fragment of LTL. At the time of writing, MTL is the most expressive logic with an NC path-checking algorithm, and UTL is the most expressive fragment of LTL with a more efficient path-checking algorithm than for full LTL (subject to standard complexity-theoretic assumptions). We then establish a connection between LTL path checking and planar circuits, which we exploit to show that any further progress in determining the precise complexity of LTL path checking would immediately entail more efficient evaluation algorithms than are known for a certain class of planar circuits. The connection further implies that the complexity of LTL path checking depends on the Boolean connectives allowed: adding Boolean exclusive or yields a temporal logic with P-complete path-checking problem.

1 Introduction

One of the most fundamental problems in the fields of testing and verification is the *path-checking problem*: determine whether a given observation[1] of a system satisfies a given specification drawn from a fixed ambient logic. The complexity of this problem plays a key role in the design and analysis of offline monitoring and runtime verification procedures [6,12]. The path-checking problem also appears in testing [1] and in Monte-Carlo-based probabilistic verification [14].

Although the problem is simply stated, determining its precise complexity can prove to be quite challenging. The case of LTL was investigated more than a decade ago [5,13], and at the time is was conjectured that the straightforward polynomial-time dynamic-programming algorithm is not optimal.[2] And indeed, using reductions to planar circuits and tree-contraction algorithms, it was recently proved [9] that LTL path checking allows an efficient parallel algorithm

---

* Full version of the paper is available as [3].
[1] In this paper, all observations (paths, traces, words, etc.) considered are finite.
[2] The best known lower bound for LTL path checking is NC^1, which crudely arises from the NC^1-hardness of mere Boolean formula evaluation.

J. Esparza et al. (Eds.): ICALP 2014, Part II, LNCS 8573, pp. 86–97, 2014.
© Springer-Verlag Berlin Heidelberg 2014

and lies in NC—in fact, in $AC^1[logDCFL]$. (This seminal result was rewarded by the ICALP 2009 best-paper award.) More recently, this work was extended to a very restricted metric extension of LTL, in which only temporal operators of the form $U_{\leq b}$ are allowed [10].

In this paper, we give an algorithm for full Metric Temporal Logic (MTL) with the same complexity—$AC^1[logDCFL]$—known algorithm for LTL.

We reprise the strategy, introduced in [9], to represent temporal operators using a special class of planar monotone circuits, together with a generic algorithm [4] as a subroutine to evaluate those circuits. Such circuits have a very special form, which led the authors of [9] to ask whether the complexity of the path-checking algorithm can be improved by devising specialised circuit-evaluation algorithms. In this paper, we present evidence to the contrary, by showing that the evaluation of circuits drawn from a class of planar circuits studied in [11] is reducible to LTL path checking; any further progress in determining the precise complexity of the latter would therefore immediately entail more efficient evaluation algorithms than are known for this class of planar circuits. It is worth pointing out that augmenting this class of planar circuits with NOT gates makes the evaluation problem P-complete [7]. It follows that the complexity of path checking is sensitive to non-monotone connectives, as allowing Boolean exclusive-or in formulae enables the evaluation of circuits from this augmented class, and is therefore itself P-complete.

An examination of the algorithmic constructions of [9] shows that the most intricate parts arise in handling the Until operator. In this paper, we show that the removal of binary operators from the logic, yielding Unary Temporal Logic (UTL), leads to a much simpler path-checking problem, enabling us to devise an AC^1 algorithm for UTL path checking.

At the time of writing, our results provide (i) the most expressive known extension of LTL with an NC path-checking algorithm (MTL), (ii) the simplest known extension of LTL with a strictly harder path-checking problem (LTL + Xor), and (iii) the most expressive known fragment of LTL with a strictly more efficient path-checking algorithm than for full LTL (UTL).[3]

2 Preliminaries

We denote Boolean true and false by $\top$ and $\bot$, respectively. The set $\{\bot, \top\}$ is denoted by $\mathbb{B}$. A vector $v \in \mathbb{B}^n$ is **downward monotone** if $v(i+1) = \top \implies v(i) = \top$. It is **upward monotone** if $v(i-1) = \top \implies v(i) = \top$. A vector is **monotone** if it is upward or downward monotone. The set of monotone vectors is denoted by $\mathbb{M}$.

Temporal Logics: Let AP be a set of atomic propositions, $p \in$ AP and $I \subseteq \mathbb{R}_{\geq 0}$ be an interval with endpoints in $\mathbb{N} \cup \{\infty\}$. The formulae of **Metric Temporal Logic** (MTL) are defined recursively as follows.

$$\varphi = p \mid \neg p \mid \varphi \wedge \varphi \mid \varphi \vee \varphi \mid X_I\,\varphi \mid Y_I\,\varphi \mid \varphi\,U_I\,\varphi \mid \varphi\,S_I\,\varphi \mid \varphi\,R_I\,\varphi \mid \varphi\,T_I\,\varphi$$

---

[3] Subject to standard complexity-theoretic assumptions.

All logics and results presented in this paper apply to temporal logics with past temporal operators. Note that negation is applied only to atomic propositions. Other operators are expressible using the following semantic equalities: $F_I \varphi = \top \, U_I \, \varphi$, $G_I \varphi = \neg \, F_I \, \neg \varphi$, $\varphi \, R_I \, \psi = \neg(\neg \varphi \, U_I \, \neg \psi)$ and $\varphi \, T_I \, \psi = \neg(\neg \varphi \, S_I \, \neg \psi)$. **Linear Temporal Logic** (LTL) is the subset of MTL in which I is always $[0, \infty)$ (and is omitted). The fragment UTL of LTL consists of all Boolean connectives and unary (X, F, G) temporal operators and their past duals.

A **trace** π over AP of length n is a function $\pi : \{1, \ldots, n\} \times AP \to \mathbb{B}$ assigning a truth value to every $p \in AP$ at every index. We identify $p \in AP$ with a vector in $\mathbb{B}^n$ and use $p(i) = \top$ if $\pi(i, p) = \top$. The proposition that is true only in the interval $[i, j]$ and false otherwise is denoted by $\chi_{i,j}$, i.e., $\chi_{i,j}(k) = \top$ if $i \leq k \leq j$ and $\chi_{i,j}(k) = \bot$ otherwise. To evaluate MTL formulae on π, we further associate with π a sequence of strictly-increasing **timestamps** $t_1 < \ldots < t_n$.

Given an MTL formula φ and index $1 \leq i \leq n$, the satisfaction relation $\pi, i \models \varphi$ is defined recursively as follows.

$$
\begin{aligned}
&\pi, i \models p && \text{if } p(i) = \top \\
&\pi, i \models \varphi_1 \wedge \varphi_2 && \text{if } \pi, i \models \varphi_1 \text{ and } \pi, i \models \varphi_2 \\
&\pi, i \models \varphi_1 \vee \varphi_2 && \text{if } \pi, i \models \varphi_1 \text{ or } \pi, i \models \varphi_2 \\
&\pi, i \models X_I \varphi && \text{if } i + 1 < n \wedge t_{i+1} - t_i \in I \wedge \pi, i+1 \models \varphi \\
&\pi, i \models Y_I \varphi && \text{if } i > 1 \text{ and } t_i - t_{i-1} \in I \text{ and } \pi, i-1 \models \varphi
\end{aligned}
$$

$$
\pi, i \models \varphi_1 U_I \varphi_2 \text{ if } \exists j \, . \, (i \leq j \leq n) \wedge \begin{pmatrix} \pi, j \models \varphi_2 \\ t_j - t_i \in I \\ \forall k \, . \, i \leq k < j \implies \pi, k \models \varphi_1 \end{pmatrix}
$$

$$
\pi, i \models \varphi_1 S_I \varphi_2 \text{ if } \exists j \, . \, (i \geq j \geq 1) \wedge \begin{pmatrix} \pi, j \models \varphi_2 \\ t_i - t_j \in I \\ \forall k \, . \, i \geq k > j \implies \pi, k \models \varphi_1 \end{pmatrix}
$$

This paper studies the complexity of evaluating a given formula on a given trace.

Definition 1. *The* **path-checking problem** *for logic $\mathcal{L}$ is to determine, given a trace π and a formula φ of $\mathcal{L}$, whether $\pi, 1 \models \varphi$.*

Let φ be an MTL formula. Working from the smallest subformulae and using the above definitions to tabulate the values $\pi, i \models \psi$ for every i and subformula ψ yields a polynomial dynamic-programming algorithm evaluating φ on π.

Theorem 1 ([13]). *The path-checking problem for* MTL *is in* P.

Given a trace π and formula φ, we represent the value of φ on π as the vector $v \in \mathbb{B}^n$ such that $v(i) = \top$ if and only if $\pi, i \models \varphi$. We further represent LTL temporal operators as functions over vectors written in infix notation. For example, $U : \mathbb{B}^n \times \mathbb{B}^n \to \mathbb{B}^n$ is a function such that $(p \, U \, q)(i) = \top$ if and only if there is $i \leq j \leq n$ such that $q(j) = \top$ and $p(k) = \top$ for all $i \leq k < j$.

A **formula context** $\varphi(X)$ is a formula with one occurrence of a proposition replaced by a variable X. If $\psi(X)$ is another formula context then $(\varphi \circ \psi)(X)$ is the context obtained by substituting $\psi(X)$ for X in $\varphi(X)$. If $q \in AP$ is a proposition then $\varphi(q)$ is obtained by substituting q for X. For example, $((p\mathrm{U}X)\circ(X\mathrm{S}q))(r) = (p\mathrm{U}(X\mathrm{S}q))(r) = p\mathrm{U}(r\mathrm{S}q)$. Composing formula contexts increases the size linearly as a formula context contain only one occurrence of X.

Circuits: A **Boolean circuit** (C, δ) consists of a set of **gates** C and a **predecessor** function $\delta : C \to \mathcal{P}(C)$. The type of a gate is either OR, AND, NOT, ID, ONE or ZERO. If c is of type τ and $\delta(c) = \{c_1, \ldots, c_n\}$ then we write $c = (\tau, c_1, \ldots, c_n)$. If $d \in \delta(c)$ then we say c **depends** on d or that there is a **wire** from d to c. The ONE and ZERO gates provide constants inputs. A gate is an **input** gate if it does not have a predecessor. A gate is an **output** gate if it is not a predecessor of any other gate. A circuit is **monotone** if it has no NOT gates. It is **planar** if the underlying DAG is planar. In this paper, all edges (wires) are straight-line segment and so a **planar embedding** is induced by a function $\gamma : C \to \mathbb{R}^2$ assigning a point in the plane to every gate.

A circuit is **layered** if it can be partitioned into **layers** $C_0, \ldots, C_n$ such that each wire goes from C_i to C_{i+1} for some i. Thus, C_0 contains only input gates. A layered circuit is **stratified** if all input gates appear in C_0. A circuit is **upward planar** if there is a planar embedding such that every edge monotonically increases in the upward direction—the direction of the evaluation of C. A circuit is **upward layered (stratified)** if it is both upward planar and layered (stratified). Each layer C_i of an upward-layered circuit consists of gates $\alpha_{i,j}$ in the left-to-right ordering. Each $\alpha_{i,j}$ depends on a contiguous block $\alpha_{i-1,l}, \ldots, \alpha_{i-1,r}$ layer below and the wires do not cross: if $\alpha_{i,j}$ depends on $\alpha_{i-1,q}$ and $\alpha_{i,k}$ depends on $\alpha_{i-1,r}$ then $j \leq k \iff q \leq r$. Fig. 3 shows upward stratified monotone circuits.

Given a circuit with one output gate, the **circuit value problem**, abbreviated as **CVP**, is the problem of determining the value of the output gate.

Complexity Classes: The class logDCFL consists of problems that are logspace many-one reducible to deterministic context-free languages. Equivalently, it is the class of problems decidable by a deterministic logspace Turing machine equipped with a stack and terminating in polynomial time. The circuit class AC^i for $i \in \mathbb{N}$ consists of problems decidable by polynomial-size unbounded fan-in circuits of depth $\log^i$. All circuits in this paper are **uniform**—can be generated by a deterministic logspace Turing machine. Given a problem S and a complexity class C, we write $S \in \mathsf{AC}^1[C]$ if there is a family of AC^1 circuits with additional unbounded fan-in C-oracle gates that decide S. It is known that

$$\mathsf{L} \subseteq \mathsf{logDCFL} \subseteq \mathsf{AC}^1 \subseteq \mathsf{AC}^1[\mathsf{logDCFL}] \subseteq \mathsf{AC}^2 \subseteq \cdots \subseteq \mathsf{AC}^i \subseteq \mathsf{AC}^{i+1} \subseteq \cdots \subseteq \mathsf{P}$$

and that CVP for upward-stratified circuits is P-complete [7], CVP for monotone upward-stratified circuits is in logDCFL [4] and that CVP for monotone upward-layered circuits is in $\mathsf{AC}^1[\mathsf{logDCFL}]$ [11].

Tree Contraction: Let $T = (V, E)$ be a binary tree, the tree contraction algorithm [8] reduces T to a single node using a sequence of tree contraction steps. Let $l \in T$ be a leaf, p be its parent and s its sibling[4]. A tree contraction step

---

[4] If l does not have a sibling then we take s to be a fresh node.

collapses the triple (l, p, s) into a single node. Formally, a new tree $T' = (V', E')$ is obtained from T as follows: $V' = V \setminus \{l, p\}$

$$E' = \begin{cases} E \setminus \{(p, l), (p, s)\} & \text{if } p \text{ is the root of } T \\ (E \setminus \{(p, l), (p, s), (q, p)\}) \cup \{q, s\} & \text{otherwise } (q \text{ is the parent of } p) \end{cases}$$

Note that a contraction step is local and hence multiple non-interfering contractions can be performed in parallel. A tree contraction algorithm using only $\lceil \log n \rceil$ parallel steps exists [8]. Further, this algorithm can be implemented in AC^1.

Let φ be an LTL formula and π a trace. A tree contraction algorithm evaluating φ on π was given in [9]. The tree T used in [9] is the parse tree of φ. The leaves of T correspond to the atomic propositions and the internal nodes to Boolean or temporal operators. Each contraction step (l, p, s) partially evaluates the operator associated with p.

For example, suppose that the formula rooted at p is $\psi \cup q$ where q is a proposition. Even if the value of ψ is unknown, we can still make some inferences. E.g., if $q(i) = \top$ then $(\psi \cup q)(i) = \top$. If the last value $q(|\pi|) = \bot$ then $(\psi \cup q)(|\pi|) = \bot$ and so on. The contraction step removes the nodes for ψ and $\cup$ and then labels the node s by the partial evaluation of the function $(X \cup q) \circ \psi$. It was shown in [9] how to represent, manipulate and evaluate these functions efficiently. When a subformula ψ is fully collapsed into a single node then the associated function is fully evaluated and the node is labelled by the constant $(\psi(1), \ldots, \psi(|\pi|)) \in \mathbb{B}^{|\pi|}$. The contraction algorithm eventually reduces the tree into a single node, which is labelled by $(\varphi(1), \ldots, \varphi(|\pi|)) \in \mathbb{B}^{|\pi|}$.

In general, a tree-contraction algorithm can evaluate a function f on a tree; each contraction step partially evaluating f on a subtree. In this paper, the evaluation is done as follows. Let $\mathcal{C}$ be the set of constants and $\mathcal{F}$ be a collection, closed under composition, of admissible functions $f : \mathcal{C} \to \mathcal{C}$.

- A constant $c_v \in \mathcal{C}$ is attached to every leaf v of T. The values of c_v for the initial leaves are given as a part of the input.
- A function $f_v \in \mathcal{F}$ is attached to every node v of T. Initially, f_v is the identity function.
- A tree contraction of (l, p, s) first builds $f' \in \mathcal{F}$ (depending on c_l and p) implementing the partial evaluation on p. Let $f'' = f_p \circ f'$. If s is a leaf then c_s is replaced by $f''(c_s)$. Otherwise, f_s is replaced by $f'' \circ f_s$.

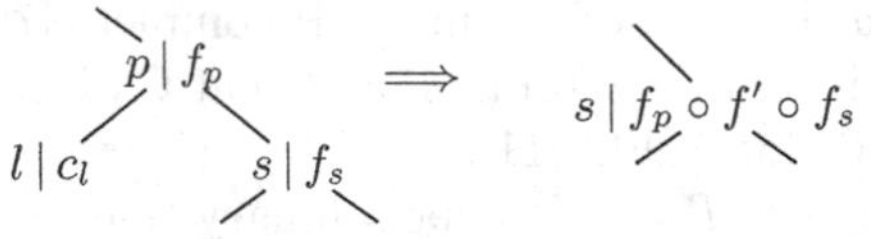

Fig. 1. An example of a tree contraction step

The output of the algorithm is the constant attached to the single remaining node. If each contraction step and admissible functions are in the complexity class C then, by [8], the contraction algorithm calculating c_{root} is in $\mathsf{AC}^1[C]$.

A tree contraction algorithm for LTL path checking [9] runs in $\mathsf{AC}^1[\mathsf{logDCFL}]$. Constants $\mathcal{C} = \mathbb{B}^n$ denote the truth values of propositions and subformulae. Functions $\mathcal{F}$ are represented by upward stratified circuits with n input and n output gates (**transducer circuits**), which are closed under composition [9] and their evaluation and composition is in $\mathsf{logDCFL}$ [2]. For a fixed $s \in \mathbb{B}^n$, [9] gives transducer circuits for $s \wedge x, s \vee x, s\,\mathsf{U}\,x$, and $s\,\mathsf{R}\,x$ as the functions of $x \in \mathbb{B}^n$. In Section 4, we give transducer circuits for MTL temporal operators.

3 Reduction from Upward Layered CVP to LTL Path Checking

Given an upward layered monotone circuit C with n gates and m wires we show how to build an LTL formula φ over at most $2n$ propositions and a trace π of length $|\pi| \leq m$ such that C evaluates to $\top$ if and only if $\pi \models \varphi$.

Denote the layers of C by $C_0, \ldots, C_k$ and the size of each C_i by n_i. Let $\alpha_{i,j}$ be the gates in C_i in the left-to-right order in the upward planar embedding of C. For each layer, we partition the trace into blocks—each of which stores the outputs of a gate in the layer. Fig. 2 shows a valid partitioning. In the figure, gate a occupies block $[1,1]$, gate e occupies $[3,5]$, gate g occupies $[1,7]$, etc.

In general, a valid partitioning consists of a trace π and intervals $v(i,j)$ associated with each gate $\alpha_{i,j}$ such that $v(i,j)$ overlaps precisely with the blocks of the gates the gate $\alpha_{i,j}$ depends on. Formally,

- intervals $v(i,1), v(i,2), \ldots, v(i,n_i)$ are disjoint and partition $[1,|\pi|]$ for every i,
- if $\alpha_{i+1,j}$ depends on $\alpha_{i,p}, \alpha_{i,p+1}, \ldots, \alpha_{i,q}$ then $v(i+1,j) \subseteq \cup_{r=p,\ldots,q} v(i,r)$ and $v(i+1,j)$ overlaps with each $v(i,r)$ for $p \leq r \leq q$,

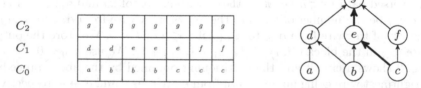

Fig. 2. An upward layered circuit (on the right) with its partition (on the left). The path π for the gate labelled e is highlighted.

Suppose we are given a valid partitioning. Then for $i > 0$ and every $1 \leq j \leq n_i$ we build a formula context $\varphi_{i,j}$ mimicking the evaluation of the gate $\alpha_{i,j}$.

For example, suppose that the gate e in Fig. 2 is an OR gate and the values of the block in the first layer is $r = (a,b,b,b,c,c,c) \in \mathbb{B}^7$ for some $a,b,c \in \mathbb{B}$. Recall that $(\varphi\,\mathsf{U}\,\psi)(i) = \psi(i) \vee (\varphi(i) \wedge (\varphi\,\mathsf{U}\,\psi)(i+1))$. Hence, if $\varphi(i) = \bot$ then $(\varphi\,\mathsf{U}\,\psi)(i) = \psi(i)$ and if $\varphi(i) = \top$ then $(\varphi\,\mathsf{U}\,\psi)(i) = \psi(i) \vee (\varphi\,\mathsf{U}\,\psi)(i+1)$. Further recall that $\chi_{i,j}$ is a proposition that is true on $[i,j]$ and false otherwise. Hence, $(\chi_{3,4}\,\mathsf{U}\,r)(1) = a, (\chi_{3,4}\,\mathsf{U}\,r)(2) = b$ and $(\chi_{3,4}\,\mathsf{U}\,r)(5,6,7) = c$. Also, $(\chi_{3,4}\,\mathsf{U}\,r)(4) = r(4) \vee (\chi_{3,4}\,\mathsf{U}\,r)(5) = b \vee c$. Finally, $(\chi_{3,4}\,\mathsf{U}\,r)(3) = r(3) \vee (\chi_{3,4}\,\mathsf{U}$

$r)(4) = b \vee (b \vee c) = b \vee c$. So $\chi_{3,4} \cup r = (a, b, b \vee c, b \vee c, c, c, c)$. Performing a similar calculation backwards, we get $\chi_{4,5} \, S \, (\chi_{3,4} \cup r) = (a, b, b \vee c, b \vee c, b \vee c, c, c)$ which gives the value of block e in Fig. 2 and leaves other blocks unchanged.

Denote the type of $\alpha_{i,j}$ by τ and the left and the right endpoint of $v(i, j)$ by l and r, respectively. Then $\varphi_{i,j}$ is constructed as follows:
  - If $\tau = $ ONE then $\varphi_{i,j}(X) = \chi_{l,r} \vee X$.
  - If $\tau = $ ZERO then $\varphi_{i,j}(X) = (\neg \chi_{l,r}) \wedge X$.
  - If $\tau = $ ID then $\varphi_{i,j}(X) = X$.
  - If $\tau = $ OR then $\varphi_{i,j}(X) = \chi_{l+1,r} \, S \, (\chi_{l,r-1} \cup X)$.
  - If $\tau = $ AND then $\varphi_{i,j}(X) = \chi_{l+1,r} \, T \, (\chi_{l,r-1} \, R \, X)$.

It can be shown that the formula context $\varphi_{i,j}$ updates the block $v(i, j)$ and leaves the other blocks unchanged. Hence, the formula context $\psi_i(X) = \varphi_{i,1} \circ \varphi_{i,2} \circ \cdots \circ \varphi_{i,n_i}$ evaluates the i-th layer C_i of C.

Formally, for each layer C_i let $r_i \in \mathbb{B}^n$ be a proposition such that $r_i(k) = \top$ if $k \in v(i, j)$ for some j and $\alpha_{i,j}$ evaluates to $\top$ and $r_i(k) = \bot$, otherwise. Then, the formula $\varphi = (\psi_k \circ \psi_{k-1} \circ \cdots \circ \psi_1)(r_0)$ computes the output of the circuit.

Lemma 1. *Let ψ_i, φ be as above. Then $\psi_i(r_{i-1}) = r_i$ and $\varphi(r_0)(1) = \top$ if and only if C evaluates to $\top$. Moreover, φ can be built in L.*

Finally, we show how to devise $v(i, j)$'s – the partitioning of the trace. Without loss of generality, connecting to a gate in the previous layer if necessary, we assume that all ONE and ZERO gates not in C_0 have at least one predecessor.

Given a gate $\alpha_{i,j}$ there is unique rightmost gate in the layer C_{i+1} that $\alpha_{i,j}$ is connected to by a wire. Now, start at $\alpha_{i,j}$ and take the rightmost wires until the sink is reached. Denote the traversed path by π_u. Similarly, there is unique rightmost gate in the layer C_{i-1} that $\alpha_{i,j}$ is connected to by a wire. Start at $\alpha_{i,j}$ and take the rightmost wires going down until a gate in C_0 is reached. Denote the traversed path by π_d. Let π be the concatenation of π_d and π_u. (See Fig. 2)

Let $k_{i,j}$ be the number of wires to the left of π. A wire from $\alpha_{i,j}$ to $\alpha_{i+1,k}$ is to the left of the wire from $\alpha_{i,a}$ to $\alpha_{i+1,b}$ if $j < a$ or $k < b$. We store the output of gate $\alpha_{i,j}$ in the block $v(i, j) := [k_{i,j-1} + 1, k_{i,j} + 1]$. We use $k_{i,0} = 0$.

Fig. 2 shows a circuit and the partitioning obtained by the above procedure. The rightmost wire going up and down from e are $e \to g$ and $c \to e$, respectively. Thus, $\pi_u = e \to g$ and $\pi_d = c \to e$. The path $\pi = c \to e \to g$ is highlighted in the figure. Four wires $a \to d, b \to d, b \to e, d \to g$ are to the left of π. We associate the block $[3, 5]$ with gate e. All blocks, grouped by layers, are shown in Fig. 2.

The following lemma summarises the important properties of $k_{i,j}$'s.

Lemma 2. *Let $k_{i,j}$'s and $v(i, j)$'s be as above. Then the following hold:*
  - $k_{i,j-1} < k_{i,j}$ *for every i and j,*
  - $k_{i,n_i} = k_{j,n_j}$ *for every i and j,*
  - $k_{i,n_i} \leq m$ *for every i,*
  - *for every i and $j = 1, \ldots, n_i$ the intervals $v(i, j)$'s partition $[1, k_{i,n_i}]$,*
  - *if $\alpha_{i+1,j}$ depends on $\alpha_{i,p}, \alpha_{i,p+1}, \ldots, \alpha_{i,q}$ then $v(i + 1, j) \subseteq \cup_{r=p,\ldots,q} v(i, r)$ and $v(i + 1, j)$ overlaps with each $v(i, r)$ for $p \leq r \leq q$,*
  - *each $k_{i,j}$ can be computed in L.*

This finishes the reduction from upward-layered CVP to LTL path checking. It was shown in [9] that the latter is in $\mathsf{AC}^1[\mathsf{logDCFL}]$. Therefore:

Theorem 2. *The CVP for upward-layered monotone circuits is in* $\mathsf{AC}^1[\mathsf{logDCFL}]$.

An alternative proof of Theorem 2 already appeared in [11]. Moreover, the relationship shows that any improvement in LTL path checking would entail an improvement in the evaluation of upward-layered monotone circuits.

The above reduction assumes the monotonicity of the input circuit. However, if the target logic LTL is extended to include binary exclusive or (xor) as a connective, then evaluating NOT gates becomes possible using $\varphi_{i,j}(X) = \chi_{l,r} \oplus X$ as a formula context for NOT gate $\alpha_{i,j}$. Noting that CVP is P-complete for general (non-monotone) upward stratified circuits [7], we have the following:

Theorem 3. LTL + Xor *path checking is* P-*complete*.

Thus, the complexity of LTL path checking depends on the monotonicity of the Boolean connectives present in the formula.

4   MTL Path Checking is Efficiently Parallelisable

We now show how the tree-contraction method of [9] extends to full MTL; giving an $\mathsf{AC}^1[\mathsf{logDCFL}]$ path-checking algorithm for MTL. By [9], summarised in Section 2, it suffices to give upward stratified transducer circuits for U_I and its duals.

Let π be the input trace with (floating-point) timestamps $t_1, \ldots, t_n$. Fix an interval I and consider the U_I operator. We now describe a dynamic-programming approach that yields planar circuits calculating $(\psi_1 \, \mathsf{U}_I \, \psi_2)(i)$. For $i \neq j$ the values $(\psi_1 \, \mathsf{U}_I \, \psi_2)(i)$ and $(\psi_1 \, \mathsf{U}_I \, \psi_2)(j)$ depend on the values of subformulae in some future intervals. In general, these intervals overlap and so naive constructions of transducer circuits are not planar. See Fig. 3 for the kind of circuits we build.

Recall, that the tree contraction is applied only to a leaf, its parent and its sibling. Let $s \in \mathbb{B}^n$ be a vector. We need to construct only circuits for $s \, \mathsf{U}_I \, \varphi$ and $\varphi \, \mathsf{U}_I \, s$ for known s. First consider the case $s \, \mathsf{U}_I \, \varphi$. (see left part of Fig. 3)

For index $1 \leq i \leq n$ the formula $(s \, \mathsf{U}_I \, \varphi)(i)$ is true if there is $j \geq i$ such that $t_j \in t_i + I$ and $\varphi(j) = \top$ and $s(k) = \top$ for all $i \leq k < j$. So let $T_i = \{j \mid t_j \in t_i + I\}$ be the set of indices in $t_i + I$. If $T_i = \emptyset$ then $(s \, \mathsf{U}_I \, \varphi)(i) = \bot$.

Otherwise, let $\mathrm{first}(i) = \min T_i$ and $\mathrm{last}(i) = \max T_i$ be the first and the last index in the interval $t_i + I$, respectively. So $(s \, \mathsf{U}_I \, \varphi)(i)$ is true if there exists $\mathrm{first}(i) \leq j \leq \mathrm{last}(i)$ such that $\varphi(j) = \top$ and $s(k) = \top$ for all $i \leq k < j$.

Now, the value of s is known. So let $\mathrm{seg}(i) = \min\{j \mid j \geq i \wedge s(j) = \bot\}$ be the first index no smaller than i such that $s(j)$ evaluates to false, i.e., $s(j)$ is true from i to $\mathrm{seg}(i) - 1$. Thus, $(s \, \mathsf{U}_I \, \varphi)(i)$ is true if there exists $\mathrm{first}(i) \leq j \leq \mathrm{last}(i)$ such that $\varphi(j) = \top$ and $j \leq \mathrm{seg}(i)$. So take $L_i = \mathrm{first}(i)$ and $R_i = \min(\mathrm{last}(i), \mathrm{seg}(i))$. Then $(s \, \mathsf{U}_I \, \varphi)(i)$ is true if $\bigvee_{L_i \leq j \leq R_i} \varphi(j)$ is true.

To build the circuits, we formalise the intuition from the left half of Fig. 3. The circuit C consists of internal gates $d_{p,q}$ and output gates o_i for each $1 \leq i \leq n$.

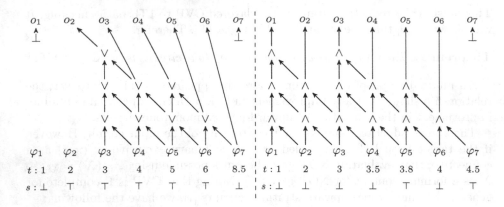

Fig. 3. Transducer circuits for $s\,\mathrm{U}_{[1,5]}\,\varphi$ and $\varphi\,\mathrm{U}_{[1,5]}\,s$. The first line below the circuits are timestamps, the second row are values of s. Note that different timestamps and s are used in the two examples. The inputs and the outputs of the circuits are denoted φ_i and o_i respectively.

Each internal gate $d_{p,q}$ calculates $\varphi_p \vee \cdots \vee \varphi_q$. Precisely, $d_{p,q}$ is present in the circuit if there is an i such that $L_i \le p \le q \le R_i$. If $p = q$ then $l(d_{p,q}) = (\mathrm{ID}, \varphi_p)$. Otherwise, $l(d_{p,q}) = (\mathrm{OR}, d_{p,q-1}, d_{p+1,q})$.

For the output gates, we define o_i so that $o_i = \bigvee_{L_i \le j \le R_i} \varphi(j) = (s\,\mathrm{U}_I\,\varphi)(i)$. Specifically, if $T_i = \emptyset$ then we set $l(o_i) = \bot$, otherwise, $l(o_i) = (\mathrm{ID}, d_{L_i, R_i})$.

An embedding $\gamma : C \to \mathbb{R}^2$ for the circuit C is $\gamma(o_i) = (i, n)$, $\gamma(\varphi_i) = (i, 0)$ and $\gamma(d_{p,q}) = (p, q - p + 1)$. Observe that $L_i \le L_{i+1}$ and $R_i \le R_{i+1}$. Hence, it cannot happen that $L_i < L_j \le R_j < R_i$ for some i and j. So the intervals may overlap but never is one properly contained in another. This ensures that the embedding is planar.

Finally, note that it is possible to compute L_i and R_i for every i in logarithmic space. Hence, the circuit construction can be carried out in logarithmic space.

Lemma 3. *Let p be any proposition. For each i, set the input φ_i of the circuit to $p(i)$. Then for each j, the value of o_j is true if and only if $(s\,\mathrm{U}_I\,p)(j)$ is true.*

We now give an analogous derivation and circuit construction for $\varphi\,\mathrm{U}_I\,s$. See the right side of Fig. 3 for an example of a resulting circuit.

For index $1 \le i \le n$ the formula $(\varphi\,\mathrm{U}_I\,s)(i)$ is true if there exists $j \ge i$ such that $t_j \in t_i + I$ and $s(j) = \top$ and $\varphi(k) = \top$ for all $i \le k < j$. Since s is known, we choose the first possible j. So let $\mathrm{limit}(i) = \min\{j \mid \mathrm{first}(i) \le j \le \mathrm{last}(i) \wedge s(j) = \top\}$ be the first j in the interval $t_i + I$ such that $s(j)$ is true.

If there is no such index then $(\varphi\,\mathrm{U}_I\,s)(i) = \bot$. Otherwise, $(\varphi\,\mathrm{U}_I\,s)(i)$ is true if $\varphi(k) = \top$ for all $i \le k < \mathrm{limit}(i)$. That is, $(\varphi\,\mathrm{U}_I\,s)(i) = \bigwedge_{i \le j < \mathrm{limit}(i)} \varphi_j$.

Now, the circuit C (see right half of Fig. 3) consists of gates $c_{p,q}$ calculating $\varphi_p \wedge \cdots \wedge \varphi_q$ and output gates o_i for $i = 1 \ldots n$. The gate $c_{p,q}$ is present in C if there is i such that $i \le p \le q < \mathrm{limit}(i)$. If $p = q$ then $l(c_{p,q}) = (\mathrm{ID}, \varphi_p)$. Otherwise, $l(c_{p,q}) = (\mathrm{AND}, c_{p,q-1}, c_{p+1,q})$.

For output, we set o_i so that $o_i = \bigwedge_{i \leq j < \text{limit}(i)} \varphi_j = (\varphi U_I s)(i)$. If $\text{limit}(i) = \infty$ then $l(o_i) = \bot$, if $\text{limit}(i) = i$ then $l(o_i) = \top$ and else $l(o_i) = (\text{ID}, c_{i,\text{limit}(i)-1})$.

The embedding $\gamma : C \to \mathbb{R}^2$ of the circuit C is the same as above, $\gamma(o_i) = (i, n)$, $\gamma(\varphi_i) = (i, 0)$ and $\gamma(c_{p,q}) = (p, q - p + 1)$. Since, $i < j$ implies $\text{limit}(i) \leq \text{limit}(j)$, the embedding is planar.

This finishes the construction of circuits for U_I. Circuits for the dual operators of U_I are obtained either by dualising OR and AND gates (Release operator), by performing the construction backwards in time (Since operator) or both (Trigger operator). Therefore,

Theorem 4. MTL *path checking is in* $\mathsf{AC}^1[\mathsf{logDCFL}]$.

A considerably weaker result appeared in [10], where the authors gave circuits and an $\mathsf{AC}^1[\mathsf{logDCFL}]$ algorithm only for a fragment of MTL interpreted over traces with integral timestamps $t_i = i$ and intervals of the form $[0, a]$ for $a \in \mathbb{N}$.

5   UTL

The most complicated circuits in the LTL path-checking algorithm [9] correspond to $s \, U \, \psi$ and $\psi \, U \, s$ formulae. As in the case of MTL, the circuits are also not uniform but depend on s. In this section, we devise an AC^1 tree-contraction algorithm for UTL—the fragment of LTL obtained by omitting binary temporal operators. The algorithm works even if XOR is allowed and is based on the analysis of functions arising in the tree contraction algorithm applied to UTL formulae. First consider the future-only fragment of UTL.

Let $p \in \mathbb{B}^n$ be any proposition. If $p(i) = \bot$ for every i then $(Fp)(i) = \bot$ for every i. Otherwise, let i be the largest index such that $p(i) = \top$. Then, $(Fp)(j) = \top$ for all $j \leq i$. By construction, $p(k) = \bot$ for all $k > i$. Hence, $(Fp)(k) = \bot$ for all $k > i$. Thus, Fp is downward monotone and depends only on the largest i with $p(i) = \top$. In particular, only $n + 1$ possible values exist for Fp.

Similarly, let t be the largest index such that $p(t) = \bot$. Then $p(j) = \top$ for all $j > t$. Hence $(Gp)(j) = \top$ for all $j > t$. Since $p(t) = \bot$ we have $(Gp)(k) = \bot$ for all $k \leq t$. Thus, Gp is upward monotone and depends only on the largest t with $p(t) = \bot$. In particular, only $n + 1$ possible values exist for Gp.

So for any formula ψ the value of $F \circ \psi$ or $G \circ \psi$ is a monotone vector—of which there are only $2n$ many. Hence for any formula context $\varphi(X)$, the formula contexts $\varphi \circ (FX)$ and $\varphi \circ (GX)$ can be represented as $g \circ F$ or $g \circ G$ where $g : \mathbb{M} \to \mathbb{B}^n$ is a *function with monotone domain*. Since $|\mathbb{M}| = O(n)$, enumerating all outputs of g explicitly requires only $|g| = O(n^2)$ space. Similar results hold for the past equivalents of G and F.

Now, Boolean operators are applied componentwise and obey the usual identities: $\bot \wedge p = \bot, \top \wedge p = p, \bot \vee p = p, \top \vee p = \top, \bot \oplus p = p$ and $\top \oplus p = \neg p$. Therefore, to represent partial evaluation of conjunction $(p \wedge X, x \wedge X)$, disjunction $(p \vee X, X \vee p)$ and xor $(p \oplus X, X \oplus p)$ it suffices to keep track whether each component is $\bot, \top$ or equal to the original or the negation of the value in X.

Furthermore, Next (Xp) and Yesterday (Yp) temporal operators shift p by 1 and -1, respectively. Let m be the size of the input formula. The last two paragraphs motivate the definition of filters: let $v \in \{\bot, \top, \text{ID}, \text{NOT}\}^n$ and $k \in [-m, m]$ satisfy $v(i) \in \mathbb{B}$ if $i + k \notin \{1, \ldots, n\}$. Then a **filter with offset k and pattern** v is the function $f_{v,k} : \mathbb{B}^n \to \mathbb{B}^n$ such that

$$f_{v,k}(p)(i) = \begin{cases} \bot & \text{if } v(i) = \bot \\ \top & \text{if } v(i) = \top \\ p(i + k) & \text{if } v(i) = \text{ID} \\ \neg p(i + k) & \text{if } v(i) = \text{NOT} \end{cases}$$

The identity function as well as the partial evaluation of conjunction, disjunction, and xor are expressible as filters with offset 0. Temporal operators Next and Yesterday are identity filters with offsets 1 and -1, respectively. Note that filters are closed under composition.

Storing v explicitly and k in unary requires $O(n + |\varphi|)$ bits per filter. By fully expanding the definition, we can evaluate and compose two filters in AC^0. Moreover, if $g : \mathbb{M} \to \mathbb{B}^n$ is a function with monotone domain then $(f_{v,k} \circ g) : \mathbb{M} \to \mathbb{B}^n$ is also a function with monotone domain and the composition in AC^0.

Lemma 4. *There are uniform AC^0 circuits calculating $f_{v,k} \circ f_{v',k'}$ and $f_{v,k}(p)(i)$ and $f_{v,k} \circ g$ and $\mathrm{F} \circ g$ and $\mathrm{G} \circ g$, where f's are filters and g is a function with monotone domain.*

We represent the functions arising in the tree-contraction algorithm as follows. If the contracted subtree S does not contain F or G operators then it is representable by a filter. If it contains F or G then let T be the first such occurrence. Then the segment from the leaves to T is representable by a filter and the segment above T is representable by a function with monotone domain. Thus, the function h associated with S can be represented as:

$$h = \begin{cases} \text{filter} & \text{no temporal operator} \\ f \circ T \circ \text{filter} & T \text{ is the first temporal operator; } f : \mathbb{M} \to \mathbb{B}^n \end{cases}$$

Now, if the contracted node is a Boolean connective, X or Y then we calculate $f_{v,k} \circ h$ for an appropriate filter. If the contracted node is F or G then we calculate $\mathrm{F} \circ h$ or $\mathrm{G} \circ h$. In either case, the resulting function is representable using the above format. Moreover, by Lemma 4, the composition is in AC^0. Hence, the complexity of the tree contraction algorithm is $\mathsf{AC}^1[\mathsf{AC}^0] = \mathsf{AC}^1$.

Theorem 5. UTL *path checking is in* AC^1.

Same results apply to past temporal operators. Note that the construction works also when the negation is applied to arbitrary subformulae, and not only to propositions. Also note that $F_{[a,\infty)}p$ is downward monotone and the corresponding circuits are constructible in logarithmic space. Therefore, the above arguments apply to the more powerful logic $\mathsf{UTL}_\geq$ obtained by allowing $F_{[a,\infty)}$ and $G_{[b,\infty)}$ operators. To the best of our knowledge, $\mathsf{UTL}_\geq$ is the most expressive and powerful fragment of LTL with a sub-$\mathsf{AC}^1[\mathsf{logDCFL}]$ path-checking problem.

6   Conclusion

The results obtained in this paper shed further light on the complexity landscape of temporal-logic path-checking problems. Several open questions however remain, the main one being to determine the precise complexity of LTL path checking. In particular, there has been no progress on the trivial NC^1 lower bound over the past ten years. Furthermore, might it be possible to separate the complexity of LTL and MTL, or of these logics and their future-only fragment?

Acknowledgments. This research was financially supported by EPSRC.

References

1. Artho, C., Barringer, H., Goldberg, A., Havelund, K., Khurshid, S., Lowry, M., Pasareanu, C., Rosu, G., Sen, K., Visser, W., Washington, R.: Combining test case generation and runtime verification. Theoretical Computer Science 336(2-3) (2005)
2. Barrington, D.A.M., Lu, C.-J., Miltersen, P.B., Skyum, S.: On monotone planar circuits. In: Proceedings of Fourteenth Annual IEEE Conference on Computational Complexity (1999)
3. Bundala, D., Ouaknine, J.: On the complexity of path checking of temporal logics (full version). CoRR, abs/1312.7603
4. Chakraborty, T., Datta, S.: One-input-face MPCVP is hard for L, but in LogDCFL. In: Arun-Kumar, S., Garg, N. (eds.) FSTTCS 2006. LNCS, vol. 4337, pp. 57–68. Springer, Heidelberg (2006)
5. Demri, S., Schnoebelen, P.: The complexity of propositional linear temporal logics in simple cases. Information and Computation 174(1), 84–103 (2002)
6. Finkbeiner, B., Sipma, H.: Checking finite traces using alternating automata. Formal Methods in System Design 24(2), 101–127 (2004)
7. Goldschlager, L.M.: The monotone and planar circuit value problems are log space complete for P. SIGACT News 9 (July 1977)
8. Karp, R.M., Ramachandran, V.: Parallel algorithms for shared-memory machines. In: Handbook of Theoretical Computer Science, Volume A: Algorithms and Complexity (A) (1990)
9. Kuhtz, L., Finkbeiner, B.: LTL path checking is efficiently parallelizable. In: Albers, S., Marchetti-Spaccamela, A., Matias, Y., Nikoletseas, S., Thomas, W. (eds.) ICALP 2009, Part II. LNCS, vol. 5556, pp. 235–246. Springer, Heidelberg (2009)
10. Kuhtz, L., Finkbeiner, B.: Efficient parallel path checking for linear-time temporal logic with past and bounds. Logical Methods in Computer Science 8(4) (2012)
11. Limaye, N., Mahajan, M., Sarma, J.M.N.: Evaluating monotone circuits on cylinders, planes and tori. In: Durand, B., Thomas, W. (eds.) STACS 2006. LNCS, vol. 3884, pp. 660–671. Springer, Heidelberg (2006)
12. Maler, O., Nickovic, D.: Monitoring temporal properties of continuous signals. In: Lakhnech, Y., Yovine, S. (eds.) FORMATS 2004 and FTRTFT 2004. LNCS, vol. 3253, pp. 152–166. Springer, Heidelberg (2004)
13. Markey, N., Schnoebelen, P.: Model checking a path (Preliminary report). In: Amadio, R.M., Lugiez, D. (eds.) CONCUR 2003. LNCS, vol. 2761, pp. 251–265. Springer, Heidelberg (2003)
14. Younes, H.L.S., Simmons, R.G.: Probabilistic verification of discrete event systems using acceptance sampling. In: Brinksma, E., Larsen, K.G. (eds.) CAV 2002. LNCS, vol. 2404, pp. 223–235. Springer, Heidelberg (2002)

Parameterised Linearisability

Andrea Cerone[1], Alexey Gotsman[1], and Hongseok Yang[2]

[1] IMDEA Software Institute, Madrid, Spain
[2] University of Oxford, England

Abstract Many concurrent libraries are parameterised, meaning that they implement generic algorithms that take another library as a parameter. In such cases, the standard way of stating the correctness of concurrent libraries via linearisability is inapplicable. We generalise linearisability to parameterised libraries and investigate subtle trade-offs between the assumptions that such libraries can make about their environment and the conditions that linearisability has to impose on different types of interactions with it. We prove that the resulting parameterised linearisability is closed under instantiating parameter libraries and composing several non-interacting libraries, and furthermore implies observational refinement. These results allow modularising the reasoning about concurrent programs using parameterised libraries and confirm the appropriateness of the proposed definitions. We illustrate the applicability of our results by proving the correctness of a parameterised library implementing flat combining.

1 Introduction

Concurrent libraries encapsulate high-performance concurrent algorithms and data structures behind a well-defined interface, providing a set of methods for clients to call. Many such libraries [6,7,13] are *parameterised*, meaning that they implement generic algorithms that take another library as a parameter and use it to implement more complex functionality (we give a concrete example in §2). Reasoning about the correctness of parameterised libraries is challenging, as it requires considering all possible libraries that they can take as parameters.

Correctness of concurrent libraries is usually stated using *linearisability* [8], which fixes a certain correspondence between the *concrete* library implementation and a (possibly simpler) *abstract* library, whose behaviour the concrete one is supposed to simulate. For example, a high-performance concurrent stack that allows multiple push and pop operations to access the data structure at the same time may be specified by an abstract library where each operation takes effect atomically. However, linearisability considers only *ground* libraries, where all of the library implementation is given, and is thus inapplicable to parameterised ones. In this paper we propose a notion of *parameterised linearisability* (§3 and §4) that lifts this limitation. The key idea is to take into account not only interactions of a library with its client, but also with its parameter library, with the two types of interactions being subject to different conditions.

A challenge we have to deal with while generalising linearisability in this way is that parameterised libraries are often correct only under some assumptions about the context in which they are used. Thus, a parameterised library may assume that the library it takes as a parameter is *encapsulated*, meaning that clients cannot call its methods directly. A parameterised library may also accept as a parameter only libraries satisfying certain properties. For this reason, we actually present three notions of parameterised

J. Esparza et al. (Eds.): ICALP 2014, Part II, LNCS 8573, pp. 98–109, 2014.
© Springer-Verlag Berlin Heidelberg 2014

linearisability, appropriate for different situations: a general one, which does not make any assumptions about the client or the parameter library, a notion appropriate for the case when the parameter library is encapsulated, and *up-to linearisability*, which allows making assumptions about the parameter library. These notions differ in subtle ways: we find that there is a trade-off between the assumptions that parameterised libraries make about their environment and the conditions that a notion of linearisability has to impose on different types of interactions with it.

We prove that the proposed notions of parameterised linearisability are *contextual* (§5), i.e., closed under parameter instantiation. This includes the case when the parameter library is itself parameterised. On the other hand, when the parameter is an ordinary ground library, this result allows us to derive the classical linearisability of the instantiated library from our notion for the parameterised one. We also prove that parameterised linearisability is *compositional* (§5): if several non-interacting libraries are linearisable, so is their composition. Finally, we show that parameterised linearisability implies *observational refinement* (§6): the behaviours of any complete program using a concrete parameterised library can be reproduced if the program uses a corresponding abstract one instead. All these results allow modularising the reasoning about concurrent programs using parameterised libraries: contextuality and compositionality break the reasoning about complex parameterised libraries into that about individual libraries from which they are constructed; observational refinement then lifts this to complete programs, including clients. The properties of parameterised linearisability we establish also serve to confirm the appropriateness of the proposed definitions.

We illustrate the applicability of our results by proving the up-to linearisability of flat combining [6] (§4), a generic algorithm for converting hard-to-parallelise sequential data structures into concurrent ones.

Due to space constraints, we defer the proofs of most theorems to [1, §B].

2   Parameterised Libraries

We consider *parameterised libraries* (or simply libraries) L, which provide some *public methods* to their *clients*. The latter are multi-threaded programs that can call the methods in parallel. In §4 and §6 we introduce a particular syntax for libraries and clients; for now it suffices to treat them abstractly. Our libraries are called parameterised because we allow their method implementations to call *abstract methods*, whose implementation is left unspecified. Abstract methods are meant to be implemented by another library provided by L's client, which we call the *parameter library* of L.

We identify methods by names from a set $\mathcal{M}$, ranged over by m, and threads by identifiers from a set $\mathcal{T}$, ranged over by t. For the sake of simplicity, we assume that methods take a single integer as a parameter and always return an integer. We annotate libraries with types as in $L : M \to M'$, where $M, M' \subseteq \mathcal{M}$ give the sets of abstract and public methods of L, respectively. If $M = \emptyset$ we call L a *ground library*. The sets M and M' do not have to be disjoint: methods in $M \cap M'$ may be called by L's clients, but their implementation is inherited from the one given by the parameter library.

Example: Flat Combining. Flat combining [6] is a recent synchronisation paradigm, which can be viewed [14] as a parameterised library $\mathsf{FC} : \{m_i\}_{i=1}^n \to \{\mathsf{do\_}m_i\}_{i=1}^n$ for a given set of methods $\{m_i\}_{i=1}^n$. In Figure 1 we show a pseudocode of its implementation, which simplifies the original one in ways orthogonal to our goals. FC takes a library, whose methods m_i are meant to be executed sequentially, and efficiently turns it into a library with methods do_m_i that can be called concurrently.

As usual, this is achieved by means of mutual exclusion, implemented using a lock, but in a way that is more sophisticated than just acquiring it before calling a method m_i. A thread executing do_m_i first publishes the operation it would like to execute and its parameter in its entry of the requests array. It then spins, trying to acquire the global lock. Having acquired a lock, the thread becomes a *combiner*: it performs the operations requested by all threads, stored in requests, by calling methods m_i of the parameter library and writing the values returned into the retval field of the corresponding entries in requests. Each spinning thread periodically checks this field and stops if some other thread has performed the operation it requested (for simplicity, we assume that nil

```
LOCK lock;
struct{op,param,retval} requests[NThread];

do_mᵢ(int z):
  requests[mytid()].op = i;
  requests[mytid()].param = z;
  requests[mytid()].retval = nil;
  do:
    if (lock.tryacquire()):
      for (t = 0; t < NThread; t++):
        if (requests[t].retval == nil):
          int j = requests[t].op;
          int w = requests[t].param;
          requests[t].retval = mⱼ(w);
      lock.release();
  while (requests[mytid()].retval == nil);
  return requests[mytid()].retval;
```

Fig. 1. Flat combining: implementation FC

is a special value that is never returned by any method). This algorithm benefits from cache locality when the combiner executes several operations in sequence, and thus yields good performance even for hard-to-parallelise data structures, such as stacks and queues.

In this paper, we develop a framework for specifying and verifying parameterised concurrent libraries. For flat combining, our framework suggests using an **abstract** library $FC^\sharp$: $\{m_i\}_{i=1}^n \to \{do\_m_i\}_{i=1}^n$ in Figure 2 as a specification for the **concrete** library in Figure 1. $FC^\sharp$ specifies the expected behaviour of flat combining by using the naive mutual exclusion. Showing that the implementation satisfies this specification in our framework amounts to proving that it is related to $FC^\sharp$ by *parameterised linearisability*, which we present next.

```
LOCK lock;
do_mᵢ(int z):
  lock.acquire();
  int retval = mᵢ(z);
  lock.release();
  return retval;
```

Fig. 2. Flat combining: specification $FC^\sharp$

3   Histories and Parameterised Linearisability

Histories. Informally, for a concrete library (such as the one in Figure 1) to be correct with respect to an abstract one (such as the one in Figure 2), the two should interact with their environment—the client and the parameter library—in similar ways. In this paper, we assume that different libraries and their clients access disjoint portions of memory, and thus interactions between them are limited to passing parameters and return values at method calls and returns. This is a standard assumption [8], which we believe can be relaxed using existing techniques [5]; see §7 for discussion. We record interactions of a parameterised library $L : M \to M'$ with its environment using *histories* (Definition 1 below), which are certain sequences of **actions** of the form

$$\mathsf{Act} ::= (t, \mathsf{call?}\, m'(z)) \mid (t, \mathsf{ret!}\, m'(z)) \mid (t, \mathsf{call!}\, m(z)) \mid (t, \mathsf{ret?}\, m(z)),$$

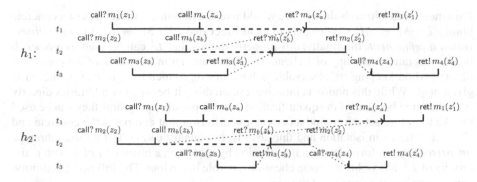

Fig. 4. Illustration of histories and parameterised linearisability. A solid line represents a thread executing the code of the parameterised library, and a dashed one, the parameter library.

where $t \in \mathcal{T}$ is the thread performing the action, $m' \in M'$ or $m \in M$ is the method involved, and $z \in \mathbb{Z}$ is the method parameter or a return value.

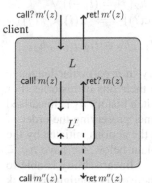

Fig. 3. Interactions of a library L with its client and parameter library L'

We illustrate the meaning of the actions in Figure 3: call? and ret! describe the client invoking public methods m' of the parameterised library L, and call! and ret? the library L invoking implementations of abstract methods m provided by a parameter library L'. We denote the sets of actions corresponding to interactions with these two entities by ClAct and AbsAct, respectively. In the spirit of the opponent-proponent distinction in game semantics [9,11], we annotate actions by ! or ? depending on whether the action was initiated by L or by an external entity, and we denote the corresponding sets of actions by Act! and Act?. We also use sets ActCall?, ActRet!, ActCall! and ActRet? with the expected meaning. Clients can also call methods $m'' \in M \cap M'$ directly, as represented by the dashed lines in the figure. Since such interactions do not involve the library L, we do not include them into Act. Histories are finite sequences of actions with invocations of abstract methods properly nested inside those of public ones.

Definition 1 (Histories). *A **history** $h : M \to M'$ is a finite sequence of actions such that for every t, the projection of h to t's actions is a prefix of a sequence generated by the grammar SHist below, where $m \in M$ and $m' \in M'$:*

SHist $::= \varepsilon \mid (t, \mathsf{call?}\, m'(z))\, \mathsf{IntSHist}\, (t, \mathsf{ret!}\, m'(z')) \mid \mathsf{SHist}\, \mathsf{SHist}$
IntSHist $::= \varepsilon \mid (t, \mathsf{call!}\, m(z))\, (t, \mathsf{ret?}\, m(z')) \mid \mathsf{IntSHist}\, \mathsf{IntSHist}$

We denote the set of histories by Hist. See Figure 4 for examples. In this paper, we focus on safety properties of libraries and thus let histories be finite. This assumption is also taken by the classical notion of linearisability [8] and can be relaxed as described in [4] (§7). For a history h and $A \subseteq$ Act, we let $h|_A$ be the projection of h onto actions in A and we denote the i-th action in h by $h(i)$.

Parameterised Linearisability. We would like the notion of correctness of a concrete library $L : M \to M'$ with respect to an abstract one $L^\sharp : M \to M'$ to imply *observational refinement*. Informally, this property means that $L^\sharp$ can be used to replace L in any program (consisting of a client, the library and an instantiation of the parameter library) while keeping its observable behaviours reproducible; a formal definition is given in §6. While this notion is intuitive, establishing it between two libraries directly is challenging because of the quantification over all possible programs they can be used by. We therefore set out to find a correctness criterion that compares the concrete and abstract libraries in isolation and thus avoids this quantification. For ground libraries, *linearisability* [8] formulates such a criterion by matching a history h_1 of L with a history h_2 of $L^\sharp$ that yields the same client-observable behaviour. The following definition generalises it to parameterised libraries.

Definition 2 (Parameterised linearisability: general case). *A history $h_1 : M \to M'$ is* **linearised** *by another one $h_2 : M \to M'$, written $h_1 \sqsubseteq h_2$, if there exists a permutation $\pi : \mathbb{N} \to \mathbb{N}$ such that*

$$\forall i.\, h_1(i) = h_2(\pi(i)) \wedge (\forall j.\, i < j \wedge ((\exists t.\, h_1(i) = (t, -) \wedge h_2(j) = (t, -)) \vee$$
$$(h_1(i) \in \mathsf{Act!} \wedge h_1(j) \in \mathsf{Act?})) \implies \pi(i) < \pi(j)).$$

For sets of histories H_1, H_2 we let $H_1 \sqsubseteq H_2 \iff \forall h_1 \in H_1.\, \exists h_2 \in H_2.\, h_1 \sqsubseteq h_2$.

In §4 we show how to generate all histories of a library in a particular language and define linearisability on libraries by the $\sqsubseteq$ relation on their sets of histories. For now we explain the above abstract definition. According to it, a history h_1 is linearised by a history h_2 when the latter is a permutation of the former preserving the order of actions within threads and the precedence relation between the actions initiated by the library and those initiated by its environment. As we explain below, we have $h_1 \sqsubseteq h_2$ for the histories h_1, h_2 in Figure 4. Hence, parameterised linearisability is able to match a history of a concurrent library with a simpler one where every contiguous block of library execution (e.g., the one between $(t_1, \mathsf{call?}\, m_1(z_1))$ and $(t_1, \mathsf{call!}\, m_a(z_a))$) is executed without interleaving with other such blocks. On the other hand, $h_2 \not\sqsubseteq h_1$, since $(t_1, \mathsf{call!}\, m_a(z_a))$ precedes $(t_3, \mathsf{call?}\, m_3(z_3))$ in h_2, but not in h_1.

When $h_1, h_2 : \emptyset \to M'$, i.e., these are histories of a ground library and thus contain only call? and ret! actions, Definition 2 coincides with a variant of the classical linearisability [8], which requires preserving the order between ret! and call? actions. For example, Definition 2 requires preserving the order between $(t_2, \mathsf{ret!}\, m_2(z_2'))$ and $(t_3, \mathsf{call?}\, m_4(z_4))$ in h_1 from Figure 4 (shown by a diagonal arrow). This requirement is needed for linearisability to imply observational refinement: informally, during the interval of time between $(t_2, \mathsf{ret!}\, m_2(z_2'))$ and $(t_3, \mathsf{call?}\, m_4(z_4))$ in an execution of a program producing h_1, both threads t_2 and t_3 execute pieces of client code, which can communicate via the client memory. To preserve the behaviour of the client when replacing the concrete library in the program by an abstract one in observational refinement, this communication must not be affected, and, for this, the abstract library has to admit a history in which the order between the above actions is preserved.

When $h_1, h_2 : M \to M'$ correspond to a non-ground parameterised library, i.e., $M \neq \emptyset$, a similar situation arises with communication between the methods of the parameter library executing in different threads. For this reason, our generalisation of linearisability requires preserving the order between call! and ret? actions, such as

$(t_2, \text{call}!\ m_b(z_b))$ and $(t_1, \text{ret}?\ m_a(z'_a))$ in Figure 4; this requirement is dual to the one considered in classical linearisability. It is not enough, however. Definition 2 also requires preserving the order between call! and call?, as well as ret! and ret? actions, e.g., $(t_3, \text{ret}!\ m_3(z'_3))$ and $(t_2, \text{ret}?\ m_b(z'_b))$ in Figure 4. In the case when $M \cap M' \neq \emptyset$, this is also required to validate observational refinement. For example, during the interval of time between $(t_3, \text{ret}!\ m_3(z'_3))$ and $(t_2, \text{ret}?\ m_b(z'_b))$ in an execution producing h_1, the client code in thread t_3 can call a method $m'_b \in M \cap M'$ of the parameter library (cf. the dashed arrows in Figure 3). The code of the method m'_b executed by t_3 can then communicate with that of the method m_b executed by t_2, and to preserve this communication, we need to preserve the order between $(t_3, \text{ret}!\ m_3(z'_3))$ and $(t_2, \text{ret}?\ m_b(z'_b))$.

In §5 and §6 we prove that the above notion of linearisability indeed validates observational refinement. If the library $L : M \to M'$ producing the histories h_1, h_2 in Definition 2 is such that $M \cap M' = \emptyset$, then the client cannot directly call methods of its parameter library, and, as we show, parameterised linearisability can be weakened without invalidating observational refinement.

Definition 3 (Parameterised linearisability: encapsulated case). *For $h_1, h_2 : M \to M'$ with $M \cap M' = \emptyset$ we let $h_1 \sqsubseteq_e h_2$ if there exists a permutation $\pi : \mathbb{N} \to \mathbb{N}$ such that*

$$\forall i.\ h_1(i) = h_2(\pi(i)) \wedge (\forall j.\ i < j \wedge ((\exists t.\ h_1(i) = (t, -) \wedge h_2(j) = (t, -)) \vee$$
$$(h_1(i), h_1(j)) \in (\text{ActRet}! \times \text{ActCall}?) \cup (\text{ActCall}! \times \text{ActRet}?)) \implies \pi(i) < \pi(j)).$$

Since this definition does not take into account the order between $(t_1, \text{call}!\ m_a(z_a))$ and $(t_3, \text{call}?\ m_3(z_3))$ in h_2 from Figure 4, we have $h_2 \sqsubseteq_e h_1$ even though $h_2 \not\sqsubseteq h_1$.

Definitions 2 and 3 do not make any assumptions about the implementation of the parameter library. However, sometimes the correctness of a parameterised library can only be established under certain assumptions about the behaviour of its parameter. In particular, this is the case for the flat combining library from §2. In its implementation FC from Figure 1, a request by a thread t to execute a method m_i of the parameter library can be fulfilled by another thread t' who happens to act as a combiner; in contrast, the specification FC$^\sharp$ in Figure 2 pretends that m_i is executed in the requesting thread. Thus, FC and FC$^\sharp$ will behave differently if we supply as their parameter a library whose methods depend on the identifiers of executing threads (e.g., with m_i implemented as "return mytid()"). As a consequence, FC does not simulate FC$^\sharp$. On the other hand, this will be the case if we restrict ourselves to parameter libraries whose behaviour is independent of thread identifiers. The following version of parameterised linearisability allows us to use such assumptions, formulated as closure properties on histories of interactions between a parameterised library and its parameter. Given a history h, let $\overline{h}$ be the history obtained by swapping ! and ? actions in h.

Definition 4 (Up-to linearisability). *For $h_1, h_2 : M \to M'$ such that $M \cap M' = \emptyset$ and a binary relation $\mathcal{R}$ on histories of type $\emptyset \to M$, we say that h_1 is **linearised by** h_2 **up to** $\mathcal{R}$, written $h_1 \sqsubseteq_\mathcal{R} h_2$, if $(h_1|_{\text{ClAct}}) \sqsubseteq (h_2|_{\text{ClAct}})$ and $\overline{(h_1|_{\text{AbsAct}})}\ \mathcal{R}\ \overline{(h_2|_{\text{AbsAct}})}$.*

For flat combining, a suitable relation $\mathcal{R}_t$ relates two histories if one can be obtained from the other by replacing thread identifiers of some pairs of a call and a corresponding (if any) return action. There are other useful choices of $\mathcal{R}$, such as equivalence up to commuting abstract method invocations [7].

So far we have defined our notions of linearisability abstractly, on sets of histories. We next introduce a language for parameterised libraries and show how to generate sets of histories of a library in this language. This lets us lift the notion of linearisability to libraries and prove that FC in Figure 1 is indeed linearised up to $\mathcal{R}_t$ by $FC^\sharp$ in Figure 2.

4   Lifting Linearisability to Libraries

Library Syntax. We use the following language to define libraries:

$$L ::= \langle \text{public} : B; \text{private} : B \rangle \qquad B ::= \varepsilon \mid (m \Leftarrow C); B \mid (\text{abstract } m); B$$
$$C ::= c \mid m() \mid C; C \mid \text{if}(E) \text{ then } C \text{ else } C \mid \text{while}(E) \, C$$

A parameterised library L is a collection of methods, some implemented by commands C and others declared as abstract, meant to be implemented by a parameter library. Methods can be public or private, with only the former made available to clients. In §5 and §6 we extend the language to complete programs, consisting of a multithreaded client using a parameterised library with its parameter instantiated. In particular, we introduce private methods here to define parameter library instantiation in §5.

In commands, c ranges over **primitive commands** from a set PComm, and E over expressions, whose set we leave unspecified. The command $m()$ invokes the method m; it does not mention its parameter or return value, since, as we explain below, these are passed via dedicated thread-local memory locations. We consider only well-formed libraries where a method is declared at most once and every method called is declared. We identify libraries up to the order of method declarations and α-renaming of private non-abstract methods. For a library $L = \langle \text{public} : B_{\text{pub}}; \text{private} : B_{\text{pvt}} \rangle$ we have $L :$ $\text{Abs}(L) \to \text{Pub}(L)$, where $\text{Pub}(L)$ is the set of methods declared in B_{pub}, and $\text{Abs}(L)$ of those declared as abstract in B_{pub} or B_{pvt}.

Linearisability on Libraries and the Semantics Idea. We now show how to generate the set of histories $[\![L]\!] \in 2^{\text{Hist}}$ of a library L. Then we let a library L_1 be **linearised by** a library L_2, written $L_1 \sqsubseteq L_2$, if $[\![L_1]\!] \sqsubseteq [\![L_2]\!]$; similarly for $\sqsubseteq_e$ and $\sqsubseteq_{\mathcal{R}}$.

We actually generate all library *traces*, which, unlike histories, also record its internal actions. Let us extend the set of actions Act with elements of the forms (t, c) for $c \in$ PComm, $(t, \text{call } m(z))$ and $(t, \text{ret } m(z))$, leading to a set TrAct. The latter two kinds of actions correspond to calls and returns between methods implemented inside the library. A *trace* τ is a finite sequence of elements in TrAct; we let Traces $=$ TrAct*.

The denotation $[\![L]\!]$ of a library $L : M \to M'$ includes the histories extracted from traces that L produces in any possible environment, i.e., assuming that client threads perform any sequences of calls to methods in M' with arbitrary parameter values and that abstract methods in M return arbitrary values. The definition of $[\![L]\!]$ follows the intuitive semantics of our programming language. An impatient reader can skip it on first reading and jump directly to Theorem 1 at the end of this section.

Heaps and Primitive Command Semantics. Let Locs be the set of memory locations. As we noted in §3, we impose a standard restriction that different libraries and their clients access different sets of memory locations, except the ones used for method parameter passing. Formally, we assume that each library L is associated with a set of its locations $\text{Locs}_L \subseteq \text{Locs}$. The state of L is thus given by a **heap** $\sigma \in \text{Locs}_L \to \mathbb{Z}$. We assume a special subset of locations $\{\text{arg}_t\}_{t \in \mathcal{T}}$ belonging to every Locs_L, which we use to pass parameters and return values for method invocations in thread t.

We assume that the execution of primitive commands and the evaluation of expressions are atomic. The semantics of a primitive command $c \in$ PComm used by a

Traces of commands $(\!|C|\!)_t : (\mathcal{M} \times \mathcal{T} \to 2^{\mathsf{Traces}}) \to 2^{\mathsf{Traces}}$

$(\!|c|\!)_t \eta = \{(t, c)\}$ $(\!|C_1; C_2|\!)_t \eta = \{\tau_1 \tau_2 \mid \tau_1 \in (\!|C_1|\!)_t \eta \wedge \tau_2 \in (\!|C_2|\!)_t \eta\}$

$(\!|\mathsf{if}(E) \text{ then } C_1 \text{ else } C_2|\!)_t \eta = (t, \mathsf{assume}(E))\, ((\!|C_1|\!)_t \eta) \cup (t, \mathsf{assume}(!E))((\!|C_2|\!)_t \eta)$

$(\!|\mathsf{while}(E)\ C|\!)_t \eta = ((t, \mathsf{assume}(E))([\![C]\!]\eta))^*(t, \mathsf{assume}(!E))$

$(\!|m()|\!)_t \eta = \begin{cases} \{(t, \mathsf{call!}\ m(z))\,\tau\,(t, \mathsf{ret?}\ m(z')) \mid \tau \in \eta(m, t) \wedge z, z' \in \mathbb{Z}\}, & \text{if } m \in M \\ \{(t, \mathsf{call}\ m(z))\,\tau\,(t, \mathsf{ret}\ m(z')) \mid \tau \in \eta(m, t) \wedge z, z' \in \mathbb{Z}\}, & \text{otherwise} \end{cases}$

Traces of library bodies

$\mathcal{F} : (\mathcal{M} \times \mathcal{T} \to 2^{\mathsf{Traces}}) \to (\mathcal{M} \times \mathcal{T} \to 2^{\mathsf{Traces}})$ | $(\!|B|\!) : \mathcal{M} \times \mathcal{T} \to 2^{\mathsf{Traces}}$

$(\mathcal{F}(\eta))(m, t) = \begin{cases} \eta(m, t) \cup ((\!|C|\!)_t \eta), & \text{if } (m \Leftarrow C) \text{ appears in } L \\ \{\varepsilon\}, & \text{if } m \in M \\ \emptyset, & \text{otherwise} \end{cases}$ | $(\!|B_{\mathsf{pub}}; B_{\mathsf{pvt}}|\!) = \mathsf{lfp}(\mathcal{F})$

Traces of libraries $(\!|L : M \to M'|\!) : 2^{\mathsf{Traces}}$

$$(\!|L|\!) = \mathsf{prefix}\left(\bigcup_{k>0} \Big\|_{t=1}^{k} \left(\bigcup_{\substack{z,z' \in \mathbb{Z} \\ m \in M' \setminus M}} (t, \mathsf{call?}\ m(z))\,((\!|B_{\mathsf{pub}}; B_{\mathsf{pvt}}|\!)(m, t))\,(t, \mathsf{ret!}\ m(z'))\right)^*\right)$$

Fig. 5. Possible traces of a library $L = \langle \mathsf{public} : B_{\mathsf{pub}}; \mathsf{private} : B_{\mathsf{pvt}} \rangle : M \to M'$. Here $\big\|_{t=1}^{k} T_t$ denotes the set of all interleavings of traces from the sets $T_1, \ldots, T_k$.

$\sigma \leadsto_{\mathsf{call}\ m(z),t}^{L} \sigma'$ iff $\sigma' = \sigma, \sigma(\mathsf{arg}_t) = z$ $\sigma \leadsto_{\mathsf{ret}\ m(z),t}^{L} \sigma'$ iff $\sigma' = \sigma, \sigma(\mathsf{arg}_t) = z$

$\sigma \leadsto_{\mathsf{call?}\ m(z),t}^{L} \sigma'$ iff $\sigma' = \sigma[\mathsf{arg}_t \mapsto z]$ $\sigma \leadsto_{\mathsf{ret!}\ m(z),t}^{L} \sigma'$ iff $\sigma' = \sigma, \sigma(\mathsf{arg}_t) = z$

$\sigma \leadsto_{\mathsf{call!}\ m(z),t}^{L} \sigma'$ iff $\sigma' = \sigma, \sigma(\mathsf{arg}_t) = z$ $\sigma \leadsto_{\mathsf{ret?}\ m(z),t}^{L} \sigma'$ iff $\sigma' = \sigma[\mathsf{arg}_t \mapsto z]$

Fig. 6. Transformers for calls and returns to, from and inside a library L

library L is defined by a family of transformers $\{\leadsto_{c,t}^{L}\}_{t \in \mathcal{T}}$, where $\leadsto_{c,t}^{L} \subseteq (\mathsf{Locs}_L \to \mathbb{Z}) \times (\mathsf{Locs}_L \to \mathbb{Z})$ describes how c affects the state of the library. The fact that the transformers are defined on locations from Locs_L formalises our assumption that L accesses only these locations. We assume that the transformers satisfy some standard properties [15], deferred to [1, §A] due to space constraints. To define the semantics of expressions, we assume that for each E the set PComm contains a special command $\mathsf{assume}(E)$, used only in defining the semantics, that allows the computation to proceed only if E is non-zero: $\sigma \leadsto_{\mathsf{assume}(E),t}^{L} \sigma'$ iff $\sigma' = \sigma$ and E is non-zero in σ.

Library Denotations. The set of traces of a library is generated in two stages. First, we generate a superset $(\!|L|\!) \subseteq 2^{\mathsf{Traces}}$ of traces produced by L, defined in Figure 5. If we think of commands as control-flow graphs, these traces contain interleavings of all possible paths through the control-flow graphs of L's methods, invoked in an arbitrary sequence. We then select those traces in $(\!|L|\!)$ that correspond to valid executions starting in a given heap using a predicate $[\![\tau]\!]_L : (\mathsf{Locs}_L \to \mathbb{Z}) \to \{\mathsf{true}, \mathsf{false}\}$. We define $[\![\cdot]\!]_L$ by generalising $\leadsto$ to calls and returns as shown in Figure 6 and letting

$[\![\varepsilon]\!]_L \sigma = \mathsf{true}; \quad [\![(t, a)\,\tau]\!]_L \sigma = \text{if } (\exists \sigma'.\ \sigma \leadsto_{a,t}^{L} \sigma' \wedge [\![\tau]\!]_L \sigma' = \mathsf{true}) \text{ then true else false}.$

Finally, we let the set of histories $[\![L]\!]$ of a library L consist of those obtained from traces representing its valid executions from a heap with all locations set to 0:

$$[\![L]\!] = \mathsf{history}(\{\tau \in (\!|L|\!) \mid [\![\tau]\!]_L (\lambda x \in \mathsf{Locs}_L.\ 0) = \mathsf{true}\}),$$

where history projects to actions in Act.

Theorem 1 (Correctness of flat combining). *For the libraries* FC *in Figure 1 and* FC$^\sharp$ *in Figure 2 and the relation* $\mathcal{R}_t$ *from* §3 *we have* FC $\sqsubseteq_{\mathcal{R}_t}$ FC$^\sharp$.

PROOF SKETCH. Consider $h \in [\![FC]\!]$. In such a history, any invocation of an abstract method $(t, \text{call! } m_i(z_i)) (t, \text{ret? } m_i(z_i'))$ happens within the execution of the corresponding wrapper method $(t', \text{call? } \text{do\_}m_i(z_i)) (t', \text{ret! } \text{do\_}m_i(z_i'))$ (or just $(t', \text{call? } \text{do\_}m_i(z_i))$ if the execution of the method is uncompleted in h), though not necessarily in the same thread. This correspondence is one-to-one, as different invocations of abstract methods correspond to different requests to perform them. Furthermore, abstract methods in h are executed sequentially. We then construct a history h' by replacing every abstract method call $(t, \text{call! } m_i(z_i)) (t, \text{ret? } m_i(z_i'))$ in $h|_{\text{AbsAct}}$ by

$$(t', \text{call? } \text{do\_}m_i(z_i)) (t', \text{call! } m_i(z_i)) (t', \text{ret? } m_i(z_i')) (t', \text{ret! } \text{do\_}m_i(z_i')),$$

where t' is the thread identifier of the corresponding wrapper method invocation (similarly for uncompleted invocations). It is easy to see that $\overline{(h|_{\text{AbsAct}})} \; \mathcal{R}_t \; \overline{(h'|_{\text{AbsAct}})}$ and $h' \in [\![FC^\sharp]\!]$. Since the execution of an abstract method in h happens within the execution of the corresponding wrapper method, we also have $(h|_{\text{ClAct}}) \sqsubseteq (h'|_{\text{ClAct}})$.   □

5   Instantiating Library Parameters and Contextuality

We now define how library parameters are instantiated and show that our notions of linearisability are preserved under such instantiations. To this end, we introduce a partial operation ∘ on libraries of §4: informally, for $L_1 : M \to M'$ and $L_2 : M' \to M''$ the library $L_2 \circ L_1 : M \to M''$ is obtained by instantiating abstract methods in L_2 with their implementations from L_1. Note that L_1 can itself have abstract methods M, which are left unimplemented in $L_2 \circ L_1$. Since we assume that different libraries operate in disjoint address spaces, for ∘ to be defined we require that the sets of locations of L_1 and L_2 be disjoint, with the exception of those used for method parameter passing. To avoid name clashes, we also require that public non-abstract methods of L_2 not be declared as abstract in L_1 (private non-abstract methods are not an issue, since we identify libraries up to their α-renaming); this also disallows recursion between L_2 and L_1.

Definition 5 (Parameter library instantiation). *Consider* $L_1 : M \to M'$ *and* $L_2 : M' \to M''$ *such that* $(M'' \setminus M') \cap M = \emptyset$ *and* $\text{Locs}_{L_1} \cap \text{Locs}_{L_2} = \{\text{arg}_t\}_{t \in \mathcal{T}}$. *Then* $L_2 \circ L_1 : M \to M''$ *is the library with* $\text{Locs}_{L_2 \circ L_1} = \text{Locs}_{L_1} \cup \text{Locs}_{L_2}$ *obtained by erasing the declarations for methods in* M' *from* L_2, *reclassifying the methods from* $M' \setminus M''$ *in* L_1 *as private, and concatenating the method declarations of the resulting two libraries. We write* $(L_2 \circ L_1)\!\downarrow$ *when* $L_2 \circ L_1$ *is defined.*

We now show that the notions of parameterised linearisability we proposed are *contextual*, i.e., closed under library instantiations. This property is useful in that it allows us to break the reasoning about a complex library into that about individual libraries from which it is constructed. As we show in §6, contextuality also helps us establish observational refinement.

Theorem 2 (Contextuality of parameterised linearisability: general case). *For* $L_1, L_2 : M \to M'$ *such that* $L_1 \sqsubseteq L_2$:

   (i) $\forall L : M'' \to M. (L_1 \circ L)\!\downarrow \wedge (L_2 \circ L)\!\downarrow \implies L_1 \circ L \sqsubseteq L_2 \circ L.$
   (ii) $\forall L : M' \to M''. (L \circ L_1)\!\downarrow \wedge (L \circ L_2)\!\downarrow \implies L \circ L_1 \sqsubseteq L \circ L_2.$

Theorem 3 (Contextuality of parameterised linearisability: encapsulated case). *For* $L_1, L_2 : M \to M'$ *such that* $M \cap M' = \emptyset$ *and* $L_1 \sqsubseteq_e L_2$:

(i) $\forall L : M'' \to M. (L_1 \circ L)\!\downarrow \wedge (L_2 \circ L)\!\downarrow \implies L_1 \circ L \sqsubseteq_e L_2 \circ L.$

(ii) $\forall L : M' \to M''. (L \circ L_1)\!\downarrow \wedge (L \circ L_2)\!\downarrow \implies L \circ L_1 \sqsubseteq_e L \circ L_2.$

The restriction on method names in Definition 5 ensures that the library compositions in Theorem 3 have no public abstract methods and can thus be compared by $\sqsubseteq_e$. Note that if L is ground, then so are $L_1 \circ L$ and $L_2 \circ L$. In this case, Theorems 2(i) and 3(i) allow us to establish classical linearisability from parameterised one.

Stating the contextuality of $\sqsubseteq_{\mathcal{R}}$ is more subtle. The relationship $L_1 \sqsubseteq_{\mathcal{R}} L_2$ allows the use of abstract methods by L_1 and L_2 to differ according to $\mathcal{R}$. As a consequence, for a non-ground parameter library L, their use by $L_1 \circ L$ and $L_2 \circ L$ may also differ according to another relation $\mathcal{G}$. We now introduce a property of L ensuring that a change in L's interactions with its client according to $\mathcal{R}$ (the *rely*) leads to a change in L's interactions with its abstract methods according to $\mathcal{G}$ (the *guarantee*).

Definition 6 (Rely-guarantee Closure). *Let* $\mathcal{R}, \mathcal{G}$ *be relations between histories of type* $\emptyset \to M'$ *and* $\emptyset \to M$, *respectively. A library* $L : M \to M'$ *is* $\binom{\mathcal{R}}{\mathcal{G}}$-*closed if for all* $h \in [\![L]\!]$ *and* $h' : \emptyset \to M'$ *we have*

$$(h|_{\mathsf{ClAct}}) \; \mathcal{R} \; h' \implies \exists h'' \in [\![L]\!]. \, (h''|_{\mathsf{ClAct}} = h') \wedge \overline{(h|_{\mathsf{AbsAct}})} \; \mathcal{G} \; \overline{(h''|_{\mathsf{AbsAct}})}.$$

Due to space constraints, we state contextuality of $\sqsubseteq_{\mathcal{R}}$ only for the case in which library parameters do not have public abstract methods. A more general statement which relaxes this assumption is given in [1, §B].

Theorem 4 (Contextuality of linearisability up to $\mathcal{R}$). *For* $L_1, L_2 : M \to M'$ *such that* $M \cap M' = \emptyset$ *and a relation* $\mathcal{R}$ *such that* $L_1 \sqsubseteq_{\mathcal{R}} L_2$:

(i) $\forall L : M'' \to M. \forall \mathcal{G}. M'' \cap M = \emptyset \wedge (L \text{ is } \binom{\mathcal{R}}{\mathcal{G}}\text{-closed}) \wedge$
$(L_1 \circ L)\!\downarrow \wedge (L_2 \circ L)\!\downarrow \implies L_1 \circ L \sqsubseteq_{\mathcal{G}} L_2 \circ L.$

(ii) $\forall L : M' \to M''. (L \circ L_1)\!\downarrow \wedge (L \circ L_2)\!\downarrow \implies L \circ L_1 \sqsubseteq_{\mathcal{R}} L \circ L_2.$

When L in Theorem 4(i) is ground, $\mathcal{G}$ becomes irrelevant. In this case we say that L is $\mathcal{R}$-*closed* if it is $\binom{\mathcal{R}}{\{(\varepsilon,\varepsilon)\}}$-closed. Hence, from Theorems 1 and 4(i) we get that for any $\mathcal{R}_t$-closed (§3) library L we have $\mathsf{FC} \circ L \sqsubseteq \mathsf{FC}^{\sharp} \circ L$: instantiating flat combining with a library insensitive to thread identifiers, e.g., a sequential stack or a queue, yields a concurrent library linearisable in the classical sense.

Given two libraries $L_1 : M_1 \to M_1'$ and $L_2 : M_2 \to M_2'$ that do not interact, i.e., $(M_1 \cup M_1') \cap (M_2 \cup M_2') = \emptyset$, we may wish to compose them by merging their method declarations into a library $L_1 \uplus L_2 : M_1 \uplus M_2 \to M_1' \uplus M_2'$, as originally proposed in [8]. Our notions of linearisability are also closed under this composition.

Theorem 5 (Compositionality of parameterised linearisability). *For* $L_1, L_1' : M_1 \to M_1'$ *and* $L_2, L_2' : M_2 \to M_2'$ *such that* $(M_1 \cup M_1') \cap (M_2 \cup M_2') = \emptyset$:

(i) $L_1 \sqsubseteq L_1' \wedge L_2 \sqsubseteq L_2' \implies L_1 \uplus L_2 \sqsubseteq L_1' \uplus L_2'.$

(ii) $L_1 \sqsubseteq_e L_1' \wedge L_2 \sqsubseteq_e L_2' \implies L_1 \uplus L_2 \sqsubseteq_e L_1' \uplus L_2'.$

(iii) $\forall \mathcal{R}, \mathcal{G}. L_1 \sqsubseteq_{\mathcal{R}} L_1' \wedge L_2 \sqsubseteq_{\mathcal{G}} L_2' \implies L_1 \uplus L_2 \sqsubseteq_{\mathcal{R} \otimes \mathcal{G}} L_1' \uplus L_2',$ *where* $\mathcal{R} \otimes \mathcal{G}$ *relates histories if their projections to* M_1 *actions are related by* $\mathcal{R}$ *and the projections to* M_2 *actions are related by* $\mathcal{G}$.

6 Clients and Observational Refinement

A **program** P has the form let L in $C_1 \parallel \ldots \parallel C_n$, where $L : \emptyset \to M$ is a ground library and $C_1 \parallel \ldots \parallel C_n$ is a client such that $C_1, \ldots, C_n$ call only methods in M, written $(C_1 \parallel \ldots \parallel C_n) : M$. Using the contextuality results from §5, we now show that our notions of linearisability imply observational refinement for such programs.

The semantics of a program P is given by the set of its traces $[\![P]\!] \in 2^{\mathsf{Traces}}$, which include actions (t, c) recording the execution of primitive commands c by client threads C_t and the library L, as well as $(t, \mathsf{call}\, m(z))$ and $(t, \mathsf{ret}\, m(z))$ actions corresponding to the former invoking methods of the latter. The semantics $[\![P]\!]$ is defined similarly to that of libraries in §4. In particular, we assume that client threads C_t access only locations in a set $\mathsf{Locs}_{\mathsf{client}}$ such that $\mathsf{Locs}_{\mathsf{client}} \cap \mathsf{Locs}_L = \{\mathsf{arg}_t\}_{t \in \mathcal{T}}$ for any L. Due to space constraints, we defer the definition of $[\![P]\!]$ to [1, §A]. We define the **observable behaviour** $\mathsf{obs}(\tau)$ of a trace $\tau \in [\![P]\!]$ as its projection to client actions, i.e., those outside method invocations, and lift obs to sets of traces as expected.

Definition 7 (Observational refinement). *For $L_1, L_2 : M \to M'$ we say that L_1 **observationally refines** L_2, written $L_1 \sqsubseteq_{\mathsf{obs}} L_2$, if for any ground library $L : \emptyset \to M$ and client $(C_1 \parallel \ldots \parallel C_n) : M'$ we have*

$$\mathsf{obs}([\![\mathsf{let}\, (L_1 \circ L)\, \mathsf{in}\, C_1 \parallel \ldots \parallel C_n]\!]) \subseteq \mathsf{obs}([\![\mathsf{let}\, (L_2 \circ L)\, \mathsf{in}\, C_1 \parallel \ldots \parallel C_n]\!]).$$

*For a binary relation $\mathcal{R}$ on histories we say that L_1 **observationally refines** L_2 **up to** $\mathcal{R}$, written $L_1 \sqsubseteq^{\mathcal{R}}_{\mathsf{obs}} L_2$, if the above is true under the assumption that L is $\mathcal{R}$-closed.*

Thus, $L_1 \sqsubseteq_{\mathsf{obs}} L_2$ means that L_1 can be replaced by L_2 in any program that uses it while keeping observable behaviours reproducible. This allows us to check a property of a program using L_1 (e.g., the flat combining implementation in Figure 1) by checking this property on a program with L_1 replaced by a possibly simpler L_2 (e.g., the flat combining specification in Figure 2). Using Theorems 2–4, we can show that our notions of linearisability validate observational refinement.

Theorem 6 (Observational refinement). *For any libraries $L_1, L_2 : M \to M'$:*

(i) $L_1 \sqsubseteq L_2 \implies L_1 \sqsubseteq_{\mathsf{obs}} L_2$.

(ii) $M \cap M' = \emptyset \wedge L_1 \sqsubseteq_{\mathsf{e}} L_2 \implies L_1 \sqsubseteq_{\mathsf{obs}} L_2$.

(iii) $\forall \mathcal{R}.\ M \cap M' = \emptyset \wedge L_1 \sqsubseteq_{\mathcal{R}} L_2 \implies L_1 \sqsubseteq^{\mathcal{R}}_{\mathsf{obs}} L_2$.

7 Related Work

Linearisability has recently been extended to handle liveness properties, ownership transfer and weak memory models [4,5,10]. Most of these extensions have exploited the connection between linearisability and observational refinement [2]. The same methodology is adopted in the present work, but for studying two previously unexplored topics: parameterised libraries and the impact that common restrictions on their contexts have on the definition of linearisability. We believe that our results are compatible with the existing ones and can thus be extended to cover liveness and ownership transfer [4,5].

Our work shares techniques with game semantics of concurrent programming languages [12,3] and Jeffrey and Rathke's semantics of concurrent objects [11] (in particular, we use the ? and ! notation from the latter). The proofs of our contextuality theorems rely on the fact that library denotations satisfy certain closure properties related to $\sqsubseteq, \sqsubseteq_{\mathsf{e}}$

and $\sqsubseteq_{\mathcal{R}}$, which are similar to those exploited in these prior works. However, there are two important differences. First, prior work has not studied common restrictions on library contexts (such as the encapsulation and closure conditions in Definitions 3 and 4) and the induced stronger notions of refinement between libraries, the two key topics of this paper. Second, prior works have considered all higher-order functions, while our parameterised libraries are limited to second order. Our motivation for constraining the setting in this way is to use a simple semantics and study the key issues involved in linearisability of parameterised libraries without using sophisticated machinery from game semantics, such as justification pointers and views [9], designed for accurately modelling higher-order features. However, it is definitely a promising direction to look for appropriate notions of linearisability for full higher-order concurrent libraries by combining the ideas from this paper with those from game semantics.

Turon et al. proposed CaReSL [14], a logic that allows proving observational refinements between higher-order concurrent programs directly, without going via linearisability. Their work is complimentary to ours: it provides efficient proof techniques, whereas we identify obligations to prove, independent of a particular proof system.

Acknowledgements. We thank Thomas Dinsdale-Young and Ilya Sergey for comments that helped improve the paper. This work was supported by the EU FET project ADVENT.

References

1. Cerone, A., Gotsman, A., Yang, H.: Parameterised linearisability (extended version), http://software.imdea.org/~gotsman/
2. Filipovic, I., O'Hearn, P.W., Rinetzky, N., Yang, H.: Abstraction for concurrent objects. Theor. Comput. Sci. 411(51-52) (2010)
3. Ghica, D.R., Murawski, A.S.: Angelic semantics of fine-grained concurrency. Ann. Pure Appl. Logic 151(2-3) (2008)
4. Gotsman, A., Yang, H.: Liveness-preserving atomicity abstraction. In: Aceto, L., Henzinger, M., Sgall, J. (eds.) ICALP 2011, Part II. LNCS, vol. 6756, pp. 453–465. Springer, Heidelberg (2011)
5. Gotsman, A., Yang, H.: Linearizability with ownership transfer. Logical Methods in Computer Science 9 (2013)
6. Hendler, D., Incze, I., Shavit, N., Tzafrir, M.: Flat combining and the synchronization-parallelism tradeoff. In: SPAA (2010)
7. Herlihy, M., Koskinen, E.: Transactional boosting: a methodology for highly-concurrent transactional objects. In: PPOPP (2008)
8. Herlihy, M., Wing, J.M.: Linearizability: A correctness condition for concurrent objects. ACM Trans. Program. Lang. Syst. 12(3) (1990)
9. Hyland, J.M.E., Luke Ong, C.-H.: On full abstraction for PCF: I, II, and III. Inf. Comput. 163(2) (2000)
10. Jagadeesan, R., Petri, G., Pitcher, C., Riely, J.: Quarantining weakness. In: Felleisen, M., Gardner, P. (eds.) ESOP 2013. LNCS, vol. 7792, pp. 492–511. Springer, Heidelberg (2013)
11. Jeffrey, A., Rathke, J.: A fully abstract testing semantics for concurrent objects. Theor. Comput. Sci. 338(1-3) (2005)
12. Laird, J.: A game semantics of idealized CSP. ENTCS 45 (2001)
13. Russo, C.V.: The Joins Concurrency Library. In: Hanus, M. (ed.) PADL 2007. LNCS, vol. 4354, pp. 260–274. Springer, Heidelberg (2007)
14. Turon, A., Dreyer, D., Birkedal, L.: Unifying refinement and Hoare-style reasoning in a logic for higher-order concurrency. In: ICFP (2013)
15. Yang, H., O'Hearn, P.W.: A semantic basis for local reasoning. In: Nielsen, M., Engberg, U. (eds.) FOSSACS 2002. LNCS, vol. 2303, pp. 402–416. Springer, Heidelberg (2002)

Games with a Weak Adversary[*,**]

Krishnendu Chatterjee[1] and Laurent Doyen[2]

[1] IST Austria
[2] LSV, ENS Cachan & CNRS, France

Abstract. We consider multi-player graph games with partial-observation and parity objective. While the decision problem for three-player games with a coalition of the first and second players against the third player is undecidable in general, we present a decidability result for partial-observation games where the first and third player are in a coalition against the second player, thus where the second player is adversarial but weaker due to partial-observation. We establish tight complexity bounds in the case where player 1 is less informed than player 2, namely 2-EXPTIME-completeness for parity objectives. The symmetric case of player 1 more informed than player 2 is much more complicated, and we show that already in the case where player 1 has perfect observation, memory of size non-elementary is necessary in general for reachability objectives, and the problem is decidable for safety and reachability objectives. From our results we derive new complexity results for partial-observation stochastic games.

1 Introduction

Games on Graphs. Games played on graphs are central in several important problems in computer science, such as reactive synthesis [21,22], verification of open systems [2], and many others. The game is played by several players on a finite-state graph, with a set of angelic (existential) players and a set of demonic (universal) players as follows: the game starts at an initial state, and given the current state, the successor state is determined by the choice of moves of the players. The outcome of the game is a *play*, which is an infinite sequence of states in the graph. A *strategy* is a transducer to resolve choices in a game for a player that given a finite prefix of the play specifies the next move. Given an objective (the desired set of behaviors or plays), the goal of the existential players is to ensure the play belongs to the objective irrespective of the strategies of the universal players. In verification and control of reactive systems an objective is typically an ω-regular set of paths. The class of ω-regular languages, that extends classical regular languages to infinite strings, provides a robust specification language to express all commonly used specifications, and parity objectives are a canonical way to define such ω-regular specifications [27]. Thus games on graphs with parity objectives provide a general framework for analysis of reactive systems.

---

[*] This research was partly supported by Austrian Science Fund (FWF) Grant No P23499- N23, FWF NFN Grant No S11407-N23 (RiSE), ERC Start grant (279307: Graph Games), Microsoft Faculty Fellowship Award, and European project Cassting (FP7-601148).
[**] Fuller version: [1].

Perfect vs Partial Observation. Many results about games on graphs make the hypothesis of *perfect observation* (i.e., players have perfect or complete observation about the state of the game). In this setting, due to determinacy (or switching of the strategy quantifiers for existential and universal players) [17], the questions expressed by an arbitrary alternation of quantifiers reduce to a single alternation, and thus are equivalent to solving two-player games (all the existential players against all the universal players). However, the assumption of perfect observation is often not realistic in practice. For example in the control of physical systems, digital sensors with finite precision provide partial information to the controller about the system state [12,14]. Similarly, in a concurrent system the modules expose partial interfaces and have access to the public variables of the other processes, but not to their private variables [25,2]. Such situations are better modeled in the more general framework of *partial-observation* games [24,25,26].

Partial-Observation Games. Since partial-observation games are not determined, unlike the perfect-observation setting, the multi-player game problems do not reduce to the case of two-player games. Typically, multi-player partial-observation games are studied in the following setting: a set of partial-observation existential players, against a perfect-observation universal player, such as for distributed synthesis [21,13,23]. The problem of deciding if the existential players can ensure a reachability (or a safety) objective is undecidable in general, even for two existential players [20,21]. However, if the information of the existential players form a chain (i.e., existential player 1 more informed than existential player 2, existential player 2 more informed than existential player 3, and so on), then the problem is decidable [21,16,18].

Games with a Weak Adversary. One aspect of multi-player games that has been largely ignored is the presence of weaker universal players that do not have perfect observation. However, it is natural in the analysis of composite reactive systems that some universal players represent components that do not have access to all variables of the system. In this work we consider games where adversarial players can have partial observation. If there are two existential (resp., two universal) players with incomparable partial observation, then the undecidability results follows from [20,21]; and if the information of the existential (resp., universal) players form a chain, then they can be reduced to one partial-observation existential (resp., universal) player. We consider the following case of partial-observation games: one partial-observation existential player (player 1), one partial-observation universal player (player 2), one perfect-observation existential player (player 3), and one perfect-observation universal player (player 4). Roughly, having more partial-observation players in general leads to undecidability, and having more perfect-observation players reduces to two perfect-observation players. We first present our results and then discuss two applications of the model.

Results. Our main results are as follows:
1. *Player 1 less informed.* We first consider the case when player 1 is less informed than player 2. We establish the following results: (i) a 2-EXPTIME upper bound for parity objectives and a 2-EXPTIME lower bound for reachability objectives (i.e., we establish 2-EXPTIME-completeness); (ii) an EXPSPACE upper bound for parity objectives when player 1 is blind (has only one observation), and EXPSPACE lower bound for reachability objectives even when both player 1 and player 2 are

Table 1. Complexity of qualitative analysis (almost-sure winning) for partial-observation stochastic games with partial observation for player 1 with reachability and parity objectives. Player 2 has either perfect observation or more information than player 1(new results boldfaced). For positive winning, all entries other than the first (randomized strategies for player 1 and perfect observation for player 2) remain the same, and the complexity for the first entry for positive winning is PTIME-complete.

	Reachability		Parity		Parity	
Player 2	Finite- or infinite-memory strategies		Infinite-memory strategies		Finite-memory strategies	
Player 1	Perfect	More informed	Perfect	More informed	Perfect	More informed
Randomized	EXP-c [9]	EXP-c [4]	Undec. [3,8]	Undec. [3,8]	EXP-c [10]	**2EXP**
Pure	EXP-c [7]	**2EXP-c**	Undec. [3]	Undec. [3]	EXP-c [10]	**2EXP-c**

blind. In all these cases, if the objective can be ensured then the upper bound on memory requirement of winning strategies is at most doubly exponential.

2. *Player 1 more informed.* We consider the case when player 1 can be more informed as compared to player 2, and show that even when player 1 has perfect observation there is a non-elementary lower bound on the memory required by winning strategies. This result is also in sharp contrast with distributed games, where if only one player has partial observation then the upper bound on memory of winning strategies is exponential.

Applications. We discuss two applications of our results: the sequential synthesis problem, and new complexity results for partial-observation *stochastic* games.

1. The sequential synthesis problem consists of a set of partially implemented modules, where first a set of modules needs to be refined, followed by a refinement of some modules by an external source, and then the remaining modules are refined so that the composite open reactive system satisfies a specification. Given the first two refinements cannot access all private variables, we have a four-player game where the first refinement corresponds to player 1, the second refinement to player 2, the third refinement to player 3, and player 4 is the environment.

2. In partial-observation stochastic games, there are two partial-observation players (one existential and one universal) playing in the presence of uncertainty in the transition function (i.e., stochastic transition function). The qualitative analysis question is to decide the existence of a strategy for the existential player to ensure the parity objective with probability 1 (or with positive probability) against all strategies of the universal player. The witness strategy can be randomized or deterministic (pure). While the qualitative problem is undecidable, the practically relevant restriction to finite-memory pure strategies reduces to the four-player game problem. Moreover, for finite-memory strategies, the decision problem for randomized strategies reduces to the pure-strategy question [7]. By the results we establish in this paper, new decidability and complexity results are obtained for the qualitative analysis of partial-observation stochastic games with player 2 partially informed but more informed than player 1. The complexity results for almost-sure winning are summarized in Table 1. Surprisingly for reachability objectives, whether player 2 is perfectly informed or more informed than player 1 does not change the complexity for randomized strategies, but it results in an exponential increase in the complexity for pure strategies.

2   Definitions

We first consider three-player (non-stochastic) games with parity objectives and we establish new complexity results in Section 3 that we later extend to four-player games in Section 5. We also present the related model of two-player stochastic games for which our contribution implies new complexity results.

Three-player games. Given alphabets A_i of actions for player i ($i = 1, 2, 3$), a *three-player game* is a tuple $G = \langle Q, q_0, \delta \rangle$ where:

- Q is a finite set of states with $q_0 \in Q$ the initial state; and
- $\delta : Q \times A_1 \times A_2 \times A_3 \to Q$ is a deterministic transition function that, given a current state q, and actions $a_1 \in A_1$, $a_2 \in A_2$, $a_3 \in A_3$ of the players, gives the successor state $q' = \delta(q, a_1, a_2, a_3)$.

The games we consider are sometimes called *concurrent* because all three players need to choose simultaneously an action to determine a successor state. The special class of *turn-based* games corresponds to the case where in every state, one player has the turn and his sole action determines the successor state. In our framework, a turn-based state for player 1 is a state $q \in Q$ such that $\delta(q, a_1, a_2, a_3) = \delta(q, a_1, a_2', a_3')$ for all $a_1 \in A_1$, $a_2, a_2' \in A_2$, and $a_3, a_3' \in A_3$. We define analogously turn-based states for player 2 and player 3. A game is turn-based if every state of G is turn-based (for some player). The class of two-player games is obtained when A_3 is a singleton. In a game G, given $s \subseteq Q$, $a_1 \in A_1$, $a_2 \in A_2$, let $\text{post}^G(s, a_1, a_2, -) = \{q' \in Q \mid \exists q \in s \cdot \exists a_3 \in A_3 : q' = \delta(q, a_1, a_2, a_3)\}$.

Observations. For $i = 1, 2, 3$, a set $\mathcal{O}_i \subseteq 2^Q$ of *observations* (for player i) is a partition of Q (i.e., $\mathcal{O}_i$ is a set of non-empty and non-overlapping subsets of Q, and their union covers Q). Let $\text{obs}_i : Q \to \mathcal{O}_i$ be the function that assigns to each state $q \in Q$ the (unique) observation for player i that contains q, i.e. such that $q \in \text{obs}_i(q)$. The functions obs_i are extended to sequences $\rho = q_0 \ldots q_n$ of states in the natural way, namely $\text{obs}_i(\rho) = \text{obs}_i(q_0) \ldots \text{obs}_i(q_n)$. We say that player i is *blind* if $\mathcal{O}_i = \{Q\}$, that is player i has only one observation; player i has *perfect information* if $\mathcal{O}_i = \{\{q\} \mid q \in Q\}$, that is player i can distinguish each state; and player 1 is *less informed* than player 2 (we also say player 2 is more informed) if for all $o_2 \in \mathcal{O}_2$, there exists $o_1 \in \mathcal{O}_1$ such that $o_2 \subseteq o_1$.

Strategies. For $i = 1, 2, 3$, let Σ_i be the set of *strategies* $\sigma_i : \mathcal{O}_i^+ \to A_i$ of player i that, given a sequence of past observations, give an action for player i. Equivalently, we sometimes view a strategy of player i as a function $\sigma_i : Q^+ \to A_i$ satisfying $\sigma_i(\rho) = \sigma_i(\rho')$ for all $\rho, \rho' \in Q^+$ such that $\text{obs}_i(\rho) = \text{obs}_i(\rho')$, and say that σ_i is *observation-based*.

Outcome. Given strategies $\sigma_i \in \Sigma_i$ ($i = 1, 2, 3$) in G, the *outcome play* from a state q_0 is the infinite sequence $\rho_{q_0}^{\sigma_1, \sigma_2, \sigma_3} = q_0 q_1 \ldots$ such that for all $j \geq 0$, we have $q_{j+1} = \delta(q_j, a_1^j, a_2^j, a_3^j)$ where $a_i^j = \sigma_i(q_0 \ldots q_j)$ (for $i = 1, 2, 3$).

Objectives. An *objective* is a set $\alpha \subseteq Q^\omega$ of infinite sequences of states. A play ρ *satisfies* the objective α if $\rho \in \alpha$. An objective α is *visible* for player i if for all $\rho, \rho' \in Q^\omega$, if $\rho \in \alpha$ and $\mathrm{obs}_i(\rho) = \mathrm{obs}_i(\rho')$, then $\rho' \in \alpha$. We consider the following objectives:

- *Reachability.* Given a set $\mathcal{T} \subseteq Q$ of target states, the *reachability* objective $\mathrm{Reach}(\mathcal{T})$ requires that a state in $\mathcal{T}$ be visited at least once, that is, $\mathrm{Reach}(\mathcal{T}) = \{\rho = q_0 q_1 \cdots \mid \exists k \geq 0 : q_k \in \mathcal{T}\}$.
- *Safety.* Given a set $\mathcal{T} \subseteq Q$ of target states, the *safety* objective $\mathrm{Safe}(\mathcal{T})$ requires that only states in $\mathcal{T}$ be visited, that is, $\mathrm{Safe}(\mathcal{T}) = \{\rho = q_0 q_1 \cdots \mid \forall k \geq 0 : q_k \in \mathcal{T}\}$.
- *Parity.* For a play $\rho = q_0 q_1 \ldots$ we denote by $\mathrm{Inf}(\rho)$ the set of states that occur infinitely often in ρ, that is, $\mathrm{Inf}(\rho) = \{q \in Q \mid \forall k \geq 0 \cdot \exists n \geq k : q_n = q\}$. For $d \in \mathbb{N}$, let $p : Q \to \{0, 1, \ldots, d\}$ be a priority function, which maps each state to a nonnegative integer priority. The parity objective $\mathrm{Parity}(p)$ requires that the minimum priority occurring infinitely often be even. Formally, $\mathrm{Parity}(p) = \{\rho \mid \min\{p(q) \mid q \in \mathrm{Inf}(\rho)\}$ is even$\}$. Parity objectives are a canonical way to express ω-regular objectives [27]. If the priority function is constant over observations of player i, that is for all observations $\gamma \in \mathcal{O}_i$ we have $p(q) = p(q')$ for all $q, q' \in \gamma$, then the parity objective $\mathrm{Parity}(p)$ is visible for player i.

Decision problem. Given a game $G = \langle Q, q_0, \delta \rangle$ and an objective $\alpha \subseteq Q^\omega$, the *three-player decision problem* is to decide if $\exists \sigma_1 \in \Sigma_1 \cdot \forall \sigma_2 \in \Sigma_2 \cdot \exists \sigma_3 \in \Sigma_3 : \rho_{q_0}^{\sigma_1, \sigma_2, \sigma_3} \in \alpha$.

The results for the three-player decision problem have implications for decision problems on partial-observation stochastic games that we formally define below.

Two-player partial-observation stochastic games. Given alphabet A_i of actions, and set $\mathcal{O}_i$ of observations (for player $i \in \{1, 2\}$), a *two-player partial-observation stochastic game* (for brevity, two-player stochastic game) is a tuple $G = \langle Q, q_0, \delta \rangle$ where Q is a finite set of states, $q_0 \in Q$ is the initial state, and $\delta : Q \times A_1 \times A_2 \to \mathcal{D}(Q)$ is a probabilistic transition where $\mathcal{D}(Q)$ is the set of probability distributions $\kappa : Q \to [0, 1]$ on Q, such that $\sum_{q \in Q} \kappa(q) = 1$. Given a current state q and actions a, b for the players, the transition probability to a successor state q' is $\delta(q, a, b)(q')$. Observation-based strategies are defined as for three-player games. An *outcome play* from a state q_0 under strategies σ_1, σ_2 is an infinite sequence $\rho = q_0 a_0 b_0 q_1 \ldots$ such that $a_i = \sigma_1(q_0 \ldots q_i)$, $b_i = \sigma_2(q_0 \ldots q_i)$, and $\delta(q_i, a_i, b_i)(q_{i+1}) > 0$ for all $i \geq 0$.

Qualitative analysis. Given an objective α that is Borel measurable (all Borel sets in the Cantor topology and all objectives considered in this paper are measurable [15]), a strategy σ_1 for player 1 is *almost-sure winning* (resp., *positive winning*) for the objective α from q_0 if for all observation-based strategies σ_2 for player 2, we have $\mathrm{Pr}_{q_0}^{\sigma_1, \sigma_2}(\alpha) = 1$ (resp., $\mathrm{Pr}_{q_0}^{\sigma_1, \sigma_2}(\alpha) > 0$) where $\mathrm{Pr}_{q_0}^{\sigma_1, \sigma_2}(\cdot)$ is the unique probability measure induced by the natural probability measure on finite prefixes of plays (i.e., the product of the transition probabilities in the prefix).

3   Three-Player Games with Player 1 Less Informed

We consider the three-player (non-stochastic) games defined in Section 2. We show that for reachability and parity objectives the three-player decision problem is decidable when player 1 is less informed than player 2. The problem is EXPSPACE-complete when player 1 is blind, and 2-EXPTIME-complete in general.

Remark 1. Observe that once the strategies of the first two players are fixed we obtain a graph, and in graphs perfect-information coincides with blind for construction of a path (see [6, Lemma 2] that counting strategies that count the number of steps are sufficient which can be ensured by a player with no information). Hence without loss of generality we consider that player 3 has perfect observation, and drop the observation for player 3.

Theorem 1 (Upper Bounds). *Given a three-player game $G = \langle Q, q_0, \delta \rangle$ with player 1 less informed than player 2 and a parity objective α, the problem of deciding whether $\exists \sigma_1 \in \Sigma_1 \cdot \forall \sigma_2 \in \Sigma_2 \cdot \exists \sigma_3 \in \Sigma_3 : \rho_{q_0}^{\sigma_1, \sigma_2, \sigma_3} \in \alpha$ can be solved in 2-EXPTIME. If player 1 is blind, then the problem can be solved in EXPSPACE.*

Proof. The proof is by a reduction of the decision problem for three-player games to a decision problem for partial-observation two-player games with the same objective. We present the reduction for parity objectives that are visible for player 2 (defined by priority functions that are constant over observations of player 2). The general case of not necessarily visible parity objectives can be solved using a reduction to visible objectives, as in [6, Section 3].

Given a three-player game $G = \langle Q, q_0, \delta \rangle$ over alphabet of actions A_i ($i = 1, 2, 3$), and observations $\mathcal{O}_1, \mathcal{O}_2 \subseteq 2^Q$ for player 1 and player 2, with player 1 less informed than player 2, we construct a two-player game $H = \langle Q_H, \{q_0\}, \delta_H \rangle$ over alphabet of actions A'_i ($i = 1, 2$), and observations $\mathcal{O}'_1 \subseteq 2^{Q_H}$ and perfect observation for player 2, where (intuitive explanations follow):

- $Q_H = \{s \in 2^Q \mid s \neq \varnothing \wedge \exists o_2 \in \mathcal{O}_2 : s \subseteq o_2\}$;
- $A'_1 = A_1 \times (2^Q \times A_2 \to \mathcal{O}_2)$, and $A'_2 = A_2$;
- $\mathcal{O}'_1 = \{\{s \in Q_H \mid s \subseteq o_1\} \mid o_1 \in \mathcal{O}_1\}$, and let $\mathrm{obs}'_1 : Q_H \to \mathcal{O}'_1$ be the corresponding observation function;
- $\delta_H(s, (a_1, f), a_2) = \mathsf{post}^G(s, a_1, a_2, -) \cap f(s, a_2)$.

Intuitively, the state space Q_H is the set of knowledges of player 2 about the current state in G, i.e., the sets of states compatible with an observation of player 2. Along a play in H, the knowledge of player 2 is updated to represent the set of possible current states in which the game G can be. In H player 2 has perfect observation and the role of player 1 in the game H is to simulate the actions of both player 1 and player 3 in G. Since player 2 fixes his strategy before player 3 in G, the simulation should not let player 2 know player-3's action, but only the observation that player 2 will actually see while playing the game. The actions of player 1 in H are pairs $(a_1, f) \in A'_1$ where a_1 is a simple action of player 1 in G, and f gives the observation $f(s, a_2)$ received by player 2 after the response of player 3 to the action a_2 of player 2 when the knowledge of player 2 is s. In H, player 1 has partial observation, as he cannot distinguish knowledges of player 2 that belong to the same observation of player 1 in G. The transition relation updates the knowledges of player 2 as expected. Note that $|\mathcal{O}_1| = |\mathcal{O}'_1|$, and therefore if player 1 is blind in G then he is blind in H as well.

Given a visible parity objective $\alpha = \mathsf{Parity}(p)$ where $p : Q \to \{0, 1, \ldots, d\}$ is constant over observations of player 2, let $\alpha' = \mathsf{Parity}(p')$ where $p'(s) = p(q)$ for all $q \in s$ and $s \in Q_H$. Note that the function p' is well defined since s is a subset of an observation of player 2 and thus $p(q) = p(q')$ for all $q, q' \in s$. However, the parity objective $\alpha' = \text{Parity}(p')$ may not be visible to player 1 in G. We establish that given witness strategies in G we can construct witness strategies in H and vice-versa, and the details of the strategy constructions are presented in [1].   □

Theorem 2 (Lower Bounds). *Given a three-player game $G = \langle Q, q_0, \delta \rangle$ with player 1 less informed than player 2 and a reachability objective α, the problem of deciding whether $\exists \sigma_1 \in \Sigma_1 \cdot \forall \sigma_2 \in \Sigma_2 \cdot \exists \sigma_3 \in \Sigma_3 : \rho_{q_0}^{\sigma_1, \sigma_2, \sigma_3} \in \alpha$ is 2-EXPTIME-hard. If player 1 is blind (and even when player 2 is also blind), then the problem is EXPSPACE-hard.*

Proof. The proof of 2-EXPTIME-hardness is obtained by a polynomial-time reduction of the membership problem for exponential-space *alternating* Turing machines to the three-player problem. The same reduction for the special case of exponential-space *nondeterministic* Turing machines shows EXPSPACE-hardness when player 1 is blind (because our reduction yields a game in which player 1 is blind when we start from a nondeterministic Turing machine). The membership problem for Turing machines is to decide, given a Turing machine M and a finite word w, whether M accepts w. The membership problem is 2-EXPTIME-complete for exponential-space alternating Turing machines, and EXPSPACE-complete for exponential-space nondeterministic Turing machines [19].

An alternating Turing machine is a tuple $M = \langle Q_\vee, Q_\wedge, \Sigma, \Gamma, \Delta, q_0, q_{acc}, q_{rej} \rangle$ where the state space $Q = Q_\vee \cup Q_\wedge$ consists of the set $Q_\vee$ of or-states, and the set $Q_\wedge$ of and-states. The input alphabet is Σ, the tape alphabet is $\Gamma = \Sigma \cup \{\#\}$ where $\#$ is the blank symbol. The initial state is q_0, the accepting state is q_{acc}, and the rejecting state is q_{rej}. The transition relation is $\Delta \subseteq Q \times \Gamma \times Q \times \Gamma \times \{-1, 1\}$, where a transition $(q, \gamma, q', \gamma', d) \in \Delta$ intuitively means that, given the machine is in state q, and the symbol under the tape head is γ, the machine can move to state q', replace the symbol under the tape head by γ', and move the tape head to the neighbor cell in direction d. A configuration c of M is a sequence $c \in (\Gamma \cup (Q \times \Gamma))^\omega$ with exactly one symbol in $Q \times \Gamma$, which indicates the current state of the machine and the position of the tape head. The initial configuration of M on $w = a_0 a_1 \dots a_n$ is $c_0 = (q_0, a_0) \cdot a_1 \cdot a_2 \cdots a_n \cdot \#^\omega$. Given the initial configuration of M on w, it is routine to define the execution trees of M where at least one successor of each configuration in an or-state, and all successors of the configurations in an and-state are present (and we assume that all branches reach either q_{acc} or q_{rej}), and to say that M accepts w if all branches of some execution tree reach q_{acc}. Note that $Q_\wedge = \varnothing$ for nondeterministic Turing machines, and in that case the execution tree reduces to a single path. A Turing machine M uses exponential space if for all words w, all configurations in the execution of M on w contain at most $2^{O(|w|)}$ non-blank symbols.

We present the key steps of our reduction from alternating Turing machines. Given a Turing machine M and a word w, we construct a three-player game with reachability objective in which player 1 and player 2 have to simulate the execution of M on w, and player 1 has to announce the successive configurations and transitions of the machine along the execution. Player 1 announces configurations one symbol at a time, thus the alphabet of player 1 is $A_1 = \Gamma \cup (Q \times \Gamma) \cup \Delta$. In an initialization phase, the transition relation of the game forces player 1 to announce the initial configuration

c_0 (this can be done with $O(n)$ states in the game, where $n = |w|$). Then, the game proceeds to a loop where player 1 keeps announcing symbols of configurations. At all times along the execution, some finite information is stored in the finite state space of the game: a window of the last three symbols z_1, z_2, z_3 announced by player 1, as well as the last symbol head $\in Q \times \Gamma$ announced by player 1 (that indicates the current machine state and the position of the tape head). After the initialization phase, we should have $z_1 = z_2 = z_3 = \#$ and head $= (q_0, a_0)$. When player 1 has announced a full configuration, he moves to a state of the game where either player 1 or player 2 has to announce a transition of the machine: for head $= (p, a)$, if $p \in Q_\vee$, then player 1 chooses the next transition, and if $p \in Q_\wedge$, then player 2 chooses. Note that the transitions chosen by player 2 are visible to player 1 and this is the only information that player 1 observes. Hence player 1 is less informed than player 2, and both player 1 and player 2 are blind when the machine is nondeterministic. If a transition $(q, \gamma, q', \gamma', d)$ is chosen by player i, and either $p \neq q$ or $a \neq \gamma$, then player i loses (i.e., a sink state is reached to let player 1 lose, and the target state of the reachability objective is reached to let player 2 lose). If at some point player 1 announces a symbol (p, a) with $p = q_{acc}$, then player 1 wins the game.

The role of player 2 is to check that player 1 faithfully simulates the execution of the Turing machine, and correctly announces the configurations. After every announcement of a symbol by player 1, the game offers the possibility to player 2 to compare this symbol with the symbol at the same position in the next configuration. We say that player 2 *checks* (and whether player 2 checks or not is not visible to player 1), and the checked symbol is stored as z_2. Note that player 2 can be blind to check because player 2 fixes his strategy after player 1. The window z_1, z_2, z_3 stored in the state space of the game provides enough information to update the middle cell z_2 in the next configuration, and it allows the game to verify the check of player 2. However, the distance (in number of steps) between the same position in two consecutive configurations is exponential (say 2^n for simplicity), and the state space of the game is not large enough to check that such a distance exists between the two symbols compared by player 2. We use player 3 to check that player 2 makes a comparison at the correct position. When player 2 decides to check, he has to count from 0 to 2^n by announcing after every symbol of player 1 a sequence of n bits, initially all zeros (again, this can be enforced by the structure of the game with $O(n)$ states). It is then the responsibility of player 3 to check that player 2 counts correctly. To check this, player 3 can at any time choose a bit position $p \in \{0, \ldots, n-1\}$ and store the bit value b_p announced by player 2 at position p. The value of b_p and p is not visible to player 2. While player 2 announces the bits $b_{p+1}, \ldots, b_{n-1}$ at position $p+1, \ldots, n-1$, the finite state of the game is used to flip the value of b_p if all bits $b_{p+1}, \ldots, b_{n-1}$ are equal to 1, hence updating b_p to the value of the p-th bit in what should be the next announcement of player 2. In the next bit sequence announced by player 2, the p-th bit is compared with b_p. If they match, then the game goes to a sink state (as player 2 has faithfully counted), and if they differ then the game goes to the target state (as player 2 is caught cheating). It can be shown that this can be enforced by the structure of the game with $O(n^2)$ states, that is $O(n)$ states for each value of p. As before, whether player 3 checks or not is not visible to player 2.

Note that the checks of player 2 and player 3 are one-shot: the game will be over (either in a sink or target state) when the check is finished. This is enough to ensure a faithful simulation by player 1, and a faithful counting by player 2, because (1) partial observation allows to hide to a player the time when a check occurs, and (2) player 2 fixes his strategy after player 1 (and player 3 after player 2), thus they can decide to run a check exactly when player 1 (or player 2) is not faithful. This ensures that player 1 does not win if he does not simulate the execution of M on w, and that player 2 does not win if he does not count correctly.

Hence this reduction ensures that M accepts w if and only if the answer to the three-player game problem is YES, where the reachability objective is satisfied if player 1 eventually announces that the machine has reached q_{acc} (that is if M accepts w), or if player 2 cheats in counting, which can be detected by player 3.   □

4   Three-Player Games with Player 1 Perfect

When player 2 is less informed than player 1, we show that three-player games get much more complicated (even in the special case where player 1 has perfect information). We note that for reachability objectives, the three-player decision problem is equivalent to the qualitative analysis of positive winning in two-player stochastic games, and we show that the techniques developed in the analysis of two-player stochastic games can be extended to solve the three-player decision problem with safety objectives as well.

For reachability objectives, the three-player decision problem is equivalent to the problem of positive winning in two-player stochastic games where the third player is replaced by a probabilistic choice over the action set with uniform probability. Intuitively, after player 1 and player 2 fixed their strategy, the fact that player 3 can construct a (finite) path to the target set is equivalent to the fact that such a path has positive probability when the choices of player 3 are replaced by uniform probabilistic transitions. Given a three-player game $G = \langle Q, q_0, \delta \rangle$, let $\mathsf{Uniform}(G) = \langle Q, q_0, \delta' \rangle$ be the two-player partial-observation *stochastic* game (with same state space, action sets, and observations for player 1 and player 2) where $\delta'(q, a_1, a_2)(q') = \frac{|\{a_3 | \delta(q, a_1, a_2, a_3) = q'\}|}{|A_3|}$ for all $a_1 \in A_1$, $a_2 \in A_2$, and $q, q' \in Q$. Formally, the equivalence result is presented in Lemma 1, and the equivalence holds for all three-player games (not restricted to three-player games where player 1 has perfect information). However, we will use Lemma 1 to establish results for three-player games where player 1 has perfect information.

Lemma 1. *Given a three-player game G and a reachability objective α, the answer to the three-player decision problem for $\langle G, \alpha \rangle$ is YES if and only if player 1 is positive winning for α in the two-player partial-observation stochastic game $\mathsf{Uniform}(G)$.*

Reachability objectives. Even in the special case where player 1 has perfect information, and for reachability objectives, non-elementary memory is necessary in general for player 1 to win in three-player games. This result follows from Lemma 1 and from the result of [7, Example 4.2 Journal version] showing that non-elementary memory is necessary to win with positive probability in two-player stochastic games. It also follows from Lemma 1 and the result of [7, Corollary 4.9 Journal version] that the three-player decision problem for reachability games is decidable. The decidability result can be extended to safety objectives [1].

Theorem 3. *When player 1 has perfect information, the three-player decision problem is decidable for both reachability and safety games, and for reachability games memory of size non-elementary is necessary in general for player 1.*

5   Four-Player Games

We show that the results presented for three-player games extend to games with four players (the fourth player is universal and perfectly informed). The definition of four-player games and related notions is a straightforward extension of Section 2.

In a four-player game with player 1 less informed than player 2, and perfect information for both player 3 and player 4, consider the *four-player decision problem* which is to decide if $\exists \sigma_1 \in \Sigma_1 \cdot \forall \sigma_2 \in \Sigma_2 \cdot \exists \sigma_3 \in \Sigma_3 \cdot \forall \sigma_4 \in \Sigma_4 : \rho_{q_0}^{\sigma_1, \sigma_2, \sigma_3, \sigma_4} \in \alpha$ for a parity objective α (also see [1, Remark 2] for further discussion). Since player 3 and player 4 have perfect information, we assume without loss of generality that the game is turn-based for them, that is there is a partition of the state space Q into two sets Q_3 and Q_4 (where $Q = Q_3 \cup Q_4$) such that the transition function is the union of $\delta_3 : Q_3 \times A_1 \times A_2 \times A_3 \to Q$ and $\delta_4 : Q_4 \times A_1 \times A_2 \times A_4 \to Q$. Strategies and outcomes are defined analogously to three-player games. A strategy of player $i \in \{3, 4\}$ is of the form $\sigma_i : Q^* \cdot Q_i \to A_i$.

We present a polynomial reduction of the problem for four-player games to solving a three-player game with the first player less informed than the second player [1]. Hardness follows from the special case of three-player games.

Theorem 4. *The four-player decision problem with player 1 less informed than player 2, and perfect information for both player 3 and player 4 is 2-EXPTIME-complete for parity objectives.*

6   Applications

We now discuss applications of our results in the context of synthesis and qualitative analysis of two-player partial-observation stochastic games.

Sequential Synthesis. The *sequential synthesis* problem consists of an open system of partially implemented modules (with possible non-determinism or choices) $M_1, M_2, \ldots, M_n$ that need to be refined (i.e., the choices determined by strategies) such that the composite system after refinement satisfy a specification. The system is open in the sense that after the refinement the composite system is reactive and interact with an environment. Consider the problem where first a set $M_1, \ldots, M_k$ of modules are refined, then a set $M_{k+1}, \ldots, M_\ell$ are refined by an external implementor, and finally the remaining set of modules are refined. In other words, the modules are refined sequentially: first a set of modules whose refinement can be controlled, then a set of modules whose refinement cannot be controlled as they are implemented externally, and finally the remaining set of modules. If the refinements of modules $M_1, \ldots, M_\ell$ do not have access to private variables of the remaining modules we obtain a partial-observation game with four players: the first (existential) player corresponds to the refinement of modules $M_1, \ldots, M_k$, the second (universal) player corresponds to the refinement of modules $M_{k+1}, \ldots, M_\ell$, the third (existential) player corresponds to the refinement of the remaining modules, and the fourth (adversarial) player is the environment. If the second player has access to all the variables visible to the first player, then player 1 is less informed.

Two-Player Partial-observation Stochastic Games. Our results for four-player games imply new complexity results for two-player stochastic games. For qualitative analysis (positive and almost-sure winning) under finite-memory strategies for the players the following reduction has been established in [10, Lemma 1] (see Lemma 2.1 of the arxiv version): the probabilistic transition function can be replaced by a turn-based gadget consisting of two perfect-observation players, one angelic (existential) and one demonic (universal). The turn-based gadget is the same as used for perfect-observation stochastic games [5,11]. In [10], only the special case of perfect observation for player 2 was considered, and hence the problem reduced to three-player games where only player 1 has partial observation and the other two players have perfect observation. In case where player 2 has partial observation, the reduction of [10] requires two perfect-observation players, and gives the problem of four-player games (with perfect observation for player 3 and player 4). Hence when player 1 is less informed, we obtain a 2-EXPTIME upper bound from Theorem 4, and obtain a 2-EXPTIME lower bound from Theorem 2 and Lemma 1 (see [1] for lower bound for almost-sure winning). Thus we obtain the following result.

Theorem 5. *The qualitative analysis problems (almost-sure and positive winning) for two-player partial-observation stochastic parity games where player 1 is less informed than player 2, under finite-memory strategies for both players, are 2-EXPTIME-complete.*

Remark 2. Note that the lower bounds for Theorem 5 are established for reachability objectives. Moreover, it was shown in [7, Section 5] that for qualitative analysis of two-player partial-observation stochastic games with reachability objectives, finite-memory strategies suffice, i.e., if there is a strategy to ensure almost-sure (resp., positive) winning, then there is a finite-memory strategy. Thus the results of Theorem 5 hold for reachability objectives even without the restriction of finite-memory strategies, and it extends the result of [7, Theorem 1] which showed EXPTIME-completeness for reachability objectives when player 2 has perfect observation.

References

1. ArXiv (2014), Full version http://arxiv.org/abs/1404.5453
2. Alur, R., Henzinger, T.A., Kupferman, O.: Alternating-time temporal logic. Journal of the ACM 49, 672–713 (2002)
3. Baier, C., Bertrand, N., Größer, M.: On decision problems for probabilistic Büchi automata. In: Amadio, R.M. (ed.) FoSSaCS 2008. LNCS, vol. 4962, pp. 287–301. Springer, Heidelberg (2008)

4. Bertrand, N., Genest, B., Gimbert, H.: Qualitative determinacy and decidability of stochastic games with signals. In: Proc. of LICS, pp. 319–328 (2009)
5. Chatterjee, K.: Stochastic ω-Regular Games. PhD thesis, UC Berkeley (2007)
6. Chatterjee, K., Doyen, L.: The complexity of partial-observation parity games. In: Fermüller, C.G., Voronkov, A. (eds.) LPAR-17. LNCS, vol. 6397, pp. 1–14. Springer, Heidelberg (2010)
7. Chatterjee, K., Doyen, L.: Partial-observation stochastic games: How to win when belief fails. In: Proc. of LICS 2012; Journal version ACM ToCL, pp. 175–184. IEEE (2012)
8. Chatterjee, K., Doyen, L., Gimbert, H., Henzinger, T.A.: Randomness for free. In: Hliněný, P., Kučera, A. (eds.) MFCS 2010. LNCS, vol. 6281, pp. 246–257. Springer, Heidelberg (2010)
9. Chatterjee, K., Doyen, L., Henzinger, T.A., Raskin, J.-F.: Algorithms for omega-regular games of incomplete information. Logical Methods in Computer Science 3(3:4) (2007)
10. Chatterjee, K., Doyen, L., Nain, S., Vardi, M.Y.: The complexity of partial-observation stochastic parity games with finite-memory strategies. In: Muscholl, A. (ed.) FoSSaCS 2014. LNCS, vol. 8412, pp. 242–257. Springer, Heidelberg (2014)
11. Chatterjee, K., Jurdziński, M., Henzinger, T.A.: Simple stochastic parity games. In: Baaz, M., Makowsky, J.A. (eds.) CSL 2003. LNCS, vol. 2803, pp. 100–113. Springer, Heidelberg (2003)
12. De Wulf, M., Doyen, L., Raskin, J.-F.: A lattice theory for solving games of imperfect information. In: Hespanha, J.P., Tiwari, A. (eds.) HSCC 2006. LNCS, vol. 3927, pp. 153–168. Springer, Heidelberg (2006)
13. Finkbeiner, B., Schewe, S.: Coordination logic. In: Dawar, A., Veith, H. (eds.) CSL 2010. LNCS, vol. 6247, pp. 305–319. Springer, Heidelberg (2010)
14. Henzinger, T.A., Kopke, P.W.: Discrete-time control for rectangular hybrid automata. Theor. Comp. Science 221, 369–392 (1999)
15. Kechris, A.: Classical Descriptive Set Theory. Springer (1995)
16. Madhusudan, P., Thiagarajan, P.S.: Distributed controller synthesis for local specifications. In: Orejas, F., Spirakis, P.G., van Leeuwen, J. (eds.) ICALP 2001. LNCS, vol. 2076, pp. 396–407. Springer, Heidelberg (2001)
17. Martin, D.A.: Borel determinacy. Annals of Mathematics 102(2), 363–371 (1975)
18. Mohalik, S., Walukiewicz, I.: Distributed games. In: Pandya, P.K., Radhakrishnan, J. (eds.) FSTTCS 2003. LNCS, vol. 2914, pp. 338–351. Springer, Heidelberg (2003)
19. Papadimitriou, C.H.: Computational complexity. Addison-Wesley (1994)
20. Peterson, G.L., Reif, J.H.: Multiple-person alternation. In: FOCS, pp. 348–363 (1979)
21. Pnueli, A., Rosner, R.: On the synthesis of a reactive module. In: Proc. of POPL, pp. 179–190. ACM Press (1989)
22. Ramadge, P.J., Wonham, W.M.: Supervisory control of a class of discrete-event processes. SIAM Journal of Control and Optimization 25(1), 206–230 (1987)
23. Ramanujam, R., Simon, S.: A communication based model for games of imperfect information. In: Gastin, P., Laroussinie, F. (eds.) CONCUR 2010. LNCS, vol. 6269, pp. 509–523. Springer, Heidelberg (2010)
24. Reif, J.H.: Universal games of incomplete information. In: Proc. of STOC, pp. 288–308 (1979)
25. Reif, J.H.: The complexity of two-player games of incomplete information. JCSS 29, –301 (1984)
26. Reif, J.H., Peterson, G.L.: A dynamic logic of multiprocessing with incomplete information. In: Proc. of POPL, pp. 193–202. ACM (1980)
27. Thomas, W.: Languages, automata, and logic. In: Handbook of Formal Languages. Beyond Words, vol. 3, ch. 7, pp. 389–455. Springer (1997)

The Complexity of Ergodic Mean-payoff Games [*],[†]

Krishnendu Chatterjee and Rasmus Ibsen-Jensen

IST Austria

Abstract. We study two-player (zero-sum) concurrent mean-payoff games played on a finite-state graph. We focus on the important sub-class of ergodic games where all states are visited infinitely often with probability 1. The algorithmic study of ergodic games was initiated in a seminal work of Hoffman and Karp in 1966, but all basic complexity questions have remained unresolved. Our main results for ergodic games are as follows: We establish (1) an optimal exponential bound on the patience of stationary strategies (where patience of a distribution is the inverse of the smallest positive probability and represents a complexity measure of a stationary strategy); (2) the approximation problem lies in FNP; (3) the approximation problem is at least as hard as the decision problem for simple stochastic games (for which NP ∩ coNP is the long-standing best known bound). We present a variant of the strategy-iteration algorithm by Hoffman and Karp; show that both our algorithm and the classical value-iteration algorithm can approximate the value in exponential time; and identify a subclass where the value-iteration algorithm is a FPTAS. We also show that the exact value can be expressed in the existential theory of the reals, and establish square-root sum hardness for a related class of games.

1 Introduction

Concurrent Games. Concurrent games are played over finite-state graphs by two players (Player 1 and Player 2) for an infinite number of rounds. In every round, both players simultaneously choose moves (or actions), and the current state and the joint moves determine a probability distribution over the successor states. The outcome of the game (or a *play*) is an infinite sequence of states and action pairs. Concurrent games were introduced in a seminal work by Shapley [21], and they are the most well-studied game models in stochastic graph games, with many important special cases.

Mean-payoff (Limit-average) Objectives. The most fundamental objective for concurrent games is the *limit-average* (or mean-payoff) objective, where a reward is associated to every transition and the payoff of a play is the limit-inferior (or limit-superior) average of the rewards of the play. The original work of Shapley [21] considered *discounted* sum objectives (or games that stop with probability 1); and the class of concurrent games with limit-average objectives (or games that have zero stop probabilities) was introduced by Gillette in [14]. The Player-1 *value* val(s) of the game at a state s is the supremum value of the expectation that Player 1 can guarantee for the limit-average objective against all strategies of Player 2. The games are zero-sum, so the objective of

---

[*] The research was partly supported by FWF Grant No P 23499-N23, FWF NFN Grant No S11407-N23 (RiSE), ERC Start grant (279307: Graph Games), and Microsoft faculty fellows award.

[†] Full version available at [1].

J. Esparza et al. (Eds.): ICALP 2014, Part II, LNCS 8573, pp. 122–133, 2014.
© Springer-Verlag Berlin Heidelberg 2014

Player 2 is the opposite. The study of concurrent mean-payoff games and its sub-classes have received huge attention over the last decades, both for mathematical results as well as algorithmic studies. Some key celebrated results are as follows: (1) the existence of values (or determinacy or equivalence of switching of strategy quantifiers for the players as in von-Neumann's min-max theorem) for concurrent discounted games was established in [21]; (2) the existence of values for the celebrated game of Big-Match was established in [4]; and (3) developing on the results of [4] and on Puiseux series [3] the existence of values for concurrent mean-payoff games was established in [19].

Sub-classes. The general class of concurrent mean-payoff games is notoriously difficult for algorithmic analysis. The current best known solution for general concurrent mean-payoff games is achieved by a reduction to the theory of the reals over addition and multiplication with three quantifier alternations [7] (also see [16] for a better reduction for constant state spaces). The strategies that are required in general for concurrent mean-payoff games are infinite-memory strategies that depend in a complex way on the history of the game [19,4], and analysis of such strategies make the algorithmic study complicated. Hence several sub-classes of concurrent mean-payoff games have been studied algorithmically both in terms of restrictions of the graph structure and restrictions of the objective. The three prominent restrictions in terms of the graph structure are as follows: (1) *Ergodic games (aka irreducible games)* where every state is visited infinitely often almost-surely. (2) *Turn-based stochastic games*, where in each state at most one player can choose between multiple moves. (3) *Deterministic games*, where the transition functions are deterministic. The most well-studied restriction in terms of objective is the *reachability* objectives. A reachability objective consists of a set U of *terminal* states (absorbing or sink states that are states with only self-loops), such that the set U is exactly the set of states where out-going transitions are assigned reward 1 and all other transitions are assigned reward 0. For all these sub-classes, except deterministic mean-payoff games (that is ergodic mean-payoff games, concurrent reachability games, and turn-based stochastic mean-payoff games) *stationary* strategies are sufficient, where a stationary strategy is independent of the past history of the game and depends only on the current state.

An Example. Consider the ergodic mean-payoff game shown in Figure 1. All transitions other than the dashed edges have probability 1, and each dashed edge has probability $1/2$. The transitions are annotated with the rewards. The stationary optimal strategy for both players is to play the first action (a_1 and b_1 for Player 1 and Player 2, respectively) with probability $4 - 2 \cdot \sqrt{3}$ in state s, and this ensures that the value is $\sqrt{3}$.

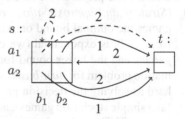

Fig. 1. Example game G

Previous Results. The decision problem of whether the value of the game at a state is at least a given threshold for turn-based stochastic reachability games (and also turn-based mean-payoff games with deterministic transition function) lie in NP ∩ coNP [8,23]. They are among the rare and intriguing combinatorial problems that lie in NP ∩ coNP, but not known to be in PTIME. The existence of polynomial-time algorithms for the above decision questions are long-standing open problems. The algorithmic solution for turn-based games that is most efficient in practice is the *strategy-iteration* algorithm, where the algorithm iterates over local improvement of strategies which is then established to converge to a globally optimal strategy. For ergodic games, Hoffman and Karp [17] presented a strategy-iteration algorithm and established that stationary strategies are sufficient. For concurrent reachability games, again stationary strategies are sufficient (for ϵ-optimal strategies, for all $\epsilon > 0$) [12,9]; the decision problem is in PSPACE and *square-root sum* hard [10].

Key Intriguing Complexity Questions. There are several key intriguing open questions related to the complexity of the various sub-classes of concurrent mean-payoff games. Some of them are: (1) Does there exist a sub-class of concurrent mean-payoff games where the approximation problem is simpler than the exact decision problem, e.g., the decision problem is square-root sum hard, but the approximation problem can be solved in FNP? (2) There is no convergence result associated with the two classical algorithms, namely the strategy-iteration algorithm [17], and the value-iteration algorithm, for ergodic games; and is it possible to establish convergence for them for approximating the values. (3) The complexity of a stationary strategy is described by its *patience* which is the inverse of the minimum non-zero probability assigned to a move [12], and no bound is known for the patience of stationary strategies for ergodic games.

Our Results. The study of the ergodic games was initiated in the seminal work of Hoffman and Karp [17], and most of the complexity questions (related to computational-, strategy-, and algorithmic-complexity) have remained open. In this work we focus on the complexity of simple generalizations of ergodic games (that subsume ergodic games). Ergodic games form a very important sub-class of concurrent games subsuming the special cases of uni-chain Markov decision processes and uni-chain turn-based stochastic games (that have been studied in great depth in the literature with numerous applications, see [13,20]). We consider generalizations of ergodic games called *sure* ergodic games where all plays are guaranteed to reach an ergodic component (a sub-game that is ergodic); and *almost-sure* ergodic games where with probability 1 an ergodic component is reached. Every ergodic game is sure ergodic, and every sure ergodic game is almost-sure ergodic. Intuitively the generalizations allow us to consider that after a finite prefix an ergodic component is reached.

1. *(Strategy and approximation complexity).* We show that for almost-sure ergodic games the optimal bound on patience required for ϵ-optimal stationary strategies, for $\epsilon > 0$, is exponential (we establish the upper bound for almost-sure ergodic games, and the lower bound for ergodic games). We then show that the approximation problem for *turn-based* stochastic ergodic mean-payoff games is at least as hard as solving the decision problem for turn-based stochastic reachability games (aka simple stochastic games); and finally show that the approximation problem belongs to FNP for almost-sure ergodic games. Observe that our results imply that improving our FNP-bound for the approximation problem to polynomial time would require solving the long-standing open question of whether the decision problem of turn-based stochastic reachability games can be solved in polynomial time.

2. *(Algorithm).* We present a variant of the Hoffman-Karp algorithm and show that for all ϵ-approximation (for $\epsilon > 0$) our algorithm converges with in exponential number of iterations for almost-sure ergodic games. We analyze the value-iteration algorithm for ergodic games and show that for all $\epsilon > 0$, the value-iteration algorithm requires at most $O(\underline{H} \cdot W \cdot \epsilon^{-1} \cdot \log(\epsilon^{-1}))$ iterations, where $\underline{H}$ is the upper

bound on the expected hitting time of state pairs that Player 1 can ensure and W is the maximal reward value. We show that $\underline{H}$ is at most $n \cdot (\delta_{\min})^{-n}$, where n is the number of states of the game, and $\delta_{\min}$ the smallest positive transition probability. Thus our result establishes an exponential upper bound for the value-iteration algorithm for approximation. This result is in sharp contrast to concurrent reachability games where the value-iteration algorithm requires double exponentially many steps [15]. Observe that we have a polynomial-time approximation scheme if $\underline{H}$ is polynomial and the numbers W and ϵ are represented in unary. Thus we identify a subclass of ergodic games where the value-iteration algorithm is polynomial (see Remark 2 for further details).

3. *(Complexity of the decision problem for the exact value)*. We show that the exact decision problem for almost-sure ergodic games can be expressed in the existential theory of the reals (in contrast to general concurrent mean-payoff games where quantifier alternations are required). Finally, we show that the exact decision problem for sure ergodic games is square-root sum hard.

Technical Contribution and Remarks. Our main result is establishing the optimal bound of exponential patience for ϵ-optimal stationary strategies, for $\epsilon > 0$, in almost-sure ergodic games. Our result is in sharp contrast to the optimal bound of double-exponential patience for concurrent reachability games [15], and also the double-exponential iterations required by the strategy-iteration and the value-iteration algorithms for concurrent reachability games [15]. Our upper bound on the exponential patience is achieved by a coupling argument. While coupling argument is a well-established tool in probability theory, to the best of our knowledge the argument has not been used for concurrent mean-payoff games before. Our lower bound example constructs a family of ergodic mean-payoff games where exponential patience is required. Our results provide a complete picture for almost-sure and sure ergodic games (subsuming ergodic games) in terms of strategy-, computational-, and algorithmic-complexity; and present answers to some of the key intriguing open questions related to the computational complexity of concurrent mean-payoff games.

Comparison with Results for Shapley Games. For Shapley (concurrent discounted) games, the exact decision problem is square-root sum hard [11], and the fact that the approximation problem is in FNP is straight-forward to prove (for details, see [18, Lemma 6, Section 1.10]). The more interesting and challenging question is whether the approximation problem can be solved in PPAD. The PPAD complexity for the approximation problem for Shapley games was established in [11]; and the PPAD complexity arguments use the existence of unique (Banach) fixpoint (due to contraction mapping) and the fact that weak approximation implies strong approximation. A PPAD complexity result for the class of ergodic games (in particular, whether weak approximation implies strong approximation) is a subject for future work. Another interesting direction of future work would be to extend our results for concurrent games where the values of all states are very close together; and for this class of games existence of near optimal stationary strategies was established in [5].

2 Definitions

Probability Distributions. For a finite set A, a *probability distribution* on A is a function $\delta \colon A \to [0, 1]$ such that $\sum_{a \in A} \delta(a) = 1$. We denote the set of probability distributions on A by $\mathcal{D}(A)$. Given a distribution $\delta \in \mathcal{D}(A)$, we denote by $\mathrm{Supp}(\delta) = \{x \in A \mid \delta(x) > 0\}$ the *support* of the distribution δ.

Concurrent Game Structures. A *concurrent stochastic game structure* $G = (S, A, \Gamma_1, \Gamma_2, \delta)$ consists of:

- A finite state space S and a finite set A of actions (or moves).
- Two move assignments $\Gamma_1, \Gamma_2 \colon S \to 2^A \setminus \emptyset$. For $i \in \{1, 2\}$, assignment Γ_i associates with each state $s \in S$ the non-empty set $\Gamma_i(s) \subseteq A$ of moves available to Player i at state s.
- A probabilistic transition function $\delta \colon S \times A \times A \to \mathcal{D}(S)$, which for every $s \in S$ and $a_1 \in \Gamma_1(s)$ and $a_2 \in \Gamma_2(s)$, gives a probability distribution $\delta(s, a_1, a_2) \in \mathcal{D}(S)$ for the successor state.

We denote by $\delta_{\min}$ the minimum non-zero transition probability, i.e., $\delta_{\min} = \min_{s, t \in S} \min_{a_1 \in \Gamma_1(s), a_2 \in \Gamma_2(s)} \{\delta(s, a_1, a_2)(t) \mid \delta(s, a_1, a_2)(t) > 0\}$. We denote by n the number of states (i.e., $n = |S|$), and by m the maximal number of actions available for a player at a state (i.e., $m = \max_{s \in S} \max\{|\Gamma_1(s)|, |\Gamma_2(s)|\}$). We denote by r the number of *random* states where the transition function is not deterministic, i.e., $r = |\{s \in S \mid \exists a_1 \in \Gamma_1(s), a_2 \in \Gamma_2(s).|\mathrm{Supp}(\delta(s, a_1, a_2))| \geq 2\}|$.

Plays. At every state $s \in S$, Player 1 chooses a move $a_1 \in \Gamma_1(s)$, and simultaneously and independently Player 2 chooses a move $a_2 \in \Gamma_2(s)$. The game then proceeds to the successor state t with probability $\delta(s, a_1, a_2)(t)$, for all $t \in S$. A *path* or a *play* of G is an infinite sequence $\pi = ((s_0, a_1^0, a_2^0), (s_1, a_1^1, a_2^1), (s_2, a_1^2, a_2^2) \ldots)$ of states and action pairs such that for all $k \geq 0$ we have (i) $a_1^k \in \Gamma_1(s_k)$ and $a_2^k \in \Gamma_2(s_k)$; and (ii) $s_{k+1} \in \mathrm{Supp}(\delta(s_k, a_1^k, a_2^k))$. Let Π be the set of all paths.

Strategies. A *strategy* for a player is a recipe that describes how to extend prefixes of a play. Formally, a strategy for Player $i \in \{1, 2\}$ is a mapping $\sigma_i \colon (S \times A \times A)^* \times S \to \mathcal{D}(A)$ that associates with every finite sequence $x \in (S \times A \times A)^*$ of state and action pairs, and the current state s in S, representing the past history of the game, a probability distribution $\sigma_i(x \cdot s)$ used to select the next move. The strategy σ_i can prescribe only moves that are available to Player i; that is, for all sequences $x \in (S \times A \times A)^*$ and states $s \in S$, we require that $\mathrm{Supp}(\sigma_i(x \cdot s)) \subseteq \Gamma_i(s)$. We denote by Σ_i the set of all strategies for Player $i \in \{1, 2\}$. Once the starting state s and the strategies σ_1 and σ_2 for the two players have been chosen, then we have a random walk $\pi_s^{\sigma_1, \sigma_2}$ for which the probabilities of events are uniquely defined [22], where an *event* $\mathcal{A} \subseteq \Pi$ is a measurable set of paths. For an event $\mathcal{A} \subseteq \Pi$, we denote by $\mathrm{Pr}_s^{\sigma_1, \sigma_2}(\mathcal{A})$ the probability that a path belongs to $\mathcal{A}$ when the game starts from s and the players use the strategies σ_1 and σ_2; and denote $\mathbb{E}_s^{\sigma_1, \sigma_2}[\cdot]$ as the associated expectation measure. We consider in particular stationary and positional strategies. A strategy σ_i is *stationary* (or memoryless) if it is independent of the history but only depends on the current state, i.e., for all $x, x' \in (S \times A \times A)^*$ and all $s \in S$, we have $\sigma_i(x \cdot s) = \sigma_i(x' \cdot s)$, and thus can be expressed as a function $\sigma_i \colon S \to \mathcal{D}(A)$. For stationary strategies, the complexity of the strategy is described by the *patience* of the strategy, which is the inverse of the minimum non-zero probability assigned to an action [12]. Formally, for a stationary strategy $\sigma_i \colon S \to \mathcal{D}(A)$ for Player i, the patience is $\max_{s \in S} \max_{a \in \Gamma_i(s)} \{\frac{1}{\sigma_i(s)(a)} \mid \sigma_i(s)(a) > 0\}$. A strategy is *pure (deterministic)* if it does not use randomization, i.e., for any history there is always some unique action a that is played with probability 1.

A pure stationary strategy σ_i is also called a *positional* strategy, and represented as a function $\sigma_i : S \to A$. We call a pair of strategies $(\sigma_1, \sigma_2) \in \Sigma_1 \times \Sigma_2$ a *strategy profile*.

The Mean-payoff Function. In this work we consider maximizing *limit-average* (or mean-payoff) functions for Player 1, and the objective of Player 2 is opposite (i.e., the games are zero-sum). We consider concurrent games with a reward function $R : S \times A \times A \to [0,1]$ that assigns a reward value $0 \le R(s, a_1, a_2) \le 1$ for all $s \in S$, $a_1 \in \Gamma_1(s)$, and $a_2 \in \Gamma_2(s)$. For a path $\pi = ((s_0, a_1^0, a_2^0), (s_1, a_1^1, a_2^1), \dots)$, the limit-inferior average (resp. limit-superior average) is defined as follows: $\mathsf{LimInfAvg}(\pi) = \liminf_{n\to\infty} \frac{1}{n} \cdot \sum_{i=0}^{n-1} R(s_i, a_1^i, a_2^i)$ (resp. $\mathsf{LimSupAvg}(\pi) = \limsup_{n\to\infty} \frac{1}{n} \cdot \sum_{i=0}^{n-1} R(s_i, a_1^i, a_2^i)$). For brevity we denote concurrent games with mean-payoff functions as CMPGs (concurrent mean-payoff games).

Values and ϵ-optimal Strategies. Given a CMPG G and a reward function R, the *lower value* $\underline{v}_s$ (resp. the *upper value* $\overline{v}_s$) at a state s is defined as follows:

$$\underline{v}_s = \sup_{\sigma_1 \in \Sigma_1} \inf_{\sigma_2 \in \Sigma_2} \mathbb{E}_s^{\sigma_1, \sigma_2}[\mathsf{LimInfAvg}]; \qquad \overline{v}_s = \inf_{\sigma_2 \in \Sigma_2} \sup_{\sigma_1 \in \Sigma_1} \mathbb{E}_s^{\sigma_1, \sigma_2}[\mathsf{LimSupAvg}].$$

The celebrated result of Mertens and Neyman [19] shows that the upper and lower value coincide and gives the *value* of the game denoted as v_s. For $\epsilon \ge 0$, a strategy σ_1 for Player 1 is ϵ-*optimal* if we have $v_s - \epsilon \le \inf_{\sigma_2 \in \Sigma_2} \mathbb{E}_s^{\sigma_1, \sigma_2}[\mathsf{LimInfAvg}]$. An *optimal* strategy is a 0-optimal strategy.

Game Classes. We consider the following special classes of CMPGs.

1. *Variants of ergodic CMPGs.* Given a CMPG G, a set C of states in G is called an *ergodic component*, if for all states $s, t \in C$, for all strategy profiles (σ_1, σ_2), if we start at s, then t is visited infinitely often with probability 1 in the random walk $\pi_s^{\sigma_1, \sigma_2}$. A CMPG is *ergodic* if the set S of states is an ergodic component. A CMPG is *sure ergodic* if for all strategy profiles (σ_1, σ_2) and for all start states s, ergodic components are reached certainly (all plays reach some ergodic component). A CMPG is *almost-sure ergodic* if for all strategy profiles (σ_1, σ_2) and for all start states s, ergodic components are reached with probability 1. Observe that every ergodic CMPG is also a sure ergodic CMPG, and every sure ergodic CMPG is also an almost-sure ergodic CMPG.

2. *Turn-based stochastic games, MDPs and SSGs.* A game structure G is *turn-based stochastic* if at every state at most one player can choose among multiple moves; that is, for every state $s \in S$ there exists at most one $i \in \{1, 2\}$ with $|\Gamma_i(s)| > 1$. A game structure is a Player-2 *Markov decision process (MDP)* if for all $s \in S$ we have $|\Gamma_1(s)| = 1$, i.e., only Player 2 has choice of actions in the game, and Player-1 MDPs are defined analogously. A *simple stochastic game (SSG)* [8] is an almost-sure ergodic turn-based stochastic game with reachability objective.

Remark 1. The results of Hoffman and Karp [17] established that for ergodic CMPGs *optimal stationary* strategies exist (for both players). Also, for an ergodic CMPG the value for every state is the same, which is called the value of the game. The result for existence of optimal stationary strategies easily extends to almost-sure ergodic CMPGs.

Value and the Approximation Problem. Given a CMPG G, a state s of G, and a rational threshold λ, the *value* problem is the decision problem that asks whether v_s is at most λ. Given a CMPG G, a state s of G, and a tolerance $\epsilon > 0$, the *approximation* problem asks to compute an interval of length ϵ such that the value v_s lies in the interval. We present the formal definition of the decision version of the approximation problem in Section 3.2. In the following sections we consider the value problem and the approximation problem for almost-sure ergodic, sure ergodic, and ergodic games.

3   Complexity of Approximation for Almost-Sure Ergodic Games

In this section we present three results for almost-sure ergodic games: (1) First we establish (in Section 3.1) an optimal exponential bound on the patience of ϵ-optimal stationary strategies, for all $\epsilon > 0$. (2) Second we show (in Section 3.2) that the approximation problem (even for turn-based stochastic ergodic mean-payoff games) is at least as hard as solving the value problem for SSGs. (3) Finally, we show (in Section 3.2) that the approximation problem lies in FNP.

3.1   Strategy Complexity

In this section we present results related to ϵ-optimal stationary strategies for almost-sure ergodic CMPGs, that on one hand establishes an optimal exponential bound for patience, and on the other hand is used to establish the complexity of approximation of values in Section 3.2. The results of this section is also used in the algorithmic analysis in Section 3.3. We start with the notion of q-rounded strategies.

The Classes of q-rounded Distributions and Strategies. For $q \in \mathbb{N}$, a distribution d over a finite set Z is a *q-rounded distribution* if for all $z \in Z$ we have that $d(z) = \frac{p}{q}$ for some number $p \in \mathbb{N}$. A stationary strategy σ is a *q-rounded strategy*, if for all states s the distribution $\sigma(s)$ is a q-rounded distribution.

Patience. Observe that the patience of a q-rounded strategy is at most q. We show that for almost-sure ergodic CMPGs for all $\epsilon > 0$ there are q-rounded ϵ-optimal strategies, where q is $\lceil 4 \cdot \epsilon^{-1} \cdot m \cdot n^2 \cdot (\delta_{\min})^{-r} \rceil$. This immediately implies an exponential upper bound on the patience. We start with a lemma related to the probability of reaching states that are guaranteed to be reached with positive probability.

Lemma 1. *Given a CMPG G, let s be a state in G, and T be a set of states such that for all strategy profiles the set T is reachable (with positive probability) from s. For all strategy profiles the probability to reach T from s in n steps is at least $(\delta_{\min})^r$ (where r is the number of random states).*

Variation Distance. We use a coupling argument in our proofs and this requires the definition of variation distance of two probability distributions. Given a finite set Z, and two distributions d_1 and d_2 over Z, the *variation distance* of the distributions is $\mathrm{var}(d_1, d_2) = \frac{1}{2} \cdot \sum_{z \in Z} |d_1(z) - d_2(z)|$.

Coupling and Coupling Lemma. Let Z be a finite set. For distributions d_1 and d_2 over the finite set Z, a *coupling* ω is a distribution over $Z \times Z$, such that for all $z \in Z$ we have $\sum_{z' \in Z} \omega(z, z') = d_1(z)$ and also for all $z' \in Z$ we have $\sum_{z \in Z} \omega(z, z') = d_2(z')$. We only use the second part of coupling lemma [2] which is stated as follows:

- **(Coupling Lemma).** For a pair of distributions d_1 and d_2, there exists a coupling ω of d_1 and d_2, such that for a random variable (X, Y) from the distribution ω, we have that $\mathrm{var}(d_1, d_2) = \Pr[X \neq Y]$.

We show that in almost-sure ergodic CMPGs strategies that play with probabilities "close" to what is played by an optimal strategy also achieve values that are "close" to the values achieved by the optimal strategy.

Lemma 2. *Consider an almost-sure ergodic CMPG and let $\epsilon > 0$ be a real number. Let σ_1 be an optimal stationary strategy for Player 1. Let σ_1' be a stationary strategy for Player 1 s.t. $\sigma_1'(s)(a) \in [\sigma_1(s)(a) - \frac{1}{q}; \sigma_1(s)(a) + \frac{1}{q}]$, where $q = 4 \cdot \epsilon^{-1} \cdot m \cdot n^2 \cdot (\delta_{\min})^{-r}$, for all states s and actions $a \in \Gamma_1(s)$. Then the strategy σ_1' is an ϵ-optimal strategy.*

Proof. First observe that we can consider $\epsilon \leq 1$, because as the rewards are in the interval $[0, 1]$ any strategy is an ϵ-optimal strategy for $\epsilon \geq 1$. We will present the proof when the play starts in an ergodic component; and the details of the other case is in [1]. We show that σ_1' guarantees a mean-payoff within ϵ of the mean-payoff guaranteed by σ_1, thus implying the statement. Let σ_2 be a positional best response strategy against σ_1'. Our proof is based on a novel *coupling* argument. For any state s, it is clear that the variation distance between $\sigma_1'(s)$ and $\sigma_1(s)$ is at most $\frac{|\Gamma_1(s)|}{2 \cdot q}$, by definition of $\sigma_1'(s)$. For a state s, let d_1^s be the distribution over states defined as follows: for $t \in S$ we have $d_1^s(t) = \sum_{a_1 \in \Gamma_1(s)} \sum_{a_2 \in \Gamma_2(s)} \delta(s, a_1, a_2)(t) \cdot \sigma_1(s)(a_1) \cdot \sigma_2(s)(a_2)$. Define d_2^s similarly using $\sigma_1'(s)$ instead of $\sigma_1(s)$. Then d_1^s and d_2^s also have a variation distance of at most $\frac{|\Gamma_1(s)|}{2 \cdot q} \leq \frac{m}{2 \cdot q}$. Let s_0 be the start state, and $P = \pi_{s_0}^{\sigma_1, \sigma_2}$ be the random walk from s_0, where Player 1 follows σ_1 and Player 2 follows σ_2. Also let $P' = \pi_{s_0}^{\sigma_1', \sigma_2}$ be the similar defined walk, except that Player 1 follows σ_1' instead of σ_1. Let X^i be the random variable indicating the i-th state of P, and let Y^i be the similar defined random variable in P'.

Consider that s_0 is part of an ergodic component. Irrespective of the strategy profile, all states of the ergodic component are visited infinitely often almost-surely (by definition of an ergodic component). Hence, we can apply Lemma 1 and obtain that we require at most $n \cdot (\delta_{\min})^r = \frac{\epsilon \cdot q}{4 \cdot n \cdot m}$ steps in expectation to get from one state of the component to any other state of the component.

Coupling argument. We now construct a coupling argument. We define the coupling using induction. First observe that $X^0 = Y^0 = s_0$ (the starting state). For $i, j \in \mathbb{N}$, let $a_{i,j} \geq 0$ be the smallest number such that $X^{i+1} = Y^{j+1+a_{i,j}}$. By the preceding we know that $a_{i,j}$ exists for all i, j with probability 1 and $a_{i,j} \leq \frac{\epsilon \cdot q}{4 \cdot n \cdot m}$ in expectation. The coupling is done as follows: (1) (Base case): Couple X^0 and Y^0. We have that $X^0 = Y^0$; (2) (Inductive case): (i) if X^i is coupled to Y^j and $X^i = Y^j = s_i$, then also couple X^{i+1} and Y^{j+1} such that $\Pr[X^{i+1} \neq Y^{j+1}] = \text{var}(d_1^s, d_2^s)$ (using coupling lemma); (ii) if X^i is coupled to Y^j, but $X^i \neq Y^j$, then $X^{i+1} = Y^{j+1+a_{i,j}} = s_{i+1}$ and X^{i+1} is coupled to $Y^{j+1+a_{i,j}}$, and we couple X^{i+2} and $Y^{j+2+a_{i,j}}$ such that $\Pr[X^{i+2} \neq Y^{j+2+a_{i,j}}] = \text{var}(d_1^{s_{i+1}}, d_2^{s_{i+1}})$ (using coupling lemma). Notice that all X^i are coupled to some Y^j almost-surely; and moreover in expectation $\frac{j}{i}$ is bounded as follows: $\frac{j}{i} \leq 1 + \frac{m}{2 \cdot q} \cdot \frac{\epsilon \cdot q}{4 \cdot n \cdot m} = 1 + \frac{\epsilon}{8 \cdot n}$. The expression can be understood as follows: consider X^i being coupled to Y^j. With probability at most $\frac{m}{2 \cdot q}$ they differ. In that case X^{i+1} is coupled to $Y^{j+1+a_{i,j}}$. Otherwise X^{i+1} is coupled to Y^{j+1}. By using our bound on $a_{i,j}$ we get the desired expression. For a state s, let f_s (resp. f_s') denote the limit-average frequency of s given σ_1 (resp. σ_1') and σ_2. Then it follows easily that for every state s, we have $|f_s - f_s'| \leq \frac{\epsilon}{8 \cdot n}$. The formal argument is as follows: for every state s, consider the reward function R_s that assigns reward 1 to all transitions from s and 0 otherwise; and then it is clear that the difference of the mean-payoffs of P and P' is maximized if the mean-payoff of P is 1 under R_s and the rewards of the steps of P' that are not coupled to P are 0. In that case the mean-payoff of P' under R_s is at least $\frac{1}{1+\frac{\epsilon}{8 \cdot n}} > 1 - \frac{\epsilon}{8 \cdot n}$ (since $1 > 1 - \left(\frac{\epsilon}{8 \cdot n}\right)^2 = (1 + \frac{\epsilon}{8 \cdot n})(1 - \frac{\epsilon}{8 \cdot n})$) in expectation and thus the difference between the mean-payoff of P and the mean-payoff of P' under R_s is at most $\frac{\epsilon}{8 \cdot n}$ in expectation. The mean-payoff value if Player 1 follows a stationary strategy σ_1^1 and Player 2 follows a stationary strategy σ_2^1, such that the frequencies of the states encountered is f_s^1, is $\sum_{s \in S} \sum_{a_1 \in \Gamma_1(s)} \sum_{a_2 \in \Gamma_2(s)} f_s^1 \cdot \sigma_1^1(s)(a_1) \cdot \sigma_2^1(s)(a_2) \cdot R(s, a_1, a_2)$. Thus the differences in mean-payoff value when Player 1 follows σ_1 (resp. σ_1') and Player 2 follows the positional strategy σ_2, which plays action a_2^s in state s, is $\sum_{s \in S} \sum_{a_1 \in \Gamma_1(s)} \left(f_s \cdot \sigma_1(s)(a_1) - f_s' \cdot \sigma_1'(s)(a_1)\right) \cdot R(s, a_1, a_2^s)$. Since $|f_s - f_s'| \leq \frac{\epsilon}{8 \cdot n}$ (by the preceding argument) and $|\sigma_1(s)(a_1) - \sigma_1'(s)(a_1)| \leq \frac{1}{q}$ for all $s \in S$ and $a_1 \in \Gamma_1(s)$ (by definition), we have $\sum_{s \in S} \sum_{a_1 \in \Gamma_1(s)} \left(f_s \cdot \sigma_1(s)(a_1) - f_s' \cdot \sigma_1'(s)(a_1)\right) \cdot R(s, a_1, a_2^s) \leq \frac{\epsilon}{2}$. The desired result follows.   □

We show that for every integer $q' \geq \ell$, for every distribution over ℓ elements, there exists a q'-rounded distribution "close" to it. Together with Lemma 2 it shows the existence of q'-rounded ϵ-optimal strategies, for every integer q' greater than the q defined in Lemma 2.

Lemma 3. *Let d_1 be a distribution over a finite set Z of size ℓ. Then for all integers $q \geq \ell$ there exists a q-rounded distribution d_2 over Z, such that $|d_1(z) - d_2(z)| < \frac{1}{q}$.*

Corollary 1. *For all almost-sure ergodic CMPGs, for all $\epsilon > 0$, there exists an ϵ-optimal, q'-rounded strategy σ_1 for Player 1, for all integers $q' \geq q$, where $q = 4 \cdot \epsilon^{-1} \cdot m \cdot n^2 \cdot (\delta_{\min})^{-r}$.*

Exponential Lower Bound on Patience. We present a family of ergodic CMPGs where the lower bound on patience is exponential in r for every $1/48$-optimal strategies (details in [1]).

Theorem 1 (Strategy Complexity). *The following assertions hold:*

1. *(Upper bound). For almost-sure ergodic CMPGs, for all $\epsilon > 0$, there exists an ϵ-optimal strategy of patience at most $\lceil 4 \cdot \epsilon^{-1} \cdot m \cdot n^2 \cdot (\delta_{\min})^{-r} \rceil$.*
2. *(Lower bound). There exists a family of ergodic CMPGs $G_n^{\delta_{\min}}$, for each odd $n \geq 9$ and $0 < \delta_{\min} < \frac{1}{2 \cdot n}$, such that $n = r + 5$ and any $\frac{1}{48}$-optimal strategy in $G_n^{\delta_{\min}}$ has patience at least $\frac{1}{2} \cdot (\delta_{\min})^{-r/4}$.*

3.2   Approximation Complexity

We establish the approximation complexity for almost-sure ergodic CMPGs.

Hardness of Approximation. We present a polynomial reduction from the value problem for SSGs to the problem of approximation of values for turn-based stochastic ergodic mean-payoff games (TEMPGs) (details in [1]).

Approximation Decision Problem. Given an almost-sure ergodic CMPG G (with rational transition probabilities given in binary), a state s, an $\epsilon > 0$ (in binary), and a rational number λ (in binary), the promise problem PROMVALERG (i) accepts if the value of s is at least λ, (ii) rejects if the value of s is at most $\lambda - \epsilon$, and (iii) if the value is in the interval $(\lambda - \epsilon; \lambda)$, then it may both accept or reject.

Theorem 2 (Approximation Complexity). *For almost-sure ergodic CMPGs, the following assertions hold:*

1. *(Upper bound). The problem* PROMVALERG *is in* FNP.
2. *(Hardness). The problem of finding the value of a state in an SSG is polynomial time Turing reducible to the problem* PROMVALERG, *even for turn-based stochastic ergodic mean-payoff games (TEMPGs).*

3.3   Strategy-Iteration Algorithm for Almost-Sure Ergodic CMPGs

The classic algorithm for solving ergodic CMPGs was given by Hoffman and Karp [17]. We present a variant of the algorithm, and show that for every $\epsilon > 0$ it runs in exponential time for ϵ approximation. Our algorithm is a variant of the original algorithm, where instead of all stationary strategies we iterate over q-rounded strategies, depending on $\epsilon > 0$. We refer our algorithm as VARHOFFMANKARP and show the following result (details in [1]).

Theorem 3. *For an almost-sure ergodic CMPG, for all $\epsilon > 0$,* VARHOFFMANKARP *correctly computes an ϵ-optimal strategy, and (i) requires at most $O\left(\left(\epsilon^{-1} \cdot m \cdot n^2 \cdot (\delta_{\min})^{-r}\right)^{n \cdot m}\right)$ iterations, and each iteration requires at most $O(2^{\text{POLY}(m)} \cdot \text{POLY}(n, \log(\epsilon^{-1}), \log(\delta_{\min}^{-1})))$ time; and (ii) requires polynomial space.*

4   Analysis of the Value-Iteration Algorithm

We show that the classical value-iteration algorithm requires at most exponentially many steps to approximate the value of ergodic concurrent mean-payoff games (ECMPGs).

Notations. Given an ECMPG G, let v^* denote the value of the game (recall that all states in an ECMPG have the same value). Let $v_s^T = \sup_{\sigma_1 \in \Sigma_1} \inf_{\sigma_2 \in \Sigma_2} \mathbb{E}_s^{\sigma_1, \sigma_2}[\text{Avg}_T]$ denote the value function for the objective Avg_T, i.e., playing the game for T steps. For an ECMPG G we call the game with the objective Avg_T as G_T. A strategy σ_1 is optimal for the objective Avg_T if $v_s^T = \inf_{\sigma_2 \in \Sigma_2} \mathbb{E}_s^{\sigma_1, \sigma_2}[\text{Avg}_T]$, and a strategy σ_2 is optimal for the objective Avg_T if $v_s^T = \sup_{\sigma_1 \in \Sigma_1} \mathbb{E}_s^{\sigma_1, \sigma_2}[\text{Avg}_T]$. The function v_s^T is computed iteratively in T and is refered to as the *value-iteration* algorithm. It is well-known that $v_s = \liminf_{T \to \infty} v_s^T = \limsup_{T \to \infty} v_s^T$ [19]. We first establish a result that shows that for all T there exist s and s' such that v^* is bounded by v_s^T and $v_{s'}^T$.

Lemma 4. *For all ECMPGs G and for all $T > 0$, there exists a pair of states s', s, such that $v_{s'}^T \le v^* \le v_s^T$.*

Proof Overview: The proof is by contradiction, that is, we assume that for all s we have $v_s^T < v^*$ (the other case follows from the same game where the players have exchanged roles). The idea is that we can consider plays of G, defined by an optimal strategy for the objective LimInfAvg for Player 1 in G and a strategy for Player 2 that plays an optimal strategy for objective Avg_T in G_T for T steps and then starts over. We then split the plays into sub-plays of length T. Since for all s we have $v_s^T < v^*$ and because Player 2 plays optimally in the sub-plays, in every segment of length T the expected mean-payoff is strictly less than v^*. But then also the expected mean-payoff of the plays is strictly less than v^*. This contradicts that Player 1 played optimally (which ensures that the expected mean-payoff is at least v^*).

The Numbers $\underline{H}$ and $\overline{H}$. Given an ECMPG G, strategies σ_1 and σ_2 for the players, and two states s and t, let $H_{s,t}^{\sigma_1,\sigma_2}$ denote the expected hitting time from s to t, given the strategies. Let $H_{\sigma_1} = \sup_{\sigma_2 \in \Sigma_2} \max_{s,t \in S} H_{s,t}^{\sigma_1,\sigma_2}$; and $\underline{H} = \inf_{\sigma_1 \in \Sigma_1} H_{\sigma_1}$ and $\overline{H} = \sup_{\sigma_1 \in \Sigma_1} H_{\sigma_1}$. Intuitively, $\underline{H}$ is the minimum expected hitting time between all state pairs that Player 1 can ensure against all strategies of Player 2.

Lemma 5. *For all ECMPGs G we have $\underline{H} \leq \overline{H} \leq n \cdot (\delta_{\min})^{-r}$.*

We now present our result for the bounds required for approximation by the value-iteration algorithm.

Theorem 4. *For all ECMPGs, for all $0 < \epsilon < 1$, and all $T \geq 4 \cdot \underline{H} \cdot c \cdot \log c$, for $c = 2 \cdot \epsilon^{-1}$, we have that $v^* - \epsilon \leq \min_s v_s^T \leq v^* \leq \max_s v_s^T \leq v^* + \epsilon$.*

Proof. (Sketch). Let $T \geq 4 \cdot \underline{H} \cdot c \cdot \log c$, for $c = 2 \cdot \epsilon^{-1}$. Also, let $c' = 4 \cdot \underline{H} \cdot \log c$ and $T' = T - c'$. By Lemma 4 we have $\min_s v_s^T \leq v^* \leq \max_s v_s^T$. We now argue that $v^* - \epsilon \leq \min_s v_s^T$, and then $\max_s v_s^T \leq v^* + \epsilon$ follows by considering the game where the players have exchanged roles. Let s' be some state in $\arg\min_{s'} v_{s'}^T$ and let s'' be some state such that $v^* \leq v_{s''}^{T'}$ (such a state exists by Lemma 4). Let σ_1' be an optimal strategy for the objective $\text{Avg}_{T'}$ in $G_{T'}$, and let σ_1^* be a strategy that ensures that the hitting time from s' to s'' is at most $2 \cdot \underline{H}$, i.e., $H_{\sigma_1^*} \leq 2 \cdot \underline{H}$ (such a strategy exists by definition of $\underline{H}$). Let σ_1 be the strategy for Player 1 that plays as σ_1^* until s'' is reached, and then switches to σ_1'. We show that σ_1 ensures that $v_{s'}^T$ is at least $v^* - \epsilon$. $\square$

Remark 2. Theorem 4 presents the bound for value-iteration when the rewards are in $[0, 1]$. If the rewards are in $[0, W]$, for some positive integer W, then for ϵ-approximation we first divide all rewards by W, and then apply results of Theorem 4 for ϵ/W-approximation. We have shown that in the worst case $\underline{H}$ is at most $n \cdot (\delta_{\min})^{-r}$. If $\underline{H}, W, \epsilon^{-1}$ are bounded by a polynomial, then the value-iteration algorithm requires polynomial-time to approximate; and hence if $\underline{H}$ and W are bounded by polynomial, then the value-iteration algorithm is a FPTAS. In particular, if either (i) r is constant and $(\delta_{\min})^{-1}$ is bounded by a polynomial, or (ii) $(\delta_{\min})^{-1}$ is bounded by a constant and r is logarithmic in n, then $\underline{H}$ is polynomial; and if W is polynomial as well, then the value-iteration algorithm is a FPTAS. There could also be other cases where $\underline{H}$ is polynomial, and then the value-iteration is a pseudo-polynomial time algorithm for constant-factor approximation.

5 Exact Value Problem for Almost-Sure Ergodic Games

We present two results for the exact value problem: (1) First we show that for almost-sure ergodic CMPGs the exact value can be expressed in the existential theory of the reals (for details about the existential theory see [6]). This is achieved by first showing that the fixpoint of the Hoffman-Karp algorithm can be expressed in the existential theory; and then combining it with a sentence in the existential theory for reachability games. (2) We establish that the value problem for sure ergodic CMPGs is square-root sum hard (using techniques similiar to [11]) generalizing the example shown in Figure 1.

Theorem 5. *(1) The value problem for almost-sure ergodic CMPGs can be expressed in the existential theory of the reals. (2) The value problem for sure ergodic CMPGs is square-root sum hard.*

References

1. ArXiv CoRR (2014), Full version http://arxiv.org/abs/1404.5734
2. Aldous, D.: Random walks on finite groups and rapidly mixing Markov chains. Lecture Notes in Mathematics, vol. 986, pp. 243–297. Springer, Berlin (1983)
3. Bewley, T., Kohlberg, E.: The asymptotic behavior of stochastic games. Math. Op. Res. (1) (1976)
4. Blackwell, D., Ferguson, T.: The big match. AMS 39, 159–163 (1968)
5. Boros, E., Elbassioni, K., Gurvich, V., Makino, K.: A potential reduction algorithm for two-person zero-sum limiting average payoff stochastic games. RUTCOR Research Report 13-2012 (2012)
6. Canny, J.F.: Some algebraic and geometric computations in PSPACE. In: STOC, pp. 460–467 (1988)
7. Chatterjee, K., Majumdar, R., Henzinger, T.A.: Stochastic limit-average games are in EXP-TIME. Int. J. Game Theory 37(2), 219–234 (2008)
8. Condon, A.: The complexity of stochastic games. I&C 96(2), 203–224 (1992)
9. de Alfaro, L., Majumdar, R.: Quantitative solution of omega-regular games. In: STOC 2001, pp. 675–683. ACM Press (2001)
10. Etessami, K., Yannakakis, M.: Recursive concurrent stochastic games. Logical Methods in Computer Science 4(4) (2008)
11. Etessami, K., Yannakakis, M.: On the complexity of nash equilibria and other fixed points. SIAM J. Comput. 39(6), 2531–2597 (2010)
12. Everett, H.: Recursive games. In: CTG. AMS, vol. 39, pp. 47–78 (1957)
13. Filar, J., Vrieze, K.: Competitive Markov Decision Processes. Springer (1997)
14. Gillette, D.: Stochastic games with zero stop probabilitites. In: CTG, pp. 179–188. Princeton University Press (1957)
15. Hansen, K.A., Ibsen-Jensen, R., Miltersen, P.B.: The complexity of solving reachability games using value and strategy iteration. In: Kulikov, A., Vereshchagin, N. (eds.) CSR 2011. LNCS, vol. 6651, pp. 77–90. Springer, Heidelberg (2011)
16. Hansen, K.A., Koucký, M., Lauritzen, N., Miltersen, P.B., Tsigaridas, E.P.: Exact algorithms for solving stochastic games: extended abstract. In: STOC, pp. 205–214 (2011)
17. Hoffman, A.J., Karp, R.M.: On nonterminating stochastic games. Management Science 12(5), 359–370 (1966)
18. Ibsen-Jensen, R.: Strategy complexity of two-player, zero-sum games. PhD thesis, Aarhus University (2013)
19. Mertens, J., Neyman, A.: Stochastic games. IJGT 10, 53–66 (1981)
20. Puterman, M.: Markov Decision Processes. John Wiley and Sons (1994)
21. Shapley, L.: Stochastic games. PNAS 39, 1095–1100 (1953)
22. Vardi, M.: Automatic verification of probabilistic concurrent finite-state systems. In: FOCS 1985, pp. 327–338. IEEE Computer Society Press (1985)
23. Zwick, U., Paterson, M.: The complexity of mean payoff games on graphs. Theoretical Computer Science 158, 343–359 (1996)

Toward a Structure Theory of Regular Infinitary Trace Languages

Namit Chaturvedi[*]

RWTH Aachen University, Lehrstuhl für Informatik 7, Aachen, Germany
chaturvedi@automata.rwth-aachen.de

Abstract. The family of regular languages of infinite words is structured into a hierarchy where each level is characterized by a class of
deterministic ω-automata – the class of deterministic Büchi automata
being the most prominent among them. In this paper, we analyze the
situation of regular languages of infinite Mazurkiewicz traces that model
non-terminating, concurrent behaviors of distributed systems. Here, a
corresponding classification is still missing. We introduce the model of
"synchronization-aware asynchronous automata", which allows us to initiate a classification of regular infinitary trace languages in a form that
is in nice correspondence to the case of ω-regular word languages.

1 Introduction

In the theory of ω-regular word languages, a natural classification is induced
by various forms of deterministic ω-automata. The three fundamental cases are
given by (a) deterministic Muller automata, capturing the class of ω-regular word
languages; (b) deterministic Büchi automata, capturing recurrence properties
of infinite words; and (c) weak automata, capturing reachability properties of
infinite words. In this paper, we concentrate on the first two automata models,
on which fundamental facts can be summarized as follows (see e.g. [8]):

1. A language is deterministically Büchi recognizable if and only if it can be
   expressed as $\lim(K) := \{\alpha \in \Sigma^\omega \mid \alpha$ has infinitely many prefixes in $K\}$ for
   some regular language $K \subseteq \Sigma^*$.
2. An ω-regular language is deterministically Büchi recognizable if and only if
   this language is recognized by a Muller automaton whose acceptance component is closed under supersets.
3. The class of Boolean combinations of deterministically Büchi recognizable
   languages coincides with the class of Muller recognizable languages.

We consider the question of defining corresponding classes in the framework
of Mazurkiewicz traces [4] that model infinite, concurrent behaviors of a finite
set of interacting processes. The concept of "ω-regular trace language" can be
introduced in close correspondence to the case of ω-regular word languages,

---

[*] Supported by the DFG Research Training Group-1298 AlgoSyn and the CASSTING
    Project funded by the European Commission's 7th Research Framework Programme.

J. Esparza et al. (Eds.): ICALP 2014, Part II, LNCS 8573, pp. 134–145, 2014.

for example[1], in terms of finite partially-commutative monoids, asynchronous automata, concurrent regular expressions, or MSO logic.

However, it is remarkable that there does not yet exist a definition of Büchi automaton over traces that allows for results analogous to any of the items 1–3 above. The objective of the present paper is to fill this gap, while making sure at all times that the corresponding definitions and results for word languages emerge from our study as special cases.

Muscholl [7] took a major step toward establishing such structural results by introducing a parameterized lim operator for trace languages. She showed that the class of Boolean combinations of parameterized lim-languages is precisely the class of ω-regular trace languages, and also characterized the class of linearizations of these parameterized languages in terms of "I-diamond" Büchi (word) automata with "extended" acceptance condition. The respective family of I-diamond automata characterizing Boolean combinations of linearizations of reachability languages (where an infinite trace is in the language if it contains a certain finite prefix) is studied in [2]. However, I-diamond word automata do not offer a proper modeling of concurrency as realized over traces.

We introduce a new concept of asynchronous automata, viz. *synchronization-aware asynchronous automata* (over traces rather than their linearizations). These, when equipped with Büchi and Muller acceptance conditions, establish not only item 1, but also items 2 and 3 above. At the same time, the synchronization-aware Muller automata are equivalent in expressive power to the standard deterministic asynchronous Muller automata for infinitary trace languages. Thus we provide a new framework that prepares – at least in important parts – a structure theory for ω-regular trace languages that is compatible with that of deterministic ω-automata over words.

Synchronization-aware automata are "aware" of the fact that during a run over an infinite trace, the set of processes may be partitioned in a manner that each part is minimal and, after a finite prefix, a process belonging to one part never interacts directly or indirectly with a process belonging to another part. The processes infer this partition by observing their infinitely recurring interactions. Although infinite traces induce such partitions in all asynchronous automata, current models cannot perform such inferencing.

Another aspect of infinite runs is that while some processes may remain *live* ad infinitum, others may halt after finitely many steps. However, the set of live processes can be explicitly coded in the Büchi acceptance condition since this directly corresponds to Muscholl's parameterized lim operation mentioned above.

By combining both these aspects, we obtain the family of synchronization-aware Büchi automata corresponding to item 1 above (see Thm. 13). We also introduce synchronization-aware Muller automata recognizing precisely the class of ω-regular trace languages (see Thm. 18). Finally, Theorems 20 and 21 respectively demonstrate a characterization à la item 2 and the equivalence result of item 3. We conclude with a discussion of a number of open problems.

---

[1] We refer the reader to [4] for a comprehensive survey of early results.

2 Preliminaries

2.1 Finite and Infinite Traces

Over a finite alphabet Σ, let $D \subseteq \Sigma^2$ be a binary, reflexive, and symmetric *dependence relation*. We also refer to the corresponding *independence relation* $I = \Sigma^2 \setminus D$, and to the *independence alphabet* (Σ, I). Given an independence alphabet, a *finite trace* is an isomorphism class of directed acyclic graphs $t = [V, \lessdot, \lambda]$ where V is a finite set of events; $\lambda \colon V \to \Sigma$ is a labeling function; and for events $e, e' \in V \colon \lambda(e) D \lambda(e') \Leftrightarrow e \lessdot e'$ or $e' \lessdot e$ or $e = e'$. The *concatenation* of two finite traces $t_1 = [V_1, \lessdot_1, \lambda_1]$ and $t_2 = [V_2, \lessdot_2, \lambda_2]$ is given by $t_1 \odot t_2 = [V_1 \uplus V_2, \lessdot', \lambda_1 \uplus \lambda_2]$, where $\lessdot' = \lessdot_1 \uplus \lessdot_2 \uplus \{(e_1, e_2) \in V_1 \times V_2 \mid \lambda_1(e_1) D \lambda_2(e_2)\}$. We denote the set of all finite traces over an alphabet (Σ, I) with $\mathbb{M}(\Sigma, I)$.

For convenience, we work with "simplified" traces $t = [V, \lessdot, \lambda]$ where we remove all edges that may be inferred from others, i.e. by $\lessdot$ we mean $\lessdot \setminus \lessdot^2$ (see Fig. 1a). We also refer to the partial order $<$ obtained from the transitive closure of this edge relation; and define relations $\leq$, $\gtrdot$, $\geq$, and $>$ in the natural manner. We use the abbreviation $e \in t$ to convey $t = [V, \lessdot, \lambda]$ and $e \in V$.

An *infinite trace* is a directed acyclic graph $\theta = [V, \lessdot, \lambda]$ where V is a countable set of events, and λ and $\lessdot$ are like above except $\lessdot$ satisfies an additional requirement, namely, for each $e \in \theta$, the set $\{e' \in \theta \mid e' \leq e\}$ is finite. Denote the set of all infinite traces with $\mathbb{R}(\Sigma, I)$. For traces $t \in \mathbb{M}(\Sigma, I), \theta \in \mathbb{R}(\Sigma, I)$, we refer to sets $\mathsf{alph}(t)$, $\mathsf{alph}(\theta)$ of letters occurring in them, and to the set $\mathsf{alphinf}(\theta)$ of letters occurring infinitely often in θ.

We say t_1 is a *prefix* of t_2, i.e. $t_1 \sqsubseteq t_2 \colon\Leftrightarrow \exists t' \colon t_2 = t_1 \odot t'$, and $t_1 \sqsubset t_2$ iff $t_1 \sqsubseteq t_2$ and $t_1 \neq t_2$. We also refer to prefixes t of some $\theta \in \mathbb{R}(\Sigma, I)$ in a similar way. If $E \subseteq t$ is a set of events, then $t[E] = [V', \lessdot', \lambda']$ is a prefix of t with the set $V' := \{f \in t \mid f \leq e \text{ for some } e \in E\}$ and $\lessdot'$ and λ' are obtained by restricting the corresponding entities in t to V'. The least upper bound of two traces t_1, t_2, whenever it exists, denoted $t_1 \sqcup t_2$ is the smallest trace s such that $t_1 \sqsubseteq s \wedge t_2 \sqsubseteq s$. Similarly, if it exists, the greatest lower bound of t_1 and t_2, denoted $t_1 \sqcap t_2$, is the largest trace s such that $s \sqsubseteq t_1 \wedge s \sqsubseteq t_2$.

2.2 Asynchronous Transition Systems

We refer to a deterministic asynchronous automaton as a pair $\mathfrak{A} = (\mathfrak{T}, \mathcal{F})$, where $\mathfrak{T}$ is a deterministic asynchronous transition system and $\mathcal{F}$ is an appropriate acceptance condition. We discuss these components separately.

For a fixed alphabet (Σ, I), an asynchronous transition system consists of a set $\mathcal{P}$ of *processes*, a mapping $\mathsf{dom} \colon \Sigma \to 2^{\mathcal{P}}$ assigning the *domain* of each letter such that $\bigcup_{a \in \Sigma} \mathsf{dom}(a) = \mathcal{P}$ and $a \mathbin{I} b \Leftrightarrow \mathsf{dom}(a) \cap \mathsf{dom}(b) = \emptyset$. Naturally, for $\Sigma' \subseteq \Sigma$, we also refer to $\mathsf{dom}(\Sigma') := \bigcup_{a \in \Sigma'} \mathsf{dom}(a)$. Moreover for an event $e \in t$, we refer to $\mathsf{dom}(e)$ instead of referring to $\mathsf{dom}(\lambda(e))$. Similarly, for $E \subseteq t$.

Processes p have sets X_p of *local p-states*. Introducing a symbol $\$ \notin \bigcup_{p \in \mathcal{P}} X_p$, for a set $P \subseteq \mathcal{P}$ the set X_P of *P-states* is a defined as $X_P := \{(x_p)_{p \in \mathcal{P}} \mid x_i \in X_{p_i} \text{ if } p_i \in P, \text{ otherwise } x_i = \$\}$. We find it convenient to assume an order over

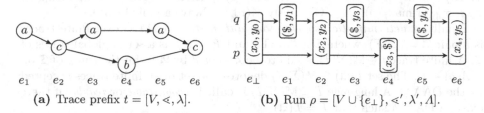

(a) Trace prefix $t = [V, <, \lambda]$.   **(b)** Run $\rho = [V \cup \{e_\perp\}, <', \lambda', \Lambda]$.

Fig. 1. For $\Sigma = \{a, b, c\}$, $a \, I \, b$, a finite trace (prefix) $t \in \mathbb{M}(\Sigma, I)$ and the run ρ of an ATS, with $\mathsf{dom}(a) = \{q\}, \mathsf{dom}(b) = \{p\}$, and $\mathsf{dom}(c) = \{p, q\}$

$\mathcal{P}$ and view a P-state as a tuple. So we refer to a *state* as a tuple $\pi \in X_P$ for some $P \subseteq \mathcal{P}$. A state is a *global state* if $P = \mathcal{P}$. We always distinguish between a $\{p\}$-state π and a local p-state x; and for a state π, define the *p-state in* π as $\pi_{|p} := x_p \in X_p \cup \{\$\}$, and similarly the *$P$-state* $\pi_{|P}$ *in* π. Also, $\mathsf{dom}(\pi) := \{p \in \mathcal{P} \mid \pi_{|p} \neq \$\}$. Finally, we denote the set of all states $X_{2^\mathcal{P}} := \bigcup_{P \subseteq \mathcal{P}} X_P$.

We now define a *deterministic asynchronous transition system* (an *ATS*) as a tuple $\mathfrak{T} = ((X_p)_{p \in \mathcal{P}}, (\delta_a)_{a \in \Sigma}, \pi_0)$, where X_p are sets of local p-states; transition functions $\delta_a \colon X_{\mathsf{dom}(a)} \to X_{\mathsf{dom}(a)}$ define how processes jointly perform state transitions on input letters a; and $\pi_0 \in X_\mathcal{P}$ is the global initial state of $\mathfrak{T}$.

Given a trace $t = [V, <, \lambda] \in \mathbb{M}(\Sigma, I)$, or $\theta = [V, <, \lambda] \in \mathbb{R}(\Sigma, I)$, we define the corresponding *run* $\rho = [V', <', \lambda', \Lambda]$ of $\mathfrak{T}$ on the trace where $V' := V \cup \{e_\perp\}$ contains a fictional, minimum event $e_\perp$. The relation $<'$ is identical to the edge relation $<$, except that $e_\perp$ is the unique minimum event.

During the run ρ of an ATS $\mathfrak{T}$ over a trace, each process p makes state transitions on events $e \in \mathsf{dom}^{-1}(p)$. Each such event may be called a *p-event* as well as a *P-event* where $P = \mathsf{dom}(e)$. All p-events in the run are totally ordered, and this order $<'_p$ can be defined with the help of the order $<$ of the trace. The *maximum p-event in* ρ according to the ordering $<'_p$ is denoted as $\max_p(\rho) \geq e_\perp$. If it exists, the p-predecessor f of an event e is denoted by $f <'_p e$. The labeling λ' is defined similarly except $\lambda'(e_\perp) := \epsilon$; and $\Lambda \colon V' \to X_{2^\mathcal{P}}$ is defined inductively:

- $\Lambda(e_\perp) := (\pi_0)$,
- for any $e >' e_\perp$, if 1. $a = \lambda(e)$, and 2. for $e_p <'_p e$, if $x_p = \Lambda(e_p)_{|p}$ are the most recent p-states just before e, then $\Lambda(e) := (y_p)_{p \in \mathcal{P}}$, where the local-state $y_p = \delta_a((x_p)_{p \in \mathsf{dom}(e)})_{|p}$ if $p \in \mathsf{dom}(e)$, $y_p = \$$ otherwise.

Fig. 1 shows the labeled events of a trace and the corresponding run; but λ' is omitted in ρ for readability. The processes are assumed to be totally ordered, hence the representation of states as tuples. Note that, in Fig. 1b, the edges are shown as per the relations $<'_p, p \in \mathcal{P}$. Importantly, although $e_\perp <' e_2$ and $e_\perp <'_p e_2$, it is not the case that $e_\perp <' e_2$.

Analogous to trace prefixes, we refer to run prefixes, and to prefixes $\rho[e], \rho[E]$ for $e \in \rho$ and $E \subseteq \rho$ respectively. For $e \in \rho$, we also refer to the label $\Lambda(e)$ as the *state of $\mathfrak{T}$ at* e. Similarly, if ρ is a finite run, then the *state of $\mathfrak{T}$ at* ρ is given by $\Lambda(\rho) = (x_p)_{p \in \mathcal{P}}$ where $x_p := \Lambda(\max_p(\rho))_{|p}$ is the p-state of $\mathfrak{T}$ at $\max_p(\rho)$; $x_p = \pi_{0|p}$ if $\max_p(\rho) = e_\perp$. Obviously, $\Lambda(\rho)$ is always a global state.

Finally, a *deterministic asynchronous automaton* (a *DAA*) over finite traces is a pair $\mathfrak{A} = (\mathfrak{T}, F)$, where $\mathfrak{T}$ is an ATS and $F \subseteq X_{\mathcal{P}}$ is a set of global states of $\mathfrak{T}$. A finite trace $t \in \mathbb{M}(\Sigma, I)$ is said to be *accepted* by $\mathfrak{A}$ if, for the run ρ of $\mathfrak{T}$ on t, $\Lambda(\rho) \in F$. The set $L(\mathfrak{A}) \subseteq \mathbb{M}(\Sigma, I)$ denotes the set of all finite traces accepted by the DAA $\mathfrak{A}$. A language $T \subseteq \mathbb{M}(\Sigma, I)$ is called *recognizable* or *regular* if there exists a DAA $\mathfrak{A}$ such that $T = L(\mathfrak{A})$.

2.3 Regular Infinitary Languages

The definition of regular languages of infinite traces, ω-*regular trace languages*, was first provided by Gastin-Petit using monoid morphisms [5]. We use as definition, a characterization of the same family in terms of deterministic asynchronous (cellular) Muller automata [3,7]. The notion of acceptance of an infinite trace $\theta \in \mathbb{R}(\Sigma, I)$ by an ATS $\mathfrak{T}$ is defined by referring to the *local infinity sets* $\mathsf{Inf}_p(\rho)$ of local p-states that occur infinitely often during the run ρ of $\mathfrak{T}$ over θ, with

$$\mathsf{Inf}_p(\rho) := \begin{cases} \left\{ x \in X_p \mid \exists^\infty e \in \rho : \Lambda(e)_{|p} = x \right\} & \text{if } p \in \mathsf{dom}(\mathsf{alphinf}(\theta)), \\ \left\{ x \in X_p \middle| \begin{array}{l} \exists e \in \rho : e = \max_p(\rho) \\ \text{and } \Lambda(e)_{|p} = x \end{array} \right\} & \text{otherwise.} \end{cases}$$

Let $\mathcal{F} = \{F_1, F_2, \dots\}$ be a table where each $F_i = (F_i^p)_{p \in \mathcal{P}}$ is a tuple of sets of local states of the processes. A *deterministic asynchronous Büchi automaton* (a *DABA*) is a pair $\mathfrak{A} = (\mathfrak{T}, \mathcal{F})$. A DABA is said to accept a trace $\theta \in \mathbb{R}(\Sigma, I)$ if, on the run ρ of $\mathfrak{A}$ on θ, there exists a tuple $F_i \in \mathcal{F}$ such that for each process p, $F_i^p \subseteq \mathsf{Inf}_p(\rho)$ [5,3]. A *deterministic asynchronous Muller automaton* (a *DAMA*) is a pair $\mathfrak{A} = (\mathfrak{T}, \mathcal{F})$, and is said to accept a trace θ if there exists a tuple $F_i \in \mathcal{F}$ such that for each process p, $F_i^p = \mathsf{Inf}_p(\rho)$ [3].

Definition 1. *A language $\Theta \subseteq \mathbb{R}(\Sigma, I)$ is said to be a* regular infinitary language *(or an ω-regular trace language) if it is recognized by a DAMA.*

Definition 2 ([3]). *For a language $T \subseteq \mathbb{M}(\Sigma, I)$ finite traces, the* infinitary limit *of T, denoted $\mathsf{lim}(T)$, is the language containing traces $\theta \in \mathbb{R}(\Sigma, I)$ such that there exists a sequence $(t_i)_{i \in \mathbb{N}}, t_i \in T$ satisfying $t_i \sqsubset t_{i+1}$ and $\bigsqcup_{i \in \mathbb{N}} t_i = \theta$.*

Fig. 2 illustrates the definition of $\mathsf{lim}(T)$ with the help of an infinite run of an asynchronous automaton recognizing T. Fig. 2a illustrates an induced run if the trace $\theta \notin \mathsf{lim}(T)$, whereas Fig. 2b illustrates the contrary.

Muscholl studies infinitary limits that are parameterized by a set of letters. This set governs which letters from the alphabet must occur infinitely often in the traces, and which letters may not. Recalling the dependence relation $D \subseteq \Sigma^2$, for a set $A \subseteq \Sigma$ we define $D(A) := \{a \in \Sigma \mid \exists b \in A : aDb\}$.

Definition 3 ([7]). *For $T \subseteq \mathbb{M}(\Sigma, I)$ and some $A \subseteq \Sigma$, the A-infinitary limit of T is defined as $\mathsf{lim}_A(T) := \{\theta \in \mathsf{lim}(T) \mid D(\mathsf{alphinf}(\theta)) = D(A)\}$.*

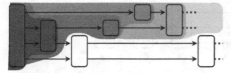 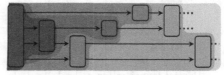

(a) $t_i \in T$ and $t_i \sqsubset t_{i+1}$, but $\theta \notin \lim(T)$ (b) $\theta \in \lim(T)$ since each event is eventu-
since $\bigsqcup_{i \in \mathbb{N}} t_i \neq \theta$.   ally covered by an accepting prefix.

Fig. 2. Illustrating Def. 2. Shaded regions constitute sequences of accepting runs.

Definition 4 ([7]). *An ω-regular trace language is called a* deterministic trace language *if it can be expressed as a finite union $\bigcup_i \lim_{A_i}(T_i)$ for regular trace languages $T_i \subseteq \mathrm{M}(\Sigma, I)$ and sets $A_i \subseteq \Sigma$.*

Clearly, the language $\lim(T)$ is a deterministic trace language since $\lim(T) = \bigcup_{A \subseteq \Sigma} \lim_A(T)$. However, not every deterministic trace language can be expressed in the form $\lim(T)$ for any T.

It is still open whether there exists a DABA recognizing the language $\lim(T)$ for any given regular trace language $T \subseteq \mathrm{M}(\Sigma, I)$. Furthermore, there exist deterministic trace languages that are not accepted by any DABA [7]. In this regard the term "deterministic trace language" [7] is not well motivated, since it has no equivalent in any of the classes of deterministic asynchronous ω-automata known so far. The results of this paper justify this term by providing a matching class of deterministic, "synchronization-aware" Büchi automata.

2.4   Secondaries and Frontiers

During a run ρ of an ATS, the processes can be thought of as "possessing and updating information" regarding other processes [6]. If ρ is finite and $p, q \in \mathcal{P}$, the *first-hand information* that p has about q at ρ, denoted by $\mathsf{latest}_{p \to q}(\rho)$, is the maximal q-event in the prefix $\rho[\max_p(\rho)]$. Trivially, $\mathsf{latest}_{p \to p}(\rho) = \max_p(\rho)$. Similarly, for $p, q, r \in \mathcal{P}$, the *second-hand information* that p has about r via q at ρ, denoted by $\mathsf{latest}_{p \to q \to r}(\rho)$, is the maximal r-event in the prefix $\rho[\mathsf{latest}_{p \to q}(\rho)]$. Trivially, $\mathsf{latest}_{p \to p \to q}(\rho) = \mathsf{latest}_{p \to q}(\rho)$.

The *primary information* of p at ρ is defined as the ordered set $\mathsf{Pri}_p(\rho) := \{\mathsf{latest}_{p \to q}(\rho) \mid q \in \mathcal{P}\}$. The *secondary information* of p at ρ is given by the set $\mathsf{Sec}_p(\rho) := \{\mathsf{latest}_{p \to q \to r}(\rho) \mid q, r \in \mathcal{P}\}$. It is easy to see that on the one hand $\mathsf{Pri}_p(\rho) \subseteq \mathsf{Sec}_p(\rho)$, on the other hand the events of $\mathsf{Sec}_p(\rho)$ may be ordered as per the partial order $<$ of ρ. This gives us a view of the *secondary graph* of p at ρ, which we identify with secondary information itself. In this paper, we are mainly interested in secondary information of the form $\mathsf{Sec}_p(\rho[e])$ for $p \in \mathrm{dom}(e)$. Since, $\mathsf{Sec}_p(\rho[e]) = \mathsf{Sec}_q(\rho[e])$ for all $p, q \in \mathrm{dom}(e)$, for convenience we denote this information simply as $\mathsf{Sec}(e)$.

While referring to finite runs ρ over finite traces, or over finite prefixes of infinite traces, it is useful to refer to their maximum p-events as a set. Define

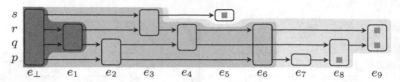

Partial frontiers for ρ: $\{e_5\}$, $\{e_9\}$, $\{e_5, e_9\}$, $\{e_8, e_9\}$, and $\{e_5, e_8, e_9\}$.
At e_4, $\rho_\sqcap = \rho[e_1]$; and at e_9, $\rho_\sqcap = \rho[e_6]$. Note that $e_5 \notin \rho[e_9]$.

Fig. 3. Partial frontiers (see below); and illustration of Lemma 6 (see Ex. 7)

frontier of ρ as $H_\rho := \{e \in \rho \mid \exists p \in \mathcal{P}, e = \max_p(\rho)\}$. Any set $H \subseteq H_\rho$ is called a *partial frontier* if it is upward closed with respect to the $<$ order over the events. E.g., the set $\{e_5, e_8\}$ in Fig. 3 is not a partial frontier of ρ since it is not an upward closed subset of the frontier $\{e_5, e_8, e_9\}$.

Finally, for event $e \in \rho$, define the *top of e in ρ* as $\top_\rho(e) := \{f \in \rho \mid e \leq f \wedge \exists p \in \mathcal{P}: f = \max_p(\rho)\}$. Of course for any $e_1, \ldots, e_n \in \rho$, $\bigcup_{i=1}^{n} \top_\rho(e_i)$ is a partial frontier of ρ. If $\Lambda(\rho)$ is the global state of an automaton, and if H is a (partial) frontier of ρ, then we define $\Lambda(H) := \Lambda(\rho)_{|\mathsf{dom}(H)}$. Roughly speaking, identifying a reasonable set of partial frontiers is necessary and sufficient for computing the global state at the end of a finite run.

3 A New Model of Asynchronous Automata

Any infinite run ρ of an ATS $\mathfrak{T}$ over a trace $\theta \in \mathbb{R}(\Sigma, I)$ yields a partition $\Psi = (P_1, \ldots, P_n)$ of set $\mathcal{P}$ of processes such that each part $P_i \subseteq \mathcal{P}$ is minimal, and after finite prefixes $\rho_i \sqsubset \rho$, the processes $p \in P_i$ no longer interact directly or indirectly with another process $p' \in P_j, i \neq j$. We wish to obtain a family of ATS's where each process can infer during a run the part to which it belongs. Owing to space restrictions, we present a concise discussion here, and refer the reader to [1] for details and for proofs of all the claims made in this section.

3.1 Degrees of Synchronization

For an ATS $\mathfrak{T}$ and a run ρ of $\mathfrak{T}$ over any trace, we associate with each event $e \in \rho$ a measure (cf. Def. 5) of how much information is exchanged among the processes in $\mathsf{dom}(e)$. We use sets $P \subseteq \mathcal{P}$ of processes as the gauge for this measure.

Definition 5. *For a run ρ of an ATS and an event $e \in \rho$, let the* secondary update at e *be the set* $\mathcal{U}_e := \{g \in \rho[e] \mid \exists p, q, r \in \mathcal{P}, \exists f_p \lessdot_p e : g = \mathsf{latest}_{p \to q \to r}(f_p) \neq \mathsf{latest}_{p \to q \to r}(e)\}$. *Then, the* degree of synchronization at e *is defined as as the set* $\mathsf{ds}(e) := \bigcup_{g \in \mathcal{U}_e} \mathsf{dom}(\top_{\rho[e]}(g))$. *By default,* $\mathsf{ds}(e_\perp) := \mathcal{P}$.

The set $\mathsf{ds}(e)$ implies that there must exist prefixes $\rho' \sqsubseteq \rho[e]$ with partial frontiers H, $\mathsf{dom}(H) = \mathsf{ds}(e)$, such that for some process $p \in \mathsf{dom}(e)$ with a predecessor $f_p \lessdot_p e$, $H \not\subseteq \rho[f_p]$. The following lemma illustrates this point, and demonstrates the importance of the set $\mathcal{U}_e$.

Lemma 6. *For $e \in \rho$, $e > e_\perp$, let $\rho_\sqcap := \prod_{f_p <_p e} \rho[f_p]$ be the greatest lower bound of all its p-prefixes. For every prefix $\rho' \sqsubseteq \rho[e]$ with $\rho' \not\sqsubseteq \rho_\sqcap$, there exist $H \subseteq \rho'$ and $U \subseteq \mathcal{U}_e$ such that 1. H is a partial frontier in ρ' with $\mathsf{dom}(H) = \mathsf{ds}(e)$; and 2. $\bigcup_{g \in U} \top_{\rho'}(g) = H$.*

Example 7. Referring to Fig. 3, at e_4, we have $e_2 <_q e_4$ and $e_3 <_r e_4$. Then, $\mathsf{ds}(e_4) = \mathcal{P}$ because $\mathcal{U}_{e_4} = \{e_\perp, e_1, e_2, e_3\}$. For instance $e_\perp = \mathsf{latest}_{q \to r \to s}(e_2) \neq \mathsf{latest}_{q \to r \to s}(e_4)$. Since $\rho_\sqcap = \rho[e_1]$, we have four possibilities of ρ', viz. $\rho'_1 = \rho[e_4]$, $\rho'_2 = \rho[e_2, e_3]$, $\rho'_3 = \rho[e_3]$, and $\rho'_4 = \rho[e_2]$. For ρ'_4, $H = \{e_\perp, e_1, e_2\}$ and we can choose $U = e_\perp \subseteq \mathcal{U}_{e_4}$. Symmetrically for ρ'_3. Also verify that, for ρ'_2, $H = U = \{e_2, e_3\}$; and for ρ'_1, $H = \{e_2, e_3, e_4\}$ and $U = \{e_\perp\}$.

Considering e_9 next, we have $e_8 <_q e_9$, $e_6 <_r e_9$, and $\mathcal{U}_{e_9} = \{e_2, e_4, e_6, e_8\}$. For instance, $e_2 = \mathsf{latest}_{r \to q \to p}(e_6) \neq \mathsf{latest}_{r \to q \to p}(e_9) = e_8$. Clearly, $\mathsf{ds}(e_9) = \{p, q, r\}$. And since $\rho_\sqcap = \rho[e_6]$, we have three possibilities of $\rho' \sqsubseteq \rho[e_9]$ s.t. $\rho' \not\sqsubseteq \rho_\sqcap$, the most interesting one being $\rho' = \rho[e_7]$. Now $H = \{e_4, e_6, e_7\}$ is the partial frontier of $\rho[e_7]$ with $\mathsf{dom}(H) = \mathsf{ds}(e_9)$, so we choose $U = \{e_2\} \subseteq \mathcal{U}_{e_9}$. ⊠

Remark 8. If $\mathcal{M}_e$ is the set of the (mutually concurrent) minimal events of $\mathcal{U}_e$, then it suffices to always consider $U = \mathcal{M}_e$ in Lemma 6.

Why we are interested in precisely these frontiers will be clear from Lemma 9 and Remark 10 below. Presently, with respect to the partial frontiers H that are revealed by Lemma 6 at an event e, we refer to the set Y_e of states $\Lambda(H)$ as the *yield at* e. Clearly, for each $\pi_1, \pi_2 \in Y_e$: $\mathsf{dom}(\pi_1) = \mathsf{dom}(\pi_2) = \mathsf{ds}(e)$. We say that the yield Y_e is *bigger* than yield Y_f if $\mathsf{ds}(f) \subsetneq \mathsf{ds}(e)$.

Lemma 9. *For an infinite run ρ and $p \in \mathcal{P}$, if $p \in \mathsf{dom}(\mathsf{alphinf}(\rho))$ then there exists a unique maximal $P \subseteq \mathcal{P}$ such that $\exists^\infty e \in \rho : p \in \mathsf{dom}(e) \wedge \mathsf{ds}(e) = P$.*

We call the set P from Lemma 9 the *max-degree of p-synchronizations in* ρ, denoted by $\lceil \mathsf{ds}_p(\rho) \rceil$. For processes $p \notin \mathsf{dom}(\mathsf{alphinf}(\rho))$ that eventually halt, we define $\lceil \mathsf{ds}_p(\rho) \rceil := \{p\}$ regardless of the value of $\mathsf{ds}(\mathsf{max}_p(\rho))$ The following remark follows immediately from Lemma 9, and demonstrates the "symmetric" nature of max-degree of synchronizations.

Remark 10. For an infinite run ρ and $p, q \in \mathcal{P}$, either $\lceil \mathsf{ds}_p(\rho) \rceil = \lceil \mathsf{ds}_q(\rho) \rceil$ or $\lceil \mathsf{ds}_p(\rho) \rceil \cap \lceil \mathsf{ds}_q(\rho) \rceil = \emptyset$.

In particular, for each part $P_i \in \Psi : q \in P_i \Leftrightarrow \lceil \mathsf{ds}_q(\rho) \rceil = P_i$. This concretizes our observation that every run ρ induces a partition Ψ of the set of states, where each part is minimal.

Definition 11. *A synchronization-aware transition system (an SATS) is a pair $(\mathfrak{T}, \mathcal{D})$ where $\mathfrak{T} = ((X_p)_{p \in \mathcal{P}}, (\delta_a)_{a \in \Sigma}, \pi_0)$ is an ATS and $\mathcal{D} = (\mathcal{D}_p)_{p \in \mathcal{P}}$ is a collection of mappings $\mathcal{D}_p : X_p \to 2^{\mathcal{P}}$ such that 1. $\mathcal{D}_p(\pi_{0|p}) = \mathcal{P}$, and 2. for every run ρ of $\mathfrak{T}$ and every event $e \in \rho$, if $\Lambda(e) = \pi$ and $p \in \mathsf{dom}(e)$ then $\mathsf{ds}(e) = P \Leftrightarrow \mathcal{D}_p(\pi_{|p}) = P$.*

This definition implies that the local p-states of an SATS always match the degrees of synchronization of events where they occur. It is easy to see that property 2 therein is in fact decidable, whence the definition is "syntactic".

3.2   Synchronization-aware Asynchronous Büchi Automata

A set $X \subseteq X_p$ of local p-states is called *homosynchronous* if for all local p-states $x, y \in X$: $\mathcal{D}_p(x) = \mathcal{D}_p(y)$. For an infinite run ρ of an SATS, we define the homosynchronous *maximal local infinity sets* $\lceil \mathsf{Inf}_p(\rho) \rceil$ as follows.

$$
\lceil \mathsf{Inf}_p(\rho) \rceil := \begin{cases} \left\{ x \in X_p \middle| \begin{array}{l} \mathcal{D}_p(x) = \lceil \mathsf{ds}_p(\rho) \rceil \text{ and} \\ \exists^\infty e \in \rho : \Lambda(e)_{|p} = x \end{array} \right\} & \text{if } p \in \mathsf{dom}(\mathsf{alphinf}(\theta)), \\[4ex] \left\{ x \in X_p \middle| \begin{array}{l} \exists e \in \rho : e = \mathsf{max}_p(\rho) \\ \text{and } \Lambda(e)_{|p} = x \end{array} \right\} & \text{otherwise.} \end{cases}
$$

Definition 12. *A deterministic, synchronization-aware asynchronous Büchi automaton (a D-SABA) is a tuple* $\mathfrak{A} = (\mathfrak{T}, \mathcal{D}, \mathcal{F})$, *where* $(\mathfrak{T}, \mathcal{D})$ *is an SATS, and the acceptance table* $\mathcal{F} = \{(Q_1, F_1), \ldots (Q_k, F_k)\}$ *is such that each* $Q_i \subseteq \mathcal{P}$ *and* $F_i = (F_i^p)_{p \in \mathcal{P}}$ *is a tuple of homosynchronous sets* F_i^p. *A D-SABA* $\mathfrak{A}$ *accepts a trace* $\theta \in \mathbb{R}(\Sigma, I)$ *if, for the run* ρ *of* $\mathfrak{A}$ *on* θ, *there exists a pair* $(Q_i, F_i) \in \mathcal{F}$ *s.t.* $\mathsf{dom}(\mathsf{alphinf}(\theta)) = Q_i$ *and for each process* $p \in \mathcal{P}$: $F_i^p \cap \lceil \mathsf{Inf}_p(\rho) \rceil \neq \emptyset$.

The above definition essentially requires that processes p ignore all of their infinitely occurring local p-states except those whose image under $\mathcal{D}_p$ matches the maximal degree of p-synchronizations. One of our main results is as follows.

Theorem 13. *A language* $\Theta \subseteq \mathbb{R}(\Sigma, I)$ *is recognized by a D-SABA iff* Θ *is a deterministic trace language, i.e.* Θ *can be expressed as a finite union of languages of the form* $\mathsf{lim}_A(T)$ *for regular languages* $T \subseteq \mathbb{M}(\Sigma, I)$ *and sets* $A \subseteq \Sigma$.

We prove this claim by breaking it up into Lemmas 14 and 15, and Prop. 16.

Lemma 14. *Given a regular trace language* $T \subseteq \mathbb{M}(\Sigma, I)$ *and a set* $A \subseteq \Sigma$, *there exists a D-SABA accepting* $\Theta = \mathsf{lim}_A(T)$.

To prove Lemma 14, we start with a DAA $\mathfrak{A} = (\mathfrak{T}, F)$ recognizing T and construct a D-SABA $\mathfrak{A}' = (\mathfrak{T}', \mathcal{D}, \mathcal{F})$ where $(\mathfrak{T}', \mathcal{D})$ is an SATS such that over every trace $\theta \in \mathbb{R}(\Sigma, I)$ (a) the run ρ' of $\mathfrak{T}'$ mimics the run ρ of $\mathfrak{T}$; and (b) at each event $e \in \rho'$, $\mathfrak{T}'$ computes the yield Y_e for the corresponding event $e \in \rho$.

Fig. 4 illustrates a run ρ induced by a trace $\theta \in \mathsf{lim}_A(T)$ on $\mathfrak{A}$. The shaded regions represent the partition Ψ of $\mathcal{P}$ induced by ρ. Note that $f = \mathsf{max}_s(\rho)$ and $\lceil \mathsf{ds}_s(\rho) \rceil = \{s\}$ even though $\mathsf{ds}(f) = \mathcal{P}$. It is easy to see here that all partial frontiers H' in the top region are concurrent to all partial frontiers H'' in the bottom region. This means that $H' \cup H'' \cup \{f\}$ are partial frontiers of some prefixes $t \sqsubseteq \rho$. In particular, if $\mathsf{dom}(H'), \mathsf{dom}(H'') \in \Psi$ then $H := H' \cup H'' \cup \{f\}$ is a frontier, and $\Lambda(H)$ is the global state at t.

Lemma 6 helps in retroactively computing partial frontiers. One can verify that $\mathsf{ds}(e_5) = \{p, q, r\}$ and $H'' = \{e_1, e_2, e_3\}$ is one of the partial frontiers computed at e_5. Then $\Lambda(H'')$ belongs to the yield Y_{e_5} at e_5. Similarly, at g we have $Y_g = \{\Lambda(g)\}$. Lastly, if $\pi_s = \Lambda(f)_{|\{s\}}$ is the $\{s\}$-state at f, then by "joining" the

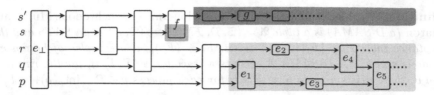

Fig. 4. Processes eventually halt, or settle in maximally interacting sets

yields Y_{e_5} and Y_g with π_s, we obtain a set Π of global states which contains the state $\Lambda(H)$ at prefix $\rho[H] \sqsubset \rho$, for $H = \{e_1, e_2, e_3, f, g\}$.

However, such computations of global states are only required at the "end" of the infinite run ρ. By joining π_s with the maximal yields that occur infinitely often (as guaranteed by Lemma 9 and Remark 10), $\mathfrak{T}'$ can compute precisely the set of global states occurring infinitely often in the run ρ of $\mathfrak{A}$.

Consequently, a local p-state of the SATS $\mathfrak{T}'$ is of the form $\overline{x} = (x, \overline{\mathsf{Sec}}, Y)$, where x is a local p-state of $\mathfrak{T}$, $\overline{\mathsf{Sec}}$ is a finite data structure to help compute the yields, and Y is a yield. $\mathfrak{T}'$ ensures that for each $e \in \rho'$ of $\mathfrak{T}'$, $\Lambda(e)_{|p} = (x, \overline{\mathsf{Sec}}, Y)$ iff for the corresponding $e \in \rho$ of $\mathfrak{T}$, $\Lambda(e)_p = x$ and $Y = Y_e$ is the yield at e.

Since the set $Q := \mathsf{dom}(A)$ of "live" processes is given, $\mathfrak{T}'$ can distinguish between the cases, e.g., that $p \in \mathsf{dom}(\mathsf{alphinf}(\theta))$ and $s \notin \mathsf{dom}(\mathsf{alphinf}(\theta))$ as shown in Fig. 4. By observing the sets $\lceil \mathsf{Inf}_p(\rho') \rceil, p \in \mathcal{P}$ in its run ρ', $\mathfrak{T}'$ can extract (a) the infinitely recurring maximal yields Y_p of $\mathfrak{T}$, from infinitely recurring maximal p-states $\overline{x}$ of live processes p; and (b) the final $\{p\}$-states π_p of $\mathfrak{T}$, from the final p-states $\overline{x}$ for processes p that halt.

Thus, $\mathfrak{T}'$ computes the set Π of global states occurring infinitely often in the run ρ of $\mathfrak{A}$. The run ρ' of $\mathfrak{T}'$ is accepting if Π has a non-empty intersection with the acceptance set F of $\mathfrak{A}$. The Büchi acceptance table $\mathcal{F} = \{(Q, F_1), \ldots (Q, F_k)\}$ is defined accordingly. For precise construction and proofs, see [1].

Lemma 15. *If $\mathfrak{A} = (\mathfrak{T}, \mathcal{D}, \mathcal{F})$ is a D-SABA with $|\mathcal{F}| = 1$ and $L(\mathfrak{A}) = \Theta$, then there exists a set $A \subseteq \Sigma$ and $T \subseteq \mathbb{M}(\Sigma, I)$ regular such that $\Theta = \lim_A(T)$.*

The proof of this lemma relies on constructing a non-deterministic asynchronous automaton recognizing the language T such that if $\mathcal{F} = \{(Q, F)\}$ then for $A := \mathsf{dom}^{-1}(Q) \setminus \mathsf{dom}^{-1}(\mathcal{P} \setminus Q)$ it holds that $\Theta = \lim_A(T)$ (cf. [1]).

Proposition 16. *The family of D-SABA-recognizable languages is closed under finite unions.*

Hence, Thm. 13 follows. Lastly, following the result established for the class of deterministic trace languages in [7], one obtains that the family of D-SABA-recognizable languages is also closed under finite intersections.

3.3 Synchronization-aware Asynchronous Muller Automata

We now define the class of synchronization-aware asynchronous Muller automata that accept precisely the ω-regular trace languages.

Definition 17. *A deterministic synchronization-aware asynchronous Muller automaton (a D-SAMA) is a tuple* $\mathfrak{A} = (\mathfrak{T}, \mathcal{D}, \mathcal{F})$, *where* $(\mathfrak{T}, \mathcal{D})$ *is an SATS and the acceptance table* $\mathcal{F} = \{F_1, \ldots F_k\}$ *is s.t.* $F_i = (F_i^p)_{p \in \mathcal{P}}$ *are tuples of homosynchronous sets* F_i^p. *A D-SAMA* $\mathfrak{A}$ *accepts a trace* $\theta \in \mathbb{R}(\Sigma, I)$ *if, for the run* ρ *of* $\mathfrak{A}$ *on* θ, *there exists a tuple* $F_i \in \mathcal{F}$ *s.t. for each process* $p \in \mathcal{P}$: $\lceil \mathsf{Inf}_p(\rho) \rceil = F_i^p$.

Theorem 18. *Any language* $\Theta \subseteq \mathbb{R}(\Sigma, I)$ *of infinite traces is recognized by a D-SAMA if and only if* Θ *is recognized by a DAMA.*

The proofs of this theorem and of the result that the family of D-SAMAs is closed under Boolean operations may be found in [1].

4   Characterization of Deterministic Büchi Recognizability

A prominent result on ω-regular word languages, due to Landweber [8], states that a language $L \subseteq \Sigma^\omega$ is deterministically Büchi recognizable iff for some (in fact, for each) deterministic Muller automaton recognizing L the acceptance component – assuming it contains only realizable sets – is closed under supersets. The stronger (bracketed) version supplies a decision procedure for Büchi recognizability of ω-regular languages. Here we present a weaker existential characterization over infinite traces. We define supersets in a manner that retains the essence of acceptance tables. Consider $F_1 = (F_1^p)_{p \in \mathcal{P}}$ and $F_2 = (F_2^p)_{p \in \mathcal{P}}$ from $\mathcal{F}$ where both F_1 and F_2 are tuples of homosynchronous sets F_1^p and $F_2^p, p \in \mathcal{P}$. We say that F_1 *is a superset of* F_2 denoted $F_1 \supseteq F_2$ if for each $p \in \mathcal{P}$, $F_1^p \supseteq F_2^p$. A table $\mathcal{F}$ is said to be *closed under supersets* if $\big((F \in \mathcal{F}) \wedge (F' \supseteq F)\big) \Rightarrow (F' \in \mathcal{F})$.

While discussing the closure under supersets, we must exempt the acceptance tuples that guarantee the halting of some processes. Let $F \in \mathcal{F}$ be a realizable acceptance tuple with $F^p = \{x\} \subseteq X_p$ for some $p \in \mathcal{P}$. Process p is guaranteed to halt during any run ρ that is accepted by referring to F only if it is the case that during two successive visits to x, p must visit another state $y \in X_p$ such that $\mathcal{D}_p(y) \not\subseteq \mathcal{D}_p(x)$. Then p must halt because otherwise, either $\lceil \mathsf{ds}_p(\rho) \rceil \supsetneq \mathcal{D}_p(x)$ or $\lceil \mathsf{Inf}_p(\rho) \rceil \supsetneq \{x\}$. Such a singleton F^p is referred to as a *finitary acceptance set*.

Definition 19. *A Muller acceptance table* $\mathcal{F}$ *is said to be* closed under supersets modulo finitary acceptance sets *if (a) whenever* $F \in \mathcal{F}$ *does not contain any finitary acceptance sets and* $F' \supseteq F$, *then* $F' \in \mathcal{F}$; *and (b) whenever* $F \in \mathcal{F}$ *contains a finitary acceptance set* F^p *and* $F' \supseteq F$ *with* $F'^p = F^p$, *then* $F' \in \mathcal{F}$.

Theorem 20. *A language* Θ *is recognized by a D-SABA* $\mathfrak{B} = (\mathfrak{T}', \mathcal{D}', \mathcal{F}')$ *if and only if* Θ *is recognized by a D-SAMA* $\mathfrak{A} = (\mathfrak{T}, \mathcal{D}, \mathcal{F})$ *whose acceptance table* $\mathcal{F}$ *is closed under supersets modulo finitary acceptance sets.*

As mentioned previously, every ω-regular trace language can be written as a finite Boolean combination of A-infinitary limit languages [3]. Our results allow us to state an equivalent claim by referring to classes of automata.

Theorem 21. *For any language* $\Theta \subseteq \mathbb{R}(\Sigma, I)$ *of infinite traces,* Θ *is D-SAMA recognizable if and only if* Θ *can be expressed as a finite Boolean combination of D-SABA recognizable languages.*

5   Conclusion

We introduced synchronization-aware asynchronous transition systems that allow us to define for the first time the family of deterministic Büchi automata that matches the expressive power of the lim operator for trace languages. Not only is this definition a generalization of that for the word case but, more importantly, the corresponding languages are closed under finite unions and intersections – analogous to the deterministically Büchi recognizable word languages. In this sense, our results have further justified Muscholl's definition of "deterministic trace languages" as finite unions of parameterized lim-languages. Finally, we have also characterized deterministically Büchi recognizable trace languages in terms of recognition via a special subset of deterministic Muller automata.

The results of this paper uncover a clear path for completing a structure theory of regular infinitary trace languages. In ongoing work, we address the issue of weak recognizability, leading to a definition of weak D-SAMA's recognizing the languages that can be expressed as Boolean combinations of reachability trace languages. A next step is concerned with conceivable characterization of these weak trace languages as those that are recognized by both D-SABA's and D-SAcBA's (the latter equipped with the co-Büchi acceptance condition). Finally, it would be interesting to establish decidability of membership in each of these subclasses, for instance, by showing a strong Landweber theorem as indicated at the beginning of Section 4.

Acknowledgement. I am grateful to Wolfgang Thomas for encouragement and numerous suggestions for improving this paper, to Marcus Gelderie and Christof Löding for many fruitful discussions. I also thank anonymous referees of a previous conference to which a prior version of this paper was submitted.

References

1. Chaturvedi, N.: Languages of infinite traces and deterministic asynchronous automata. Technical Report AIB-2014-04, RWTH Aachen University (2014)
2. Chaturvedi, N., Gelderie, M.: Weak ω-Regular Trace Languages. arXiv.org, CoRR abs/1402.3199 (2014)
3. Diekert, V., Muscholl, A.: Deterministic asynchronous automata for infinite traces. Acta Informatica 31(4), 379–397 (1994)
4. Diekert, V., Rozenberg, G. (eds.): The Book of Traces. World Scientific Publishing Co., Inc., River Edge (1995)
5. Gastin, P., Petit, A.: Asynchronous cellular automata for infinite traces. In: Kuich, W. (ed.) ICALP 1992. LNCS, vol. 623, pp. 583–594. Springer, Heidelberg (1992)
6. Madhavan, M.: Automata on distributed alphabets. In: D'Souza, D., Shankar, P. (eds.) Modern Applications of Automata Theory. IISc Research Monographs Series, vol. 2, pp. 257–288. World Scientific (May 2012)
7. Muscholl, A.: Über die Erkennbarkeit unendlicher Spuren. PhD thesis (1994)
8. Perrin, D., Pin, J.-É.: Automata and Infinite Words. In: Infinite Words: Automata, Semigroups, Logic and Games. Pure and Applied Mathematics, vol. 141. Elsevier (2004)

Unary Pushdown Automata
and Straight-Line Programs*

Dmitry Chistikov and Rupak Majumdar

Max Planck Institute for Software Systems (MPI-SWS)
Kaiserslautern and Saarbrücken, Germany
{dch,rupak}@mpi-sws.org

Abstract. We consider decision problems for deterministic pushdown automata over the unary alphabet (udpda, for short). Udpda are a simple computation model that accept exactly the unary regular languages, but can be exponentially more succinct than finite-state automata. We complete the complexity landscape for udpda by showing that emptiness (and thus universality) is **P**-hard, equivalence and compressed membership problems are **P**-complete, and inclusion is **coNP**-complete. Our upper bounds are based on a *translation theorem* between udpda and straight-line programs over the binary alphabet (SLPs). We show that the characteristic sequence of any udpda can be represented as a pair of SLPs—one for the prefix, one for the lasso—that have size linear in the size of the udpda and can be computed in polynomial time. Hence, decision problems on udpda are reduced to decision problems on SLPs. Conversely, any SLP can be converted in logarithmic space into a udpda, and this forms the basis for our lower bound proofs. We show **coNP**-hardness of the ordered matching problem for SLPs, from which we derive **coNP**-hardness for inclusion. In addition, we complete the complexity landscape for unary nondeterministic pushdown automata by showing that the universality problem is $\Pi_2\mathbf{P}$-hard, using a new class of integer expressions. Our techniques have applications beyond udpda. We show that our results imply $\Pi_2\mathbf{P}$-completeness for a natural fragment of Presburger arithmetic and **coNP** lower bounds for compressed matching problems with one-character wildcards.

1 Introduction

Any model of computation comes with a set of fundamental decision questions: emptiness (does a machine accept some input?), universality (does it accept all inputs?), inclusion (are all inputs accepted by one machine also accepted by another?), and equivalence (do two machines accept exactly the same inputs?). The theoretical computer science community has a fairly good understanding of the precise complexity of these problems for most "classical" models, such as finite and pushdown automata, with only a few prominent open questions (e. g., the precise complexity of equivalence for deterministic pushdown automata).

---

* The full version of the paper is available at http://arxiv.org/abs/1403.0509.

J. Esparza et al. (Eds.): ICALP 2014, Part II, LNCS 8573, pp. 146–157, 2014.

In this paper, we study a simple class of machines: deterministic pushdown automata working on unary alphabets (unary dpda, or *udpda* for short). A classic theorem of Ginsburg and Rice [7] shows that they accept exactly the unary regular languages, albeit with potentially exponential succinctness when compared to finite automata. However, the precise complexity of most basic decision problems for udpda has remained open.

Our first and main contribution is that we close the complexity picture for these devices. We show that emptiness is already **P**-hard for udpda (even when the stack is bounded by a linear function of the number of states) and thus **P**-complete. By closure under complementation, it follows that universality is **P**-complete as well. Our main technical construction shows equivalence is in **P** (and so **P**-complete). Somewhat unexpectedly, inclusion is **coNP**-complete. In addition, we study the *compressed membership* problem: given a udpda over the alphabet $\{a\}$ and a number n in binary, is a^n in the language? We show that this problem is **P**-complete too.

A natural attempt at a decision procedure for equivalence or compressed membership would go through translations to finite automata (since udpda only accept regular languages, such a translation is possible). Unfortunately, these automata can be exponentially larger than the udpda and, as we demonstrate, such algorithms are not optimal. Instead, our approach establishes a connection to *straight-line programs* (*SLPs*) on binary words—a well-studied model for word compression (see, e.g., Lohrey [20]). An SLP $\mathcal{P}$ is a context-free grammar generating a single word, denoted $\mathrm{eval}(\mathcal{P})$, over $\{0,1\}$. Our main construction is a translation theorem: for any udpda, we construct in polynomial time two SLPs $\mathcal{P}'$ and $\mathcal{P}''$ such that the infinite sequence $\mathrm{eval}(\mathcal{P}') \cdot \mathrm{eval}(\mathcal{P}'')^\omega \in \{0,1\}^\omega$ is the characteristic sequence of the language of the udpda (for any $i \geq 0$, its ith element is 1 iff a^i is in the language). With this construction, decision problems on udpda reduce to decision problems on compressed words. Conversely, we show that from any pair $(\mathcal{P}', \mathcal{P}'')$ of SLPs one can compute, in logarithmic space, a udpda accepting the language with characteristic sequence $\mathrm{eval}(\mathcal{P}') \cdot \mathrm{eval}(\mathcal{P}'')^\omega$. Thus, lower bounds for computational complexity of decision problems for udpda may be obtained from the corresponding lower bounds for SLPs. Indeed, we show **coNP**-hardness of inclusion via **coNP**-hardness of the *ordered matching* problem for compressed words (i.e., is $\mathrm{eval}(\mathcal{P}_1) \leq \mathrm{eval}(\mathcal{P}_2)$ letter-by-letter, where the alphabet comes with an ordering $\leq$), a problem of independent interest.

As a second contribution, we complete the complexity picture for unary *nondeterministic* pushdown automata (unpda, for short). For unpda, the precise complexity of most decision problems was already known [14]. The remaining open question was the precise complexity of the universality problem, and we show that it is $\Pi_2\mathbf{P}$-hard (membership in $\Pi_2\mathbf{P}$ was shown earlier by Huynh [14]). An equivalent question was left open in Kopczyński and To [18] in 2010, but the question was posed as early as 1976 by Hunt III, Rosenkrantz, and Szymanski [12, Open Problem 2], where it was asked whether the problem was in **NP** or **PSPACE** or outside both. Huynh's $\Pi_2\mathbf{P}$-completeness result for equivalence [14] showed, in particular, that universality was in **PSPACE**, and our

$\Pi_2\mathbf{P}$-hardness result reveals that membership in $\mathbf{NP}$ is unlikely under usual complexity assumptions. As a corollary, we characterize the complexity of the $\forall_{\text{bounded}} \exists^*$-fragment of Presburger arithmetic, where the universal quantifier ranges over numbers at most exponential in the size of the formula.

To show $\Pi_2\mathbf{P}$-hardness, we show hardness of the universality problem for a class of integer expressions. Several decision problems of this form, with the set of operations $\{+, \cup\}$, were studied in the classic paper of Stockmeyer and Meyer [30], and we show that checking universality of expressions over $\{+, \cup, \times 2, \times \mathbb{N}\}$ is $\Pi_2\mathbf{P}$-complete (the upper bound follows from Huynh [14]).

Related Work. Table 1 provides the current complexity picture, including the results in this paper. Results on general alphabets are mostly classical and included for comparison. Note that the complexity landscape for udpda differs from those for unpda, dpda, and finite automata. Upper bounds for emptiness and universality are classical, and the lower bounds for emptiness are originally by Jones and Laaser [17] and Goldschlager [9]. In the nondeterministic unary case, $\mathbf{NP}$-completeness of compressed membership is from Huynh [14], rediscovered later by Plandowski and Rytter [25]. The $\mathbf{PSPACE}$-completeness of the compressed membership problem for binary pushdown automata (see definition in Section 6) is by Lohrey [22].

The main remaining open question is the precise complexity of the equivalence problem for dpda. It was shown decidable by Sénizergues [28] and primitive recursive by Stirling [29] and Jančar [15], but only $\mathbf{P}$-hardness (from emptiness) is currently known. Recently, the equivalence question for dpda when the stack alphabet is unary was shown to be $\mathbf{NL}$-complete by Böhm, Göller, and Jančar [4]. From this, it is easy to show that emptiness and universality are also $\mathbf{NL}$-complete. Compressed membership, however, remains $\mathbf{PSPACE}$-complete (see Caussinus et al. [5] and Lohrey [21]), and inclusion is already undecidable (see Valiant [31]). When we restrict dpda to both unary input and unary stack alphabet, all five decision problems are $\mathbf{L}$-complete.

We discuss corollaries of our results and other related work in Section 6.

2   Preliminaries

Pushdown Automata. A *unary pushdown automaton* (*unpda*) over the alphabet $\{a\}$ is a finite structure $\mathcal{A} = (Q, \Gamma, \bot, q_0, F, \delta)$, with Q a set of (control) states, Γ a stack alphabet, $\bot \in \Gamma$ a bottom-of-the-stack symbol, $q_0 \in Q$ an initial state, $F \subseteq Q$ a set of final states, and $\delta \subseteq (Q \times (\{a\} \cup \{\varepsilon\}) \times \Gamma) \times (Q \times \Gamma^*)$ a set of transitions with the property that, for every $(q_1, \sigma, \gamma, q_2, s) \in \delta$, either $\gamma \neq \bot$ and $s \in (\Gamma \setminus \{\bot\})^*$, or $\gamma = \bot$ and $s \in (\Gamma \setminus \{\bot\})^* \bot$. Here and everywhere below ε denotes the empty word.

The semantics of unpda is defined in the following standard way. The set of *configurations* of $\mathcal{A}$ is $Q \times (\Gamma \setminus \{\bot\})^* \bot$. Suppose (q_1, s_1) and (q_2, s_2) are configurations; we write $(q_1, s_1) \vdash_\sigma (q_2, s_2)$ and say that a *move* to (q_2, s_2) is available to $\mathcal{A}$ at (q_1, s_1) iff there exists a transition $(q_1, \sigma, \gamma, q_2, s) \in \delta$ such that $s_1 = \gamma s'$ and $s_2 = s s'$ for some $s' \in \Gamma^*$. A unary pushdown automaton is called

Table 1. Complexity of decision problems for pushdown automata

	unary		binary	
	dpda	npda	dpda	npda
Emptiness	$\mathbf{P}^l$	$\mathbf{P}$	$\mathbf{P}$	$\mathbf{P}$
Universality	$\mathbf{P}^l$	$\Pi_2\mathbf{P}^l$	$\mathbf{P}$	undecidable
Equivalence	$\mathbf{P}^{u,l}$	$\Pi_2\mathbf{P}$	$\mathbf{P}$.. pr.rec.	undecidable
Inclusion	$\mathbf{coNP}^{u,l}$	$\Pi_2\mathbf{P}$	undecidable	undecidable
Compressed membership	$\mathbf{P}^{u,l}$	$\mathbf{NP}$	$\mathbf{PSPACE}$	$\mathbf{PSPACE}$

Legend: "dpda" and "npda" stand for deterministic and *possibly* nondeterministic pushdown automata, respectively; "unary" and "binary" refer to their input alphabets. Names of complexity classes stand for completeness with respect to logarithmic-space reductions; abbreviation "pr.rec." stands for "primitive recursive". Superscripts u and l denote new upper and lower bounds shown in this paper.

deterministic, shortened to *udpda*, if at every configuration at most one move is available.

A word $w \in \{a\}^*$ is *accepted* by $\mathcal{A}$ if there exists a configuration (q_k, s_k) with $q_k \in F$ and a sequence of moves $(q_i, s_i) \vdash_{\sigma_i} (q_{i+1}, s_{i+1})$, $i = 0, \ldots, k-1$, such that $s_0 = \bot$ and $\sigma_0 \ldots \sigma_{k-1} = w$; that is, the acceptance is by final state. The *language* of $\mathcal{A}$, denoted $L(\mathcal{A})$, is the set of all words $w \in \{a\}^*$ accepted by $\mathcal{A}$.

We define the *size* of a unary pushdown automaton $\mathcal{A}$ as $|Q| \cdot |\Gamma|$, provided that for all transitions $(q_1, \sigma, \gamma, q_2, s) \in \delta$ the length of the word s is at most 2 (see also [24]). While this definition is better suited for deterministic rather than nondeterministic automata, it already suffices for the purposes of Section 5, where we handle unpda, because it is always the case that $|\delta| \leq 2 |Q|^2 |\Gamma|^4$.

Decision Problems. We consider the following decision problems: *emptiness* $(L(\mathcal{A}) =^? \emptyset)$, *universality* $(L(\mathcal{A}) =^? \{a\}^*)$, *equivalence* $(L(\mathcal{A}_1) =^? L(\mathcal{A}_2))$, and *inclusion* $(L(\mathcal{A}_1) \subseteq^? L(\mathcal{A}_2))$. The *compressed membership* problem for unary pushdown automata is associated with the question $a^n \in^? L(\mathcal{A})$, with n given in binary as part of the input. In complexity statements, hardness is with respect to logarithmic-space reductions.

Straight-Line Programs. A *straight-line program* [20], or an *SLP*, over an alphabet Σ is a context-free grammar that generates a single word; in other words, it is a tuple $\mathcal{P} = (S, \Sigma, \Delta, \pi)$, where Σ and Δ are disjoint sets of *terminal* and *nonterminal* symbols (*terminals* and *nonterminals*), $S \in \Delta$ is the *axiom*, and the function $\pi \colon \Delta \to (\Sigma \cup \Delta)^*$ defines a set of *productions* written as "$N \to w$", $w = \pi(N)$, and satisfies the property that the relation $\{(N, D) \mid N \to w \text{ and } D \text{ occurs in } w\}$ is acyclic. An SLP $\mathcal{P}$ is said to *generate* a (unique) word $w \in \Sigma^*$, denoted eval$(\mathcal{P})$, which is the result of applying substitutions π to S.

An SLP is said to be in *Chomsky normal form* if for all productions $N \to w$ it holds that either $w \in \Sigma$ or $w \in \Delta^2$. The *size* of an SLP is the number of nonterminals in its Chomsky normal form.

3   Indicator Pairs and the Translation Theorem

We say that a pair of SLPs $(\mathcal{P}', \mathcal{P}'')$ over an alphabet Σ *generates* a sequence $c \in \Sigma^\omega$ if $\mathrm{eval}(\mathcal{P}') \cdot (\mathrm{eval}(\mathcal{P}''))^\omega = c$. We call an infinite sequence $c \in \{0,1\}^\omega$, $c = c_0 c_1 c_2 \ldots$, the *characteristic sequence* of a unary language $L \subseteq \{a\}^*$ if, for all $i \geq 0$, it holds that c_i is 1 if $a^i \in L$ and 0 otherwise. One may note that the characteristic sequence is eventually periodic if and only if L is regular.

Definition 1. *A pair of straight-line programs $(\mathcal{P}', \mathcal{P}'')$ over $\{0,1\}$ is called an* indicator pair *for a unary language $L \subseteq \{a\}^*$ if it generates the characteristic sequence of L.*

A unary language can have several different indicator pairs. Indicator pairs form a descriptional system for unary languages, with the *size* of $(\mathcal{P}', \mathcal{P}'')$ defined as the sum of sizes of $\mathcal{P}'$ and $\mathcal{P}''$. The following translation theorem shows that udpda and indicator pairs are polynomially equivalent representations for unary regular languages. We remark that the theorem does not give a normal form for udpda because of the non-uniqueness of indicator pairs.

Theorem 1 (translation theorem). *For a unary language $L \subseteq \{a\}^*$:*

(1) *if there exists a udpda $\mathcal{A}$ of size m with $L(\mathcal{A}) = L$, then there exists an indicator pair for L of size $O(m)$;*

(2) *if there exists an indicator pair for L of size m, then there exists a udpda $\mathcal{A}$ of size $O(m)$ with $L(\mathcal{A}) = L$.*

Both translations can be performed by polynomial-time algorithms, the second of which works in logarithmic space.

Proof idea. We only discuss part 1, which presents the main technical challenge. The starting point is the simple observation that a udpda $\mathcal{A}$ has a single infinite computation, provided that the input tape supplies $\mathcal{A}$ with as many input symbols a as it needs to consume. Along this computation, *events* of two types are encountered: $\mathcal{A}$ can consume a symbol from the input and can enter a final state.

The crucial technical task is to construct inductively, using dynamic programming, straight-line programs that record these events along finite computational segments. These segments are of two types: first, between matching push and pop moves ("procedure calls") and, second, from some starting point until a move pops the symbol that has been on top of the stack at that point ("exits from current context"). Loops are detected, and infinite computations are associated with pairs of SLPs: in such a pair, one SLP records the initial segment, or prefix of the computation, and the other SLP records events within the loop.

After constructing these SLPs, it remains to transform the computational "history", or *transcript*, associated with the initial configuration of $\mathcal{A}$ into the characteristic sequence. This transformation can easily be performed in polynomial time, without expanding SLPs into the words that they generate. The result is an indicator pair for $\mathcal{A}$.   □

Any (possibly nondeterministic) logarithmic-space translation from udpda to indicator pairs would imply $\mathbf{NL} = \mathbf{P}$. This is because the emptiness problem for udpda ($\mathbf{P}$-hard, see Proposition 1 in Section 4 below) reduces to checking if at least one of the SLPs in an indicator pair generates a word containing a 1, and this can be checked in $\mathbf{NL}$.

Note that going from indicator pairs to udpda is useful for obtaining lower bounds on the computational complexity of decision problems for udpda (Theorems 2 and 5). For this purpose, it suffices to model just a single SLP, but taking into account the whole pair is interesting from the point of view of descriptional complexity (see also Section 6).

4   Decision Problems for UDPDA

4.1   Compressed Membership and Equivalence

For an SLP $\mathcal{P}$, by $|\mathcal{P}|$ we denote the length of the word $\text{eval}(\mathcal{P})$, and by $\mathcal{P}[n]$ the nth symbol of $\text{eval}(\mathcal{P})$, counting from 0 (that is, $0 \leq n \leq |\mathcal{P}| - 1$). We write $\mathcal{P}_1 \equiv \mathcal{P}_2$ if and only if $\text{eval}(\mathcal{P}_1) = \text{eval}(\mathcal{P}_2)$.

The following SLP-QUERY problem is known to be $\mathbf{P}$-complete (see Lifshits and Lohrey [19]): given an SLP $\mathcal{P}$ over $\{0, 1\}$ and a number n in binary, decide whether $\mathcal{P}[n] = 1$. The problem SLP-EQUIVALENCE is only known to be in $\mathbf{P}$ (see, e. g., Lohrey [20]): given two SLPs $\mathcal{P}_1$, $\mathcal{P}_2$, decide whether $\mathcal{P}_1 \equiv \mathcal{P}_2$.

Theorem 2. UDPDA-COMPRESSED-MEMBERSHIP *is $\mathbf{P}$-complete.*

Proof. The upper bound follows from Theorem 1. Indeed, given a udpda $\mathcal{A}$ and a number n, first construct an indicator pair $(\mathcal{P}', \mathcal{P}'')$ for $L(\mathcal{A})$. Now compute $|\mathcal{P}'|$ and $|\mathcal{P}''|$ and then decide if $n \leq |\mathcal{P}'| - 1$. If so, the answer is given by $\mathcal{P}'[n]$, otherwise by $\mathcal{P}''[r]$, where $r = (n - |\mathcal{P}'|) \bmod |\mathcal{P}''|$ and in both cases 1 is interpreted as "yes" and 0 as "no".

To prove the lower bound, we reduce from the SLP-QUERY problem. Take an instance with an SLP $\mathcal{P}$ and a number n in binary. By transforming the pair $(\mathcal{P}, \mathcal{P}_0)$, with $\mathcal{P}_0$ any fixed SLP over $\{0, 1\}$, into a udpda $\mathcal{A}$ using part 2 of Theorem 1, this problem is reduced, in logspace, to whether $a^n \in L(\mathcal{A})$. $\square$

The following proposition can be shown by a reduction from the monotone circuit value problem (for hardness) and polynomial-time algorithms for emptiness of pushdown automata and complementation of deterministic pushdown automata.

Proposition 1. UDPDA-EMPTINESS, UDPDA-UNIVERSALITY *are $\mathbf{P}$-complete.*

We now extend this result to the general equivalence problem for udpda.

Theorem 3. UDPDA-EQUIVALENCE *is $\mathbf{P}$-complete.*

Proof idea. Hardness follows from Proposition 1. We show how Theorem 1 can be used to prove the upper bound: given udpda $\mathcal{A}_1$ and $\mathcal{A}_2$, first construct indicator pairs $(\mathcal{P}_1', \mathcal{P}_1'')$ and $(\mathcal{P}_2', \mathcal{P}_2'')$ for $L(\mathcal{A}_1)$ and $L(\mathcal{A}_2)$, respectively. Now reduce the problem of whether $L(\mathcal{A}_1) = L(\mathcal{A}_2)$ to SLP-EQUIVALENCE. The key observation is that an eventually periodic sequence that has periods $|\mathcal{P}_1''|$ and $|\mathcal{P}_2''|$ also has period $t = \gcd(|\mathcal{P}_1''|, |\mathcal{P}_2''|)$. Therefore, it suffices to check that, first, the initial segments of the generated sequences match and, second, that $\mathcal{P}_1''$ and $\mathcal{P}_2''$ generate powers of the same word up to a certain circular shift.   □

4.2 Inclusion

A natural idea for handling the inclusion problem for udpda would be to extend the result of Theorem 3, that is, to tackle inclusion similarly to equivalence. This raises the problem of comparing the words generated by two SLPs in the componentwise sense with respect to the order $0 \leq 1$. To the best of our knowledge, this problem has not been studied previously, so we deal with it separately. As it turns out, here one cannot hope for an efficient algorithm unless $\mathbf{P} = \mathbf{NP}$.

Let us define the following family of problems, parameterized by partial order R on the alphabet of size at least 2, and denoted SLP-COMPONENTWISE-R. The input is a pair of SLPs $\mathcal{P}_1$, $\mathcal{P}_2$ over an alphabet partially ordered by R, generating words of equal length. The output is "yes" iff for all i, $0 \leq i < |\mathcal{P}_1|$, the relation $R(\mathcal{P}_1[i], \mathcal{P}_2[i])$ holds. By SLP-COMPONENTWISE-$(0 \leq 1)$ we mean a special case of this problem with R the partial order on $\{0, 1\}$ given by $0 \leq 0$, $0 \leq 1$, $1 \leq 1$.

Theorem 4. SLP-COMPONENTWISE-R *is* **coNP**-*complete if R is not the equality relation (that is, if $R(a, b)$ holds for some $a \neq b$), and in* $\mathbf{P}$ *otherwise.*

Proof idea. The technical part is to show **coNP**-hardness of SLP-COMPONENTWISE-$(0 \leq 1)$ by a reduction from the complement of SUBSET-SUM. We use so-called Lohrey words [22, Theorem 5.2]: given a vector $w = (w_1, \ldots, w_n)$, a natural t, and the question of whether there exists an $x = (x_1, \ldots, x_n) \in \{0, 1\}^n$ such that $x \cdot w = t$, where $x \cdot w = \sum_{i=1}^{n} x_i w_i$, it is possible to construct in logarithmic space two SLPs that generate words $W_1 = \prod_{x \in \{0,1\}^n} a^{x \cdot w} b a^{s - x \cdot w}$ and $W_2 = (a^t b a^{s-t})^{2^n}$, where $s = (1, \ldots, 1) \cdot w$ and the product in W_1 enumerates the xs in the lexicographic order. Now W_1 and W_2 share a symbol b in some position iff the original instance of SUBSET-SUM is a yes-instance. Substituting 0 for a and 1 for b in the first SLP, and 0 for b and 1 for a in the second SLP brings us to (the complement of) SLP-COMPONENTWISE-$(0 \leq 1)$.   □

Remark. An alternative reduction can be derived from Bertoni, Choffrut, and Radicioni [3, Lemma 3]. A corollary of Theorem 4 on a problem of matching for compressed partial words is demonstrated in Section 6.

Theorem 5. UDPDA-INCLUSION *is* **coNP**-*complete.*

Proof idea. We rely on Theorem 1: hardness is by a reduction from SLP-COMPONENTWISE-$(0 \leq 1)$ using part 2, and membership in **coNP** depends on part 1 and follows from the fact that $L(\mathcal{A}_1) \not\subseteq L(\mathcal{A}_2)$ if and only if there exists an n such that $a^n \in L(\mathcal{A}_2) \setminus L(\mathcal{A}_1)$ and, moreover, $n \leq 2^{O(m)}$, where m is the size of the input. The upper bound on n follows from the translation to deterministic finite automata (see discussion in Section 6 or Pighizzini [24, Theorem 8]).   □

5   Universality of UNPDA

In this section we settle the complexity status of the universality problem for unary, possibly nondeterministic pushdown automata. While $\Pi_2\mathbf{P}$-completeness of equivalence and inclusion is shown by Huynh [14], it has been unknown whether the universality problem is also $\Pi_2\mathbf{P}$-hard.

For convenience of notation, we use an auxiliary descriptional system. Define *integer expressions* over the set of operations $\{+, \cup, \times 2, \times \mathbb{N}\}$ inductively: the base case is a non-negative integer n, written in binary, and the inductive step is associated with binary operations $+$, $\cup$, and unary operations $\times 2$, $\times \mathbb{N}$. To each expression E we associate a set of non-negative integers $S(E)$: $S(n) = \{n\}$, $S(E_1 + E_2) = \{s_1 + s_2 \colon s_1 \in S(E_1), s_2 \in S(E_2)\}$, $S(E_1 \cup E_2) = S(E_1) \cup S(E_2)$, $S(E \times 2) = S(E + E)$, $S(E \times \mathbb{N}) = \{sk \colon s \in S(E), k = 0, 1, 2, \dots\}$.

Expressions E_1 and E_2 are called *equivalent* iff $S(E_1) = S(E_2)$; an expression E is *universal* iff it is equivalent to $1 \times \mathbb{N}$. The problem of deciding universality is denoted by INTEGER-$\{+, \cup, \times 2, \times \mathbb{N}\}$-EXPRESSION-UNIVERSALITY.

Decision problems for integer expressions have been studied for more than 40 years: Stockmeyer and Meyer [30] showed that for expressions over $\{+, \cup\}$ compressed membership is **NP**-complete and equivalence is $\Pi_2\mathbf{P}$-complete (universality is, of course, trivial). For recent results on such problems with operations from $\{+, \cup, \cap, \times, ^-\}$, see McKenzie and Wagner [23] and Glaßer et al. [8].

Lemma 1. INTEGER-$\{+, \cup, \times 2, \times \mathbb{N}\}$-EXPRESSION-UNIVERSALITY *is* $\Pi_2\mathbf{P}$*-hard.*

Proof idea. The reduction is from the GENERALIZED-SUBSET-SUM problem [1, Lemma 6.2], which is defined as follows. The input consists of two vectors of naturals, u and v, and a natural t, and the problem is to decide whether for all $y \in \{0, 1\}^m$ there exists an $x \in \{0, 1\}^n$ such that $x \cdot u + y \cdot v = t$, where the middle dot $\cdot$ denotes the inner product. Let M be a big enough number, and consider the integer expression E defined by $E = E' \cup E''$, where

$$E' = (2^m M + 1 \times \mathbb{N}) \cup (M \times \mathbb{N} + ([0, t-1] \cup [t+1, M-1])),$$

$$E'' = \sum_{j=1}^{m}(0 \cup (2^{j-1}M + v_j)) + \sum_{i=1}^{n}(0 \cup u_i),$$

and segments $[a, b]$ are given by expressions of size $O(\log(b - a))$. Then E is universal iff the input is a yes-instance of GENERALIZED-SUBSET-SUM.   □

Remark. With *circuits* instead of formulae (see also [23] and [8]) we would not need doubling. Furthermore, we only use $\times \mathbb{N}$ on fixed numbers, so instead we could use any feature for expressing an arithmetic progression with fixed common difference.

Theorem 6. UNARY-PDA-UNIVERSALITY *is* $\Pi_2\mathbf{P}$*-complete.*

Hardness is by a reduction from INTEGER-$\{+, \cup, \times 2, \times \mathbb{N}\}$-EXPRESSION-UNIVER-SALITY, and membership in $\Pi_2\mathbf{P}$ follows from Huynh [14].

Corollary 1. *Universality, equivalence, and inclusion are* $\Pi_2 P$-*complete for (possibly nondeterministic) unary pushdown automata, unary context-free grammars, and integer expressions over* $\{+, \cup, \times 2, \times \mathbb{N}\}$.

Another consequence of Theorem 6 is that deciding equality of a (not necessarily unary) context-free language, given as a context-free grammar, to any fixed context-free language L_0 that contains an infinite regular subset, is $\Pi_2 P$-hard and, if $L_0 \subseteq \{a\}^*$, $\Pi_2 P$-complete. The lower bound is by a reduction due to Hunt III, Rosenkrantz, and Szymanski [12, Theorem 3.8], who show that deciding equivalence to $\{a\}^*$ reduces to deciding equivalence to any such L_0. The reduction is shown to be polynomial-time, but is easily seen to be logarithmic-space as well. The upper bound for the unary case is by Huynh [14]; in the general case, the problem can be undecidable.

6   Corollaries and Discussion

Descriptional Complexity Aspects of UDPDA. Theorem 1 can be used to obtain several results on descriptional complexity aspects of udpda proved earlier by Pighizzini [24]. He shows how to transform a udpda of size m into an equivalent deterministic finite automaton (DFA) with at most 2^m states [24, Theorem 8] and into an equivalent context-free grammar in Chomsky normal form (CNF) with at most $2m + 1$ nonterminals [24, Theorem 12]. In our construction m gets multiplied by a small constant, but the advantage is that we now see (the slightly weaker variants of) these results as easy corollaries of a single underlying theorem. Indeed, using an indicator pair $(\mathcal{P}', \mathcal{P}'')$ for L, it is straightforward to construct a DFA of size $|\text{eval}(\mathcal{P}')| + |\text{eval}(\mathcal{P}'')|$ accepting L, as well as to transform the pair into a CFG in CNF that generates L and has at most thrice the size of $(\mathcal{P}', \mathcal{P}'')$.

Another result which follows, even more directly, from ours is a lower bound on the size of udpda accepting a specific language L_1 [24, Theorem 15]. To obtain this lower bound, Pighizzini employs a known lower bound on the SLP-size of the word $W = W[0] \ldots W[K-1] \in \{0,1\}^K$ such that $a^n \in L_1$ iff $W[n \bmod K] = 1$. To this end, a udpda $\mathcal{A}$ accepting L_1 is intersected (we are glossing over some technicalities here) with a small deterministic finite automaton that "captures" the end of the word W. The obtained udpda, which only accepts a^K, is transformed into an equivalent context-free grammar. It is then possible to use the structure of the grammar to transform it into an SLP that produces W (note that such a transformation in general is **NP**-hard). While the proof produces from a udpda for L_1 a related SLP with a polynomial blowup, this construction depends crucially on the structure of the language L_1, so it is difficult to generalize the argument to *all* udpda and thus obtain Theorem 1. Our proof of Theorem 1 therefore follows a very different path.

Relationship to Presburger Arithmetic. An alternative way to prove the upper bound in Theorem 5 is via Presburger arithmetic, using the observation that there is a poly-time computable existential Presburger formula that expresses the membership of a word a^n in $L(\neg\mathcal{A}_1)$ and $L(\mathcal{A}_2)$. This technique distills the arguments used by Huynh [13,14] to show that the compressed membership problem for unary pushdown automata is in **NP**. It is used in a purified form by Plandowski and Rytter [25, Theorems 4 and 8], who developed a much shorter proof of the same fact (apparently unaware of the previous proof). The same idea was later rediscovered and used in a combination with Presburger arithmetic by Verma, Seidl, and Schwentick [32, Theorem 4].

Yet another application of this technique provides an alternative proof of the $\Pi_2\mathbf{P}$ upper bound for unpda universality, equivalence, and inclusion (Theorem 6 and Corollary 1). Indeed, we can use the same approach as for inclusion of udpda; the only difference is that there is no polynomial-time complementation, so another level of quantifier alternation is introduced. The proof known to date, due to Huynh [14], involves reproving Parikh's theorem and is more than 10 pages long. Reduction to Presburger formulae produces a much simpler proof.

Also, our $\Pi_2\mathbf{P}$-hardness result for unpda shows that the $\forall_{\text{bounded}} \exists^*$-fragment of Presburger arithmetic is $\Pi_2\mathbf{P}$-complete, where the variable bound by the universal quantifier is at most exponential in the size of the formula. The upper bound holds because the $\exists^*$-fragment is **NP**-complete, see von zur Gathen and Sieveking [33]. In comparison, the $\forall \exists^*$-fragment, without restrictions on the domain of the universally quantified variable, requires co-nondeterministic $2^{n^{\Omega(1)}}$ time, see Grädel [10]. Previously known fragments that are complete for the second level of the polynomial hierarchy involve alternation depth 3 and a fixed number of quantifiers, as in Grädel [11] and Schöning [27]. Also note that the $\forall^s \exists^t$-fragment is **coNP**-complete for all fixed $s \geq 1$ and $t \geq 2$, see Grädel [11].

Problems Involving Compressed Words. Recall Theorem 4: given two SLPs, it is **coNP**-complete to compare the generated words componentwise with respect to any partial order different from equality. As a corollary, we get tight complexity bounds for SLP equivalence in the presence of wildcards or, equivalently, compressed matching in the well-known model of partial words (see, e. g., Fischer and Paterson [6] and Berstel and Boasson [2]). Consider the problem SLP-PARTIAL-WORD-MATCHING: the input is a pair of SLPs $\mathcal{P}_1, \mathcal{P}_2$ over the alphabet $\{a, b, ?\}$, generating words of equal length, and the output is "yes" iff for every i, $0 \leq i < |\mathcal{P}_1|$, either $\mathcal{P}_1[i] = \mathcal{P}_2[i]$ or at least one of $\mathcal{P}_1[i]$ and $\mathcal{P}_2[i]$ is ? (a *hole*, or a single-character *wildcard*).

Schmidt-Schauß [26] defines a problem equivalent to SLP-PARTIAL-WORD-MATCHING, along with another related problem, where one needs to find occurrences of eval($\mathcal{P}_1$) in eval($\mathcal{P}_2$) (as in pattern matching), $\mathcal{P}_2$ is known to contain no holes, and two symbols match iff they are equal or at least one of them is a hole. For this related problem, he develops a polynomial-time algorithm that finds (a representation of) all matching occurrences and operates under the assumption that the number of holes in eval($\mathcal{P}_1$) is polynomial in the size of the input. He also points out that no solution for (the general case of) SLP-PARTIAL-WORD-MATCHING is known—unless a polynomial upper bound on the number of ?s in eval($\mathcal{P}_1$) and eval($\mathcal{P}_2$) is given. Our next proposition shows that such a solution is not possible unless $\mathbf{P} = \mathbf{NP}$. It is an easy consequence of Theorem 4.

Proposition 2. SLP-PARTIAL-WORD-MATCHING *is* coNP-*complete.*

Proof. Membership in **coNP** is obvious, and the hardness is by a reduction from SLP-COMPONENTWISE-$(0 \leq 1)$. Given a pair of SLPs $\mathcal{P}_1$, $\mathcal{P}_2$ over $\{0, 1\}$, substitute ? for 0 and a for 1 in $\mathcal{P}_1$, and b for 0 and ? for 1 in $\mathcal{P}_2$. The resulting pair of SLPs over $\{a, b, ?\}$ is a yes-instance of SLP-PARTIAL-WORD-MATCHING iff the original pair is a yes-instance of SLP-COMPONENTWISE-$(0 \leq 1)$.   □

The wide class of *compressed membership* problems (deciding eval$(\mathcal{P}) \in L$) is first introduced in Plandowski and Rytter [25] and further studied and discussed in, e.g., Jeż [16] and Lohrey [20]. In the case of words over the unary alphabet, $w \in \{a\}^*$, expressing w with an SLP is poly-time equivalent to representing it with its length $|w|$ written in binary. An easy corollary of Theorem 2 is that deciding $w \in L(\mathcal{A})$, where $\mathcal{A}$ is a (not necessarily unary) deterministic pushdown automaton and $w = a^n$ with n given in binary, is **P**-complete.

Acknowledgements. We thank Rayna Dimitrova, Joshua Dunfield, Rose Hoberman, Patrick Totzke, and the anonymous reviewers for comments.

References

1. Berman, P., Karpinski, M., Larmore, L.L., Plandowski, W., Rytter, W.: On the complexity of pattern matching for highly compressed two-dimensional texts. JCSS 65(2), 332–350 (2002)
2. Berstel, J., Boasson, L.: Partial words and a theorem of Fine and Wilf. TCS 218(1), 135–141 (1999)
3. Bertoni, A., Choffrut, C., Radicioni, R.: Literal shuffle of compressed words. In: Ausiello, G., Karhumäki, J., Mauri, G., Ong, L. (eds.) IFIP TCS 2008. IFIP, vol. 273, pp. 87–100. Springer, Boston (2008)
4. Böhm, S., Göller, S., Jančar, P.: Equivalence of deterministic one-counter automata is **NL**-complete. In: Boneh, D., Roughgarden, T., Feigenbaum, J. (eds.) STOC'13, pp. 131–140. ACM (2013)
5. Caussinus, H., McKenzie, P., Thérien, D., Vollmer, H.: Nondeterministic **NC**^{1} computation. JCSS 57(2), 200–212 (1998)
6. Fischer, M.J., Paterson, M.S.: String-matching and other products. In: Karp, R. (ed.) SIAM-AMS Proceedings, vol. 7, pp. 113–125. AMS (1974)
7. Ginsburg, S., Rice, H.G.: Two families of languages related to ALGOL. Journal of the ACM 9(3), 350–371 (1962)
8. Glaßer, C., Herr, K., Reitwießner, C., Travers, S., Waldherr, M.: Equivalence problems for circuits over sets of natural numbers. Theory of Computing Systems 46(1), 80–103 (2010)
9. Goldschlager, L.M.: ε-productions in context-free grammars. Acta Informatica 16(3), 303–308 (1981)
10. Grädel, E.: Dominoes and the complexity of subclasses of logical theories. Annals of Pure and Applied Logic 43(1), 1–30 (1989)
11. Grädel, E.: Subclasses of Presburger arithmetic and the polynomial-time hierarchy. TCS 56(3), 289–301 (1988)
12. Hunt III, H.B., Rosenkrantz, D.J., Szymanski, T.G.: On the equivalence, containment, and covering problems for the regular and context-free languages. JCSS 12(2), 222–268 (1976)

13. Huynh, D.T.: Commutative grammars: the complexity of uniform word problems. Information and Control 57, 21–39 (1983)
14. Huynh, D.T.: Deciding the inequivalence of context-free grammars with 1-letter terminal alphabet is Σ_2^p-complete. TCS 33(2-3), 305–326 (1984)
15. Jančar, P.: Decidability of DPDA language equivalence via first-order grammars. In: LICS 2012, pp. 415–424. IEEE Computer Society (2012)
16. Jeż, A.: The complexity of compressed membership problems for finite automata. Theory of Computing Systems, 1–34 (2013)
17. Jones, N.D., Laaser, W.T.: Complete problems for deterministic polynomial time. TCS 3(2), 105–117 (1976)
18. Kopczyński, E., To, A.W.: Parikh images of grammars: complexity and applications. In: LICS 2010, pp. 80–89. IEEE Computer Society (2010)
19. Lifshits, Y., Lohrey, M.: Querying and embedding compressed texts. In: Královič, R., Urzyczyn, P. (eds.) MFCS 2006. LNCS, vol. 4162, pp. 681–692. Springer, Heidelberg (2006)
20. Lohrey, M.: Algorithmics on SLP-compressed strings: a survey. Groups Complexity Cryptology 4(2), 241–299 (2012)
21. Lohrey, M.: Leaf languages and string compression. Information and Computation 209(6), 951–965 (2011)
22. Lohrey, M.: Word problems and membership problems on compressed words. SIAM Journal on Computing 35(5), 1210–1240 (2006)
23. McKenzie, P., Wagner, K.W.: The complexity of membership problems for circuits over sets of natural numbers. Computational Complexity 16(3), 211–244 (2007)
24. Pighizzini, G.: Deterministic pushdown automata and unary languages. International Journal of Foundations of Computer Science 20(4), 629–645 (2009)
25. Plandowski, W., Rytter, W.: Complexity of language recognition problems for compressed words. In: Karhumäki, J., Maurer, H., Păun, G., Rozenberg, G. (eds.) Jewels are Forever, pp. 262–272. Springer (1999)
26. Schmidt-Schauß, M.: Matching of compressed patterns with character variables. In: Tiwari, A. (ed.) RTA 2012. LIPIcs, vol. 15, pp. 272–287. Dagstuhl (2012)
27. Schöning, U.: Complexity of Presburger arithmetic with fixed quantifier dimension. Theory of Computing Systems 30(4), 423–428 (1997)
28. Sénizergues, G.: $L(A) = L(B)$? A simplified decidability proof. TCS 281(1-2), 555–608 (2002)
29. Stirling, C.: Deciding DPDA equivalence is primitive recursive. In: Widmayer, P., Eidenbenz, S., Triguero, F., Morales, L., Conejo, R., Hennessy, M. (eds.) ICALP 2002. LNCS, vol. 2380, pp. 821–832. Springer, Heidelberg (2002)
30. Stockmeyer, L.J., Meyer, A.R.: Word problems requiring exponential time: Preliminary report. In: STOC 1973, pp. 1–9. ACM (1973)
31. Valiant, L.: Decision procedures for families of deterministic pushdown automata. PhD thesis. University of Warwick (1973)
32. Verma, K.N., Seidl, H., Schwentick, T.: On the complexity of equational Horn clauses. In: Nieuwenhuis, R. (ed.) CADE 2005. LNCS (LNAI), vol. 3632, pp. 337–352. Springer, Heidelberg (2005)
33. Von zur Gathen, J., Sieveking, M.: A bound on solutions of linear integer equalities and inequalities. Proceedings of the AMS 72(1), 155–158 (1978)

Robustness against Power is PSpace-complete*

Egor Derevenetc[1,2] and Roland Meyer[2]

[1] Fraunhofer ITWM, Germany
[2] University of Kaiserslautern, Germany

Abstract. Power is a RISC architecture developed by IBM, Freescale, and several other companies and implemented in a series of POWER processors. The architecture features a relaxed memory model providing very weak guarantees with respect to the ordering and atomicity of memory accesses.

Due to these weaknesses, some programs that are correct under sequential consistency (SC) show undesirable effects when run under Power. We say that these programs are not robust against the Power memory model. Formally, a program is robust if every computation under Power has the same data and control dependencies as some SC computation.

Our contribution is a decision procedure for robustness of concurrent programs against the Power memory model. It is based on three ideas. First, we reformulate robustness in terms of the acyclicity of a happens-before relation. Second, we prove that among the computations with cyclic happens-before relation there is one in a certain normal form. Finally, we reduce the existence of such a normal-form computation to a language emptiness problem. Altogether, this yields a PSPACE algorithm for checking robustness against Power. We complement it by a matching lower bound to show PSPACE-completeness.

1   Introduction

To execute code as fast as possible, modern processors reorder operations. For example, Intel x86/x86-64 and SPARC processors implement the Total Store Ordering (TSO) memory model [14] which allows write buffering: store operations in each thread can be queued and get executed on memory later. Processors can also execute independent instructions out of program order as soon as the input data and computational units are available for them. This is an inherent feature of the POWER and ARM microprocessors [13]. Moreover, Power and ARM memory models, unlike TSO, do not guarantee store atomicity: one write can become visible to different threads at different times. They only ensure that all threads see stores to the same memory location in the same order; stores to different memory locations can be seen in different order by different threads.

All these optimizations are usually designed so that a single-threaded program has the illusion that its instructions are executed in program order. The picture changes in the presence of concurrency. Concurrent programs are often assumed to have sequentially consistent (SC) semantics [11]: each thread executes its

---

* The full version of this paper is available online [9].

J. Esparza et al. (Eds.): ICALP 2014, Part II, LNCS 8573, pp. 158–170, 2014.
© Springer-Verlag Berlin Heidelberg 2014

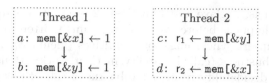

Fig. 1. Message Passing (MP) program [15]. By $\&x$ and $\&y$ we denote the addresses of the variables x and y. Initially, $x = y = 0$. The first thread writes a message into x and sets flag variable y, signifying that the message is written. The second thread reads the flag and, if it is set, expects to see the message written to x by the first thread.

operations in program order, stores become visible immediately to all threads. Concurrent programs may observe a difference from SC when run on a modern processor with a weak memory model. To see this, consider the MP program in Figure 1. SC and TSO forbid the situation where $r_1 > r_2$ upon termination of both threads. However, this is possible on Power: instruction c can read the value written by b, whereas d reads the initial value.

We call a program *not robust* against Power [16,6,7,2,5,8,4], if it exhibits non-SC behaviors when executed under the Power memory model. More formally, a program is robust if all its Power computations have the same data and control dependencies as the computations under SC. That is, for every Power computation there is a sequentially consistent computation which executes the same instructions, all loads read from the same stores in both computations, and stores to the same address happen in the same order. Robust programs produce the same results on Power and SC architectures, which means verification results for SC remain valid for the weak memory model.

We present an algorithm for deciding robustness against Power. This is the first decidability result for this architecture and, more generally, the first decidability result for a non-store-atomic memory model. We obtain the algorithm in the following steps. First, we reformulate robustness in terms of acyclicity of happens-before, using the result by Shasha and Snir [16]. Second, we show that among the computations with cyclic happens-before relation there is always one in a certain normal form. Next, we prove that the set of all normal-form computations can be generated by a multiheaded automaton — an automaton model developed recently in the context of robustness [8]. Finally, to check cyclicity of the happens-before relation we intersect this automaton with regular languages. The program is robust iff the intersection is empty. This reduces robustness to language emptiness for multiheaded automata. The algorithm works in space polynomial in the size of the program. We obtain a matching lower bound by a reduction of SC-reachability to robustness, similar to [5].

Related Work. The happens-before relation was formulated by Lamport [10]. Shasha and Snir [16] have shown that a computation violates sequential consistency iff it has a cyclic happens-before relation. Burckhardt and Musuvathi [6] proposed the first algorithm for detecting non-robustness against TSO based on monitoring SC computations. Burnim et al. [7] pointed out a mistake in the definition of TSO used in [6] and described monitoring algorithms for the TSO and

PSO memory models. Alglave and Maranget [2] presented a tool to statically over-approximate happens-before cycles in programs written in x86 and Power assembly, and to insert synchronization primitives (memory fences and syncs) as required for robustness (called stability in their work). Bouajjani et al. [5] obtained the first decidability result for robustness: robustness against TSO is PSPACE-complete for finite-state programs. In [4] they presented a reduction of robustness against TSO to SC reachability working for general programs and an algorithm for optimal fence insertion.

The Power architecture has attracted considerable recent attention. Alglave et al. [3] give an overview of the numerous publications devoted to defining its semantics. We highlight two Power models: the operational model by Sarkar et al. [15] and the axiomatic one by Mador-Haim et al. [12]. These models were extensively tested against the architecture and were proven to be equivalent [12]. Nevertheless, the operational model is known to forbid certain behaviors that are possible on real hardware[1] and in the axiomatic model[2] [3]. Fortunately, there is a suggestion for a fix: in Section 4.5 of [15] one should read *from a coherence-order-earlier write* instead of *from a different write* (two occurrences). Then, the operational model is believed to tightly over-approximate Power [1]. In the present paper we stick to the corrected operational model from [15].

Finally, we would like to note that ARM has a memory model very similar to that of Power. The differences and similarities are highlighted by Maranget et al. in [13,3]. This fact promises a relatively easy transfer of the proof techniques used in the present paper to the ARM memory model.

2   Programming Model

We define programs and their semantics in terms of automata. An *automaton* is a tuple $A = (S, \Sigma, \Delta, s_0, F)$, where S is a set of states, Σ is an alphabet, $\Delta \subseteq S \times (\Sigma \cup \{\varepsilon\}) \times S$ is a set of transitions, $s_0 \in S$ is an initial state, and $F \subseteq S$ is a set of final states. We call the automaton *finite* if S and Σ are finite. We write $s_1 \xrightarrow{a} s_2$ if $t = (s_1, a, s_2) \in \Delta$ and denote $\mathsf{src}(t) := s_1$, $\mathsf{dst}(t) := s_2$, $\mathsf{lab}(t) := a$. The *language* of the automaton is $\mathcal{L}(A) := \{\sigma \in \Sigma^* \mid s_0 \xrightarrow{\sigma} s \text{ for some } s \in F\}$. For a sequence $\sigma = a_1 \ldots a_n \in \Sigma^*$ we define $|\sigma| := n$, $\sigma[i] := a_i$, $\mathsf{first}(\sigma) := a_1$, and $\mathsf{last}(\sigma) := a_n$. We use $\cdot$ for concatenation, $\downarrow$ for projection, and ε for the empty sequence. Given $\alpha \in \Sigma^*$ and $a, b \in \alpha$, we write $a <_\alpha b$ if $\alpha = \alpha_1 \cdot a \cdot \alpha_2 \cdot b \cdot \alpha_3$. Given a function $f : X \to Y$, $x' \in X$, and $y' \in Y$, we define $f' = f[x' \hookleftarrow y']$ by $f'(x) := f(x)$ for $x \in X \setminus \{x'\}$ and $f'(x') := y'$.

A program is a finite sequence of threads: $\mathcal{P} = \mathcal{T}_1 \ldots \mathcal{T}_n$. A *thread* is an automaton $\mathcal{T}_{\mathsf{tid}} = (Q_{\mathsf{tid}}, \mathsf{CMD}, \mathcal{I}_{\mathsf{tid}}, q_{0\mathsf{tid}}, Q_{\mathsf{tid}})$ with a finite set of control states Q_{tid}, all of them being final, initial state $q_{0\mathsf{tid}}$, and a set of transitions $\mathcal{I}_{\mathsf{tid}}$ called *instructions* and labeled with *commands* CMD defined below. Each thread has an id from $\mathsf{TID} := [1..|\mathcal{P}|]$.

---

[1] http://diy.inria.fr/cats/pldi-power/#lessvs
[2] http://diy.inria.fr/cats/cav-power/

Let DOM = ADDR be a finite domain of values and addresses containing the value 0. Let REG be a finite set of registers that take values from DOM. Commands CMD include loads, stores, local assignments, and conditionals (`assume`): The set

$$\langle cmd \rangle ::= \langle reg \rangle \leftarrow \text{mem}[\langle expr \rangle] \mid \text{mem}[\langle expr \rangle] \leftarrow \langle expr \rangle$$
$$\mid \langle reg \rangle \leftarrow \langle expr \rangle \mid \text{assume}(\langle expr \rangle)$$

of expressions EXPR is defined over constants from DOM, registers from REG, and (unspecified) functions FUN over $DOM \cup \{\bot\}$. We assume that these functions return $\bot$ iff any of the arguments is $\bot$.

2.1 Power Semantics

We briefly recall the corrected model from [15]. The state of a running program consists of the runtime states of threads and the state of a storage subsystem.

The runtime state of a thread includes information about the instructions being executed by the thread. In order to start executing an instruction, the thread must *fetch* it. The thread can fetch any instruction whose source control state is equal to the destination state of the last fetched instruction. Then, the thread must perform any computation required by the semantics of this instruction. For example, for a load the thread must compute the address being accessed, then read the value at this address, and place it into the target register. The last step of executing an instruction is *committing* it. Committing an instruction requires committing all its *dependencies*. For example, before committing a load the thread must commit all its *address dependencies* — the instructions which define the values of registers used in the address expression — and *control dependencies* — the program-order-earlier (fetched earlier than the load) conditional instructions. Moreover, all loads and stores accessing the same address must be committed in the order in which they were fetched.

The storage subsystem keeps track, for each address, of the global ordering of stores to this address — the *coherence order* — and the last store to this address *propagated* to each thread. When a thread commits a store, this store is assigned a position in the coherence order which we identify by a rational number — the *coherence key*. We choose rational numbers (rather than naturals) to be able to insert a store between any two stores in the coherence order. The key must be greater than the coherence key of the last store to the same address propagated to this thread. The committed store is immediately propagated to its own thread. At some point later this store can be propagated to any other thread, as long as it is coherence-order-later (has a greater coherence key) than the last store to the same address propagated to that thread. When a thread loads a value from a certain address, it gets the value written by the last store to this address propagated to the thread. A thread can also forward the value being written by a not yet committed store to a later load reading the same address. This situation is called an *early read*.

An important property of Power is that it maintains the illusion of sequential consistency for single-threaded programs. This means that reorderings on the thread level must not lead to situations when, e.g., a program-order-later load reads a coherence-order-earlier store than the one read by a program-order-earlier load from the same address. In [15] these restrictions are enforced by the mechanism of restarting operations. We put these conditions into the requirements on final states of the running program instead.

To keep the paper readable, we omit the description of Power synchronization instructions: `sync`, `lwsync`, `isync`. All constructions in the paper can be consistently extended to support them with the final result continuing to hold.

Formally, we define the semantics of program $\mathcal{P}$ on Power by a *Power automaton* $Z(\mathcal{P}) := (S_Z, \mathsf{E}, \Delta_Z, s_{0Z}, F_Z)$. Here, E is a set of labels called *events* that we define together with the transitions.

State Space. A state of the Power automaton is a pair $s_Z = (\mathsf{ts}, s_Y) \in S_Z$ with runtime thread states $\mathsf{ts}\colon \mathsf{TID} \to S_X$ and storage subsystem state $s_Y \in S_Y$.

A runtime thread state $s_X = (\mathsf{fetched}, \mathsf{committed}, \mathsf{loaded}) \in S_X$ includes a finite sequence of fetched instructions $\mathsf{fetched} \in \mathcal{I}^*$, a set of indices of committed instructions $\mathsf{committed} \subseteq [1..|\mathsf{fetched}|]$, and a function giving the store read by a load $\mathsf{loaded}\colon [1..|\mathsf{fetched}|] \to \{\bot\} \cup \{\mathsf{init}_a \mid a \in \mathsf{ADDR}\} \cup \mathsf{TID} \times \mathbb{N}$. We use init_a to denote the initial store of value 0 to address a. The initial state of a running thread is $s_{0X} := (\varepsilon, \emptyset, \lambda i.\bot)$.

A state of the storage subsystem $s_Y = (\mathsf{co}, \mathsf{prop}) \in S_Y$ includes a mapping from a store instruction (its thread id and index in the list of fetched instructions) to its position in the coherence order $\mathsf{co}\colon \mathsf{TID} \times \mathbb{N} \cup \{\mathsf{init}_a \mid a \in \mathsf{ADDR}\} \to \mathbb{Q}$, and a mapping from a thread id and an address to the last store to this address propagated to this thread $\mathsf{prop}\colon \mathsf{TID} \times \mathsf{ADDR} \to \{\mathsf{init}_a \mid a \in \mathsf{ADDR}\} \cup \mathsf{TID} \times \mathbb{N}$. The initial state of the storage subsystem is $s_{0Y} := (\lambda\mathsf{tid}.\lambda i.0, \lambda\mathsf{tid}.\lambda a.\mathsf{init}_a)$.

The initial state of automaton $Z(\mathcal{P})$ is $s_{0Z} := (\lambda\mathsf{tid}.s_{0X}, s_{0Y})$.

Transition Relation. Fix a state $s_Z = (\mathsf{ts}, s_Y)$ with $s_Y = (\mathsf{co}, \mathsf{prop})$ and a thread id $\mathsf{tid} \in \mathsf{TID}$ with runtime state $\mathsf{ts}(\mathsf{tid}) = (\mathsf{fetched}, \mathsf{committed}, \mathsf{loaded})$.

Let $\mathsf{eval}(\mathsf{tid}, i, e)$ return the value in DOM of expression e in the i'th fetched instruction of thread tid, or $\bot$ when the value is undefined. Let $\mathsf{addr}(\mathsf{tid}, i)$ and $\mathsf{val}(\mathsf{tid}, i)$ return the values of the address and value arguments of the i'th fetched instruction of thread tid. We use the special value $\top$ if the instruction has no such arguments. The expressions $\mathsf{addrdep}(\mathsf{tid}, i)$, $\mathsf{datadep}(\mathsf{tid}, i)$, $\mathsf{ctrldep}(\mathsf{tid}, i)$ denote the sets of indices of instructions in thread tid being respectively address, data, and control dependencies of the i'th instruction.

Let $\mathcal{T}_{\mathsf{tid}} = (Q_{\mathsf{tid}}, \mathsf{CMD}, \mathcal{I}_{\mathsf{tid}}, q_{0\mathsf{tid}}, Q_{\mathsf{tid}}) \in \mathcal{P}$. The transition relation Δ_Z is the smallest relation defined by the rules below:

POW-FETCH. Consider $\mathsf{instr} \in \mathcal{I}_{\mathsf{tid}}$ with $\mathsf{src}(\mathsf{instr}) = \mathsf{dst}(\mathsf{last}(\mathsf{fetched}))$ or $\mathsf{src}(\mathsf{instr}) = q_{0\mathsf{tid}}$ if $\mathsf{fetched} = \varepsilon$, then:

$$(\mathsf{ts}, s_Y) \xrightarrow{(\mathsf{fetch}, \mathsf{tid}, \mathsf{instr})} (\mathsf{ts}[\mathsf{tid} \hookleftarrow (\mathsf{fetched} \cdot \mathsf{instr}, \mathsf{committed}, \mathsf{loaded})], s_Y).$$

POW-LOAD. If $\mathsf{fetched}[i]$ is a load, $\mathsf{loaded}[i] = \bot$, $a = \mathsf{addr}(\mathsf{tid}, i) \neq \bot$, then:

$$(\mathsf{ts}, s_Y) \xrightarrow{(\mathsf{load}, \mathsf{tid}, i, a)} (\mathsf{ts}[\mathsf{tid} \hookleftarrow (\mathsf{fetched}, \mathsf{committed}, \mathsf{loaded}[i \hookleftarrow \mathsf{prop}(\mathsf{tid}, a)])], s_Y).$$

POW-EARLY. Let fetched$[i]$ be a load, loaded$[i] = \perp$, and a $=$ addr(tid, i) $\neq$ $\perp$. Let $i' \in [1..i-1]$ be the greatest index such that fetched$[i']$ is a store with a$' =$ addr(tid, i') $\in \{$a$, \perp\}$. If a$' \neq \perp$, val(tid, i') $\neq \perp$, $i' \notin$ committed, then:

$(\text{ts}, s_Y) \xrightarrow{(\text{load},\text{tid},i,\text{a})} (\text{ts}[\text{tid} \leftarrow (\text{fetched}, \text{committed}, \text{loaded}[i \leftarrow (\text{tid}, i')])], s_Y)$.

POW-COMMIT. Consider $i \in [1..|\text{fetched}|] \setminus \text{committed}$ where fetched$[i]$ is not a store. Assume addrdep(tid, i) $\cup$ datadep(tid, i) $\cup$ ctrldep(tid, i) $\subseteq$ committed. Assume a $=$ addr(tid, i) $\neq \perp$, v $=$ val(tid, i) $\neq \perp$. If a $\neq \top$, assume $\{i' \in [1..i-1] \mid$ addr(tid, i') $\in \{$a$, \perp\}\} \subseteq$ committed. In case fetched$[i]$ is a load, assume loaded$[i] \neq \perp$. In case fetched$[i]$ is an `assume()`, assume v $\neq 0$. Then:

$(\text{ts}, s_Y) \xrightarrow{(\text{commit},\text{tid},i)} (\text{ts}[\text{tid} \leftarrow (\text{fetched}, \text{committed} \cup \{i\}, \text{loaded})], s_Y)$.

POW-STORE. Assume all the preconditions from the previous rule hold, but fetched$[i]$ is a store. Choose a coherence key k $\in \mathbb{Q}$ such that there is no tid$' \in$ TID, $i' \in \mathbb{N}$ for which co(tid$', i'$) $=$ k. Then:

$(\text{ts}, s_Y) \xrightarrow{(\text{commit},\text{tid},i,\text{k},\text{a})} (\text{ts}[\text{tid} \leftarrow (\text{fetched}, \text{committed} \cup \{i\}, \text{loaded})], s_Y')$,

where $s_Y' := (\text{co}[(\text{tid}, i) \leftarrow \text{k}], \text{prop})$.

Additionally, this transition is immediately followed by a POW-PROP transition propagating the store to the thread where it was committed.

POW-PROP. Consider tid$' \in$ TID, $i' \in \mathbb{N}$ with co(tid$', i'$) $\neq \perp$. Let a $=$ addr(tid$', i'$). Assume co(prop(tid, a)) $<$ co(tid$', i'$). Then:

$(\text{ts}, s_Y) \xrightarrow{(\text{prop},\text{tid},\text{tid}',i',\text{a})} (\text{ts}, (\text{co}, \text{prop}[(\text{tid}, \text{a}) \leftarrow (\text{tid}', i')]))$.

Final States. The set of final states $F_Z \subseteq S_Z$ consists of all states $s_Z = (\text{ts}, (\text{co}, \text{prop})) \in S_Z$, such that for each tid $\in$ TID, ts$[\text{tid}] = (\text{fetched}, \text{committed}, \text{loaded})$ the following holds:

FIN-COMM. All instructions are committed: committed $= [1..|\text{fetched}|]$.

FIN-LD. Loads agree with the coherence order. Let fetched$[i]$ be a load, and fetched$[i']$ be an earlier load to the same address: $i' < i$, addr(tid, i) $=$ addr(tid, i'). Then co(loaded$[i']$) $\leq$ co(loaded$[i]$).

FIN-LD-ST. Loads and stores in the same thread agree with the coherence order. Let fetched$[i]$ be a load, let fetched$[i']$ be an earlier store to the same address: $i' < i$, addr(tid, i) $=$ addr(tid, i'). Then co(tid, i') $\leq$ co(loaded$[i]$).

The set of all *Power computations of program* $\mathcal{P}$ is $\mathsf{C}_{\text{power}}(\mathcal{P}) := \mathcal{L}(Z(\mathcal{P}))$. The set of all *SC computations of the program* $\mathsf{C}_{\text{sc}}(\mathcal{P}) \subseteq \mathsf{C}_{\text{power}}(\mathcal{P})$ includes only those computations where each instruction is executed atomically, and stores are immediately propagated to all threads.

Example 1. $\sigma_{MP} = \text{fetch}(a) \cdot \text{commit}(a) \cdot \text{prop}(a, 1) \cdot \text{fetch}(b) \cdot \text{commit}(b) \cdot \text{prop}(b, 1) \cdot \text{prop}(b, 2) \cdot \text{fetch}(c) \cdot \text{fetch}(d) \cdot \text{load}(c) \cdot \text{load}(d) \cdot \text{commit}(d) \cdot \text{commit}(c)$ is a feasible Power computation of program MP in Figure 1 (we simplified the events by removing information unimportant for this example). Load c reads value 1 written by store b, because b is propagated to thread 2 before the load(c) event. Store a is never propagated to thread 2, consequently, d reads the initial value 0.

3   Robustness

Intuitively, a *trace* $T(\sigma)$ abstracts a program computation σ to the dataflow and control-flow relations between instructions. Formally, the trace of σ is a directed graph $T(\sigma) := (V, \to_{po}, \to_{co}, \to_{src}, \to_{cf})$ with nodes V and four kinds of arcs. The nodes are instructions together with their thread identifiers and fetch indices (in order to distinguish instructions executed in different threads and the same instruction executed multiple times in the same thread): $V \subseteq \{\text{init}_a \mid a \in \text{ADDR}\} \cup \bigcup_{\text{tid}\in\text{TID}} \{\text{tid}\} \times \mathbb{N} \times \mathcal{I}_{\text{tid}}$. The *program order* $\to_{po}$ is the order in which instructions were fetched in each thread. The *coherence order* $\to_{co}$ gives the global ordering of stores to each address. The *source order* $\to_{src}$ shows the store from which a load took its value. The *conflict order* $\to_{cf}$ shows, for a load, the stores to the same address following the store the load took its value from. We define the *happens-before* relation as $\to_{hb} := \to_{po} \cup \to_{co} \cup \to_{src} \cup \to_{cf}$.

We also need address $\to_{addr}$ and data $\to_{data}$ dependence relations (defined as expected based on addrdep and datadep). Since $\to_{po}$ includes all the information from the fetched component of a thread state, $\to_{addr}$ and $\to_{data}$ can be reconstructed from $\to_{po}$ by inspecting the instructions labeling the nodes. They are therefore not included in the trace explicitly.

The *robustness problem* is, given a program $\mathcal{P}$, to check whether the set of all traces under Power is a subset of all traces under SC: $T_{\text{power}}(\mathcal{P}) \subseteq T_{\text{sc}}(\mathcal{P})$, where $T_{\text{mm}}(\mathcal{P}) := \{T(\sigma) \mid \sigma \in \mathsf{C}_{\text{mm}}(\mathcal{P})\}$ for mm $\in \{\text{power}, \text{sc}\}$.

Shasha and Snir have shown that a trace belongs to an SC computation iff its happens-before relation is acyclic:

Lemma 1 ([16]). *A program $\mathcal{P}$ is robust against Power iff there is no trace $T \in T_{\text{power}}(\mathcal{P})$ with cyclic $\to_{hb}$.*

Example 2. The trace of computation σ_{MP} (Figure 2) has a cyclic happens-before relation. By Lemma 1, this means that the program is not robust. Indeed, in no SC computation load d can read 0 whereas c has read 1.

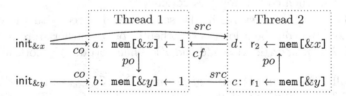

Fig. 2. Trace of computation σ_{MP} from Example 1

4   Normal-Form Computations

We say that a computation $\tau \in \mathsf{C}_{\text{power}}(\mathcal{P})$ is *in normal form of degree n* if there is a partitioning $\tau = \tau_1 \cdots \tau_n$, such that

NF-A $(\tau_2 \cdots \tau_n)\!\downarrow\!\text{fetch} = \varepsilon$.
NF-B For $j \in \{1,2\}$ let e_j, e'_j be events related to instruction (tid_j, i_j). If $e_1, e_2 \in \tau_s$ and $e'_1, e'_2 \in \tau_{s'}$, then $e_1 <_{\tau_s} e_2$ iff $e'_1 <_{\tau_{s'}} e'_2$.

With NF-A, all fetch events occur in τ_1. With NF-B, the different parts of the computation have the same ordering of related events. In the rest of this section we prove the following theorem:

Theorem 1. *A program is robust iff it has no normal-form computation of degree $|\mathcal{P}| + 3$ with cyclic happens-before relation.*

Consider $\sigma \in C_{mm}(\mathcal{P})$. By $\sigma \setminus (tid, i)$ we denote the computation obtained from σ by deleting all events related to the i'th fetched instruction in thread tid.

Lemma 2. *Consider a non-empty computation $\sigma \in C_{power}(\mathcal{P})$. Then there is a (tid_x, i_x), such that $\sigma' = \sigma \setminus (tid_x, i_x)$ satisfies $|\sigma'| < |\sigma|$ and $\sigma' \in C_{power}(\mathcal{P})$.*

Proof. Consider the last fetched instruction in each thread. If among such instructions there is a non-store instruction, delete it: its result cannot be used by any other instruction. If all these instructions are stores, delete the one, on which (1) no load or store depends via $(\to_{src} \cup \to_{data})^+ \cdot \to_{addr}$, and (2) no condition depends via $(\to_{src} \cup \to_{data})^+$.

Towards a contradiction, assume there is no such store. Consider the last fetched (store) instruction in a thread tid_1: (tid_1, i_1). Case 1: there is a load or a store (tid_2, i'_2) whose address depends on (tid_1, i_1). Case 2: there is a condition (tid_2, i'_2) whose value depends on (tid_1, i_1). Consider the last fetched instruction in thread tid_2: (tid_2, i_2). It must be a store, and it must have been committed after (tid_1, i_1): a store can only be committed after all loads and stores fetched before it have their addresses determined (Case 1) and after all preceding conditions are committed (Case 2).

Continuing the reasoning, for any last fetched instruction in a thread (tid_j, i_j) there is a last instruction in a different thread (tid_{j+1}, i_{j+1}) which must have been committed later. Taking into account finiteness of the number of threads, we get a contradiction. □

Fix a program $\mathcal{P}$. Consider a shortest Power computation $\alpha \in C_{power}(\mathcal{P})$ with cyclic $\to_{hb}$. Let (tid_x, i_x) be the instruction determined by Lemma 2. Let $\alpha := \alpha_1 \cdot x_1 \cdot \alpha_2 \cdot x_2 \cdots \alpha_n$, where $\{x_1 \ldots x_{n-1}\}$ are the events related to the i_x'th instruction fetched in thread tid_x. Then $\alpha \setminus (tid_x, i_x) := \alpha' := \alpha_1 \cdot \alpha_2 \cdots \alpha_n$. Since α' is shorter than α, its $\to_{hb}$ is acyclic. Therefore, there is a computation $\beta \in C_{sc}(\mathcal{P})$ with $T(\beta) = T(\alpha')$.

The computations β and α' consist of the same fetch, load, and commit events: fetch events are determined by $\to_{po}$; address component a of load and store commit events is determined by $\to_{addr}$, $\to_{data}$ (derivable from $\to_{po}$), and $\to_{src}$; since $\to_{co}$ is the same for both computations, we can assume that matching store commit events have the same value of coherence key k. Notably, β can have more propagate events than α' as the Power semantics does not guarantee that all stores are propagated to all threads. Now we reorder the events in each part α_j of α in the way they follow in β. This gives the computation $\gamma := \beta{\downarrow}\alpha_1 \cdot x_1 \cdot \beta{\downarrow}\alpha_2 \cdot x_2 \cdots \beta{\downarrow}\alpha_n$.

Lemma 3. $\gamma \in C_{power}(\mathcal{P})$ *and* $T(\gamma) = T(\alpha)$.

Wlog we may assume that all fetch events of α are located within $\alpha_1 \cdot x_1$: every thread can always first fetch all instructions and in the rest of the computation only execute them; such a reordering does not change the trace. Also, note that the maximal number of events an instruction can generate is $|\mathcal{P}| + 2$. This bound is achieved by a store that is fetched, committed, and propagated to all threads. Then the following lemma holds; together with Lemma 1 it proves Theorem 1.

Lemma 4. *Computation γ is in normal form of degree $|\mathcal{P}| + 3$.*

Example 3. Consider $\alpha := \mathsf{fetch}(c) \cdot \mathsf{fetch}(d) \cdot \mathsf{fetch}(a) \cdot \cancel{\mathsf{fetch}(b)} \cdot \mathsf{commit}(a) \cdot \mathsf{prop}(a, 1) \cdot \cancel{\mathsf{commit}(b)} \cdot \cancel{\mathsf{prop}(b, 1)} \cdot \cancel{\mathsf{prop}(b, 2)} \cdot \mathsf{load}(c) \cdot \mathsf{load}(d) \cdot \mathsf{commit}(d) \cdot \mathsf{commit}(c)$, which is σ_{MP} with fetch events moved to the front. We cancel the x_i events (crossed out) related to store instruction b, as b is the last instruction of thread 1 and no address depends on it (we could also cancel the events of d instead). Therefore, $\alpha_1 := \mathsf{fetch}(c) \cdot \mathsf{fetch}(d) \cdot \mathsf{fetch}(a)$, $\alpha_2 := \mathsf{commit}(a) \cdot \mathsf{prop}(a, 1)$, $\alpha_3 := \alpha_4 := \varepsilon$, $\alpha_5 := \mathsf{load}(c) \cdot \mathsf{load}(d) \cdot \mathsf{commit}(d) \cdot \mathsf{commit}(c)$, and $\alpha' := \alpha_1 \cdot \alpha_2 \cdot \alpha_3 \cdot \alpha_4 \cdot \alpha_5$. The trace of α' is the trace of α (Figure 2) with node b and adjacent arcs removed, and a source arc from init_y to c added. The SC computation with the same trace is $\beta := \mathsf{fetch}(c) \cdot \mathsf{load}(c) \cdot \mathsf{commit}(c) \cdot \mathsf{fetch}(d) \cdot \mathsf{load}(d) \cdot \mathsf{commit}(d) \cdot \mathsf{fetch}(a) \cdot \mathsf{commit}(a) \cdot \mathsf{prop}(a, 1) \cdot \mathsf{prop}(a, 2)$. The normal-form computation is $\gamma := \beta {\downarrow} \alpha_1 \cdot x_1 \cdots \beta {\downarrow} \alpha_5 = (\mathsf{fetch}(c) \cdot \mathsf{fetch}(d) \cdot \mathsf{fetch}(a)) \cdot \mathsf{fetch}(b) \cdot (\mathsf{commit}(a) \cdot \mathsf{prop}(a, 1)) \cdot \mathsf{commit}(b) \cdot \mathsf{prop}(b, 1) \cdot \mathsf{prop}(b, 2) \cdot (\mathsf{load}(c) \cdot \mathsf{commit}(c) \cdot \mathsf{load}(d) \cdot \mathsf{commit}(d))$. It is feasible and has the same trace as α and σ_{MP} (Figure 2).

5   From Normal-Form Computations to Emptiness

We now reduce robustness to language emptiness. First, we define a multiheaded automaton capable of generating all normal-form computations of a program. Next, we intersect it with regular languages that check cyclicity of the happens-before relation. Altogether, the program is robust iff the intersection is empty.

5.1   Generating Normal-Form Computations

To generate all normal-form computations, we use multiheaded automata [8]. A multiheaded automaton generates a computation $\sigma_1 \ldots \sigma_n$ by simultaneously generating its parts σ_i. The automaton has a head for each part, and transitions define the head producing an event. Formally, an *n-headed automaton over Σ* is an automaton over an extended alphabet: $A = (S, [1..n] \times \Sigma, \Delta, s_0, F)$. The *language* is $\mathcal{L}(A) := \{\mathsf{second}(\sigma {\downarrow} (\{1\} \times \Sigma) \cdots \sigma {\downarrow} (\{n\} \times \Sigma)) \mid s_0 \xrightarrow{\sigma} s \in F\}$, where $\mathsf{second}((a_1, b_1) \cdots (a_m, b_m)) := b_1 \cdots b_m$. Multiheaded automata are closed under regular intersection, and language emptiness is NL-complete [8].

We generate all normal-form computations of program $\mathcal{P}$ with the n-headed automaton $M(\mathcal{P}) := (S_M, \mathsf{E}, \Delta_M, s_{0M}, F_M)$, where $n := |\mathcal{P}| + 3$. The automaton generates all events related to a single instruction in one shot, but, possibly, in different parts of the computation. All fetch events are generated in the first part. To generate them, the automaton stores the destination state of the last fetched instruction in each thread (component ctrl-state of the automaton state).

Each instruction can only read the last value written to a register. Therefore, the automaton only needs to remember |REG| register values per thread (component reg-value). However, an instruction cannot be executed until the values of all registers that it reads become known. To obey this restriction, the automaton memorizes the part of the computation in which the register value gets computed (reg-comp-head). For example, while handling an assignment $r_1 \leftarrow r_1 + r_2$, the automaton learns that the new value of r_1 is the sum of the current values of r_1 and r_2. It also remembers that this value is available no earlier than the current values of r_1 and r_2 are computed. Similarly, the automaton remembers the parts of the computation in which the addresses of load and store instructions become known (addr-comp-head), and certain kinds of instructions get committed (reg-comm-head, assume-comm-head, addr-comm-head).

The automaton has to keep a separate memory state for each thread and for each part of the computation. The memory state of a thread in a part is updated when a store instruction gets propagated to this thread in this part. When a load instruction is handled, the automaton chooses a part where the load event takes place and uses the memory state of that part. Besides the memory valuation (mem-value), the memory state includes coherence keys (last-key) to guarantee that the generated computation respects the coherence order.

When starting the computation, the automaton non-deterministically guesses the memory valuations and coherence keys for all parts of the computation (except the first one). Upon termination, the automaton checks that the parts of the computation generated by each head fit together at the concatenation points. This ensures the overall computation is valid for the program. The trick is to remember the guess of the initial memory valuations and coherence keys in immutable components of the automaton state (mem-value$_g$, last-key$_g$). The final states require that the current memory state in part h of the computation coincides with the guessed initial state in part h + 1.

We now formally define the transition rules for assignments and loads. The remaining rules are given in [9]. Fix a state s_M and consider a thread tid $\in$ TID in control state ctrl-state(tid) = q_1 and an instruction instr = $q_1 \xrightarrow{\text{cmd}} q_2 \in \mathcal{I}_{\text{tid}}$. The automaton uses three indices from HEAD := $[1..n]$. Index $h_1 := 1$ denotes the part where the automaton generates fetch events. Index $h_2 \in$ HEAD refers to the part in which the computation of the instruction takes place. There are constraints on this index. The instruction has to be fetched, $h_2 \geq h_1$, and the computation can only complete when the value of each register r read in cmd has been computed: $h_2 \geq$ reg-comp-head(tid, r). Finally, index $h_3 \in$ HEAD determines the part of the computation where the instruction is committed. An instruction has to be computed to be committed: $h_3 \geq h_2$. Moreover, the last assignment to each register r read in cmd has to be committed before cmd can be committed itself: $h_3 \geq$ reg-comm-head(tid, r). The instruction count is incremented with each instruction: let $i :=$ instr-count(tid) + 1, then instr-count$' :=$ instr-count[tid $\leftarrow i$]. We use primed variables for the new values of state components. We overload eval(tid, e) to mean the value of expression e for the valuation of registers defined by λr.reg-value(tid, r).

MH-ASSIGN. For an assignment $\mathsf{cmd} = \mathsf{r} \leftarrow e_\mathsf{v}$, let $\mathsf{v} := \mathsf{eval}(\mathsf{tid}, e_\mathsf{v})$ be the value. We update the register $\mathsf{reg\text{-}value}' := \mathsf{reg\text{-}value}[(\mathsf{tid}, \mathsf{r}) \leftarrow \mathsf{v}]$ and store the part in which the value has been computed $\mathsf{reg\text{-}comp\text{-}head}' := \mathsf{reg\text{-}comp\text{-}head}[(\mathsf{tid}, \mathsf{r}) \leftarrow h_2]$. We also keep the part where it has been committed $\mathsf{reg\text{-}comm\text{-}head}' := \mathsf{reg\text{-}comm\text{-}head}[(\mathsf{tid}, \mathsf{r}) \leftarrow h_3]$. The transition is labeled by $\lambda := (h_1, \mathsf{fetch}, \mathsf{tid}, \mathsf{instr}) \cdot (h_3, \mathsf{commit}, \mathsf{tid}, i)$, which means it is actually decomposed into two transitions.

MH-LOAD. Let $\mathsf{cmd} = \mathsf{r} \leftarrow \mathsf{mem}[e_\mathsf{a}]$ and $\mathsf{a} := \mathsf{eval}(\mathsf{tid}, e_\mathsf{a})$. All preceding accesses to this address have to be committed before the load can be committed: $h_3 \geq \mathsf{addr\text{-}comm\text{-}head}(\mathsf{tid}, \mathsf{a})$. The value stems either from memory or from an early read. In the former case, we check that there are no pending stores $\mathsf{early\text{-}mem\text{-}value}(\mathsf{tid}, \mathsf{a}) = \perp$ and set $\mathsf{v} := \mathsf{mem\text{-}value}(\mathsf{tid}, \mathsf{a}, h_2)$. In the latter case, we find a pending store and make sure there is no later store with an undetermined address: $\mathsf{v} := \mathsf{early\text{-}mem\text{-}value}(\mathsf{tid}, \mathsf{a}, h_2)$ with $\mathsf{v} \neq \top$. We modify the register information as for assignments. We update the index of the leftmost part of the computation where all addresses are determined: $\mathsf{addr\text{-}comp\text{-}head}' := \mathsf{addr\text{-}comp\text{-}head}[\mathsf{tid} \leftarrow \max\{\mathsf{addr\text{-}comp\text{-}head}(\mathsf{tid}), h_2\}]$. We also remember the position of the last commit to the current address: $\mathsf{addr\text{-}comm\text{-}head}' := \mathsf{addr\text{-}comm\text{-}head}[(\mathsf{tid}, \mathsf{a}) \leftarrow h_3]$. The transition label is $\lambda := (h_1, \mathsf{fetch}, \mathsf{tid}, \mathsf{instr}) \cdot (h_2, \mathsf{load}, \mathsf{tid}, i, \mathsf{a}) \cdot (h_3, \mathsf{commit}, \mathsf{tid}, i)$.

The set of final states F_M consists of all states with $\mathsf{mem\text{-}value}(\mathsf{tid}, \mathsf{a}, h) = \mathsf{mem\text{-}value}_g(\mathsf{tid}, \mathsf{a}, h+1)$ and $\mathsf{last\text{-}key}(\mathsf{tid}, \mathsf{a}, h) = \mathsf{last\text{-}key}_g(\mathsf{tid}, \mathsf{a}, h+1)$.

Lemma 5. $\{\tau \in C_{power}(\mathcal{P}) \mid \tau \text{ is in normal form of degree } n\} \subseteq \mathcal{L}(M(\mathcal{P}))$ and $\mathcal{L}(M(\mathcal{P})) \subseteq C_{power}(\mathcal{P})$.

5.2 Checking Cyclicity of the Happens-Before Relation

We call a happens-before cycle *beautiful*, if it has the following form:

$$(\mathsf{tid}_1, i_1, \mathsf{instr}_1) \rightarrow_{po}{}^* (\mathsf{tid}_1, i_1', \mathsf{instr}_1') \rightarrow_{hop} \cdots$$
$$\rightarrow_{hop} (\mathsf{tid}_n, i_n, \mathsf{instr}_n) \rightarrow_{po}{}^* (\mathsf{tid}_n, i_n', \mathsf{instr}_n') \rightarrow_{hop} (\mathsf{tid}_1, i_1, \mathsf{instr}_1).$$

Here, $\rightarrow_{hop} := (\rightarrow_{co} \cup \rightarrow_{src} \cup \rightarrow_{cf})$ and $\mathsf{tid}_k \neq \mathsf{tid}_l$ for $k \neq l$. We call $\theta := \mathsf{tid}_1 \ldots \mathsf{tid}_n$ the *profile* of the cycle.

Example 4. The happens-before cycle shown in Figure 2 is beautiful.

Lemma 6 ([8]). *A computation $\tau \in C_{power}(\mathcal{P})$ has a happens-before cycle iff it has a beautiful happens-before cycle.*

Given a cycle profile θ, we define the automaton $M'(\mathcal{P}, \theta)$ as a modification of $M(\mathcal{P})$ that marks one event in each thread $\mathsf{tid}_j \in \theta$ with enter (identifying $(\mathsf{tid}_j, i_j, *)$) and a later (or the same) event with leave (identifying $(\mathsf{tid}_j, i_j', *)$, $i_j \leq i_j'$). Note that $M(\mathcal{P})$ generates the events in program order, which ensures $(\mathsf{tid}_j, i_j, *) \rightarrow_{po}{}^* (\mathsf{tid}_j, i_j', *)$. To check $(\mathsf{tid}_j, i_j', *) \rightarrow_{hop} (\mathsf{tid}_{j+1}, i_{j+1}, *)$, we use an intersection with a regular language $H^{\mathsf{tid}_j, \mathsf{tid}_{j+1}}$.

Lemma 7. *Program $\mathcal{P}$ has a beautiful cycle with profile $\theta = tid_1 \ldots tid_n$ iff*

$$M'(\mathcal{P}, \theta) \cap H^{tid_1, tid_2} \cap \ldots \cap H^{tid_n, tid_1} \neq \emptyset.$$

Automaton $M(\mathcal{P})$ is infinite-state. To ensure $M'(\mathcal{P}, \theta)$ has finitely many states, we note that the instruction indices are irrelevant for the detection of happens-before cycles (instr-count can be dropped), and that the number of different coherence keys that must be stored in the state at any moment is polynomial in the size of $\mathcal{P}$. Together with the observation that emptiness is in NL, we obtain a PSPACE upper bound for robustness. The lower bound is by a reduction of SC-reachability similar to [5].

Theorem 2. *Robustness against Power is PSPACE-complete.*

Acknowledgements. The authors thank Parosh Aziz Abdulla, Jade Alglave, Mohamed Faouzi Atig, Ahmed Bouajjani, and Carl Leonardsson for helpful discussions on Power and the reviewers for suggestions on the presentation. The authors were supported by Fraunhofer ITWM and the DFG project R2M2.

References

1. Alglave, J.: Personal communication (October 2013)
2. Alglave, J., Maranget, L.: Stability in weak memory models. In: Gopalakrishnan, G., Qadeer, S. (eds.) CAV 2011. LNCS, vol. 6806, pp. 50–66. Springer, Heidelberg (2011)
3. Alglave, J., Maranget, L., Tautschnig, M.: Herding cats. ACM TOPLAS (to appear, 2014)
4. Bouajjani, A., Derevenetc, E., Meyer, R.: Checking and enforcing robustness against TSO. In: Felleisen, M., Gardner, P. (eds.) ESOP 2013. LNCS, vol. 7792, pp. 533–553. Springer, Heidelberg (2013)
5. Bouajjani, A., Meyer, R., Möhlmann, E.: Deciding robustness against total store ordering. In: Aceto, L., Henzinger, M., Sgall, J. (eds.) ICALP 2011, Part II. LNCS, vol. 6756, pp. 428–440. Springer, Heidelberg (2011)
6. Burckhardt, S., Musuvathi, M.: Effective program verification for relaxed memory models. In: Gupta, A., Malik, S. (eds.) CAV 2008. LNCS, vol. 5123, pp. 107–120. Springer, Heidelberg (2008)
7. Burnim, J., Sen, K., Stergiou, C.: Sound and complete monitoring of sequential consistency for relaxed memory models. In: Abdulla, P.A., Leino, K.R.M. (eds.) TACAS 2011. LNCS, vol. 6605, pp. 11–25. Springer, Heidelberg (2011)
8. Calin, G., Derevenetc, E., Majumdar, R., Meyer, R.: A theory of partitioned global address spaces. In: FSTTCS. LIPIcs, vol. 24, pp. 127–139 (2013)
9. Derevenetc, E., Meyer, R.: Robustness against Power is PSPACE-complete. CoRR, abs/1404.7092 (2014), http://arxiv.org/abs/1404.7092
10. Lamport, L.: Time, clocks, and the ordering of events in a distributed system. CACM 21(7), 558–565 (1978)
11. Lamport, L.: How to make a multiprocessor computer that correctly executes multiprocess programs. IEEE Transactions on Computers 28(9), 690–691 (1979)
12. Mador-Haim, S., et al.: An axiomatic memory model for POWER multiprocessors. In: Madhusudan, P., Seshia, S.A. (eds.) CAV 2012. LNCS, vol. 7358, pp. 495–512. Springer, Heidelberg (2012)

13. Maranget, L., Sarkar, S., Sewell, P.: A tutorial introduction to the ARM and POWER relaxed memory models,
https://www.cl.cam.ac.uk/~pes20/ppc-supplemental/test7.pdf
14. Owens, S., Sarkar, S., Sewell, P.: A better x86 memory model: x86-TSO (extended version). Technical Report CL-TR-745, University of Cambridge (2009)
15. Sarkar, S., Sewell, P., Alglave, J., Maranget, L., Williams, D.: Understanding POWER multiprocessors. In: PLDI, pp. 175–186. ACM (2011)
16. Shasha, D., Snir, M.: Efficient and correct execution of parallel programs that share memory. TOPLAS 10(2), 282–312 (1988)

A Nivat Theorem for Weighted Timed Automata and Weighted Relative Distance Logic

Manfred Droste and Vitaly Perevoshchikov*

Universität Leipzig, Institut für Informatik,
04109 Leipzig, Germany
{droste,perev}@informatik.uni-leipzig.de

Abstract. Weighted timed automata (WTA) model quantitative aspects of real-time systems like continuous consumption of memory, power or financial resources. They accept quantitative timed languages where every timed word is mapped to a value, e.g., a real number. In this paper, we prove a Nivat theorem for WTA which states that recognizable quantitative timed languages are exactly those which can be obtained from recognizable boolean timed languages with the help of several simple operations. We also introduce a weighted extension of relative distance logic developed by Wilke, and we show that our weighted relative distance logic and WTA are equally expressive. The proof of this result can be derived from our Nivat theorem and Wilke's theorem for relative distance logic. Since the proof of our Nivat theorem is constructive, the translation process from logic to automata and vice versa is also constructive. This leads to decidability results for weighted relative distance logic.

Keywords: Weighted timed automata, linearly priced timed automata, average behavior, discounting, Nivat's theorem, quantitative logic.

1 Introduction

Timed automata introduced by Alur and Dill [1] are a prominent model for real-time systems. Timed automata form finite representations of infinite-state automata for which various fundamental results from the theory of finite-state automata can be transferred to the timed setting. Although time has a quantitative nature, the questions asked in the theory of timed automata are of a qualitative kind. On the other side, quantitative aspects of systems, e.g., costs, probabilities and energy consumption can be modelled using weighted automata, i.e., classical nondeterministic automata with a transition weight function. The behaviors of weighted automata can be considered as quantitative languages (also known as formal power series) where every word carries a value. Semiring-weighted automata have been extensively studied in the literature (cf. [6, 17, 20] and the handbook of weighted automata [12]).

Weighted extensions of timed automata are of much interest for the real-time community, since weighted timed automata (WTA) can model continuous time-dependent consumption of resources. In the literature, various models of WTA were considered,

---

* Supported by DFG Graduiertenkolleg 1763 (QuantLA).

J. Esparza et al. (Eds.): ICALP 2014, Part II, LNCS 8573, pp. 171–182, 2014.
© Springer-Verlag Berlin Heidelberg 2014

e.g., linearly priced timed automata [3, 4, 21], multi-weighted timed automata with knapsack-problem objective [22], and WTA with measures like average, reward-cost ratio [7, 8] and discounting [2, 18, 19]. In [24, 25], WTA over semirings were studied with respect to classical automata-theoretic questions. However, various models, e.g., WTA with average and discounting measures as well as multi-weighted automata cannot be defined using semirings. For the latter situations, only several algorithmic problems were handled. But many questions whether the results known from the theories of timed and weighted automata also hold for WTA remain open. Moreover, there is no unified framework for WTA.

The main goal of this paper is to build a bridge between the theories of WTA and timed automata. First, we develop a general model of *timed valuation monoids* for WTA. Recall that Nivat's theorem [23] is one of the fundamental characterizations of rational transductions and establishes a connection between rational transductions and rational languages. Our first main result is an extension of Nivat's theorem to WTA over timed valuation monoids. By Nivat's theorem for semiring-weighted automata described recently in [13], recognizable quantitative languages are exactly those which can be constructed from recognizable languages using operations like morphisms and intersections. The proof of this result requires the fact that finite automata are determinizable. However, timed automata do not enjoy this property. Nevertheless, for idempotent timed valuation monoids which model all mentioned examples of WTA, we do not need determinization. In this case, our Nivat theorem for WTA is similar to the one for weighted automata. In the non-idempotent case, we give an example showing that this statement does not hold true. But in this case we can establish a connection between recognizable quantitative timed languages and sequentially, deterministically or unambiguously recognizable timed languages.

As an application of our Nivat theorem, we provide a characterization of recognizable quantitative timed languages by means of quantitative logics. The classical Büchi-Elgot theorem [9] was extended to both weighted [10, 11, 14] and timed settings [26, 27]. In [24, 25], a semiring-weighted extension of Wilke's relative distance logic [26, 27] was considered. Here, we develop a different weighted version of relative distance logic based on our notion of timed valuation monoids. In our second main result, we show that this logic and WTA have the same expressive power. For the proof of this result, we use a new proof technique and our Nivat theorem to derive our result from the corresponding result for unweighted logic [26, 27]. Since the proof of our Nivat theorem is constructive, the translation process from weighted relative distance logic to WTA and vice versa is constructive. This leads to decidability results for weighted relative distance logic. In particular, based on the results of [3, 4, 21], we show the decidability of several weighted extensions of the satisfiability problem for our logic.

2 Timed Automata

An *alphabet* is a non-empty finite set. Let Σ be a non-empty set. A *finite word* over Σ is a finite sequence $a_1...a_n$ where $n \geq 0$ and $a_1, ..., a_n \in \Sigma$. If $n \geq 1$, then we say that w is *non-empty*. Let Σ^+ denote the set of all non-empty words over Σ. Let $\mathbb{R}_{\geq 0}$ denote the set of all non-negative real numbers. A *finite timed word* over Σ is a finite word over

$\Sigma \times \mathbb{R}_{\geq 0}$, i.e., a finite sequence $w = (a_1, t_1)...(a_n, t_n)$ where $n \geq 0, a_1, ..., a_n \in \Sigma$ and $t_1, ..., t_n \in \mathbb{R}_{\geq 0}$. Let $|w| = n$ and $\langle w \rangle = t_1 + ... + t_n$ and let $\mathbb{T}\Sigma^+ = (\Sigma \times \mathbb{R}_{\geq 0})^+$, the set of all non-empty finite timed words. Any set $\mathcal{L} \subseteq \mathbb{T}\Sigma^+$ of timed words is called a *timed language*.

Let C be a finite set of *clock variables* ranging over $\mathbb{R}_{\geq 0}$. A *clock constraint* over C is either TRUE or (if C is non-empty) a finite conjunction of formulas of the form $x \bowtie c$ where $x \in C, c \in \mathbb{N}$ and $\bowtie \in \{<, \leq, =, \geq, >\}$. Let $\Phi(C)$ denote the set of all clock constraints over C. A *clock valuation* over C is a mapping $\nu : C \to \mathbb{R}_{\geq 0}$ which assigns a value to each clock variable. Let $\mathbb{R}_{\geq 0}^C$ be the set of all clock valuations over C. The *satisfaction relation* $\models \subseteq \mathbb{R}_{\geq 0}^C \times \Phi(C)$ is defined as usual. Now let $\nu \in \mathbb{R}_{\geq 0}^C, t \in \mathbb{R}_{\geq 0}$ and $\Lambda \subseteq C$. Let $\nu + t$ denote the clock valuation $\nu' \in \mathbb{R}_{\geq 0}^C$ such that $\nu'(x) = \nu(x) + t$ for all $x \in C$. Let $\nu[\Lambda := 0]$ denote the clock valuation $\nu' \in \mathbb{R}_{\geq 0}^C$ such that $\nu'(x) = 0$ for all $x \in \Lambda$ and $\nu'(x) = \nu(x)$ for all $x \notin \Lambda$.

Definition 2.1. *Let Σ be an alphabet. A* timed automaton *over Σ is a tuple $\mathcal{A} = (L, C, I, E, F)$ such that L is a finite set of* locations, *C is a finite set of* clocks, *$I, F \subseteq L$ are sets of* initial *resp.* final *locations and $E \subseteq L \times \Sigma \times \Phi(C) \times 2^C \times L$ is a finite set of* edges.

For an edge $e = (\ell, a, \phi, \Lambda, \ell')$, let label$(e) = a$ be the *label* of e. A *run* of $\mathcal{A}$ is a finite sequence

$$\rho = (\ell_0, \nu_0) \xrightarrow{t_1, e_1} (\ell_1, \nu_1) \xrightarrow{t_2, e_2} ... \xrightarrow{t_n, e_n} (\ell_n, \nu_n) \tag{1}$$

where $n \geq 1, \ell_0, \ell_1, ..., \ell_n \in L, \nu_0, \nu_1, ..., \nu_n \in \mathbb{R}_{\geq 0}^C, t_1, ..., t_n \in \mathbb{R}_{\geq 0}$ and $e_1, ..., e_n \in E$ satisfy the following conditions: $\ell_0 \in I, \nu_0(x) = 0$ for all $x \in C$, $\ell_n \in F$ and, for all $1 \leq i \leq n, e_i = (\ell_{i-1}, a_i, \phi_i, \Lambda_i, \ell_i)$ for some $a_i \in \Sigma, \phi_i \in \Phi(C)$ and $\Lambda_i \subseteq C$ such that $\nu_{i-1} + t_i \models \phi_i$ and $\nu_i = (\nu_{i-1} + t_i)[\Lambda_i := 0]$. The *label* of ρ is the timed word label$(\rho) = ($label$(e_1), t_1)...($label$(e_n), t_n) \in \mathbb{T}\Sigma^+$. For any timed word $w \in \mathbb{T}\Sigma^+$, let Run$_{\mathcal{A}}(w)$ denote the set of all runs ρ of $\mathcal{A}$ such that label$(\rho) = w$. Let $\mathcal{L}(\mathcal{A}) = \{w \in \mathbb{T}\Sigma^+ \mid \text{Run}_{\mathcal{A}}(w) \neq \emptyset\}$. We say that an arbitrary timed language $\mathcal{L} \subseteq \mathbb{T}\Sigma^+$ is *recognizable* if there exists a timed automaton $\mathcal{A}$ over Σ such that $\mathcal{L}(\mathcal{A}) = \mathcal{L}$. We say that a timed automaton $\mathcal{A} = (L, C, I, E, F)$ is *unambiguous* if $|\text{Run}_{\mathcal{A}}(w)| \leq 1$ for all $w \in \mathbb{T}\Sigma^+$. We call $\mathcal{A}$ *deterministic* if $|I| = 1$ and, for all $e_1 = (\ell, a, \phi_1, \Lambda_1, \ell_1) \in E$ and $e_2 = (\ell, a, \phi_2, \Lambda_2, \ell_2) \in E$ with $e_1 \neq e_2$, there exists no clock valuation $\nu \in \mathbb{R}_{\geq 0}^C$ with $\nu \models \phi_1 \wedge \phi_2$. We call $\mathcal{A}$ *sequential* if $|I| = 1$ and, for all $e_1 = (\ell, a, \phi_1, \Lambda_1, \ell_1) \in E$ and $e_2 = (\ell, a, \phi_2, \Lambda_2, \ell_2) \in E$, we have $e_1 = e_2$; this property can be viewed as a strong form of determinism. Based on these notions, we can define *sequentially recognizable, deterministically recognizable* and *unambiguously recognizable* timed languages.

3 Weighted Timed Automata

In this section, we introduce a general model of weighted timed automata (WTA) over *timed valuation monoids*. We will show that our new model covers a variety of situations known from the literature: linearly priced timed automata [3, 4, 21] and WTA with the measures like average [7, 8] and discounting [2, 18, 19].

A *timed valuation monoid* is a tuple $\mathsf{M} = (M, +, \mathrm{val}, \mathbb{0})$ where $(M, +, \mathbb{0})$ is a commutative monoid and $\mathrm{val} : \mathbb{T}(M \times M)^+ \to M$ is a *timed valuation function*. We will say that M is the *domain* of M. We say that M is *idempotent* if $+$ is idempotent, i.e., $m + m = m$ for all $m \in M$.

Let Σ be an alphabet and $\mathsf{M} = (M, +, \mathrm{val}, \mathbb{0})$ a timed valuation monoid. A *weighted timed automaton* (WTA) over Σ and M is a tuple $\mathcal{A} = (L, C, I, E, F, \mathrm{wt})$ where (L, C, I, E, F) is a timed automaton over Σ and $\mathrm{wt} : L \cup E \to M$ is a *weight function*. Let ρ be a run of $\mathcal{A}$ of the form (1). Let $\mathrm{wt}^\sharp(\rho) \in \mathbb{T}(M \times M)^+$ be the timed word $(u_1, t_1)...(u_n, t_n)$ where, for all $1 \leq i \leq n$, $u_i = (\mathrm{wt}(\ell_{i-1}), \mathrm{wt}(e_i))$. Then, the *weight* of ρ is defined as $\mathrm{wt}_\mathcal{A}(\rho) = \mathrm{val}(\mathrm{wt}^\sharp(\rho)) \in M$. The *behavior* of $\mathcal{A}$ is the mapping $||\mathcal{A}|| : \mathbb{T}\Sigma^+ \to M$ defined by $||\mathcal{A}||(w) = \sum(\mathrm{wt}_\mathcal{A}(\rho) \mid \rho \in \mathrm{Run}_\mathcal{A}(w))$ for all $w \in \mathbb{T}\Sigma^+$. A *quantitative timed language* (QTL) over M is a mapping $\mathbb{L} : \mathbb{T}\Sigma^+ \to M$. We say that $\mathbb{L}$ is *recognizable* if there exists a WTA $\mathcal{A}$ over Σ and M such that $\mathbb{L} = ||\mathcal{A}||$.

Example 3.1. All of the subsequent WTA model the property that staying in a location invokes costs depending on the length of the stay; the subsequent transition also invokes costs but happens instantaneously. We assume that, for all $x \in \mathbb{R} \cup \{\infty\}$, $x \cdot \infty = \infty \cdot x = \infty$ and $x + \infty = \infty + x = \infty$.

(a) *Linearly priced timed automata* were considered in [3, 4, 21]. We can describe this model by the timed valuation monoid $\mathsf{M}^{\mathrm{sum}} = (\mathbb{R} \cup \{\infty\}, \min, \mathrm{val}^{\mathrm{sum}}, \infty)$ where $\mathrm{val}^{\mathrm{sum}}$ is defined by $\mathrm{val}^{\mathrm{sum}}(v) = \sum_{i=1}^n (m_i \cdot t_i + m_i')$ for all $v = ((m_1, m_1'), t_1)...((m_n, m_n'), t_n) \in \mathbb{T}(M \times M)^+$.

(b) The situation of the average behavior for WTA considered in [7, 8] can be described by means of the timed valuation monoid $\mathsf{M}^{\mathrm{avg}} = (\mathbb{R} \cup \{\infty\}, \min, \mathrm{val}^{\mathrm{avg}}, \infty)$ where $\mathrm{val}^{\mathrm{avg}}$ is defined as follows. Let $v = ((m_1, m_1'), t_1)...((m_n, m_n'), t_n) \in \mathbb{T}(M \times M)^+$. If $\langle v \rangle > 0$, then we let $\mathrm{val}^{\mathrm{avg}}(v) = \frac{\sum_{i=1}^n (m_i \cdot t_i + m_i')}{\sum_{i=1}^n t_i}$. If $\langle v \rangle = 0$, $m_1 = ... = m_n \in \mathbb{R}$ and $m_1' = ... = m_n' = 0$, then we put $\mathrm{val}^{\mathrm{avg}}(v) = m_1$. Otherwise, we put $\mathrm{val}^{\mathrm{avg}}(v) = \infty$.

(c) The model of WTA with the discounting measure was investigated in [2, 18, 19]. These WTA can be considered as WTA over the timed valuation monoid $\mathsf{M}^{\mathrm{disc}\lambda} = (\mathbb{R} \cup \{\infty\}, \min, \mathrm{val}^{\mathrm{disc}\lambda}, \infty)$ where $0 < \lambda < 1$ is a *discounting factor* and $\mathrm{val}^{\mathrm{disc}\lambda}$ is defined for all $v = ((m_1, m_1'), t_1)...((m_n, m_n'), t_n) \in \mathbb{T}(M \times M)^+$ by $\mathrm{val}^{\mathrm{disc}\lambda}(v) = \sum_{i=1}^n \lambda^{t_1 + ... + t_{i-1}} \cdot \left(\int_0^{t_i} m_i \cdot \lambda^\tau d\tau + \lambda^{t_i} \cdot m_i' \right)$.

Note that the timed valuation monoids $\mathsf{M}^{\mathrm{sum}}$, $\mathsf{M}^{\mathrm{avg}}$ and $\mathsf{M}^{\mathrm{disc}\lambda}$ are idempotent.

4 Closure Properties

In this section, we consider several closure properties of recognizable quantitative timed languages which we will use for the proof of our Nivat theorem and which could be of independent interest. For lack of space, we will omit the proofs.

Let Σ be a set, Γ an alphabet and $h : \Gamma \to \Sigma$ a mapping. For a timed word $v = (\gamma_1, t_1)...(\gamma_n, t_n) \in \mathbb{T}\Gamma^+$, we let $h(v) = (h(\gamma_1), t_1)...(h(\gamma_n), t_n) \in \mathbb{T}\Sigma^+$. Then, for a QTL $r : \mathbb{T}\Gamma^+ \to M$ over M, we define the QTL $h(r) : \mathbb{T}\Sigma^+ \to M$ over M by $h(r)(w) = \sum(r(v) \mid v \in \mathbb{T}\Gamma^+$ and $h(v) = w)$ for all $w \in \mathbb{T}\Sigma^+$. Observe that for any

$w \in \mathbb{T}\Sigma^+$ there are only finitely many $v \in \mathbb{T}\Gamma^+$ with $h(v) = w$, hence the sum exists in $(M, +)$.

Lemma 4.1. *Let* Σ, Γ *be alphabets,* $\mathbb{M} = (M, +, \mathrm{val}, \mathbb{0})$ *a timed valuation monoid and* $h : \Gamma \to \Sigma$ *a mapping. If* $r : \mathbb{T}\Gamma^+ \to M$ *is a recognizable QTL over* $\mathbb{M}$, *then the QTL* $h(r)$ *is also recognizable.*

For the proof of this lemma, we use a similar construction as in [16], Lemma 1.

Let $g : \Sigma \to M \times M$ be a mapping. We denote by $\mathrm{val} \circ g : \mathbb{T}\Sigma^+ \to M$ the QTL over $\mathbb{M}$ defined for all $w \in \mathbb{T}\Sigma^+$ by $(\mathrm{val} \circ g)(w) = \mathrm{val}(g(w))$. We say that a timed valuation monoid $\mathbb{M} = (M, +, \mathrm{val}, \mathbb{0})$ is *location-independent* if, for any $v = ((m_1, m_1'), t_1)...((m_n, m_n'), t_n) \in \mathbb{T}(M \times M)^+$ and $v' = ((k_1, k_1'), t_1)...((k_n, k_n'), t_n) \in \mathbb{T}(M \times M)^+$ with $m_i' = k_i'$ for all $1 \leq i \leq n$, we have $\mathrm{val}(v) = \mathrm{val}(v')$.

Lemma 4.2. *Let* Σ *be an alphabet,* $\mathbb{M} = (M, +, \mathrm{val}, \mathbb{0})$ *a timed valuation monoid and* $g : \Sigma \to M \times M$ *a mapping. Then,* $\mathrm{val} \circ g$ *is unambiguously recognizable. If* $\mathbb{M}$ *is location-independent, then* $\mathrm{val} \circ g$ *is sequentially recognizable.*

However, in general, $\mathrm{val} \circ g$ is not deterministically recognizable (and hence not sequentially recognizable). Let $\Sigma = \{a, b\}$ and $\mathbb{M} = \mathbb{M}^{\mathrm{sum}}$ as in Example 3.1 (a). Let $g(a) = (1, 0)$ and $g(b) = (2, 0)$. Then, one can show that $\mathrm{val} \circ g$ is not deterministically recognizable. Let $\mathcal{L} \subseteq \mathbb{T}\Sigma^+$ be a timed language and $r : \mathbb{T}\Sigma^+ \to M$ a QTL over $\mathbb{M}$. The *intersection* $(r \cap \mathcal{L}) : \mathbb{T}\Sigma^+ \to M$ is the QTL over $\mathbb{M}$ defined by $(r \cap \mathcal{L})(w) = r(w)$ if $w \in \mathcal{L}$ and $(r \cap \mathcal{L})(w) = \mathbb{0}$ if $w \in \mathbb{T}\Sigma^+ \setminus \mathcal{L}$.

Example 4.3. As opposed to weighted untimed automata, recognizable quantitative timed languages are not closed under the intersection with recognizable timed languages. Let Σ be a singleton alphabet and $\mathcal{L}$ a recognizable timed language over Σ which is not unambiguously recognizable. Wilke [26] showed that such a language exists. Consider the non-idempotent and location-independent timed valuation monoid $\mathbb{M} = (\mathbb{N}, +, \mathrm{val}, 0)$ where $+$ is the usual addition of natural numbers and $\mathrm{val}(v) = m_1' \cdot ... \cdot m_n'$ for all $v = ((m_1, m_1'), t_1)...((m_n, m_n'), t_n) \in \mathbb{T}(\mathbb{N} \times \mathbb{N})^+$. Let the QTL $r : \mathbb{T}\Sigma^+ \to \mathbb{N}$ over $\mathbb{M}$ be defined by $r(w) = 1$ for all $w \in \mathbb{T}\Sigma^+$. Then, r is recognizable but $r \cap \mathcal{L}$ is not recognizable.

Nevertheless, the intersection enjoys the following closure properties.

Lemma 4.4. *Let* Σ *be an alphabet,* $\mathbb{M} = (M, +, \mathrm{val}, \mathbb{0})$ *a timed valuation monoid,* $\mathcal{L} \subseteq \mathbb{T}\Sigma^+$ *a recognizable timed language and* $r : \mathbb{T}\Sigma^+ \to M$ *a recognizable QTL over* $\mathbb{M}$. *If* $\mathbb{M}$ *is idempotent, then* $r \cap \mathcal{L}$ *is recognizable. If* $\mathcal{L}$ *is unambiguously recognizable, then* $r \cap \mathcal{L}$ *is recognizable. If* $\mathcal{L}, r$ *are unambiguously (deterministically, sequentially, respectively) recognizable, then* $r \cap \mathcal{L}$ *is also unambiguously (deterministically, sequentially, respectively) recognizable.*

For the proof, we use a kind of product construction for timed automata.

5   A Nivat Theorem for Weighted Timed Automata

Nivat's theorem [23] (see also [5], Theorem 4.1) is one of the fundamental characterizations of rational transductions and establishes a connection between rational transductions and rational languages. A version for semiring-weighted automata was given
in [13]; this shows a connection between recognizable quantitative and qualitative languages. In this chapter, we prove a Nivat-like theorem for recognizable quantitative
timed languages.

Let Σ be an alphabet and $\mathbb{M} = (M, +, \mathrm{val}, \mathbb{0})$ a timed valuation monoid. Let
$\mathrm{REC}(\Sigma, \mathbb{M})$ denote the collection of all QTL recognizable by a WTA over Σ and $\mathbb{M}$.
Let $\mathcal{N}(\Sigma, \mathbb{M})$ (with $\mathcal{N}$ standing for Nivat) denote the set of all QTL $\mathbb{L} : \mathbb{T}\Sigma^+ \to M$
over $\mathbb{M}$ such that there exist an alphabet Γ, mappings $h : \Gamma \to \Sigma$ and $g : \Gamma \to M \times M$
and a recognizable timed language $\mathcal{L} \subseteq \mathbb{T}\Sigma^+$ such that $\mathbb{L} = h((\mathrm{val} \circ g) \cap \mathcal{L})$. Let the
collection $\mathcal{N}^{\mathrm{SEQ}}(\Sigma, \mathbb{M})$ be defined like $\mathcal{N}(\Sigma, \mathbb{M})$ with the only difference that $\mathcal{L}$ is sequentially recognizable. The collections $\mathcal{N}^{\mathrm{UNAMB}}(\Sigma, \mathbb{M})$ and $\mathcal{N}^{\mathrm{DET}}(\Sigma, \mathbb{M})$ are defined
similarly using unambiguously resp. deterministically recognizable timed languages.

Our Nivat theorem for weighted timed automata is the following.

Theorem 5.1. *Let Σ be an alphabet and $\mathbb{M}$ a timed valuation monoid. Then,*
$\mathrm{REC}(\Sigma, \mathbb{M}) = \mathcal{N}^{\mathrm{SEQ}}(\Sigma, \mathbb{M}) = \mathcal{N}^{\mathrm{DET}}(\Sigma, \mathbb{M}) = \mathcal{N}^{\mathrm{UNAMB}}(\Sigma, \mathbb{M}) \subseteq \mathcal{N}(\Sigma, \mathbb{M})$.
If $\mathbb{M}$ is idempotent, then $\mathrm{REC}(\Sigma, \mathbb{M}) = \mathcal{N}(\Sigma, \mathbb{M})$.

As opposed to the result of [13] for weighted untimed automata, the equality
$\mathrm{REC}(\Sigma, \mathbb{M}) = \mathcal{N}(\Sigma, \mathbb{M})$ does not always hold: let Σ, $\mathbb{M}$, $\mathcal{L}$ and r be defined as in
Example 4.3. Then, one can show that $r \cap \mathcal{L} \in \mathcal{N}(\Sigma, \mathbb{M}) \setminus \mathrm{REC}(\Sigma, \mathbb{M})$.

The proof of Theorem 5.1 is based on the closure properties of WTA (cf. Sect. 4)
and the following lemma.

Lemma 5.2. *Let Σ be an alphabet and $\mathbb{M}$ a timed valuation monoid. Then,*
$\mathrm{REC}(\Sigma, \mathbb{M}) \subseteq \mathcal{N}^{\mathrm{SEQ}}(\Sigma, \mathbb{M})$.

Proof (Sketch). Let $\mathcal{A} = (L, C, I, E, F, \mathrm{wt})$ be a WTA over Σ and $\mathbb{M}$. Let $\Gamma = E$. We
define the mappings $h : \Gamma \to \Sigma$ and $g : \Gamma \to M \times M$ for all $\gamma = (\ell, a, \phi, \Lambda, \ell') \in \Gamma$
by $h(\gamma) = a$ and $g(\gamma) = (\mathrm{wt}(\ell), \mathrm{wt}(\gamma))$. Let $\mathcal{L}$ be the set of all timed words $w =
(\gamma_1, \tau_1)...(\gamma_n, \tau_n)$ such that there exists a run ρ of $\mathcal{A}$ of the form (1) with $\gamma_i = e_i$ and
$\tau_i = t_i$ for all $1 \leq i \leq n$. It can be shown that $\mathcal{L}$ is sequentially recognizable and
$\|\mathcal{A}\| = h((\mathrm{val} \circ g) \cap \mathcal{L}) \in \mathcal{N}^{\mathrm{SEQ}}(\Sigma, \mathbb{M})$.   □

Let Σ be an alphabet and $\mathbb{M}$ a timed valuation monoid with the domain M. Let
$\mathcal{H}^{\mathrm{UNAMB}}(\Sigma, \mathbb{M})$ denote the collection of all QTL $\mathbb{L} : \mathbb{T}\Sigma^+ \to M$ over $\mathbb{M}$ such that there
exist an alphabet Γ, a mapping $h : \Gamma \to \Sigma$ and an unambiguously recognizable
QTL $r : \mathbb{T}\Gamma^+ \to M$ over $\mathbb{M}$ such that $\mathbb{L} = h(r)$. The collections $\mathcal{H}^{\mathrm{SEQ}}(\Sigma, \mathbb{M})$ and
$\mathcal{H}^{\mathrm{DET}}(\Sigma, \mathbb{M})$ are defined like $\mathcal{H}^{\mathrm{UNAMB}}(\Sigma, \mathbb{M})$ with the only difference that r is sequentially resp. deterministically recognizable.

As a corollary from Theorem 5.1, we establish the following connections between
recognizable and unambiguously, sequentially and deterministically recognizable QTL.
For the proof of this corollary, we apply Theorem 5.1 and closure properties of WTA
considered in Sect. 4.

Corollary 5.3. *Let Σ be an alphabet and $\mathbb{M}$ a timed valuation monoid. Then,* $\mathcal{H}^{\text{SEQ}}(\Sigma, \mathbb{M}) = \mathcal{H}^{\text{DET}}(\Sigma, \mathbb{M}) \subseteq \mathcal{H}^{\text{UNAMB}}(\Sigma, \mathbb{M}) = \text{REC}(\Sigma, \mathbb{M})$. *If $\mathbb{M}$ is location-independent, then* $\mathcal{H}^{\text{SEQ}}(\Sigma, \mathbb{M}) = \text{REC}(\Sigma, \mathbb{M})$.

However, the equality $\mathcal{H}^{\text{SEQ}}(\Sigma, \mathbb{M}) = \text{REC}(\Sigma, \mathbb{M})$ does not always hold. Let $\Sigma = \{a, b\}$ and $\mathbb{M} = \mathbb{M}^{\text{sum}}$ be the timed valuation monoid as in Example 3.1 (a); note that $\mathbb{M}$ is not location-independent. Consider the QTL $\mathbb{L} : \mathbb{T}\Sigma^+ \to M$ over $\mathbb{M}$ defined for all $w = (a_1, t_1)...(a_n, t_n)$ by $\mathbb{L}(w) = t_1$ if $a_1 = a$ and $\mathbb{L}(w) = 2 \cdot t_1$ otherwise. We can show that $\mathbb{L} \in \text{REC}(\Sigma, \mathbb{M}) \setminus \mathcal{H}^{\text{SEQ}}(\Sigma, \mathbb{M})$.

6 Weighted Relative Distance Logic

In this section, we develop a weighted relative distance logic. Relative distance logic on finite and infinite timed words was introduced by Wilke in [26, 27]. It was shown that restricted relative distance logic and timed automata have the same expressive power. Here, we will derive a weighted version of this result for finite timed words. We will show that the proof of our result can be deduced from Wilke's result and our Nivat theorem for WTA.

We fix a countable set V_1 of *first-order variables* and a countable set V_2 of *second-order variables* such that $V_1 \cap V_2 = \emptyset$. Let $V = V_1 \cup V_2$.

6.1 Relative Distance Logic

Let Σ be an alphabet. The set $\text{RDL}(\Sigma)$ of *relative distance formulas* over Σ is defined by the grammar:

$$\varphi ::= P_a(x) \mid x \leq y \mid X(x) \mid d_{\leftarrow}^{\bowtie c}(X, x) \mid \neg\varphi \mid \varphi \vee \varphi \mid \exists x.\varphi \mid \exists X.\varphi$$

where $a \in \Sigma$, $x, y \in V_1$, $X \in V_2$, $\bowtie \in \{<, \leq, =, \geq, >\}$ and $c \in \mathbb{N}$. The formulas of the form $d_{\leftarrow}^{\bowtie c}(X, x)$ are called *past formulas*.

Let $w = (a_1, t_1)...(a_n, t_n) \in \mathbb{T}\Sigma^+$ be a timed word. For every $1 \leq i \leq n$, let $\langle w \rangle_i = t_1 + ... + t_i$. The *domain* of w is the set $\text{dom}(w) = \{1, ..., n\}$ of *positions* of w. Let $y \in \text{dom}(w)$, $Y \subseteq \text{dom}(w)$, $\bowtie \in \{<, \leq, =, \geq, >\}$ and $c \in \mathbb{N}$. Then, we write $d_{\leftarrow}^{\bowtie c, w}(Y, y)$ iff either there exists a position $z \in Y$ such that $z < y$ and, for the greatest such position z, $\langle w \rangle_y - \langle w \rangle_z \bowtie c$, or there exists no position $z \in Y$ with $z < y$, and $\langle w \rangle_y \bowtie c$. A *w-assignment* is a mapping $\sigma : V \to \text{dom}(w) \cup 2^{\text{dom}(w)}$ such that $\sigma(V_1) \subseteq \text{dom}(w)$ and $\sigma(V_2) \subseteq 2^{\text{dom}(w)}$. We define the *update* $\sigma[x/i]$ to be the w-assignment such that $\sigma[x/i](x) = i$ and $\sigma[x/i](y) = \sigma(y)$ for all $y \in V \setminus \{x\}$. Similarly, for $X \in V_2$ and $I \subseteq \text{dom}(w)$, we define the update $\sigma[X/I]$. Let $\varphi \in \text{RDL}(\Sigma)$ and σ be a w-assignment. The definition that the pair (w, σ) *satisfies* the formula φ, written $(w, \sigma) \models \varphi$, is given inductively on the structure of φ as usual for MSO logic where, for the new formulas $d_{\leftarrow}^{\bowtie c}(X, x)$, we put $(w, \sigma) \models d_{\leftarrow}^{\bowtie c}(X, x)$ iff $d_{\leftarrow}^{\bowtie c, w}(\sigma(X), \sigma(x))$.

A formula $\varphi \in \text{RDL}(\Sigma)$ is called a *sentence* if every variable occurring in φ is bound by a quantifier. Note that, for a sentence $\varphi \in \text{RDL}(\Sigma)$, the relation $(w, \sigma) \models \varphi$ does not depend on σ, i.e., for any w-assignments σ_1, σ_2, $(w, \sigma_1) \models \varphi$ iff $(w, \sigma_2) \models \varphi$. Then, we will write $w \models \varphi$. For a sentence $\varphi \in \text{RDL}(\Sigma)$, let $\mathcal{L}(\varphi) = \{w \in \mathbb{T}\Sigma^+ \mid w \models \varphi\}$, the timed language *defined* by φ. Let $\Delta \subseteq \text{RDL}(\Sigma)$. We say that a timed language $\mathcal{L} \subseteq \mathbb{T}\Sigma^+$ is Δ-*definable* if there exists a sentence $\varphi \in \Delta$ such that $\mathcal{L}(\varphi) = \mathcal{L}$.

Let $\mathcal{V} = \{X_1, ..., X_m\} \subseteq V$ with $|\mathcal{V}| = m$. For $\varphi \in \text{RDL}(\Sigma)$, let $\exists\mathcal{V}.\varphi$ denote the formula $\exists X_1. \ ... \ \exists X_m.\varphi$. For a formula $\varphi \in \text{RDL}(\Sigma)$, let $\mathcal{D}(\varphi) \subseteq V_2$ denote the set of all variables X for which there exist $x \in V_1, \bowtie \in \{<, \leq, =, \geq, >\}$ and $c \in \mathbb{N}$ such that $d^{\bowtie c}_{\leftarrow}(X, x)$ is a subformula of φ. Let $\text{RDL}^{\leftarrow}(\Sigma) \subseteq \text{RDL}(\Sigma)$ denote the set of all formulas φ where quantification of second-order variables is applied only to variables not in $\mathcal{D}(\varphi)$. We denote by $\exists\text{RDL}^{\leftarrow}(\Sigma) \subseteq \text{RDL}(\Sigma)$ the set of all sentences of the form $\exists\mathcal{D}(\varphi).\varphi$.

Theorem 6.1 (Wilke [27]). *Let Σ be an alphabet and $\mathcal{L} \subseteq \mathbb{T}\Sigma^+$ a timed language. Then, $\mathcal{L}$ is recognizable iff $\mathcal{L}$ is $\exists\text{RDL}^{\leftarrow}(\Sigma)$-definable.*

6.2 Weighted Relative Distance Logic

In this subsection, we consider a weighted version of relative distance logic. For untimed words, weighted MSO logic over semirings was defined in [10]. A weighted MSO logic over (untimed) product valuation monoids was considered in [14]. We will use a similar approach to define the syntax and the semantics of our weighted relative distance logic. In [14], valuation monoids were augmented with a product operation and a unit element to define the semantics of weighted formulas. Here, we proceed in a similar way and consider timed *product* valuation monoids.

A *timed product valuation monoid* (timed pv-monoid) $\mathsf{M} = (M, +, \text{val}, \diamond, \mathbb{0}, \mathbb{1})$ is a timed valuation monoid $(M, +, \text{val}, \mathbb{0})$ equipped with a multiplication $\diamond : M \times M \to M$ and a unit $\mathbb{1} \in M$ such that $m \diamond \mathbb{1} = \mathbb{1} \diamond m = m$ and $m \diamond \mathbb{0} = \mathbb{0} \diamond m = \mathbb{0}$ for all $m \in M$, $\text{val}(((\mathbb{1}, \mathbb{1}), t_1), ..., ((\mathbb{1}, \mathbb{1}), t_n)) = \mathbb{1}$ for all $n \geq 1$ and all $t_1, ..., t_n \in \mathbb{R}_{\geq 0}$, and $\text{val}(((m_1, m'_1), t_1)...((m_n, m'_n), t_n)) = \mathbb{0}$ whenever $m'_i = \mathbb{0}$ for some $1 \leq i \leq n$. We say that M is *idempotent* if $+$ is idempotent.

Example 6.2. If we augment the timed valuation monoids M^{sum}, M^{avg} and $\mathsf{M}^{\text{disc}\lambda}$ from Example 3.1 with the multiplication $\diamond = +$ and the unit $\mathbb{1} = 0$, then we obtain the timed pv-monoids $\mathsf{M}^{\text{sum}}_0$, $\mathsf{M}^{\text{avg}}_0$ and $\mathsf{M}^{\text{disc}\lambda}_0$. Note that these timed pv-monoids are idempotent.

Motivated by the examples, for the clarity of presentation, we restrict ourselves to idempotent timed pv-monoids.

Let Σ be an alphabet and $\mathsf{M} = (M, +, \text{val}, \diamond, \mathbb{0}, \mathbb{1})$ a timed pv-monoid. The set $\text{wRDL}(\Sigma, \mathsf{M})$ of formulas of *weighted relative distance logic* over Σ and M is defined by the grammar

$$\varphi ::= \mathbb{B}.\beta \mid m \mid \varphi \vee \varphi \mid \varphi \wedge \varphi \mid \exists x.\varphi \mid \forall x.(\varphi, \varphi) \mid \exists X.\varphi$$

where $\beta \in \text{RDL}^{\leftarrow}(\Sigma)$, $m \in M$, $x \in V_1$ and $X \in V_2$; the notation $\mathbb{B}.\beta$ indicates that here β will be interpreted in a quantitative way.

Let $\mathbb{T}\Sigma^+_V$ denote the set of all pairs (w, σ) where $w \in \mathbb{T}\Sigma^+$ and σ is a w-assignment. For $\varphi \in \text{wRDL}(\Sigma, \mathsf{M})$, the *semantics* of φ is the mapping $[\![\varphi]\!] : \mathbb{T}\Sigma^+_V \to M$ defined for all $(w, \sigma) \in \mathbb{T}\Sigma^+_V$ with $w = (a_1, t_1)...(a_n, t_n)$ inductively on the structure of φ as shown in Table 1. Here, $x \in V_1$, $X \in V_2$, $\beta \in \text{RDL}^{\leftarrow}(\Sigma)$, $m \in M$ and $\varphi, \varphi_1, \varphi_2 \in \text{wRDL}(\Sigma, \mathsf{M})$.

Table 1. The semantics of weighted relative distance logic

$$[\![\mathbb{B}.\beta]\!](w,\sigma) = \begin{cases} \mathbb{1}, & \text{if } (w,\sigma) \models \beta, \\ \mathbb{0}, & \text{otherwise} \end{cases}$$

$$[\![m]\!](w,\sigma) = m$$

$$[\![\varphi_1 \vee \varphi_2]\!](w,\sigma) = [\![\varphi_1]\!](w,\sigma) + [\![\varphi_2]\!](w,\sigma)$$

$$[\![\varphi_2 \wedge \varphi_2]\!](w,\sigma) = [\![\varphi_1]\!](w,\sigma) \diamond [\![\varphi_2]\!](w,\sigma)$$

$$[\![\exists x.\varphi]\!](w,\sigma) = \sum_{i \in \mathrm{dom}(w)} [\![\varphi]\!](w,\sigma[x/i])$$

$$[\![\exists X.\varphi]\!](w,\sigma) = \sum_{I \subseteq \mathrm{dom}(w)} [\![\varphi]\!](w,\sigma[X/I])$$

$$[\![\forall x.(\varphi_1,\varphi_2)]\!](w,\sigma) = \mathrm{val}[(([\![\varphi_1]\!](w,\sigma[x/i]), [\![\varphi_2]\!](w,\sigma[x/i])), t_i)_{i \in \mathrm{dom}(w)}]$$

Remark 6.3. In [24, 25], Quaas introduced a weighted version of relative distance logic over a semiring $\mathbb{S} = (S, +, \cdot, \mathbb{0}, \mathbb{1})$ and a family of functions $\mathcal{F} \subseteq S^{\mathbb{R}_{\geq 0}}$ where elements of S model discrete weights and functions $f \in \mathcal{F}$ model continuous weights. If $\mathcal{F}$ is a one-parametric family of functions $(f_s)_{s \in S}$, then our weighted logic incorporates the logic of Quaas over $\mathbb{S}$ and $\mathcal{F}$. However, for more complicated timed valuation functions (like average and discounting) we must have formulas which combine both discrete and continuous weights. Therefore, we use the formulas $\forall x.(\varphi_1, \varphi_2)$. Our approach also extends the idea of [14] to define the semantics of formulas with a first-order universal quantifier using the valuation function.

Example 6.4. Let $\Sigma = \{a, b\}$ and let $C(a), C(b) \in \mathbb{R}$ be the *continuous costs* of a, b and $D(a), D(b) \in \mathbb{R}$ the *discrete costs*. Given a timed word $w = (\gamma_1, t_1)...(\gamma_n, t_n) \in \mathbb{T}\Sigma^+$, the *average cost* of w is defined as $A(w) = \frac{\sum_{i=1}^n (C(\gamma_i) \cdot t_i + D(\gamma_i))}{\sum_{i=1}^n t_i}$. Let $\mathsf{M}_0^{\mathrm{avg}}$ be defined as in Example 6.2. For $U \in \{C, D\}$, let $\varphi_U(x) = (P_a(x) \wedge U(a)) \vee (P_b(x) \wedge U(b))$. Consider the $\mathrm{wRDL}(\Sigma, \mathsf{M}_0^{\mathrm{avg}})$-sentence $\varphi = \forall x.(\varphi_C(x), \varphi_D(x))$. Then, for all $w \in \mathbb{T}\Sigma^+$, we have: $[\![\varphi]\!](w) = A(w)$.

A sentence $\varphi \in \mathrm{wRDL}(\Sigma, \mathsf{M})$ is defined as usual as a formula without free variables. Then, for every sentence $\varphi \in \mathrm{wRDL}(\Sigma, \mathsf{M})$, every timed word $w \in \mathbb{T}\Sigma^+$ and every w-assignment σ, the value $[\![\varphi]\!](w, \sigma)$ does not depend on σ. Hence, we can consider the semantics of φ as a quantitative timed language $[\![\varphi]\!] : \mathbb{T}\Sigma^+ \to M$ over M.

Similarly to the results of [10], in general weighted relative distance logic and WTA are not expressively equivalent. We can show that the QTL $\mathbb{L} : \mathbb{T}\Sigma^+ \to \mathbb{R} \cup \{\infty\}$ with $\mathbb{L}(w) = |w|^2$ is not recognizable over the timed valuation monoid $\mathsf{M}^{\mathrm{sum}}$. But this QTL is defined by the $\mathrm{wRDL}(\Sigma, \mathsf{M}_0^{\mathrm{sum}})$-sentence $\forall x.(0, \forall y.(0, 1))$.

Nevertheless, there is a syntactically restricted fragment of weighted relative distance logic which is expressively equivalent to WTA. Let Σ be an alphabet and $\mathsf{M} = (M, +, \mathrm{val}, \diamond, \mathbb{0}, \mathbb{1})$ an idempotent timed pv-monoid. A formula $\varphi \in \mathrm{wRDL}(\Sigma, \mathsf{M})$ is called *almost boolean* if it is built from boolean formulas $\mathbb{B}.\beta \in \mathrm{RDL}^{\leftarrow}(\Sigma, \mathsf{M})$ and constants $m \in M$ using disjunctions and conjunctions. We say that a formula φ is *syntactically restricted* if whenever it contains a subformula $\forall x.(\varphi_1, \varphi_2)$, then φ_1, φ_2 are almost boolean; whenever it contains a subformula $\varphi_1 \wedge \varphi_2$, then either φ_1, φ_2 are almost boolean or $\varphi_1 = \mathbb{B}.\varphi'$ or $\varphi_2 = \mathbb{B}.\varphi'$ with $\varphi' \in \mathrm{RDL}^{\leftarrow}(\Sigma)$; every constant $m \in M$ is in the scope of a first-order universal quantifier. Let $\mathrm{DEF}^{\mathrm{res}}(\Sigma, \mathsf{M})$ denote the collection of all QTL $\mathbb{L} : \mathbb{T}\Sigma^+ \to M$ over M such that $\mathbb{L} = [\![\varphi]\!]$ for some syntactically restricted $\mathrm{wRDL}(\Sigma, \mathsf{M})$-sentence φ.

Our main result for weighted relative distance logic is the following theorem.

Theorem 6.5. *Let Σ be an alphabet and M an idempotent timed pv-monoid. Then,* $\mathrm{DEF}^{\mathrm{res}}(\Sigma, \mathsf{M}) = \mathrm{REC}(\Sigma, \mathsf{M})$.

Now we give a sketch of the proof of this theorem. Let $\mathcal{N}^{\exists\text{RDL}^{\leftarrow}}(\Sigma, \mathbb{M})$ denote the collection of all QTL $\mathbb{L} : \mathbb{T}\Sigma^+ \to M$ over $\mathbb{M}$ such that there exist an alphabet Γ, mappings $h : \Gamma \to \Sigma$, $g : \Gamma \to M \times M$ and a $\exists\text{RDL}^{\leftarrow}(\Gamma)$-definable timed language $\mathcal{L}$ such that $\mathbb{L} = h((\text{val} \circ g) \cap \mathcal{L})$. For the proof of Theorem 6.5, we establish a Nivat-like characterization of definable QTL.

Theorem 6.6. *Let Σ be an alphabet and $\mathbb{M}$ an idempotent timed pv-monoid. Then,* $\mathcal{N}^{\exists\text{RDL}^{\leftarrow}}(\Sigma, \mathbb{M}) = \text{DEF}^{\text{res}}(\Sigma, \mathbb{M})$.

Proof (Sketch). To show the inclusion $\subseteq$, let $\mathbb{L} = h((\text{val} \circ g) \cap \mathcal{L})$ where Γ, h, g and $\mathcal{L}$ are as in the definition of $\mathcal{N}^{\exists\text{RDL}^{\leftarrow}}(\Sigma, \mathbb{M})$. Let β be a $\exists\text{RDL}^{\leftarrow}(\Sigma)$-sentence defining $\mathcal{L}$. We introduce a family $\mathcal{V} = (X_\gamma)_{\gamma \in \Gamma}$ of second-order variables not occurring in β. We replace each predicate $P_\gamma(x)$ with $\gamma \in \Gamma$ occurring in β by the formula $P_{h(\gamma)}(x) \wedge X_\gamma(x)$; so we obtain a formula $\beta' \in \exists\text{RDL}^{\leftarrow}(\Sigma)$. Assume that $\beta' = \exists\mathcal{D}(\beta'').\beta''$ with $\beta'' \in \text{RDL}^{\leftarrow}(\Sigma)$. We construct a formula Part $\in \text{RDL}^{\leftarrow}(\Sigma)$ which demands that the variables $\mathcal{V}$ form a partition of the domain, and a formula $H \in \text{RDL}^{\leftarrow}(\Sigma)$ which demands that, whenever a position of a word belongs to X_γ, then this position is labelled by $h(\gamma)$. Then, the following syntactically restricted wRDL$(\Sigma, \mathbb{M})$-sentence defines $\mathcal{L}$:

$$\exists(\mathcal{V} \cup \mathcal{D}(\beta'')).[\mathbb{B}.(\beta'' \wedge \text{Part} \wedge H) \wedge \forall x.(\bigvee_{\gamma \in \Gamma} \mathbb{B}.X_\gamma(x) \wedge g_1(\gamma), \bigvee_{\gamma \in \Gamma} \mathbb{B}.X_\gamma(x) \wedge g_2(\gamma))]$$

where, for $i \in \{1, 2\}$, g_i is the projection of g to the i-th coordinate.

To show the inclusion $\supseteq$, we introduce *canonical* wRDL$(\Sigma, \mathbb{M})$-*sentences* which are of the form $\varphi = \exists\mathcal{V}.\forall y.(\bigvee_{i=1}^k \mathbb{B}.\beta_i \wedge m_i, \bigvee_{i=1}^k \mathbb{B}.\beta_i \wedge m'_i)$ where $\mathcal{V}$ is a set of variables, $m_1, ..., m_k, m'_1, ..., m'_k \in M$ and $\beta_1, ..., \beta_k \in \text{RDL}^{\leftarrow}(\Sigma)$ are such that, for every timed word $w \in \mathbb{T}\Sigma^+$ and every w-assignment σ, there exists exactly one $i \in \{1, ..., k\}$ such that $(w, \sigma) \models \beta_i$. By structural induction every syntactically-restricted sentence can be transformed into a canonical one. It remains to prove that, for a canonical sentence φ as above, $[\![\varphi]\!] \in \mathcal{N}^{\exists\text{RDL}^{\leftarrow}}(\Sigma, \mathbb{M})$. Let $M^1_\varphi = \{m_1, ..., m_k\}$ and $M^2_\varphi = \{m'_1, ..., m'_k\}$. We put $\Gamma = \Sigma \times M^1_\varphi \times M^2_\varphi$. Let $h : \Gamma \to \Sigma$ be the projection to the first coordinate. Let $g : \Gamma \to M \times M$ be the projection to $M^1_\varphi \times M^2_\varphi$. Then we can construct a $\exists\text{RDL}^{\leftarrow}(\Gamma)$-sentence β of the form $\exists\mathcal{V}.\forall y.\beta'$ such that $[\![\varphi]\!] = h((\text{val} \circ g) \cap \mathcal{L}(\beta))$. $\square$

Then, our Theorem 6.5 follows from Theorem 6.6, the Nivat Theorem 5.1 and Wilke's Theorem 6.1.

Remark 6.7. We can also follow the approach of [10] to prove our Theorem 6.5. Compared to this way, our new proof technique has the following advantages. The proof idea of [10] involves technical details like Büchi's encodings of assignments and a bulky logical description of accepting runs of timed automata. In our new proof, these details are taken care of by Wilke's proof for unweighted relative distance logic.

Let Σ be an alphabet, $\mathbb{M}^{\text{sum}}$ the timed valuation monoid as in Example 3.1(a) and $\mathcal{A}$ a WTA over Σ and $\mathbb{M}$. As it was shown in [3, 4, 21], $\inf\{\|\mathcal{A}\|(w) \mid w \in \mathbb{T}\Sigma^+\}$ is computable. This result and our Theorem 6.5 imply decidability results for weighted relative distance logic.

- Let $\mathbb{M}^{\text{sum}}_0$ be the timed pv-monoid as in Example 6.2. It is decidable, given an alphabet Σ, a syntactically restricted sentence $\varphi \in \text{wRDL}(\Sigma, \mathbb{M}^{\text{sum}})$ with constants from $\mathbb{Q}$ and a threshold $\theta \in \mathbb{Q}$, whether there exists $w \in \mathbb{T}\Sigma^+$ with $[\![\varphi]\!](w) < \theta$.

- Let $\mathsf{M}_0^{\mathrm{avg}}$ be the timed pv-monoid as in Example 6.2. It is decidable, given an alphabet Σ, a syntactically restricted sentence $\varphi \in \mathrm{WRDL}(\Sigma, \mathsf{M}^{\mathrm{avg}})$ with constants from $\mathbb{Q}$ and a threshold $\theta \in \mathbb{Q}$, whether there exists $w \in \mathbb{T}\Sigma^+$ with $\langle w \rangle > 0$ and $[\![\varphi]\!](w) < \theta$.

7   Conclusion and Future Work

In this paper, we proved a version of Nivat's theorem for weighted timed automata on finite words which states a connection between the quantitative and qualitative behaviors of timed automata. We also considered several applications of this theorem. Using this theorem, we studied the relations between sequential, unambiguous and non-deterministic WTA. We also introduced a weighted version of Wilke's relative distance logic and established a Büchi-like result for this logic, i.e., we showed the equivalence between restricted weighted relative distance logic and WTA. Using our Nivat theorem, we deduced this from Wilke's result.

Because of space constraints, we did not present in this paper the following results. As in [14], for timed pv-monoid with additional properties there are larger fragments of weighted relative-distance logic which are still expressively equivalent to WTA. For the simplicity of presentation, we restricted ourselves to idempotent timed pv-monoids. However, we also obtained a more complicated result for non-idempotent timed pv-monoids. In [24, 25], for weighted relative distance logic over non-idempotent semi-rings, a strong restriction on the use of a first-order universal quantification was done. Surprisingly, in our result we could avoid this restriction.

Our future work concerns the following directions. Ongoing research will extend the currently obtained results to ω-infinite words. This work should be further extended to the *multi-weighted* setting for WTA, e.g., the optimal reward-cost ratio [7, 8] or the optimal consumption of several resources where some resources must be restricted [22]. A logical characterization of untimed multi-weighted automata was given in [15]. It could be also interesting to compare for the weighted and unweighted cases the complexity of translations between logic and automata. We believe that our Nivat theorem will be helpful for this.

References

[1] Alur, R., Dill, D.L.: A theory of timed automata. Theoretical Computer Science 126(2), 183–235 (1994)

[2] Alur, R., Triverdi, A.: Relating average and discounted costs for quantitative analysis of timed systems. In: EMSOFT 2011, pp. 165–174. IEEE (2011)

[3] Alur, R., La Torre, S., Pappas, G.J.: Optimal paths in weighted timed automata. In: Di Benedetto, M.D., Sangiovanni-Vincentelli, A.L. (eds.) HSCC 2001. LNCS, vol. 2034, pp. 49–62. Springer, Heidelberg (2001)

[4] Behrmann, G., Fehnker, A., Hune, T., Larsen, K.G., Petterson, P., Romijn, J., Vaandrager, F.: Minimum-cost reachability for priced timed automata. In: Di Benedetto, M.D., Sangiovanni-Vincentelli, A.L. (eds.) HSCC 2001. LNCS, vol. 2034, pp. 147–161. Springer, Heidelberg (2001)

[5] Berstel, J.: Transductions and Context-Free Languages. Teubner Studienbücher: Informatik. Teubner, Stuttgart (1979)

[6] Berstel, J., Reutenauer, C.: Rational Series and Their Languages. EATCS Monographs on Theoretical Computer Science, vol. 12. Springer (1988)

[7] Bouyer, P., Brinksma, E., Larsen, K.G.: Staying alive as cheaply as possible. In: Alur, R., Pappas, G.J. (eds.) HSCC 2004. LNCS, vol. 2993, pp. 203–218. Springer, Heidelberg (2004)

[8] Bouyer, P., Brinksma, E., Larsen, K.G.: Optimal infinite scheduling for multi-priced timed automata. Formal Methods in System Design 32, 3–23 (2008)

[9] Büchi, J.R.: Weak second-order arithmetic and finite automata. Z. Math. Logik und Grundl. Math. 6, 66–92 (1960)

[10] Droste, M., Gastin, P.: Weighted automata and weighted logics. Theoret. Comp. Sci. 380 (1-2), 69–86 (2007)

[11] Droste, M., Gastin, P.: Weighted automata and weighted logics. In: Droste, M., Kuich, W., Vogler, H. (eds.) [12], ch. 5

[12] Droste, M., Kuich, W., Vogler, H. (eds.): Handbook of Weighted Automata. EATCS Monographs on Theoretical Computer Science. Springer (2009)

[13] Droste, M., Kuske, D.: Weighted automata. In: Pin, J.-E. (ed.) Handbook: "Automata: from Mathematics to Applications". European Mathematical Society (to appear)

[14] Droste, M., Meinecke, I.: Weighted automata and weighted MSO logics for average and long-time behaviors. Inf. Comput. 220-221, 44–59 (2012)

[15] Droste, M., Perevoshchikov, V.: Multi-weighted automata and MSO logic. In: Bulatov, A.A., Shur, A.M. (eds.) CSR 2013. LNCS, vol. 7913, pp. 418–430. Springer, Heidelberg (2013)

[16] Droste, M., Vogler, H.: Kleene and Büchi theorems for weighted automata and multi-valued logic over arbitrary bounded lattices. In: Gao, Y., Lu, H., Seki, S., Yu, S. (eds.) DLT 2010. LNCS, vol. 6224, pp. 160–172. Springer, Heidelberg (2010)

[17] Eilenberg, S.: Automata, Languages and Machines, vol. A. Academic Press, New York (1974)

[18] Fahrenberg, U., Larsen, K.G.: Discount-optimal infinite runs in priced timed automata. Electr. Notes Theor. Comput. Sci. 239, 179–191 (2009)

[19] Fahrenberg, U., Larsen, K.G.: Discounting in time. Electr. Notes Theor. Comput. Sci. 253, 25–31 (2009)

[20] Kuich, W., Salomaa, A.: Semirings, Automata and Languages. EATCS Monographs on Theoretical Computer Science, vol. 5. Springer (1986)

[21] Larsen, K.G., Behrmann, G., Brinksma, E., Fehnker, A., Hune, T., Pettersson, P., Romijn, J.: As cheap as possible: Efficient cost-optimal reachability for priced timed automata. In: Berry, G., Comon, H., Finkel, A. (eds.) CAV 2001. LNCS, vol. 2102, pp. 493–505. Springer, Heidelberg (2001)

[22] Larsen, K.G., Rasmussen, J.I.: Optimal conditional reachability for multi-priced timed automata. In: Sassone, V. (ed.) FOSSACS 2005. LNCS, vol. 3441, pp. 234–249. Springer, Heidelberg (2005)

[23] Nivat, M.: Transductions des langages de Chomsky. Ann. de L'Inst. Fourier 18, 339–456 (1968)

[24] Quaas, K.: Kleene-Schützenberger and Büchi theorems for weighted timed automata. PhD thesis, Universität Leipzig (2010)

[25] Quaas, K.: MSO Logics for weighted timed automata. Formal Methods in System Design 38(3), 193–222 (2011)

[26] Wilke, T.: Automaten und Logiken zur Beschreibung zeitabhängiger Systeme. PhD thesis, Christian-Albrecht-Universität Kiel (1994)

[27] Wilke, T.: Specifying timed state sequences in powerful decidable logics and timed automata. In: Langmaack, H., de Roever, W.-P., Vytopil, J. (eds.) FTRTFT 1994 and ProCoS 1994. LNCS, vol. 863, pp. 694–715. Springer, Heidelberg (1994)

Computability in Anonymous Networks: Revocable vs. Irrecovable Outputs*

Yuval Emek[1], Jochen Seidel[2], and Roger Wattenhofer[2]

[1] Faculty of Industrial Engineering and Management, Technion, Haifa, Israel
yemek@ie.technion.ac.il
[2] Distributed Computing, ETH Zürich, Zürich, Switzerland
{seidelj,wattenhofer}@ethz.ch

Abstract. What can be computed in an anonymous network, where nodes are not equipped with unique identifiers? It turns out that the answer to this question depends on the commitment of the nodes to their first computed output value: Two classes of problems solvable in anonymous networks are defined, where in the first class nodes are allowed to revoke their outputs and in the second class they are not. These two classes are then related to the class of all centrally solvable network problems, observing that the three classes form a strict linear hierarchy, and for several classic and/or characteristic problems in distributed computing, we determine the exact class to which they belong.

Does this hierarchy exhibit complete problems? We answer this question in the affirmative by introducing the concept of a distributed oracle, thus establishing a more fine grained classification for distributed computability which we apply to the classic/characteristic problems. Among our findings is the observation that the three classes are characterized by the three pillars of distributed computing, namely, local symmetry breaking, coordination, and leader election.

1 Introduction

We study computability in networks, referred to hereafter as *distributed computability*. Distributed computability is equivalent to classic centralized (Turing Machine) computability when the nodes are equipped with unique (comparable) identifiers. However, as Anglin noticed in her seminal work [3], distributed computability becomes fascinating in *anonymous* networks, where nodes do not have unique IDs. What can be computed with deterministic algorithms merely depends on the topology of the network, and it is well known that problems like maximal independent set can be solved in an anonymous network only if the nodes are allowed to toss coins. We therefore consider the distributed computability of *randomized* algorithms running in anonymous networks. Notice that

---

* Due to space limitations most proofs are omitted or replaced by proof sketches in this extended abstract. Also most results obtained in Section 4 are left out. We refer the interested reader to the full version which is available at http://disco.ethz.ch/publications/ICALP2014-revocability-full.pdf.

J. Esparza et al. (Eds.): ICALP 2014, Part II, LNCS 8573, pp. 183–195, 2014.
© Springer-Verlag Berlin Heidelberg 2014

in the scope of this paper, we do not impose any limitations on the complexity resources (time, message/memory size, . . .), however, like in classic sequential computability theory, we do require a correct result after a finite amount of time.

Apart from its theoretical interest, the study of anonymous networks is motivated by various real-world scenarios. For example, the nodes may be indistinguishable due to their fabrication in a large-scale industrial process [5], in which equipping every node with a unique identifier (serial number) is not economically feasible. In other cases nodes may not wish to reveal their unique identity out of privacy and security concerns [24].

1.1 Setting

Distributed Problems. We consider simple (undirected, loop-free and no parallel edges) connected finite graphs G, and denote the node and edge sets of a graph G by $V(G)$ and $E(G)$ or V and E if G is clear from the context. A function $f : V(G) \to L$ is called a *labeling* of the graph G, and we refer to the set L as the set of *values* that f assigns to nodes in G. A *distributed problem* Π is a set of three-tuples (G, i, o), where G is a graph as described above, and i and o are *input labels* and *output labels* for G. For every problem there are two sets $I(\Pi)$ and $O(\Pi)$ denoting the *input values* and *output values* of Π, i.e., the values that the labels i and o assign, correspondingly. Such a three-tuple $(G, i, o) \in \Pi$ is called a *(solved) instance* of Π. An *input instance* of Π is a two-tuple (G, i) for which there exists a *valid output* o satisfying $(G, i, o) \in \Pi$, and we also write $(G, i) \in \Pi$ for input instances of the problem. We restrict ourselves to problems that are solvable in a centralized setting.

Randomized Anonymous Algorithms. Our definition of how distributed algorithms work follows the convention of [30] for synchronized network systems (message passing) with simultaneous starting times. Nodes execute the same randomized and uniform algorithm in synchronous *rounds*, and in each round we allow each node access to finitely many random bits. Every node v knows its degree $\deg(v)$ and can distinguish between its neighbors $\Gamma(v)$ (by means of a bijection $\{1, \ldots, \deg(v)\} \to \Gamma(v)$, cf. the *port model*). In each round every node sends and receives a message of unbounded, yet finite, size to and from each individual neighbor. To ease our discussion every node v is equipped with one *input register* holding some problem-dependent input value and one *output register*. The output register initially contains a special symbol ε indicating v *is not ready* to return an output. Any value $x \neq \varepsilon$ contained in v's output register is interpreted as v being *ready* to return its output and we say that v *has output* x. A global *configuration* in which all nodes are ready is called a *ready configuration*. When algorithm $\mathcal{A}$ is in a ready configuration, we define $\mathcal{A}$'s output $o_{\mathcal{A}} : V(G) \to O$ by setting $o_{\mathcal{A}}(v)$ to be the content of node v's output register. In the following, we consider two different notions of *output revocability*.

Definition (Output Revocability). An algorithm is referred to as a *write-once* algorithm if every node is restricted to write to its output register at most once. If this restriction is lifted, then we call it a *rewrite* algorithm.

In other words, in a rewrite algorithm a node may revoke its output, e.g., by writing ε to its output register. While every execution of a write-once algorithm reaches at most one ready configuration, during the execution of a rewrite algorithm many ready configurations can occur. Note that the converse does not hold: an algorithm that is guaranteed to reach at most one ready configuration is not necessarily a write-once algorithm. In the existing literature, algorithms are typically considered to be write-once algorithms.

Definition (Correctness). Fix some problem Π and an algorithm $\mathcal{A}$. A ready configuration of $\mathcal{A}$ when invoked on an input instance $(G, i) \in \Pi$ is said to be *valid* if the output $o_{\mathcal{A}}$ of $\mathcal{A}$ in this configuration is a valid output for (G, i). Algorithm $\mathcal{A}$ is said to *solve* Π if it satisfies the following two conditions for every input instance $(G, i) \in \Pi$:
1. A ready configuration is reached within finite time with probability 1.
2. Every ready configuration that can occur with a positive probability is valid.

The aforementioned definition of correctness requires that all occurring ready configurations will be correct (i.e., correspond to a valid output). In Section 2 we show that our definition of correctness is robust to certain changes. Notice that in the scope of this paper, we do not require that an algorithm *terminates* in order to be correct. However, the algorithms designed throughout the paper do terminate, and the general transformation techniques we present (i.e., compilers/simulations) can be designed to ensure termination if the algorithms to which the transformation is applied terminate.

The choice of output revocability has a significant impact on the problems that an algorithm can solve. In the following the terms WO-algorithms and RW-algorithms will thus be used to denominate write-once and rewrite algorithms running in an anonymous network, respectively; RW and WO refer to the *classes* of distributed problems solvable by these two types of algorithms. Lastly, we denote by CF the class of distributed problems that are solvable in a centralized setting (by a Turing machine), bearing in mind that this class essentially includes every *computable function* on graphs. The distinction of these classes is justified by the following observation. The full version of this paper contains a straightforward proof.

Observation. *The classes of distributed problems satisfy* WO $\subset$ RW $\subset$ CF *(in the strict sense).*

1.2 Our Contribution

What can be computed in anonymous networks? As it turns out the effect output revocability has on the distributed computability of anonymous networks is remarkable. A total of 21 problems, including some of the most fundamental problems in distributed computing, are classified according to the exact class to which they belong (Section 4).

Does the hierarchy we present exhibit complete problems? To answer this question we introduce the notion of accessing an oracle in a distributed setting and show that this notion is sound (Section 3). As the first stepping stone in this effort we show that the classes WO and RW are robust against two modifications to the aforementioned correctness condition (Section 2). In the full version of this paper each of our 21 problems is then classified according to its *hardness* or *completeness* for the three classes, thus obtaining a deeper understanding of the intrinsic properties of these problems. For reasons of brevity this extended abstract only gives a brief overview of those results (Section 4). Surprisingly, the WO, RW, and CF classes turn out to capture exactly the three *pillars* of distributed computing, namely, local symmetry breaking, coordination, and leader election, respectively.

1.3   Related Work

The history of distributed computability starts with the work of Angluin [3] proving that randomization does not help to elect a leader in anonymous networks. Later, it was shown that electing a leader in an anonymous ring network is possible if the size n of the ring is known [25], in fact, a $(2 - \varepsilon)$-approximation of n is enough [1], not only in the special case of a ring but in general networks [34]. It turns out that all these results, and many similar ones, come almost for free once our characterization for the class RW (established in the full version) is available.

There is a line of work that concentrates on *deterministic* distributed algorithms for problems in CF, in particular if some parameters of the topology of the graph (for instance, its size) are known, e.g. [35,10]. Deterministic algorithms are interesting to investigate even if the graph is restricted to a ring [18,13], and also assignments of not necessarily unique identifiers were studied in this context [31].

Another line of research studies computability in anonymous (directed) networks in connection with termination. Not unlike us it is argued that termination in distributed systems is an issue that is not directly evident, since one may be interested in systems where nodes terminate independently of others. The strongest anonymous model considered in [11] is equivalent to deterministic write-once algorithms with knowledge of an upper bound to the network size. When no prior knowledge is assumed the class of solvable problems can be fully characterized using *local views*[1] and recursive functions [14]. Extending their approach, in the context of the current paper an individual node executing a RW-algorithm can never be entirely sure about termination. We show that the class RW lies *between* the two classes WO (local termination) and CF (global termination).

Output revocability should not be confused with the concept of *eventual correctness*, where the network eventually converges to a correct output. For example, *self-stabilizing algorithms* [15] allow the system to return an incorrect output for a finite amount of time, thus allowing a fault-tolerant algorithm to recover from errors. With randomization, self-stabilizing leader election is possible on general graphs [16], hence with randomization every CF-problem is eventually solvable in an anonymous network. In our terminology eventual correctness

---

[1] Local views are only discussed in the full version of this paper.

could be viewed as requiring that some ready configuration, not necessarily the first one, is stable[2] and valid. We require though that an output is returned after finite time and that every output returned by the network is correct, but we do allow the network to revoke partial outputs. The problems solvable by self-stabilizing algorithms in directed graphs can be characterized by *fibrations* [12] similar to our characterization for RW that is presented in the full paper. The notion of eventual correctness is also used in the scope of population protocols [5] in which nodes are modeled by finite state machines, see [9] for an overview. In a clique network, the predicates a population protocol can solve are exactly those expressible in first-order Presburger arithmetic [5,6,7], whereas in bounded-degree graphs a Turing machine with linearly bounded space can be simulated [4]. It was also studied how the correctness condition for population protocols affects solvability of the CONSENSUS problem [8].

Apart from these results, not much is known about distributed computability (in contrast to distributed complexity). However, there are surprising connections between complexity and computability, which go beyond us borrowing the terms hardness and completeness. Regarding network algorithms, in the last 30 years, a lot of research went into the question how fast a particular problem can be computed by the network.

Naor and Stockmeyer [33] introduced the notion of locally checkable labelings (essentially an apply-once oracle) in identified networks and ask the question how a constant-time deterministic algorithm can decide whether the labeling represents a correct solution to a given problem. Follow-up work looked at the bit complexity required to solve problems [26,21] and a problem hierarchy depending on the size of checkable labelings was suggested [23], also for anonymous networks. Our work also yields a characterization for the decision problems in RW. However, we do not restrict the run-time to be constant and allow randomization for symmetry breaking. Pruning algorithms [27] that build a solution gradually in a write-once fashion were inspired by the same line of research, in an effort to remove the necessity of global knowledge about the graph. While our algorithms are required to give a correct output in every execution [19,22] study the notion of (p, q)-decidable decision problems (an anonymous randomized algorithm may return a wrong output with constant probability) and find a hierarchy among the solvable problem-classes depending on the success probabilities. If a randomized algorithm is allowed to fail (Monte-Carlo algorithm), then a leader can be elected [32] with high probability (w.h.p, i.e., with probability $1 - n^{-c}$ for any c). Hence any CF-problem can be solved in an anonymous network w.h.p. In contrast to that, we require a correct output with probability 1.

Non-deterministic algorithms running in an anonymous setting can fully determine the structure of the radius t-ball around itself in [20], and thus solve exactly the decision problems that are closed under so-called *t-homomorphisms*. In our model only the local view can be retrieved. It may thus be surprising

---

[2] A configuration is said to be *stable* if the nodes no longer revoke their outputs, see Section 2.

that RW-algorithms can solve exactly the problems that such non-deterministic constant-time algorithms can solve in a single round.

2   Notions of Correctness

Our definition of a correct algorithm requires every ready configuration that occurs throughout an execution to be valid. For WO-algorithms this requirement is superfluous since its execution will reach at most one ready configuration. However, RW-algorithms may invalidate or change a ready configuration after it occurred. One may therefore wonder if strengthening the definition by allowing only one durable ready configuration makes the class of solvable problems strictly smaller. On the other hand one may be tempted to weaken this definition, in hope to capture a larger class of problems by requiring only the first occurring ready configuration to be correct. Perhaps surprisingly we show that these two variants have no effect and are equivalent to the current definition of correctness. This equivalence will play a key role when we reason about RW-algorithms in the next section which covers distributed oracles.

Definition (Sustainable Correctness). A ready configuration is said to be *stable*, if the nodes no longer revoke their outputs. Algorithm $\mathcal{A}$ is said to *sustainably solve* a problem Π if it satisfies the following two conditions for every input instance $(G, i) \in \Pi$:
1. A ready configuration is reached within finite time with probability 1.
2. The first ready configuration that occurs is valid and stable.

Definition (Loose Correctness). Algorithm $\mathcal{A}$ is said to *loosely solve* a problem Π if it satisfies the following two conditions for every input instance $(G, i) \in \Pi$:
1. A ready configuration is reached within finite time with probability 1.
2. The first ready configuration that occurs is valid.

The class *Sustainable-RW* (respectively, *Loose-RW*) consists of every distributed problem that can be sustainably solved (resp., loosely solved) by a RW-algorithm. Since sustainable correctness (resp., loose correctness) is a restriction (resp., a relaxation) of correctness as defined in Section 1.1, we conclude that Sustainable-RW $\subseteq$ RW $\subseteq$ Loose-RW. Note that the corresponding classes *Sustainable-WO* and *Loose-WO* for WO-algorithms are equal to the class WO due to the write-once restriction of these algorithms. The following theorem states that also for RW-algorithms the three classes are, in fact, equal.

Theorem 1. *The classes of problems solvable by RW-algorithms under the three different notions of correctness satisfy Sustainable-RW = RW = Loose-RW.*

The proof of Theorem 1 relies on a *sustainability compiler* that takes a RW-algorithm $\mathcal{A}$ that loosely solves a problem Π and transforms it into a RW-algorithm $\hat{\mathcal{A}}$ that sustainably solves this problem. At the heart of the compiler lies the concept of *inhibiting messages*, i.e., a refinement of a simple concept referred to as *safe broadcast* in which information is broadcast throughout the whole network and no ready configuration is reached before all nodes have received the information. Specifically, the inhibiting messages ensure that the first ready configuration reached by algorithm $\hat{\mathcal{A}}$ is stable. We refer to the appended full version of this paper for the details of the sustainability compiler and its underlying inhibiting message technique.

3   Distributed Oracles

In this section, we introduce the concepts of hardness and completeness, which are central to this work and allow us to gain a deeper understanding how the computability classes relate to each other. To that end, we introduce the notion of an oracle working in a distributed setting.

Definition (Algorithm with access to a Π-oracle). Consider some problem Π. A C-algorithm, C $\in$ {WO, RW}, with *access to a Π-oracle* is a distributed C-algorithm in which every node v is equipped with a designated *oracle input register* and a designated *oracle output register*. Given some $r \geq 1$, let $\tilde{i}(v)$ be the content of v's oracle input register in round r and let $\tilde{o}(v)$ be the content of v's oracle output register in round $r + 1$. If $(G, \tilde{i})$ is an input instance of Π, then it is guaranteed that $\tilde{o}$ is a valid output for $(G, \tilde{i})$. No assumptions are made on the operation of the algorithm if $(G, \tilde{i}) \notin \Pi$.

While applying the oracle in every round of the algorithm may seem powerful, allowing the distributed algorithm to arbitrarily choose the rounds in which the oracle is applied may require some sort of global coordination, which is not necessarily possible. In comparison, a weaker definition of "accessing an oracle" would be to allow application of the oracle only once in round 1. This distinction does not make a difference for problems Π without inputs ($|I(\Pi)| = 1$), e.g., for graph theoretic problems like coloring, maximal independent set, or determining the diameter, because the oracle is always applied on the same input instance. It does however affect problems that do receive inputs ($|I(\Pi)| \geq 2$), e.g., CONSENSUS or logical AND and OR.

As stated above, based on the oracle concept, we will soon introduce the notion of hard and complete problems for the hierarchy of problem classes. This notion would be ill-defined if accessing an oracle to a problem $\Pi_C \in$ C could enhance the computational power of a C-algorithm. We ensure that the notion of an algorithm with access to an oracle is sound in the following theorem. Note that the statement of the theorem does not mention the case C = CF, since the soundness of oracles for centralized models is well understood and in any case, beyond the scope of the current paper.

Theorem 2 (Soundness). *If a problem Π is solvable by a C-algorithm, C $\in$ {RW, WO}, accessing an oracle to a problem $\Pi_C \in$ C, then Π can also be solved by a C-algorithm that does not access any oracle.*

The key to proving this theorem is to show that in a C-algorithm $\mathcal{A}^a$ that solves a problem Π with *access* to a Π_C-oracle, $\Pi_C \in$ C, one can *replace* the oracle access by simulating a C-algorithm $\mathcal{A}^r$ that solves Π_C without any oracle access. This turns out to be a non-trivial task especially for RW-algorithms since a node v simulating $\mathcal{A}^r$ cannot know for sure that the output returned by $\mathcal{A}^r$ will not be revoked later on, i.e., whether it can be safely used for the execution of $\mathcal{A}^a$. In other words, node v does not know when such a result is valid so that the execution of $\mathcal{A}^a$ can continue based on this result (as if it was returned by the Π_C-oracle). The technique we present to resolve this issue for RW-algorithms is based on Theorem 1. Since the sustainability compiler (discussed in detail in the full paper) works independently of the algorithm's access to an oracle, the arguments to establish Theorem 1 can be repeated to yield the following.

Lemma 1. *Fix some problem Π'. Let $\mathcal{A}$ be a RW-algorithm with access to a Π'-oracle loosely solving a problem Π and let $\hat{\mathcal{A}}$ be the RW-algorithm with access to a Π'-oracle obtained by applying the sustainability compiler to $\mathcal{A}$. Then $\hat{\mathcal{A}}$ sustainably solves Π with an access to a Π'-oracle.*

The ability to transform any RW-algorithm to ensure sustainable correctness plays a key role in the proof of Theorem 2. Recall that our goal is to replace the access to a Π_C-oracle of a C-algorithm $\mathcal{A}^a$ by a C-algorithm $\mathcal{A}^r$ solving Π_C without any oracle access. In other words, the crux is to show how a C-algorithm $\mathcal{A}$ can interleave the execution of algorithm $\mathcal{A}^a$ with an invocation of $\mathcal{A}^r$ in every round in a correct manner, without any additional knowledge of the run-time of $\mathcal{A}^r$ or properties of the underlying network. As noted before, in the case $C = RW$, algorithm $\mathcal{A}$ faces the issue that an output returned to a node v by $\mathcal{A}^r$ may not be part of a ready configuration and thus it is not clear whether v should use this value as an output of the Π_C-oracle that $\mathcal{A}^a$ invoked. Theorem 1 however relieves $\mathcal{A}$ from the burden of dealing with more than one ready configuration of $\mathcal{A}^r$, whereas Lemma 1 does the same with $\mathcal{A}^a$. Therefore, $\mathcal{A}$ is left with the task of determining when $\mathcal{A}^r$ and $\mathcal{A}^a$ have reached a ready configuration.

In the full version of this paper we show how this can be accomplished by carefully dividing the simulation into phases of a predetermined length and re-cycling previously used random bits. Assuming that Theorem 2 is established we introduce the concept of hard problems by borrowing the terminology from sequential complexity theory.

Definition (Hardness). For two classes $B \supseteq C$, a problem Π is said to be B-*hard* with respect to C, denoted by $\Pi \in B\text{-}hard_C$, if for every problem $\Pi_B \in B$, there exists a C-algorithm that solves Π_B with access to a Π-oracle. We say that Π is *complete* in B with respect to C, denoted by $\Pi \in B\text{-}complete_C$, if additionally Π itself is contained in B.

Following our notational convention, we would refer to an $\mathcal{NP}$-hard problem as being $\mathcal{NP}\text{-}hard_P$. For example, the problem of electing a leader is well known to be CF-$hard_{WO}$ since once a leader is available, this leader can assign unique identifiers to all other nodes and solve the problem centrally. Our definition yields the three hardness classes CF-$hard_{RW}$, CF-$hard_{WO}$ and RW-$hard_{WO}$, allowing us to study how algorithms running in anonymous networks relate to centralized algorithms as well as how the two output revocability notions relate among each other. By definition, every CF-*hard*$_{\text{WO}}$ problem is both CF-*hard*$_{\text{RW}}$ and RW-*hard*$_{\text{WO}}$; in Section 5 we present a proof sketch for the following theorem, which states that the converse direction is also true. A thorough proof appears in the full version.

Theorem 3. *It holds that* CF-*hard*$_{\text{WO}}$ = CF-*hard*$_{\text{RW}}$ ∩ RW-*hard*$_{\text{WO}}$.

4   Problem Zoo

We study the computability and hardness of 21 problems in our setting, and develop different proof techniques to tackle this tedious task. In this extended abstract we confine ourselves to summarize the fruits of our effort in Figure 1. Exemplarily we also present the hardness result for logical OR which is necessary for the sketched proof of Theorem 3 in Section 5.

Overview of Problems. We briefly explain the problems listed in Figure 1.

- LEADER-ELECTION: all but one node output "NOT LEADER", while a single node outputs "LEADER".
- UNIQUENESS: determine whether all nodes have a unique input value.
- IDs: without any input, every node must return a unique identifier.
- α-SIZE-APX: determine a value $\tilde{n}$ such that $n \leq \tilde{n} \leq \alpha \cdot n$, where n is the number of nodes in the network.
- MIN-CUT: determine a partition of the network inducing a minimum cut as well as the size of this cut.
- MIN-CUT-VALUE: determine the size of a minimum cut.
- MIN-CUT-PARTITION: determine a partition of the network inducing a minimum cut.
- DIAMETER: determine the diameter D of the network.

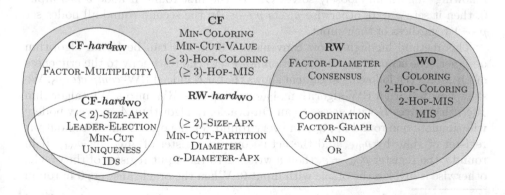

Fig. 1. Classes CF, RW and WO, and the respective hardness classes

- α-DIAMETER-APX: determine a value $\tilde{d}$ such that $D \leq \tilde{d} \leq \alpha \cdot D$.
- MIN-COLORING: color the graph with the minimum number of colors.
- COLORING: determine *some* coloring of the graph.
- k-HOP-COLORING: color the graph so that the color of every node v differs from the color of every other node in its k-hop neighborhood.
- MIS: determine a maximal independent set.
- k-HOP-MIS: find a maximal subset S of the nodes so that the distance between every two nodes in S is greater than k.
- CONSENSUS: nodes return the same value x which is at least one node's input.
- COORDINATION: determine whether all nodes have the same input.
- AND: nodes have input 0 or 1 and have to return the logical AND of all inputs.
- OR: nodes have input 0 or 1 and have to return the logical OR of all inputs.
- FACTOR-GRAPH: agree on a mapping f inducing a factor F of the network graph. Each node v returns F and its corresponding node $f(v)$ in F.
- FACTOR-DIAM: determine the diameter of a factor graph of the network.
- FACTOR-MULTIPLICITY: determine the multiplicity of the smallest factor of the network.

The last three problems on this list require the notion of graph factors[3] which is introduced in the full version of this paper. Connections from distributed computability to graph factors were witnessed before, for example in [3]. For some problems on the list, namely k-HOP-MIS, k-HOP-COLORING, and α-SIZE-APX, computability and/or hardness depends on the choice of k and α, respectively.

Of course for many problems on this list it is known whether they are contained in WO or CF \ WO. For example the well studied symmetry breaking tasks MIS or COLORING with $\Delta + 1$ colors (where Δ denotes the maximum degree of a node in the graph) are known to be in WO [29,2,28]. The work [17] presents WO-algorithms for each of the two problems 2-HOP-MIS and 2-HOP-COLORING. An example of a previously known hardness result is that an approximation α-SIZE-APX with $\alpha < 2$ is sufficient to find unique identifiers with a WO-algorithm [34].

Example (Logical OR). Denote by ρ the output register of a node v. The following "algorithm" loosely solves OR. In the first round if node v has input 0, then it sets $\rho \leftarrow 0$, otherwise v sets $\rho \leftarrow \varepsilon$. In the second round all nodes set $\rho \leftarrow 1$ regardless of their input.

This method highlights how convenient Theorem 1 can be for an algorithm designer. The straight-forward solution however is no testimony to the crudeness of OR, since the following argument shows that it is indeed RW-*complete*$_{\text{WO}}$. We show how to turn a RW-algorithm $\mathcal{A}_{\text{RW}}$ solving $\Pi \in$ RW into a WO-algorithm $\mathcal{A}_{\text{WO}}$ that solves Π with access to an OR-oracle. In algorithm $\mathcal{A}_{\text{WO}}$ every node v will simulate one round of $\mathcal{A}_{\text{RW}}$ in every round; we denote v's simulated output register of $\mathcal{A}_{\text{RW}}$ by ρ_{RW}, and the actual output register of $\mathcal{A}_{\text{WO}}$ by ρ_{WO}. If in round r the register $\rho_{\text{RW}} = \varepsilon$, then v writes 1 to the input register of the oracle, otherwise it invokes the oracle with input 0. When the oracle answers 0 in round

---

[3] In the distributed computing literature, the concept of graph factors was also referred to as covering graphs and graph lifts.

$r + 1$, the network was in a ready configuration in round r and v sets ρ_{WO} to the value contained in ρ_{RW} in round r.

5 Proof of Theorem 3

In this section we only present a sketch for the proof of Theorem 3; a comprehensive proof is presented in the appended full paper. Our proof is based on the techniques introduced in Section 2 and utilizes the aforementioned completeness result for OR. Theorem 3 states that if a problem Π is both CF-$hard_{RW}$ and RW-$hard_{WO}$, then it is also CF-$hard_{WO}$. Let $\Pi \in$ CF-$hard_{RW} \cap$ RW-$hard_{WO}$ be a problem satisfying the premise. Denote by $\mathcal{A}_{LE}$ a RW-algorithm solving LEADER-ELECTION with an access to a Π-oracle, and by $\mathcal{A}_{OR}$ a WO-algorithm solving OR with an access to a Π-oracle respectively.

The idea is to design a WO-algorithm $\mathcal{A}$ solving LEADER-ELECTION with access to a Π-oracle by simulating one execution of $\mathcal{A}_{LE}$ and multiple executions of $\mathcal{A}_{OR}$, where the task of the latter is to determine whether the former has reached a ready configuration. That is, for every simulated round r of algorithm $\mathcal{A}_{LE}$ a corresponding simulation $\mathcal{A}_{OR}$, called the *fork* $[r]$ of $\mathcal{A}_{OR}$, is initiated. The input to fork $[r]$ is 0 if v was ready in round r under $\mathcal{A}_{LE}$ (v observes that from the simulated outcome of $\mathcal{A}_{LE}$'s round r); the input is 1 otherwise. Since in $\mathcal{A}$ the Π-oracle can only be accessed once in every round, algorithm $\mathcal{A}$ uses a careful mechanism to schedule disjoint accesses by the simulated execution of $\mathcal{A}_{LE}$ and all forks to this scarce resource; we refer to the full version for the details.

The logic of OR guarantees that fork $[r]$ of $\mathcal{A}_{OR}$ has output 0 if and only if round r under $\mathcal{A}_{LE}$'s simulation is in a ready configuration. Since $\mathcal{A}_{OR}$ is a WO-algorithm, node v can immediately rely on a returned 0 value to conclude that this indeed happened. Employing Lemma 1, one can assume that $\mathcal{A}_{LE}$ sustainably solves the leader election problem, thus ensuring that the output returned in v's simulated round r of $\mathcal{A}_{LE}$ yields a correct output for LEADER-ELECTION. This establishes Theorem 3.

References

1. Abrahamson, K., Adler, A., Higham, L., Kirkpatrick, D.: Probabilistic solitude verification on a ring. In: PODC (1986)
2. Alon, N., Babai, L., Itai, A.: A fast and simple randomized parallel algorithm for the maximal independent set problem. Journal of Algorithms 7(4), 567–583 (1986)
3. Angluin, D.: Local and global properties in networks of processors (extended abstract). In: Theory of Computing (1980)
4. Angluin, D., Aspnes, J., Chan, M., Fischer, M.J., Jiang, H., Peralta, R.: Stably computable properties of network graphs. In: Prasanna, V.K., Iyengar, S.S., Spirakis, P.G., Welsh, M. (eds.) DCOSS 2005. LNCS, vol. 3560, pp. 63–74. Springer, Heidelberg (2005)
5. Angluin, D., Aspnes, J., Diamadi, Z., Fischer, M.J., Peralta, R.: Computation in networks of passively mobile finite-state sensors. In: PODC (2004)

194   Y. Emek, J. Seidel, and R. Wattenhofer

6. Angluin, D., Aspnes, J., Eisenstat, D.: Stably computable predicates are semilinear. In: PODC (2006)
7. Angluin, D., Aspnes, J., Eisenstat, D., Ruppert, E.: The computational power of population protocols. Distributed Computing 20, 279–304 (2007)
8. Angluin, D., Fischer, M.J., Jiang, H.: Stabilizing consensus in mobile networks. In: Gibbons, P.B., Abdelzaher, T., Aspnes, J., Rao, R. (eds.) DCOSS 2006. LNCS, vol. 4026, pp. 37–50. Springer, Heidelberg (2006)
9. Aspnes, J., Ruppert, E.: An introduction to population protocols. In: Garbinato, B., Miranda, H., Rodrigues, L. (eds.) MiNEMA (2009)
10. Boldi, P., Vigna, S.: Computing anonymously with arbitrary knowledge. In: PODC (1999)
11. Boldi, P., Vigna, S.: An effective characterization of computability in anonymous networks. In: Welch, J.L. (ed.) DISC 2001. LNCS, vol. 2180, pp. 33–47. Springer, Heidelberg (2001)
12. Boldi, P., Vigna, S.: Universal dynamic synchronous self-stabilization. Distributed Computing 15(3), 137–153 (2002)
13. Chalopin, J., Das, S., Santoro, N.: Groupings and pairings in anonymous networks. In: Dolev, S. (ed.) DISC 2006. LNCS, vol. 4167, pp. 105–119. Springer, Heidelberg (2006)
14. Chalopin, J., Godard, E., Métivier, Y.: Local terminations and distributed computability in anonymous networks. In: Taubenfeld, G. (ed.) DISC 2008. LNCS, vol. 5218, pp. 47–62. Springer, Heidelberg (2008)
15. Dolev, S.: Self-Stabilization (2000)
16. Dolev, S., Israeli, A., Moran, S.: Uniform dynamic self-stabilizing leader election. In: Toueg, S., Spirakis, P.G., Kirousis, L. (eds.) WDAG 1991. LNCS, vol. 579, pp. 167–180. Springer, Heidelberg (1992)
17. Emek, Y., Wattenhofer, R.: Stone age distributed computing. In: PODC (2013)
18. Flocchini, P., Kranakis, E., Krizanc, D., Luccio, F.L., Santoro, N.: Sorting and election in anonymous asynchronous rings. J. Parallel Distrib. Comput. 64(2), 254–265 (2004)
19. Fraigniaud, P., Korman, A., Peleg, D.: Local distributed decision. In: FOCS (October 2011)
20. Fraigniaud, P., Halldórsson, M.M., Korman, A.: On the impact of identifiers on local decision. In: Baldoni, R., Flocchini, P., Binoy, R. (eds.) OPODIS 2012. LNCS, vol. 7702, pp. 224–238. Springer, Heidelberg (2012)
21. Fraigniaud, P., Ilcinkas, D., Pelc, A.: Oracle size: A new measure of difficulty for communication tasks. In: PODC (2006)
22. Fraigniaud, P., Korman, A., Parter, M., Peleg, D.: Randomized distributed decision. In: Aguilera, M.K. (ed.) DISC 2012. LNCS, vol. 7611, pp. 371–385. Springer, Heidelberg (2012)
23. Göös, M., Suomela, J.: Locally checkable proofs. In: PODC (2011)
24. Guerraoui, R., Ruppert, E.: What can be implemented anonymously? In: Fraigniaud, P. (ed.) DISC 2005. LNCS, vol. 3724, pp. 244–259. Springer, Heidelberg (2005)
25. Itai, A., Rodeh, M.: Symmetry breaking in distributive networks. In: FOCS (1981)
26. Korman, A., Kutten, S., Peleg, D.: Proof labeling schemes. In: PODC (2005)
27. Korman, A., Sereni, J.S., Viennot, L.: Toward more localized local algorithms: removing assumptions concerning global knowledge. In: PODC (2011)
28. Linial, N.: Locality in distributed graph algorithms. SIAM Journal on Computing 21(1), 193–201 (1992)

29. Luby, M.: A simple parallel algorithm for the maximal independent set problem. In: Theory of Computing (1985)
30. Lynch, N.A.: Distributed Algorithms (1996)
31. Mavronicolas, M., Michael, L., Spirakis, P.: Computing on a partially eponymous ring. In: Shvartsman, A. (ed.) OPODIS 2006. LNCS, vol. 4305, pp. 380–394. Springer, Heidelberg (2006)
32. Métivier, Y., Robson, J.M., Zemmari, A.: Analysis of fully distributed splitting and naming probabilistic procedures and applications. In: Moscibroda, T., Rescigno, A.A. (eds.) SIROCCO 2013. LNCS, vol. 8179, pp. 153–164. Springer, Heidelberg (2013)
33. Naor, M., Stockmeyer, L.: What can be computed locally? SIAM Journal on Computing 24(6), 1259–1277 (1995)
34. Schieber, B., Snir, M.: Calling names on nameless networks. In: PODC (1989)
35. Yamashita, M., Kameda, T.: Computing on anonymous networks: Part i-characterizing the solvable cases. IEEE Trans. Parallel Distrib. Syst. 7(1), 69–89 (1996)

Coalgebraic Weak Bisimulation from Recursive Equations over Monads

Sergey Goncharov[1] and Dirk Pattinson[2]

[1] Department of Computer Science, FAU Erlangen-Nürnberg, Germany
[2] Research School of Computer Science, Australian National University, Australia

Abstract. Strong bisimulation for labelled transition systems is one of the most fundamental equivalences in process algebra, and has been generalised to numerous classes of systems that exhibit richer transition behaviour. Nearly all of the ensuing notions are instances of the more general notion of *coalgebraic bisimulation*. Weak bisimulation, however, has so far been much less amenable to a coalgebraic treatment. Here we attempt to close this gap by giving a coalgebraic treatment of (parametrized) weak equivalences, including weak bisimulation. Our analysis requires that the functor defining the transition type of the system is based on a suitable order-enriched monad, which allows us to capture weak equivalences by least fixpoints of recursive equations. Our notion is in agreement with existing notions of weak bisimulations for labelled transition systems, probabilistic and weighted systems, and simple Segala systems.

1 Introduction

Both strong and weak bisimulations are fundamental equivalences in process algebra [13]. Both have been adapted to systems with richer behaviour such as probabilistic and weighted transition systems. For each class of systems, strong bisimulation is defined in a similar way which is explained by universal coalgebra where strong bisimulation is recovered as a canonical equivalence that parametrically depends on the type of system [16]. Weak bisimulations are much more difficult to analyse even for labelled transition systems (LTS), and much less canonical in status (e.g. branching and delay bisimulations [21]).

We present a unified, coalgebraic treatment of various types of weak bisimulation. An important special (and motivating) case of our definition is *probabilistic weak bisimulation* of Baier and Herrmanns [2]. Unlike labelled transition systems, probabilistic weak bisimulation needs to account for *point-to-set* transitions, while *point-to-point* transitions, as for labelled transition systems, do not suffice: Every LTS with a transition relation $\rightarrow$ induces an LTS with a *weak transition relation* $\Rightarrow$ and weak bisimulation for the original system is strong bisimulation of the transformed one. This approach fails in the probabilistic case, as weak point-to-point transitions no longer form a probability distribution: in a system where $x \xrightarrow{a(0.5)} y$ and $x \xrightarrow{\tau(0.5)} x$, we obtain $x \xRightarrow{a(1)} y$ as the probability that x evolves to y along a trace of the form $\tau^* \cdot a \cdot \tau^*$ is clearly one, but also $x \xRightarrow{\tau(1)} x$ as the system will also evolve from x to x along τ^* also

J. Esparza et al. (Eds.): ICALP 2014, Part II, LNCS 8573, pp. 196–207, 2014.

with probability one (by simply doing nothing). Crucially, both events are not independent. This is resolved by relating states to state sets along transition sequences, and the probability $P(x, \Lambda, S)$ of x evolving to a state in S along a trace in Λ is the probability of the event that contains all execution sequences leading from x to S via Λ, called *total probability* in *op. cit.* By re-formulating this idea axiomatically, we show that it is applicable to a large class of systems, specifically coalgebras of the form $X \to T(X \times A)$ where T is enriched over directed complete partial orders with least element (pointed dcpos) and non-strict maps. Not surprisingly, similar (but stronger) assumptions also play a prominent role in coalgebraic trace semantics [8], and have two ramifications: the fact that the functor T that describes the branching behaviour extends to a monad allows us to consider *transition sequences*, and order-enrichment permits us to compute the cumulative effect of (sets of) transition sequences recursively using Kleene's fixpoint theorem. Our construction is parametric in an *observation pattern* that can be varied to obtain e.g. weak and delay bisimulation. We demonstrate by example that our definition generalises concrete definitions of probabilistic and weak weighted and probabilistic bisimulation found in the literature [2, 5, 18, 17].

A special role in our model is played by the operation of binary join, which is a continuous operation of the monad. We show that if it is also *algebraic* in the sense of Plotkin and Power [15], which holds in the case of LTS, then weak bisimulation can be recovered as a strong bisimulation for a system of the same type, thus reestablishing Milner's weak transition construction. In the probabilistic case, for which join is unsurprisingly nonalgebraic, we show that weak bisimulation arises as strong bisimulation of a system based on the continuation monad.

2 Preliminaries

We use basic notions of category theory and coalgebra, see e.g. [16] for an overview. For a functor $F : \mathbf{Set} \to \mathbf{Set}$, an F-*coalgebra* is a pair (X, f) with $f : X \to TX$. Coalgebras form a category where the morphisms between (X, f) and (Y, g) are functions $\phi : X \to Y$ with $g \circ \phi = F\phi \circ f$. A relation $E \subseteq X \times X$ is a *kernel bisimulation* on (X, f) if there is an F-coalgebra (Z, h) and two morphisms $\phi : (X, f) \to (Z, h)$ and $\psi : (X, g) \to (Z, h)$ such that $E = \mathsf{Ker}(\phi) = \{(x, y) \in X \times X \mid \phi(x) = \psi(y)\}$ is the kernel of ϕ. Clearly, kernel bisimulations are equivalence relations, and we only consider kernel bisimulations in what follows. Kernel bisimulation agrees with Aczel-Mendler bisimulation (and its variants) in case F preserves pullbacks weakly but is mathematically better behaved in case F does not. It also agrees (in all cases) with the notion of behavioural equivalence: a thorough comparison is provided in [20].

We take *monads* (on sets) as given by their extension form, i.e. as Kleisli triples $\mathbb{T} = (T, \eta, -^{\dagger})$ where $T : \mathbf{Set} \to \mathbf{Set}$ is a functor, $\eta_X : X \to TX$ is a map for all sets X and $f^{\dagger} : TX \to TY$ is a map for all $f : X \to TY$ subject to the equations $f^{\dagger}\eta_X = f$, $\eta_X^{\dagger} = \mathrm{id}_{TX}$ and $(f^{\dagger}g)^{\dagger} = f^{\dagger}g^{\dagger}$ for all sets X and all f, g of appropriate type. Throughout, we write T for the underlying functor of a monad $\mathbb{T}$. The *Kleisli category* induced by a monad $\mathbb{T}$ has sets as objects, but Kleisli-morphisms between X and Y are functions $f : X \to TY$ with Kleisli composition $g \circ f = g^\dagger \circ f$ where $g^\dagger \circ f$ is function composition in **Set** and η_X is the identity at X. We use Haskell-style do-notation to manipulate monad terms: for any $p \in TX$ and $q : X \to TY$ we write do $x \leftarrow p; q(x)$ to denote $q^\dagger(p) \in TY$; if $p \in T(X \times Y)$ we write do $\langle x, y \rangle \leftarrow p; q(x, y)$.

In the sequel, we consider (among other examples) monads induced by semirings: A *semiring* is a structure $(R, +, \cdot, 0, 1)$ such that $(R, +, 0)$ is a commutative monoid, $(R, \cdot, 1)$ is a monoid and multiplication distributes over addition, i.e. $x \cdot (y + z) = x \cdot y + x \cdot z$ and $(y + z) \cdot x = y \cdot x + z \cdot x$. A *positively ordered semiring* is a semiring $(R, +, \cdot, 0, 1, \leq)$ equipped with a partial order $\leq$ that is positive $(0 \leq r$ for all $r \in R)$ and compatible with the ring structure $(x \leq y$ implies that $x \square z \leq y \square z$ and $z \square x \leq z \square y$ for all $x, y, z \in R$ and $\square \in \{+, \cdot\})$. A *continuous semiring* is a positively ordered semiring where every directed set $D \subseteq R$ has a least upper bound $\sup D \in R$ that is compatible with the ring structure $(r \square \sup D = \sup\{r \square d \mid d \in D\}$ and $\sup D \square r = \sup\{d \square r \mid d \in D\}$ for all directed sets $D \subseteq R$, all $r \in R)$ and $\square \in \{+, \cdot\}$. Every continuous semiring R is a *complete semiring*, i.e. has infinite sums given by $\sum_{i \in I} r_i = \sup\{\sum_{i \in J} r_i \mid J \subseteq I \text{ finite}\}$. We refer to [7] for details. If R is complete, the functor $T_R X = X \to R$ extends to a monad $\mathbb{T}_R$, called the *complete semimodule monad* (c.f. [9]) with $\eta_X(x)(y) = 1$ if $x = y$ and $\eta_X(x)(y) = 0$, otherwise, and $f^\dagger(\phi)(y) = \sum_{x \in X} \phi(x) \cdot f(x)(y)$ for $f : X \to T_R Y$. Note if R is continuous then all $T_R X$ are pointed dcpos under the pointwise ordering of R and the same applies to Kleisli homsets, i.e. the set of Kleisli-maps of type $X \to TY$.

3   Examples

We illustrate our generic approach to weak bisimulation by means of the following examples. For all examples, strong bisimulation is well understood and known to coincide with kernel bisimulation. As we will see later, the same is true for weak bisimulation, introduced in the next section.

Labelled Transition Systems. We consider the monad $\mathbb{T}_Q$ where $Q = \{0, 1\}$ is the boolean semiring. Clearly $T_Q \cong \mathcal{P}$ where $\mathcal{P}$ is the covariant powerset functor. A *labelled transition system* can now be described as a coalgebra $(X, f : X \to T_Q(X \times A))$. It is well known that bisimulation equivalences on labelled transition systems coincide with kernel bisimulations as introduced in the previous section.

Probabilistic Systems. Consider the monad $\mathbb{T}_{[0,\infty]}$ induced by the complete semiring of non-negative real numbers, extended with infinity. Various types of probabilistic systems arise as sub-classes of systems of type $(X, f : X \to T_{[0,\infty]}(X \times A))$. For *reactive systems*, one postulates $\sum_{y \in X} f(x)(y, a) \in \{0, 1\}$ for all $x \in X$ and all $a \in A$. *Generative systems* satisfy $\sum_{(y,a) \in Y \times A} f(x)(y, a) \in \{0, 1\}$ for all $x \in X$, and *fully probabilistic systems* satisfy $\sum_{(y,a) \in X \times A} f(x)(y, a) = 1$ for all $x \in X$. We refer to [3] for a detailed analysis of various types of probabilistic systems in coalgebraic terms. It is known that probabilistic bisimulation equivalence [10] and kernel bisimulations agree [6]. Our justification of viewing these various types of probabilistic systems as $[0, \infty]$ weighted transition systems comes from the fact that kernel bisimulations are reflected by embeddings:

Lemma 1. *Let $\kappa : F \to G$ be a monic natural transformation between two set-functors F and G and (X, f) be an F-coalgebra. Then kernel bisimulations on the F-coalgebra (X, f) agree with kernel bisimulations on the G-coalgebra $(X, \kappa_X \circ f)$.*

Integer Weighted Transition Systems. Weighted transition systems, much like probabilistic systems, arise as coalgebras for the functor $FX = T_{\mathbb{N} \cup \{\infty\}}(X \times A)$ where $\mathbb{N} \cup \{\infty\}$ is the (complete) semiring of natural numbers extended with ∞ and the usual arithmetic operations. In an (integer) weighted transition system, every labelled transition comes with a weight, and we can write $x \xrightarrow{a(n)} y$ if $f(x)(y, a) = n$. In process algebra, weights represent different ways in which the same transition can be derived syntactically, e.g. $a.0 + a.0 \xrightarrow{a(2)} 0$, according to the reduction of the term on the left and right, respectively. The ensuing (strong) notion of equivalence has been studied in [1] and shown to be coalgebraic.

The three examples above are a special instance of semiring-weighted transition systems, studied for instance in [11]. This is not the case for systems that combine probability and non-determinism.

Non-Deterministic Probabilistic Systems. As we have motivated in the introduction, a coalgebraic analysis of weak bisimulation hinges on the ability to sequence transitions, i.e. the fact that the functor F defining the concrete shape of a transition system $(X, f : X \to F(X \times A))$ extends to a monad. The naive combination of probability and non-determinism, i.e. considering the functor $F = \mathcal{P} \circ \mathcal{D}$ where $\mathcal{D}(X)$ is the set of finitely distributed probability distributions does not extend to a monad [22]. One solution, discussed in *op.cit.* and elaborated in [9] is to restrict to *convex* sets of valuations. Informally, we use monad $\mathcal{C}_0\mathcal{M}$ (a variant of the $\mathcal{C}\mathcal{M}$ monad from [9]), encompassing two semiring structures, for probability and non-determinism, and the former distributes over the latter, i.e. $a +_p (b+c) = (a +_p b) + (a +_p c)$ where $+$ is nondeterministic choice and $+_p$ is probabilistic choice (choose 'left' with probability p and 'right' with probability $1 - p$). Concretely, for the underlying functor $\mathcal{C}_0\mathcal{M}$ of the monad $\mathcal{C}_0\mathcal{M}$, $\mathcal{C}_0\mathcal{M}X$ is the set of nonempty *convex* sets of finite valuations over $[0, \infty)$, i.e. finitely supported maps to $[0, \infty)$, containing the trivial valuation identically equal to 0. A set S is convex if $\sum_i r_i \cdot \xi_i \in S$ whenever all $\xi_i \in S$ and $\sum_i r_i = 1$. Our definition deviates slightly from [9] in that we require that $\mathcal{C}_0\mathcal{M}X$ contains the zero valuation, whereas in *op.cit.* (and also in [4]) this condition is used to restrict the class of systems to which the theory is applied.

4   Weak Bisimulation, Coalgebraically

Capturing weak bisimulation for transition systems $(X, f : X \to T(X \times A))$ coalgebraically, where A is a set of labels that we keep fixed throughout, amounts to two requirements: first, T needs to extend to a monad which enables us to sequence transitions. Second, we need to be able to compute the cumulative effect of transitions which requires the monad to be enriched over the category of directed complete partial orders (and non-strict morphisms).

Definition 2 (Completely Ordered Monads). A monad $\mathbb{T}$ is *completely ordered* if its Kleisli category is enriched over the category $\mathbf{DCPO}_\perp$ of directed-complete partial orders with least element (pointed dcpos) and continuous maps: every hom-set $\mathbf{Set}(X, TY)$ is a pointed dcpo and Kleisli composition is continuous, i.e. the joins $f^\dagger \circ (\bigsqcup_i g_i) = \bigsqcup_i f^\dagger \circ g_i$ and $(\bigsqcup_i f_i)^\dagger \circ g = \bigsqcup_i f_i^\dagger \circ g$ exist and are equal whenever the join on the left hand side is taken over a directed set. A *continuous operation* of arity n on a completely ordered monad is a natural transformation $\alpha : T^n \to T$ for which every component α_X is Scott-continuous.

The diligent reader will have noticed that the same type of enrichment is also required in the coalgebraic treatment of trace semantics [8]. This is by no means a surprise, as the observable effect of weak transitions are precisely given in terms of (sets of) traces.

Often, these sets are defined in terms of weak transitions of the form $\xrightarrow{\tau}^* \cdot \xrightarrow{a} \cdot \xrightarrow{\tau}^*$. We think of weak transitions as transitions along trace sets closed under Brzozowski derivatives which enables us to recursively decompose a weak transition into a (standard) transition, followed by a weak transition.

Definition 3 (Observation Pattern). An *observation pattern* over a set A of labels is a subset $B \subseteq \mathcal{P}(A^*)$ that is closed under Brzozowski derivatives, i.e. $b/a = \{w \in A^* \mid aw \in b\} \in B$ for all $b \in B$ and all $a \in A$.

Different observation patterns capture different notions of weak bisimulation, and B stands for both bisimulation and closure under Brzozowski derivatives.

Example 4 (Observation Patterns). Let A contain a silent action τ.

(i) the *strong pattern* over A is given by $B = \{\{a\} \mid a \in A\} \cup \{\emptyset, \{\epsilon\}\}$.

(ii) the *weak pattern* over A is given by $B = \{\hat{a} \mid a \in A\}$ where $\hat{\tau} = \tau^*$ and $\hat{a} = \tau^* \cdot a \cdot \tau^*$ for $a \neq \tau$.

(iii) The *delay pattern* is $B = \{\tau^* a \mid a \in A \setminus \{\tau\}\} \cup \{\tau^*\}$.

It is immediate that all are closed under Brzozowski derivatives.

Given an observation pattern that determines the notion of traces, our definition of weak bisimulation relies on the fact that the cumulative effect of transitions can be computed recursively. This is ensured by enrichment, and we have the following (see Section 2 for the do-notation):

Lemma 5. *Suppose B is an observation pattern over A, $\mathbb{T}$ is a completely ordered monad and $\oplus : T^2 \to T$ is continuous. Then the equation*

$$f_h^B(x)(b) = \begin{cases} \eta(h(x)) & \text{if } \epsilon \in b \\ \perp & \text{otherwise} \end{cases} \oplus \text{do} \langle y, a \rangle \leftarrow f(x); f_h^B(y)(b/a) \qquad (\bigstar)$$

has a unique least solution $f_h^B : X \to (TY)^B$ for all $f : X \to T(X \times A)$ and all $h : X \to Y$.

Lemma (5) follows from Kleene's fixpoint theorem [23] using order-enrichment. The central notion of our paper can now be given as follows:

Definition 6. Suppose that $\mathbb{T}$ is a completely ordered monad with a continuous operation $\oplus$, B is an observation pattern over A and let $f : X \to T(X \times A)$. An equivalence relation $E \subseteq X \times X$ is a B-$\oplus$-*bisimulation* if $E \subseteq \mathsf{Ker}(f_\pi^B)$ where $\pi : X \to X/E$ is the canonical projection (and f_π^B is the unique least solution of ($\bigstar$)). We often elide the continuous operation, and say that $x, x' \in X$ are B-bisimilar, if they are related by a B-bisimulation.

Some remarks are in order before we show that the above definition agrees with various notions of weak bisimulation studied in the literature.

Remark 7.   (i) Intuitively, the requirement $E \subseteq \mathsf{Ker}(f_\pi^B)$ expresses that any two E-related states x and x' have the same cumulative behaviour under all trace sets in B, *provided* that E-related states are not distinguished. In other words, a state $[x]_E$ of the quotient of the original system exhibits the same behaviour with respect to all trace sets in B, as the representative x of $[x]_E$. This intuition is made precise in Section 6 where we show how B-bisimulation can be recovered as strong bisimulation (and hence quotients can be constructed).

(ii) The definition of weak bisimulation above caters for systems of the form $(X, f : X \to T(X \times A))$, i.e. we implicitly consider the labels as part of the observable behaviour, or as 'output'. The role of labels appears to be reversed when computing the cumulative effect of transitions via the function $f_\pi^B : X \to T(X/E)^B$. This apparent reversal of roles is due to the fact that every element of B is a set of traces. Accordingly, the function application $f_h^B(x)(b)$ represents the totality of behaviour that can be observed along traces in b, starting from x, and trace sets are now 'input'.

As a slogan, B-bisimilarity is a B-bisimulation:

Lemma 8. *Let $(E_i)_{i \in I}$ be a family of B-bisimulation equivalences on $(X, f : X \to T(X \times A)$. Then so is the transitive closure of $\bigcup_{i \in I} E_i$.*

5   Examples, Revisited

We demonstrate that B-bisimulation agrees with the known (and expected) notion of weak bisimulation for the examples in Section 3. To instantiate the general definition to coalgebras of the form $X \to T(X \times A)$, we need to verify that the monad $\mathbb{T}$ is completely ordered. This is the case for complete semimodule monads over continuous semirings.

Lemma 9. *Let R be a continuous semiring. Then the monad $\mathbb{T}_R$ is completely ordered, and both join $\sqcup$ and semiring sum $+$ are continuous operations on T.*

This lemma in particular ensures that B-bisimulation is meaningful for transitions systems weighted in a complete semiring, and in particular for labelled, probabilistic and integer-weighted systems.

Labelled Transitions Systems. As in Section 3, labelled transition systems are coalgebras for the functor $FX = \mathcal{P}(X \times A)$. For an F-coalgebra (X, f), Equation ($\bigstar$) stipulates that

$$f_h(x)(b) = \{ h(x) \mid \epsilon \in b \} \cup \bigcup_{x \xrightarrow{a} y} f_h(y)(b/a)$$

where $x \xrightarrow{a} y$ iff $\langle y, a \rangle \in f(x)$. By Kleene's fixpoint theorem, the least solution is

$$f_h(x)(b) = \left\{ h(x_k) \mid x \xrightarrow{a_1} x_1 \xrightarrow{a_2} \dots \xrightarrow{a_k} x_k, \ a_1 \cdots a_k \in b \right\}.$$

If $\tau \in A$, B is the weak pattern and $E \subseteq X \times X$ is an equivalence, this gives

$$[x']_E \in f_\pi^B(x)(\hat{a}) \quad \text{iff} \quad x \xRightarrow{\hat{a}} x'$$

where $x \xRightarrow{\hat{a}} x'$ if there are $(y_1, a_1), \dots, (y_n, a_n)$ such that $x \xrightarrow{a_1} y_1 \xrightarrow{a_2} \cdots \xrightarrow{a_n} y_n = x'$ and $a_1 \cdots a_n \in \hat{a}$. By Definition 6, E is a B-bisimulation if for any $\langle x, y \rangle \in E$, $\{[x']_E \mid x \xRightarrow{\hat{a}} x'\} = \{[y']_E \mid y \xRightarrow{\hat{a}} y'\}$ for any $a \in A$ (including τ). The latter is easily shown to be equivalent to the standard notion of weak bisimulation equivalence. By analogous reasoning one readily recovers delay bisimulation equivalences from the delay pattern.

Probabilistic Systems. Fully probabilistic system (Section 3) are coalgebras of type $(X, f : X \to T_{[0,\infty]}(X \times A))$, where $\mathbb{T}_{[0,\infty]}$ is the complete semimodule monad induced by $[0, \infty]$ and additionally satisfy $\sum_{(y,a) \in X \times A} f(x)(y, a) = 1$ for all $x \in X$. In [2], an equivalence relation $E \subseteq X \times X$ is a weak bisimulation, if

$$P(x, \hat{a}, [y]_E) = P(x', \hat{a}, [y]_E)$$

for all $a \in A, y \in X$ and $(x, x') \in E$. Here $\hat{a}$ is given as in Example 4 and $P(x, \Lambda, C)$ is the *total probability* of the system evolving from state x to a state in C via a trace in $\Lambda \subseteq A^*$. *Op.cit.* states that total probabilities satisfy the recursive equations: $P(x, \Lambda, C) = 1$ if $\epsilon \in C$ and $x \in \Lambda$, and

$$P(x, \Lambda, C) = \sum_{(y,a) \in X \times A} f(x)(y, a) \cdot P(y, \Lambda/a, C)$$

otherwise. In fact, total probabilities are the *least* solution (with respect to the pointwise order on $[0, \infty]$) of the recursive equations above.

Lemma 10. *Let $(X, f : X \to T_{[0,\infty]}(X \times A))$ be a fully probabilistic system, B an observation pattern over A and $E \subseteq X \times X$ an equivalence relation. If $\pi : X \to X/E$ is the canonical projection, then $P(x, b, [y]_E) = f_\pi^B(x)(b)([y]_E)$ for all $x, y \in X$ and all $b \in B$, using $\sqcup$ as continuous operation.*

As a corollary, we obtain that weak bisimulation of fully probabilistic systems is a special case of B-bisimulation for the weak pattern.

Weighted Transition Systems. Weighted transition systems are technically similar to probabilistic systems as they also appear as coalgebras for a (complete) semimodule monad, but without any restriction on the sum of weights. The associated notion of *weak resource bisimulation* is described syntactically in [5]. Abstracting from the concrete syntax and taking weighted transition systems as primitive, we are faced with a situation that is reminiscent of the probabilistic case: a weak resource bisimulation equivalence on a weighted transition system $(X, f : X \to T_{\mathbb{N} \cup \infty}(X \times A))$ is an equivalence relation $E \subseteq X \times X$ such that xEy and $a \in A$ implies that $W(x, \Lambda, C) = W(y, \Lambda, C)$ for all equivalence classes

$C \in X/E$ and all Λ that are of the form $\tau^* a \tau^*$ for $a \neq \tau$ and τ^*. Here $W(x, \Lambda, C)$ is the *total weight*, i.e. the maximal number of possibilities in which x can evolve into a state in C via a path from Λ. Total weights can be understood as (weighted) sums over all *independent* paths that lead from x into C via a trace in Λ, where two paths are independent if neither is a prefix of the other. Analogously to the probabilistic case, these weights are given by the least solution of the recursive equations

$$W(x, \Lambda, C) = \begin{cases} 1 & \epsilon \in \Lambda, x \in C \\ 0 & \text{otherwise} \end{cases} \sqcup \sum_{(y,a) \in X \times A} f(x)(y, a) \cdot W(y, \Lambda/a, C)$$

and represent the total number of possibilities in which a process x can evolve into a process in C along a trace in Λ. For example, we have that $W(0 + \tau.0 + \tau.\tau.0, \tau^*, \{0\}) = 3$ representing the three different possibilities in which the given process can become inert along a τ-trace, and $W(x, \tau^*, z) = 6$ for the triangle-shaped system $x \xrightarrow{\tau(2)} y$, $x \xrightarrow{\tau(2)} z$ and $y \xrightarrow{\tau(2)} z$. It is routine to check that $W(x, b, [x']_E) = f_\pi^B(x)(b)([x']_E)$. Unlike the probabilistic case, the number of different ways in which processes may evolve is strictly additive. For the weak pattern, B-bisimulation is therefore the semantic manifestation of weak resource bisimulation advocated in [5].

Probability and Nondeterminism. Systems that combine probabilistic and nondeterministic behaviour arise as coalgebras of type $(X, f : X \to C_0 M(X \times A))$ where $C_0 \mathcal{M}$ is the monad from Section 3. Systems of this type capture so-called Segala systems. Here we stick to simple Segala systems, which are colagebras of type $\mathcal{P}(\mathcal{D} \times A)$ and for which the ensuing notion of weak probabilistic bisimulation was introduced in [18]. These systems extend probabilistic systems by additionally allowing non-deterministic transitions. As was essentially elaborated in [4], every simple Segala system embeds into a coalgebra $(X, f : X \to C_0 M(X \times A))$.

Completing a simple Segala system to a coalgebra over $C_0 \mathcal{M}$ amounts to forming *convex* sets of valuations; convexity arises from probabilistic choice as follows: given non-deterministic transitions $x \to \xi$ and $x \to \zeta$, where ξ and ζ are valuations over $X \times A$ induces a transition $x \to \xi +_p \zeta$ where $+_p$ is probabilistic choice. Following [22], one way to understand this is to also consider non-deterministic choice $+$ and to observe that

$$\xi + \zeta = (\xi + \zeta + \xi) +_p (\xi + \zeta + \zeta) = (\xi + \zeta) +_p (\xi + \zeta) + (\xi +_p \zeta) = \xi + \zeta + (\xi +_p \zeta)$$

by the axioms $\xi + \xi = \xi +_p \xi = \xi$, $(\xi + \zeta) +_p \theta = (\xi +_p \theta) + (\zeta +_p \theta)$, the last one describing the interaction between probabilistic and non-deterministic choice.

We argue that B-bisimulation where B is the weak observation pattern agrees with the notion from [18, 17]. We make a forward reference to Theorem 16 which shows that B-bisimulation for (X, f) amounts to strong bisimulation for (X, f_{id}^B). In other words, weak bisimilarity can be recovered from strong bisimilarity for the system whose transitions are weak transitions of the original system. Solving the recursive equation for f_{id}^B (where B is the weak pattern and we use the notation of Example 4) we can write $x \xRightarrow{\hat{a}} \xi$ if $\xi \in f_{\text{id}}^B(x)(a)$. Intuitively, this represents that x can evolve along a trace in $\hat{a}$ to the valuation ξ, interleaving probabilistic and nondeterministic steps. We then obtain that an equivalence relation $E \subseteq X \times X$ is a B-bisimulation if, whenever $(x, y) \in E$ and $x \overset{\hat{a}}{\Longrightarrow} \xi$, there exists ζ such that $y \overset{\hat{a}}{\Longrightarrow} \zeta$ and ξ and ζ are 'equivalent up to E', that is, $(F\pi)\xi = (F\pi)\zeta$ where $\pi : X \to X/E$ is the projection and $FX = [0, \infty)^X$. More concretely, the weak relation $\Rightarrow\, \in X \times B \times [0, \infty)^X$ is obtained by ($\bigstar$) and is the least solution of the following system:

$$x \overset{\hat{\tau}}{\Rightarrow} \delta_x$$

$$x \overset{\hat{a}}{\Rightarrow} \zeta \quad \text{iff} \quad \exists \xi \in f(x). \ \zeta \in \left\{ \sum\nolimits_{y \in X} \xi(y, a) \cdot \theta_y^\tau + \xi(y, \tau) \cdot \theta_y^a \mid \forall y.\, y \overset{\hat{b}}{\Rightarrow} \theta_y^b \right\}$$

where $x \overset{\hat{b}}{\Rightarrow} \zeta$ ($b \in \{a, \tau\}$) abbreviates $\langle x, b, \zeta \rangle \in \Rightarrow$; $\delta_y(y') = 1$ if $y = y'$ and $\delta_y(y') = 0$ otherwise; and scalar multiplication and summation act on valuations pointwise. Kleene's fixpoint theorem underlying Lemma 5 ensures that the relation $\Rightarrow$ can be calculated iteratively, i.e. $\Rightarrow\, = \bigcup_i \Rightarrow_i$ where the $\Rightarrow_i$ replace $\Rightarrow$ in the above recursive equations in the obvious way, hence making them recurrent. Then $x \overset{\hat{b}}{\Rightarrow} \zeta$ iff there is i such that $x \overset{\hat{b}}{\Rightarrow}_i \zeta$. The resulting definition in terms of weak transitions $\Rightarrow_i$ matches weak probabilistic bisimulation from [18, 17]. Note that convexity of the monad precisely ensures that ξ in the recursive clause above for $x \overset{\hat{a}}{\Rightarrow} \zeta$ represents a *combined step* of the underlying Segala system, which by definition, is exactly a convex combination of ordinary probabilistic transitions.

6 Weak Bisimulation as Strong Bisimulation

Milner's weak transition construction characterises weak bisimilarity as bisimilarity for a (modified) system whose transitions are the weak transitions of the original system. This construction does not transfer to the general case, witnessed by the case of (fully) probabilistic systems. The pivotal role is played by the continuous operation $\oplus$ that determines B-bisimulation. We show that Milner's construction generalises if $\oplus$ is *algebraic* and present a variation of the construction if algebraicity fails. An *algebraic operation* of arity n on a monad $\mathbb{T}$ (e.g. [15]) is a natural transformation $\alpha : T^n \to T$ such that $\alpha_Y \circ (f^\dagger)^n = f^\dagger \circ \alpha_X$ for all $f : X \to TY$. Algebraic operations are automatically continuous:

Lemma 11. *Algebraic operations of completely ordered monads are continuous.*

Example 12 (Algebraic Operations). Semiring summation $+$ is algebraic on continuous semimodule monads. If the underlying semiring is idempotent, e.g. the boolean semiring, summation coincides with the join operation $\sqcup$ which is therefore also algebraic. The bottom element $\bot$ is a nullary algebraic operation (constant). The join operation is algebraic on the monad $\mathcal{C}_0\mathcal{M}$ from Section 3. The join operation $\sqcup$ is generally not algebraic for free (complete) semimodule monads unless the semiring is idempotent.

Algebraicity of $\oplus$ allows us to lift Milner's construction to the coalgebraic case: B-bisimulations coincide with kernel bisimulations for a modified system of the *same transition type*. This instantiates to labelled transition systems, as $\sqcup$ is algebraic on the semimodule monad induced by the boolean semiring. We show this using a sequence of lemmas, the first asserting that algebraic operations commute over fixpoints.

Lemma 13. *Suppose $h : X \to Y$ and $u : Y \to Z$. Given a coalgebra $(X, f : X \to T(X \times A))$ we have that $f_{uh}^B = T^B u \circ f_h^B$ if $\oplus$ is algebraic.*

Similarly, sans algebraicity, B-bisimulations commute with morphisms.

Lemma 14. *Let $h : X \to Y$ be a morphism from $(X, f : X \to T(X \times A))$ to $(Y, g : Y \to T(Y \times A))$. Then $g_u^B \circ h = f_{uh}^B$ for all $u : Y \to Z$.*

Consequently, kernel bisimulations are B-bisimulations:

Corollary 15. *Let $h : X \to Y$ be a morphism of coalgebras $(X, f : X \to T(X \times A))$ and $(Y, g : Y \to T(Y \times A))$. Then $\mathsf{Ker}\, h \subseteq \mathsf{Ker}\, f_h^B$.*

Lemma 13 shows that for monads equipped with an algebraic operation $\oplus$ (such as the monad defining) labelled transition systems, we can recover B-bisimilarity as strong bisimilarity of a transformed system.

Theorem 16. *Provided $\oplus$ is algebraic, E is a B-bisimulation on a monad-type coalgebra (X, f) iff E is a kernel bisimulation equivalence on (X, f_{id}^B).*

If $\oplus$ is not algebraic it can still be possible to recover B-bisimulation as a kernel bisimulation for a system of a different type. For probabilistic systems this was done in [19]. Here, we obtain a similar result in a more conceptual way using the *continuous continuation monad* $\mathbb{T}$, which is obtained from the standard continuation monad [14] by restricting to continuous functions: the functorial part of $\mathbb{T}$ is $TX = (X \to D) \to_c D$ where $\to_c$ it the continuous function space, D is a directed-complete partial order, and $(X \to D)$ is ordered pointwise.

Lemma 17. *For a pointed dcpo D, $TX = (X \to D) \to_c D$ extends to a submonad $\mathbb{T}$ of the corresponding continuation monad, $\mathbb{T}$ is completely ordered, and every $\oplus : T^2 \to T$, given pointwise, i.e. $(p \oplus q)(c) = p(c) \oplus q(c)$, is algebraic.*

The following lemma is the B-bisimulation analogue of Lemma 1 and is the main technical tool for reducing B-bisimulation to kernel bisimulation.

Lemma 18. *Let $(X, f : X \to T(X \times A))$ be a coalgebra and $\kappa : T \to \widehat{T}$ an injective monad morphism. If $\widehat{\oplus}$ is an algebraic operation on $\widehat{T}$ such that $\widehat{\oplus} \circ \kappa^2 = \kappa \circ \oplus$ then B-$\oplus$-bisimulation equivalences on (X, f) and B-$\widehat{\oplus}$-bisimulation equivalences on $(X, \kappa f)$ agree.*

We use Lemma 18 as follows. Given a complete semimodule monad $\mathbb{T}$ over a (complete) semiring R, we embed TX into $\widehat{T}X = (X \to T1) \to_c T1$ (where $T1 = R$) by mapping $p \in TX$ to the function $\lambda c . X \to T1.c^\dagger(p)$. This embedding is injective, and the conditions of Lemma 18 are fulfilled with $\oplus = \sqcup$ and $\widehat{\oplus}$ the pointwise extension of $\oplus$ (which is algebraic by Lemma 17). This gives:

Theorem 19. *Let $\mathbb{T}$ be a continuous semimodule monad over a continuous semiring R. Let $(X, f : X \to (X \times A))$ be a coalgebra and let $\oplus$ be the join on R. Then E is a B-bisimulation equivalence on (X, f) iff it is a bisimulation equivalence on $(X, (\kappa_X \circ f)^B_{\mathrm{id}} : X \to (X \times A \to_c R) \to R)$.*

In summary, Milner's weak transition construction generalises to the coalgebraic case if $\sqcup$ is algebraic, and lifts to a different transition type for semirings.

7   Conclusions and Related Work

We have presented a generic definition, and basic structural properties, of weak bisimulation in a general, coalgebraic framework. We use coalgebraic methods and enriched monads, similar to the coalgebraic treatment of trace semantics [8]. Our definition applies uniformity to labelled transition systems, probabilistic and weighted systems, and to Segala systems from [18]. Most of our results, including the notions of B-bisimulation as a solution of the recursive equation ($\bigstar$), easily transfer to categories other than **Set**. An important conceptual contribution is the fact that algebraicity allows to generalise Milner's weak transition construction to the coalgebraic setting (Theorem 16), recovering B-bisimulation as kernel bisimulation for a (modified) system of the same transition type. We also provide an alternative for cases where this fails (Theorem 19).

Related Work. Results similar to ours are presented both in [4] and [12]. Brengos [4] uses a remarkably similar tool set (order-enriched monads) but in a substantially different way: Given a system of type $T(F + -)$ with $\mathbb{T}$ order-enriched, the monad structure on $\mathbb{T}$ extends to $T(F + -)$, and saturation w.r.t. internal transitions is achieved by iterating the obtained monad in a way resembling the weak transition construction for LTS. Examples include labelled transition systems and (simple) Segala systems. For both underlying monads, join is algebraic, so that both examples are covered by our lifting Theorem 16. Fully probabilistic systems, for which algebraicity fails, are not treated in [4]. Miculan and Peresotti [12] also approach weak bisimulation by solving recurrence relations, but only treat (continuous) semimodule monads and do not account for (simple) Segala systems. Our treatment covers all examples considered in both [4] and [12], and additionally identifies the pivotal role of algebraicity in the generalisation of Milner's construction. Sokolova et.al. [19] are concerned with probabilistic systems only and reduce probabilistic weak bisimulation to strong (kernel) bisimulation for a system of type $(- \times A \to 2) \to [0, 1]$. This is similar to our Theorem 19, which establishes an analogous transformation (to a system of type $(- \times A \to [0, \infty]) \to [0, \infty]$) by a rather more high-level argument.

Future Work. We plan to investigate to what extent our treatment extends to coalgebras $X \to T(X + FX)$ for a monad $\mathbb{T}$ (the branching type) and a functor F (the transition type) and are interested in both a logical and an equational characterisation of B-bisimulation, and in algorithms to compute B-bisimilarity.

References

[1] Aceto, L., Ingolfsdottir, A., Sack, J.: Resource bisimilarity and graded bisimilarity coincide. Information Processing Letters 111(2), 68–76 (2010)

[2] Baier, C., Hermanns, H.: Weak bisimulation for fully probabilistic processes. In: Grumberg, O. (ed.) CAV 1997. LNCS, vol. 1254, pp. 119–130. Springer, Heidelberg (1997)

[3] Bartels, F., Sokolova, A., de Vink, E.: A hierarchy of probabilistic system types. In: Coalgebraic Methods in Computer Science. ENTCS, vol. 82. Elsevier (2003)

[4] Brengos, T.: Weak bisimulation for coalgebras over order enriched monads (2013), http://arxiv.org/abs/1310.3656

[5] Corradini, F., De Nicola, R., Labella, A.: Graded modalities and resource bisimulation. In: Pandu Rangan, C., Raman, V., Ramanujam, R. (eds.) FST TCS 1999. LNCS, vol. 1738, pp. 381–393. Springer, Heidelberg (1999)

[6] de Vink, E., Rutten, J.: Bisimulation for probabilistic transition systems: a coalgebraic approach. Theoretical Computer Science 221(1-2), 271–293 (1999)

[7] Droste, M., Kuich, W.: Semirings and formal power series. In: Droste, M., Kuich, W., Vogler, H. (eds.) Handbook of Weighted Automata. Monographs in Theoretical Computer Science, pp. 3–28. Springer (2009)

[8] Hasuo, I., Jacobs, B., Sokolova, A.: Generic trace semantics via coinduction. Logical Methods in Comp. Sci. (2007)

[9] Jacobs, B.: Coalgebraic trace semantics for combined possibilitistic and probabilistic systems. Electr. Notes Theor. Comput. Sci. 203(5), 131–152 (2008)

[10] Larsen, K.G., Skou, A.: Bisimulation through probabilistic testing. Information and Computation 94(1), 1–28 (1991)

[11] Latella, D., Massink, M., de Vink, E.P.: Bisimulation of labeled state-to-function transition systems of stochastic process languages. In: Golas, U., Soboll, T. (eds.) Proc. ACCAT 2012. EPTCS, vol. 93, pp. 23–43 (2012)

[12] Miculan, M., Peressotti, M.: Weak bisimulations for labelled transition systems weighted over semirings (2013), http://arxiv.org/abs/1310.4106

[13] Milner, R.: Communication and concurrency. Prentice-Hall (1989)

[14] Moggi, E.: A modular approach to denotational semantics. In: Curien, P.-L., Pitt, D.H., Pitts, A.M., Poigné, A., Rydeheard, D.E., Abramsky, S. (eds.) CTCS 1991. LNCS, vol. 530, pp. 138–139. Springer, Heidelberg (1991)

[15] Plotkin, G., Power, J.: Notions of computation determine monads. In: Nielsen, M., Engberg, U. (eds.) FOSSACS 2002. LNCS, vol. 2303, pp. 342–356. Springer, Heidelberg (2002)

[16] Rutten, J.: Universal Coalgebra: A theory of systems. Theoret. Comput. Sci. 249(1), 3–80 (2000)

[17] Segala, R.: Modelling and Verification of Randomized Distributed Real-Time Systems. PhD thesis, Massachusetts Institute of Technology (1995)

[18] Segala, R., Lynch, N.A.: Probabilistic simulations for probabilistic processes. In: Jonsson, B., Parrow, J. (eds.) CONCUR 1994. LNCS, vol. 836, pp. 481–496. Springer, Heidelberg (1994)

[19] Sokolova, A., de Vink, E.P., Woracek, H.: Coalgebraic weak bisimulation for action-type systems. Sci. Ann. Comp. Sci. 19, 93–144 (2009)

[20] Staton, S.: Relating coalgebraic notions of bisimulation. Logical Methods in Computer Science 7(1) (2011)

[21] van Glabbeek, R.J., Weijland, W.P.: Branching time and abstraction in bisimulation semantics. J. ACM 43(3), 555–600 (1996)

[22] Varacca, D., Winskel, G.: Distributing probability over non-determinism. Math. Struct. Comput. Sci. 16, 87–113 (2006)

[23] Winskel, G.: The Formal Semantics of Programming Languages. MIT Press, Cambridge (1993)

Piecewise Boolean Algebras and Their Domains

Chris Heunen*

University of Oxford, Department of Computer Science, United Kingdom
`heunen@cs.ox.ac.uk`

Abstract. We characterise piecewise Boolean domains, that is, those domains that arise as Boolean subalgebras of a piecewise Boolean algebra. This leads to equivalent descriptions of the category of piecewise Boolean algebras: either as piecewise Boolean domains equipped with an orientation, or as full structure sheaves on piecewise Boolean domains.

1 Introduction

Boolean algebras embody the logical calculus of observations. But in many applications it does not make sense to consider any two observations simultaneously. For a simple example, can you really verify that "there is a polar bear in the Arctic" and "there is a penguin in Antarctica", when you cannot be in both places at once? This leads to the notion of a *piecewise Boolean algebra*[1], which is roughly a Boolean algebra where only certain pairs of elements have a conjunction.

You could say that the issue in the above example is merely caused by a constructive interpretation. But it is a real, practical concern in *quantum logic*, where the laws of nature forbid jointly observing certain pairs (the famous example being to measure position and momentum), and piecewise Boolean algebras consequently play a starring role [3–6].

Another cause of incompatible observations relates to *partiality*. Some (observations of) computations might not yet have returned a result, but nevertheless already give some partial information. It might not make sense to compare two partial observations, whereas the completed observations would be perfectly compatible. Partiality is also at play in quantum theory, where measurements can be fine-grained, so that the course-grained version only gives partial information. This leads to *domain theory* [7, 8].

This paper brings the two topics, domain theory and quantum logic, together. The main construction sends a piecewise Boolean algebra P to the collection $\mathrm{Sub}(P)$ of its compatible parts, *i.e.* its Boolean subalgebras. This well-known construction [1, 3–5, 9–13] assigns a domain $\mathrm{Sub}(P)$ to a piecewise Boolean algebrac P. Our main result is a characterisation of the domains of the form $\mathrm{Sub}(P)$, called *piecewise Boolean domains*; it turns out they are the so-called algebraic L-domains whose bottom two rungs satisfy some extra properties. This

---

* Supported by EPSRC Fellowship EP/L002388/1.

[1] Née *partial* Boolean algebra; recent authors use *piecewise* to avoid 'partial complete Boolean algebra' [1]. Incidentally, this is the structure Boole originally studied [2].

J. Esparza et al. (Eds.): ICALP 2014, Part II, LNCS 8573, pp. 208–219, 2014.
© Springer-Verlag Berlin Heidelberg 2014

gives an alternative description of piecewise Boolean algebras, that is more concise, amenable to domain theoretic techniques, and addresses open questions [11, Problems 1 and 2]. Colloquially, it shows that to reconstruct the whole, it suffices to know how the parts fit together, without having to know the internal structure of the parts.

Commutative rings, such as Boolean algebras, can be reconstructed from their Zariski spectrum together with the structure sheaf over that spectrum [14, V.3]. Analogously, we prove that a piecewise Boolean algebra can be reconstructed from its piecewise Boolean domain together with the structure sheaf over that domain. (Equivalently, we could use the Stone dual of the structure sheaf.) We prove a categorical equivalence between piecewise Boolean algebras, and piecewise Boolean domains with a subobject-preserving functor valued in Boolean algebras. We call the latter objects *piecewise Boolean diagrams*.

There is a beautiful microcosm principle at play in the reconstruction of a piecewise Boolean diagram from a piecewise Boolean domain: piecewise Boolean diagrams are really structure-preserving functors from a piecewise Boolean domain into the category of Boolean algebras. The piecewise Boolean diagram is almost completely determined by the piecewise Boolean domain, but some choices have to be made. We condense those choices into an *orientation*, that fixes a choice between two possibilities on each atom of a piecewise Boolean domain. Finally, we prove that the category of piecewise Boolean algebras is equivalent to the category of oriented piecewise Boolean domains.

We proceed as follows. Section 2 recalls the basics of piecewise Boolean algebras, after which Section 3 introduces piecewise Boolean domains and proves they are precisely those domains of the form $\mathrm{Sub}(P)$. This characterisation is simplified further in Section 4. Section 5 proves the equivalence between piecewise Boolean algebras and piecewise Boolean diagrams, and Section 6 reduces from piecewise Boolean diagrams to oriented piecewise Boolean domains. Finally, Section 7 concludes with directions for future work. For example, it would be interesting to explore connections to other work [15, 16].

2   Piecewise Boolean Algebras

Definition 1. *A* piecewise Boolean algebra *consists of a set P with:*

- *a reflexive and symmetric binary* (commeasurability) *relation $\odot \subseteq P \times P$;*
- *elements $0, 1 \in P$;*
- *a (total) unary operation $\neg \colon P \to P$;*
- *(partial) binary operations $\wedge, \vee \colon \odot \to P$;*

such that every set $A \subseteq P$ of pairwise commeasurable elements is contained in a set $B \subseteq P$, whose elements are also pairwise commeasurable, and on which the above operations determine a Boolean algebra structure.

A morphism *of piecewise Boolean algebras is a function that preserves commeasurability and all the algebraic structure, whenever defined. Piecewise Boolean algebras and their morphisms form a category* **PBool**.

A piecewise Boolean algebra in which every two elements are commeasurable is just a Boolean algebra. Given a piecewise Boolean algebra P, we write $\mathrm{Sub}(P)$ for the collection of its commeasurable subalgebras, ordered by inclusion. (The maximal elements of $\mathrm{Sub}(P)$ are also called *blocks*, see [6, Section 1.4].) In fact, Sub is a functor **PBool** → **Poset** to the category of partially ordered sets and monotone functions, acting on morphisms by direct image. If P is a piecewise Boolean algebra, $\mathrm{Sub}(P)$ is called its *piecewise Boolean domain*.

We now list two main results about piecewise Boolean algebras and their domains. First, we can reconstruct P from $\mathrm{Sub}(P)$ up to isomorphism.

Theorem 2 ([1]). *Any piecewise Boolean algebra P is a colimit of* $\mathrm{Sub}(P)$. □

Boolean algebras are precisely objects of the ind-completion of the category of finite Boolean algebras [14, VI.2.3], defining Boolean algebras as colimits of diagrams of finite Boolean algebras. The previous theorem extends this to piecewise Boolean algebras. Second, $\mathrm{Sub}(P)$ determines P up to isomorphism.

Theorem 3 ([11]). *If P and P' are piecewise Boolean algebras and $\varphi\colon \mathrm{Sub}(P) \to \mathrm{Sub}(P')$ is an isomorphism, then there is an isomorphism $f\colon P \to P'$ with $\varphi = \mathrm{Sub}(f)$. Moreover, f is unique iff atoms of* $\mathrm{Sub}(P)$ *are not maximal.* □

However, the functor Sub is not an equivalence. It is not faithful: see the above theorem. Neither is it full: not every monotone function $\mathrm{Sub}(P) \to \mathrm{Sub}(P')$ preserves atoms. Nevertheless, the previous two theorems show that the functor Sub is almost an equivalence. Later, we will upgrade the functor Sub to an equivalence. But first we investigate posets of the form $\mathrm{Sub}(P)$.

3   Piecewise Boolean Domains

This section characterises piecewise Boolean domains in terms of finite partition lattices, which we will characterise further in the next section. Recall that an element x of a poset P is *compact* when, if $x \leq \bigvee D$ for a directed subset $D \subseteq P$ with a supremum, then $x \leq y$ for some $y \in D$. Write $K(P)$ for the partially ordered set of compact elements of P.

Definition 4. *A poset is called a* piecewise Boolean domain *when:*

(1) it has directed suprema;
(2) it has nonempty infima;
(3) each element is the directed supremum of compact ones;
(4) the downset of each compact element is dual to a finite partition lattice.

Posets satisfying properties (1)–(3) are also known as *Scott domains* [17].

Proposition 5. *If P is a piecewise Boolean algebra, $\mathrm{Sub}(P)$ is a piecewise Boolean domain.*

Proof. If $B_i \in \mathrm{Sub}(P)$, then also $\bigcap B_i \in \mathrm{Sub}(P)$, giving nonempty infima. If $\{B_i\}$ is a directed family of elements of $\mathrm{Sub}(P)$, then $\bigcup B_i$ is a Boolean algebra, which is the supremum in $\mathrm{Sub}(P)$. To show that every element is the directed supremum of compact ones, it therefore suffices to show that the compact elements are the finite Boolean subalgebras of P. But this is easily verified. Finally, the downset of any compact element is pairwise commeasurable, hence a finite Boolean algebra, and it is dual to a finite partition lattice. [12, 13]. $\square$

We now set out to prove that any piecewise Boolean domain L is of the form $\mathrm{Sub}(P)$ for some piecewise Boolean algebra P. The first step is to show L gives rise to a functor $L \to \mathbf{Bool}$ that preserves the structure of L. For $x \in L$, we write $\mathrm{Sub}(x)$ for the principal ideal of x.

Remark 6. Both occurrences of Sub are instances of a more general scheme. If $\mathbf{C}$ is a category with epi-mono factorizations, we write $\mathrm{Sub} \colon \mathbf{C} \to \mathbf{Poset}$ for the covariant subobject functor. It acts as direct image on morphisms $f \colon x \to y$, that is, a subobject $m \colon \bullet \rightarrowtail x$ gets mapped to the image $f[m] \colon \mathrm{Im}(f \circ m) \rightarrowtail y$. If $\mathbf{C}$ is a poset, then $\mathrm{Sub}(x)$ is just the principal ideal of x, and functoriality just means that $\mathrm{Sub}(x) \subseteq \mathrm{Sub}(y)$ when $x \leq y$. If $\mathbf{C} = \mathbf{Bool}$, then $\mathrm{Sub}(B)$ is the lattice of Boolean subalgebras of B, and the direct image $f[A]$ of a Boolean subalgebra A under a homomorphism $f \colon B \to B'$ is a Boolean subalgebra of B'. By slight abuse of notation, if $\mathbf{C}$ is the category $\mathbf{PBool}$, we let $\mathrm{Sub}(P)$ be the poset of Boolean subalgebras of P (instead of piecewise Boolean subalgebras), as before. The action on morphisms by direct image is then still well-defined.

Lemma 7. *Let L be a piecewise Boolean domain.*

(a) For each $x \in L$ there is a Boolean algebra $F(x)$ with $\mathrm{Sub}(F(x)) \cong \mathrm{Sub}(x)$.
(b) There is a functor $F \colon L \to \mathbf{Bool}$ and a natural isomorphism $\mathrm{Sub} \circ F \cong \mathrm{Sub}$.

Proof. Properties (1) and (2) make L into an L-domain [8, Theorem 2.9]. Adding property (3) makes L into an algebraic L-domain [8, Section 2.2]. It follows that every downset is an algebraic lattice [8, Corollary 1.7 and Proposition 2.8], and in fact that $\bigcup_x K(\mathrm{Sub}(x)) = K(L)$ [8, Proposition 1.6]. Finally, property (4) ensures that every downset satisfies the following property: it is an algebraic lattice, and each compact element in it is dual to a finite partition lattice. Therefore every downset is the lattice of Boolean subalgebras of some Boolean algebra [12], establishing (a).

Towards (b), define $\varphi_{x,y}$ for $x \leq y \in L$ as the following composition.

$$
\begin{array}{ccc}
\mathrm{Sub}(x) & \xleftarrow{\ \cong\ } & \mathrm{Sub}(F(x)) \\
{\scriptstyle \mathrm{Sub}(x \leq y)}\downarrow & & \downarrow{\scriptstyle \varphi_{x,y}} \\
\mathrm{Sub}(y) & \xrightarrow[\cong]{} & \mathrm{Sub}(F(y))
\end{array}
$$

Because $\mathrm{Sub}(x \leq y)$ is a monomorphism of complete lattices [8, Proposition 2.8], so is $\varphi_{x,y}$. Now, $\mathrm{Sub}(\varphi_{x,y})(\mathrm{Sub}(F(x))) \in \mathrm{Sub}(\mathrm{Sub}(F(y)))$; that is, the direct image of $\varphi_{x,y}$ is downward closed in $\mathrm{Sub}(F(y))$. So, by construction, the direct image of $\varphi_{x,y}$ is $\mathrm{Sub}(B)$, where $B = \varphi_{x,y}(F(x))$. Hence $\varphi_{x,y}$ factors as an isomorphism $\psi\colon \mathrm{Sub}(F(x)) \to \mathrm{Sub}(B)$ followed by an inclusion $\mathrm{Sub}(B) \subseteq \mathrm{Sub}(F(y))$. By [12, Theorem 4] or [13, Corollary 2], there is an isomorphism $f\colon F(x) \to B$ such that $\psi = \mathrm{Sub}(f)$. Also, $B \in \mathrm{Sub}(B) \subseteq \mathrm{Sub}(F(y))$, so B is a Boolean subalgebra of $F(y)$. That is, there is an inclusion $g\colon B \hookrightarrow F(y)$ such that $\mathrm{Sub}(g)$ is the inclusion $\mathrm{Sub}(B) \subseteq \mathrm{Sub}(F(y))$. Thus $F(x \leq y) := g \circ f\colon F(x) \rightarrowtail F(y)$ is a monomorphism of Boolean algebras that satisfies $\mathrm{Sub}(F(x \leq y)) = \varphi_{x,y}$. If $|F(x)| \neq 4$, then $F(x \leq y)$ is in fact the unique such map [12, Lemma 5], and in this case it follows that $F(y \leq z) \circ F(x \leq y) = F(x \leq z)$.

Next, we will adjust $F(x \leq y)$ for $|F(x)| = 4$ if need be, to ensure functoriality of F. Let x be an atom of L. If x is maximal, there is nothing to do. Otherwise choose y covering x. Select one of the two possible $F(x < y)$ inducing $\varphi_{x,y}$. Now, for any $y' > x$ such that $z = y \vee y'$ exists we need to choose $F(x < y')$ making the following diagram commute.

$$
\begin{array}{ccc}
F(x) & \xrightarrow{\;F(x < y)\;} & F(y) \\
{\scriptstyle F(x < y')}\Big\downarrow & & \Big\downarrow{\scriptstyle F(y < z)} \\
F(y') & \xrightarrow[\;F(y' < z)\;]{} & F(z)
\end{array}
\qquad (*)
$$

Let us write α_z for the isomorphism $\mathrm{Sub}(F(z)) \to \mathrm{Sub}(z)$. Next, notice that $X := F(y < z) \circ F(x < y)[F(x)] = \varphi_{x,z}(F(x)) = \alpha_z(x) \subseteq F(z)$, and similarly $Y := F(y' < z)[F(y')] = \varphi_{y',z}(F(y')) = \alpha_z(y') \subseteq F(z)$; because $x < y'$ hence $X \subseteq Y$, and there is a unique $F(x < y')$ making the diagram commute. Moreover $\mathrm{Sub}(F(x < y')) = \varphi_{x,y'}$. Thus F is functorial, and the isomorphisms $\mathrm{Sub} \circ F \cong \mathrm{Sub}$ are natural by construction. This proves part (b). $\qquad\square$

We say a functor $F\colon L \to \mathbf{Bool}$ *preserves subobjects* when there is a natural isomorphism $\mathrm{Sub} \circ F \cong \mathrm{Sub}$.

Next, we show that the data contained in the functor $L \to \mathbf{Bool}$ can equivalently be packaged as a piecewise Boolean algebra by taking its colimit.

Lemma 8. *Let L be a piecewise Boolean domain, let F be the functor of Lemma 7, and let the piecewise Boolean algebra P be the colimit of F in* **PBool***.*

(a) Maximal elements of L correspond bijectively to maximal elements of $\mathrm{Sub}(P)$.
(b) The colimit maps $F(x) \to P$ are injective.

Proof. In general, colimits of piecewise Boolean algebras are hard to compute (see [1, Theorem 2], and also [18]). But injectivity of $F(x \leq y)$ makes it manageable. Namely, $P = \coprod_{x \in L} F(x)/\sim$, where $\sim$ is the smallest equivalence relation satisfying $b \sim F(x \leq y)(b)$ when $x \leq y$ and $b \in F(x)$. That is, $F(x_1) \ni b_1 \sim b_n \in F(x_n)$ means there are $x_2, \ldots, x_{n-1} \in L$ with $x_1 \geq x_2 \leq x_3 \geq x_4 \leq x_5 \geq \cdots \geq x_{n-1} \leq x_n$, and $b_i \in F(x_i)$ for $i = 2, \ldots, n-1$ that satisfy

$b_{i+1} = F(x_i \leq x_{i+1})(b_i)$ for even i and $b_i = F(x_{i+1} \leq x_i)(x_{i+1})$ for odd i. Let us write $p_x \colon F(x) \to P$ for the colimiting maps $p_x(a) = [a]_\sim$.

If x_1 and x_n are maximal, then without loss of generality we may assume that x_i is maximal for odd i and that $x_{i+1} = x_i \wedge x_{i+2}$ for odd i. By the naturality of Lemma 7(b), this means that the subalgebra $F(x_2)$ of $F(x_1)$ and $F(x_3)$ is identified. So, by injectivity of $F(x \leq y)$, the only way the entire algebra $F(x_1)$ can be identified with $F(x_n)$ is when $x_1 = \ldots = x_n$.

Define a function $f \colon \mathrm{Max}(L) \to \mathrm{Max}(\mathrm{Sub}(P))$ by $f(x) = p_x[F(x)] = [F(x)]_\sim$. The discussion above shows that f is injective. Any $B \in \mathrm{Sub}(P)$ is commeasurable, and hence there is $x \in L$ such that $B \subseteq [F(x)]_\sim$. If B is maximal, then we must have $B = f(x)$. Thus f is well-defined, and surjective. This proves (a).

For part (b), let $x \in L$. It follows from Zorn's Lemma and property (1) that x is below some maximal $y \in L$. By part (a), then p_y is injective. Therefore $p_x = p_y \circ F(x \leq y)$ is injective, too. $\qquad\square$

We are now ready to prove our main result.

Theorem 9. *Any piecewise Boolean domain is isomorphic to* $\mathrm{Sub}(P)$ *for a piecewise Boolean algebra* P.

Proof. Let L be a piecewise Boolean domain. Fix a functor F as in Lemma 7, and its piecewise Boolean algebra colimit $p_x \colon F(x) \to P$ as in Lemma 8. Define $f \colon L \to \mathrm{Sub}(P)$ as $f(x) = p_x[F(x)]$.

We first prove that f is surjective. Any $B \in \mathrm{Sub}(P)$ is commeasurable, so B is a Boolean subalgebra of $p_y[F(y)]$ for some $y \in L$. Hence $p_y^{-1}(B) \in \mathrm{Sub}(F(y))$. Because F preserves subobjects, $p_y^{-1}(B) = F(x \leq y)[F(x)]$ for some $y \leq x$. Then:

$$f(x) = p_x[F(x)] = p_y \circ F(x \leq y)[F(x)] = p_y[p_y^{-1}(B)] = B.$$

Next we prove that f is injective by exhibiting a left-inverse $g \colon \mathrm{Sub}(P) \to L$. Set $g(B) = \bigwedge\{x \in L \mid B \subseteq f(x)\}$. Note that $g(f(x)) = \bigwedge\{y \mid [F(x)]_\sim \subseteq [F(y)]_\sim\} \leq x$. Now, if $y \leq x$ then $[F(y)]_\sim = p_y[F(y)] = p_x \circ F(y \leq x)[F(y)] \subseteq [F(x)]_\sim$. Hence if also $[F(x)]_\sim \subseteq [F(y)]_\sim$, then $F(y \leq x)$ is an isomorphism, and $x = y$. So $g(f(x)) = x$.

Clearly $g(B) \leq g(C)$ when $B \subseteq C$, so $f(x) \subseteq f(y)$ implies $x \leq y$. Conversely, if $x \leq y$, then $f(x) = p_x[F(x)] = p_y[F(x \leq y)[F(x)]] \subseteq p_y[F(y)] = f(y)$. Thus f is an order isomorphism $\mathrm{Sub}(P) \cong L$. $\qquad\square$

4   Partition Lattices

There exist many characterisations of finite partition lattices [19–25]. We now summarise one of them that we will use to reformulate condition (4). In a partition lattice, the intervals $[p, 1]$ for atoms p are again partition lattices. This leads to the following result. For terminology, recall that a finite lattice is *(upper) semimodular* when x covers $x \wedge y$ implies that $x \vee y$ covers y, that a *geometric lattice* is a finite atomistic semimodular lattice, and that an element x of a lattice is called *modular* if $a \vee (x \wedge y) = (a \vee x) \wedge y$ for all $a \leq y$.

Theorem 10 ([24, 25]). *Suppose L is a geometric lattice with a modular coatom, and the interval $[p, 1]$ is a partition lattice of height $n - 1$ for all atoms p. If $n \leq 4$, assume that L has $\binom{n}{2}$ atoms. Then L is a partition lattice of height n. Conversely, a partition lattice of height n satisfies these requirements.* ☐

Let us call a lattice *cogeometric* when it is dual to a geometric lattice; this is equivalent to being finite, lower semimodular, and coatomistic. We can now simplify condition (4), showing that piecewise Boolean domains are domains that are determined entirely by their behaviour on the bottom three rungs.

Proposition 11. *A poset is a piecewise Boolean domain precisely if it meets conditions (1)–(3) and*

(4') the downset of a compact element is cogeometric and has a modular atom;
(4") each element of height $n \leq 3$ covers exactly $\binom{n+1}{2}$ elements.

Proof. We show that we may replace condition (4) in Definition 4 by (4') and (4"). Observe that a dual lattice having a modular coatom is equivalent to the lattice itself having a modular atom. Assuming condition (4) and $x \in K(L)$, then $\mathrm{Sub}(x)$ is dual to a finite partition lattice, so that condition (4') is satisfied. For $\mathrm{ht}(x) \leq 4$, condition (4") is verified by computing the partition lattices of height up to three, see Figure 1.

Conversely, assume (4') and (4"). Then the downset of each compact element is finite, so that compact elements have finite height. Hence condition (4) follows by induction on the height by Theorem 10. ☐

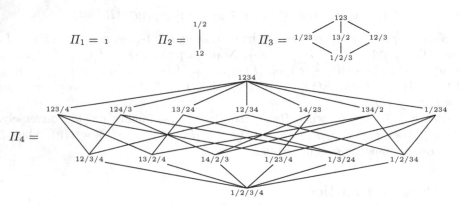

Fig. 1. The partition lattices of height up to three

5   Piecewise Boolean Diagrams

Definition 12. *A piecewise Boolean diagram is a subobject-preserving functor from a piecewise Boolean domain to **Bool**. A morphism of piecewise Boolean diagrams from $F: L \to$ **Bool** to $F': L' \to$ **Bool** consists of a morphism $\varphi: L \to L'$*

of posets and a natural transformation $\eta\colon F \Rightarrow F' \circ \varphi$. *Piecewise Boolean diagrams and their morphisms form a category* **PBoolD**. *Composition is given by* $(\psi, \theta) \circ (\varphi, \eta) = (\psi \circ \varphi, \theta\varphi \cdot \eta)$, *and identies are* $(\mathrm{id}, \mathrm{Id})$.

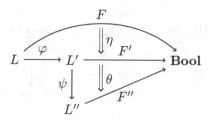

Notice that, because F preserves subobjects, also $\mathrm{Sub}(\varphi(x)) = \varphi[\mathrm{Sub}(x)]$, so that φ preserves directed suprema.

The functor Sub extends from piecewise Boolean domains to piecewise Boolean diagrams as follows.

Proposition 13. *There is a functor* $\mathrm{Spec}\colon$ **PBool** $\to$ **PBoolD** *defined as follows. On objects* $P \in$ **PBool**, *define* $\mathrm{Spec}(P)\colon \mathrm{Sub}(P) \to$ **Bool** *by* $B \mapsto B$. *On morphisms* $f\colon P \to P'$, *define* $\mathrm{Spec}(f)_B = f\!\restriction_B\colon B \to f[B]$. □

There is also a functor in the other direction. We will prove that the two functors in fact form an equivalence.

Proposition 14. *There is a functor* $\mathrm{colim}\colon$ **PBoolD** $\to$ **PBool** *defined as follows. On objects* $F\colon L \to$ **Bool**, *let* $\mathrm{colim}(F)$ *be the colimit* $p_x\colon F(x) \to \coprod F(x)/ \sim$. *On morphisms* $(\varphi, \eta)\colon F \to F'$, *let* $\mathrm{colim}(\varphi, \eta)$ *be the morphism* $\mathrm{colim}(F) \to \mathrm{colim}(F')$ *induced by the cocone* $p'_{\varphi(x)} \circ \eta_x\colon F(x) \to \mathrm{colim}(F')$. □

Theorem 15. *The functors* Spec *and* colim *form an equivalence between the category of piecewise Boolean algebras and the category of piecewise Boolean diagrams.*

Proof. If $P \in$ **PBool**, then $\mathrm{colim}(\mathrm{Spec}(P)) \cong P$ by Theorem 2. The isomorphism $P \cong \mathrm{colim}(\mathrm{Spec}(P))$ is given by $b \mapsto [b]_\sim$. If $f\colon P \to P'$, unrolling definitions shows that $\mathrm{colim}(\mathrm{Spec}(f))$ sends $[b]_\sim$ to $[f(b)]_\sim$. Therefore $\mathrm{colim} \circ \mathrm{Spec}$ is naturally isomorphic to the identity.

For a diagram $F\colon L \to$ **Bool**, fix $P = \mathrm{colim}(F)$. Set $\varphi\colon L \to \mathrm{Sub}(P)$ by $x \mapsto p_x[F(x)]$, and $\eta_x = p_x\colon F(x) \to p_x[F(x)]$. This is a well-defined isomorphism $(\varphi, \eta)\colon F \to \mathrm{Spec}(\mathrm{colim}(F))$ by Lemma 8. If $(\psi, \varepsilon)\colon F \to F'$, then $(\psi', \varepsilon') = \mathrm{Spec}(\mathrm{colim}(\psi, \varepsilon))$ consists of $\psi'\colon \mathrm{Sub}(\mathrm{colim}(F)) \to \mathrm{Sub}(\mathrm{colim}(F'))$ given by $\psi'(B) = [\bigcup_{b \in B \cap F(x)} \varepsilon_x(b)]_\sim$, and $\varepsilon'_B\colon B \to [\varepsilon[B]]_\sim$ given by $\varepsilon'_B(b) = [\varepsilon_x(b)]_\sim$ when $b \in F(x)$. It follows that

$$\psi' \circ \varphi(x) = [\varepsilon_x[F(x)]]_\sim = \varphi' \circ \psi(x),$$
$$(\eta'\psi \cdot \varepsilon)_x(b) = [\varepsilon_x(b)]_\sim = (\varepsilon'\varphi \cdot \eta)_x(b),$$

whence $(\varphi', \eta') \circ (\psi, \varepsilon) = (\psi', \varepsilon') \circ (\varphi, \eta)$, and $\mathrm{Spec} \circ \mathrm{colim}$ is naturally isomorphic to the identity. □

6   Orientation

We have lifted the functor Sub, that is full nor faithful, to an equivalence.

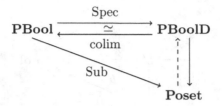

However, the cost was to add the full structure sheaf to $\mathrm{Sub}(P)$. In this section we reduce to minimal extra structure on a piecewise Boolean domain instead of the full structure sheaf. In other words: we want to find a converse to the forgetful functor, dashed in the diagram above. Lemma 7 goes towards such a functor, on the level of objects. However, notice that its proof required making some arbitrary choices. We will now fix these choices to obtain a functor.

Proposition 16. *Let L be a piecewise Boolean domain. If $x \in L$ is not an atom or 0, we may fix $F(x)$ to be the power set of the set of modular atoms in $\mathrm{Sub}(x)$ in Lemma 7(a).*

Proof. If x has at least four, it follows from a lattice-theoretic characterisation of partition lattices by Sachs [19, Theorem 14] that $\mathrm{Sub}(x)$ is dually isomorphic to the lattice of partitions of $\{$modular coatoms in $\mathrm{Sub}(x)^{\mathrm{op}}\}$.

For x of height two or three we may explicitly compute which coatoms of Π_n are modular. Notice that the element $y = 12/34$ is not modular in Π_4 (see Figure 1); taking $x = 13/2/4$ and $z = 13/24$ gives $x \vee (y \wedge z) = x \neq z = (x \vee y) \wedge z$. Similarly, $13/24$ and $14/23$ are not modular. But $123/4$, $124/3$, $134/2$, $234/1$ are modular elements. Hence Π_4 has 4 modular coatoms. Similarly, one can check that all 3 coatoms in Π_3 are modular. □

Definition 17. *An* orientation *of a piecewise Boolean domain L consists of a pointed four-element Boolean algebra $b_a \in F(a)$ for each atom $a \in L$. A morphism of oriented piecewise Boolean domains consists of a monotone function $\varphi \colon L \to L'$ satisfying*

- *if $a \in L$ is an atom, then either $\varphi(a)$ is an atom or $\varphi(a) = 0$,*
- *if a is a modular atom in $\mathrm{Sub}(x)$, then $\varphi(a)$ is modular in $\mathrm{Sub}(\varphi(x))$,*

and a map $\eta_a \colon F(a) \to F'(\varphi(a))$ satisfying $\eta_a(b_a) = b'_{\varphi(a)}$ for atoms $a \in L$ for which $\varphi(a)$ is a nonmaximal atom. The resulting category is denoted **OPBoolD**.

Proposition 18. *The functor $\mathrm{Sub} \colon$ **PBool** $\to$ **Poset** extends to orientations as follows. On objects, the orientation is given by $F(B) = B$. The point b_B is the unique element of $\mathrm{At}(C) \cap B$ for an atom B covered by C, and 0 if B is maximal. A morphism $\varphi = \mathrm{Sub}(f)$ extends to orientations by $\eta_B = f\restriction_B \colon B \to f[B]$.*

Proof. First of all, notice that this is well-defined on objects. If $B \in \mathrm{At}(\mathrm{Sub}(P))$ is covered by $C \in \mathrm{Sub}(P)$, say $B = \{0, x, \neg x, 1\}$ for $x \in P$, then precisely one of x and $\neg x$ must be an atom in C (and the other one a coatom). Also, this does not depend on C.

We have to show it is also well-defined on morphisms $f \colon P \to P'$. If B is an atom, say $B = \{0, x, \neg x, 1\}$, then $\varphi(B) = f[B] = \{0, f(x), \neg f(x), 1\}$ is clearly either an atom or $\{0, 1\}$. If $f[B]$ is a nonmaximal atom, then $f[C]$ covers $f[B]$ for some $C \in \mathrm{Sub}(P)$ covering B, so $f(b_B) = b'_{f[B]}$ by construction. Now suppose B is modular in $\mathrm{Sub}(D)$. Let $A' \subseteq C' \in \mathrm{Sub}(f[D])$; then $A' = f[A]$ and $C' = f[C]$ for some $A, C \in \mathrm{Sub}(D)$, namely $A = f^{-1}(A') \cap D$. Since $A \vee C$ is generated by $A \cup C$, we have $f[A \vee C] = f[A] \vee f[C]$ by [26, Proposition 2.4.4]. We may assume $B \cap C = \{0, 1\}$, for if $B \subseteq C$ then $f[C] \subseteq f[A] \vee f[B] = f[A \vee B]$ and $f[B]$ is modular in $\mathrm{Sub}(f[D])$. Of course always $f[B \cap C] \subseteq f[B] \cap f[C]$. Hence $A' \vee (f[B] \cap C') = f[A \vee B] \cap f[C] \supseteq f[(A \vee B) \cap C] = f[A \vee (B \cap C)] = f[A] \vee f[B \cap C] = f[A]$. Because $A \subseteq C$, the reverse inclusion also holds, and $f[B]$ is modular in $\mathrm{Sub}(f[D])$.

Finally, this extension is clearly functorial. $\qquad\square$

It follows that the forgetful functor **PBoolD** $\to$ **Poset** also extends to orientations as a functor **PBoolD** $\to$ **OPBoolD**.

Lemma 19. *An oriented piecewise Boolean domain* (L, F, b) *extends uniquely to a piecewise Boolean diagram* $F \colon L \to$ **Bool** *where* $F(a \leq x)(b_a)$ *is an atom if* x *covers an atom* $a \in L$.

Proof. It suffices to show that the requirement in the statement fixes the choice of maps $F(a \leq y)$ for atoms $a \in L$ in Lemma 7(b) in a well-defined way. Pick any y covering a, and fix $F(a < y)$ to be the map that sends b_a to an atom in $F(y)$. By diagram $(*)$, then $F(a < y')$ maps b_a to an atom for any $y' > a$ for which $z = y \vee y'$ exists (because Theorem 15 lets us assume that $F = \mathrm{Spec}(P)$ for some piecewise Boolean algebra P). Hence $F(a < y)$ does not depend on the choice of y. $\qquad\square$

Lemma 20. *A morphism of oriented piecewise Boolean domains extends uniquely to a morphism of piecewise Boolean diagrams.*

Proof. We have to extend a map $\eta_a \colon F(a) \to F'(\varphi(a))$, that is only defined on atoms $a \in L$, to a natural transformation $\eta_x \colon F(x) \to F'(\varphi(x))$. Let $x \in L$ be nonzero, and let $b' \in F(x)$. Then there is an atom $a \leq x$ and an element $b \in F(a)$ such that $F(a \leq x)(b) = b'$. Define $\eta_x(b) = F'(\varphi(a) \leq \varphi(x))(b')$. Because a and b are unique unless $b' \in \{0, 1\}$, this is a well-defined function. Moreover, it is natural by construction. Therefore it is also automatically unique.

We have to show η_x is a homomorphism of Boolean algebras. It clearly preserves 0 and $\neg$, so it suffices to show that it preserves $\wedge$. Let $b \neq b' \in F(x)$, say $b \in F(a)$ and $b' \in F(a')$ for atoms $a, a' \leq x$. By naturality, we may assume that $x = a \vee a'$. Hence x and $\varphi(x)$ have height 2, and $F(x)$ and $F'(\varphi(x))$ have 8 elements. We can now distinguish four cases, depending on whether $b = b_a$ and

$b' = b_{a'}$ or not. In each case it is easy to see that $\eta_x(b \wedge b') = \eta_x(b) \wedge \eta_x(b')$. For example, if $b = b_a$ and $b' = b_{a'}$, then they are distinct atoms in $F(x)$, so $b \wedge b' = 0$. But $\eta_x(b) = b'_a$ and $\eta_x(b') = b'_{a'}$ are distinct atoms in $F(\varphi(x))$, so $\eta_x(b) \wedge \eta_x(b') = 0$, too.   □

It follows that morphisms of oriented piecewise Boolean domains preserve directed suprema.

Theorem 21. *There is a functor* **OPBoolD** $\to$ **PBoolD** *that, together with the forgetful functor, forms an isomorphism of categories.*

Proof. Lemmas 19 and 20 define the functor on objects and morphisms, respectively; it is functorial by construction. Extending an oriented piecewise Boolean domain to a piecewise Boolean diagram and then restricting again to an oriented piecewise Boolean domain leads back to the original. Conversely, starting with a piecewise Boolean diagram, restricting it to an oriented piecewise Boolean domain, and then extending, leads back to the original piecewise Boolean diagram by unicity. Hence this is an isomorphism of categories.   □

7 Future Work

We conclude by listing several directions for future research.

- Many examples of piecewise Boolean algebras come from *orthomodular lattices* [1, 6]. These are precisely the piecewise Boolean algebras that are *transitive* and *joined*: the union $\leq$ of the orders on each commeasurable subalgebra is a transitive relation, and every two elements have a least upper bound with respect to $\leq$ [6, 1.4.22]; see also [4, 5]. An isomorphism of piecewise Boolean algebras between orthomodular lattices is in fact an isomorphism of orthomodular lattices.[2] Reformulating these properties in terms of piecewise Boolean domains would extend our results to orthomodular lattices.
- The introduction discussed the analogy between piecewise Boolean diagrams on a piecewise Boolean domains and structure sheaves on a Zariski spectrum. The latter form a topos and hence come with an internal logic [9]. However, piecewise Boolean domains are not (pointless) topological spaces. Can we formalise a notion of "skew sheaf" over piecewise Boolean domains so that it still makes sense to perform logic in the resulting "skew topos"?
- An obvious question is whether our results extend to piecewise *complete* Boolean algebras.
- Although there are many characterisations of finite partition lattices, there is no known equivalence between the category of finite partition lattices and the category of finite sets. For concreteness' sake, it would be very satisfying to explicate the maps $\varphi_{x,y}$ in Lemma 7.
- Any C*-algebra A gives rise to a piecewise Boolean algebra P. In fact, $\mathrm{Sub}(P)$ determines A up to isomorphism of Jordan algebras [28, 29]. Can our results be used to give an equivalent description of Jordan C*-algebras?

---

[2] This was observed in Sarah Cannon's MSc thesis [27], which prompted this work.

References

1. van den Berg, B., Heunen, C.: Noncommutativity as a colimit. Appl. Cat. Struct. 20(4), 393–414 (2012)
2. Hailperin, T.: Boole's algebra isn't Boolean algebra. Math. Mag. 54(4) (1981)
3. Hughes, R.I.G.: Omnibus review. J. Symb. Logic 50(2), 558–566 (1985)
4. Finch, P.E.: On the structure of quantum logic. J. Symb. Logic 34(2) (1969)
5. Gudder, S.P.: Partial algebraic structures associated with orthomodular posets. Pacific J. Math. 41(3) (1972)
6. Kalmbach, G.: Orthomodular Lattices. Acad. Pr (1983)
7. Abramsky, S., Jung, A.: Domain Theory. In: Handbook of Logic in Comp. Sci., vol. 3. Clarendon Press (1994)
8. Jung, A.: Cartesian closed categories of domains. PhD thesis, Tech. Hochsch. Darmstadt (1988)
9. Heunen, C., Landsman, N.P., Spitters, B.: A topos for algebraic quantum theory. Comm. Math. Phys. 291, 63–110 (2009)
10. Döring, A., Barbosa, R.S.: Unsharp values, domains and topoi. In: Quantum Field Theory and Gravity, pp. 65–96. Birkhäuser (2011)
11. Harding, J., Navara, M.: Subalgebras of orthomodular lattices. Order 28, 549–563 (2011)
12. Grätzer, G., Koh, K.M., Makkai, M.: On the lattice of subalgebras of a Boolean algebra. Proc. Amer. Math. Soc. 36, 87–92 (1972)
13. Sachs, D.: The lattice of subalgebras of a Boolean algebra. Can. J. Math. 14, 451–460 (1962)
14. Johnstone, P.T.: Stone spaces. Cambridge Studies in Advanced Mathematics, vol. 3. Cambridge Univ. Pr. (1982)
15. Laird, J.: Locally Boolean domains. Theor. Comp. Sci. 342(1), 132–148 (2005)
16. Abramsky, S., Vickers, S.: Quantales, observational logic and process semantics. Math. Struct. Comp. Sci. 3, 161–227 (1993)
17. Scott, D.S.: Domains for denotational semantics. In: Nielsen, M., Schmidt, E.M. (eds.) ICALP 1982. LNCS, vol. 140, pp. 577–613. Springer, Heidelberg (1982)
18. Haimo, F.: Some limits of Boolean algebras. Proc. Amer. Math. Soc. 2(4), 566–576 (1951)
19. Sachs, D.: Partition and modulated lattices. Pacific J. Math. 11(1), 325–345 (1961)
20. Sasaki, U., Fujiwara, S.: The characterization of partition lattices. J. Sci. Hiroshima Univ (A) 15, 189–201 (1952)
21. Ore, O.: Theory of equivalence relations. Duke Math. J. 9(3), 573–627 (1942)
22. Firby, P.A.: Lattices and compactifications I, II. Proc. London Math. Soc. 27, 22–60 (1973)
23. Aigner, M.: Uniformität des Verbandes der Partitionen. Math. Ann. 207, 1–22 (1974)
24. Stonesifer, J.R., Bogart, K.P.: Characterizations of partition lattices. Alg. Univ. 19, 92–98 (1984)
25. Yoon, Y.J.: Characterizations of partition lattices. Bull. Korean Math. Soc. 31(2), 237–242 (1994)
26. Koppelberg, S.: Handbook of Boolean algebras, vol. 1. North-Holland (1989)
27. Cannon, S.: The spectral presheaf of an orthomodular lattice. Master's thesis, Univ. Oxford (2013)
28. Harding, J., Döring, A.: Abelian subalgebras and the Jordan structure of a von Neumann algebra. Houston J. Math (2014)
29. Hamhalter, J.: Isomorphisms of ordered structures of abelian C*-subalgebras of C*-algebras. J. Math. Anal. Appl. 383, 391–399 (2011)

Between Linearizability and Quiescent Consistency*
Quantitative Quiescent Consistency

Radha Jagadeesan and James Riely

DePaul University

Abstract Linearizability is the de facto correctness criterion for concurrent data structures. Unfortunately, linearizability imposes a performance penalty which scales linearly in the number of contending threads. Quiescent consistency is an alternative criterion which guarantees that a concurrent data structure behaves correctly when accessed sequentially. Yet quiescent consistency says very little about executions that have any contention.

We define quantitative quiescent consistency (QQC), a relaxation of linearizability where the degree of relaxation is proportional to the degree of contention. When quiescent, no relaxation is allowed, and therefore QQC refines quiescent consistency, unlike other proposed relaxations of linearizability. We show that high performance counters and stacks designed to satisfy quiescent consistency continue to satisfy QQC. The precise assumptions under which QQC holds provides fresh insight on these structures. To demonstrate the robustness of QQC, we provide three natural characterizations and prove compositionality.

1 Introduction

This paper defines *Quantitative Quiescent Consistency (QQC)* as a criterion that lies between linearizability [10] and quiescent consistency [3], [11], [17]. The following example should give some intuition about these criteria.

Example 1.1. Consider a counter object with a single getAndIncrement method. The counter's sequential behavior can be defined as a set of strings such as $[^+\ ]_0^+\ \{^+\ \}_1^+\ (^+\ )_2^+$ where $[^+$ denotes an invocation (or call) of the method and $]_i^+$ denotes the response (or return) with value i. Suppose each invocation is initiated by a different thread.

A concurrent execution may have overlapping method invocations. For example, in $(^+\ [^+\ ]_0^+\ \{^+\ \}_1^+\ )_2^+$ the execution of $(^+\ )_2^+$ overlaps with both $[^+\ ]_0^+$ and $\{^+\ \}_1^+$, whereas $[^+\ ]_0^+$ finishes executing before $\{^+\ \}_1^+$ begins. Consider the following four executions.

$$(^+\ [^+\ ]_0^+\ \{^+\ \}_1^+\ )_2^+ \qquad (^+\ \{^+\ \}_1^+\ [^+\ ]_0^+\ )_2^+ \qquad [^+\ (^+\ )_2^+\ \{^+\ \}_1^+\ ]_0^+ \qquad [^+\ (^+\ )_2^+\ ]_0^+\ \{^+\ \}_1^+$$

Linearizability states roughly that *every* response-to-invocation order in a concurrent execution must be consistent with the sequential specification. Thus, the first execution is linearizable, since the response of $[^+\ ]_0^+$ precedes the invocation of $\{^+\ \}_1^+$ in the specification. However, none of the other executions is linearizable. For example, the response of $\{^+\ \}_1^+$ precedes the invocation of $[^+\ ]_0^+$ in the second execution.

---

* Research supported by NSF 0916741.

The full version of this paper is available at http://arxiv.org/abs/1402.4043.

J. Esparza et al. (Eds.): ICALP 2014, Part II, LNCS 8573, pp. 220–231, 2014.

Linearizability can also be understood in terms the *linearization point* of a method execution, which must occur between the invocation and response. From this perspective, the first execution above is linearizable because we can find a sequence of linearization points that agrees with the specification; this requires only that the linearization point of $(^+\ )_2^+$ follow that of $\{^+\ \}_1^+$. No such sequence of linearization points exists for the two other executions.

Quiescent consistency is similar to linearizability, except that the response-to-invocation order must be respected only across a quiescent point, that is, a point with no open method calls. The first three executions above are quiescently consistent because there are no non-trivial quiescent points. The last execution fails to be quiescently consistent since the order from $(^+\ )_2^+$ to $\{^+\ \}_1^+$ is not preserved in the specification.

We define *Quantitative Quiescent Consistency (QQC)* to require that the number of response-to-invocation pairs that are out-of-order at any point be bounded by the number of open calls that might be ordered later in the specification. We also give a *counting characterization* of QQC, which requires that if a response matches the ith method call in the specification, then it must be preceded by at least i invocations.

The first two executions above are QQC; however, the last two are not. In the second execution, the open call to $(^+\ )_2^+$ justifies the return of $\{^+\ \}_1^+$ before $[^+\ ]_0^+$ since $(^+\ )_2^+$ occurs after $\{^+\ \}_1^+$ in the specification. However, in the third execution, the return of $(^+\ )_2^+$ before $\{^+\ \}_1^+$ cannot be justified only by the call to $[^+\ ]_0^+$ since $[^+\ ]_0^+$ occurs earlier in the specification. Following the counting characterization sketched above, the third execution fails since $(^+\ )_2^+$ is the third method call in the specification trace, but the response of $(^+\ )_2^+$ is only preceded by two invocations: $[^+$ and $(^+$. □

Quiescent consistency is too coarse to be of much use in reasoning about concurrent executions. For example, a sequence of interlocking calls never reaches a quiescent point; therefore it is trivially quiescently consistent. This includes obviously correct executions, such as $[^+\ (^+\ ]_0^+\ [^+\ )_1^+\ (^+\ ]_2^+\ [^+\ )_3^+\ (^+\ ]_4^+\ [^+ \cdots$, nearly correct executions, such as $[^+\ (^+\ ]_1^+\ [^+\ )_0^+\ (^+\ ]_3^+\ [^+\ )_2^+\ (^+\ ]_5^+\ [^+ \cdots$, and also ridiculous executions, such as $[^+\ (^+\ ]_{1074}^+\ [^+\ )_{17}^+\ (^+\ ]_{2344}^+\ [^+\ )_3^+\ (^+ \cdots$.

Linearizability has proven quite useful in reasoning about concurrent executions; however, it fundamentally constrains efficiency in a multicore setting: Dwork, Herlihy, and Waarts [6] show that if many threads concurrently access a linearizable counter, there must be either a location with high contention or an execution path that accesses many shared variables. Shavit [14] argues that the performance penalty of linearizable data structures is increasingly unacceptable in the multicore age. This observation has lead to a recent renewal of interest in nonlinearizable data structures. As a simple example, consider the following counter implementation: a simplified version of the counting networks of Aspnes, Herlihy, and Shavit[3].

```
class Counter<N:Int> {
    field b:[0..N-1] = 0;                    // 1 balancer
    field c:Int[]    = [0, 1, ..., N-1];     // N counters
    method getAndIncrement():Int {
        val i:[0..N-1];
        atomic { i = b; b++; }
        atomic { val v = c[i]; c[i] += N; return v; } } }
```

The N-Counter has two fields: a *balancer* b and an array c of N integer counters. There are two atomic actions in the code: The first reads and updates the balancer, setting the local index variable i. The second reads and updates the ith counter. Although the balancer has high contention in our simplified implementation, the counters do not; balancers that avoid high contention are described in [3].

Example 1.2. The N-Counter behaves like a sequential counter if calls to getAnd-Increment are sequentialized. To see this, consider a 2-Counter, with initial state $\langle b = 0, c = [0, 1] \rangle$. In a series of sequential calls, the state progresses as follows, where we show the execution of the first atomic with the invocation and the second atomic with the response. The execution $[^+ \,]_0^+ \, \{^+ \, \}_1^+ \, (^+ \,)_2^+$ can be elaborated as follows.

$$\langle b = 0, c = [0, 1] \rangle \xrightarrow{[^+} \langle b = 1, c = [0, 1] \rangle \xrightarrow{]_0^+} \langle b = 1, c = [2, 1] \rangle$$
$$\xrightarrow{\{^+} \langle b = 0, c = [2, 1] \rangle \xrightarrow{\}_1^+} \langle b = 0, c = [2, 3] \rangle$$
$$\xrightarrow{(^+} \langle b = 1, c = [2, 3] \rangle \xrightarrow{)_2^+} \langle b = 1, c = [4, 3] \rangle$$

When there is concurrent access, the 2-Counter allows nonlinearizable executions, such as $(^+ \, \{^+ \, \}_1^+ \, [^+ \,]_0^+ \,)_2^+$.

$$\langle b = 0, c = [0, 1] \rangle \xrightarrow{(^+} \langle b = 1, c = [0, 1] \rangle$$
$$\xrightarrow{\{^+} \langle b = 0, c = [0, 1] \rangle \xrightarrow{\}_1^+} \langle b = 0, c = [0, 3] \rangle$$
$$\xrightarrow{[^+} \langle b = 1, c = [0, 3] \rangle \xrightarrow{]_0^+} \langle b = 1, c = [2, 3] \rangle$$
$$\xrightarrow{)_2^+} \langle b = 1, c = [4, 3] \rangle$$

With a sequence of interlocking calls, it is also possible for the N-Counter to execute as $[^+ \, (^+ \,]_1^+ \, [^+ \,)_0^+ \, (^+ \,]_3^+ \, [^+ \,)_2^+ \, (^+ \,]_5^+ \, [^+ \cdots$, producing an infinite sequence of values that are just slightly out of order. Using the results of this paper, one can conclude that with a maximum of two open calls, the value returned by getAndIncrement will be "off" by no more than 2, but this does not follow from quiescent consistency.   □

Our results are related to those of [2], [3], [5], [16]. In particular, Aspnes, Herlihy, and Shavit[3] prove that in any *quiescent* state (with no call that has not returned), such a counter has a "step-property", indicating the shape of c. Between $\}_1^+$ and $]_0^+$ in the second displayed execution of Example 1.2, the states with $c = [0, 3]$ do *not* have the step property, since the two adjacent counters differ by more than 1.

Aspnes, Herlihy, and Shavit[3] imply that the step property is related to quiescent consistency, but they do not provide a formal definition. It appears that they have in mind is something like the following: An execution is *weakly quiescent consistent* if any uninterrupted subsequence of *sequential* calls (single calls separated by quiescent points) is a subtrace of a specification trace.

The situation is delicate: Although the increment-only counters of [3] are quiescently consistent in the sense we defined in Example 1.1 (indeed, they are QQC), the increment-decrement counters of [2], [5], [16] are only *weakly* quiescent consistent. Indeed, the theorems proven in [16] state only that, at a quiescent point, a variant of the step property holds. They state nothing about the actual values read from the individual counters. Instead, our definition requires that a quiescently consistent execution be a permutation of *some* specification trace, even if it has no nontrivial quiescent points.

Example 1.3. Consider an extension of the 2-Counter with decrementAndGet.

```
method decrementAndGet():Int {
    val i:[0..N-1];
    atomic { i = b-1; b--; }
    atomic { c[i] -= N; return c[i]; } }
```

The execution $[^+ \{^+ (^- <^- >_{-2}]^+_{-2} \}^+_1)^-_1$ is possible, although this is not a permutation of any specification trace. The execution proceeds as follows.

$$\langle b = 0, c = [0,1]\rangle \xrightarrow{[^+} \langle b = 1, c = [0,1] \rangle \xrightarrow{\{^+} \langle b = 0, c = [0,1]\rangle$$
$$\xrightarrow{(^-} \langle b = 1, c = [0,1] \rangle \xrightarrow{<^-} \langle b = 0, c = [0,1]\rangle$$
$$\xrightarrow{>_{-2}} \langle b = 0, c = [-2,1]\rangle \xrightarrow{]^+_{-2}} \langle b = 0, c = [0,1]\rangle$$
$$\xrightarrow{\}^+_1} \langle b = 0, c = [0,3] \rangle \xrightarrow{)^-_1} \langle b = 0, c = [0,1]\rangle \qquad \square$$

It is important to emphasize that this increment-decrement counter is not even quiescently consistent according to our definition. There is no hope that it could satisfy any stronger criterion.

Of course counters are not the only data structures of interest. In the full paper, we treat concurrent stacks in detail. We define a simplified N-Stack below; the full, tree-based data structure is defined in Shavit and Touitou[16].

```
class Stack<N:Int> {
    field b:[0..N-1] = 0;                        // 1 balancer
    field s:Stack[]  = [[], [], ..., []];  // N stacks of values
    method push(x:Object):Unit {
        val i:[0..N-1];
        atomic { i = b; b++; }
        atomic { val v = s[i].push(x); return v; } }
    method pop():Object {
        val i:[0..N-1];
        atomic { i = b-1; b--; }
        atomic { val v = s[i].pop(); return v; } } }
```

The trace given in Example 1.3 for the increment-decrement counter is also a trace of the stack, where we interpret + as push and - as pop. Whereas this is a nonsense execution for a counter, it is a linearizable execution of a stack: simply choose the linearization points so that each push occurs immediately before the corresponding pop. Nonetheless, the N-Stack is only *weakly* quiescent consistent in general.

Example 1.4. The N-Stack generates the execution $[^+_a]^+ (^+_b)^+ \{^+_c <^- >_a^- \}^+$ as follows.

$$\langle b = 0, s = [[\,], [\,]] \rangle \xrightarrow{[^+_a} \langle b = 1, s = [[\,], [\,]] \rangle \xrightarrow{]^+} \langle b = 1, s = [[a], [\,]] \rangle$$
$$\xrightarrow{(^+_b} \langle b = 0, s = [[a], [\,]] \rangle \xrightarrow{)^+} \langle b = 0, s = [[a], [b]]\rangle$$
$$\xrightarrow{\{^+_c} \langle b = 1, s = [[a], [b]]\rangle$$
$$\xrightarrow{<^-} \langle b = 0, s = [[a], [b]]\rangle \xrightarrow{>_a^-} \langle b = 0, s = [[\,], [b]] \rangle$$
$$\xrightarrow{\}^+} \langle b = 0, s = [[c], [b]]\rangle$$

However, this specification is not quiescently consistent with any stack execution: There is a quiescent point after each of the first two pushes; therefore it is impossible to pop a before b. This execution is possible even when there are several pushes beforehand. $\square$

In the case of the N-Stack, a simple *local* constraint can be imposed in order to establish quiescent consistency: intuitively, we require that no pop *overtakes* a push on the same stack s[i]. In the full paper, we show that the stack is actually *QQC* under this constraint, and therefore quiescently consistent. We also prove that the elimination-tree stacks of Shavit and Touitou [16] are QQC. The increment-only counters of [3] are also QQC; the proofs for the tree-based increment-only counter follow the structure of the proofs for the elimination-tree stacks. (We have not found a *local* constraint under which the increment-decrement counter is quiescently consistent.) Our correctness result is much stronger than that of [16], which only proves *weak* quiescent consistency.

The preliminary version of Shavit and Touitou's paper [15] suggests an upcoming definition ε-*linearizability*, "a variant of linearizability that captures the notion of 'almostness' by allowing a certain fraction of concurrent operations to be out-of-order." This thread was picked up by Afek, Korland, and Yanovsky[1] and improved by Henzinger, Kirsch, Payer, Sezgin, and Sokolova[9]. As defined in [9], the idea is to define a cost metric on relaxations of strings and to bound the relaxation cost for the specification trace that matches an execution. This relaxation-based approach has been used to validate several novel concurrent data structures [1], [7]. With the exception of the increment-only counter validated in [1], all of these data structures intentionally violate quiescent consistency. In Section 4, we show that this approach in incomparable to QQC.

With QQC, the maximal degradation depends upon the amount of concurrent access, whereas in the relaxation-based approach it does not. Thus, QQC "degrades gracefully" as concurrency increases. In particular, a QQC data structure that is accessed sequentially will exactly obey the sequential specification, whereas a data structure validated against the relaxation-based approach may not.

In the rest of the paper, we formalize QQC and study its properties. Our contributions are as follows.
– We define linearizability (Section 2), quiescent consistency (Section 3) and QQC (Section 4) in terms of partial orders over events with duration. As in Example 1.1, the definitions are given in terms of the order from response to invocation.
– For sequential specifications, we provide alternative characterizations of linearizability, quiescent consistency and QQC in terms of the number of invocations that precede a response. For linearizability, this approach can be found in [4].
– We provide an alternative characterization of QQC in terms of a proxy that controls access to the underlying sequential data structure. The proxy adds a form of *speculation* to the flat combining technique of Hendler, Incze, Shavit, and Tzafrir[8]. This characterization can be seen as a language generator, rather than an accepter.
– Like linearizability and quiescent consistency [11], QQC is non-blocking and compositional. Like quiescent consistency and unlike linearizability, a QQC execution may not respect program order, and therefore QQC is incomparable to sequential consistency [12]. We prove that QQC is compositional for sequential specifications, in the sense of Herlihy and Wing[10].
– We show that QQC is useful for reasoning about data structures in the literature. In the full paper, we prove that the elimination tree stacks of Shavit and Touitou[16] are QQC, as long as no pop overtakes a push on the same stack.

2 Linearizability

A *trace* is a labelled partial order with polarity and bracketing. We use ? and ! to denote polarities. The polarity indicates whether an event in the partial order is a call/input (?) or a return/output (!). Bracketing matches each return with the particular call that precedes it. Let p–t range over traces and let a, b range over *names*, which form the carrier set of the partial order. We introduce notation over traces as needed.

Intuitively, linearizability requires that the response-to-invocation order in an execution be respected by a specification trace. To show that s'' is linearizable, it suffices to do the following

- Choose a specification trace t.
- Choose an *extension* s' of s'' that closes the open calls in s''. We say that s' *extends* s'' if (1) if s'' is a prefix of s', and (2) all of the new events in $s' - s''$ are ordered after all events of *opposite polarity* in s'' (that is, calls after returns and returns after calls). Let extensions(s'') be the set of extensions of s''.
- Choose a renaming $s =_\alpha s'$ such that $s =_\pi t$. Here $=_\alpha$ denotes equivalence up to renaming and $=_\pi$ denotes equivalence up to permutation. This establishes that s' is a permutation of t. The names are witness to the permutation.
- Show that for every response $a^!$ and invocation $b^?$, if $a^!$ precedes $b^?$ in s ($a^! \Rightarrow_s b^?$), then the same must be true in t ($a^! \Rightarrow_t b^?$).

This definition differs from the traditional one in several small details, enumerated in the full paper. In particular, we allow $s' \in$ extensions(s'') to include calls that are not in s'', in addition to returns. We can refactor the definition slightly to pull it into the shape used to define quiescent consistency and QQC.

Definition 2.1. For traces s, t, we write $s \sqsubseteq_{\text{lin}} t$ if $s =_\pi t$ and for every prefix $p \leq_{\text{pre}} s$

$$\forall a^! \in p.\ \forall b^? \in s - p.\ (a^! \Rightarrow_s b^?) \text{ implies } (a^! \Rightarrow_t b^?).$$

Then $(s'' \sqsubseteq_{\text{lin}} t) \triangleq (\exists s' \in \text{extensions}(s'').\ \exists s =_\alpha s'.\ s \sqsubseteq_{\text{lin}} t)$,

and $(S \sqsubseteq_{\text{lin}} T) \triangleq (\forall s'' \in S.\ \exists t \in T.\ s'' \sqsubseteq_{\text{lin}} t).$ □

This characterization of linearizability requires that we look at every way to *cut* the trace s into a prefix p and suffix $s - p$. We then look at the return events in p and the call events in $s - p$ and ensure that the order of events *crossing the cut* is respected in t. The definitions are equivalent since we quantify over all possible cuts.

Consider the counter specification from Example 1.1: $[^+]_0^+ \{^+ \}_1^+ (^+)_2^+$. The trace $\{^+ [^+ \}_1^+ (^+]_0^+)_2^+$ is linearizable. The interesting cut is $\{^+ [^+ \}_1^+$ which requires only that $\{^+ \}_1^+$ precede $(^+)_2^+$ in the specification. By the same reasoning, $\{^+ (^+ \}_1^+ [^+)_2^+]_0^+$, is not linearizable, since it requires that $\{^+ \}_1^+$ precede $[^+]_0^+$.

Given a sequential specification, a trace is linearizable if every return is preceded by the calls that come before it in specification order. This holds for *operational* traces, in which all events of opposite polarity are ordered. Operational traces correspond to those generated by a standard interleaving semantics. Define $s \leq_\pi t$ to mean that s is a subtrace of a permutation of t: $(s \leq_\pi t) \triangleq (\exists s'.\ s \subseteq s' =_\pi t)$.

Theorem 2.2. Let t be a sequential trace with name order $(a_1^?, a_1^!, a_2^?, a_2^!, \ldots, a_n^?, a_n^!)$. Let s be an operational trace such that $s \leq_\pi t$. Then

$$s \sqsubseteq_{\text{lin}} t \quad \text{iff} \quad \forall a_j^! \in s.\ \{a_1^?, \ldots, a_j^?\} \subseteq \{a_i^? \mid a_i^? \Rightarrow_s a_j^!\} \qquad □$$

3   Quiescent Consistency

Let open(s) be the set of calls in s that have no matching return. We say that trace s is *quiescent* if open(s) $= \emptyset$. This notion of quiescence does not require that there be no active thread, but only that there be no open calls. Thus, this notion of quiescence is compatible with libraries that maintain their own thread pools.

The definition of quiescent consistency is similar to Definition 2.1 of linearizability. The difference lies in the quantifier for the prefix p: Whereas linearizability quantifies over *every* prefix, quiescent consistency only quantifies over *quiescent* prefixes.

Definition 3.1. We write $s \sqsubseteq_{\mathsf{qc}} t$ if $s =_\pi t$ and for any *quiescent* prefix $p \leq_{\mathsf{pre}} s$

$$\forall a^! \in p. \ \forall b^? \in s - p. \ (a^! \Rightarrow_s b^?) \text{ implies } (a^! \Rightarrow_t b^?).$$ □

($\sqsubseteq_{\mathsf{qc}}$) is defined similarly to ($\sqsubseteq_{\mathsf{lin}}$). Again let us revisit the counter specification from Example 1.1: $[^+ \ ]_0^+ \ \{^+ \ \}_1^+ \ (^+ \ )_2^+$. This notion of quiescent consistency places some constraints on the system even when it has no nontrivial quiescent points. For example, the execution $[^+ \ \{^+ \ (^+ \ )_3^+ \ \}_1^+ \ ]_0^+$ is not quiescently consistent with the given specification, since it is not a permutation. If one extends the execution to $[^+ \ \{^+ \ (^+ \ )_3^+ \ \}_1^+ \ ]_0^+ \ <^+ \ >_2^+$ and attempts to matches it against the specification $[^+ \ ]_0^+ \ \{^+ \ \}_1^+ \ <^+ \ >_2^+ \ (^+ \ )_3^+$, quiescent consistency continues to fail: In the quiescent prefix $[^+ \ \{^+ \ (^+ \ )_3^+ \ \}_1^+ \ ]_0^+$, the order across the cut from $)_3^+$ to $<^+$ is not preserved in the specification.

For linearizability, only responses need be included in the extensions of a trace. The same does not hold for quiescent consistency. For example, since $(^+ \ \{^+ \ \}_1^+ \ [^+ \ ]_0^+ \ )_2^+$ is quiescently consistent, its prefix $(^+ \ \{^+ \ \}_1^+$ should also be quiescently consistent. However, there is no specification trace that can be matched that does not include $[^+ \ ]_0^+$. Therefore, it does not suffice merely to close the open call by adding $)_2^+$; we must also include $[^+$ and $]_0^+$.

We now give a counting characterization of quiescent consistency. Define $u \Mapsto_s v$ to mean that $u \Rightarrow_s v$ and there is no quiescent cut that separates u and v.

Theorem 3.2. Let t be a sequential trace with name order $(a_1^?, a_1^!, a_2^?, a_2^!, \ldots, a_n^?, a_n^!)$. Let s be an operational trace such that $s \leq_\pi t$. Then

$$s \sqsubseteq_{\mathsf{qc}} t \quad \text{iff} \quad \forall a_j^! \in s. \ |\{a_1^?, \ldots, a_j^?\}| \leq |\{a_i^? \mid a_i^? \Rightarrow_s a_j^!\} \cup \{a_i^? \mid a_j^! \Mapsto_s a_i^?\}| \quad □$$

If $a_j^!$, the jth return in t, occurs in s, then there must be at least j calls contained in two sets: (1) the calls that precede $a_j^!$ in s, and (2) the calls that follow $a_j^!$ in s but are "quiescently concurrent" — that is, not separated by a quiescent point.

4   Quantitative Quiescent Consistency

We provide three characterizations of QQC and prove their equivalence. (1) Definition 4.1 defines QQC in the style that we have defined linearizability and quiescent consistency, from response to invocation. (2) Theorem 4.3 provides a *counting characterization* of QQC, which requires that if a response matches the ith method call in the specification, then it must be preceded by at least i invocations. (3) Theorem 4.4 provides an operational characterization of QQC as a proxy between the concurrent world and an underlying sequential data structure.

To develop some intuition for the what is allowed by QQC, we give some examples using the 2-Counter from the introduction. First we note that the capability given by an open call can be used repeatedly, as in $(^+ \ [^+ \ ]_1^+ \ \{^+ \ \}_0^+ \ [^+ \ ]_3^+ \ \{^+ \ \}_2^+ \ [^+ \ ]_5^+ \ \{^+ \ \}_4^+ \ )_6^+$. The open call $(^+$ enables the inversion of $\{^+ \ \}_0^+$ with $[^+ \ ]_1^+$ and also of $\{^+ \ \}_2^+$ with $[^+ \ ]_3^+$.

Alternatively, multiple open calls may be accumulated to create a trace with events that are arbitrarily far off, as in $(^+ \ [^+ \ ]_1^+ \ (^+ \ [^+ \ ]_3^+ \ (^+ \ [^+ \ ]_5^+ \ (^+ \ [^+ \ ]_7^+ \ [^+ \ ]_0^+ \ )_2^+ \ )_4^+ \ )_6^+ \ )_8^+$. Note that $[^+ \ ]_0^+$ *follows* $[^+ \ ]_7^+$ in this execution! It is worth emphasizing that the order between these actions is observable to the outside: a single thread can call getAndIncrement and get 7, then subsequently call getAndIncrement and get 0. Such behaviors are a hallmark of nonlinearizable data structures. In general, an N-Counter can give results that are $k \times N$ off of the expected value, where k is the maximum number of open calls and N is the width of the counter. There is no way to bound the behavior of this counter, as in [9], without also bounding the amount of concurrency, as in [1].

It is also possible for open calls to overlap in nontrivial ways. The trace $(^+ \ [^+ \ ]_1^+ \ \{^+ \ [^+ \ ]_0^+ \ )_3^+ \ (^+ \ )_2^+ \ \}_4^+$ is QQC. Here, the first $(^+$ justifies the out-of-order execution of $[^+ \ ]_1^+$ and $[^+ \ ]_0^+$. The subsequent $\{^+$ justifies an inversion of the previous justifier, namely $(^+ \ )_3^+$ and $(^+ \ )_2^+$. A similar example is $\{^+ \ (^+ \ )_1^+ \ (^+ \ [^+ \ ]_0^+ \ )_3^+ \ [^+ \ ]_2^+ \ \}_4^+$.

Finally, we note that the stack execution $\{_c^+ \ [^- \ ]_a^- \ (_a^+ \ )^+ \ \}^+$ is QQC with respect to the specification $(_a^+ \ )^+ \ [^- \ ]_a^- \ \{_c^+ \ \}^+$. This follows from exactly the kind of reasoning that we have done for the counter. For the counter this simply means that we are seeing an integer value early, but for a stack holding pointers, it means that we can potentially see a pointer before it has been allocated! To prevent such executions, causality can be specified as a relation from calls to returns, consistent with specification order: A trace is *causal* if it respects the specified causality relation. We have elided causality from the definition of QQC because it is orthogonal and can be enforced independently.

Linearizability requires that for *every* cut, *all* response-to-invocation order crossing the cut must be respected in the specification. Quiescent consistency limits attention to *quiescent* cuts. QQC restores the quantification over every cut, but relaxes the requirement to match all response-to-invocation order crossing the cut. When checking response-to-invocation pairs across the cut, QQC allows some invocations to be ignored. How many?

One constraint comes from our desire to refine quiescent consistency. For quiescent cuts, we cannot drop any invocations, since quiescent consistency does not. As a first attempt at a definition, we may take the number of dropped invocations at any cut to be bounded by $|\text{open}(p)|$. This criterion would allow both of the traces $(^+ \ \{^+ \ \}_1^+ \ [^+ \ ]_0^+ \ )_2^+$ and $[^+ \ (^+ \ )_2^+ \ \{^+ \ \}_1^+ \ ]_0^+$ in Example 1.1. In each case, the interesting cut splits the trace in half, with one open call and one completed. In the first trace, we can ignore $[^+$ in the suffix, and in the second trace, we can ignore $\{^+$ in the suffix; thus, both are allowed. However, in the second trace, the first call completed is two steps in the future, even though there is only one concurrent action. In the first trace this does not happen. The difference can be seen by looking not only at the number of open calls, but also at *which* calls are open. In the first trace we have $(^+$ before $\}_1^+$, and in the second, we have $[^+$ before $)_2^+$. We say that $(^+$ is *early* for $\}_1^+$, since it does not precede $\}_1^+$ in the specification, whereas $[^+$ is not early for $)_2^+$, since it *does* precede $)_2^+$. We restrict our attention to calls that are both open and early with respect to the response of interest.

Given a specification t and a response $a^! \in t$, none of the actions in the t-down-closure of $a^!$ could possibly be early for $a^!$; any other action could be. Thus, the actions in $\mathsf{open}(p) - (\downarrow_t a^!)$ are both open and early for $a^!$. This leads us to the following definition. (In the full paper, we show that for sequential specifications, we can swap the quantifiers $(\exists r)$ and $(\forall a^!)$, pulling out the existential.)

Definition 4.1. We write $s \sqsubseteq_{\mathsf{qqc}} t$ if $s =_\pi t$ and for any prefix $p \leq_{\mathsf{pre}} s$

$$\forall a^! \in p.\ \exists r \subseteq s.\ |r| \leq |\mathsf{open}(p) - (\downarrow_t a^!)|.$$
$$\forall b^? \in ((s - p) - r).\ (a^! \Rightarrow_s b^?) \text{ implies } (a^! \Rightarrow_t b^?). \qquad \square$$

As before, $(\lesssim_{\mathsf{qqc}})$ is defined by analogy to $(\lesssim_{\mathsf{lin}})$.

Theorem 4.2. $(\lesssim_{\mathsf{lin}}) \subset (\lesssim_{\mathsf{qqc}}) \subset (\lesssim_{\mathsf{qc}})$. $\qquad \square$

Given the subtlety of Definition 4.1, it may be surprising that QQC has the following simple characterization for sequential specifications.

Theorem 4.3. Let t be a sequential trace with name order $(a_1^?, a_1^!, a_2^?, a_2^!, \ldots, a_n^?, a_n^!)$. Let s be an operational trace such that $s \leq_\pi t$. Then

$$s \lesssim_{\mathsf{qqc}} t \quad \text{iff} \quad \forall a_j^! \in s.\ |\{a_1^?, \ldots, a_j^?\}| \leq |\{a_i^? \mid a_i^? \Rightarrow_s a_j^!\}| \qquad \square$$

This characterization provides a simple method for calculating whether a trace is QQC. For example, the trace $\{^+\ (^+\ )_1^+\ (^+\ [^+\ ]_0^+\ )_3^+\ [^+\ ]_2^+\ \}_4^+$ is QQC since $)_1^+$ is preceded by two calls, $]_0^+$, $)_3^+$ by four, and $]_2^+$, $\}_4^+$ by five. The trace $\{^+\ (^+\ )_1^+\ (^+\ )_3^+\ [^+\ ]_0^+\ [^+\ ]_2^+\ \}_4^+$ is not QQC since $)_3^+$ is only preceded by three calls, yet it is the fourth call in the specification.

Our third characterization of QQC describes how QQC affects an arbitrary sequential data structure, using a *proxy* that generates QQC traces from an underlying sequential implementation. This characterization of QQC incorporates *speculation* into flat combining [8]. We push the obligation to predict the future into the underlying sequential object, with must conform to the following interface.

```
interface Object {
    method run(i:Invocation):Response;
    method predict():Invocation;  }
```

The run method passes invocations to the underlying sequential structure and returns the appropriate response. The predict method is an oracle that guesses the invocations that are to come in the future. It is the use of predict that makes our code speculative.

The code for the proxy is given in Figure 1. Communication between the implementation threads and the underlying Object is mediated by two maps. When a thread would like to interact with the Object, it creates a semaphore, registers it in called and waits. Upon awakening, the thread removes the result from returned and returns.

The Object is serviced by a single *proxy* thread which loops forever making one of two nondeterministic choices. The proxy keeps two private maps. Upon receiving an invocation in called, the proxy moves the invocation from called to received. Rather than executing the received invocation, the proxy asks the oracle to predict an arbitrary invocation i and executes that instead, placing the result in executed. Once a invocation is both received and executed, it may become returned.

At the beginning of this section, we noted that the stack execution $\{_c^+\ [^-\ ]_a^-\ (_a^+\ )^+\ \}^+$ is QQC with respect to the specification $(_a^+\ )^+\ [^-\ ]_a^-\ \{_c^+\ \}^+$. How can such a trace possibly

```
class QQCProxy<o:Object> {
  field called:ThreadSafeMultiMap<Invocation,Semaphore> = [];
  field returned:ThreadSafeMap    <Semaphore, Response>  = [];
  method run(i:Invocation):Response { // proxy for external access to o
    val m:Semaphore = [];
    called.add(i, m);
    m.wait();
    return returned.remove(m); }
  thread { // single thread to interact with o
    val received:MultiMap<Invocation,Semaphore> = [];
    val executed:MultiMap<Invocation,Response>  = [];
    repeatedly choose {
      choice if called.notEmpty() {
        received.add(called.removeAny());
        val i:Invocation = o.predict();
        val r:Response    = o.run(i);
        executed.add(i, r); }
      choice if exists i in received.keys() intersect executed.keys() {
        val m:Semaphore = received.remove(i);
        val r:Response  = executed.remove(i);
        returned.add(m, r);
        m.signal(); } } } }
```

Fig. 1. QQC Proxy

be generated? The execution of the proxy proceeds as follows. Upon receipt of $\{_c^+$, the proxy executes $(_a^+$, storing response $)^+$. Upon receipt of $[^-$, the proxy executes $[^-$, storing response $]_a^-$. At this point $[^- \,]_a^-$ can return. Upon receipt of $(_a^+$, the proxy executes $\{_c^+$, storing response $\}^+$. At this point both $(_a^+ \,)^+$ and $\{_c^+ \,\}^+$ can return.

Such noncausal behaviors can be eliminated by requiring when a pop is executed, a corresponding push must have been received. The prior execution is invalidated since $(_a^+ \,)^+$ is not received when $[^- \,]_a^-$ returns. However, nonlinearizable behaviors are still allowed. For example $\{_c^+ \, [_a^+ \,]^+ \, (_b^+ \,)^+ \, \}^+ \, [^- \,]_a^- \, (^- \,)_b^-$ is generating by predicting $(_b^+ \,)^+$.

Theorem 4.4. The concurrent proxy is sound for QQC with respect to the underlying `Object`. *It is also complete for operational traces.*   □

In the full paper, we show that the elimination-tree stack of [16] and increment-only counter of [3] are QQC. The characterizations of QQC also allow us to predict the QQC behavior of other data structures, such as a queues, even if no implementation is known. The following examples, from Sezgin[13], allow a useful comparison with [9].

To see that QQC makes distinctions not found in [9], consider the two stack traces $\{_c^+$ $[_a^+ \,]^+ \, (_b^+ \,)^+ <^- >_a^- \, \}^+$ and $\{_c^+ \, [_a^+ \,]^+ \, (_b^+ \,)^+ \, \}^+ <^- >_a^-$. In the framework of [9], these are both 1 out-of-order (when a is popped, at least b must be above a on the stack). However, only the first is QQC.

In the other direction, the queue execution $\{_a^+ \, [_{b_1}^+ \,]^+ \, [_{b_1}^+ \,]^+ \cdots [_{b_n}^+ \,]^+ \, (_c^+ \,)^+ <^- >_c^- \, \}^+$ is QQC with respect to the queue specification $(_c^+ \,)^+ \, [_{b_1}^+ \,]^+ \, [_{b_1}^+ \,]^+ \cdots [_{b_n}^+ \,]^+ <^- >_c^- \, \{_a^+ \,\}^+$. In the framework of [9], this would be n out-of-order because at least all b_i's should be in the queue before c is inserted into the queue; the removal of c from the queue must happen when there are n elements ahead of c in the queue.

Finally, we prove compositionality for QQC. Let $\div$ denote partial order difference.

Theorem 4.5. Let t_1 and t_2 be sequential traces.

Let s, s_1 and s_2 be operational traces such that $s_1 = s \div s_2$ and $s_2 = s \div s_1$.

For $i \in \{1, 2\}$, suppose that each $s_i \sqsubseteq_{\mathsf{qqc}} t_i$.

Then there exists a sequential trace $t \in (t_1 \parallel\mkern-6mu\mid t_2)$ such that $s \sqsubseteq_{\mathsf{qqc}} t$.

PROOF SKETCH. Assume that the names in t_1 and t_2 are disjoint. Let the sequence of names in t_1 be $(a_1^?, a_1^!, \ldots, a_m^?, a_m^!)$ and sequence of names in t_2 be $(b_1^?, b_1^!, \ldots, b_n^?, b_n^!)$. Applying Theorem 4.3 to the supposition $s_1 \sqsubseteq_{\mathsf{lin}} t_1$, we have that $j \leq \left| \{a_i^? \mid a_i^? \Rightarrow_s a_j^!\} \right|$, and similarly $\ell \leq \left| \{b_k^? \mid b_k^? \Rightarrow_s b_\ell^!\} \right|$. It suffices to construct an interleaving $t \in (t_1 \parallel\mkern-6mu\mid t_2)$ such that whenever t contains a subsequence with names

$$a_j^?, a_j^!, b_k^?, b_k^!, b_{k+1}^?, b_{k+1}^!, \ldots, b_{k+x}^?, b_{k+x}^!$$

then for every $k \leq \ell \leq k+x$, we have

$$\{a_i^? \mid a_i^? \Rightarrow_s a_j^!\} \subseteq \{a_i^? \mid a_i^? \Rightarrow_s b_\ell^!\}$$

and symmetrically for subsequences $b_k^?, b_k^!, a_j^?, a_j^!, a_{j+1}^?, a_{j+1}^!, \ldots, a_{j+y}^?, a_{j+y}^!$. To demonstrate the existence of an appropriate t, it suffices to show that $\mathsf{merge}(a_1^? a_1^! \ldots a_m^? a_m^!, b_1^? b_1^! \ldots b_n^? b_n^!)$ is nonempty. By operationality, it must be the case that either (1) $a_j^! \Rightarrow_s b_\ell^!$, in which case $\{a_i^? \mid a_i^? \Rightarrow_s a_j^!\} \subseteq \{a_i^? \mid a_i^? \Rightarrow_s b_\ell^!\}$, (2) $b_\ell^! \Rightarrow_s a_j^!$, in which case $\{b_k^? \mid b_k^? \Rightarrow_s b_\ell^!\} \subseteq \{b_k^? \mid b_k^? \Rightarrow_s a_j^!\}$, or (3) $a_j^!$ and $b_\ell^!$ are unordered, in which case both conclusions hold. Therefore an appropriate t exists. $\qquad\square$

5 Conclusions

Quantitative quiescent consistency (QQC) is a correctness criterion for concurrent data structures that relaxes linearizability and refines quiescent consistency. To the best of our knowledge, it is the first such criterion to be proposed.

To show that QQC is a robust concept, we have provided three alternate characterizations: (1) in the style of linearizability, (2) counting the number of calls before a return, and (3) using speculative flat combining. We have also proven compositionality (in the style of Herlihy and Wing [10]) and, in the full paper, the correctness of data structures defined by Aspnes, Herlihy, and Shavit [3] and Shavit and Touitou [16].

In order to establish the correctness of the elimination-tree stack of [16], we had to restrict attention to traces in which no pop *overtakes* a push on the same stack. (The formalities are given in the full paper.) A related constraint appears in a footnote of [14]: "To keep things simple, pop operations should block until a matching push appears." This, however, is not strong enough to guarantee quiescent consistency as we have defined it. Our analysis provides a full account: The stack is QQC with the no-overtaking requirement and only weakly quiescently consistent without it.

There are many unanswered questions, chief among them: Is QQC useful in reasoning about client programs? Is there a verification methodology for QQC analogous to that developed for linearizability? Are there other useful data structures that can be shown to satisfy QQC?

Linearizability is, at its core, *linear*. We have defined QQC in terms of general partial orders, and yet the results reported here are stated in terms of sequential specifications. Partly we have done this so that we can relate the definition of QQC to the vast amount of existing work on linearizability. However, the general case is interesting.

Acknowledgements. Gustavo Petri participated in the early discussions motivating this work. Alexey Gotsman suggested the connection to flat combining. Ali Sezgin provided a comparison with [9]. We also thank Alan Jeffrey, Corin Pitcher and Hongseok Yang for useful discussion.

References

[1] Afek, Y., Korland, G., Yanovsky, E.: Quasi-linearizability: Relaxed consistency for improved concurrency. In: Lu, C., Masuzawa, T., Mosbah, M. (eds.) OPODIS 2010. LNCS, vol. 6490, pp. 395–410. Springer, Heidelberg (2010)

[2] Aiello, W., Busch, C., Herlihy, M., et al.: Supporting increment and decrementoperations in balancing networks. Chicago J. Theor. Comput. Sci. (2000)

[3] Aspnes, J., Herlihy, M., Shavit, N.: Counting networks. J. ACM 41(5), 1020–1048 (1994)

[4] Batty, M., Dodds, M., Gotsman, A.: Library abstraction for C/C++ concurrency. In: POPL (2013)

[5] Busch, C., Mavronicolas, M.: The strength of counting networks (abstract). In: Burns, J.E., Moses, Y. (eds.) PODC, p. 311. ACM (1996)

[6] Dwork, C., Herlihy, M., Waarts, O.: Contention in shared memory algorithms. J. ACM 44(6), 779–805 (1997)

[7] Haas, A., Lippautz, M., Henzinger, T.A., et al.: Distributed queues in shared memory. In: Conf. Computing Frontiers, p. 17. ACM (2013)

[8] Hendler, D., Incze, I., Shavit, N., Tzafrir, M.: Flat combining and the synchronization-parallelism tradeoff. In: SPAA, pp. 355–364 (2010)

[9] Henzinger, T.A., Kirsch, C.M., Payer, H., Sezgin, A., Sokolova, A.: Quantitative relaxation of concurrent data structures. In: POPL, pp. 317–328 (2013)

[10] Herlihy, M., Shavit, N.: The Art of Multiprocessor Programming. Morgan Kaufmann (2008)

[11] Herlihy, M., Wing, J.M.: Linearizability: a correctness condition for concurrent objects. ACM TOPLAS 12(3), 463–492 (1990)

[12] Lamport, L.: How to make a multiprocessor computer that correctly executes multiprocess programs. IEEE Trans. Comput. 28(9), 690–691 (1979)

[13] Sezgin, A.: Private correspondence (March 18, 2014)

[14] Shavit, N.: Data structures in the multicore age. Commun. ACM 54(3), 76–84 (2011)

[15] Shavit, N., Touitou, D.: Elimination trees and the construction of pools and stacks (preliminary version). In: SPAA, pp. 54–63 (1995)

[16] Shavit, N., Touitou, D.: Elimination trees and the construction of pools and stacks. Theory Comput. Syst. 30(6), 645–670 (1997)

[17] Shavit, N., Zemach, A.: Diffracting trees. ACM Trans. Comput. Syst. 14(4), 385–428 (1996)

Bisimulation Equivalence
of First-Order Grammars*

Petr Jančar**

Dept Comp. Sci., FEI, Techn. Univ. of Ostrava (VŠB-TUO),
17. listopadu 15, 70833 Ostrava, Czech Rep.
petr.jancar@vsb.cz

Abstract. A decidability proof for bisimulation equivalence of first-order grammars (i.e., finite sets of labelled rules for rewriting roots of first-order terms) is presented. The result, generalizing the decidability of the DPDA (deterministic pushdown automata) equivalence, is equivalent to the result achieved by Sénizergues (1998, 2005) in the framework of equational graphs, or of PDA with restricted ε-steps, but the framework of classical first-order terms seems to be particularly useful for providing a concise proof that should be understandable for a wider audience.

1 Introduction

Decision problems for semantic equivalences have been a frequent topic in computer science. Pushdown automata (PDA) constitute a well-known example; language equivalence of PDA is a standard undecidable problem, but the decidability for deterministic PDA (DPDA) is a famous result by Sénizergues [14].

In concurrency theory, logic, verification, and other areas, a finer equivalence, called *bisimulation equivalence* or *bisimilarity*, has emerged as another fundamental behavioural equivalence; on deterministic systems it essentially coincides with language equivalence. We name [1] to exemplify the first decidability results for infinite-state systems, and refer to [16] for a survey of a specific area.

One of the most involved results in the area [15] shows the decidability of bisimilarity of equational graphs with finite out-degree (or of PDA with deterministic popping ε-steps); this generalizes the result for DPDA. The recent nonelementary lower bound [2] for the problem is, in fact, TOWER-hardness in the terminology of [13], and it holds even for real-time PDA, i.e. PDA with no ε-steps. For the full above mentioned PDA the problem is even not primitive recursive, since it is Ackermann-hard [11]. In the deterministic case, the equivalence problem is known to be PTIME-hard, and has a primitive recursive upper bound shown by Stirling [17]; a finer analysis places the problem in TOWER [11]. This complexity gap is just one indication that the respective fundamental equivalence problems are far from being fully understood. Another

---

* A version with more details can be found at arxiv.org/abs/1405.7923.
** Supported by the Grant Agency of the Czech Rep., project GAČR:P202/11/0340.

J. Esparza et al. (Eds.): ICALP 2014, Part II, LNCS 8573, pp. 232–243, 2014.

such indication might be the length and the technical nature of the so far published proofs (including the unpublished [18]).

This paper is an attempt to make a further step in clarifying the main decidability proof in the mentioned area. It provides a self-contained decidability proof for bisimulation equivalence in labelled transition systems generated by *first-order grammars* (FO-grammars), which seems to be a particularly convenient formalism. The states are here first-order terms over a specified finite set of function symbols (or "nonterminals"); the transitions are induced by a finite set of labelled rules that allow to rewrite the roots of terms. This framework is equivalent to the framework of [15], as follows already from the works referred to in [4], e.g. (A concrete transformation from PDA to FO-grammars can be found, e.g., in [9].) The proof here is in principle based on the same high-level ideas as the proof in [15] but it is considerably shorter and simpler. This paper is a (self-contained) continuation of [9] where the first-order term framework was used to give a decidability proof in the deterministic case.

Some further related work is briefly discussed in the concluding remarks.

2   Preliminaries and Result

In this section we define the basic notions and state the result. Some standard definitions are restricted when we do not need the full generality.

By $\mathbb{N}$ we denote the set $\{0, 1, 2, \dots\}$ of nonnegative integers; we use $[i, j]$ to denote the set $\{i, i+1, \dots, j\}$. For a set $\mathcal{A}$, by $\mathcal{A}^*$ we denote the set of finite sequences of elements of $\mathcal{A}$, which are also called *words* (over $\mathcal{A}$). By $|w|$ we denote the *length* of $w \in \mathcal{A}^*$. By ε we denote the *empty sequence* (hence $|\varepsilon| = 0$).

LTSs. A *labelled transition system* (an LTS) is a tuple $\mathcal{L} = (\mathcal{S}, \Sigma, (\xrightarrow{a})_{a \in \Sigma})$ where $\mathcal{S}$ is a *finite or countable* set of *states*, Σ is a finite set of *actions* (or *letters*), and $\xrightarrow{a} \subseteq \mathcal{S} \times \mathcal{S}$ is a set of *a-transitions* (for each $a \in \Sigma$). Moreover, we assume *image-finiteness*, which requires that the set $\{s' \mid s \xrightarrow{a} s'\}$ is finite for each pair $s \in \mathcal{S}$, $a \in \Sigma$. We say that $\mathcal{L}$ is a *deterministic LTS* if for each pair $s \in \mathcal{S}$, $a \in \Sigma$ there is at most one s' such that $s \xrightarrow{a} s'$.

By $s \xrightarrow{w} s'$, where $w = a_1 a_2 \dots a_n \in \Sigma^*$, we denote that there is a *path* $s = s_0 \xrightarrow{a_1} s_1 \xrightarrow{a_2} \cdots \xrightarrow{a_n} s_n = s'$; if $s \xrightarrow{w} s'$, then s' is *reachable from* s, within $|w|$ steps. By $s \xrightarrow{w}$ we denote that w is *enabled by* s, i.e., $s \xrightarrow{w} s'$ for some s'. If $\mathcal{L}$ is deterministic, then $s \xrightarrow{w} s'$ or $s \xrightarrow{w}$ denotes a unique path.

(Stratified) Bisimilarity. Let $\mathcal{L} = (\mathcal{S}, \Sigma, (\xrightarrow{a})_{a \in \Sigma})$ be a given LTS. We say that a *set* $\mathcal{B} \subseteq \mathcal{S} \times \mathcal{S}$ *covers* $(s, t) \in \mathcal{S} \times \mathcal{S}$ if for any $a \in \Sigma$ and $s' \in \mathcal{S}$ such that $s \xrightarrow{a} s'$ there is $t' \in \mathcal{S}$ such that $t \xrightarrow{a} t'$ and $(s', t') \in \mathcal{B}$, and for any $a \in \Sigma$ and $t' \in \mathcal{S}$ such that $t \xrightarrow{a} t'$ there is $s' \in \mathcal{S}$ such that $s \xrightarrow{a} s'$ and $(s', t') \in \mathcal{B}$. For $\mathcal{B}, \mathcal{B}' \subseteq \mathcal{S} \times \mathcal{S}$ we say that $\mathcal{B}'$ *covers* $\mathcal{B}$ if $\mathcal{B}'$ covers each $(s, t) \in \mathcal{B}$. A set $\mathcal{B} \subseteq \mathcal{S} \times \mathcal{S}$ is a *bisimulation* if $\mathcal{B}$ covers $\mathcal{B}$. States $s, t \in \mathcal{S}$ are *bisimilar*, written $s \sim t$, if there is a bisimulation $\mathcal{B}$ containing (s, t). We note the standard fact that $\sim \subseteq \mathcal{S} \times \mathcal{S}$ is the maximal bisimulation, the union of all bisimulations.

We put $\sim_0 = \mathcal{S} \times \mathcal{S}$. For $k \in \mathbb{N}$, $\sim_{k+1} \subseteq \mathcal{S} \times \mathcal{S}$ is the set of all pairs covered by $\sim_k$. We easily verify that $\sim$ and $\sim_k$ are equivalence relations, and that $\sim_0 \supseteq$

$\sim_1 \supseteq \sim_2 \supseteq \cdots \cdots \supseteq \sim$. For the (first infinite) ordinal ω we put $s \sim_\omega t$ if $s \sim_k t$ for all $k \in \mathbb{N}$; hence $\sim_\omega = \bigcap_{k \in \mathbb{N}} \sim_k$. It is a standard fact that $\bigcap_{k \in \mathbb{N}} \sim_k$ is a bisimulation in any image-finite LTS, where we thus have $\sim = \sim_\omega$.

Eq-levels. Given an image-finite LTS, we attach the *equivalence level* (eq-level) to each pair of states: $\mathrm{EQLV}(s,t) = \max \{ k \in \mathbb{N} \cup \{\omega\} \mid s \sim_k t \}$.

First-order-term LTSs Informally. We focus on certain (image-finite) LTSs in which states are first-order terms; we mean standard finite terms primarily but it will turn out convenient to consider also infinite regular terms (i.e. infinite terms with finitely many pairwise different subterms). The terms are built from *variables* from a fixed countable set $\mathrm{VAR} = \{x_1, x_2, x_3, \dots\}$ and from *function symbols*, also called *(ranked) nonterminals*, from some specified finite set $\mathcal{N}$; each $A \in \mathcal{N}$ has $arity(A) \in \mathbb{N}$. An example of a (standard finite) term is $A(D(x_5, C(x_2, B)), x_5, B)$ where the arities of A, B, C, D are $3, 0, 2, 2$, respectively. Transitions are determined by a finite set of *root-rewriting* rules. An example of a "non-popping" rule is $A(x_1, x_2, x_3) \xrightarrow{a} C(D(x_3, B), x_2)$, an example of a "popping" rule is $A(x_1, x_2, x_3) \xrightarrow{b} x_1$. Each rule induces the transitions arising by applying the same substitution σ to both the left-hand side (lhs) and the right-hand side (rhs). E.g., the rule $A(x_1, x_2, x_3) \xrightarrow{a} C(D(x_3, B), x_2)$ and the substitution σ for which $\sigma(x_1) = D(x_5, C(x_2, B))$, $\sigma(x_2) = x_5$, $\sigma(x_3) = B$ (where $A(x_1, x_2, x_3)$ after applying σ becomes $A(D(x_5, C(x_2, B)), x_5, B)$) induce the transition $A(D(x_5, C(x_2, B)), x_5, B) \xrightarrow{a} C(D(B, B), x_5)$; the rule $A(x_1, x_2, x_3) \xrightarrow{b} x_1$ and σ induce $A(D(x_5, C(x_2, B)), x_5, B) \xrightarrow{b} D(x_5, C(x_2, B))$.

The Result Informally. We will show that there is an algorithm that computes $\mathrm{EQLV}(T_0, U_0)$ when given a finite set of root-rewriting rules and two terms T_0, U_0. In the rest of this section we formalize this statement, making also some conventions about our use of (finite and infinite) terms and substitutions.

Regular Terms, Presentation Size. We identify terms with their syntactic trees, and denote them by $E, F, \dots$. Thus a *term E over $\mathcal{N}$* is a rooted, ordered, finite or infinite tree where each node has a label from $\mathcal{N} \cup \mathrm{VAR}$; if the label of a node is $x_i \in \mathrm{VAR}$, then the node has no successors, and if the label is $A \in \mathcal{N}$, then it has m (immediate) successor-nodes where $m = arity(A)$. A subtree of a term (i.e. tree) E is also called a *subterm* of E. A subterm can have more (maybe infinitely many) *occurrences* in E. Each *subterm-occurrence* has its (nesting) *depth in E*, which is its (naturally defined) distance from the root of E. We also use the standard notation for terms: we write $E = x_i$ or $E = A(G_1, \dots, G_m)$ with the obvious meaning; in the latter case we have $\mathrm{ROOT}(E) = A \in \mathcal{N}$, $m = arity(A)$, and $G_1, \dots, G_m$ are the *root-successors*, i.e., the ordered subterm-occurrences with the depth 1.

A *term E is finite* if the respective tree is finite; by $\mathrm{HEIGHT}(E)$ we then mean the largest depth of a subterm in E. A (possibly infinite) *term is regular* if it has only finitely many subterms (though the subterms may be infinite and can have infinitely many occurrences). Any regular term has a natural *finite-graph presentation* (with possible cycles); by $\mathrm{PRSIZE}(E)$ (the presentation size of E) we mean the size of the smallest graph presentation of E.

In what follows, by a "term" we mean a "regular term" if we do not say explicitly that the term is finite. (We do not consider non-regular terms.) We reserve symbols E, F, G, H, and also T, U, V, W, for denoting (regular) terms.

Substitutions, Associative Composition. By $\text{TERMS}_{\mathcal{N}}$ we denote the set of all (regular) terms over a set $\mathcal{N}$ of (ranked) nonterminals. A *substitution* σ is a mapping $\sigma : \text{VAR} \to \text{TERMS}_{\mathcal{N}}$ whose *support* $\text{SUPP}(\sigma) = \{x_i \mid \sigma(x_i) \neq x_i\}$ is *finite*; we reserve the symbol σ for substitutions. By $\text{RANGE}(\sigma)$ we mean the set $\{\sigma(x_i) \mid x_i \in \text{SUPP}(\sigma)\}$. By *applying a substitution* σ to a term E we get the term $E\sigma$ that arises from E by replacing each occurrence of x_i with $\sigma(x_i)$. Hence $E = x_i$ implies $E\sigma = x_i\sigma = \sigma(x_i)$. The *composition of substitutions*, where $\sigma = \sigma_1\sigma_2$ satisfies $\sigma(x_i) = (\sigma_1(x_i))\sigma_2$, can be easily verified to be associative. We thus write simply $E\sigma_1\sigma_2$ when meaning $(E\sigma_1)\sigma_2$ or $E(\sigma_1\sigma_2)$.

First-order Grammars. A *first-order grammar*, an *FO-grammar* or just a *grammar* for short, is a tuple $\mathcal{G} = (\mathcal{N}, \Sigma, \mathcal{R})$ where $\mathcal{N}$ is a finite set of ranked *nonterminals*, viewed as function symbols with arities, Σ is a finite set of *actions* (or letters), and $\mathcal{R}$ is a finite set of *rules* of the form $A(x_1, x_2, \ldots, x_m) \xrightarrow{a} E$ where $A \in \mathcal{N}$, $arity(A) = m$, $a \in \Sigma$, and E is a *finite* term over $\mathcal{N}$ in which each occurring variable is from the set $\{x_1, x_2, \ldots, x_m\}$.

LTSs Generated by Grammars. Given $\mathcal{G} = (\mathcal{N}, \Sigma, \mathcal{R})$, by $\mathcal{L}_{\mathcal{G}}^{\text{R}}$ we denote the *(rule based)* LTS $\mathcal{L}_{\mathcal{G}}^{\text{R}} = (\text{TERMS}_{\mathcal{N}}, \mathcal{R}, (\xrightarrow{r})_{r \in \mathcal{R}})$ where each rule r of the form $A(x_1, x_2, \ldots, x_m) \xrightarrow{a} E$ induces $(A(x_1, \ldots, x_m))\sigma \xrightarrow{r} E\sigma$ for any substitution σ. (Thus also $A(x_1, \ldots, x_m) \xrightarrow{r} E$, using σ with $\text{SUPP}(\sigma) = \emptyset$.) The LTS $\mathcal{L}_{\mathcal{G}}^{\text{R}}$ is deterministic, since for each F and r there is at most one H such that $F \xrightarrow{r} H$. We note that *transitions cannot add variables*, i.e., $F \xrightarrow{w} H$ implies that each variable occurring in H also occurs in F. We also note that $F \xrightarrow{w} H$ implies $F\sigma \xrightarrow{w} H\sigma$ for any substitution σ.

Since the right-hand sides (rhs) E in the rules $A(x_1, \ldots, x_m) \xrightarrow{a} E$ are finite, all *terms reachable from a finite term* are *finite*. (It is technically convenient to have the rhs finite while including regular terms into our LTSs.)

By the *action-based* LTS we mean $\mathcal{L}_{\mathcal{G}}^{\text{A}} = (\text{TERMS}_{\mathcal{N}}, \Sigma, (\xrightarrow{a})_{a \in \Sigma})$ where each rule $A(x_1, \ldots, x_m) \xrightarrow{a} E$ induces $(A(x_1, \ldots, x_m))\sigma \xrightarrow{a} E\sigma$ (for any σ). Hence $F \xrightarrow{w} H$ in $\mathcal{L}_{\mathcal{G}}^{\text{R}}$ implies $F \xrightarrow{\text{ACT}(w)} H$ in $\mathcal{L}_{\mathcal{G}}^{\text{A}}$, where $\text{ACT}(w)$ is the naturally defined *action-image* of w. We note that $\mathcal{L}_{\mathcal{G}}^{\text{A}}$ is image-finite, and non-deterministic in general. We *complete the definition of* $\mathcal{L}_{\mathcal{G}}^{\text{A}}$ *by stipulating that*

$$\text{EQLV}(x_i, H) = 0 \text{ if } H \neq x_i \text{ (in particular, } x_i \not\sim_1 x_j \text{ for } i \neq j).$$

Remark. This reflects the fact that $x_i \neq H$ implies that $\text{EQLV}(x_i\sigma, H\sigma) = 0$ for some σ, unless the underlying grammar $\mathcal{G}$ is trivial.

In what follows we refer to the action-based LTSs $\mathcal{L}_{\mathcal{G}}^{\text{A}}$, if we do not say explicitly that we have $\mathcal{L}_{\mathcal{G}}^{\text{R}}$ in mind.

Theorem 1. *There is an algorithm that, given an FO-grammar* $\mathcal{G} = (\mathcal{N}, \Sigma, \mathcal{R})$ *and* $T_0, U_0 \in \text{TERMS}(\mathcal{N})$, *computes* $\text{EQLV}(T_0, U_0)$ *in* $\mathcal{L}_{\mathcal{G}}^{\text{A}}$.

3   Proof of Theorem 1

As a convenient tool we introduce a round-based game between Prover(she) and Refuter(he); the game is more involved than the standard bisimulation game. We start with a simple first version of the game, and then we stepwise enhance it. Refuter will be always able to force his win in finite time if the terms in the initial pair (T_0, U_0) are non-equivalent. Prover will be always able to avoid losing if $T_0 \sim U_0$, but only in the final version she will be able to force her win in finite time. Before the first game-version we observe some simple standard facts related to (stratified) bisimulation equivalence.

Expansions. Assume an LTS $\mathcal{L} = (\mathcal{S}, \Sigma, (\xrightarrow{a})_{a \in \Sigma})$. By $\mathcal{B} \triangleleft \mathcal{B}'$, where $\mathcal{B}, \mathcal{B}' \subseteq \mathcal{S} \times \mathcal{S}$, we denote that $\mathcal{B}'$ is a *minimal expansion for* $\mathcal{B}$, i.e., $\mathcal{B}'$ covers $\mathcal{B}$ and no proper subset of $\mathcal{B}'$ covers $\mathcal{B}$; this also implies that for each $(s', t') \in \mathcal{B}'$ there is $(s, t) \in \mathcal{B}$ such that $s \xrightarrow{a} s'$ and $t \xrightarrow{a} t'$ for some $a \in \Sigma$. We note that $\emptyset \triangleleft \emptyset$, and if s, t are dead (not enabling any action), then $\{(s, t)\} \triangleleft \emptyset$.

For any $k \in \mathbb{N}$ we have $k < \omega$ and we stipulate $\omega - k = \omega + k = \omega$. We also stipulate $\min \emptyset = \omega$, and define $\mathrm{MINEQL}(\mathcal{B}) = \min\{\mathrm{EQLV}(s, t) \mid (s, t) \in \mathcal{B}\}$.

Proposition 2.
(1) If $\mathrm{MINEQL}(\mathcal{B}) = 0$ then there is no $\mathcal{B}'$ such that $\mathcal{B} \triangleleft \mathcal{B}'$.
(2) If $\mathcal{B} \triangleleft \mathcal{B}'$ and $\mathrm{MINEQL}(\mathcal{B}) < \omega$, then $\mathrm{MINEQL}(\mathcal{B}) > \mathrm{MINEQL}(\mathcal{B}')$.
(3) If $\mathrm{MINEQL}(\mathcal{B}) > 0$ then there is $\mathcal{B}'$ such that $\mathcal{B} \triangleleft \mathcal{B}'$ and $\mathrm{MINEQL}(\mathcal{B}') \geq \mathrm{MINEQL}(\mathcal{B}) - 1$. (In particular, if $\mathcal{B} \subseteq \sim$ then $\mathcal{B} \triangleleft \mathcal{B}'$ for some $\mathcal{B}' \subseteq \sim$.)
(4) For $k \in \mathbb{N}$ we have $s \sim_k t$ iff there is a sequence $\{(s, t)\} \triangleleft \mathcal{B}_1 \triangleleft \mathcal{B}_2 \triangleleft \cdots \triangleleft \mathcal{B}_k$.

Prover-Refuter Game (First Version). A play starts with a grammar $\mathcal{G} = (\mathcal{N}, \Sigma, \mathcal{R})$ and an *initial pair* (T_0, U_0) of terms. For $i = 0, 1, 2 \ldots$, the $(i+1)$-th *round* of the play starts with some specified pair (T_i, U_i) and proceeds as follows:

1. Prover chooses $k > 0$ and some $\mathcal{B}_j \subseteq \mathrm{TERMS}_{\mathcal{N}} \times \mathrm{TERMS}_{\mathcal{N}}$ for $j = 1, 2 \ldots, k$ and shows that $\mathcal{B}_0 \triangleleft \mathcal{B}_1 \triangleleft \mathcal{B}_2 \triangleleft \ldots \triangleleft \mathcal{B}_k$ where $\mathcal{B}_0 = \{(T_i, U_i)\}$.
   If this is impossible (i.e., if $T_i \not\sim_1 U_i$), then Refuter wins.
2. Refuter chooses a pair (T_i', U_i') in $\mathcal{B}_k \smallsetminus \bigcup_{j=0}^{k-1} \mathcal{B}_j$. If this is impossible, i.e. if $\mathcal{B}_k \subseteq \bigcup_{j=0}^{k-1} \mathcal{B}_j$ (which includes the case $\mathcal{B}_k = \emptyset$), then Prover wins. (In this case $T_i \sim U_i$, as follows by using Prop. 2.)
3. The pair $(T_{i+1}, U_{i+1}) = (T_i', U_i')$ is taken for starting the $(i+2)$-th round.

We say that *Refuter* uses the *least-eqlevel strategy*, if he always chooses (T_i', U_i') so that $\mathrm{EQLV}(T_i', U_i') = \mathrm{MINEQL}(\mathcal{B}_k)$; in this case $\mathrm{EQLV}(T_i', U_i') < \mathrm{MINEQL}(\bigcup_{j=0}^{k-1} \mathcal{B}_j)$, and thus $\mathrm{EQLV}(T_i', U_i') < \mathrm{EQLV}(T_i, U_i) - (k-1)$, unless $T \sim U$ for all $(T, U) \in \bigcup_{j=0}^{k} \mathcal{B}_j$. We easily observe the following facts.

Proposition 3. *Let $\mathrm{EQLV}(T_0, U_0) = e \in \mathbb{N} \cup \{\omega\}$.*
1. If $e < \omega$, then Refuter wins within $e+1$ rounds by the least-eqlevel strategy.
2. Prover can guarantee that she will not lose within e rounds.

We note that $\sim_k$ is decidable for each fixed k, and there is an algorithm that, given $\mathcal{G}$, T_0, U_0, outputs $\mathrm{EQLv}(T_0, U_0)$ if $T_0 \not\sim U_0$, and does not halt if $T_0 \sim U_0$.

Prover's Additional Tool. A challenge is to add sound possibilities to Prover to enable her to force her win in finite time if $T_0 \sim U_0$. We allow Prover to claim a win when she can (soundly) demonstrate, in some $(i{+}1)$-th round, that either Refuter has not used the least-eqlevel strategy or $T_0 \sim U_0$. This new abstract rule does not change Prop. 3. A simple instance is a *repeat*: if $\{T_i, U_i\} = \{T_j, U_j\}$ for some $j < i$, then Prover can claim her win.

We further assume that Prover wins when a repeat appears, and we look at more involved options for Prover, enabling to "balance", i.e., to replace T_i', U_i' with T_{i+1}, U_{i+1} that are "closer" to each other, while not changing the eq-level if Refuter uses the least eq-level strategy. Before formulating the second version of the game, we clarify the crucial underlying facts. First a trivial one:

Proposition 4. *Assume an LTS $\mathcal{L}$. If $\mathrm{EQLv}(s, t) = k$ and $\mathrm{EQLv}(s, s') > k$, then $\mathrm{EQLv}(s', t) = k$ (since $s' \sim_k s \sim_k t$ and $s' \sim_{k+1} s \not\sim_{k+1} t$).*

Congruence, the Crux of Balancing. Assume a grammar $\mathcal{G} = (\mathcal{N}, \Sigma, \mathcal{R})$. We put $\sigma \sim_k \sigma'$ (for two substitutions σ, σ') if $\sigma(x_i) \sim_k \sigma'(x_i)$ for each $x_i \in \mathrm{VAR}$, and we define $\mathrm{EQLv}(\sigma, \sigma') = \max \{ k \in \mathbb{N} \cup \{\omega\} \mid \sigma \sim_k \sigma' \}$. We now note that $\sim_k$ and $\sim$ are congruences (which is obvious, e.g. by induction on k):

Proposition 5.
(1) If $E \sim_k F$, then $E\sigma \sim_k F\sigma$; hence $\mathrm{EQLv}(E, F) \le \mathrm{EQLv}(E\sigma, F\sigma)$.
(2) If $\sigma \sim_k \sigma'$, then $E\sigma \sim_k E\sigma'$; hence $\mathrm{EQLv}(\sigma, \sigma') \le \mathrm{EQLv}(E\sigma, E\sigma')$.

We illustrate how Prover can use the above facts. Suppose the $(i{+}1)$-th round starts with (T_i, U_i) and Refuter chooses (T_i', U_i') in $\mathcal{B}_k$ (we refer to the notation in the game definition). We thus have $T_i \xrightarrow{u_1} T_i'$, $U_i \xrightarrow{u_2} U_i'$ in $\mathcal{L}_\mathcal{G}^R$, for some $u_1, u_2 \in \mathcal{R}^*$, where $|u_1| = |u_2| = k$ (and $\mathrm{ACT}(u_1) = \mathrm{ACT}(u_2)$). Suppose that $T_i \xrightarrow{u_1} T_i'$ is *not a shortest path* from T_i to T_i' (in $\mathcal{L}_\mathcal{G}^R$). Then we have $T_i \xrightarrow{v_1} T_i'$ for some $v_1 \in \mathcal{R}^*$ where $|v_1| < |u_1|$. Since $\{(T_i, U_i)\} \lhd \mathcal{B}_1 \lhd \mathcal{B}_2 \lhd \cdots \lhd \mathcal{B}_k$, we must have $U_i \xrightarrow{v_2} U''$ for some U'' and some $v_2 \in \mathcal{R}^*$ such that $|v_1| = |v_2|$ (and $\mathrm{ACT}(v_1) = \mathrm{ACT}(v_2)$) and $(T_i', U'') \in \bigcup_{j=0}^{k-1} \mathcal{B}_j$. (Hence $\mathrm{EQLv}(T_i', U'') > \mathrm{EQLv}(T_i', U_i')$ when $T_i' \not\sim U_i'$ and Refuter uses the least-eqlevel strategy.)

Therefore Refuter "cannot protest" when Prover puts $(T_{i+1}, U_{i+1}) = (U'', U_i')$ instead of $(T_{i+1}, U_{i+1}) = (T_i', U_i')$, since $\mathrm{EQLv}(T_i', U_i') = \mathrm{EQLv}(U'', U_i')$ if Refuter uses the least-eqlevel strategy. We note that U'', U_i' are close to each other in the sense that they are both reachable within k steps from one "pivot term", namely U_i. We have $U_i \rightsquigarrow_k (U'', U_i')$, where generally we define

$$W \rightsquigarrow_k (T, U) \Leftrightarrow_{\mathrm{df}} W \xrightarrow{v_1} T, W \xrightarrow{v_2} U \text{ for some } v_1, v_2 \text{ of length at most } k. \quad (1)$$

We give Prover also other possibilities how to replace T_i' or U_i':

Prover-Refuter Game (Second Version). The only change w.r.t. the first game-version is in the point 3:

3. Prover creates (T_{i+1}, U_{i+1}) for the start of the $(i+2)$-th round:
   Either she puts $(T_{i+1}, U_{i+1}) = (T_i', U_i')$, thus making *no change*, or she can use one of the following options if available:

   i/ *Left-balancing*: Prover presents T_i' as $G\sigma$ for some *finite term* G and some substitution σ, where for each $V \in \text{RANGE}(\sigma)$ she finds V' such that $(V, V') \in \bigcup_{j=0}^{k-1} \mathcal{B}_j$. She defines σ' with $\text{SUPP}(\sigma') = \text{SUPP}(\sigma)$ as follows: if $\sigma(x_\ell) = V$, then $\sigma'(x_\ell) = V'$, where (V, V') is an above found pair. Finally she puts $(T_{i+1}, U_{i+1}) = (G\sigma', U_i')$.

   ii/ *Right-balancing*: Symmetrically, Prover presents U_i' as $G\sigma$, finds all appropriate pairs (V', V) in $\bigcup_{j=0}^{k-1} \mathcal{B}_j$, and puts $(T_{i+1}, U_{i+1}) = (T_i', G\sigma')$.

Our previous illustration was a special case: we had $T_i' = G\sigma$ where $G = x_1$, $\text{SUPP}(\sigma) = \{x_1\}$ and $\sigma(x_1) = T_i'$, and we replaced $G\sigma$ with $G\sigma'$ where $\sigma'(x_1) = U''$ (and thus $(\sigma(x_i), \sigma'(x_i)) \in \bigcup_{j=0}^{k-1} \mathcal{B}_j$ for all $x_i \in \text{SUPP}(\sigma) = \text{SUPP}(\sigma') = \{x_1\}$).

We can easily verify that Prop. 3 holds also for the second game-version. The crucial point is that $\text{EQLV}(T_{i+1}, U_{i+1}) = \text{EQLV}(T_i', U_i')$ when Refuter uses the least-eqlevel strategy (this is based on Prop. 5(2) and Prop. 4).

When doing a left-balancing, replacing $(T_i', U_i') = (G\sigma, U_i')$ with the *bal-result* $(T_{i+1}, U_{i+1}) = (G\sigma', U_i')$, we might not have $U_i \leadsto_k (T_{i+1}, U_{i+1})$, but we surely have $U_i \overset{\text{L}:d}{\leadsto}_k (T_{i+1}, U_{i+1})$, for $d = \text{HEIGHT}(G)$, where we generally define

$$W \overset{\text{L}:d}{\leadsto}_k (T, U) \Leftrightarrow_{\text{df}} \text{ there is a finite term } G \text{ and } \sigma \text{ such that} \qquad (2)$$

$T = G\sigma$, $\text{HEIGHT}(G) \leq d$, and U and all $V \in \text{RANGE}(\sigma)$ are reachable from W within k steps; here L signals that we allow a special head, of height at most d, in the left-hand component. Symmetrically we define $W \overset{\text{R}:d}{\leadsto}_k (T, U)$, where R refers to the right-hand component.

Two Remaining Steps in the Proof of Theorem 1. We recall that a play gives rise to a sequence $(T_0, U_0), (T_1, U_1), (T_2, U_2), \ldots$ of pairs of terms that are the starting pairs for the rounds $1, 2, 3, \ldots$, respectively. We show that in the case $T_0 \sim U_0$ Prover can force a certain potentially infinite (n, g)-subsequence of $(T_1, U_1), (T_2, U_2), \ldots$ (by Lemma 6), and then we bound the lengths of eqlevel-decreasing (n, g)-sequences, which is used in the final game-version.

Eqlevel-decreasing Sequences, and (n, g)-sequences. A *sequence* $(V_1, W_1), (V_2, W_2), (V_3, W_3), \ldots$ is *eqlevel-decreasing* if $\omega > \text{EQLV}(V_1, W_1) > \text{EQLV}(V_2, W_2) > \cdots$. In this case the sequence must be finite, and our requirement $V_1 \nsim W_1$ implies that its length is bounded by $1 + \text{EQLV}(V_1, W_1)$.

Given a pair (n, g) where $n \in \mathbb{N}$ and $g : \mathbb{N}_+ \to \mathbb{N}_+$ is a nondecreasing function (where $\mathbb{N}_+ = \{1, 2, \ldots\}$), a (finite or infinite) sequence of pairs of terms is an (n, g)-*sequence* if it can be presented as $(E_1\sigma, F_1\sigma), (E_2\sigma, F_2\sigma), (E_3\sigma, F_3\sigma), \ldots$ for a substitution σ with $|\text{SUPP}(\sigma)| \leq n$, where $\text{PRSIZE}(E_j, F_j) \leq g(j)$ for $j = 1, 2, \ldots$. (We put $\text{PRSIZE}(E, F) = \text{PRSIZE}(E) + \text{PRSIZE}(F)$, say.) Thus the growth of the (regular) "head-terms" E_j, F_j is bounded by the function g, while at most n fixed "tail-subterms" (of unrestricted size) suffice for this presentation.

Lemma 6. *There are n, g determined by (in fact, computable from) grammar $\mathcal{G}$ such that Prover can force for any initial $T_0 \sim U_0$ that she either wins or the sequence $(T_1, U_1), (T_2, U_2), \ldots$ has an infinite (n, g)-subsequence.*

Proof. Before showing a strategy of Prover that proves the claim, we introduce some technical notions related to a given grammar $\mathcal{G} = (\mathcal{N}, \Sigma, \mathcal{R})$; in our notation we assume that $arity(A) = m$ for all $A \in \mathcal{N}$.

If $A(x_1, \ldots, x_m) \xrightarrow{w} x_i$ in $\mathcal{L}_\mathcal{G}^R$, then we call $w \in \mathcal{R}^*$ an (A, i)-*sink word*. We assume that for each pair $A \in \mathcal{N}$, $i \in [1, m]$ there is a fixed shortest (A, i)-sink word $w_{[A,i]}$, and we put $M_0 = 1 + \max\{|w_{[A,i]}|; A \in \mathcal{N}, i \in [1, m]\}$. The words $w_{[A,i]}$ can be found and (exponentially bounded) M_0 can be computed by standard dynamic programming approach; the grammar can be easily transformed to the required form if there are no (A, i)-sink words for some A, i.

A *path* $V \xrightarrow{u}$ in $\mathcal{L}_\mathcal{G}^R$ is *root-performable*, if $A(x_1, \ldots, x_m) \xrightarrow{u}$ where $A = \text{ROOT}(V)$ (in which case u is enabled by any term with the root A). A *path* $V \xrightarrow{w}$ in $\mathcal{L}_\mathcal{G}^R$ is a *non-sink segment*, a *non-sink* for short, if $|w| = M_0$ and $V \xrightarrow{w}$ is root-performable. (For each root-successor V' in V we thus have $V \xrightarrow{v} V'$ for some v shorter than w.)

A *path* $T \xrightarrow{u} T'$ in $\mathcal{L}_\mathcal{G}^R$ is *sinking* if it contains no non-sink, i.e., for any partition $u = u_1 u_2 u_3$ with $|u_2| = M_0$ we have $u_2 = u_2' u_2''$ ($u_2' \neq \varepsilon$) where $T \xrightarrow{u_1} V \xrightarrow{u_2'} V' \xrightarrow{u_2'' u_3} T'$ and V' is a root-successor in V. Hence if $T \xrightarrow{u} T'$ is sinking, then it can be written $T \xrightarrow{u_1} V \xrightarrow{u_2} T'$ where $|u_2| < M_0$ and V is a subterm of T in depth at least $|u| \div M_0$. (By $\div$ we denote integer division.)

Finally we consider a shortest path $T \xrightarrow{u} T'$ from T to T' that is not sinking. It can be written $T \xrightarrow{u_1} V \xrightarrow{u_2} V' \xrightarrow{u_3} T'$ where $V \xrightarrow{u_2} V'$ is the last non-sink. Since $V \xrightarrow{u_2 u_3} T'$ is root-performable (if a prefix of $u_2 u_3$ exposes a root-successor in V, then $V \xrightarrow{u_2 u_3} T'$ is not a shortest path from V to T', which contradicts the assumption that $T \xrightarrow{u} T'$ is shortest), for $A = \text{ROOT}(V)$ we have $A(x_1, \ldots, x_m) \xrightarrow{u_2 u_3} G$ (and thus $V = (A(x_1, \ldots, x_m))\sigma \xrightarrow{u_2 u_3} T' = G\sigma$ where $\text{RANGE}(\sigma)$ consists of the root-successors in V), and this path is sinking after the first step (since we took the last non-sink). Obviously, $\text{HEIGHT}(G) \leq M_0'$, for some M_0' computable from $\mathcal{G}$. We now take M_1 such that $(M_1 \div M_0) > M_0'$, and show Prover's strategy in the $(i+1)$-th round, starting with $T_i \sim U_i$:

i/ Prover chooses $k = M_1$ and $\{(T_i, U_i)\} = \mathcal{B}_0 \lhd \mathcal{B}_1 \lhd \mathcal{B}_2 \lhd \ldots \lhd \mathcal{B}_{M_1}$ where $\mathcal{B}_j \subseteq \sim$ (for all $j \in [1, M_1]$). Refuter chooses (T_i', U_i') and we get the paths $T_i \xrightarrow{u_1} T_i'$, $U_i \xrightarrow{u_2} U_i'$ in $\mathcal{L}_\mathcal{G}^R$ (where $|u_1| = |u_2| = M_1$, $\text{ACT}(u_1) = \text{ACT}(u_2)$).

ii/ If $T_i \xrightarrow{u_1} T_i'$ is not shortest or contains a non-sink, and *Prover did not do a right-balancing in the (previous) i-th round*, then she makes a left-balancing, replacing $(T_i', U_i') = (G\sigma, U_i')$ with $(T_{i+1}, U_{i+1}) = (G\sigma', U_i')$, for some head G with the smallest possible height. (We know that $\text{HEIGHT}(G) \leq M_0'$.)

iii/ If ii/ did not apply, and $U_i \xrightarrow{u_2} U_i'$ is not shortest or contains a non-sink, and Prover did not do a left-balancing in the i-th round, then she makes a right-balancing, symmetrically to ii/.

iv/ If none of ii/, iii/ applied, Prover puts $(T_{i+1}, U_{i+1}) = (T_i', U_i')$.

We recall that each bal-result (T_{i+1}, U_{i+1}) has its pivot W, where $W \overset{\text{L}:\, M_0'}{\leadsto} M_1$
(T_{i+1}, U_{i+1}) or $W \overset{\text{R}:\, M_0'}{\leadsto} M_1 (T_{i+1}, U_{i+1})$ (recall the definition (2)).

If Prover balances in the $(i+1)$-th round, e.g., she does a left-balancing with
pivot W, hence $W \overset{\text{L}:\, M_0'}{\leadsto} M_1 (T_{i+1}, U_{i+1}) = (G\sigma', U_{i+1})$, and she cannot balance
in the $(i+2)$-th round, since the respective path $T_{i+1} = G\sigma' \overset{u}{\longrightarrow} T_{i+1}'$ is sinking
(and shortest), then $W \leadsto_{2M_1} (T_{i+2}, U_{i+2})$ (we have chosen M_1 large enough so
that the sinking path $G\sigma' \overset{u}{\longrightarrow} T_{i+1}'$ "erases" the special head G and exposes
some $\sigma'(x_\ell)$ that is reachable from W within M_1 steps).

We now explore an infinite play from $T_0 \sim U_0$ where Prover uses the above
strategy. There are infinitely many balancings; otherwise from some round on we
would have constant sinking on both sides, which necessarily leads to a repeat
since our terms are regular. We denote the respective pivots $W_1, W_2, \ldots$, and
easily verify that for each j we have a path $W_j \overset{w_j}{\longrightarrow} W_{j+1}$ (in $\mathcal{L}_{\mathcal{G}}^{\text{R}}$) of the form
$W_j \overset{v_1}{\longrightarrow} V \overset{v_2}{\longrightarrow} V' \overset{v_3}{\longrightarrow} W_{j+1}$ where $|v_1|, |v_3|$ are bounded (surely by $2M_1$)
and V' is a subterm of V; though v_2 can be sometimes long, we can assume
it to be sinking. Suppose that there is a term V such that the "pivot path"
$W_1 \overset{w_1}{\longrightarrow} W_2 \overset{w_2}{\longrightarrow} W_3 \overset{w_3}{\longrightarrow} \cdots$ visits subterms of V infinitely often. Then the pivots
W_j are infinitely often boundedly reachable from a subterm of V, as follows from
the above form of paths $W_j \overset{w_j}{\longrightarrow} W_{j+1}$. In this case one pivot reappears infinitely
often but there are boundedly many bal-results for one pivot — a repeat.

In some segment $W_j \overset{w_j}{\longrightarrow} W_{j+1}$ we thus have V, hence $W_j \overset{w_j'}{\longrightarrow} V \overset{w_j''}{\longrightarrow} W_{j+1}$,
such that its subterms are never visited by the path after; this implies that the
infinite path $V \overset{w_j''}{\longrightarrow} \overset{w_{j+1}}{\longrightarrow} \overset{w_{j+2}}{\longrightarrow} \cdots$ is root-performable; for $A = \text{ROOT}(V)$ we have
$A(x_1, \ldots, x_m) \overset{w_j''}{\longrightarrow} G_1 \overset{w_{j+1}}{\longrightarrow} G_2 \overset{w_{j+2}}{\longrightarrow} G_3 \overset{w_{j+3}}{\longrightarrow} \cdots$. Hence $V = (A(x_1, \ldots, x_m))\sigma'$
and $W_{j+\ell} = G_\ell \sigma'$ for σ' whose range consists of the root-successors in V. We
note that $\text{HEIGHT}(G_\ell)$ can only boundedly grow (with growing ℓ).

We are interested in the bal-results related to $G_1\sigma', G_2\sigma', G_3\sigma', \ldots$. Since the
bal-result (T, U) related to $G_j\sigma'$ satisfies $G_j\sigma' \overset{\text{L}:\, M_0'}{\leadsto} M_1 (T, U)$ or $G_j\sigma' \overset{\text{R}:\, M_0'}{\leadsto} M_1$
(T, U) (recall (2)), it is useful to rather write $V = F\sigma$ for a finite F in
which each branch has length M_1 if it is not a complete branch of V, and
where $\text{RANGE}(\sigma')$ consists of the subterms of V with depth M_1. Then we
rewrite $G_1\sigma', G_2\sigma', G_3\sigma', \ldots$ as $H_1\sigma, H_2\sigma, H_3\sigma, \ldots$ (where each occurrence of
$x_i \in \text{SUPP}(\sigma)$ in H_ℓ has depth at least M_1), and we now easily derive that the
respective bal-results can be written as $(E_1\sigma, F_1\sigma), (E_2\sigma, F_2\sigma), (E_3\sigma, F_3\sigma), \ldots$,
namely as an (n, g)-sequence for n, g determined by the grammar $\mathcal{G}$.   □

Lemma 9 bounds the lengths of eqlevel-decreasing (n, g)-sequences. Before its
proof we show some useful facts and convenient notions, assuming a grammar
$\mathcal{G} = (\mathcal{N}, \Sigma, \mathcal{R})$. We first recall that $\text{EQLV}(E, F) \leq \text{EQLV}(E\sigma, F\sigma)$, and note:

Proposition 7. *If* $\text{EQLV}(E, F) = k < e = \text{EQLV}(E\sigma, F\sigma)$ *(where* $e \in \mathbb{N} \cup \{\omega\}$*)*
then there are $x_i \in \text{SUPP}(\sigma)$, $H \neq x_i$, *and* $w \in \Sigma^*$, *where* $|w| \leq k$, *such that*
$E \overset{w}{\longrightarrow} x_i$, $F \overset{w}{\longrightarrow} H$ *or* $E \overset{w}{\longrightarrow} H$, $F \overset{w}{\longrightarrow} x_i$, *and* $\sigma(x_i) \sim_{e-k} H\sigma$.

Proof. We take $\{(E\sigma, F\sigma)\} = \mathcal{B}_0 \lhd \mathcal{B}_1 \lhd \cdots \lhd \mathcal{B}_{k+1}$, so that $\text{MinEqL}(\mathcal{B}_j) = e - j$. When trying to make $\{(E, F)\} = \mathcal{B}'_0 \lhd \mathcal{B}'_1 \lhd \cdots \lhd \mathcal{B}'_{k+1}$, by replacing σ with the empty-support substitution in the pairs in $\mathcal{B}_j$, we must get (x_i, H), $H \neq x_i$, somewhere in $\bigcup_{j=0}^{k} \mathcal{B}'_j$, otherwise we proved $E \sim_{k+1} F$. $\square$

By $\{(x_i, H)\}$ we denote the substitution that (only) replaces x_i with H. Hence $\{(x_i, H)\}\sigma$ is the substitution arising from σ by replacing $\sigma(x_i)$ with $H\sigma$. We note that the ("limit" regular) term $H' = H\{(x_i, H)\}\{(x_i, H)\}\cdots$ is well defined; a graph presentation of H' arises from a graph of H by redirecting each arc leading to x_i (if there is any) towards the root.

By $\sigma_{[-x_i]}$ we denote the substitution arising from σ by removing x_i from the support (if it is there), i.e., $\sigma_{[-x_i]}(x_i) = x_i$ and $\sigma_{[-x_i]}(x_j) = \sigma(x_j)$ for $j \neq i$. Since x_i does not occur in the above H', we have $H'\sigma = H'\sigma_{[-x_i]}$. Prop. 5(2) can be repeatedly used to show the following fact, referring to the above H':

Proposition 8. *If $\sigma(x_i) \sim_k H\sigma$ and $H \neq x_i$, then $\sigma \sim_k \{(x_i, H')\}\sigma_{[-x_i]}$.*

For any $\mathcal{B} \subseteq \text{Terms}_{\mathcal{N}} \times \text{Terms}_{\mathcal{N}}$ we put $\text{MaxEqL}(\mathcal{B}) = \max\{\text{EqLv}(E, F) \mid (E, F) \in \mathcal{B}\}$, stipulating $\max \emptyset = 0$. ($\text{MinEqL}(\mathcal{B})$ has been already defined.) For any $b \in \mathbb{N}$, we put $\text{SIZE}_{\leq b} = \{(E, F) \mid \text{PrSize}(E, F) \leq b\}$. For each $b \in \mathbb{N}$ we define the following *finite* number: $\text{MEL}_b = \text{MaxEqL}(\text{SIZE}_{\leq b} \cap \not\sim)$.

For any $n \in \mathbb{N}$ and $g : \mathbb{N}_+ \to \mathbb{N}_+$ we define $\ell_{n,g} \in \mathbb{N}$ by the following recursive definition: $\ell_{0,g} = 1 + \text{MEL}_{g(1)}$, $\ell_{n+1,g} = 1 + \text{MEL}_{g(1)} + \ell_{n,g'}$ where

$$g'(j) = g(1 + \text{MEL}_{g(1)} + j) + 2 \cdot (g(1) + \text{MEL}_{g(1)} \cdot \text{StepInc}) \text{ for all } j \in \mathbb{N}_+. \quad (3)$$

StepInc (step-increase) is the maximal size of the right-hand sides of the rules of $\mathcal{G}$; hence $F \xrightarrow{w} G$ implies $\text{PrSize}(G) \leq \text{PrSize}(F) + |w| \cdot \text{StepInc}$.

Lemma 9. *Any eqlevel-decreasing (n, g)-sequence has length at most $\ell_{n,g}$.*

Proof. By induction on n. Assume an eqlevel-decreasing (n, g)-sequence $(E_1\sigma, F_1\sigma), (E_2\sigma, F_2\sigma), \ldots, (E_\ell\sigma, F_\ell\sigma)$; recall that we also require $E_1\sigma \not\sim F_1\sigma$. Hence $E_1 \not\sim F_1$, and we have $\text{EqLv}(E_1, F_1) \leq \text{MEL}_{g(1)}$. If $n = 0$, then $(E_1, F_1) = (E_1\sigma, F_1\sigma)$, and thus $\ell \leq 1 + \text{EqLv}(E_1, F_1) \leq 1 + \text{MEL}_{g(1)} = \ell_{0,g}$.

If $\text{EqLv}(E_1, F_1) = k < e = \text{EqLv}(E_1\sigma, F_1\sigma)$ then $\sigma \sim_{e-k} \{(x_i, H')\}\sigma_{[-x_i]}$ for some $x_i \in \text{supp}(\sigma)$ and some H' with $\text{PrSize}(H') \leq g(1) + \text{MEL}_{g(1)} \cdot \text{StepInc}$; this can be easily derived from Prop. 7 and 8.

We now put $(shift)$ $\mathtt{s} = 1 + \text{MEL}_{g(1)}$; hence $\mathtt{s} > \text{EqLv}(E_1, F_1) = k$, and thus $\text{EqLv}(E_{\mathtt{s}+1}\sigma, F_{\mathtt{s}+1}\sigma) < e - k$. For $j = 1, 2, \ldots, \ell - \mathtt{s}$ we define $(E'_j, F'_j) = (E_{\mathtt{s}+j}\{(x_i, H')\}, F_{\mathtt{s}+j}\{(x_i, H')\})$. We thus have $\text{EqLv}(E'_j\sigma_{[-x_i]}, F'_j\sigma_{[-x_i]}) = \text{EqLv}(E_{\mathtt{s}+j}\sigma, F_{\mathtt{s}+j}\sigma)$ (by Prop. 4). Hence $\ell - \mathtt{s} \leq \ell_{n-1,g'}$ by the induction hypothesis, for g' defined by (3). $\square$

If $T_0 \sim U_0$, then Prover can force a potentially infinite (n, g)-sequence for certain n, g. She could claim a win after creating an (n, g)-sequence longer than $\ell_{n,g}$, if she could demonstrate $\ell_{n,g}$. Inspecting the above, we can verify that for computing $\ell_{n,g}$ for concrete n, g it suffices to know $\text{SIZE}_{\leq B} \cap \not\sim$ for a sufficiently large

$B \in \mathbb{N}$. The final idea is that Prover guesses B and $\mathcal{C} \subseteq \text{SIZE}_{\leq B}$, demonstrating the finite eq-levels of all pairs in $\mathcal{C}$. She claims that $\text{REST} = \text{SIZE}_{\leq B} \setminus \mathcal{C}$ is a subset of $\sim$, and she computes $\ell_{n,g}^{\mathcal{C}}$ by the recursive procedure for $\ell_{n,g}$ above, but consistently with her choice of $\mathcal{C}$ and the claim that $\text{REST} \subseteq \sim$. Mimicking the proof of Lemma 9, we derive the next lemma, which leads to the final game-version.

Lemma 10. *Any eqlevel-decreasing (n, g)-sequence starting with a pair whose eq-level is less than $\text{MINEQL}(\text{REST})$ has length at most $\ell_{n,g}^{\mathcal{C}}$.*

Prover-Refuter Game (Third Version). We separate $\mathcal{G}$ from the initial pair, now denoted E_0, F_0, to stress that the initial phase depends on $\mathcal{G}$ only.

i) A grammar $\mathcal{G} = (\mathcal{N}, \Sigma, \mathcal{R})$ is given.
ii) Prover provides some finite set $\mathcal{C} \subseteq \text{TERMS}_{\mathcal{N}} \times \text{TERMS}_{\mathcal{N}}$, some $n \in \mathbb{N}$, a sequence of increasing values denoted $g(1), g(2), \ldots, g(i)$ for some $i \in \mathbb{N}$, and some $B \in \mathbb{N}$ such that $\mathcal{C} \subseteq \text{SIZE}_{\leq B}$. For each pair $(E, F) \in \mathcal{C}$ Prover provides $e \in \mathbb{N}$ and demonstrates that $\text{EQLV}(E, F) = e$ (recall that $\sim_k$ is decidable for each $k \in \mathbb{N}$); thus $\mathcal{C} \subseteq \not\sim$. Prover now computes $\ell_{n,g}^{\mathcal{C}}$ (which fails when B is not sufficiently large).
iii) An initial pair (E_0, F_0) is given.
iv) For $\text{REST} = \text{SIZE}_{\leq B} \setminus \mathcal{C}$, Refuter chooses (T_0, U_0) from $\{(E_0, F_0)\} \cup \text{REST}$ (with the least eq-level when using the least-eqlevel strategy).
v) Now a play of the second game-version starts with (T_0, U_0). A new feature is that Prover can claim her win when she shows that $(T_1, U_1), (T_2, U_2), \ldots$ contains an (n, g)-subsequence that is longer than $\ell_{n,g}^{\mathcal{C}}$.

The least-eqlevel strategy guarantees Refuter's win for $E_0 \not\sim F_0$ (by Lemma 10). On the other hand, Prover can correctly guess $\mathcal{C} = \text{SIZE}_{\leq B} \cap \not\sim$ for B that is sufficient for computing (the real) $\ell_{n,g}$ (related to n, g that are guaranteed for $\mathcal{G}$ by Lemma 6), and she can force her win when $E_0 \sim F_0$.

Since a winning strategy of Prover (for any $\mathcal{G}, E_0, F_0$ where $E_0 \sim F_0$) is finitely presentable and effectively verifiable, a proof of Theorem 1 is now clear.

Concluding Remarks. Further work is needed to fully understand the discussed problems. Even the case of BPA processes, generated by real-time PDA with a single control-state, is not quite clear. Here the bisimilarity problem is EXPTIME-hard [12] and in 2-EXPTIME [3] (proven explicitly in [10]); for the subclass of normed BPA the problem is polynomial [7] (see [5] for the best published upper bound). Another issue is the precise decidability border. This was also studied in [8]; in our context, allowing (nondeterministic) popping ε-rules $A(x_1, \ldots, x_m) \xrightarrow{\varepsilon} x_i$ in FO-grammars leads to undecidability of bisimilarity. This aspect has been very recently refined, using branching bisimilarity [19]. Y. Fu and Q. Yin [6] also announced a result that would translate in our setting as the decidability of branching bisimilarity of FO-grammars with popping ε-rules (while "pushing" ε-rules lead to undecidability by adapting the constructions in [8]). Such result should be also achievable by the following adaptation of our Prover-Refuter game: Prover is required to tell, whenever she presents a new (sub)term V, if V is equivalent with some root-successor V' in V, and she must also extend the respective sets $\mathcal{B}_j$ with such claimed pairs (V, V'). Her later choices must be consistent, which essentially allows us to proceed analogously to the case studied in this paper.

References

1. Baeten, J., Bergstra, J., Klop, J.: Decidability of bisimulation equivalence for processes generating context-free languages. J. ACM 40(3), 653–682 (1993)
2. Benedikt, M., Göller, S., Kiefer, S., Murawski, A.S.: Bisimilarity of pushdown automata is nonelementary. In: Proc. LICS 2013, pp. 488–498. IEEE Computer Society Press (2013)
3. Burkart, O., Caucal, D., Steffen, B.: An elementary bisimulation decision procedure for arbitrary context-free processes. In: Hájek, P., Wiedermann, J. (eds.) MFCS 1995. LNCS, vol. 969, pp. 423–433. Springer, Heidelberg (1995)
4. Courcelle, B.: Recursive applicative program schemes. In: Handbook of Theoretical Computer Science, vol. B, pp. 459–492. Elsevier, MIT Press (1990)
5. Czerwiński, W., Lasota, S.: Fast equivalence-checking for normed context-free processes. In: Proc. FSTTCS 2010. LIPIcs, vol. 8. Schloss Dagstuhl - Leibniz-Zentrum für Informatik (2010)
6. Fu, Y., Yin, Q.: Dividing line between decidable PDA's and undecidable ones. CoRR abs/1404.7015 (2014)
7. Hirshfeld, Y., Jerrum, M., Moller, F.: A polynomial algorithm for deciding bisimilarity of normed context-free processes. Theor. Comput. Sci. 158, 143–159 (1996)
8. Jančar, P., Srba, J.: Undecidability of bisimilarity by Defender's forcing. J. ACM 55(1) (2008)
9. Jančar, P.: Decidability of DPDA language equivalence via first-order grammars. In: Proc. LICS 2012, pp. 415–424. IEEE Computer Society (2012)
10. Jančar, P.: Bisimilarity on basic process algebra is in 2-ExpTime (an explicit proof). Logical Methods in Computer Science 9(1) (2013)
11. Jančar, P.: Equivalences of pushdown systems are hard. In: Muscholl, A. (ed.) FOSSACS 2014. LNCS, vol. 8412, pp. 1–28. Springer, Heidelberg (2014)
12. Kiefer, S.: BPA bisimilarity is EXPTIME-hard. Inf. Proc. Letters 113(4), 101–106 (2013)
13. Schmitz, S.: Complexity hierarchies beyond elementary. CoRR abs/1312.5686 (2013)
14. Sénizergues, G.: L(A)=L(B)? Decidability results from complete formal systems. Theor. Comput. Sci. 251(1–2), 1–166 (2001)
15. Sénizergues, G.: The bisimulation problem for equational graphs of finite outdegree. SIAM J.Comput. 34(5), 1025–1106 (2005), presented at FOCS 1998
16. Srba, J.: Roadmap of infinite results. In: Current Trends In Theoretical Computer Science, The Challenge of the New Century, vol. 2, pp. 337–350. World Scientific Publishing Co. (2004), http://users-cs.au.dk/srba/roadmap/
17. Stirling, C.: Deciding DPDA equivalence is primitive recursive. In: Widmayer, P., Triguero, F., Morales, R., Hennessy, M., Eidenbenz, S., Conejo, R. (eds.) ICALP 2002. LNCS, vol. 2380, pp. 821–832. Springer, Heidelberg (2002)
18. Stirling, C.: Decidability of bisimulation equivalence for pushdown processes (2000), available at the author's web-page
19. Yin, Q., Fu, Y., He, C., Huang, M., Tao, X.: Branching bisimilarity checking for PRS. CoRR abs/1402.0050 (2014), Accepted to ICALP 2014

Context Unification is in PSPACE[*]

Artur Jeż[1,2]

[1] Max Planck Institute für Informatik, Saarbrücken, Germany
[2] Institute of Computer Science, University of Wrocław, Wrocław, Poland

Abstract. Contexts are terms with one 'hole', i.e. a place in which we can substitute an argument. In context unification we are given an equation over terms with variables representing contexts and ask about the satisfiability of this equation. Context unification at the same time is subsumed by a second-order unification, which is undecidable, and subsumes satisfiability of word equations, which is in PSPACE. We show that context unification is in PSPACE, so as word equations. For both problems NP is still the best known lower-bound.

This result is obtained by an extension of the recompression technique, recently developed by the author and used in particular to obtain a new PSPACE algorithm for satisfiability of word equations, to context unification. The recompression is based on applying simple compression rules (replacing pairs of neighbouring function symbols), which are (conceptually) applied on the solution of the context equation and modifying the equation in a way so that such compression steps can be performed directly on the equation, without the knowledge of the actual solution.

Keywords: Context unification; Second order unification; Term rewriting.

1 Introduction

Context Unification. In context unification we solve equations over context terms, let us recall the appropriate notions: A *ground context* is a ground term (i.e. function symbols have fixed arity and a symbol of arity k has exactly k children) with exactly one occurrence of a special constant Ω (a *hole*), which represents a missing argument. Ground contexts can be applied to ground terms, which results in a replacement of the hole by the given ground term; similarly we define a composition of two ground contexts, which is again a ground context. We build context terms using using function symbols, variables (which shall denote ground terms) and context variables (which shall denote ground contexts). A context equation is an equation between two such terms. A solution of a context equation assigns to each context variable a ground context (over the given input signature) and to each variable a ground term (over the same signature) such that both sides of the equation evaluate to the same (ground) term. The

---

[*] Supported by the Humboldt Foundation Postdoctoral grant. The full version of this paper is available at http://arxiv.org/abs/1310.4367.

J. Esparza et al. (Eds.): ICALP 2014, Part II, LNCS 8573, pp. 244–255, 2014.

corresponding decision problem is known as the *context unification*. It was introduced by Comon [1,2] and independently by Schmidt-Schauß [10] and found many applications in diverse fields.

The problem gained considerable attention and there was a large body of work [2,3,5,6,7,11,12,13] focused on context unification and several partial results were obtained. This popularity is fueled by the fact that this is the only known natural problem which is on one hand subsumed by second order unification, which is undecidable in many restricted cases, and subsumes satisfiability of word equations, which is decidable (in PSPACE). To be more precise: in second-order unification the argument of the second-order variable X can be used unbounded number of times in the substitution term for X. Hence context unification imposes a semantic restriction on the substitutions, which makes the proofs of undecidability of second order unification inapplicable. On the other hand, when the underlying signature contains only unary functions and constants, the context equation becomes a word equation. Whether algorithms for word equations extend to context unification, remained an open problem.

In this paper we show that context unification can be nondeterministically decided in polynomial space. The presented algorithm ContextEqSat is an extension of the recompression-based algorithm for word equations recently developed by the author [4].

Theorem 1. ContextEqSat *non-deterministically verifies the satisfiability of a context equation. It stores equation of length* $\mathcal{O}(nk)$ *and uses additional* $\mathcal{O}(n^2 k^2)$ *memory, where n is the size of the input equation while k is the maximal arity of symbols used in the equation.*

Corollary 1. *Context unification is in* PSPACE.

The best known lower bound for context unification is NP, which holds already for systems of word equations over one-letter signature.

Recompression. The connection between compression and unification was first observed by Plandowski and Rytter [9], who showed that each length-minimal solution (of size N) of the word equation (of size n) has a poly$(n, \log N)$ description (in terms of LZ77); this work gave no upper-bound on N, though. This connection was further exploited by Plandowski [8], whose PSPACE algorithm works on compressed representation of the word equation (and uses some finely tuned word factorisations to process such an equation).

The recompression method, which was introduced recently by the author [4], further exploits the connection between compression and word equations. It employs simple compression steps to compress a solution of a word equation. The crucial trick is that those compression steps are performed directly on the equation, without unfolding the implicit word. In order to make such compression steps applicable, the variables in the word equation are modified a bit. The advantage over the earlier solution by Plandowski is the apparent simplicity of both the compression steps and the modifications applied to the variables.

It is known that compression-based techniques are applicable also to context unification [14]. This paper extends the recompression method to terms in full generality. The first step is to devise a set of operations that guarantee a reduction of the term size by a constant factor. Since typically in a term at least half of symbols' occurrences are at leaves, one of our rules 'absorbs' constants into their parents: we replace a term $f(t_1, \ldots, t_{i-1}, c, t_{i+1}, \ldots, t_m)$ (where c is a constant) with $f'(t_1, \ldots, t_{i-1}, t_{i+1}, \ldots, t_m)$, where f' is a fresh function symbol (i.e. not in signature Σ). Applying such compression rule to every constant present at the beginning of the phase reduces the size of the term by a constant factor, assuming that there are no long 'chains', i.e. sequences of function symbols of arity 1. However, chains are very similar to strings and so it is natural to take the (known) compression rules used by string recompression [4]. There are two such rules: one replaces maximal chains a^ℓ with a single fresh symbol a_ℓ, the other replaces a chain ab, where $a \neq b$, with a fresh letter c.

It can be shown that such compression rules reduce the size of the tree by a constant factor. It remains to show, how to modify the equation so that the compression of the solution of the equation is performed directly on the equation. The modification boils down to replacement of variable X with $a(X)$ or $X(b)$, where a is the label of the topmost node in substitution for X and b is the one above the Ω (i.e. above the 'hole').

All our operations do not introduce new context variables, though they *can* introduce new variables (i.e. ones denoting closed terms). Still it can be shown that at any time there are at most kn variables, where k is the maximal arity of symbols in the input equation. The variable replacements can introduce $\mathcal{O}(1)$ new symbols per variable and context variable, thus $\mathcal{O}(kn)$ new symbols are introduced in one phase. On the other hand, the replacement rules guarantee that the size of the context equation is decreased by a constant factor (for appropriate nondeterministic choices). Thus the size of the equation remains $\mathcal{O}(nk)$.

2   Labelled Trees and Their Compression

Labelled Trees. We deal with rooted, ordered trees. Nodes are labelled with a ranked alphabet Σ, i.e. each $a \in \Sigma$ has a fixed arity $\mathrm{ar}(f)$. A tree is *well-formed* if a node labelled with f has exactly $\mathrm{ar}(f)$ children. Unless explicitly written, we consider well-formed trees (i.e. *ground terms* over Σ).

A set of labels Σ may be infinite and it is growing during the run of our algorithm. Without loss of generality, if Σ contains a label of arity k it also contains at least one label of arity k', for each $k' \leq k$. Initially Σ may be restricted to symbols used in the equation (and perhaps one additional binary symbol). We call the labels from Σ *letters*, letters of arity 1 *unary letters* and letters of arity 0 *constants*. Γ (perhaps with some subscripts) is used for some subalphabet of Σ, say, letters used in some particular tree or letters of arity at least 1, etc.

We replace with new letters fragments of a tree that are not necessarily well-formed. Thus a *subtree* is not necessarily well-formed, but in such a case we explicitly mention it. A *pattern* is a tree (perhaps not well-formed) in which a node labelled with f has *at most* $\mathrm{ar}(f)$ children; since we imagine a pattern as a part of a term with some of the subterms removed, the $0 \leq m \leq \mathrm{ar}(f)$ children of f in the pattern are numbered $1 \leq i_1 < \cdots < i_m \leq \mathrm{ar}(f)$ to denote which children of f are those in a 'real term'. A subpattern of a tree t is any subtree of t which is a pattern; we often consider individual *occurrences* of subpatterns of a tree t within t. In this terminology, our algorithm replaces occurrences of subpattern p of t in t by pattern p' (the subtrees rooted in children which are omitted in p are attached in p' in the same order, details are given later).

A *chain* is a pattern that consists only of unary letters. We consider 2-chains, so consisting only of two unary letters (usually different) and a-chains, which consists solely of letters a. We treat chains as strings and write them in the string notation and 'concatenate' them: for two chains s and s' the ss' denotes the chain obtained by attaching the top node in s to the bottom one in s. A chain t' that is a subpattern of t is a *chain subpattern* of t, an occurrence of a chain subpattern a^ℓ is a-*maximal* if it cannot be extended by a up nor down.

Local Compression of Trees. We perform three types of compressions on a tree t, all of them replace subpatterns by a single letter:

a-**chain Compression.** For a unary letter a we replace each a-maximal chain subpattern a^ℓ (where $\ell > 1$) by a fresh unary letter a_ℓ (making the father/child of a^ℓ the father/child of a_ℓ).

a, b **Pair Compression.** For two unary letters a and b we replace each 2-chain subpattern ab with a fresh unary letter c.

(f, i, c) **Leaf Compression.** For a constant c and letter f of arity $\mathrm{ar}(f) = m \geq i \geq 1$, we replace each subtree $f(t_1, \ldots, t_{i-1}, c, t_{i+1}, \ldots, t_m)$ with $f'(t_1, \ldots, t_{i-1}, t_{i+1}, \ldots, t_m)$ where f' is a fresh letter of arity $m - 1$.

To make the compression effective, we apply several compression steps in parallel: consider the a-maximal chain compression. As occurrences of a-maximal and b-maximal chain subpatterns do not overlap (it does not matter whether $a = b$ or not), we can perform a-maximal chain compression for $a \in \Gamma_1$ in parallel (as long as the letters that are used to replace the chains are not taken from Γ_1). We call the resulting procedure TreeChainComp(Γ_1, t) or simply *chain compression*, when Γ_1 and t are clear from the context.

Algorithm 1. TreeLeafComp($\Gamma_{\geq 1}, \Gamma_0, t$):

1: **for** $f \in \Gamma_{\geq 1}, 0 < i_1 < i_2 < \cdots < i_\ell \leq \mathrm{ar}(f) =: m, (a_1, a_2, \ldots, a_\ell) \in \Gamma_0^\ell$ **do**
2:     replace each $f(t_1, \ldots, t_m)$ by $f'(t_1, \ldots, t_{i_1-1}, t_{i_1+1}, \ldots, t_{i_\ell-1}, t_{i_\ell+1}, \ldots, t_m)$
     where $t_{i_j} = a_j$ for $1 \leq j \leq \ell$ and $t_i \notin \Gamma_0$ for $i \notin \{i_1, \ldots, i_\ell\}$

For $\Gamma_{\geq 1}$ and Γ_0 that consist of letters of arity at least 1 and 0, respectively, we would like to perform all possible (f, i, a) leaf compressions for $f \in \Gamma_{\geq 1}$, $i \leq \mathrm{ar}(f)$ and $a \in \Gamma_0$, at the same time. To avoid the ambiguity we introduce the $(f, i_1, a_1, i_2, a_2, \ldots, i_\ell, a_\ell)$ leaf compression, which replaces each subtree $f(t_1, \ldots, t_m)$ where $t_{i_\ell} = a_\ell$ and $t_i \notin \Gamma_0$ for $i \notin \{i_1, \ldots, i_\ell\}$ with $f'(t_1, \ldots, t_{i_1-1}, t_{i_1+1}, \ldots, t_{i_\ell-1}, t_{i_\ell+1}, \ldots, t_m)$. Note that in this way for a given node there is exactly one $(f, i_1, a_1, i_2, a_2, \ldots, i_\ell, a_\ell)$ compression that is applicable to this node, which allows making several such compressions in parallel (as long as we do not try to compress also the letters introduced during the compression). When $\Gamma_{\geq 1}$, Γ_0, and t are clear from the context, we simply call this operation *leaf compression* or TreeLeafComp.

In case of pair compression we cannot make parallel compressions, as 2-chains may overlap. However, parallel a, b pair compressions are possible when we take a and b from disjoint subalphabets Γ_1 and Γ_2. Those subalphabets are usually a partition of letters present in some tree and so we call them a *partition*, even if we do not explicitly say of what. In this case for each unary letter we

---
Algorithm 2. TreeComp(t)

1: $\Gamma \leftarrow$ unary letters in t
2: $t \leftarrow$ TreeChainComp(Γ, t)
3: $\Gamma \leftarrow$ unary letters in t
4: guess partition of Γ into $\Gamma_1 \uplus \Gamma_2$
5: $t \leftarrow$ TreePartitionComp(Γ_1, Γ_2, t)
6: $\Gamma_0 \leftarrow$ constants in t
7: $\Gamma_{\geq 1} \leftarrow$ letters of arity ≥ 1 in t
8: $t \leftarrow$ TreeLeafComp($\Gamma_{\geq 1}, \Gamma_0, t$)

---

can tell whether it should be the parent node or the child node in the compression step and the result does not depend on the order of the considered pairs, as long as new letters are outside $\Gamma_1 \cup \Gamma_2$. The obtained procedure is called TreePartitionComp(Γ_1, Γ_2, t), when t is clear from the context we refer to it simply as Γ_1, Γ_2 compression (we list the Γ_1 and Γ_2 to stress the dependency of the procedure on them).

The final compression procedure TreeComp applies all subprocedures once and it shrinks the tree by a constant factor.

Theorem 2. *Let* $t' = $ TreeComp(t), *then* $|t'| < 3|t|/4$ *for some partition* Γ_1, Γ_2.

3   Context Unification

By Ω we denote a special constant outside Σ; no letter added to Σ is Ω. A *ground context* is a ground $(\Sigma \cup \{\Omega\})$-term t, where $\text{ar}(\Omega) = 0$, that has exactly one occurrence of the constant Ω.

Given a ground context s and a ground term/context t we write st for the ground term/context that is obtained from s when we replace the occurrence of Ω in s by t. In the same spirit, when a is a unary letter, we usually write at to denote $a(t)$.

Let $\mathcal{V}$ denote an infinite set of context variables $X, Y, Z, \ldots$, while $\mathcal{X}$ denotes variables $x, y, z, \ldots$. The *terms* over $\Sigma, \mathcal{X}, \mathcal{V}$ are ground terms over $\Sigma \cup \mathcal{X} \cup \mathcal{V}$ in which $\text{ar}(X) = 1$ and $\text{ar}(x) = 0$ for each $x \in \mathcal{X}$ and $X \in \mathcal{V}$. A *context equation* is an equation of the form $u = v$ where both u and v are terms.

Letters from Σ occurring in a context equation are *explicit letters* and those occurrences are *explicit*.

A substitution replaces variables with ground terms and context variables with ground contexts. It is extended to terms and context terms in a natural way: let us just note that $S(X(t))$ is equal to $S(X)S(t)$. A substitution S is a *solution* of a context equation $u = v$ if $S(u) = S(v)$. A solution S is *size-minimal*, if for every solution S' it holds that $|S(u)| \leq |S'(u)|$; it is non-empty if $S(X) \neq \Omega$ for each $X \in \mathcal{X}$ occurring in $u = v$; we are interested only in non-empty solutions. This is not restricting, as for the input instance we can guess, which context variable is Ω and remove all such variables.

We recall the bound on *exponent of periodicity* for context equations.

Lemma 1 ([12]). *Let S be a size-minimal solution of a context equation $u = v$. Suppose that $S(X)$ (or $S(x)$) can be written as $t s^m t'$, where t, s, t' are ground context terms (t' is a ground term, respectively). Then $m = 2^{\mathcal{O}(|u|+|v|)}$.*

For a ground term $S(u)$ and an occurrence of a letter a in it we say that this occurrence *comes from* u if it was obtained from the letter in u and that it *comes from* X (or x) if it was obtained from $S(X)$ (or $S(x)$, respectively).

Solutions may use letters from Σ that do not occur in the equation, however, we can eliminate almost all such letters.

Lemma 2. *Consider a context equation $u = v$ over a signature Σ and let $\Gamma \subseteq \Sigma$ be the set of letters used in $u = v$. Then*

1. *If Σ has no constant then $u = v$ is not satisfiable.*
2. *If Σ has only constant and unary letters then $u = v$ reduces to a word equation.*
3. *If Σ has both constant and letter of arity greater than 1 then $u = v$ has a solution over Σ iff it has a solution over $\Gamma \cup c \cup f$, where c, f are letters, of arity 0 and 2, respectively. Moreover, if there is a constant $c' \in \Gamma$ then we can take $c = c'$ and if there is an f' of arity 2 in Γ then we can take $f = f'$.*

The proof follows by a simple substitution argument: if letter f is used in the solution but not in the equation then in the solution f can be replaced with another letter of the same arity.

The usage of Lemma 2 is as follows: the case without constants in Σ is trivial, the case with only unary letters and constants reduces to simpler problem of word equations. In the remaining nontrivial case we can remove from Σ every letter that is not used in the equation (except perhaps 2 letters). In particular, we eliminate infinite signatures and make sure that the arity of letters in Σ is bounded by the size of the equation.

4 Compression of Non-crossing Subpatterns

In this section we adapt the tree compression from Section 2 to the case when the tree is given implicitly, as a solution of a context equation. To this end we identify cases, in which performing such a compression is easy and those in which it is hard and show how to make the compression in the easy cases. In the next section we present how to transform the difficult cases to the easy ones.

We first formalise the notions about correctness of the nondeterministic procedures transforming the equation. A (nondeterministic) procedure is *sound*, when given an unsatisfiable word equation $u = v$ it cannot transform it to a satisfiable one, regardless of the nondeterministic choices; such a procedure is *complete*, if given a satisfiable equation $u = v$ for some nondeterministic choices it returns a satisfiable equation $u' = v'$. A composition of sound (complete) procedures is also sound (complete).

Non-crossing Partitions. Suppose that we want to perform the Γ_1, Γ_2 compression on the equation $u = v$ with a solution S. i.e. we want to replace each occurrence of a chain subpattern $ab \in \Gamma_1\Gamma_2$ with a fresh unary letter c. Such replacement is easy, when the occurrence of ab subpattern comes from the equation or from $S(X)$ (or $S(x)$) for some context variable X (or a variable x, respectively): in the former case we modify the equation by replacing each subpattern ab with c, in the latter the modification is done implicitly (i.e. we replace the subpattern ab in $S(X)$ or $S(x)$ with c). The problematic part is with the ab chain subpattern that is of neither of those forms, as they 'cross' between $S(X)$ (or $S(x)$) and some letter outside this $S(X)$ (or $S(x)$). This is formalised as follows: For an equation $u = v$ and a non-empty substitution S we say that an occurrence of a chain subpattern ab in $S(u)$ (or $S(v)$) is *explicit in S* if the occurrences of both a and b come from explicit letters a and b in $u = v$; *implicit in S* if the occurrences of both a and b come from the same occurrence of $S(x)$ (or $S(X)$); *crossing in S* otherwise. We say that ab is a *crossing pair* in S if they have at least one crossing occurrence in S; otherwise ab is a *non-crossing pair* (in S). A partition Γ_1, Γ_2 of Γ is *non-crossing* (in S) if there is no crossing pair $ab \in \Gamma_1\Gamma_2$ (in S); otherwise it is *crossing* (in S). Unless explicitly written, we consider only crossing/noncrossing pairs ab in which $a \neq b$.

The notions of a crossing chain subpattern can be defined in a more operational way: for a non-empty substitution S by the *first letter* of $S(X)$ ($S(x)$) we denote the topmost-letter in $S(X)$ ($S(x)$, respectively), by the *last letter* of $S(X)$ we denote the function symbol that is the father of Ω in $S(X)$. Then ab is crossing in non-empty S iff one of the following conditions holds for some context variables X, Y (or a context variable X and a variable y):

(CP1) aX (or ax) is a chain subpattern in $u = v$ and b is the first letter of $S(X)$ (or $S(x)$, respectively) *or*

(CP2) Xb is a chain subpattern in $u = v$ and a is the last letter of $S(X)$ *or*

(CP3) XY (or Xy) is a chain subpattern in $u = v$, a is the last letter of $S(X)$ and b the first letter of $S(Y)$ ($S(y)$, respectively).

When a partition Γ_1, Γ_2 is non-crossing in a solution S, we can simulate the TreePartitionComp($\Gamma_1, \Gamma_2, S(u)$) on $u = v$ simply be performing the Γ_1, Γ_2 compression on the explicit letters in the equation. To be more precise we treat the equation $u = v$ as a term over $\Sigma \cup \mathcal{X} \cup \mathcal{V} \cup \{=\}$ (imagine u and v as children of the root labelled with '=', which has arity 2) and apply the Γ_1, Γ_2 pair compression on this tree, we call this operation PartitionComp($\Gamma_1, \Gamma_2, 'u = v'$).

Lemma 3. PartitionComp($\Gamma_1, \Gamma_2, 'u = v'$) *is sound. If $u = v$ has a solution S such that Γ_1, Γ_2 is a non-crossing partition in S then it is complete: the returned equation $u' = v'$ has a solution S' such that $S'(u') = $ TreePartitionComp($\Gamma_1, \Gamma_2, S(u)$).*

The occurrences of ab that come from explicit letters are compressed, the ones that come from $S(X)$ and $S(x)$ are compressed by changing the solution and there are no other possibilities, by assumption that ab is noncrossing.

Non-crossing a-maximal Chains and Their Compression. Suppose that we want to perform the a-maximal chain compression on $u = v$ which has a solution S, i.e. all occurrences of a-maximal chains are to be replaced. Such replacement is easy, when the chain subpattern comes from the equation or from $S(X)$ (or $S(x)$) for some context variable X (or a variable x, respectively). The problematic part is with the occurrences that are of neither of those forms, as they 'cross' between $S(X)$ (or $S(x)$) and another subtree. This is formalised as follows: For an equation $u = v$ and a substitution S we say that an occurrence of an a-maximal chain subpattern a^ℓ in $S(u)$ (or $S(v)$) is *explicit in S* if this occurrence comes wholly from u (or v); *implicit in S* if this occurrence comes wholly from a single occurrence of $S(X)$ or $S(x)$; *crossing in S* otherwise. We say that a has a *crossing chain* if there is at least one occurrence of a crossing a-maximal chain subpattern. Otherwise, a *has no crossing chain*. It is easy to show that a has a crossing chain if and only if aa is a crossing pair, in particular the classification (CP1)–(CP3) applies.

When no unary letter (from Γ) has a crossing chain to simulate the chain compression on the context equation we perform the TreeChainComp on the explicit letters, treating the context equation as a tree, similarly as in the case of the Γ_1, Γ_2 compression; we refer to this algorithm as $\mathsf{ChainComp}(\Gamma, `u = v\,')$.

Lemma 4. $\mathsf{ChainComp}(\Gamma, `u = v\,')$ *is sound. If $u = v$ has a solution S such that no letter in Γ has a crossing chain then it is complete: the returned equation $u' = v'$ has a solution S' such that $S'(u') = \mathsf{TreeChainComp}(\Gamma, S(u))$.*

Non-crossing Father-Leaf Pairs and Their Compression. Suppose now that given a context equation $u = v$ with a solution S we would like to perform leaf compression on $S(u)$ and $S(v)$. This is easy if each occurrence of respective subpattern comes either from explicit letters in $u = v$ or wholly from $S(X)$ (or $S(x)$) then we treat $u = v$ as a tree and perform leaf compression on it. Consider a subpattern consisting of f with a child a on some position $i \leq \mathrm{ar}(f)$. For an equation $u = v$ and a substitution S we say that an occurrence of such a subpattern is *explicit in S* if both the occurrence of f and a come from explicit letters in u (or v); *implicit in S* if both the occurrence of f and a come from a single occurence of $S(X)$ or $S(x)$; *crossing in S* otherwise. Then (f, a) is a *crossing parent-leaf pair* in $u = v$ in S if it has at least one crossing occurrence in $u = v$ in S. Otherwise it is *noncrossing* in S.

Then (f, a) is a crossing father-leaf pair in S if and only if one of the following holds for some context variable X and variable y

(CFL 1) f with a son x is a subpattern in $u = v$ and $S(x) = a$ *or*
(CFL 2) Xa is a subpattern in $u = v$ and the last letter of $S(X)$ is f *or*
(CFL 3) Xy is a subpattern in $u = v$, $S(y) = a$ and f is the last letter of $S(X)$.

When there is no crossing father-leaf pair (f, a) for $f \in \Gamma_{\geq 1}$ and $a \in \Gamma_0$ then to simulate leaf compression on $S(u)$ and $S(v)$ it is enough to perform it on $u = v$, treating it as a tree. This procedure is called LeafComp$(\Gamma_{\geq 1}, \Gamma_0, 'u = v')$.

Lemma 5. LeafComp *is sound. If $u = v$ has a solution S such that there is no crossing father-leaf pair (f, a) with $f \in \Gamma_{\geq 1}$ and $a \in \Gamma_0$ in $u = v$ in S then it is complete: the returned equation $u' = v'$ has a solution S' such that $S'(u') =$ TreeLeafComp$(\Gamma_{\geq 1}, \Gamma_0, S(u))$.*

5   Uncrossing

One cannot assume that an arbitrary partition Γ_1, Γ_2 is noncrossing, nor that there are no crossing chains nor crossing father-leaf pairs. Still, for a fixed partition Γ_1, Γ_2 and a solution S we can modify the instance so that this partition becomes non-crossing in a solution S' (that corresponds to S of the original equation); similarly, given an equation $u = v$ we can turn it into an equation that has no letters with a crossing chain in a solution S'; lastly, for $\Gamma_{\geq 1}$ and Γ_0 we can modify the instance so that no father-leaf pair (f, a) with $f \in \Gamma_{\geq 1}$ and Γ_0 is crossing in S'.

Uncrossing Partitions. We first show how to turn a partition into a non-crossing one. For each of (CP1)–(CP3) we modify the instance so that ab is no longer a crossing pair:

- In (CP1) we *pop up* the letter b: we replace X (or x) with bX (bx, respectively). We also modify the solution $S(X)$ $(S(x))$ from $S(X) = bt$ $(S(x) = bt$, respectively) to $S'(X) = t$ $(S'(x) = t$, respectively). If $S'(X) = \Omega$, we remove X from the equation.
- In (CP2) we *pop down* the letter a: we replace each X with Xa. We also implicitly change $S(X) = sa\Omega$ to $S'(X) = s$. If $S'(X)$ is empty, we remove X.
- The case (CP3) is a combination of the two cases above, in which we need to pop-down from X and pop-up from Y (or y).

Popping can be performed on all $ab \in \Gamma_1\Gamma_2$ in parallel.

Lemma 6. *Let Γ_1, Γ_2 be disjoint. Then* Pop$(\Gamma_1, \Gamma_2, 'u = v')$ *is sound and complete; more precisely, if $u = v$ has a solution S then for appropriate non-deterministic choices the returned equation $u' = v'$ has a non-empty solution S' such that $S'(u') = S(u)$ and Γ_1, Γ_2 is a non-crossing partition in S'.*

Algorithm 3. Pop$(\Gamma_1, \Gamma_2, 'u = v')$

1: **for** $X \in \mathcal{V}$ **do**
2:     let a be the last letter of $S(X)$
3:     **if** $a \in \Gamma_1$ **then**
4:         replace each X by Xa
5:         **if** $S(X)$ is empty **then**
6:             replace each $X(s)$ by s
7: **for** $X \in \mathcal{V}$ or $x \in \mathcal{X}$ **do**
8:     Do a symmetric action for first letter replacing X (or x) with bX (or bx)

Uncrossing Chains. Suppose that some unary letter a has a crossing chain in a non-empty solution S. Recall that a has a crossing chain if and only if aa satisfies one of (CP1)–(CP3). Suppose that (CP2) holds. Then we can replace X with Xa throughout the equation $u = v$ (implicitly changing $S(X) = ta\Omega$ to $S(X) = t$) but it can still happen that a is the last letter of $S(X)$. So we keep popping down a until the last letter of $S(X)$ is not a, in other words we replace X with Xa^r, where $S(X) = ta^r\Omega$ and the last letter of t is not a. Then a and X no longer satisfy (CP2), as $S'(X)$ ends with a letter different than a. A symmetric action and analysis applies to (CP1), and (CP3) follows by applying the popping down for X and popping up for Y (or y). To simplify the description, for a ground term (or context) t we say that a^ℓ is the a-*prefix* of t if $t = a^\ell t'$ and the first letter of t' is not a (t' may be empty). Similarly, for a ground context t we say that b^r is a b-*suffix* of t if $t = t'b^r\Omega$ and the last letter of t' is not b (t' may be empty). CutPrefSuff(Γ_1, '$u = v$') pops up a-prefix and b-suffix down from each context variable and variable (where $a, b \in \Gamma_1$).

Lemma 7. CutPrefSuff(Γ_1, '$u = v$') *is sound and complete; more precisely, if* $u = v$ *has a non-empty solution S then for appropriate non-deterministic choices the returned $u' = v'$ has a solution S' such that $S'(u') = S(u)$ and there are no crossing chains in S'.*

Uncrossing Father-Leaf Pairs. Let (f, a) be a crossing father-leaf pair, this is because it satisfies one of (CFL1)–(CFL3).

- In (CFL1) we *pop up* the letter a from x: we replace each x with $a = S(x)$.
- In (CFL2) we *pop down* the letter f: let $S(X) = sf(t_1, \ldots, t_{i-1}, \Omega, t_{i+1}, \ldots, t_m)$. Then we replace each X with $Xf(x_1, \ldots, x_{i-1}, \Omega, x_{i+1}, \ldots, x_m)$, where $x_1, \ldots, x_{i-1}, x_{i+1}, \ldots, x_m$ are fresh variables. We also implicitly change $S(X) = sf(t_1, \ldots, t_{i-1}, \Omega, t_{i+1}, \ldots, t_m)$ to $S'(X) = s$ and add $S'(x_j) = t_j$ for $j = 1 \ldots, i-1, i+1, \ldots, m$. If $S'(X)$ is empty, we remove X from the equation.
- The third case (CFL3) is a combination of (CFL1)–(CFL2), in which we need to pop down from X and pop up from y.

This procedure is performed on all $f \in \Gamma_{\geq 1}$ and $a \in \Gamma_0$ in parallel.

Note that popping down last letters in $\Gamma_{\geq 1}$ from X is done only when it is needed: i.e. we want to make (f, i, a) leaf compression, f is the last letter of $S(X)$, its i-th child is Ω and Xa occurs in $u = v$.

Lemma 8. *Let $\Gamma_{\geq 1}$ be an alphabet without constants and Γ_0 alphabet of constants.*

Algorithm 4. GenPop($\Gamma_{\geq 1}, \Gamma_0$, '$u = v$')

1: **for** $x \in \mathcal{X}$ **do**
2:    **if** $S(x) \in \Gamma_0$ **then**
3:       replace each x in $u = v$ by $S(x)$
4: **for** $X \in \mathcal{V}$ **do**
5:    let f be the last letter of $S(X)$, $m \leftarrow \mathrm{ar}(f)$
6:    let Ω be i-th child of its father in $S(X)$
7:    **if** Xa occurs in $u = v$, $a \in \Gamma_0$, $f \in \Gamma_{\geq 1}$ **then**
8:       replace each X in $u = v$ by
      $Xf(x_1, \ldots, x_{i-1}, \Omega, x_{i+1}, \ldots, x_m)$
9:       **if** $S(X)$ is empty **then**
10:          replace each $X(u)$ by u
11: **for** new variables $x \in \mathcal{X}$ **do**
12:    **if** $S(x) \in \Gamma_0$ **then**
13:       replace each x in $u = v$ by $S(x)$

GenPop($\Gamma_{\geq 1}, \Gamma_0, \text{'}u = v\text{'}$) *is sound and complete, more precisely: if $u = v$ has a solution S then for appropriate non-deterministic choices the returned $u' = v'$ has a solution S' such that $S'(u') = S(u)$ and there is no crossing father-leaf pair (f, a) with $f \in \Gamma_{\geq 1}$ and $a \in \Gamma_0$ in S'.*

6 Main Algorithm

Our algorithm for testing the satisfiability of context equations works in *phases*, each of which is divided into two subphases. In each subphase we uncross all unary letters and then perform the chain compression; guess the Γ_1, Γ_2 partition of unary letters, uncross it and perform the Γ_1, Γ_2 compression and lastly remove all crossing father-leaf pairs and perform the leaf compression.

The first subphase ensures that the size of the (size-minimal) solution decreases by a constant factor (cf. Theorem 2), the second phase is used to make sure that the size of the equation is bounded (in some sense the second phase decreases the size of the equation, but as the equation grows in both subphases due to popping, in total we can only guarantee that the equation is of more or less the same size).

Algorithm 5. ContextEqSat($\text{'}u = v\text{'}, \Sigma$) Satisfiability of a context equation

---

1: **while** $|u| > 1$ or $|v| > 1$ **do**
2:     **for** $i \leftarrow 1..2$ **do**   ▷ One iteration to shorten the solution, one to the equation
3:         $\Gamma_1 \leftarrow$ unary letters in $u = v$ ▷ By Lemma 2
4:         CutPrefSuff($\Gamma_1, \text{'}u = v\text{'}$)   ▷ No letter has a crossing block
5:         ChainComp($\Gamma_1, \text{'}u = v\text{'}$)   ▷ Chain compression
6:         $\Gamma \leftarrow$ the set of unary in $u = v$ ▷ By Lemma 2
7:         guess partition of Γ into Γ_1 and Γ_2
8:         Pop($\Gamma_1, \Gamma_2, \text{'}u = v\text{'}$)   ▷ Γ_1, Γ_2 is a non-crossing partition
9:         PartitionComp($\Gamma_1, \Gamma_2, \text{'}u = v\text{'}$)   ▷ Γ_1, Γ_2 compression
10:        $\Gamma_0 \leftarrow$ constants in $\text{'}u = v\text{'}$ plus one fresh constant c ▷ By Lemma 2
11:        $\Gamma_{\geq 1} \leftarrow$ non-constants in $\text{'}u = v\text{'}$ plus one fresh letter f of arity 2
12:        GenPop($\Gamma_{\geq 1}, \Gamma_0, \text{'}u = v\text{'}$)   ▷ No crossing father-leaf pairs
13:        LeafComp($\Gamma_{\geq 1}, \Gamma_0, \text{'}u = v\text{'}$)   ▷ Leaf compression
14: Solve the problem naively   ▷ With sides of size 1, the problem is trivial

---

Lemma 9. ContextEqSat *is sound and complete, to be more precise for some nondeterministic choices the following conditions are satisfied:*

1. *kept equation has at most n context variables, kn variables and $\mathcal{O}(nk)$ letters;*
2. *if N is the size of the size-minimal solution at the beginning of the phase then at the end of the phase the equation has a solution of size at most $3N/4$;*
3. *the additional memory usage is at most $\mathcal{O}(k^2 n^2)$ (counted in bits);*
4. *the maximal arity of symbols in Σ does not increase during ContextEqSat.*

The intuition of the space bound is as follows: in the second subphase we treat the equation as a term and try to ensure that its size drops by one fourth, just as in the case of Theorem 2. However, in the meantime we also increased the size of the equation, as we pop the letters into the context equation (in both subphases). The number of those letters depends linearly on the number of occurrences of variables and context variables in $u = v$, which is known to be $\mathcal{O}(kn)$. Hence when $u = v$ and $u' = v'$ are the equation at the beginning and end of the phase, we have that $|u'| + |v'| \leq \frac{3}{4}(|u| + |v|) + ckn$ for some constant c. By a simple induction this shows that $|u| + |v| \leq 4ckn$.

References

1. Comon, H.: Completion of rewrite systems with membership constraints. Part I: Deduction rules. J. Symb. Comput. 25(4), 397–419 (1998)
2. Comon, H.: Completion of rewrite systems with membership constraints. Part II: Constraint solving. J. Symb. Comput. 25(4), 421–453 (1998)
3. Gascón, A., Godoy, G., Schmidt-Schauß, M., Tiwari, A.: Context unification with one context variable. J. Symb. Comput. 45(2), 173–193 (2010)
4. Jeż, A.: Recompression: a simple and powerful technique for word equations. In: Portier, N., Wilke, T. (eds.) STACS. LIPIcs, vol. 20, pp. 233–244. Schloss Dagstuhl–Leibniz-Zentrum für Informatik, Dagstuhl (2013)
5. Levy, J.: Linear second-order unification. In: Ganzinger, H. (ed.) RTA 1996. LNCS, vol. 1103, pp. 332–346. Springer, Heidelberg (1996)
6. Levy, J., Schmidt-Schauß, M., Villaret, M.: On the complexity of bounded second-order unification and stratified context unification. Logic Journal of the IGPL 19(6), 763–789 (2011)
7. Levy, J., Villaret, M.: Currying second-order unification problems. In: Tison, S. (ed.) RTA 2002. LNCS, vol. 2378, pp. 326–339. Springer, Heidelberg (2002)
8. Plandowski, W.: Satisfiability of word equations with constants is in PSPACE. J. ACM 51(3), 483–496 (2004)
9. Plandowski, W., Rytter, W.: Application of lempel-ziv encodings to the solution of word equations. In: Larsen, K.G., Skyum, S., Winskel, G. (eds.) ICALP 1998. LNCS, vol. 1443, pp. 731–742. Springer, Heidelberg (1998)
10. Schmidt-Schauß, M.: Unification of stratified second-order terms, internal Report 12/94, Johann-Wolfgang-Goethe-Universität (1994)
11. Schmidt-Schauß, M.: A decision algorithm for stratified context unification. J. Log. Comput. 12(6), 929–953 (2002)
12. Schmidt-Schauß, M., Schulz, K.U.: On the exponent of periodicity of minimal solutions of context equations. In: Nipkow, T. (ed.) RTA 1998. LNCS, vol. 1379, pp. 61–75. Springer, Heidelberg (1998)
13. Schmidt-Schauß, M., Schulz, K.U.: Solvability of context equations with two context variables is decidable. J. Symb. Comput. 33(1), 77–122 (2002)
14. Schmidt-Schauß, M., Schulz, K.U.: Decidability of bounded higher-order unification. J. Symb. Comput. 40(2), 905–954 (2005)

Monodic Fragments of Probabilistic First-Order Logic

Jean Christoph Jung[1], Carsten Lutz[1], Sergey Goncharov[2], and Lutz Schröder[2]

[1] Universität Bremen, Germany
{jeanjung,clu}@informatik.uni-bremen.de
[2] Friedrich-Alexander-Universität Erlangen-Nürnberg, Germany
{Sergey.Goncharov,Lutz.Schroeder}@fau.de

Abstract. By classical results of Abadi and Halpern, validity for probabilistic first-order logic of type 2 (ProbFO) is Π_1^2-complete and thus not recursively enumerable, and even small fragments of ProbFO are undecidable. In temporal first-order logic, which has similar computational properties, these problems have been addressed by imposing *monodicity*, that is, by allowing temporal operators to be applied only to formulas with at most one free variable. In this paper, we identify a monodic fragment of ProbFO and show that it enjoys favorable computational properties. Specifically, the valid sentences of monodic ProbFO are recursively enumerable and a slight variation of Halpern's axiom system for type-2 ProbFO on bounded domains is sound and complete for monodic ProbFO. Moreover, decidability can be obtained by restricting the FO part of monodic ProbFO to any decidable FO fragment. In some cases, which notably include the guarded fragment, our general constructions result in tight complexity bounds.

1 Introduction

Both logic and probability theory are fundamental to the formalization and solution of many important problems in computer science. While logic is a way to address the combinatorics hidden in such problems, the main use of probabilities is to capture uncertainty that arises from many different sources such as noisy or untrusted data (in database systems), a high level of abstraction (in verification), or incomplete training data (in machine learning). Unfortunately, the combination of logic and probability is notoriously difficult and involves a large number of choices and trade-offs, which has resulted in a broad spectrum of probabilistic logic formalisms to be proposed that vary greatly in spirit, semantics, and expressive power.

A natural and fundamental way to combine logic and probabilities is to enrich classical first-order logic (FO) with a probabilistic component [12,4,5]. Although reasoning in the resulting probabilistic FO logics is, of course, undecidable, they are still useful as a general and uniform 'baseline formalism' that encompasses many other probabilistic logics, much in the same way that FO provides a baseline formalism for many other logics used in computer science. However, it turns out that probabilistic FO logics are not only undecidable, but tend to be computationally even less well-behaved than classical FO. They come in essentially two versions, called type-1 and type-2 [12]. While type-1 is for reasoning about statistical probabilities, reflected in the semantics by a probability distribution over the domain of the FO structure, the purpose of type-2 is

J. Esparza et al. (Eds.): ICALP 2014, Part II, LNCS 8573, pp. 256–267, 2014.

reasoning about subjective probabilities by adopting a possible worlds semantics. In this paper, we concentrate on the latter and use 'ProbFO' to refer to Halpern's probabilistic FO logic of type-2 [12]. The disastrous computational behaviour of ProbFO was analyzed by Abadi and Halpern, who show that validity is Π_1^2-complete [1], thus outside the arithmetic and analytic hierarchies and, in particular, far from being recursively enumerable. This result holds up even when only unary predicates are admitted, and we add in this work the observation that ProbFO is still Π_1^1-hard even with only two (object) variables (and no quantification over real-valued variables, see below).

Our aim in the current work is to analyze how and how far the problematic computational properties of ProbFO can be improved. We start by observing that there is a clear semantic and computational similarity between ProbFO and temporal first-order logic (TFO). Both logics use a possible worlds semantics, and although TFO is only Π_1^1-complete, just like ProbFO it is not recursively enumerable. In the case of TFO, Hodkinson, Wolter and Zakharyaschev have given an elegant explanation of why this is the case and how better computational properties can be recovered, by introducing the *monodic* fragment of TFO that restricts temporal operators to be applied only to formulas with at most one free variable [17]. In fact, monodic TFO turns out to be recursively enumerable [23] and decidable fragments of monodic TFO can often be obtained by restricting the FO part of monodic TFO to a decidable FO fragment [18,14,16,15]. In the present work, we identify a monodic fragment of ProbFO and show that, as in the case of TFO, this recovers good computational properties. Note that the formulas of unrestricted ProbFO are obtained by combining classical FO with the language of real closed fields (including quantification over real numbers) via real-valued terms of the form $w(\varphi)$ denoting the probability that the formula φ is true. Atomic formulas in this extended language of real closed fields are called *weight formulas*. In analogy to TFO, a natural candidate for monodicity in ProbFO is to admit only weight terms $w(\varphi)$ in which φ has at most one free first-order variable. We show that this is not an effective choice since the resulting fragment of ProbFO still fails to be recursively enumerable.

We thus have to adopt stronger restrictions and define a ProbFO formula to be monodic if it contains no variables for real numbers (thus no quantification over the reals) and every weight formula in it contains at most one free (object) variable. Under this definition, we can establish a useful abstract representation of models of monodic ProbFO formulas – so-called *quasi-models* – which are essentially a collection of monadic formula types that satisfy certain integrity conditions and are associated with a system of polynomial inequalities over the reals to capture probabilities. This representation yields in a rather direct way that monodic ProbFO is recursively enumerable. Moreover, we exploit quasi-models to establish a concrete axiomatization of monodic ProbFO, a variation of a complete axiomatization of unrestricted ProbFO on finite domains of fixed size by Halpern [12] (we use unrestricted domains). Finally, quasi-models can be used to identify decidable fragments of monodic ProbFO. We show that for any FO-fragment $\mathcal{L}$ such that a slightly generalized version of satisfiability in $\mathcal{L}$ called *realizability* is decidable, monodic Prob$\mathcal{L}$ is decidable, too. In particular, we thus obtain decidability for the case where $\mathcal{L}$ is the monadic fragment of FO, the guarded fragment (GF), the two-variable fragment, and the guarded negation fragment. The finite model property transfers in the same way.

We also analyze the computational complexity of some important decidable fragments of monodic ProbFO. The naive version of our general algorithm yields a $2\text{NEXPTIME}^{\exists\mathbb{R},\mathcal{C}}$ upper bound where superscripts denote access to oracles, $\exists\mathbb{R}$ is the class of problems that reduce in polynomial time to solving systems of polynomial inequalities over the reals (recall $\text{NP} \subseteq \exists\mathbb{R} \subseteq \text{PSPACE}$), and $\mathcal{C}$ is the complexity of deciding realizability in the underlying FO fragment $\mathcal{L}$. We then propose two improvements to our algorithm. The first one consists of a more careful realizability check as known from monodic TFO, and this modification sometimes allows removing the oracle for $\mathcal{C}$. For monodic ProbGF, in particular, we obtain in this way an improved $2\text{NEXPTIME}^{\exists\mathbb{R}}$ upper bound. The second improvement applies only when $\mathcal{L}$ satisfies a certain model-theoretic property that we call *closedness under unions of types*, and it allows improving the runtime by one exponential by reducing the size of quasi-models. GF satisfies the mentioned property, and thus we obtain a tight 2EXPTIME upper bound for monodic ProbGF. We also obtain a $\text{NEXPTIME}^{\exists\mathbb{R}}$ upper bound when the arity of predicates is bounded, and a tight NEXPTIME upper bound for the case where only linear weight formulas are admitted, that is, multiplication of probabilities is disallowed. Note that the relatively high computational complexities are partly due to the fact that we aim at identifying *maximal* decidable fragments of monodic ProbFO. In fact, monodic ProbFO can be viewed as a natural generalization of the family of probabilistic description logics introduced in [20,11] and provides a principled explanation for why these logics are computationally much more well-behaved than traditional ProbFO. Conversely, the mentioned description logics can be viewed as fragments of monodic ProbFO with lower computational complexity, typically EXPTIME-complete.

Proofs are generally omitted or only sketched; a full version is available at http://www.informatik.uni-bremen.de/tdki/research/papers.html.

2   Preliminaries

Type-2 probabilistic first-order logic (ProbFO) [12] comprises two sorts: objects of the domain of discourse and the real numbers $\mathbb{R}$. Accordingly, there are *object variables* and *field variables*, the latter being used to represent probabilities. *Object terms* are object variables or *object constants*. *ProbFO-formulas* and *field terms* are defined by mutual recursion:

$$\varphi, \psi ::= R(t_1, \ldots, t_k) \mid \varphi \wedge \psi \mid \neg\varphi \mid \forall x\, \varphi(x) \mid f_1 \le f_2$$
$$f_1, f_2 ::= 0 \mid 1 \mid \mathsf{w}(\varphi) \mid f_1 + f_2 \mid f_1 \times f_2$$

where R is a k-ary predicate symbol, $t_1, \ldots, t_k$ are object terms, and formulas of the form $f_1 \le f_2$ are called *weight formulas*. Quantification is possible both over object and field variables, with field variables ranging over $\mathbb{R}$. We use $\text{ProbFO}^=$ to denote the extension of ProbFO with equality on object terms. We could admit rational constants in field terms to represent concrete probabilities, but as usual we refrain from doing so because rational constants can be eliminated by clearing denominators.

Formulas of ProbFO are interpreted in *probabilistic structures* $\mathfrak{M} = (D, W, \mu, \pi)$ that consist of a non-empty domain D, a set of worlds W, a discrete probability distribution μ over W and a function π that maps each pair (R, w) to a subset of D^k and each pair (c, w) to an element of D for each k-ary predicate symbol R, $w \in W$, and constant symbol c. Intuitively, $\mathfrak{M}$ can be viewed as a set of classical FO structures (over the same domain) with weights given by μ. A *valuation* for $\mathfrak{M}$ is a function ν that maps object variables to elements of D and field variables to real numbers. Given $\mathfrak{M}$, ν, and a world $w \in W$, the interpretation $[f]_{(\mathfrak{M},w,\nu)} \in \mathbb{R}$ of a field term f is defined in the natural way, with terms $\mathsf{w}(\varphi)$ interpreted as $[\mathsf{w}(\varphi)]_{(\mathfrak{M},w,\nu)} = \mu(\{w' \in W \mid (\mathfrak{M}, w', \nu) \models \varphi\})$. The semantics of formulas is standard. A ProbFO-sentence φ is *satisfiable* if there is a probabilistic structure $\mathfrak{M} = (D, W, \mu, \pi)$ and a world $w \in W$ such that $(\mathfrak{M}, w) \models \varphi$. A sentence φ is *valid* if $\neg\varphi$ is not satisfiable.

When we speak of (non-probabilistic) first-order logic (FO), we mean the FO fragment of ProbFO as introduced above. In particular, we mean FO without equality unless we write FO$^=$. A classical FO-structure has the form $\mathfrak{A} = (A, \pi)$ where A is a domain and π is a function as above except that its second argument (the world) is omitted.

3 Monodic ProbFO

Abadi and Halpern have shown that validity in ProbFO is Π_1^2-complete, and thus highly undecidable and far from being recursively enumerable [1]. They also show that already over vocabularies that contain *only constants*, validity is Π_∞^1-complete when equality is allowed. The lower bounds of these theorems are proved by reductions of suitable higher-order theories of integer arithmetics. We give additional evidence of the computational difficulty of ProbFO by proving (in the full version) the following orthogonal result by a reduction of recurring domino systems that is rather different in spirit from the mentioned reductions from integer arithmetic.

Theorem 1. *Validity in ProbFO is Π_1^1-hard even if quantification over field variables is disallowed and only two object variables are admitted.*

The mentioned previous results and Theorem 1 illustrate that several restrictions of ProbFO that might at first sight seem promising fail to improve the computational properties of this logic. Inspired by the good computational properties of *monodic fragments* of temporal first-order logic [17,23], we aim to define monodic fragments of ProbFO that are computationally well-behaved. In the context of temporal first-order logic, a formula is monodic when temporal operators are applied only to formulas with at most one free variable. We first show that one has to be careful when adapting this notion to ProbFO; the following result is proved by a reduction of finite validity in FO.

Theorem 2. *Validity in ProbFO is Π_1^0-hard even if only one free object variable is allowed to occur in weight formulas.*

Although a natural candidate for monodicity, the restriction formulated in Theorem 2 is thus not strong enough to regain recursive enumerability. Intuitively, this is because it is still possible to compare the probabilities of different domain elements such as in the formula $\forall x \forall r\, (\mathsf{w}(A(x)) = r \Rightarrow \exists y\, \mathsf{w}(A(y)) = r/2)$, which says that for each object x, there is an object y that has half the probability of satisfying A. To avoid this, we require a weight formula with a free object variable to have *no* other free variables (object or field). This restriction makes field variables and quantification over them mostly useless, so we disallow them altogether.

Definition 3 (Monodic ProbFO Formula). A ProbFO formula is *monodic* if it contains no field variables and every weight formula contains at most one free (object) variable.

We will see that the above definition of monodicity indeed guarantees good computational properties such as recursive enumerability of validity. In the balance, of course, we lose some expressive power, in particular the ability to relate different domain elements in terms of their probabilities. The following proposition gives explicit examples of ProbFO-formulas that cannot be expressed in monodic ProbFO. Its proof relies on Theorem 7 below, which states that every satisfiable monodic ProbFO sentence is satisfiable in a model with only finitely many worlds. In the full version, we show how to enforce infinitely many worlds using the formulas in Proposition 4.

Proposition 4. *The following formulas are not expressible in monodic ProbFO:*

1. $\mathsf{w}(P(x,y)) \sim p$ *with* P *binary,* $p \in (0,1)$*, and* $\sim \in \{<, \leq, =, \geq, >\}$*;*
2. $\mathsf{w}(A(x)) > \mathsf{w}(A(y))$ *with* A *unary.*

Formulas as in Item 1 can be used to express that any two persons who show up at a party together and both wear rings are probably married; with a formula as in Item 2, we could say that children are more likely to use a smartphone than their parents. Note that formulas such as $\mathsf{w}(\exists y\, P(x,y)) \sim p$ and $\mathsf{w}(\forall y\, P(x,y)) \sim p$, which are similar to the formulas in Item 1 but only have one free variable, *do* fall within monodic ProbFO.

The following theorem illustrates that the positive results for monodic ProbFO rely on disallowing equality; it is again proved by reduction of finite validity in FO.

Theorem 5. *Validity in monodic ProbFO$^=$ is* Π_1^0*-hard.*

4 The Quasi-Model Machinery

We introduce quasi-models, an abstraction of probabilistic structures that underlies the proofs of all positive results established in this paper. This requires some preliminaries. In the following, fix a monodic ProbFO-sentence φ_0. We denote by $\mathsf{sub}(\varphi_0)$ the set of all subformulas of φ_0 and their negations, and by $\mathsf{sub}_n(\varphi_0)$ the formulas from $\mathsf{sub}(\varphi_0)$ with precisely n free variables, for $n \in \{0, 1\}$. By $\mathsf{con}(\varphi_0)$, we denote the set of all constant symbols that occur in φ_0. Reflecting monodicity, we concentrate on formulas with at most one free variable when defining quasi-models. In particular, these formulas are from the following set, where x is a distinguished variable:

$$\mathsf{sub}_x(\varphi_0) = \mathsf{sub}_0(\varphi_0) \cup \{\psi(x), \psi(c) \mid \psi(y) \in \mathsf{sub}_1(\varphi_0), c \in \mathsf{con}(\varphi_0)\}.$$

We introduce a way to represent ProbFO formulas as FO formulas by replacing weight formulas with new predicates. Introduce a fresh nullary predicate symbol P_ψ for every weight formula $\psi \in \mathsf{sub}_0(\varphi_0)$ and a fresh unary predicate symbol P_ψ for every weight formula $\psi \in \mathsf{sub}_1(\varphi_0)$. Denote by $\overline{\varphi}$ the FO formula that is obtained from the ProbFO formula φ by replacing each weight formula $\psi()$ (resp. $\psi(x)$) that is not within the scope of another weight formula with $P_\psi()$ (resp. $P_\psi(x)$). This notation is lifted to sets of formulas in the obvious way.

A *type* is a subset t of $\mathsf{sub}_x(\varphi_0)$ such that the set of FO formulas $\overline{t}$ is a maximal satisfiable subset of $\overline{\mathsf{sub}_x(\varphi_0)}$. Intuitively, a type is a set of FO formulas with one free variable that are satisfied by a domain element in a world of a probabilistic structure; it also records the sentences true in that world, including the FO formulas with one free variable that are satisfied by constants. Two types t_1, t_2 *agree on sentences*, written $t_1 \equiv_0 t_2$, if for all sentences $\psi \in \text{sub}_x(\varphi_0)$, we have $\psi \in t_1$ iff $\psi \in t_2$.

A *world type* is a set of types that agree on sentences; it can be viewed as an abstract representation of a world in a probabilistic structure, that is, of an FO structure. For an FO structure $\mathfrak{A} = (A, \pi)$ and an element $d \in A$, define

$$\text{tp}(\mathfrak{A}, d) = \{\psi \in \text{sub}_x(\varphi_0) \mid \mathfrak{A} \models \overline{\psi}[d]\} \quad \text{and} \quad \text{tp}(\mathfrak{A}) = \{\text{tp}(\mathfrak{A}, d) \mid d \in A\}.$$

Note that $\text{tp}(\mathfrak{A}, d)$ is a type and $\text{tp}(\mathfrak{A})$ is a world type. A world type T is *realizable* if there is an FO structure $\mathfrak{A}$ such that $\text{tp}(\mathfrak{A}) = T$, that is, if the FO formula $\overline{\chi}(T)$ is satisfiable, where we define

$$\chi(T) = \bigwedge_{t \in T} \exists x \bigwedge t(x) \wedge \forall x \bigvee_{t \in T} \bigwedge t(x).$$

World types will play a central role in the definition of quasi-models, but need to be suitably enriched with (i) runs that describe the types of a single domain element in *all* worlds of a probabilistic structure and (ii) relevant conditions that have to be satisfied by the probabilities of worlds. Note that runs and world types in a sense represent orthogonal dimensions. Let Q be a set of world types. A *run through* Q is a function r that assigns to each world type $T \in Q$ a non-empty set $r(T) \subseteq T$ and is *coherent*, that is, whenever some $t \in r(T)$ contains a weight formula θ, then for all $T' \in Q$ and $t' \in r(T')$, we have $\theta \in t'$. Coherence allows us to write $\theta \in r$ to denote that for all (equivalently: some) $T \in Q$ and $t \in r(T)$, we have $\theta \in t$. A run selects a set of types for each world type instead of only a single type because each world type can represent several actual worlds, and an element might have different types in each of these worlds. A *quasi-model candidate* is a triple (T_0, Q, R) with T_0 a world type, Q a set of world types, and R a set of runs through $Q \cup \{T_0\}$ such that for all $T \in Q \cup \{T_0\}$ and $t \in T$, there is a run $r \in R$ with $t \in r(T)$. Intuitively, T_0 describes a (single) world of probability 0 while each $T \in Q$ describes worlds of positive probability. To address Point (ii) above and obtain our final quasi-model representation, we augment quasi-model candidates with a system of polynomial inequalities. It uses a variable x_T for each world type T to represent the probability of T (obtained by summing up the probabilities of all worlds of world type T) and a variable $x_{r,t,T}$ for each run r, world type T, and type $t \in T$ to describe the (summed up) probability of those worlds of world type T in which the element described by run r has type t.

Definition 6 (Quasi-Model). A quasi-model candidate (T_0, Q, R) is a *quasi-model* if every $T \in Q \cup \{T_0\}$ is realizable and the following system of polynomial inequalities $\mathcal{E}(Q, R)$ has a *positive* solution over the reals:

1. distribution on world types: $\sum_{T \in Q} x_T = 1$;

2. the probabilities of the types associated by a run $r \in R$ to a quasi-world $T \in Q$ sum up to the probability of T: $x_T = \sum_{t \in r(T)} x_{r,t,T}$;

3. runs respect weight formulas, that is, for all $f_1 \sim f_2 \in r$ with $\sim \in \{\leq, >\}$[1] we include an equation $[f_1]_r \sim [f_2]_r$ where $[f]_r$ is obtained from f by replacing each outermost term $w(\psi(x))$ with the following expression describing its probability:

$$\sum_{T \in Q} \sum_{t \in r(T), \psi(x) \in t} x_{r,t,T}.$$

---

[1] We write $f_1 > f_2 \in r$ in place of $f_1 \leq f_2 \notin r$.

Note that the field terms f_1, f_2 in Item 3 of Definition 6 can contain addition and multiplication, thus the system $\mathcal{E}(Q, R)$ need not be linear.

We say that a quasi-model candidate (or quasi model) (T_0, Q, R) *satisfies* a ProbFO sentence φ_0 if $\varphi_0 \in t$ for some $t \in T_0$. The following provides the basis for our use of quasi-models in subsequent sections.

Theorem 7. *A monodic ProbFO sentence φ_0 is satisfiable iff it is satisfied in some quasi-model. Moreover, any satisfiable monodic ProbFO sentence is satisfied in a probabilistic structure with finitely many worlds.*

In the "$\Rightarrow$" direction, we read off a quasi-model satisfying φ_0 from a probabilistic structure that satisfies φ_0. To show that the system $\mathcal{E}(Q, R)$ has a solution, the values for the variables x_T and $x_{r,t,T}$ are also read off in a straightforward way.

The "$\Leftarrow$" direction is more interesting. Let (T_0, Q, R) be a quasi-model that satisfies φ_0. Hence, every $T \in Q \cup \{T_0\}$ is realizable and $\mathcal{E}(Q, R)$ has a positive solution; we use x_T^* to denote the value of x_T in this solution and likewise for $x_{r,t,T}^*$. To construct a probabilistic structure $\mathfrak{M}$ that satisfies φ_0, it would be convenient to use the world types in Q as worlds. Since runs can associate more than one type with a world type, though, this is not sufficient. We thus need to subdivide each $T \in Q$ into several worlds, each accommodating a *single* type that a given run assigns to T. This has to be done in a careful way since we have to do this simultaneously for *all runs* while also ensuring that all types in T are realized in each of the worlds that T is subdivided into.

Let $r \in R$ and $T \in Q$. A *subdivision of T for r* is a tuple $s = (b_1, \ldots, b_n, \zeta)$ such that $b_1 < b_2 < \cdots < b_n = x_T^*$, $n = |r(T)| + 1$, and ζ is a surjective function that assigns to every b_i a type $\zeta(b_i) \in r(T)$ such that for all $t \in r(T)$ we have $\sum_{i \in [1,n], \zeta(b_i) = t}(b_i - b_{i-1}) = x_{r,t,T}^*$ where, here and in what follows, $b_0 := 0$. Intuitively, the interval $[0, x_T^*]$ represents the probability covered by all worlds of type T and we subdivide this range into the intervals $(b_i, b_{i+1}]$, with $i < n$. Elements described by the run r then have type $\zeta(b_{i+1})$ in the interval $(b_i, b_{i+1}]$. For easier reference, we say for all $p \in (0, x_T^*]$ that s *has type t at p* if $\zeta(b_i) = t$ and $p \in (b_{i-1}, b_i]$. A *subdivided run* is a pair (r, S) with r a run through Q and S a function that assigns to every $T \in Q$ a subdivision $S(T)$ of T for r. If we had only the single run r, we could use the subintervals identified by a subdivided run (r, S) as worlds. Since this is not the case, we first identify a sufficiently rich set of subdivisions which we then combine into a finer 'overall' subdivision: in the full version, we show how to define a finite set Γ of subdivided runs such that

$(*)$ for all $T \in Q$, $t \in T$, and $p \in (0, x_T^*]$, there is some $(r, S) \in \Gamma$ such that $S(T)$ has type t at p.

To define the worlds for a world type T, let $z_1 < \cdots < z_m$ be all numbers that occur in a subdivision for T in (a subdivided run from) Γ. We introduce one world of type T for every z_i and assign to it the probability $z_i - z_{i-1}$, with $z_0 := 0$. Note that the probabilities of all worlds for T sum up to x_T^*. Doing this for all world types T (and adding one world with world type T_0 and probability 0) gives us the set of worlds W for the desired probabilistic structure $\mathfrak{M}$ along with their probabilities $\mu(w)$. Since every $T \in Q \cup \{T_0\}$ is realizable, we find for each $w \in W$ an FO structure $\mathfrak{A}_w$ that realizes the world type T associated with w. The domain of $\mathfrak{M}$ is the disjoint union of the domains of all these $\mathfrak{A}_w$ (recall that we do not allow equality), and the further construction of

$\mathfrak{M}$ is detailed in the full version. Notably, $(*)$ guarantees that every $t \in T$ is realized in every world w associated with world type T.

5 Recursive Enumerability and Axiomatization

We now show that the set of valid monodic ProbFO sentences is recursively enumerable and also provide a concrete axiomatization. For the former, it suffices to provide a semi-decision procedure for unsatisfiability, based on Theorem 7. The crucial observation is that, for any input sentence φ_0, the number of quasi-model candidates (T_0, Q, R) that satisfy φ_0 is bounded. It is thus possible to construct all quasi-model candidates that satisfy φ_0 and then eliminate those that do not satisfy the system of polynomial inequalities $\mathcal{E}(Q, R)$ from Definition 6. Then, enumerate all unsatisfiable FO formulas. For each such formula ψ, eliminate all quasi-model candidates (T_0, Q, R) such that $\overline{\chi}(T) = \psi$ for some $T \in Q \cup \{T_0\}$ (since T is not realizable, (T_0, Q, R) cannot be a quasi-model). Once all quasi-model candidates have been eliminated, return with 'φ_0 is unsatisfiable'.

Theorem 8. *The set of valid monodic ProbFO sentences is recursively enumerable.*

Halpern gives an axiomatization of ProbFO for the case where probabilistic structures are restricted to a domain of bounded size [12]. We propose a variation of this axiomatization that is sound and complete for monodic ProbFO (without assuming bounded domains). Let AX_2 be the set of the following axioms:

- PC: an axiomatization of FO [7];[2]
- OF: all instances of the axioms of ordered fields (formulated in terms of $\leq$) that are well-formed formulas in monodic ProbFO;
- PW_1: $\varphi \Rightarrow (\mathsf{w}(\varphi) = 1)$ if all occurrences of predicate symbols in φ are inside the scope of $\mathsf{w}()$;
- PW_2: $\mathsf{w}(\varphi) \geq 0$;
- PW_3: $\mathsf{w}(\varphi \wedge \psi) + \mathsf{w}(\varphi \wedge \neg\psi) = \mathsf{w}(\varphi)$;
- PW_4: $\mathsf{w}(\exists x\, \varphi(x)) > 0 \Rightarrow \exists x\, \mathsf{w}(\varphi(x)) > 0$;
- RPW: from $\varphi \equiv \psi$ infer $\mathsf{w}(\varphi) = \mathsf{w}(\psi)$.

In comparison to Halpern's axiomatization, we have removed the axiom FIN_N for bounded domains of size N and added axiom PW_4. This axiom follows from Halpern's axiomatization, but is independent of the axioms that remain when FIN_N is removed – in a nutshell, its soundness over discrete measures depends on σ-additivity, while PW_3 captures only finite additivity. Moreover, as we exclude field variables, we no longer need the full axiomatization of real-closed fields but, by the Artin-Schreier Theorem [3], can make do with the axioms of ordered fields. These can be phrased as quantifier-free open formulas (e.g. $x \geq 0 \vee -x \geq 0$) and hence can be instantiated to monodic ProbFO formulas (by replacing real variables with weight terms, observing the monodicity restriction).

---

[2] Since constants can be interpreted differently in different worlds, a slight adaptation of the definition of when a term t is *substitutable* for x in the axiom $\forall x\, \varphi \Rightarrow \varphi(x/t)$ is necessary [12].

Theorem 9. AX_2 *axiomatizes validity in monodic ProbFO.*

Soundness is proved essentially as in [12]. For showing completeness, we make use of Theorem 7. We use $AX_2 \vdash \varphi$ to denote that φ can be derived in AX_2, and call a sentence φ *consistent* if $AX_2 \vdash \neg\varphi$ does not hold. By Theorem 7, it suffices to show that if a monodic ProbFO sentence φ_0 is consistent, then there is a quasi-model that satisfies φ_0. The strategy is to use consistency of φ_0 to derive a consistent sentence φ' that describes a quasi-model that satisfies φ_0. The general structure of φ' is

$$\chi(T_0) \wedge \sum_{T \in Q} \mathsf{w}(\chi(T)) = 1 \wedge \bigwedge_{T \in Q} \mathsf{w}(\chi(T)) > 0 \wedge \Psi(x_1, \dots, x_k)$$

where T_0 is a world type that contains some t with $\varphi_0 \in t$, Q is a set of world types, $\Psi(x_1, \dots, x_k)$ is a quantifier-free formula in which each free variable x_i identifies a run r_i through $Q \cup \{T_0\}$, and Ψ is a conjunction of weight formulas with weight terms of the form $\mathsf{w}(\chi(T))$ and $\mathsf{w}(\chi(T) \wedge t(x_i))$ for some $t \in T$. Intuitively, these weight terms correspond to the variables x_T and $x_{r_i,t,T}$, respectively, in $\mathcal{E}(Q, R)$; moreover, $\Psi(x_1, \dots, x_k)$ describes precisely $\mathcal{E}(Q, R)$ under this correspondence. Observe that φ' is consistent relative to OF and thus has a solution which is also a solution to $\mathcal{E}(Q, R)$.

6 Decidability and Complexity

Theorem 7 reduces satisfiability in monodic ProbFO to satisfiability in FO and solvability of systems of polynomial inequalities over the reals. In the following, we use this observation to establish decidability results for fragments of monodic ProbFO that are obtained by restricting its FO part to a decidable FO fragment such as the guarded fragment or the two-variable fragment. We also derive complexity results, which in some cases are tight. For a fragment $\mathcal{L}$ of FO, *monodic Prob$\mathcal{L}$* is the fragment of monodic ProbFO that consists of all formulas φ such that, for all $\psi \in \mathrm{sub}(\varphi)$, the FO formula $\overline{\psi}$ belongs to $\mathcal{L}$. To warm up, we start with considering the finite model property (FMP). Recall that, by Theorem 7, even full monodic ProbFO has the FMP regarding the number of worlds. Here, we thus mean the number of domain elements.

Theorem 10. *For an FO fragment $\mathcal{L}$, monodic Prob$\mathcal{L}$ has the FMP iff $\mathcal{L}$ has the FMP.*

Theorem 10 is a direct consequence of the proof of Theorem 7. In the "if"-direction of that proof, we combine FO structures that witness realizability of world types. If $\mathcal{L}$ has the finite model property, we can choose these structures to be finite. Then, the resulting probabilistic structure is also finite.

Based on quasi-models, transfer of decidability is also easy to establish. We say that *realizability is decidable in $\mathcal{L}$* if it is decidable whether a given world type T formulated in monodic Prob$\mathcal{L}$ is realizable, that is, whether the $\mathcal{L}$ formula $\overline{\chi}(T)$ is satisfiable.

Theorem 11. *If realizability is decidable in the FO fragment $\mathcal{L}$, then so is satisfiability in monodic Prob$\mathcal{L}$.*

Theorem 11 is established by the following algorithm which decides satisfiability of a given Prob$\mathcal{L}$ sentence φ_0:

1. guess a quasi-model candidate (T_0, Q, R) that satisfies φ_0;
2. verify that the system $\mathcal{E}(Q, R)$ has a positive solution in $\mathbb{R}$;
3. verify that each world type $T \in Q \cup \{T_0\}$ is realizable.

Step 1 is effective since the size of quasi-model candidates is bounded by a computable function in the size of φ_0 (analyzed in more detail below).

Theorem 11 applies for instance to the monadic fragment of FO (MonaFO), the guarded fragment (GF) [2], the guarded negation fragment (GNFO) [6], and the two-variable fragment FO_2 [10]: In all these cases, the formulas $\overline{\chi}(T)$ for checking realizability remain within the fragment, and satisfiability in all the mentioned fragments is decidable.

Corollary 12. *Let $\mathcal{L}$ be one of MonaFO, GF, GNFO, FO_2. Then satisfiability in monodic Prob$\mathcal{L}$ is decidable.*

To analyze the complexity of the algorithm from the proof of Theorem 11, first note that it suffices to guess a quasi-model candidate (T_0, Q, R) of size at most double exponential in the size of φ_0. In fact, Q contains at most double exponentially many world types T, and each T contains at most exponentially many types. While R can in principle be larger than double exponential, it suffices to include one run r for each $T \in Q \cup \{T_0\}$ and $t \in T$, such that $t \in r(T)$. Considering for example GF in which satisfiability is 2ExpTime-complete, we thus obtain a 2NExpTime$^{\exists\mathbb{R},2\text{ExpTime}}$ upper bound for satisfiability in monodic ProbGF where the superscripts indicate access to two oracles: one for solving systems of polynomial inequalities over the reals and one for realizability in GF. Recall that $\exists\mathbb{R}$ denotes the class of all problems that are reducible in polynomial time to solving the mentioned systems [22], and that $NP \subseteq \exists\mathbb{R} \subseteq PSPACE$.

For many FO fragments $\mathcal{L}$, though, we can improve on the upper bounds obtained in this direct way. First, it is helpful to not consider satisfiability of the exponential size realizability formula $\overline{\chi}(T)$ as a black box. In particular, the regular structure of $\overline{\chi}(T)$ implies that its satisfiability can be decided in time double exponential in the size of φ_0 for GF and in space exponential in the size of φ_0 for both MonaFO and FO_2 [16]. This yields a 2NExpTime$^{\exists\mathbb{R}}$ upper bound for monodic ProbGF, monodic ProbMonaFO, and monodic ProbFO_2. Second, for some FO fragments $\mathcal{L}$ the quasi-model machinery can be refined so that each quasi-model candidate has at most exponential size. The following is a sufficient condition for when this is possible.

Definition 13. *An FO fragment $\mathcal{L}$ is closed under unions of types if for each $\mathcal{L}$-sentence ψ and any two structures $\mathfrak{A}_1$ and $\mathfrak{A}_2$ that satisfy the same sentences from $sub_x(\psi)$, there is a structure $\mathfrak{B}$ such that $tp(\mathfrak{B}) = tp(\mathfrak{A}_1) \cup tp(\mathfrak{A}_2)$.*

For GF without constant symbols, closure under unions of types can be shown easily by taking disjoint unions. The following result is proved in the full version.

Theorem 14. *If $\mathcal{L}$ is closed under unions of types, then for every satisfiable monodic Prob$\mathcal{L}$ sentence φ_0, there is a quasi-model (T_0, Q, R) that satisfies φ_0 and in which no two distinct world types agree on sentences.*

As a consequence of Theorem 14, we obtain the following improved complexity bounds for monodic ProbGF.

Corollary 15. *Satisfiability in monodic ProbGF is*

(a) 2ExpTime-*complete;*
(b) *in* NExpTime$^{\exists\mathbb{R}}$ *when the arity of predicates is bounded;*
(c) NExpTime-*complete when additionally only linear weight formulas are allowed.*

For part (a), where our general machinery even yields a tight upper bound, it suffices to guess a quasi-model candidate (T_0, Q, R) of exponential size (Theorem 14); the associated system $\mathcal{E}(Q, R)$ is then also of exponential size and thus the existence of a solution can thus be checked in space exponential in the size of the input formula φ_0 since $\exists\mathbb{R} \subseteq$ PSpace. It remains to verify that every world type is realizable, in time double exponential in the size of φ_0. The lower bound is inherited from satisfiability in GF. For part (b), we can argue analogously with the difference that realizability can be checked in exponential time. For part (c), observe that in this case $\mathcal{E}(Q, R)$ is a system of linear inequalities and can thus be solved in polynomial time. The lower bound follows from the fact that the NExpTime-hard modal logic S5$_{\mathcal{ALC}}$ [8] is contained in this fragment.

Other FO fragments such as FO$_2$ and MonaFO are not closed under unions of types. Consider for example the FO$_2$ sentence $\psi = \forall x (\forall y\, R(x, y) \vee \forall y\, \neg R(y, x))$ which states that R is either the full relation or the empty relation. It does this in a slightly unorthodox way to ensure that no sentence from $\text{sub}_x(\psi)$ can distinguish the two cases. But the cases are distinguished in types because if R is full, then every type contains the formula $\forall y\, R(x, y)$ and if R is empty, then every type contains its negation. It is thus easy to show that closure under unions of types fails.

7   Conclusion

We have analyzed the reasons for the bad computational behaviour of ProbFO and we have shown that, unlike other natural restrictions that fail to establish recursive enumerability and decidability, monodicity is able to tame ProbFO computationally. We thus believe that monodic ProbFO lays a promising foundation for identifying decidable and useful probabilistic logics for computer science.

An interesting direction for further research is to enrich monodic ProbFO with additional expressive power that enables more complex and succinct statements about independence and conditioning. Note that existing decidable probabilistic first-order logics used in statistical relational learning such as Markov logic [9,21] are largely orthogonal to monodic ProbFO as they typically assume bounded domains and their main use is to encode a fixed distribution for a propositional theory (over ground instances).

Another important extension to be investigated is to combine statistical and subjective probabilities in a probabilistic FO logic. A basic version of ProbFO that combines both kinds of probability was considered by Halpern [12] under the name *type-3* ProbFO, and later refined to include stronger forms of independence and conditioning [19]. Adapting the quasi-model machinery to type-3 ProbFO and the mentioned extensions is a challenging open research objective.

We have given a tight upper complexity bound for monodic ProbGF. An open problem that remains is to determine the exact computational complexity of other relevant fragments of monodic ProbFO such as monodic ProbMonaFO, ProbFO$_2$, and negation-guarded monodic ProbFO.

Acknowledgments. This work was supported by the DFG project Probabilistic Description Logics (LU1417/1-1, SCHR1118/6-1).

References

1. Abadi, M., Halpern, J.: Decidability and expressiveness for first-order logics of probability. Inf. Comput. 112, 1–36 (1994)
2. Andréka, H., van Benthem, J., Németi, I.: Back and forth between modal logic and classical logic. Logic J. IGPL 3, 685–720 (1995)
3. Artin, E., Schreier, O.: Algebraische Konstruktion reeller Körper. Abh. Math. Sem. Univ. Hamburg 5, 85–99 (1927)
4. Bacchus, F.: Representing and reasoning with probabilistic knowledge - a logical approach to probabilities. MIT Press (1990)
5. Bacchus, F., Grove, A., Koller, D., Halpern, J.: From statistics to beliefs. In: Artificial Intelligence, AAAI 1992, pp. 602–608. AAAI Press/The MIT Press (1992)
6. Bárány, V., ten Cate, B., Segoufin, L.: Guarded negation. In: Aceto, L., Henzinger, M., Sgall, J. (eds.) ICALP 2011, Part II. LNCS, vol. 6756, pp. 356–367. Springer, Heidelberg (2011)
7. Enderton, H.B.: A mathematical introduction to logic. Academic Press (1972)
8. Gabbay, D., Kurucz, A., Wolter, F., Zakharyaschev, M.: Many-dimensional modal logics: theory and applications. Studies in Logic, vol. 148. Elsevier (2003)
9. Getoor, L., Taskar, B.: Introduction to Statistical Relational Learning. MIT Press (2007)
10. Grädel, E., Kolaitis, P., Vardi, M.: On the decision problem for two-variable first-order logic. Bull. Symb. Log. 3, 53–69 (1997)
11. Gutiérrez-Basulto, V., Jung, J., Lutz, C., Schröder, L.: A closer look at the probabilistic description logic prob-$\mathcal{EL}$. In: Artificial Intelligence, AAAI 2011. AAAI Press (2011)
12. Halpern, J.: An analysis of first-order logics of probability. Artif. Intell. 46, 311–350 (1990)
13. Harel, D.: Recurring dominoes: making the highly undecidable highly understandable. Ann. Discrete Math. 24, 51–72 (1985)
14. Hodkinson, I.: Monodic packed fragment with equality is decidable. Stud. Log. 72, 185–197 (2002)
15. Hodkinson, I.: Complexity of monodic guarded fragments over linear and real time. Ann. Pure Appl. Logic 138, 94–125 (2006)
16. Hodkinson, I., Kontchakov, R., Kurucz, A., Wolter, F., Zakharyaschev, M.: On the computational complexity of decidable fragments of first-order linear temporal logics. In: Proc. TIME-ICTL 2003, pp. 91–98. IEEE Computer Society (2003)
17. Hodkinson, I., Wolter, F., Zakharyaschev, M.: Decidable fragment of first-order temporal logics. Ann. Pure Appl. Logic 106, 85–134 (2000)
18. Hodkinson, I., Wolter, F., Zakharyaschev, M.: Monodic fragments of first-order temporal logics: 2000-2001 A.D. In: Nieuwenhuis, R., Voronkov, A. (eds.) LPAR 2001. LNCS (LNAI), vol. 2250, pp. 1–23. Springer, Heidelberg (2001)
19. Koller, D., Halpern, J.: Irrelevance and conditioning in first-order probabilistic logic. In: Proc. AAAI/IAAI 1096, pp. 569–576. AAAI Press / The MIT Press (1996)
20. Lutz, C., Schröder, L.: Probabilistic description logics for subjective uncertainty. In: Principles of Knowledge Representation and Reasoning, KR 2010. AAAI Press (2010)
21. Richardson, M., Domingos, P.: Markov logic networks. Machine Learning 62, 107–136 (2006)
22. Schaefer, M.: Complexity of some geometric and topological problems. In: Eppstein, D., Gansner, E.R. (eds.) GD 2009. LNCS, vol. 5849, pp. 334–344. Springer, Heidelberg (2010)
23. Wolter, F., Zakharyaschev, M.: Axiomatizing the monodic fragment of first-order temporal logic. Ann. Pure Appl. Logic 118, 133–145 (2002)

Stability and Complexity of Minimising Probabilistic Automata*

Stefan Kiefer and Björn Wachter

University of Oxford, UK

Abstract. We consider the state-minimisation problem for weighted and probabilistic automata. We provide a numerically stable polynomial-time minimisation algorithm for weighted automata, with guaranteed bounds on the numerical error when run with floating-point arithmetic. Our algorithm can also be used for "lossy" minimisation with bounded error. We show an application in image compression. In the second part of the paper we study the complexity of the minimisation problem for probabilistic automata. We prove that the problem is NP-hard and in PSPACE, improving a recent EXPTIME-result.

1 Introduction

Probabilistic and weighted automata were introduced in the 1960s, with many fundamental results established by Schützenberger [25] and Rabin [23]. Nowadays probabilistic automata are widely used in automated verification, natural-language processing, and machine learning.

Probabilistic automata (PAs) generalise deterministic finite automata (DFAs): The transition relation specifies, for each state q and each input letter a, a probability distribution on the successor state. Instead of a single initial state, a PA has a probability distribution over states; and instead of accepting states, a PA has an acceptance probability for each state. As a consequence, the language induced by a PA is a *probabilistic language*, i.e., a mapping $L : \Sigma^* \to [0,1]$, which assigns each word an acceptance probability. Weighted automata (WAs), in turn, generalise PAs: the numbers appearing in the specification of a WA may be arbitrary real numbers. As a consequence, a WA induces a *weighted language*, i.e., a mapping $L : \Sigma^* \to \mathbb{R}$. Loosely speaking, the weight of a word w is the sum of the weights of all accepting w-labelled paths through the WA.

Given an automaton, it is natural to ask for a small automaton that accepts the same weighted language. A small automaton is particularly desirable when further algorithms are run on the automaton, and the runtime of those algorithms depends crucially on the size of the automaton [17]. In this paper we consider the problem of minimising the number of states of a given WA or PA, while preserving its (weighted or probabilistic) language.

WAs can be minimised in polynomial time, using, e.g., the standardisation procedure of [25]. When implemented efficiently (for instance using triangular

---

* For a full version of this paper, see [19].

J. Esparza et al. (Eds.): ICALP 2014, Part II, LNCS 8573, pp. 268–279, 2014.

matrices), one obtains an $O(|\Sigma|n^3)$ minimisation algorithm, where n is the number of states. As PAs are special WAs, the same holds in principle for PAs.

There are two problems with these algorithms: (1) numerical instability, i.e., round-off errors can lead to an automaton that is not minimal and/or induces a different probabilistic language; and (2) minimising a PA using WA minimisation algorithms does not necessarily result in a PA: transition weights may, e.g., become negative. This paper deals with those two issues.

Concerning problem (1), numerical stability is crucial under two scenarios: (a) when the automaton size makes the use of exact rational arithmetic prohibitive, and thus necessitates floating-point arithmetic [17]; or (b) when exact minimisation yields an automaton that is still too large and a "lossy compression" is called for, as in image compression [15]. Besides finding a numerically stable algorithm, we aim at two further goals: First, a stable algorithm should also be efficient; i.e., it should be as fast as classical (efficient, but possibly unstable) algorithms. Second, stability should be provable, and ideally there should be easily computable error bounds. In Section 3 we provide a numerically stable $O(|\Sigma|n^3)$ algorithm for minimising WAs. The algorithm generalises the *Arnoldi iteration* [2] which is used for locating eigenvalues in numerical linear algebra. The key ingredient, leading to numerical stability and allowing us to give error bounds, is the use of special orthonormal matrices, called *Householder reflectors* [14]. To the best of the authors' knowledge, these techniques have not been previously utilised for computations on weighted automata.

Problem (2) suggests a study of the computational complexity of the *PA minimisation problem*: given a PA and $m \in \mathbb{N}$, is there an equivalent PA with m states? In the 1960s and 70s, PAs were studied extensively, see the survey [7] for references and Paz's influential textbook [22]. PAs appear in various flavours and under different names. For instance, in *stochastic sequential machines* [22] there is no fixed initial state distribution, so the semantics of a stochastic sequential machine is not a probabilistic language, but a mapping from initial distributions to probabilistic languages. This gives rise to several notions of minimality in this model [22]. In this paper we consider only PAs with an initial state distribution; equivalence means equality of probabilistic languages.

One may be tempted to think that PA minimisation is trivially in NP, by guessing the minimal PA and verifying equivalence. However, it is not clear that the minimal PA has rational transition probabilities, even if this holds for the original PA.

For DFAs, which are special PAs, an automaton is minimal (i.e., has the least number of states) if and only if all states are reachable and no two states are equivalent. However, this equivalence does in general not hold for PAs. In fact, even if a PA has the property that no state behaves like a convex combination of other states, the PA may nevertheless not be minimal. As an example, consider the PA in the middle of Figure 2 on page 276. State 3 behaves like a convex combination of states 2 and 4: state 3 can be removed by splitting its incoming arc with weight 1 in two arcs with weight 1/2 each and redirecting the new arcs to states 2 and 4. The resulting PA is equivalent and no state can be replaced by a convex combination of other states. But the PA on the right of the figure is equivalent and has even fewer states.

In Section 4 we show that the PA minimisation problem is NP-hard by a reduction from 3SAT. A step in our reduction is to show that the following problem, the *hypercube problem*, is NP-hard: given a convex polytope P within the d-dimensional unit hypercube and $m \in \mathbb{N}$, is there a convex polytope with m vertices that is nested between P and the hypercube? We then reduce the hypercube problem to PA minimisation. To the best of the authors' knowledge, no lower complexity bound for PA minimisation has been previously obtained, and there was no reduction from the hypercube problem to PA minimisation. However, towards the converse direction, the textbook [22] suggests that an algorithm for the hypercube problem could serve as a "subroutine" for a PA minimisation algorithm, leaving the decidability of both problems open. In fact, problems similar to the hypercube problem were subsequently studied in the field of computational geometry, citing PA minimisation as a motivation [26,21,11,10].

The PA minimisation problem was shown to be decidable in [20], where the authors provided an exponential reduction to the existential theory of the reals, which, in turn, is decidable in PSPACE [8,24], but not known to be PSPACE-hard. In Section 4.2 we give a polynomial-time reduction from the PA minimisation problem to the existential theory of the reals. It follows that the PA minimisation problem is in PSPACE, improving the EXPTIME result of [20].

2 Preliminaries

In the technical development that follows it is more convenient to talk about vectors and transition matrices than about states, edges, alphabet labels and weights. However, a PA "of size n" can be easily viewed as a PA with states $1, 2, \ldots, n$. We use this equivalence in pictures.

Let $\mathbb{N} = \{0, 1, 2, \ldots\}$. For $n \in \mathbb{N}$ we write $\mathbb{N}_n$ for the set $\{1, 2, \ldots, n\}$. For $m, n \in \mathbb{N}$, elements of $\mathbb{R}^m$ and $\mathbb{R}^{m \times n}$ are viewed as vectors and matrices, respectively. Vectors are row vectors by default. Let $\alpha \in \mathbb{R}^m$ and $M \in \mathbb{R}^{m \times n}$. We denote the entries by $\alpha[i]$ and $M[i, j]$ for $i \in \mathbb{N}_m$ and $j \in \mathbb{N}_n$. By $M[i, \cdot]$ we refer to the ith row of M. By $\alpha[i..j]$ for $i \leq j$ we refer to the sub-vector $(\alpha[i], \alpha[i+1], \ldots, \alpha[j])$, and similarly for matrices. We denote the transpose by α^T (a column vector) and $M^T \in \mathbb{R}^{n \times m}$. We write I_n for the $n \times n$ identity matrix. When the dimension is clear from the context, we write $e(i)$ for the vector with $e(i)[i] = 1$ and $e(i)[j] = 0$ for $j \neq i$. A vector $\alpha \in \mathbb{R}^m$ is *stochastic* if $\alpha[i] \geq 0$ for all $i \in \mathbb{N}_m$ and $\sum_{i=1}^m \alpha[i] \leq 1$. A matrix is *stochastic* if all its rows are stochastic. By $\|\cdot\| = \|\cdot\|_2$, we mean the 2-norm for vectors and matrices throughout the paper unless specified otherwise. If a matrix M is stochastic, then $\|M\| \leq \|M\|_1 \leq 1$. For a set $V \subseteq \mathbb{R}^n$, we write $\langle V \rangle$ to denote the vector space spanned by V, where we often omit the braces when denoting V. For instance, if $\alpha, \beta \in \mathbb{R}^n$, then $\langle \{\alpha, \beta\} \rangle = \langle \alpha, \beta \rangle = \{r\alpha + s\beta \mid r, s \in \mathbb{R}\}$.

An $\mathbb{R}$-*weighted automaton (WA)* $\mathcal{A} = (n, \Sigma, M, \alpha, \eta)$ consists of a size $n \in \mathbb{N}$, a finite alphabet Σ, a map $M : \Sigma \to \mathbb{R}^{n \times n}$, an initial (row) vector

$\alpha \in \mathbb{R}^n$, and a final (column) vector $\eta \in \mathbb{R}^n$. Extend M to Σ^* by setting $M(a_1 \cdots a_k) := M(a_1) \cdots M(a_k)$. The *language* $L_\mathcal{A}$ of a WA $\mathcal{A}$ is the mapping $L_\mathcal{A} : \Sigma^* \to \mathbb{R}$ with $L_\mathcal{A}(w) = \alpha M(w)\eta$. WAs $\mathcal{A}, \mathcal{B}$ over the same alphabet Σ are said to be *equivalent* if $L_\mathcal{A} = L_\mathcal{B}$. A WA $\mathcal{A}$ is *minimal* if there is no equivalent WA $\mathcal{B}$ of smaller size.

A *probabilistic automaton (PA)* $\mathcal{A} = (n, \Sigma, M, \alpha, \eta)$ is a WA, where α is stochastic, $M(a)$ is stochastic for all $a \in \Sigma$, and $\eta \in [0, 1]^n$. A PA is a *DFA* if all numbers in M, α, η are 0 or 1.

3 Stable WA Minimisation

In this section we discuss WA minimisation. In Section 3.1 we describe a WA minimisation algorithm in terms of elementary linear algebra. The presentation reminds of Brzozowski's algorithm for NFA minimisation [6].[1] WA minimisation techniques are well known, originating in [25], cf. also [4, Chapter II] and [3]. Our algorithm and its correctness proof may be of independent interest, as they appear to be particularly succinct. In Sections 3.2 and 3.3 we take further advantage of the linear algebra setting and develop a numerically stable WA minimisation algorithm.

3.1 Brzozowski-like WA Minimisation

Let $\mathcal{A} = (n, \Sigma, M, \alpha, \eta)$ be a WA. Define the *forward space* of $\mathcal{A}$ as the (row) vector space $\mathsf{F} := \langle \alpha M(w) \mid w \in \Sigma^* \rangle$. Similarly, let the *backward space* of $\mathcal{A}$ be the (column) vector space $\mathsf{B} := \langle M(w)\eta \mid w \in \Sigma^* \rangle$. Let $\overrightarrow{n} \in \mathbb{N}$ and $F \in \mathbb{R}^{\overrightarrow{n} \times n}$ such that the rows of F form a basis of F. Similarly, let $\overleftarrow{n} \in \mathbb{N}$ and $B \in \mathbb{R}^{n \times \overleftarrow{n}}$ such that the columns of B form a basis of B. Since $\mathsf{F}M(a) \subseteq \mathsf{F}$ and $M(a)\mathsf{B} \subseteq \mathsf{B}$ for all $a \in \Sigma$, there exist maps $\overrightarrow{M} : \Sigma \to \mathbb{R}^{\overrightarrow{n} \times \overrightarrow{n}}$ and $\overleftarrow{M} : \Sigma \to \mathbb{R}^{\overleftarrow{n} \times \overleftarrow{n}}$ such that

$$FM(a) = \overrightarrow{M}(a)F \quad \text{and} \quad M(a)B = B\overleftarrow{M}(a) \quad \text{for all } a \in \Sigma. \tag{1}$$

We call $(F, \overrightarrow{M})$ a *forward reduction* and $(B, \overleftarrow{M})$ a *backward reduction*. We will show that minimisation reduces to computing such reductions. By symmetry we can focus on forward reductions. We call a forward reduction $(F, \overrightarrow{M})$ *canonical* if $F[1, \cdot]$ (i.e., the first row of F) is a multiple of α, and the rows of F are orthonormal, i.e., $FF^T = I_{\overrightarrow{n}}$.

Let $\mathcal{A} = (n, \Sigma, M, \alpha, \eta)$ be a WA with forward and backward reductions $(F, \overrightarrow{M})$ and $(B, \overleftarrow{M})$, respectively. Let $\overrightarrow{\alpha} \in \mathbb{R}^{\overrightarrow{n}}$ be a row vector such that $\alpha = \overrightarrow{\alpha}F$; let $\overleftarrow{\eta} \in \mathbb{R}^{\overleftarrow{n}}$ be a column vector such that $\eta = B\overleftarrow{\eta}$. (If $(F, \overrightarrow{M})$ is canonical, we have $\overrightarrow{\alpha} = (\pm\|\alpha\|, 0, \ldots, 0)$.) Call $\overrightarrow{\mathcal{A}} := (\overrightarrow{n}, \Sigma, \overrightarrow{M}, \overrightarrow{\alpha}, F\eta)$ a *forward WA* of $\mathcal{A}$ with base F and $\overleftarrow{\mathcal{A}} := (\overleftarrow{n}, \Sigma, \overleftarrow{M}, \alpha B, \overleftarrow{\eta})$ a *backward WA* of $\mathcal{A}$ with base B. By extending (1) one can see that these automata are equivalent to $\mathcal{A}$:

---

[1] In [5] a very general Brzozowski-like minimization algorithm is presented in terms of universal algebra. One can show that it specialises to ours in the WA setting.

Proposition 1. *Let $\mathcal{A}$ be a WA. Then $L_{\mathcal{A}} = L_{\overleftarrow{\mathcal{A}}} = L_{\overleftrightarrow{\mathcal{A}}}$.*

Further, applying both constructions consecutively yields a minimal WA:

Theorem 2. *Let $\mathcal{A}$ be a WA. Let $\mathcal{A}' = \overleftrightarrow{\mathcal{A}}$ or $\mathcal{A}' = \overrightarrow{\overleftarrow{\mathcal{A}}}$. Then $\mathcal{A}'$ is minimal and equivalent to $\mathcal{A}$.*

Theorem 2 mirrors Brzozowski's NFA minimisation algorithm. We give a short proof in [19].

3.2   Numerically Stable WA Minimisation

Theorem 2 reduces the problem of minimising a WA to the problem of computing a forward and a backward reduction. In the following we focus on computing a *canonical* (see above for the definition) forward reduction $(F, \overrightarrow{M})$. Figure 1 shows a generalisation of Arnoldi's iteration [2] to multiple matrices. Arnoldi's iteration is typically used for locating eigenvalues [12]. Its generalisation to multiple matrices is novel, to the best of the authors's knowledge. Using (1) one can see that it computes a canonical forward reduction by iteratively extending a partial orthonormal basis $\{f_1, \ldots, f_j\}$ for the forward space F.

---

function ArnoldiReduction
input: $\alpha \in \mathbb{R}^n$; $M : \Sigma \to \mathbb{R}^{n \times n}$
output: canonical forward reduction $(F, \overrightarrow{M})$ with $F \in \mathbb{R}^{\vec{n} \times n}$ and $\overrightarrow{M} : \Sigma \to \mathbb{R}^{\vec{n} \times \vec{n}}$
$\qquad \ell := 0;\ j := 1;\ f_1 := \alpha/\|\alpha\|$ (or $f_1 := -\alpha/\|\alpha\|$)
$\qquad$ while $\ell < j$ do
$\qquad\qquad \ell := \ell + 1$
$\qquad\qquad$ for $a \in \Sigma$ do
$\qquad\qquad\qquad$ if $f_\ell M(a) \notin \langle f_1, \ldots, f_j \rangle$
$\qquad\qquad\qquad\qquad j := j + 1$
$\qquad\qquad\qquad\qquad$ define f_j orthonormal to $f_1, \ldots, f_{j-1}$ such that
$\qquad\qquad\qquad\qquad\qquad \langle f_1, \ldots, f_{j-1}, f_\ell M(a) \rangle = \langle f_1, \ldots, f_j \rangle$
$\qquad\qquad\qquad\qquad$ define $\overrightarrow{M}(a)[\ell, \cdot]$ such that $f_\ell M(a) = \sum_{i=1}^{j} \overrightarrow{M}(a)[\ell, i] f_i$
$\qquad\qquad\qquad\qquad\qquad$ and $\overrightarrow{M}(a)[\ell, j+1..n] = (0, \ldots, 0)$
$\qquad \vec{n} := j;$ form $F \in \mathbb{R}^{\vec{n} \times \vec{n}}$ with rows $f_1, \ldots, f_{\vec{n}}$
$\qquad$ return F and $\overrightarrow{M}(a)[1..\vec{n}, 1..\vec{n}]$ for all $a \in \Sigma$

Fig. 1. Generalised Arnoldi iteration

For efficiency, one would like to run generalised Arnoldi iteration (Figure 1) using floating-point arithmetic. This leads to *round-off errors*. The check "if $f_\ell M(a) \notin \langle f_1, \ldots, f_j \rangle$" is particularly problematic: since the vectors $f_1, \ldots, f_j$ are computed with floating-point arithmetic, we cannot expect that $f_\ell M(a)$ lies *exactly* in the vector space spanned by those vectors, even if that would be the case without round-off errors. As a consequence, we need to introduce an *error tolerance parameter* $\tau > 0$, so that the check "$f_\ell M(a) \notin \langle f_1, \ldots, f_j \rangle$" returns *true* only if $f_\ell M(a)$ has a "distance" of more than τ to the vector space

$\langle f_1, \ldots, f_j \rangle$.[2] Without such a "fuzzy" comparison the resulting automaton could even have more states than the original one. The error tolerance parameter τ causes further errors.

To assess the impact of those errors, we use the *standard model of floating-point arithmetic*, which assumes that the elementary operations $+, -, \cdot, /$ are computed exactly, up to a relative error of at most the *machine epsilon* $\varepsilon_{\text{mach}} \geq 0$. It is stated in [13, Chapter 2]: "This model is valid for most computers, and, in particular, holds for IEEE standard arithmetic." The bit length of numbers arising in a numerical computation is bounded by hardware, using suitable roundoff. So we adopt the convention of numerical linear algebra to take the number of arithmetic operations as a measure of time complexity.

The algorithm ArnoldiReduction (Figure 1) leaves open how to implement the conditional "if $f_\ell M(a) \notin \langle f_1, \ldots, f_j \rangle$", and how to compute the new basis element f_j. In [19] we propose an instantiation *HouseholderReduction* of Arnoldi-Reduction based on so-called *Householder reflectors* [14], which are special orthonormal matrices. We prove the following stability property:

Proposition 3. *Consider the algorithm* HouseholderReduction *in [19], which has the following interface:*

function HouseholderReduction
input: $\alpha \in \mathbb{R}^n$; $M : \Sigma \to \mathbb{R}^{n \times n}$; error tolerance parameter $\tau \geq 0$
output: canonical forward reduction $(F, \overrightarrow{M})$ with $F \in \mathbb{R}^{\vec{n} \times n}$ and $\overrightarrow{M} : \Sigma \to \mathbb{R}^{\vec{n} \times \vec{n}}$

We have:

1. *The number of arithmetic operations is* $O(|\Sigma| n^3)$.
2. HouseholderReduction *instantiates* ArnoldiReduction.
3. *The computed matrices satisfy the following error bound: For each* $a \in \Sigma$, *the matrix* $\mathcal{E}(a) \in \mathbb{R}^{\vec{n} \times n}$ *with* $\mathcal{E}(a) := FM(a) - \overrightarrow{M}(a)F$ *satisfies*

$$\|\mathcal{E}(a)\| \leq 2\sqrt{n}\tau + cmn^3 \varepsilon_{\text{mach}},$$

where $m > 0$ *is such that* $\|M(a)\| \leq m$ *holds for all* $a \in \Sigma$, *and* $c > 0$ *is an input-independent constant.*

The proof follows classical error-analysis techniques for QR factorisations with Householder reflectors [13, Chapter 19], but is substantially complicated by the presence of the "if" conditional and the resulting need for the τ parameter. By Proposition 3.2. HouseholderReduction computes a precise canonical forward reduction for $\varepsilon_{\text{mach}} = \tau = 0$. For positive $\varepsilon_{\text{mach}}$ and τ the error bound grows linearly in $\varepsilon_{\text{mach}}$ and τ, and with modest polynomials in the WA size n. In practice $\varepsilon_{\text{mach}}$ is very small[3], so that the term $cmn^3\varepsilon_{\text{mach}}$ can virtually be ignored.

The use of Householder reflectors is crucial to obtain the bound of Proposition 3. Let us mention a few alternative techniques, which have been used for computing certain matrix factorisations. Such factorisations (QR or LU) are

---

[2] This will be made formal in our algorithm.
[3] With IEEE double precision, e.g., it holds $\varepsilon_{\text{mach}} = 2^{-53}$ [13].

related to our algorithm. *Gaussian elimination* can also be used for WA min-imisation in time $O(|\Sigma|n^3)$, but its stability is governed by the *growth factor*, which can be exponential even with *pivoting* [13, Chapter 9], so the bound on $\|\mathcal{E}(a)\|$ in Proposition 3 would include a term of the form $2^n \varepsilon_{mach}$. The most straightforward implementation of ArnoldiReduction would use the *Classical Gram-Schmidt* process, which is highly unstable [13, Chapter 19.8]. A variant, the *Modified Gram-Schmidt* process is stable, but the error analysis is compli-cated by a possibly loss of orthogonality of the computed matrix F. The extent of that loss depends on certain condition numbers (cf. [13, Equation (19.30)]), which are hard to estimate or control in our case. In contrast, our error bound is independent of condition numbers.

Using Theorem 2 we can prove:

Theorem 4. *Consider the following algorithm:*

function HouseholderMinimisation
input: WA $\mathcal{A} = (n, \Sigma, M, \alpha, \eta)$; error tolerance parameter $\tau \geq 0$
output: minimised WA $\mathcal{A}' = (n', \Sigma, M', \alpha', \eta')$.
    compute forward reduction $(F, \overrightarrow{M})$ of $\mathcal{A}$ using HouseholderReduction
    form $\overrightarrow{\mathcal{A}} := (\overrightarrow{n}, \Sigma, \overrightarrow{M}, \overrightarrow{\alpha}, \overrightarrow{\eta})$ as the forward WA of $\mathcal{A}$ with base F
    compute backward reduction (B, M') of $\overrightarrow{\mathcal{A}}$ using HouseholderReduction
    form $\mathcal{A}' := (n', \Sigma, M', \alpha', \eta')$ as the backward WA of $\overrightarrow{\mathcal{A}}$ with base B
    return $\mathcal{A}'$

We have:

1. *The number of arithmetic operations is $O(|\Sigma|n^3)$.*
2. *For $\varepsilon_{mach} = \tau = 0$, the computed WA $\mathcal{A}'$ is minimal and equivalent to $\mathcal{A}$.*
3. *Let $\tau > 0$. Let $m > 0$ such that $\|A\| \leq m$ holds for all $A \in \{M(a), \overrightarrow{M}(a), M'(a) \mid a \in \Sigma\}$. Then for all $w \in \Sigma^*$ we have*

$$|L_{\mathcal{A}}(w) - L_{\mathcal{A}'}(w)| \leq 4|w|\|\alpha\|m^{|w|-1}\|\eta\|\sqrt{n}\tau$$
$$+ c\max\{|w|, 1\}\|\alpha\|m^{|w|}\|\eta\|n^3\varepsilon_{mach},$$

where $c > 0$ is an input-independent constant.

The algorithm computes a backward reduction by running the straightforward backward variant of HouseholderReduction. We remark that for PAs one can take $m = 1$ for the norm bound m from part 3. of the theorem (or $m = 1 + \varepsilon$ for a small ε if unfortunate roundoff errors occur). It is hard to avoid an error bound exponential in the word length $|w|$, as $|L_{\mathcal{A}}(w)|$ itself may be exponential in $|w|$ (consider a WA of size 1 with $M(a) = 2$). Theorem 4 is proved in [19].

The error bounds in Proposition 3 and Theorem 4 suggest to choose a small value for the error tolerance parameter τ. But as we have discussed, the computed WA may be non-minimal if τ is set too small or even to 0, intuitively because round-off errors may cause the algorithm to overlook minimisation opportunities. So it seems advisable to choose τ smaller (by a few orders of magnitude) than the desired bound on $\|\mathcal{E}(a)\|$, but larger (by a few orders of magnitude) than ε_{mach}.

Note that for $\varepsilon_{mach} > 0$ Theorem 4 does not provide a bound on the number of states of $\mathcal{A}'$.

To illustrate the stability issue we have experimented with minimising a PA $\mathcal{A}$ derived from Herman's protocol as in [17]. The PA has 190 states and $\Sigma = \{a\}$. When minimising with the (unstable) Classical Gram-Schmidt process, we have measured a huge error of $|L_{\mathcal{A}}(a^{190}) - L_{\mathcal{A}'}(a^{190})| \approx 10^{36}$. With the Modified Gram-Schmidt process and the method from Theorem 4 the corresponding errors were about 10^{-7}, which is in the same order as the error tolerance parameter τ.

3.3 Lossy WA Minimisation

A larger error tolerance parameter τ leads to more "aggressive" minimisation of a possibly already minimal WA. The price to pay is a shift in the language: one would expect only $L'_{\mathcal{A}}(w) \approx L_{\mathcal{A}}(w)$. Theorem 4 provides a bound on this imprecision. In this section we illustrate the trade-off between size and precision using an application in image compression.

Weighted automata can be used for image compression, as suggested by Culik et al. [15]. An image, represented as a two-dimensional matrix of grey-scale values, can be encoded as a weighted automaton where each pixel is addressed by a unique word. To obtain this automaton, the image is recursively subdivided into quadrants. There is a state for each quadrant and transitions from a quadrant to its sub-quadrants. At the level of the pixels, the automaton accepts with the correct grey-scale value.

Following this idea, we have implemented a prototype tool for image compression based on the algorithm of Theorem 4. We give details and show example pictures in [19]. This application illustrates lossy minimisation. The point is that Theorem 4 guarantees bounds on the loss.

4 The Complexity of PA Minimisation

Given a PA $\mathcal{A} = (n, \Sigma, M, \alpha, \eta)$ and $n' \in \mathbb{N}$, the *PA minimisation problem* asks whether there exists a PA $\mathcal{A}' = (n', \Sigma, M', \alpha', \eta')$ so that $\mathcal{A}$ and $\mathcal{A}'$ are equivalent. For the complexity results in this section we assume that the numbers in the description of the given PA are fractions of natural numbers represented in binary, so they are rational. In Section 4.1 we show that the minimisation problem is NP-hard. In Section 4.2 we show that the problem is in PSPACE by providing a polynomial-time reduction to the existential theory of the reals.

4.1 NP-Hardness

We will show:

Theorem 5. *The PA minimisation problem is NP-hard.*

For the proof we reduce from a geometrical problem, the *hypercube problem*, which we show to be NP-hard. Given $d \in \mathbb{N}$, a finite set $P = \{p_1, \ldots, p_k\} \subseteq [0,1]^d$

of vectors ("points") within the d-dimensional unit hypercube, and $\ell \in \mathbb{N}$, the *hypercube problem* asks whether there is a set $Q = \{q_1, \ldots, q_\ell\} \subseteq [0,1]^d$ of at most ℓ points within the hypercube such that $conv(Q) \supseteq P$, where

$$conv(Q) := \{\lambda_1 q_1 + \cdots + \lambda_\ell q_\ell \mid \lambda_1, \ldots, \lambda_\ell \geq 0, \ \lambda_1 + \cdots + \lambda_\ell = 1\}$$

denotes the convex hull of Q. Geometrically, the convex hull of P can be viewed as a convex polytope, nested inside the hypercube, which is another convex polytope. The hypercube problem asks whether a convex polytope with at most ℓ vertices can be nested in between those polytopes. The answer is trivially yes, if $\ell \geq k$ (take $Q = P$) or if $\ell \geq 2^d$ (take $Q = \{0,1\}^d$). We speak of the *restricted hypercube problem* if P contains the origin $(0, \ldots, 0)$. We prove the following:

Proposition 6. *The restricted hypercube problem can in polynomial time be reduced to the PA minimisation problem.*

Proof (sketch). Let $d \in \mathbb{N}$ and $P = \{p_1, \ldots, p_k\} \subseteq [0,1]^d$ and $\ell \in \mathbb{N}$ be an instance of the restricted hypercube problem, where $p_1 = (0, \ldots, 0)$ and $\ell \geq 1$. We construct in polynomial time a PA $\mathcal{A} = (k+1, \Sigma, M, \alpha, \eta)$ such that there is a set $Q = \{q_1, \ldots, q_\ell\} \subseteq [0,1]^d$ with $conv(Q) \supseteq P$ if and only if there is a PA $\mathcal{A}' = (\ell+1, \Sigma, M', \alpha', \eta')$ equivalent to $\mathcal{A}$. Take $\Sigma := \{a_2, \ldots, a_k\} \cup \{b_1, \ldots, b_d\}$. Set $M(a_i)[1, i] := 1$ and $M(b_s)[i, k+1] := p_i[s]$ for all $i \in \{2, \ldots, k\}$ and all $s \in \mathbb{N}_d$, and set all other entries of M to 0. Set $\alpha := e(1)$ and $\eta := e(k+1)^T$. Figure 2 shows an example of this reduction. We prove the correctness of this reduction in [19]. $\square$

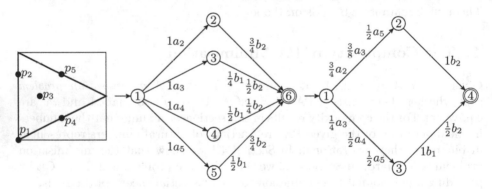

Fig. 2. Reduction from the hypercube problem to the minimisation problem. The left figure shows an instance of the hypercube problem with $d = 2$ and $P = \{p_1, \ldots, p_5\} = \{(0,0), (0, \frac{3}{4}), (\frac{1}{4}, \frac{1}{2}), (\frac{1}{2}, \frac{1}{4}), (\frac{1}{2}, \frac{3}{4})\}$. It also suggests a set $Q = \{(0,0), (0,1), (1, \frac{1}{2})\}$ with $conv(Q) \supseteq P$. The middle figure depicts the PA $\mathcal{A}$ obtained from P. The right figure depicts a minimal equivalent PA $\mathcal{A}'$, corresponding to the set Q suggested in the left figure.

Next we show that the hypercube problem is NP-hard, which together with Proposition 6 implies Theorem 5. A related problem is known[4] to be NP-hard:

Theorem 7 (Theorem 4.2 of [10]). *Given two nested convex polyhedra in three dimensions, the problem of nesting a convex polyhedron with minimum faces between the two polyhedra is NP-hard.*

Note that this NP-hardness result holds even in $d = 3$ dimensions. However, the outer polyhedron is not required to be a cube, and the problem is about minimising the number of faces rather than the number of vertices. Using a completely different technique we show:

Proposition 8. *The hypercube problem is NP-hard. This holds even for the restricted hypercube problem.*

The proof is by a reduction from 3SAT, see [19].

Remark 9. The hypercube problem is in PSPACE, by appealing to decision algorithms for $ExTh(\mathbb{R})$, the existential fragment of the first-order theory of the reals. For every fixed d the hypercube problem is[5] in P, exploiting the fact that $ExTh(\mathbb{R})$ can be decided in polynomial time, if the number of variables is fixed. (For $d = 2$ an efficient algorithm is provided in [1].) It is an open question whether the hypercube problem is in NP. It is also open whether the search for a minimum Q can be restricted to sets of points with rational coordinates (this holds for $d = 2$).

Propositions 6 and 8 together imply Theorem 5.

4.2 Reduction to the Existential Theory of the Reals

In this section we reduce the PA minimisation problem to $ExTh(\mathbb{R})$, the existential fragment of the first-order theory of the reals. A formula of $ExTh(\mathbb{R})$ is of the form $\exists x_1 \ldots \exists x_m R(x_1, \ldots, x_n)$, where $R(x_1, \ldots, x_n)$ is a boolean combination of comparisons of the form $p(x_1, \ldots, x_n) \sim 0$, where $p(x_1, \ldots, x_n)$ is a multivariate polynomial and $\sim \in \{<, >, \leq, \geq, =, \neq\}$. The validity of closed formulas ($m = n$) is decidable in PSPACE [8,24], and is not known to be PSPACE-hard.

Proposition 10. *Let $\mathcal{A}_1 = (n_1, \Sigma, M_1, \alpha_1, \eta_1)$ be a PA. A PA $\mathcal{A}_2 = (n_2, \Sigma, M_2, \alpha_2, \eta_2)$ is equivalent to $\mathcal{A}_1$ if and only if there exist matrices $\overrightarrow{M}(a) \in \mathbb{R}^{(n_1+n_2) \times (n_1+n_2)}$ for $a \in \Sigma$ and a matrix $F \in \mathbb{R}^{(n_1+n_2) \times (n_1+n_2)}$ such that $F[1, \cdot] = (\alpha_1, \alpha_2)$, and $F(\eta_1^T, -\eta_2^T)^T = (0, \ldots, 0)^T$, and*

$$F \begin{pmatrix} M_1(a) & 0 \\ 0 & M_2(a) \end{pmatrix} = \overrightarrow{M}(a)F \qquad \text{for all } a \in \Sigma.$$

The proof is in [19]. The conditions of Proposition 10 on $\mathcal{A}_2$, including that it be a PA, can be phrased in $ExTh(\mathbb{R})$. Thus it follows:

---

[4] The authors thank Joseph O'Rourke for pointing out [10].
[5] This observation is in part due to Radu Grigore.

Theorem 11. *The PA minimisation problem can be reduced in polynomial time to ExTh($\mathbb{R}$). Hence, PA minimisation is in PSPACE.*

Theorem 11 improves on a result in [20] where the minimisation problem was shown to be in EXPTIME. (More precisely, Theorem 4 of [20] states that a minimal PA can be computed in EXPSPACE, but the proof reveals that the decision problem can be solved in EXPTIME.)

5   Conclusions and Open Questions

We have developed a numerically stable and efficient algorithm for minimising WAs, based on linear algebra and Brzozowski-like automata minimisation. We have given bounds on the minimisation error in terms of both the machine epsilon and the error tolerance parameter τ.

We have shown NP-hardness for PA minimisation, and have given a polynomial-time reduction to ExTh($\mathbb{R}$). Our work leaves open the precise complexity of the PA minimisation problem. The authors do not know whether the search for a minimal PA can be restricted to PAs with rational numbers. As stated in the Remark after Proposition 8, the corresponding question is open even for the hypercube problem. If rational numbers indeed suffice, then an NP algorithm might exist that guesses the (rational numbers of the) minimal PA and checks for equivalence with the given PA. Proving PSPACE-hardness would imply PSPACE-hardness of ExTh($\mathbb{R}$), thus solving a longstanding open problem.

For comparison, the corresponding minimisation problems involving WAs (a generalisation of PAs) and DFAs (a special case of PAs) lie in P. More precisely, minimisation of WAs (with rational numbers) is in randomised NC [18], and DFA minimisation is NL-complete [9]. NFA minimisation is PSPACE-complete [16].

Acknowledgements. The authors would like to thank James Worrell, Radu Grigore, and Joseph O'Rourke for valuable discussions, and the anonymous referees for their helpful comments. Stefan Kiefer is supported by a Royal Society University Research Fellowship.

References

1. Aggarwal, A., Booth, H., O'Rourke, J., Suri, S., Yap, C.K.: Finding minimal convex nested polygons. Information and Computation 83(1), 98–110 (1989)
2. Arnoldi, W.E.: The principle of minimized iteration in the solution of the matrix eigenvalue problem. Quarterly of Applied Mathematics 9, 17–29 (1951)
3. Beimel, A., Bergadano, F., Bshouty, N.H., Kushilevitz, E., Varricchio, S.: Learning functions represented as multiplicity automata. Journal of the ACM 47(3), 506–530 (2000)
4. Berstel, J., Reutenauer, C.: Rational Series and Their Languages. Springer (1988)
5. Bonchi, F., Bonsangue, M.M., Hansen, H.H., Panangaden, P., Rutten, J.J.M.M., Silva, A.: Algebra-coalgebra duality in Brzozowski's minimization algorithm. ACM Transactions on Computational Logic (to appear)

6. Brzozowski, J.A.: Canonical regular expressions and minimal state graphs for definite events. In: Symposium on Mathematical Theory of Automata. MRI Symposia Series, vol. 12, pp. 529–561. Polytechnic Press, Polytechnic Institute of Brooklyn (1962)
7. Bukharaev, R.G.: Probabilistic automata. Journal of Soviet Mathematics 13(3), 359–386 (1980)
8. Canny, J.: Some algebraic and geometric computations in PSPACE. In: Proceedings of STOC 1988, pp. 460–467 (1988)
9. Cho, S., Huynh, D.T.: The parallel complexity of finite-state automata problems. Information and Computation 97(1), 1–22 (1992)
10. Das, G., Goodrich, M.T.: On the complexity of approximating and illuminating three-dimensional convex polyhedra. In: Sack, J.-R., Akl, S.G., Dehne, F., Santoro, N. (eds.) WADS 1995. LNCS, vol. 955, pp. 74–85. Springer, Heidelberg (1995)
11. Das, G., Joseph, D.: Minimum vertex hulls for polyhedral domains. Theoretical Computer Science 103(1), 107–135 (1992)
12. Golub, G.H., van Loan, C.F.: Matrix Computations. John Hopkins University Press (1989)
13. Higham, N.J.: Accuracy and Stability of Numerical Algorithms, 2nd edn. SIAM (2002)
14. Householder, A.S.: Unitary triangularization of a nonsymmetric matrix. Journal of the ACM 5(4), 339–342 (1958)
15. Culik II, K., Kari, J.: Image compression using weighted finite automata. Computers & Graphics 17(3), 305–313 (1993)
16. Jiang, T., Ravikumar, B.: Minimal NFA problems are hard. SIAM Journal on Computing 22(6), 1117–1141 (1993)
17. Kiefer, S., Murawski, A.S., Ouaknine, J., Wachter, B., Worrell, J.: Language equivalence for probabilistic automata. In: Gopalakrishnan, G., Qadeer, S. (eds.) CAV 2011. LNCS, vol. 6806, pp. 526–540. Springer, Heidelberg (2011)
18. Kiefer, S., Murawski, A.S., Ouaknine, J., Wachter, B., Worrell, J.: On the complexity of equivalence and minimisation for Q-weighted automata. Logical Methods in Computer Science 9(1:8), 1–22 (2013)
19. Kiefer, S., Wachter, B.: Stability and complexity of minimising probabilistic automata. Technical report, arxiv.org (2014), http://arxiv.org/abs/1404.6673
20. Mateus, P., Qiu, D., Li, L.: On the complexity of minimizing probabilistic and quantum automata. Information and Computation 218, 36–53 (2012)
21. Mitchell, J.S.B., Suri, S.: Separation and approximation of polyhedral objects. In: Proceedings of SODA, pp. 296–306 (1992)
22. Paz, A.: Introduction to probabilistic automata. Academic Press (1971)
23. Rabin, M.O.: Probabilistic automata. Information and Control 6(3), 230–245 (1963)
24. Renegar, J.: On the computational complexity and geometry of the first-order theory of the reals. Parts I–III. Journal of Symbolic Computation 13(3), 255–352 (1992)
25. Schützenberger, M.-P.: On the definition of a family of automata. Information and Control 4, 245–270 (1961)
26. Silio, C.B.: An efficient simplex coverability algorithm in E^2 with application to stochastic sequential machines. IEEE Transactions on Computers C-28(2), 109–120 (1979)

Kleene Algebra with Equations

Dexter Kozen and Konstantinos Mamouras

Computer Science Department, Cornell University, Ithaca, NY 14853-7501, USA
{kozen,mamouras}@cs.cornell.edu

Abstract. We identify sufficient conditions for the construction of free language models for systems of Kleene algebra with additional equations. The construction applies to a broad class of extensions of KA and provides a uniform approach to deductive completeness.

1 Introduction

Kleene algebra (KA) is the algebra of regular expressions. Introduced by Stephen Cole Kleene in 1956, it is fundamental and ubiquitous in computer science. It has proven useful in countless applications, from program specification and verification to the design and analysis of algorithms [1–8].

One can augment KA with Booleans in a seamless way to obtain Kleene algebra with tests (KAT). Unlike many other related logics for program verification, KAT is classically based, requiring no specialized syntax or deductive apparatus other than classical equational logic. In practice, statements in the logic are typically universal Horn formulas

$$s_1 = t_1 \to s_2 = t_2 \to \cdots \to s_n = t_n \to s = t,$$

where the conclusion $s = t$ is the main target task and the premises $s_i = t_i$ are the verification conditions needed to prove it. The conclusion $s = t$ may encode a partial correctness assertion, an equivalence between an optimized and an unoptimized version of a program, or an equivalence between a program annotated with static analysis information and the unannotated program. The verification conditions $s_i = t_i$ are typically simple properties of the underlying domain of computation that describe how atomic actions interact with atomic assertions. They may require first-order interpreted reasoning, but are proven once and for all, then abstracted to propositional form. The proof of the conclusion $s = t$ from the premises takes place at the propositional level in KAT. This methodology affords a clean separation of the theory of the domain of computation from the program restructuring operations. It is advantageous to separate the two levels of reasoning, because the full first-order theory of the domain of computation may be highly undecidable, even though we may only need small parts of it. By isolating those parts, we can often maintain decidability and deductive completeness.

A typical form of premise that arises frequently in practice is a *commutativity condition* $pb = bp$ for an action p and a test b. This captures the idea that the action p does not affect the truth of b. For example, the action p might be an

J. Esparza et al. (Eds.): ICALP 2014, Part II, LNCS 8573, pp. 280–292, 2014.

assignment $x := 3$ and b might be a test $y = 4$, where x and y are distinct variables. It is clear that the truth value of b is not affected by the action p, so it would be the same before as after. But once this is established, we no longer need to know what p and b are, but only that $pb = bp$. It follows by purely equational reasoning in KAT that $p_1 b = bp_1 \to \cdots \to p_n b = bp_n \to qb = bq$, where q is any program built from atomic actions $p_1, \ldots, p_n$.

In some instances, Horn formulas with premises of a certain form can be reduced to the equational theory without loss of deductive completeness or decision efficiency using a technique known as *elimination of hypotheses* [3, 9, 10]. One important class of premises for which this is possible are those of the form $s = 0$. The universal Horn theory restricted to premises of this form is called the *Hoare theory*, because it subsumes Hoare logic: the partial correctness assertion $\{b\}p\{c\}$ can be encoded as the equation $bp\bar{c} = 0$. Other forms that arise frequently in practice are $bp = b$, which says that the action p is not necessary if b is true, useful in optimizations to eliminate redundant actions; and $pq = qp$, which says that the atomic actions p and q can occur in either order with the same effect, useful in reasoning about concurrency. Unfortunately, KAT with general commutativity assumptions $pq = qp$ is undecidable [11].

As a case in point, the NetKAT system [8] incorporates a number of such equational premises as part of the theory, which are taken as additional axioms besides those of KAT. Proofs of deductive completeness and complexity as given in [8] required extensive adaptation of the analogous proofs for KA and KAT. Indeed, this was already the case with KAT, which was an adaptation of KA to incorporate an embedded Boolean algebra.

Although each of these instances was studied separately, there are some striking similarities. It turns out that the key to progress in all of them is the identification of a suitable class of *language models* that characterize the equational theory of the system. A language model is a structure in which expressions are interpreted as sets of elements of some monoid. The language models should form the free models for the system at hand. For KA, a language model is the regular sets of strings over a finite alphabet, elements of a free monoid; for KAT, the regular sets of guarded strings; for NetKAT, the regular sets of strings of a certain reduced form. Once a suitable class of language models can be determined, this opens the door to a systematic treatment of deductive completeness. It is also clear from previous work [8, 12–15] that the existence of coalgebraic decision algorithms also depends strongly on the existence of language models (although we do not develop this connection in this paper). The question thus presents itself: Is there a general set of criteria that admit a uniform construction of language models and that would apply in a broad range of situations and subsume previous ad hoc constructions? That is the subject of this paper.

Alas, such a grand unifying framework is unlikely, given the negative results of [11] and of §2. However, we have identified a framework that goes quite far in this direction. It applies in the case in which the additional equational axioms are monoid equations or partial monoid equations (as is the case in all the examples mentioned above) and is based on a well-studied class of rewrite systems called *inverse context-free systems* [16]. We give criteria in terms of these rewrite systems that imply the existence of free language models in a wide range of previously studied instances, as well as some new ones.

This paper is organized as follows. In §2 we present preliminary definitions and our negative result limiting the applicability of the method. In §3 we establish a connection between the classical theory of string rewriting and Kleene algebra. We recall from [16] the definition of *inverse context-free rewrite systems* and the key result that they preserve regularity. The original proof involved an automata-theoretic construction, but we show that it can be carried out axiomatically in KA. In §4 we give examples of partial and total monoid equations and give a general construction that establishes completeness in those cases. The construction is a special case of the more general results of §5, but we start with it as a conceptual first step to illustrate the ideas. However, we can already derive some interesting consequences in this special case. In §5, we establish completeness for typed monoid equations. This is the most general setting covered in this paper. We give the completeness proof along with several applications. In §6 we present conclusions, future work, and open problems.

Proofs are omitted for lack of space. A full version is available online [17].

2   Preliminaries and a Negative Result

A *Kleene algebra* (KA) is an idempotent semiring $(K, +, \cdot, ^*, 0, 1)$ with an iteration operator * satisfying

$$1 + aa^* \leq a^* \quad 1 + a^*a \leq a^* \quad ax \leq x \Rightarrow a^*x \leq x \quad xa \leq x \Rightarrow xa^* \leq x$$

where $\leq$ refers to the natural partial order on K: $a \leq b \Leftrightarrow a + b = b$. A *Kleene algebra with tests* (KAT) is a two-sorted structure $(K, B, +, \cdot, ^*, ^-, 0, 1)$ such that $(K, +, \cdot, ^*, 0, 1)$ is a KA, $(B, +, \cdot, ^-, 0, 1)$ is a Boolean algebra, and $(B, +, \cdot, 0, 1)$ is a subalgebra of $(K, +, \cdot, 0, 1)$ as an idempotent semiring.

Let Σ be a finite alphabet of symbols. The *free monoid* $(\Sigma^*, \cdot, \epsilon)$ generated by Σ is the set Σ^* of words over Σ together with the operation $\cdot$ of string concatenation and the empty string ϵ as identity. To generalize this construction, we consider a *finitely presented monoid* $M = \langle a, b, \ldots \mid u_1 \equiv u_2, v_1 \equiv v_2, \ldots \rangle$ with a finite set of generators $\Sigma = \{a, b, \ldots\}$ and a finite set of relations $R = \{(u_1, u_2), (v_1, v_2), \ldots\}$. We interchangeably write a relation as an equation $u \equiv u'$ or as a pair (u, u'). Let $\leftrightarrow_R^*$ be the smallest congruence on Σ^* that contains R. The congruence class of a string u is denoted by $[u]$. The finitely presented monoid $M = \langle \Sigma \mid R \rangle = \Sigma^*/R$ has the congruence classes $\{[u] \mid u \in \Sigma^*\}$ of $\leftrightarrow_R^*$ as its carrier. Multiplication is given by $[u] \cdot [v] \mapsto [uv]$, and the identity is $[\epsilon]$.

We define *regular expressions* over the alphabet Σ to be the terms given by the grammar $e, e_1, e_2 ::= a \in \Sigma \mid 1 \mid 0 \mid e_1 + e_2 \mid e_1; e_2 \mid e^*$. We can interpret a regular expression as a subset of a finitely presented monoid $M = \langle \Sigma \mid R \rangle$ with multiplication $\cdot$ and identity $1_M = [\epsilon]$. The function $\mathcal{R}_M$, called *language interpretation* in M, sends a regular expression to a set of elements of M:

$$\mathcal{R}_M(a) = \{[a]\} \qquad \mathcal{R}_M(e_1 + e_2) = \mathcal{R}_M(e_1) \cup \mathcal{R}_M(e_2)$$
$$\mathcal{R}_M(1) = \{1_M\} \qquad \mathcal{R}_M(e_1; e_2) = \mathcal{R}_M(e_1) \cdot \mathcal{R}_M(e_2)$$
$$\mathcal{R}_M(0) = \emptyset \qquad \mathcal{R}_M(e^*) = \bigcup_{n \geq 0} \mathcal{R}_M(e)^n$$

where $\cdot$ on sets is given by $A \cdot B = \{u \cdot v \mid u \in A, v \in B\}$, and A^n is defined inductively as $A^0 = \mathcal{R}_M(1)$ and $A^{n+1} = A^n \cdot A$. The image of the interpretation $\mathcal{R}_M$ together with the operations $\cup$, $\cdot$, $*$, $\emptyset$, $\{1_M\}$ is the *algebra of regular sets* over M, denoted by $\mathsf{Reg}\,M$. If M is the free monoid Σ^*, then $\mathcal{R}_M$ is the standard language interpretation of regular expressions.

It is known that the algebra of regular sets $\mathsf{Reg}\,\Sigma^*$ is the free Kleene algebra generated by Σ [18]. This is equivalent to the completeness of the axioms of KA for the standard language interpretation $\mathcal{R}$ of regular expressions. That is, for any two regular expressions e_1, e_2 over Σ, if $\mathcal{R}(e_1) = \mathcal{R}(e_2)$ then $\mathsf{KA} \vdash e_1 \equiv e_2$. The question then arises if this result extends to the general case of $\mathsf{Reg}\,M$ for a finitely presented monoid $M = \langle \Sigma \mid R \rangle$. We ask the question of whether $\mathcal{R}_M(e_1) = \mathcal{R}_M(e_2)$ implies provability of $e_1 \equiv e_2$ in a system of KA augmented with (at least) the equations corresponding to the relations R.

In general, the answer to the question posed in the previous paragraph is negative. That is, there exists a finitely presented monoid $M = \langle \Sigma \mid R \rangle$ such that the equational theory of $\mathsf{Reg}\,M$ is not recursively enumerable, and therefore not recursively axiomatizable. The *equational theory* of the Kleene algebra $\mathsf{Reg}\,M$ is the set of equations between regular expressions that are true in $\mathsf{Reg}\,M$ under the interpretation $\mathcal{R}_M$, i.e., the set $\{e_1 \equiv e_2 \mid \mathcal{R}_M(e_1) = \mathcal{R}_M(e_2)\}$. We show this negative result using the ideas developed in [11]. The proof specifies a way to construct effectively the monoid whose existence we claim.

Theorem 1. There exists a finitely presented monoid M such that the equational theory of $\mathsf{Reg}\,M$ is not recursively enumerable.

This negative result says that we can only hope to identify subclasses of finitely presented monoids M such that the algebra $\mathsf{Reg}\,M$ of regular sets over M is axiomatizable. The idea is to first restrict attention to those finite monoid presentations, where the equations can be oriented to give a confluent and terminating rewrite system. This allows one to consider as canonical representatives the irreducible strings of the congruence classes. Then, we focus on a subclass that allows two crucial algebraic constructions: a "descendants" automata-theoretic construction, and an "ancestors" construction, which is a homomorphism.

The proof of Theorem 1 is similar to that of [11, Theorem 4.1(ii)], but strictly speaking, neither theorem follows from the other. The theorem of [11] gives a uniform Π_2^0-lower bound when the monoid is considered part of the input, whereas Theorem 1 gives a Π_1^0-lower bound for a fixed monoid.

3 String Rewriting Systems

In this section we establish a connection between the classical theory of string rewriting systems and Kleene algebra. More specifically, we recall a result regarding the preservation of regularity: for every *inverse context-free* system R and a regular set L, the set of the R-descendants of L is also regular [16]. This result involves an automata-theoretic construction, which can be modeled in KA, because an automaton can be represented as an appropriate KA term [18].

The combinatorial arguments of the construction can then be replaced by equational reasoning in KA. As it turns out, this connection will allow us to obtain powerful completeness metatheorems in later sections.

A *string rewriting system* R over a finite alphabet Σ consists of rules $\ell \to r$, where ℓ and r are finite strings over Σ. This extends to the *one-step rewrite relation* $\to_R$, given by $x\ell y \to_R xry$, for strings x, y and rule $\ell \to r$ of R. If $x \to_R y$ then we say that y is an *R-successor* of x, and x is an *R-predecessor* of y. We write $\to_R^*$ for the reflexive-transitive closure of $\to_R$, which is called the *rewrite relation* for R. If u, v are strings for which $u \to_R^* v$ we say that v is an *R-descendant* of u, and that u is an *R-ancestor* of v. For a set of strings L:

$$\mathsf{Desc}_R(L) = \{v \mid \exists u \in L.\ u \to_R^* v\} \qquad \mathsf{Ance}_R(L) = \{u \mid \exists v \in L.\ u \to_R^* v\}$$

So, $\mathsf{Desc}_R(L)$ is the set of all the R-descendants of the strings in L, and similarly $\mathsf{Ance}_R(L)$ is the set of all R-ancestors of the strings in L. The *inverse system* R^{-1} of R is the system that results by taking a rule $r \to \ell$ for every rule $\ell \to r$ of R. If u is an R-ancestor of a string v, then u is an R^{-1}-descendant of v. Define $\leftrightarrow_R^*$ to be the smallest congruence on Σ^* that contains $\{(u, v) \mid u \to v \text{ is } R\text{-rule}\}$. The congruence class of a string u is denoted by $[u]$.

Let R be a rewrite system. We say that R is *terminating* if there is no infinite rewrite chain $x_0 \to_R x_1 \to_R x_2 \to_R \cdots$. If R has rules of the form $\ell \to r$ with $|r| < |\ell|$ then it is terminating, because every rule application strictly reduces the length of the string. A string x is called *R-irreducible* if no rule of R applies to it, that is, there is no y with $x \to_R y$. We say that R is *confluent* if $u \to_R^* x$ and $u \to_R^* y$ imply that there exists z with $x \to_R^* z$ and $y \to_R^* z$. It is said that R has the *Church-Rosser property* (we also say that "R is Church-Rosser") if for all strings x, y with $x \leftrightarrow_R^* y$ there exists z such that $x \to_R^* z$ and $y \to_R^* z$. It is a standard result that confluence and the Church-Rosser property are equivalent [16]. A system R is said to be *locally (or weakly) confluent* if for all strings u, x, y with $u \to_R x$ and $u \to_R y$, there exists a string z such that $x \to_R^* z$ and $y \to_R^* z$. If R is both locally confluent and terminating, then R is confluent [16, 19].

Suppose that R is confluent and terminating. We map each string u to the unique R-irreducible string $\mathsf{nf}_R(u)$ that results from rewriting u as much as possible. For strings u, v, it holds that $u \leftrightarrow_R^* v$ iff $\mathsf{nf}_R(u) = \mathsf{nf}_R(v)$. So, two strings are congruent iff they can be rewritten to the same R-irreducible. For every congruence class $[u]$ of $\leftrightarrow_R^*$, we choose as *canonical representative* (normal form) the R-irreducible string $\mathsf{nf}_R(u)$.

Definition 1 (Total Coalesced Product). Assume that R is confluent and terminating, and let I_R be the set of R-irreducible strings. Define the binary operation $\diamond$ on I_R, called *coalesced product*, by $u \diamond v = \mathsf{nf}_R(uv)$. We lift the operation to sets of R-irreducible strings as $A \diamond B = \{u \diamond v \mid u \in A,\ v \in B\}$.

Definition 2. Let R be an arbitrary string rewrite system. For a language $L \subseteq \Sigma^*$, we define $\mathscr{C}_R(L) = \bigcup_{u \in L}[u] = \{v \mid \exists u \in L.\ v \leftrightarrow_R^* u\}$. Assume additionally that R is confluent and terminating, so that the function nf_R is well-defined. For $L \subseteq \Sigma^*$, we define $\mathscr{G}_R(L) = \{\mathsf{nf}_R(u) \mid u \in L\}$.

Lemma 1. Let R be a confluent and terminating rewrite system over Σ.

1. $\mathscr{C}_R(L) = \bigcup\{[u] \mid u \in \mathscr{G}_R(L)\}$, for a language $L \subseteq \Sigma^*$.
2. $\mathscr{G}_R(L_1) = \mathscr{G}_R(L_2)$ iff $\mathscr{C}_R(L_1) = \mathscr{C}_R(L_2)$, for languages $L_1, L_2 \subseteq \Sigma^*$.
3. $\mathscr{C}_R(L) = \mathsf{Ance}_R(\mathsf{Desc}_R(L))$, for a language $L \subseteq \Sigma^*$.

A rewrite system R is said to *preserve regularity* if for every regular language L, the R-descendants $\mathsf{Desc}_R(L)$ form a regular set. A system R is called *inverse context-free* if it only contains rules of the form $\ell \to r$, where $|r| \leq 1$. That is, every right-hand side of a rule is either a single letter or the empty string. A classical result of the theory of string rewriting is that inverse context-free systems preserve regularity (see Chapter 4 of [16] for a detailed proof). The proof of this fact uses a construction on finite automata, which we briefly present here. We will be referring to it as the *descendants construction*. Suppose that L is a regular language, recognized by an automaton $\mathcal{A}$. The automaton is possibly nondeterministic and it may have epsilon transitions. We will describe a sequence of transformations on $\mathcal{A}$. When the sequence reaches a fixpoint, we obtain an automaton (nondeterministic with epsilon transitions) that recognizes $\mathsf{Desc}_R(L)$.

– Suppose that the system R has a rule $\ell \to a$, where a is a single letter, and $\ell = \ell_1 \ell_2 \cdots \ell_m$ is a string of length m. We assume that there is an ℓ-path from the state q_0 to the state q_n of the automaton. That is, a sequence

$$q_0 \xrightarrow{x_1} q_1 \xrightarrow{x_2} q_2 \xrightarrow{x_3} \cdots \xrightarrow{x_{n-1}} q_{n-1} \xrightarrow{x_n} q_n,$$

where each x_i is a letter or ϵ, $x_1 \cdot x_2 \cdot \ldots \cdot x_n = \ell$, and each $q_{i-1} \xrightarrow{x_i} q_i$ is a transition of the automaton. We add the transition $q_0 \xrightarrow{a} q_n$. The idea is that if the automaton accepts $x \ell y$, then it should also accept the R-descendant $x a y$.

– Similarly, suppose that the system R has a rule $\ell \to \epsilon$, where ϵ is the empty string, and that there is an ℓ-path from the state q_0 to the state q_n. Then, we add the epsilon transition $q_0 \xrightarrow{\epsilon} q_n$ to the transition table of the automaton. This process is iterated until no new transitions are added. The resulting automaton accepts exactly the set of R-descendants $\mathsf{Desc}_R(L)$.

Theorem 2. Let R be an inverse context-free rewrite system and e a regular expression whose interpretation is $L = \mathcal{R}(e)$. We can construct effectively a new regular expression $\hat{e}$ such that $\mathsf{KA}_R \vdash e \equiv \hat{e}$ and $\mathcal{R}(\hat{e}) = \mathsf{Desc}_R(L)$. KA_R is the system KA augmented with an equation $\ell \equiv r$ for every rewrite rule $\ell \to r$ of R.

Theorem 2 says that the descendants construction, which is combinatorial, can be modeled algebraically in the system of KA with some extra equations. This is a central technical result that we will use for our later theorems.

4   Completeness: (Partial) Monoid Equations

In this section we present our first completeness metatheorems, from which we can prove the existence of free language models for systems of KA with extra monoid and partial monoid equations. Our metatheorems are not only a conceptual first step towards the more general typed monoid case, which we investigate in §5, but they also allow us to obtain previously unknown completeness results. As a concrete novel application, think of the assignment statement $x := c$, where c is a constant. The action $x := c$ is idempotent, meaning that the effect of $x := c; x := c$ is the same as the effect of $x := c$. We express this fact with the monoid equation $aa \equiv a$, where a is a single letter abstraction of the assignment. KA can be augmented with any number of such idempotence equations, and our metatheorem implies the existence of a free language model (see Example 1).

Definition 3 (Language Interpretation). Let R be a confluent and terminating rewrite system. The corresponding coalesced product is $\diamond$. We define the function $\mathcal{G}_R$ that sends a regular expression to a set of R-irreducibles:

$$\mathcal{G}_R(a) = \{\mathsf{nf}_R(a)\} \qquad \mathcal{G}_R(e_1 + e_2) = \mathcal{G}_R(e_1) \cup \mathcal{G}_R(e_2)$$
$$\mathcal{G}_R(0) = \emptyset \qquad \mathcal{G}_R(e_1; e_2) = \mathcal{G}_R(e_1) \diamond \mathcal{G}_R(e_2)$$
$$\mathcal{G}_R(1) = \{\mathsf{nf}_R(\epsilon)\} \qquad \mathcal{G}_R(e^*) = \bigcup_{n \geq 0} \mathcal{G}_R(e)^{\langle n \rangle}$$

where, for a set A of R-irreducibles, $A^{\langle n \rangle}$ is defined by $A^{\langle 0 \rangle} = \mathcal{G}_R(1)$ and $A^{\langle n+1 \rangle} = A^{\langle n \rangle} \diamond A$. We also define the interpretation $\mathcal{C}_R(e) = \mathscr{C}_R(\mathcal{R}(e)) = \bigcup_{u \in \mathcal{R}(e)} [u]$.

Let R be a confluent and terminating system over Σ, and $M = \langle \Sigma \mid R \rangle$ be the corresponding monoid. For a regular expression e, we have that $\mathcal{R}_M(e) = \{[u] \mid u \in \mathcal{G}_R(e)\}$. The algebra $\mathsf{Reg}\,M$ is isomorphic to the algebra that is the image of $\mathcal{G}_R$. This implies that $\mathcal{R}_M(e_1) = \mathcal{R}_M(e_2)$ iff $\mathcal{G}_R(e_1) = \mathcal{G}_R(e_2)$. So, our investigations of completeness can be w.r.t. the interpretation $\mathcal{G}_R$.

Lemma 2. Let R be a confluent and terminating string rewrite system.
1. $\mathcal{G}_R(e) = \{\mathsf{nf}_R(u) \mid u \in \mathcal{R}(e)\} = \mathscr{G}_R(\mathcal{R}(e))$, for an expression e.
2. $\mathcal{C}_R(e) = \bigcup\{[v] \mid v \in \mathcal{G}_R(e)\}$, for an expression e.
3. $\mathcal{G}_R(e_1) = \mathcal{G}_R(e_2)$ iff $\mathcal{C}_R(e_1) = \mathcal{C}_R(e_2)$, for expressions e_1, e_2.

Definition 4 (Well-Behaved Rewrite System). Let R be a rewrite system over Σ. We say that R is *well-behaved* if it consists of finitely many rules $\ell \to r$ with $|r| = 1$ and $|\ell| > 1$, and it additionally satisfies confluence and the following property: For every letter a of the alphabet, the R-ancestors of a form a regular set $\mathcal{R}(e_a)$ for some expression e_a, so that $\mathsf{KA}_R \vdash e_a \equiv a$. Recall that KA_R is the system of KA extended with equations corresponding to the rules of R.

Intuitively, we say that R is well-behaved if it allows two important algebraic constructions. First, the special form of the rules allows the automata-theoretic descendants construction (described in §3), which can be modeled in KA, because automata can be encoded as matrices. Then, the regularity requirement for the sets of R-ancestors of single letters implies that we can apply a homomorphism to obtain all the ancestors of a regular set. We can thus "close" a regular expression under the congruence induced by R.

Theorem 3 (Completeness). *Let R be a well-behaved rewrite system over Σ. For any expressions e_1 and e_2, $\mathcal{G}_R(e_1) = \mathcal{G}_R(e_2)$ implies that $\mathsf{KA}_R \vdash e_1 \equiv e_2$.*

Example 1 (Idempotence Hypotheses). We will see how the general completeness metatheorem we have shown (Theorem 3) can be used to obtain a completeness result for the regular algebra of a simple finitely presented monoid. Consider the monoid $M = \langle a, b \mid aa \equiv a \rangle$. The rewrite system R contains only the rule $aa \to a$. In order to invoke Theorem 3 we verify that R is well-behaved:

- For the only rule $\ell = aa \to a = r$ of R, we have that $|r| = 1$ and $|\ell| > 1$.
- To show confluence of R, it is sufficient to show local confluence, since R is terminating. This is known as Newman's Lemma (see [16, 19]). We have the following critical-pair lemma: Suppose that $u \to x$ and $u \to y$. If $x = y$, we are done. If $x \neq y$, then u, x, y must be of the following forms: $u = v_1 a^{m+1} v_2 a^{n+1} v_3$, $x = v_1 a^m v_2 a^{n+1} v_3$, and $y = v_1 a^{m+1} v_2 a^n v_3$. Notice now that $x, y \to v_1 a^m v_2 a^n v_3$, which establishes local confluence.
- For the R-ancestors of the letters a and b, we see that $\mathsf{Ance}_R(b) = \{b\}$, and $\mathsf{Ance}_R(a) = \{a^i \mid i \geq 1\} = \mathcal{R}(a^+)$, where $a^+ = a; a^*$. We put $e_b = b$ and $e_a = a^+$. Clearly, $\mathsf{KA}_R \vdash e_b \equiv b$. Reasoning in KA_R: $a \leq a^+$ and $a^+ = a; a^* \leq a \Longleftarrow a; a \leq a \Longleftarrow a; a \equiv a$. We have thus shown that $\mathsf{KA}_R \vdash e_a \equiv a$.

Since the rewrite system R satisfies the conditions of Theorem 3, we get completeness of KA together with the equation $a; a \equiv a$ for the interpretation $\mathcal{R}_M$.

We would like to generalize our result in a way that allows us to designate certain strings as being *non-well-formed* or *undefined*. Any string with a non-well-formed substring has to be discarded from the interpretation. For a string $a_1 \cdots a_k$ over the alphabet, we declare it to be non-well-formed using the equation $a_1 \cdots a_k \equiv \bot$, where $\bot$ is a special "undefined" symbol not in the alphabet.

We define a *partial monoid* to be an algebraic structure $(M, \cdot, 1_M, \bot_M)$ satisfying the monoid axioms, as well as the equations $x \cdot \bot_M = \bot_M$ and $\bot_M \cdot x = \bot_M$. The identity is 1_M, and $\bot_M$ is called the *undefined element* of M. In a presentation of a partial monoid $M_\bot = \langle \Sigma \mid x_1 \equiv y_1, x_2 \equiv y_2, \ldots, z_1 \equiv \bot, z_2 \equiv \bot, \ldots \rangle$ we allow equations $x \equiv y$ between strings over Σ (call the collection of these R), as well as equations of the form $z \equiv \bot$, where z is a string over Σ ($\bot$ is not in Σ). In order to give a concrete description of the partial monoid, we consider the strings over the extended alphabet $\Sigma \cup \{\bot\}$, and the equations $R_\bot$:

$$x_i \equiv y_i \qquad z_i \equiv \bot \qquad a\bot \equiv \bot, \ \bot a \equiv \bot \ \ (a \in \Sigma) \qquad \bot\bot \equiv \bot$$

Let $\sim$ be the smallest congruence on $(\Sigma \cup \{\bot\})^*$ that contains the relations $R_\bot$. The partial monoid $M_\bot$ is the set of strings $(\Sigma \cup \{\bot\})^*$ quotiented by the congruence $\sim$, and hence equal to $\langle \Sigma \cup \{\bot\} \mid R_\bot \rangle$. The identity is the $\sim$-congruence class $[\epsilon]$, and the undefined element is the class of $[\bot]$.

Assumption 1. We collect a list of assumptions for $(\Sigma, R, R_\bot)$. First, assume that R is a confluent and terminating rewrite system over the alphabet Σ. The rewrite system $R_\bot$ extends R with rules of the form $z \to \bot$, where $z \in \Sigma^*$ and $|z| \geq 2$. Moreover, $R_\bot$ contains the rule $\bot\bot \to \bot$, as well as all the rules $a\bot \to \bot$ and $\bot a \to \bot$ for every letter $a \in \Sigma$. We further assume that $R_\bot$ is terminating, and that the *seamlessness property* is satisfied: If xzy is a string with $z \to \bot$ in $R_\bot$, then any R-successor of xzy is of the form $x'z'y'$, where $z' \to \bot$ is in $R_\bot$. Intuitively, seamlessness says that if a string contains a non-well-formed substring, then no R-rewriting can make it well-formed.

Definition 5 (Partial Coalesced Product). Let $(\Sigma, R, R_\bot)$ satisfy Assumption 1. Define the partial *coalesced product* $\diamond$ on $R_\bot$-irreducibles in Σ^*:

$$u \diamond v = \mathsf{nf}_R(uv), \text{ if } uv \not\sim \bot; \qquad u \diamond v = \text{undefined, if } uv \sim \bot.$$

The condition $uv \not\sim \perp$ is equivalent to $\mathsf{nf}_R(uv)$ not having a substring z with $z \to \perp$. We lift the coalesced product into a total operation on sets of $R_\perp$-irreducibles: $A \diamond B = \{u \diamond v \mid u \diamond v \text{ exists}, u \in A, v \in B\}$.

Definition 6 (Language Interpretation). Let $(\Sigma, R, R_\perp)$ satisfy Assumption 1. For a string u, define $[u]_\Sigma = \Sigma^* \cap [u]$. For a language $L \subseteq \Sigma^*$, put:
$$\mathscr{G}_{R_\perp}(L) = \{\mathsf{nf}_R(u) \mid u \in L\} \setminus [\perp]_\Sigma \qquad \mathscr{C}_{R_\perp}(L) = [\perp]_\Sigma \cup \bigcup_{u \in L}[u]_\Sigma$$
Now, $\mathcal{G}_{R_\perp}$ sends a regular expression to a set of $R_\perp$-irreducibles of Σ^*:
$$\begin{aligned}
\mathcal{G}_{R_\perp}(a) &= \{\mathsf{nf}_R(a)\} \setminus [\perp]_\Sigma & \mathcal{G}_{R_\perp}(e_1 + e_2) &= \mathcal{G}_{R_\perp}(e_1) \cup \mathcal{G}_{R_\perp}(e_2) \\
\mathcal{G}_{R_\perp}(0) &= \emptyset & \mathcal{G}_{R_\perp}(e_1; e_2) &= \mathcal{G}_{R_\perp}(e_1) \diamond \mathcal{G}_{R_\perp}(e_2) \\
\mathcal{G}_{R_\perp}(1) &= \{\mathsf{nf}_R(\epsilon)\} \setminus [\perp]_\Sigma & \mathcal{G}_{R_\perp}(e^*) &= \bigcup_{n \geq 0} \mathcal{G}_{R_\perp}(e)^{\langle n \rangle}
\end{aligned}$$
where $A^{\langle 0 \rangle} = \mathcal{G}_{R_\perp}(1)$ and $A^{\langle n+1 \rangle} = A^{\langle n \rangle} \diamond A$. Define $\mathcal{C}_{R_\perp}(e) = \mathscr{C}_{R_\perp}(\mathcal{R}(e))$. The interpretation $\mathcal{G}_{R_\perp}$ discards the undefined strings, but $\mathcal{C}_{R_\perp}$ adds them all in.

Definition 7 (Well-Behaved). We suppose that $(\Sigma, R, R_\perp)$ satisfies Assumption 1. We say that it is *well-behaved* if $R_\perp$ consists of finitely many rules, every rule $\ell \to r$ of R satisfies $|r| = 1$ and $|\ell| > 1$, and it satisfies the property: For every letter a of the alphabet, the R-ancestors of a form a regular set $\mathcal{R}(e_a)$ for some regular expression e_a, so that $\mathsf{KA}_R \vdash e_a \equiv a$. The empty string and the single-letter strings are $R_\perp$-irreducible. We write $\mathsf{KA}_{R_\perp}$ for the system KA_R extended with an equation $a_1; \cdots ; a_k \equiv 0$ for every rule $a_1 \cdots a_k \to \perp$ of $R_\perp$.

Theorem 4 (Completeness). *Suppose that $(\Sigma, R, R_\perp)$ is well-behaved. Then, $\mathcal{G}_{R_\perp}(e_1) = \mathcal{G}_{R_\perp}(e_2)$ implies that $\mathsf{KA}_{R_\perp} \vdash e_1 \equiv e_2$.*

5   Completeness: Typed Monoid Equations

We further generalize the partial monoid setting by assuming more structure on the strings and the rewrite system. One major difference from the partial monoid case is the introduction of a new category of primitive symbols, the subidentities, which allow the encoding of Booleans. We show how to cover several examples: plain KAT, KAT with simple Hoare hypotheses $b; p; c \equiv 0$, KAT with hypotheses $c; p \equiv c$, and NetKAT. There are even more applications which for lack of space we do not present here: commutativity equations $b; p \equiv p; b$ (test b, atomic action p), Boolean equations $b \equiv c$ (tests b, c), and so on. These examples attest to the generality and wide applicability of our technique.

Assumption 2. We collect a list of assumptions for $(P, Id, R, R_\perp)$. Let $\Sigma = P \cup Id$ be a finite alphabet, whose symbols are partitioned into a set P of *action symbols* and a set Id of *subidentities*. We write $p, q, r, \ldots$ to vary over actions symbols, $\alpha, \beta, \gamma, \ldots$ to vary over subidentities, and $a, b, c, \ldots$ to vary over arbitrary symbols of Σ. Let S be the subset of Σ^* consisting of all strings in which an action symbol p always appears surrounded by subidentities, as in $\alpha p \beta$. The set S is regular, and the corresponding regular expression is $e_S = Id \cdot (Id^* \cdot P \cdot Id)^* \cdot Id^*$. Let R be a rewrite system over Σ that includes at least the rules $\alpha \alpha \to \alpha$ for every subidentity $\alpha \in Id$, and additionally it satisfies:

(1) S is closed under $\to_R$: if $x \in S$ and $x \to_R y$ then $y \in S$. Moreover, S is closed under the inverse of $\to_R$: if $y \in S$ and $x \to_R y$ then $x \in S$. (2) For every rule $\ell \to r$ of R we have that $|\ell| > |r|$. (3) R is confluent on S: For $u, x, y \in S$, $u \to_R^* x$ and $u \to_R^* y$ imply that $x \to_R^* z$ and $y \to_R^* z$ for some $z \in S$. Now, suppose that $R_\perp$ extends R with the rules $\alpha\beta \to \perp$ for all subidentities $\alpha \neq \beta$, and possibly more rules of the form $z \to \perp$, where $z \in S$ and $|z| \geq 2$. Moreover, $R_\perp$ contains all the rules $a\perp \to \perp$, $\perp a \to \perp$ (for each $a \in \Sigma$), as well as the rule $\perp\perp \to \perp$. We assume that $R_\perp$ satisfies additionally the *seamlessness property*: For $xzy \in S$ with $z \to \perp$ in $R_\perp$, any R-successor of xzy is of the form $x'z'y'$ for some rule $z' \to \perp$ of $R_\perp$. We will use the term *irreducible* (unqualified) to mean $R_\perp$-irreducible of S. Finally, define the function cp to send every letter a of Σ to a finite subset $\mathsf{cp}(a)$ of S, called the *components* of a. For a subidentity $\alpha \in Id$, we put $\mathsf{cp}(\alpha) = \{\alpha\}$. For an action symbol $p \in P$, we put $\mathsf{cp}(p) = \{\alpha p\beta \mid \alpha, \beta \in Id\}$.

Definition 8 (Language Interpretation). Let $(P, Id, R, R_\perp)$ satisfy Assumption 2. For a string u, we put $[u]_S = S \cap [u]$. For a language $L \subseteq S$, we define:
$$\mathscr{G}_{R_\perp}(L) = \{\mathsf{nf}_R(u) \mid u \in L\} \setminus [\perp]_S \qquad \mathscr{C}_{R_\perp}(L) = [\perp]_S \cup \bigcup_{u \in L}[u]_S$$
The *coalesced product* of irreducibles, written $\diamond$, is defined as in Definition 5. The interpretation $\mathcal{G}_{R_\perp}$ sends a regular expression to a set of irreducibles:
$$\mathcal{G}_{R_\perp}(a) = \mathsf{nf}_R(\mathsf{cp}(a)) \setminus [\perp]_S \qquad \mathcal{G}_{R_\perp}(e_1 + e_2) = \mathcal{G}_{R_\perp}(e_1) \cup \mathcal{G}_{R_\perp}(e_2)$$
$$\mathcal{G}_{R_\perp}(0) = \emptyset \qquad \mathcal{G}_{R_\perp}(e_1; e_2) = \mathcal{G}_{R_\perp}(e_1) \diamond \mathcal{G}_{R_\perp}(e_2)$$
$$\mathcal{G}_{R_\perp}(1) = Id \qquad \mathcal{G}_{R_\perp}(e^*) = \bigcup_{n \geq 0} \mathcal{G}_{R_\perp}(e)^{(n)}$$
Define $\mathcal{C}_{R_\perp}(e) = \mathscr{C}_{R_\perp}(\mathcal{R}(e))$, for expressions e with $\mathcal{R}(e) \subseteq S$.

Definition 9 (Well-Behaved). Let $(P, Id, R, R_\perp)$ be a tuple satisfying Assumption 2. We say that the tuple is *well-behaved* if $R_\perp$ consists of finitely many rules, every rule $\ell \to r$ of R satisfies $|r| = 1$ and $|\ell| > 1$, and it satisfies the following property: For every letter a of the alphabet, the R-ancestors of a form a regular set $\mathcal{R}(e_a)$ for some regular expression e_a, so that $\mathsf{KA}_R \vdash e_a \equiv a$.

We define the finite collection E of *equations associated* with the well-behaved tuple $(P, Id, R, R_\perp)$ to contain: (1) an equation $x \equiv y$ for every rule $x \to y$ of R, (2) an equation $z \equiv 0$ for every rule $z \to \perp$ of $R_\perp$, as well as (3) the equation $\sum_{\alpha \in Id} \alpha \equiv 1$. We write KA_E for the system of KA augmented with the equations E. It is easy to prove in KA_E the equation $\sum_{x \in \mathsf{cp}(a)} x \equiv a$ for every letter a.

Theorem 5 (Completeness). *Let $(P, Id, R, R_\perp)$ be well-behaved, and E be the associated equations. Then, $\mathcal{G}_{R_\perp}(e_1) = \mathcal{G}_{R_\perp}(e_2)$ implies that $\mathsf{KA}_E \vdash e_1 \equiv e_2$.*

Applications. Theorem 5 gives us four completeness results as corollaries. First, we show that KAT is complete for the standard interpretation of KAT terms as sets of guarded strings. We then consider the case of KAT extended with simple Hoare hypotheses $b; p; c \equiv 0$ (tests b, c, atomic action p), and with hypotheses $c; p \equiv c$. We conclude with a completeness proof for NetKAT.

Theorem 6. *Let $\mathcal{G}_{\mathsf{KAT}}$ be the standard interpretation of KAT expressions. For any e_1 and e_2, it holds that $\mathcal{G}_{\mathsf{KAT}}(e_1) = \mathcal{G}_{\mathsf{KAT}}(e_2)$ implies $\mathsf{KAT} \vdash e_1 \equiv e_2$.*

A *simple Hoare assertion* is an expression $\{b\}p\{c\}$, where b, c are tests and p is an atomic action. It can be encoded in KAT with the equation $b; p; \neg c \equiv 0$. This equation is equivalent to the conjunction of the equations $\beta; p; \gamma \equiv 0$, where β, γ are atoms with $\beta \leq b$ and $\gamma \leq \neg c$. So, w.l.o.g. we restrict attention to assertions of the form $\beta; p; \gamma \equiv 0$, where β, γ are atoms and p is an atomic action.

Theorem 7. Let Z_h be a finite collection of strings of the form $\gamma p \delta$, where γ, δ are atoms and p is an atomic action symbol. Let W be the set of strings containing some $\gamma p \delta$ in Z_h, and H be the collection of equations $\gamma; p; \delta \equiv 0$ for every $\gamma p \delta$ in Z_h. Define the interpretation $\mathcal{G}_h$ by $\mathcal{G}_h(e) = \mathcal{G}_{\mathsf{KAT}}(e) \setminus W$. Then, $\mathcal{G}_h(e_1) = \mathcal{G}_h(e_2)$ implies $\mathsf{KAT} + H \vdash e_1 \equiv e_2$.

We consider now another class of equations of the form $c; p \equiv c$, where c is a test and p is an atomic action. We see that $c; p \equiv c$ is equivalent to the conjunction of $\gamma; p \equiv \gamma$ for $\gamma \leq c$. So, we can restrict our attention to equations of the form $\gamma; p \equiv \gamma$, where γ is an atom, and p is an atomic action.

Theorem 8. Let X be a finite set of strings of the form γp, where γ is an atom and p is an atomic action symbol, and H be the set of equations $\gamma; p \equiv \gamma$ for every γp in X. For an atomic action symbol p, define the set of atoms $A(p) = \{\gamma \mid \gamma p \in X\}$. Let $\mathcal{G}_h$ be the interpretation that differs from $\mathcal{G}_{\mathsf{KAT}}$ only for the base case of atomic action symbols: $\mathcal{G}_h(p) = A(p) \cup \{\gamma p \delta \mid \gamma \notin A(p)\}$. Then, $\mathcal{G}_h(e_1) = \mathcal{G}_h(e_2)$ implies $\mathsf{KAT} + H \vdash e_1 \equiv e_2$, for any KAT expressions e_1, e_2.

We turn to the case of NetKAT. Fix an alphabet At of atoms. For $\alpha \in At$ we introduce an action symbol p_α, and we put $P = \{p_\alpha \mid \alpha \in At\}$. Let dup be a new action symbol, and set $\Sigma = P \cup \{\mathsf{dup}\} \cup At$. NetKAT extends KA with:

$$\sum_{\alpha \in At} \alpha \equiv 1 \qquad \alpha; \mathsf{dup} \equiv \mathsf{dup}; \alpha \qquad p_\alpha \equiv p_\alpha; \alpha$$
$$\alpha; \beta \equiv 0 \ (\alpha \neq \beta) \qquad p_\alpha; p_\beta \equiv p_\beta \qquad \alpha \equiv \alpha; p_\alpha$$

The axioms imply $\alpha; \alpha \equiv \alpha; p_\alpha; \alpha \equiv \alpha; p_\alpha \equiv \alpha$, for every atom α. So, NetKAT can also be defined as an extension of KAT. The following axioms

$$\sum_{\alpha \in At} \alpha \equiv 1 \qquad \alpha; \alpha \equiv \alpha \qquad a; p_\alpha; \alpha \equiv \alpha \qquad \alpha; \mathsf{dup}; \beta \equiv 0 \ (\alpha \neq \beta)$$
$$\alpha; \beta \equiv 0 \ (\alpha \neq \beta) \qquad p_\alpha; \alpha; p_\beta \equiv p_\beta \qquad \alpha; p_\beta; \gamma \equiv 0 \ (\beta \neq \gamma)$$

give an equivalent axiomatization of NetKAT.

Theorem 9. Let At be the subidentities (atoms), and $P' = P \cup \{\mathsf{dup}\}$ be the alphabet of action symbols, where $P = \{p_\alpha \mid \alpha \in At\}$. Define R and $R_\perp$ as:

$$\alpha\alpha \to \alpha \ (\alpha \in At) \qquad \alpha p_\alpha \alpha \to \alpha \ (\alpha \in At) \qquad p_\alpha \alpha p_\beta \to p_\beta \ (\alpha, \beta \in At)$$
$$\alpha\beta \to \perp \ (\alpha \neq \beta) \qquad \alpha\mathsf{dup}\beta \to \perp \ (\alpha \neq \beta) \qquad \alpha p_\beta \gamma \to \perp \ (\beta \neq \gamma)$$

$(P', At, R, R_\perp)$ is well-behaved, and NetKAT is complete for $\mathcal{G}_{R_\perp}$.

6   Conclusion

We have identified sufficient conditions for the construction of free language models for systems of Kleene algebra with additional equations. The construction provides a uniform approach to deductive completeness and coalgebraic decision procedures. The criteria are given in terms of *inverse context-free rewrite systems* [16]. They imply the existence of free language models in a wide range of previously studied instances, including KAT [6] and NetKAT [8], as well as some new ones. We have also given a negative result that establishes a limit to the applicability of the technique.

For the future, we would like to investigate the possibility of developing a uniform approach to coalgebraic bisimulation-based decision procedures [8, 12–15]. Such decision procedures typically involve some variant of Brzozowski derivatives and are highly dependent on the existence of language models.

Acknowledgments. We thank Bjørn Grathwohl, Stathis Zachos, and the anonymous reviewers for helpful suggestions. This work was supported by the National Security Agency under award #H98230-14-C-0140.

References

1. Angus, A., Kozen, D.: Kleene algebra with tests and program schematology. Technical Report TR2001-1844, CS Department, Cornell University (July 2001)
2. Barth, A., Kozen, D.: Equational verification of cache blocking in LU decomposition using Kleene algebra with tests. Technical Report TR2002-1865, Computer Science Department, Cornell University (June 2002)
3. Cohen, E.: Hypotheses in Kleene algebra. Technical report, Bellcore (1993)
4. Cohen, E.: Lazy caching in Kleene algebra (1994)
5. Cohen, E.: Using Kleene algebra to reason about concurrency control. Technical report, Telcordia, Morristown, N.J (1994)
6. Kozen, D.: Kleene algebra with tests. Transactions on Programming Languages and Systems 19(3), 427–443 (1997)
7. Kozen, D., Patron, M.C.: Certification of compiler optimizations using Kleene algebra with tests. In: Proc. 1st Int. Conf. Comput. Logic (CL 2000), pp. 568–582 (2000)
8. Anderson, C.J., Foster, N., Guha, A., Jeannin, J.-B., Kozen, D., Schlesinger, C., Walker, D.: NetKAT: Semantic foundations for networks. In: Proceedings of POPL 2014, San Diego, California, USA, pp. 113–126. ACM (January 2014)
9. Kozen, D., Smith, F.: Kleene algebra with tests: Completeness and decidability. In: van Dalen, D., Bezem, M. (eds.) CSL 1996. LNCS, vol. 1258, pp. 244–259. Springer, Heidelberg (1997)
10. Hardin, C., Kozen, D.: On the elimination of hypotheses in Kleene algebra with tests. Technical Report TR2002-1879, CS Department, Cornell University (2002)
11. Kozen, D.: On the complexity of reasoning in Kleene algebra. Information and Computation 179, 152–162 (2002)
12. Foster, N., Kozen, D., Milano, M., Silva, A., Thompson, L.: A coalgebraic decision procedure for NetKAT. Technical Report, Computing and Information Science, Cornell University (2014), http://hdl.handle.net/1813/36255
13. Grathwohl, N.B.B., Kozen, D., Mamouras, K.: KAT + B! Technical Report, CIS, Cornell University (January 2014), http://hdl.handle.net/1813/34898
14. Rot, J., Bonsangue, M., Rutten, J.: Coalgebraic bisimulation-up-to. In: van Emde Boas, P., Groen, F.C.A., Italiano, G.F., Nawrocki, J., Sack, H. (eds.) SOFSEM 2013. LNCS, vol. 7741, pp. 369–381. Springer, Heidelberg (2013)

15. Bonchi, F., Pous, D.: Checking NFA equivalence with bisimulations up to congruence. In: Proceedings of POPL 2013, pp. 457–468. ACM (2013)
16. Book, R.V., Otto, F.: String-Rewriting Systems. Springer (1993)
17. Kozen, D., Mamouras, K.: Kleene algebra with equations. Technical Report CIS, Cornell University (February 2014), http://hdl.handle.net/1813/36202
18. Kozen, D.: A completeness theorem for Kleene algebras and the algebra of regular events. Infor. and Comput. 110(2), 366–390 (1994)
19. Baader, F., Nipkow, T.: Term Rewriting and All That. CUP (1998)

All–Instances Termination of Chase
is Undecidable

Tomasz Gogacz* and Jerzy Marcinkowski**

Institute of Computer Science,
University of Wrocław, Poland

Abstract. We show that all–instances termination of chase is undecidable. More precisely, there is no algorithm deciding, for a given set $\mathcal{T}$ consisting of Tuple Generating Dependencies (a.k.a. Datalog$^\exists$ program), whether the $\mathcal{T}$-chase on D will terminate for every finite database instance D. Our method applies to Oblivious Chase, Semi-Oblivious Chase and – after a slight modification – also for Standard Chase. This means that we give a (negative) solution to the all–instances termination problem for all version of chase that are usually considered.

The arity we need for our undecidability proof is three. We also show that the problem is EXPSPACE-hard for binary signatures, but decidability for this case is left open.

Both the proofs – for ternary and binary signatures – are easy. Once you know them.

1 Introduction

The chase procedure was defined in late 1970s and has been considered one of the most fundamental database theory algorithms since then. It has been applied to a wide spectrum of problems, for example for checking containment of queries under constraints [ASU79] or for testing implication between sets of database dependencies ([MMS79], [BV84]). A new wave of interest in this notion began when the theory of data exchange was founded ([FKPP05]), where chase is used to compute solutions to data exchange problems. This interest was further strengthened recently by the Datalog$^\pm$ program [CGL09], [CGL12].

The **basic idea of a $\mathcal{T}$-chase** is as follows. We consider a set $\mathcal{T}$ of Tuple Generating Dependencies[1], which means rules (constraints) of the form:

$$\Phi(\bar{x}, \bar{y}) \Rightarrow \exists \bar{z} \, \Psi(\bar{x}, \bar{z})$$

where Φ and Ψ are conjunctive queries[2], and where $\bar{x}$, $\bar{y}$ and $\bar{z}$ are tuples of variables. Then, for a database instance D we try – step by step – to extend D,

---

* Supported by Polish National Science Centre grant 2013/09/N/ST6/01188.
** Supported by Polish National Science Centre grant DEC-2013/09/B/ST6/01535.

[1] Such sets are also known as Datalog$^\exists$ programs, and we will use the word "program" in this sense. While chase is sometimes also defined for other types of dependencies, we only consider Tuple Generating Dependencies in this paper.

[2] Φ and Ψ are positive, without equality. Our negative results hold for single head TGDs, which means that Ψ is a single atom.

J. Esparza et al. (Eds.): ICALP 2014, Part II, LNCS 8573, pp. 293–304, 2014.
© Springer-Verlag Berlin Heidelberg 2014

by adding new elements and atoms, so that the new database satisfies the constraints from $\mathcal{T}$: *whenever* there are some elements $\bar{a}$, $\bar{b}$ in the current structure, such that $\Phi(\bar{a}, \bar{b})$ is true, a tuple $\bar{c}$ of new elements is created and new relational atoms added, to make $\Psi(\bar{a}, \bar{c})$ also true. Notice that the tuple $\bar{z}$ can be empty. In such case the TGD under consideration degenerates to a plain Datalog rule.

As it turns out, there are several possible semantics of the *whenever* above, leading to several versions of the chase procedure. The Standard Chase (also known as Restricted Chase) is a lazy version – it only adds new elements if $\Phi(\bar{a}, \bar{b})$ is true in the current structure, but $\exists \bar{z}\, \Psi(\bar{a}, \bar{z})$ is (at this point of execution) false. Oblivious and Semi-Oblivious Chase ([M09]) are eager versions. Oblivious Chase always adds one tuple $\bar{c}$ for each tuple $\bar{a}, \bar{b}$ such that $\Phi(\bar{a}, \bar{b})$ is true. Semi-Oblivious Chase always adds one tuple $\bar{c}$ for each tuple $\bar{a}$ such that $\exists \bar{y}\, \Phi(\bar{a}, \bar{y})$ is true.

It is not hard to notice that the order of execution does not matter for Oblivious and Semi-Oblivious Chase. Whatever order the candidate tuples are picked in, we will eventually get the same structure[3]. But Standard Chase is non-deterministic – different orders in which tuples are picked can eventually lead to different structures.

One more version of the procedure is Core Chase (see [DNR08]). It is again a lazy version, but a parallel one: all the rules applicable at some point are triggered at the same time. In this way the non-determinism of Standard Chase is got rid of. For reasons that we will not discuss here Core Chase is slightly more complicated than that (and not really practical – the cost of each step is DP-complete).

As we said before, the chase procedure is almost ubiquitous in database theory. This phenomenon is discussed in [DNR08]: *"the applicability of the same tool to (..) seemingly different problems is not accidental, and it is due to a deeper, tool–independent reason: to solve these problems, it suffices to exhibit a representative (database) instance U with two key properties, and the chase is an algorithm for finding such an instance."* The two key properties of the instance U, being the result of $\mathcal{T}$-chase on an a database instance D, for given set $\mathcal{T}$ of tuple generating dependencies and for given database instance D are that:

-U is a model of $\mathcal{T}$ and D;

-U is universal - there is a homomorphism from U into every model of D and $\mathcal{T}$.

But U, or **Chase**$(D, \mathcal{T})$, as we prefer to call the structure resulting from running a $\mathcal{T}$-chase on D, is in many cases only useful when it is finite, which only happens if (and only if) the chase procedure terminates. One of the applications where finiteness of Chase$(D, \mathcal{T})$ is a key issue is considered in [FKPP05], and a sufficient condition on $\mathcal{T}$, implying finiteness of Chase$(D, \mathcal{T})$ was studied in this paper, called Weak Acyclicity. Weak Acyclicity is a property of $\mathcal{T}$ alone, so it implies termination regardless of D. This reflects the fact that the typical context in which database constraints are analyzed is the static analysis context – we want to optimize $\mathcal{T}$ before knowing D. So, in particular, it is natural to want to be sure that $\mathcal{T}$–chase on D will terminate on D before knowing the D itself.

---

[3] If chase does not terminate the claim is true provided the order is fair – each tuple will be eventually picked.

Many other conditions like that were studied. For example the Stratified-Witness property ([DT03]), which is historically earlier, and stronger (i.e. narrower), than Weak Acyclicity. Then it was the Rich Acyclicity criterion, introduced in [HS07], and proved in [GO11] to imply termination of Oblivious Chase for all instances D. A condition based on stratification of rules was introduced in [DNR08]. As it turned out to only guarantee termination of the Standard Chase, another class of sets of rules – Corrected Stratified Class (CSC) was defined in [MSL09], with Oblivious Chase terminating for all instances D. Then, in [MSL09a], CSC was extended to Inductively Restricted (IR) class, and further to a whole hierarchy of classes $T[k]$, where $T[2] = IR$.

This list is by no means exhaustive – see Adrian Onet's thesis [O12] for a 35-pages long survey chapter about sufficient conditions for chase termination. What is however worth mentioning is that all the known conditions imply all-instances termination and thus none of them depends on D.

With so much effort spent on finding the sufficient conditions it is natural to ask about decidability of the all–instances termination problem itself. But surprisingly, this fundamental problem has so far remained open. Some work was done, but mostly on a related problem of chase termination for given program $\mathcal{T}$ and also **given** database instance D. It was shown to be undecidable in [DNR08] for Core Chase and Standard Chase (♠). In [M09] it was noticed that the proof of ♠ works also for Semi-oblivious and Oblivious chase. The only previous results concerning decidability of the all–instances chase termination problem can be found in [GO13], where the problem is shown to be undecidable for Core Chase (♡1) and the Standard$_\exists$ sub-version (♡2), where we ask, for given $\mathcal{T}$, whether for each database instance D **there exists** a terminating execution path of $\mathcal{T}$-Standard Chase on D (let us remind here that Standard Chase is a non-deterministic procedure). And this is again not really the most natural question as – having some $\mathcal{T}$ in mind – we want to be sure that whenever and however we run a $\mathcal{T}$-chase, it will always terminate[4].

Another result in [GO13] is undecidability of all–instances chase termination problem for sets of constraints where, apart from TGDs, a denial constraint is allowed, which is a conjunctive query Q such that when Q is proved somewhere in Chase($\mathcal{T}, D$) then the chase procedure terminates and "fails" (♣).

One more result from [M09], which can be slightly confusing, is undecidability of what is there – misleadingly – called "all-instances termination" (◇). The signature Σ of the TGDs there is a disjoint union of two sub-signatures Σ_1 and Σ_2 but only instances where the relations in Σ_2 are initially empty are allowed.

1.1 Our Contribution

The main result of this paper is:

Theorem 1. *All-instance termination of Oblivious Chase is undecidable (and r.e.-hard) for Datalog$^\exists$ programs consisting of single-head TGDs over ternary signatures.*

---

[4] The termination problem for the Standard$_\exists$ version is shown to be Π_2^0 complete in [GO13]. But the result statement there is not correct: co-r.e. completeness is claimed.

Proof of Theorem 1 is presented in Section 3. It can also be read, without any changes, as a proof of undecidability of all-instances termination of Semi-Oblivious Chase. In short Subsection 3.5 we modify the proof to show that also all-instances–all–paths Standard Chase termination is undecidable[5].

It is common knowledge that whatever can be said about TGDs over high arity signature usually remains true for binary signature, as long as multi-head TGDs are allowed. And also the other way round – one who is prepared to pay the arity cost can usually translate everything into the language of single-head TGDs. This fails however in the context of chase termination: one can easily modify our proof of Theorem 1 to get undecidability of all-instance chase termination for multi-head binary TGDs, but only for Semi-Oblivious Chase, not for Oblivious. See the full paper for more detailed discussion . In Section 4 we show:

Theorem 2. *All-instance termination of Oblivious Chase is EXPSPACE-hard for Datalog$^\exists$ programs consisting of single-head TGDs over binary signatures.*

Upper Bounds. It follows easily from Lemma 3 that the all-instances termination problem of Oblivious and Semi-Oblivious Chase is recursively enumerable, and so Theorem 1 provides matching lower bounds. But Lemma 3 is not true for all–instances–all–paths Standard Chase termination, and thus the only upper bound known for this problem is the Π_2^0 level of the Arithmetical Hierarchy. Our conjecture is that the problem is in fact also r.e., but much more insight into the structure of Standard Chase is needed in order to prove this claim.

The lower bound given by Theorem 2 is not matched by any upper bound, and we believe that the problem is undecidable. A similarity that is maybe worth being mentioned here (see also next subsection) is that Datalog programs uniform boundedness is also known to be undecidable for ternary arities but decidability was left open for the binary case [M99].

2   Techniques

It will not be too unfair to say that the proof of ♠, in [DNR08], is not complicated. The possibility of having our favorite instance D fixed gives a lot of control, and having this control it is not hard to encode a computation of a machine of one's choice as $\mathcal{T}$-chase for some program $\mathcal{T}$. The same can be said about ◇, whose proof, in [M09], is an adaptation of the proof of ♠ – the input instance over signature Σ_1 is neglected, a new instance, over Σ_2, hardwired in dedicated TGDs, is created, and then the proof from [DNR08] is applied.

Flooding Rule. The schema from [DNR08] is repeated, in a sense, in the proof of ♡2 in [GO13]. The instance D is treated as an input of some machine, and chase simulates the computation of this machine on given input. Chase terminates when the computation does. The problem are the instances D which

---

[5] This is for sake of completeness, as it was earlier shown in [GO13] that undecidability of all-instances termination of Oblivious Chase implies undecidability of all-instances–all–paths Standard Chase termination.

contain too much positive information to be understood by a Datalog$^\exists$ program as a finite input – for example instances that contain a loop, which is unavoidably seen by a program as an infinite path.

The trick used in [GO13] to make sure that chase will terminate on such unwelcome instances is the flooding rule – a technique earlier used in 1990s in the numerous papers dealing with the Datalog boundedness problem [GMSV93]. Let us illustrate it by an example:

Example. Consider the program $\mathcal{T}$:

(i) $U(x,y,z), E(z,w) \Rightarrow \exists u\, U(y,u,w)$;
(ii) $E(x,y) \rightarrow E^+(x,y)$;
(iii) $E^+(x,y), E(y,z) \Rightarrow E^+(x,z)$;
(iv) $E^+(x,x) \Rightarrow U(y,u,w)$ (flooding rule)

To see what is going on here, notice that E-atoms are never produced. Rules (ii) and (iii) compute E^+, being the (non-reflexive) transitive closure of E. Rule (i) unfolds graph E: if $\langle x,y\rangle$ is an edge in the unfolding, y is "over" an element z in E and if there is an edge $\langle z,w\rangle$ in E then a new element u must exist in the unfolding, being "over" w.

It is easy to see that whatever D we begin with, $\mathcal{T}$-Standard Chase on D has a terminating path. If E is acyclic, then rule (i) terminates for all chase variants. If E has a cycle, then rules (ii) and (iii) can prove $E^+(a,a)$ for some a, and then rule (iv) can be used to "flood" the predicate U, so that in consequence, the head of (i) will be always satisfied and (i) will never be triggered again.

But there is no hope for this trick to work for the all–instances–all–paths Standard Chase: flooding rule only terminates a Standard Chase if we can make sure it is always used early enough to prevent new elements to be born, which means that it must be us who decides what the execution order is.

Clearly, this technique also fails for the eager chase variants. $\mathcal{T}$-Oblivious Chase on D does not terminate whenever D is an instance containing an atom $U(a,b,c)$, for some c belonging to a cycle in E.

Notice also that adding a denial constraint to the constraints ($\clubsuit$) is just another way of using a flooding rule – instead of flooding the database we make the chase fail.

Drinking from the Well of Positivity. The trick we invented in this paper to replace the flooding rule is as follows. We treat the instance D as the only source of some positive facts: there are predicates which are never proved, they can only come with D.

Then the idea is that each new element a of Chase uses the path leading from D to a to run its private computation of some Turing-complete computational model. Only Datalog rules are used in this computation so we do not need to bother about termination. In order to be able to give birth to a successor a must first reach, by means of atoms created during its private computation, some atom that can only be found in D. Elements of Chase which are already too far away from this source of positivity cannot drink from it any more, dessicate, and do not produce offspring, thus causing the chase to terminate.

As we are going to see in the next Section, once one knows the above idea, the proof of Theorem 1 is easy.

3 Proof of Theorem 1

3.1 The Well of Positivity

From now on, whenever we say "chase" we mean Oblivious Chase.

Informally we say that Oblivious Chase creates one witness for each tuple satisfying the body of an existential TGD, regardless whether such a witness is already present in the current database instance or not. One of the ways how this informal statement can be formalized is to construct, for a given Datalog$^\exists$ program $\mathcal{T}$ a new such program $\mathcal{T}'$, by replacing each TGD in $\mathcal{T}$, of the form:

(i) $\Phi(\bar{x}) \Rightarrow \exists y\, \Psi(y, \bar{x})$

where (i) is the number of the rule in $\mathcal{T}$, and Φ and Ψ are conjunctive queries, by a rule:

(i') $\Phi(\bar{x}) \Rightarrow \Psi(h_i(\bar{x}), \bar{x})$

where h_i is a Skolem function. In this way Chase$(D, \mathcal{T})$ is the structure whose active domain is a subset of Herbrand universe, where the elements of D are treated as constants, and terms are built out of constants using the Skolem functions h_i, and which is a minimal model for all the rules of the program $\mathcal{T}'$. Since $\mathcal{T}'$ is a Prolog program it always has such a minimal model.

Now the question whether the $\mathcal{T}$-Oblivious Chase on D terminates is equivalent to the question whether Chase$(D, \mathcal{T})$, seen as a substructure of the Herbrand universe, contains, for each $k \in \mathbb{N}$, a term of depth at least k.

For a given signature Σ, an element a_Σ of a database instance D over Σ will be called a *well of positivity* if for each relation $R \in \Sigma$ the atom $R(a_\Sigma, a_\Sigma, \ldots a_\Sigma)$ is true in D. By D_Σ we will denote the database instance consisting of a single element, being a well of positivity.

Lemma 3. *The following conditions are equivalent for Datalog$^\exists$ program $\mathcal{T}$:*

(i) for each database instance D, $\mathcal{T}$-Oblivious Chase on D terminates;
(ii) $\mathcal{T}$- Oblivious Chase terminates on D_Σ.

This is (rephrased) Theorem 2 in [M09]. We sketch its proof for completeness.

Proof. Only the (ii)$\Rightarrow$ (i) implication needs a proof. Let us assume that there exists D such that Chase$(D, \mathcal{T})$, seen as a substructure of the Herbrand universe, contains, for each $k \in \mathbb{N}$, a term of depth at least k. What we need to prove is that also Chase$(D_\Sigma, \mathcal{T})$ does contain such a term.

So let t be a term of depth at least k in Chase$(D, \mathcal{T})$. This means that there is a derivation, in program $\mathcal{T}'$, having atoms of D in its leaves and some atom containing t in its root. When we replace all the elements of D, occurring in atoms of this derivation, by the well of positivity a_Σ, then we will get another valid derivation in program $\mathcal{T}'$, leading, instead of t, to some new term t' in Chase$(D_\Sigma, \mathcal{T})$. And the depth of t' is equal to the depth of t – the two terms only differ at the level of constants, but are equal otherwise. $\square$

3.2   The Problem to be Reduced

The undecidable problem we are going to encode is the halting problem for finite automata with three counters (3CM). More precisely, the instance of the problem Halt3CM is a triple consisting of finite set Q of states, of some initial state $q_1 \in Q$ and of a finite set Π of instructions, each of them of the following format:

if the current state is $q \in Q$,
the value of the 1st counter (is|is not) zero
and the value of the 2nd counter (is|is not) zero
then:
change the state to $q' \in Q$;
(increment|decrement|keep unchanged) the value of the 1st counter,
(increment|decrement|keep unchanged) the value of the 2nd counter,
increment the value of the third counter.

We assume here that the automaton is deterministic, which means that the part of the instruction which is after then is a function of the part occurring before then. This function is partial – if a configuration is reached with no instruction applicable then the automaton halts.

The problem, called Halt3CM, is whether, for a given 3CM M, executing the instructions of M will ever halt when started from the state q_1 and three empty counters. Since 3CM is a Minsky Machine with an additional counter, Halt3CM is of course undecidable. From now on each time we say "M halts" we mean that it halts after started from q_1 and three empty counters.

Notice that the value of the third counter is never read by the automaton, and the counter is incremented in each step. This leads to the following:

Lemma 4. *A 3CM halts if and only if the set of values of its third counter is bounded.*

From now on a 3CM $M = \langle Q, q_1, \Pi \rangle$ is fixed and we will construct a Datalog$^\exists$ program $\mathcal{T}_M$, over some signature Σ_M such that $\mathcal{T}_M$-Oblivious Chase on D_{Σ_M} terminates if and only if M halts.

3.3   Encoding the Automaton as a Conway Function

Now we will encode the computation of M as a sequence of iterations of a Conway function. This technique is by no means new, but maybe not as widely known as some other undecidable problems, so we include this subsection for completeness.

Suppose $|Q| = m$. Let $p_1 = 2$, $p_2 = 3, \ldots p_{m+3}$ be the first $m + 3$ primes and let $p = p_1 p_2 \ldots p_{m+3}$. Let $\mathbf{c}$ be a configuration of M with the state being q_i and c_1, c_2 and c_3 being respectively values of the first, second and third counter. Then by $e(\mathbf{c})$ (or *encoding of* $\mathbf{c}$) we will mean the number:

$$p_i p_{m+1}^{c_1} p_{m+2}^{c_2} p_{m+3}^{c_3}$$

Notice that if $\mathbf{c}$ is the initial configuration of M then $e(\mathbf{c}) = 2$.

For two configurations $\mathbf{c}$, $\mathbf{c}'$ of M we will say that they are consecutive when $\mathbf{c}'$ is a result of executing a single step of M in $\mathbf{c}$ or when there is no instruction that can be executed in $\mathbf{c}$ and $\mathbf{c} = \mathbf{c}'$. Now it is easy to see that:

Theorem 5. *There exist natural numbers* q_0, $q_1, \ldots q_{p-1}$, r_0, $r_1, \ldots r_{p-1}$, *such that for each two consecutive configurations* $\mathbf{c}$, $\mathbf{c}'$ *of* M, *such that* $e(\mathbf{c}) = i \mod p$ *it holds that* $e(\mathbf{c}') = \frac{q_i e(\mathbf{c})}{r_i}$.

For the proof of this theorem notice that the remainder i of $e(\mathbf{c})$ modulo p carries all the information needed for M to decide which instruction should be applied: the state is q_j if and only if i is divisible by p_j and the value of the (for example) second counter is non-zero if and only if i is divisible by p_{m+2}. It is equally easy to see that executing an instruction boils down to division (removing the old state, decrementing a counter) and multiplication (moving to a new state, incrementing a counter).

From now on the numbers q_0, $q_1, \ldots q_{p-1}$, r_0, $r_1, \ldots r_{p-1}$ provided for M by Theorem 5 are fixed. Denote by g a function that maps a natural number n to nq_i/r_i, where $n = i \mod p$. Let $\mathcal{G} = \{g^n(2) : n \in \mathbb{N}\}$ be the smallest subset of $\mathbb{N}$ which contains 2 and is closed under g. Clearly, M halts if and only if $\mathcal{G}$ is bounded. So, what remains for us to do is to construct such a Datalog$^\exists$ program $\mathcal{T}_M$ that $\mathcal{T}_M$-Oblivious Chase on D_{Σ_M} terminates if and only if $\mathcal{G}$ is bounded. Notice that it is here where the third counter is important.

3.4   The Program $\mathcal{T}_M$

Denote by $\mathcal{QR}$ the set $\{q_0, q_1, \ldots q_{p-1}, r_0, r, \ldots r_{p-1}\}$. The signature Σ_M will consist of the following relations:

- a binary relation E, which will pretend to be the successor relation on the natural numbers;
- for each $j \in \mathcal{QR}$ a binary relation E^j – only needed to keep rule (d3) short;
- a unary relation H, which will never occur in the head of any rule, so its only atom will be $H(a_\Sigma)$;
- for each $0 \le i \le p-1$ a ternary relation T^i, with $T^i_x(y, z)$ meaning something like "x thinks that $\frac{y}{z} = \frac{q_i}{r_i}$". Normally we should of course write $T(x, y, z)$ rather than $T_x(y, z)$. But we like $T_x(y, z)$ more, and it is still ternary;
- for each $0 \le i \le p-1$ a binary relation R^i, with $R^i_x(y)$ meaning something like "x thinks that $i = y \mod p$";
- a binary relation G, with $G_x(y)$ meaning "x thinks that $y \in \mathcal{G}$";
- a unary relation N, with $N(x)$ meaning that x is a natural number. N is not really needed, we only have it because otherwise the bodies of rules (d2) and (d4) would be empty, and we do not like rules with empty bodies.

Now we are ready to write the program $\mathcal{T}_M$. There is one existential rule:

(e) $G_x(y), H(y) \Rightarrow \exists z\; E(z, x)$.

Read this rule as "Once x has drunk from the well of positivity, it is allowed to give birth to a new element z."

There will be also several Datalog rules:

(d0) $E(y, y_1), E(y_1, y_2), \ldots E(y_{j-1}, y_j) \Rightarrow E^j(y, y_j)$ one rule for each $j \in \mathcal{QR}$;

(d1) $E(z, x) \Rightarrow N(z)$

Rules of the form (d2) and (d3) form a recursive definition of multiplication by addition (remember – x always thinks it equals zero):

(d2) $N(x) \Rightarrow T_x^i(x, x)$ one rule for each $0 \leq i \leq p - 1$;

(d3) $T_x^i(y, z), E^{q_i}(y, y'), E^{r_i}(z, z') \Rightarrow T_x^i(y', z')$ one rule for each $0 \leq i \leq p - 1$;

The next two rules count modulo p:

(d4) $N(x) \Rightarrow R_x^0(x)$;

(d5) $R_x^i(y), E(y, y') \Rightarrow R_x^j(y')$ whenever $j = i + 1 \mod p$;

Now, once we have all the predicates we need for the multiplications, and for remainders modulo p, we can easily write rules which will compute the set $\mathcal{G}$. First of them says – as long as x keeps assuming that it equals zero – that $2 \in \mathcal{G}$:

(d6) $E(x, y), E(y, z) \Rightarrow G_x(z)$

Second rule for G says that $\mathcal{G}$ is closed with respect to the function g:

(d7) $R_x^i(y), G_x(y), T_x^i(y, z) \Rightarrow G_x(z)$ one rule for each $0 \leq i \leq p - 1$;

Notice that the rules (d2)–(d7) form a sort of a private Datalog program for each x, and the atoms proved by such programs for different x, x' never see each other (this is reflected in our notation, which suggests that x is more than merely an argument of the predicates, but part of their names). Rule (e) creates a new element z, such that $E(z, x)$, when the program for x can prove that $G_x(y)$ for some y such that $H(y)$. But, as we said, there is no rule saying that something is in H and the only element a such that $\mathrm{Chase}(D_{\Sigma_M}, \mathcal{T}_M) \models H(a)$ is the well of positivity a_{Σ_M}. So (e) creates a new element z, such that $E(z, x)$, when the program for x can prove that $G_x(a_{\Sigma_M})$.

Now we have a lemma that Theorem 1 follows from:

Lemma 6. $\mathcal{T}_M$-*Oblivious Chase on* D_{Σ_M} *terminates if and only if* $\mathcal{G}$ *is bounded.*

We think that the lemma follows directly from the construction of $\mathcal{T}_M$. But the readers who like it more formal, are invited to read the full paper.

3.5 The Case of All-Instances-All-Paths Standard Chase Termination

For any $\mathcal{T}$ and D any structure being a result of running a $\mathcal{T}$-Standard Chase on D is a subset of (oblivious) $\mathrm{Chase}(D, \mathcal{T})$. This means that if $\mathcal{G}$ is bounded, then $\mathcal{T}_M$-Standard Chase terminates on each instance and each path. What remains to be seen is that if $\mathcal{G}$ is not bounded, then there exists D such that $\mathcal{T}_M$-Standard Chase does not terminate on some path. It is easy to see that a structure D, consisting of the well of positivity a_{Σ_M} and of some a such that $D \models E(a, a_{\Sigma_M})$, has this property.

4 Proof of Theorem 2

It is harder to prove any nontrivial lower bound for all-instances Oblivious Chase termination problem for single-head TGDs over binary signatures, than to prove undecidability in the general case. In the proof of Theorem 2 we try to repeat the idea of proof of Theorem 1, creating a new element of some E-path, for a binary E, only when some private computation, run by the last element a of the current path, terminates. But, while having arity three at our disposal, we could run many mutually non-interfering computations using the same arena, now we must construct a separate arena for each element of the E-path being built.

This arena needs to be huge enough to contain a complex computation, but on the other hand the process of the construction of the arena should never lead to an infinite chase. In other words we need to – and we think it is not immediately clear how to do it – find a binary Datalog$^\exists$ program which builds a huge (i.e. greater than exponential, with respect to the size of the program) (Oblivious) Chase, when run on a_Σ, but finally terminates.

4.1 Constructing the Arena: Chase of Exponential Depth

Let m be a fixed natural number and let $M = 2^m$. Consider the program $\mathcal{T}_b^0(m)$ consisting of the following rules:

(d0) $H(x) \Rightarrow K(x)$

(d0') $H(x) \Rightarrow C_i(x)$ \hfill (one rule for each $i \in \{0, 1, \ldots m\}$)

(e) $K(x) \Rightarrow \exists y\, R(x, y)$

(d1) $R(x, y) \Rightarrow T(y, y)$

(d2) $T(x, y), R(x', z), R(z, x), R(y', y) \Rightarrow T(x', y')$

(d3) $T(x, y), C_i(x) \Rightarrow C_{i+1}(y)$ \hfill (one rule for each $i \in \{0, 1, \ldots m - 1\}$)

(d4) $R(x, y), C_m(x) \Rightarrow K(y)$

Let now a be any element such that $H(a)$ (which means that a may be, but may not be, a well of positivity), and let D_a be a database instance containing a as a single element.

Exercise 7 *Chase*$(D_a, \mathcal{T}_b^0(m))$, *seen as a graph over predicate R, is a path of length $M + 1$, having a as its first element*

For a solution to this exercise see the full paper. **Hint:** like in Section 3 there is no rule saying that something is in H, and the only element satisfying H plays the role of the well of positivity. Also like in Section 3, Oblivious Chase produces a path (this time it is an R-path) – if an element is in K then it is "close enough" to H to be able to produce R-offspring. The predicates C_i are resources – the further we are from H the more we are running out them.

4.2   Constructing the Arena: Chase of Double Exponential Size

For fixed natural numbers m and p consider now the program $T_b^1(m, p)$ consisting of all the rules that can be obtained from the rules of $T_b^0(m)$ by replacing each occurrence of the predicate R with one of the predicates $R_1, \ldots R_p$. For example rule (e) will be replaced by p new rules while rule (d2) will be replaced by p^3 new rules. Let a and D_a be as in the previous subsection. Then the analysis of $\text{Chase}(D_a, T_b^1(m, p))$ is analogous to the analysis of $\text{Chase}(D_a, T_b^0(m))$, except that the structure we now get is a p-ary tree of depth $M + 1$ rather than a path of length $M + 1$. Notice that the same elements are created regardless if a is a well of positivity, or any element just satisfying $H(a)$.

4.3   The Encoding Lemma and How it Implies Theorem 2

Now $\text{Chase}(D_a, T_b^1(m, p))$ can be used as an arena, where we can run some computation. Let a and D_a be as before.

Lemma 8 (The encoding Lemma). *The problem:*

Given $m, p \in \mathbb{N}$ and a Datalog program T, with EDB relations H, R_1, $R_2, \ldots R_p$ and IDB relations P (binary) and G_1, G, G_2 and C (unary). Is it the case that:

$$\text{Chase}(D_a, T_b^1(m, p) \cup T) \models C(a) \quad ?$$

---

is EXSPACE-hard.
The size of the instance is here the size of the program $T_b^1(m, p) \cup T$.

For the proof of the Lemma see the full paper. Notice that $\text{Chase}(D_a, T_b^1(m, p)) \cup T)$ has the same set of elements as $\text{Chase}(D_a, T_b^1(m))$ – this is because the Datalog rules of T do not prove any atoms that could be used by $T_b^1(m, p)$.

Let now $T_b^2(m, p)$ be $T_b^1(m, p)$ with the following additional rules:

(d') $E(x, y) \Rightarrow H(y)$
(e') $C(x) \Rightarrow \exists z\, E(x, z)$.

Now, Theorem 2 follows from the next lemma, whose proof can be found in the full paper:

Lemma 9. *For a Datalog program T, as in Lemma 8, the following two conditions are equivalent:*

– $\text{Chase}(D_a, T_b^1(m, p) \cup T) \models C(a)$
– $\text{Chase}(D_\Sigma, T_b^2(m, p) \cup T$ *does not terminate.*

References

[ASU79]   Aho, A.V., Sagiv, Y., Ullman, J.D.: Efficient Optimization of a Class of Relational Expressions. ACM Transactions on Database Systems 4(4), 435–454 (1979)

[B84]   Beeri, C., Vardi, M.Y.: A proof procedure for data dependencies. Journal of the ACM (JACM) 31(4), 718–741 (1984)

[CGK08]   Calı̀, A., Gottlob, G., Kifer, M.: Taming the infinite chase: Query answering under expressive relational constraints. In: Proc. of KR, pp. 70–80 (2008)

[CGL09]   Calı̀, A., Gottlob, G., Lukasiewicz, T.: Datalog +/-: A unified approach to ontologies and integrity constraints. In: Proceedings of the 12th International Conference on Database Theory. ACM (2009)

[CGL12]   Calı̀, A., Gottlob, G., Lukasiewicz, T.: A general datalog-based framework for tractable query answering over ontologies. Web Semantics: Science, Services and Agents on the World Wide Web 14, 57–83 (2012)

[DT03]   Deutsch, A., Tannen, V.: Reformulation of XML Queries and Constraints. In: Calvanese, D., Lenzerini, M., Motwani, R. (eds.) ICDT 2003. LNCS, vol. 2572, pp. 225–238. Springer, Heidelberg (2002)

[DNR08]   Deutsch, A., Nash, A., Remmel, J.: The chase revisited. In: Proceedings of the 27th ACM SIGMOD-SIGACT-SIGART Symposium on Principles of Database Systems. ACM (2008)

[FKPP05] Fagin, R., Kolaitis, P.G., Miller, R.J., Popa, L.: Data exchange: semantics and query answering. Theoretical Computer Science 336(1), 89–124 (2005)

[G10]   Greco, S., Spezzano, F.: Chase termination: A constraints rewriting approach. Proceedings of the VLDB Endowment 3(1-2), 93–104 (2010)

[GMSV93] Gaifman, H., Mairson, H., Sagiv, Y., Vardi, M.Y.: Undecidable optimization problems for database logic programs. Journal of the ACM 40(3), 683–713 (1993)

[GO11]   Grahne, G., Onet, A.: On Conditional Chase Termination. AMW 11, 46 (2011)

[GO13]   Grahne, G., Onet, A.: Anatomy of the chase. arXiv:1303.6682 (2013)

[HS07]   Hernich, A., Schweikardt, N.: CWA-solutions for data exchange settings with target dependencies. In: Proceedings of the Twenty-sixth ACM SIGMOD-SIGACT-SIGART Symposium on Principles of Database Systems. ACM (2007)

[M99]   Marcinkowski, J.: Achilles, turtle, and undecidable boundedness problems for small DATALOG programs. SIAM Journal on Computing 29(1), 231–257 (1999)

[MMS79]   Maier, D., Mendelzon, A.O., Sagiv, Y.: Testing implications of data dependencies. ACM Transactions on Database Systems 4(4), 455–469 (1979)

[M09]   Marnette, B.: Generalized schema-mappings: from termination to tractability. In: Proceedings of the Twenty-eighth ACM SIGMOD-SIGACT-SIGART Symposium on Principles of Database Systems. ACM (2009)

[MSL09]   Meier, M., Schmidt, M., Lausen, G.: On chase termination beyond stratification. Proceedings of the VLDB Endowment 2(1), 970–981 (2009)

[MSL09a] Meier, M., Schmidt, M., Lausen, G.: On chase termination beyond stratification (technical report and erratum), http://arxiv.org/abs/0906.4228

[O13]   Onet, A.: The chase procedure and its applications in data exchange. In: Data Exchange, Integration, and Streams. Dagstuhl Follow-Ups. Schloss Dagstuhl-Leibniz-Zentrum für Informatik, Germany (2013)

Non-uniform Polytime Computation in the Infinitary Affine Lambda-Calculus

Damiano Mazza

CNRS, UMR 7030, LIPN, Université Paris 13, Sorbonne Paris Cité, France
Damiano.Mazza@lipn.univ-paris13.fr

Abstract. We give an implicit, functional characterization of the class of non-uniform polynomial time languages, based on an infinitary affine lambda-calculus and on previously defined bounded-complexity subsystems of linear (or affine) logic. The fact that the characterization is implicit means that the complexity is guaranteed by structural properties of programs rather than explicit resource bounds. As a corollary, we obtain a proof of the (already known) P-completeness of the normalization problem for the affine lambda-calculus which mimics in an interesting way Ladner's P-completeness proof of CIRCUIT VALUE (essentially, the argument giving the Cook-Levin theorem). This suggests that the relationship between affine and usual lambda-calculus is deeply similar to that between Boolean circuits and Turing machines.

1 Introduction

Loosely speaking, the aim of implicit computational complexity is to replace clocks (or other explicit resource bounds) with certificates. For example, if we consider polynomial time computation, the idea is to define a structured programming language whose programs guarantee a polynomial dependence of the runtime on the input by construction, *i.e.*, because they satisfy some syntactic condition, not because their execution is artificially stopped after a polynomial number of steps. At the same time, such a programming language must be expressive enough so that every polynomial time function may be somehow implemented. Notable early examples of such methodology are the work of Bellantoni and Cook [3], Leivant and Marion [10], and Jones [6].

We consider here the question of finding an implicit characterization of non-uniform polynomial time, *i.e.*, the class P/poly. Our approach brings together two lines of work, both based on linear logic. The first is the linear-logical take at implicit computational complexity initiated by Girard [5] and reformulated in the λ-calculus, for example, by Asperti and Roversi [2]. The second is the author's work on the infinitary affine λ-calculus [12], previously considered also by Kfoury [7] and Melliès [13].

For our present purposes, the essence of linear logic is in its resource awareness. Linear (or, more precisely, affine) types describe volatile data, which may be accessed only once. Accordingly, the linear (or affine) functional type $A \multimap B$ describes programs producing an output of type B by using their input of type A exactly (or at most) once. Persistent data is described by the type $!A$, which may be understood as volatile access

J. Esparza et al. (Eds.): ICALP 2014, Part II, LNCS 8573, pp. 305–317, 2014.

to a bottomless pile of copies of A, thus obtaining unlimited access to A. The usual functional type $A \to B$ may then be expressed by $!A \multimap B$.

In the λ-calculus, which is the prototypical functional language, affinity takes the form of forbidding duplication, which translates into an extremely simple syntactic restriction: each variable must appear at most once in a term. Our previous work [12] shows how affine λ-terms may approximate usual λ-terms arbitrarily well, in a precise topological sense which is compatible with computation (*i.e.*, reduction is continuous). In the limit, usual λ-terms are recovered by considering infinitary affine terms (thus taking quite literally the above idea of "bottomless pile"). However, the limit process (which is just the completion of a uniform space) introduces a host of infinitary terms which do not correspond to any usual λ-term. The reason is easily explained: to act as a persistent memory cell, a datum of type $!A$ must contain infinitely many *identical* copies of a datum of type A. Without further constraints, the infinitary affine λ-calculus allows memory cells whose content changes arbitrarily with each access. This is the "functional gateway" to non-uniform computation.

The technical contribution of this paper is to "tame" the non-uniformity of the unrestricted calculus $\ell\Lambda_\infty$ of [12] so as to keep it within interesting boundaries, namely those of P/poly. Let us give an informal description of what this means. Using (an adaptation of) the standard λ-calculus encodings of binary strings, we may say that a term t *decides* $L \subseteq \{0,1\}^*$ in $\ell\Lambda_\infty$ if, given $w \in \{0,1\}^*$, $t\underline{w} \to^* \underline{b}$ with $b \in \{0,1\}$ according to whether w belongs to L ($\underline{w}$ is the encoding of w and $\to^*$ is the reduction relation of the calculus). Now, t is generally infinite, but we may define a canonical sequence $\lfloor t \rfloor_n$ of approximations of t, which are finite affine terms such that $\lim \lfloor t \rfloor_n = t$. Intuitively, $\lfloor t \rfloor_n$ behaves like t in which every internal memory cell is limited to at most n accesses. We may then appeal to the continuity of reduction, by which, if we let u_n be the normal form of $\lfloor t \rfloor_n \underline{w}$, we have that $\underline{b} = \lim u_n$. But our topology is such that pieces of data like $\underline{b}$ are isolated points, so there exists $m \in \mathbb{N}$ such that $u_n = \underline{b}$ for all $n \geq m$. This means that a finite approximation of t suffices to compute $t\underline{w}$. The size of $\lfloor t \rfloor_m$ is linear in m, so the question is: How big is m? Can we relate it to $|w|$? If we can make m be polynomial in $|w|$, the language decided by t is in P/poly: we may use the $\lfloor t \rfloor_m$ as (polynomial) advice and then normalize $\lfloor t \rfloor_m \underline{w}$, which may be done in polynomial time in $|w|$ because it is a finite affine term.

There exist several λ-calculus characterizations of P based on linear logic (most notably Girard's [5] and Lafont's [9]) and the naive idea to polynomially bound m would be to reuse the recipes given therein. However, non-uniformity in the λ-calculus is extremely subtle and the approach "take your favorite λ-calculus characterization of P and add non-uniformity" does not necessarily yield P/poly. The most surprising aspect is that polytime non-uniformity seems to refuse the logical principle of contraction (expressed by the formula $!A \multimap !A \otimes !A$): in its presence, m may be exponentially big and we may therefore decide any language (an intuitive explanation is given below). This rules out Girard's approach [5]. Lafont's system [9] does not use contraction but appears to have the opposite problem: we are currently unaware of whether the expressiveness of its non-uniform version reaches P/poly.

The key to our solution is a new structural constraint on terms, which we call *parsimony*. In $\ell\Lambda_\infty$, affinity is enforced by giving a unique integer index to each occurrence of non-linear variable x: intuitively, x_i means "access to the i-th copy of the datum contained in x". In this setting, contraction (corresponding to duplication) is implemented using "Hilbert's hotel": from an infinite family $(x_i)_{i \in \mathbb{N}}$ representing an argument of type $!A$, we make two infinite families, e.g. x_{2i} and x_{2i+1}. Iterating this n times, we obtain a family whose first element is $x_{O(2^n)}$, causing the exponential growth rate of m mentioned above. A very high level description of parsimony is that, when such a reallocation of a family of occurrences is performed, the resulting families may not "waste" indices: each of them contains either finitely many x_i (i.e., it is finite), or almost all of them (i.e., it is co-finite). Parsimony therefore refuses contraction, which necessarily produces infinite co-infinite families. Instead, it allows an asymmetric form of contraction, also known as absorption, expressed by the formula $!A \multimap !A \otimes A$.

Parsimony is coupled with *stratification*, which is a staple of Girard's work [5]. Stratification partitions a program into rigid levels which may not interact and, very roughly speaking, forbids the self-reference that makes the λ-calculus Turing powerful. Alone, it guarantees termination (in elementary time, in the uniform case). Without it, parsimonious terms may diverge and the question of bounding m may not make sense.

For brevity, most of the results are given here without proof. An extended version of this paper, containing the missing proofs, is available on the author's web page.

2 The Affine Lambda-Calculus

Pre-terms. We fix two denumerably infinite disjoint sets of *linear variables*, ranged over by a, b, c, and *non-linear variables*, ranged over by x, y, z. *Patterns* and *pre-terms* are generated by the grammar

$$\mathsf{p}, \mathsf{q} ::= a \mid x \mid \mathsf{p} \otimes \mathsf{q}, \qquad t, u ::= \perp \mid a \mid x_i \mid \lambda \mathsf{p}.t \mid tu \mid t \otimes u \mid \mathbf{u},$$

where $i \in \mathbb{N}$ and $\mathbf{u}$, which we refer to as a *box*, is a finite sequence of pre-terms, which for convenience we identify with a function from $\mathbb{N}$ to pre-terms almost everywhere equal to $\perp$. We also use the explicit notation $\langle \mathbf{u}(0), \ldots, \mathbf{u}(n-1) \rangle$, in which we imply that $\mathbf{u}(i) = \perp$ for all $i \geq n$. If a variable a or x appears in a pattern p, we write $a \in \mathsf{p}$ or $x \in \mathsf{p}$. We require that, in $\mathsf{p} \otimes \mathsf{q}$, a variable cannot appear both in p and q. Free and bound variables are defined as customary, the only point worth mentioning is that if $x \in \mathsf{p}$, then *all* occurrences of the form x_i are bound in $\lambda \mathsf{p}.t$. As usual, we identify two pre-terms if they only differ in the names of their bound variables (α-equivalence).

Shallow contexts and *contexts* are defined by the following grammar:

$$S ::= \bullet \mid \lambda \mathsf{p}.S \mid St \mid tS \mid S \otimes t \mid t \otimes S \qquad C ::= S \mid \langle u_0, \ldots, C, \ldots, u_n \rangle,$$

where $t, u_1, \ldots, u_n$ are arbitrary pre-terms. As usual, we denote by $C[t]$ the term obtained by substituting t to $\bullet$ in the context C. We say that u is a *subterm* of t, and we write $u \sqsubseteq t$, if there exists a context C such that $t = C[u]$.

We will find useful to see pre-terms as labelled trees. Intuitively, this is done in the obvious way: a pre-term t induces a function $t : \mathbb{N}^* \longrightarrow \Sigma$, where $\Sigma := \{\perp, a, x_i, \lambda \mathsf{p}, @, \otimes, !\}$ and $\mathbb{N}^*$ is the set of finite sequences of natural numbers, ranged over by α and with the empty sequence denoted by ϵ. Sequences of arbitrary integers are needed because of boxes. The symbol $t[\alpha]$ denotes the kind of constructor at position α in t: $\perp$, a variable (a or x_i), an abstraction (λ), an application (@), a tensor ($\otimes$) or a box (!). Also, when $u \sqsubseteq t$, we say that u occurs at position α if u is rooted at position α in t. Note that, when the position α does not exist in t, we assume that $t[\alpha] = \perp$.

Terms and Reduction. A *term* is a pre-term t such that:

- every linear variable and occurrence of non-linear variable appears at most once in t (*i.e.*, if $x_i, x_j \sqsubseteq t$ occur at different positions, then $i \neq j$);
- whenever $u \sqsubseteq t$, the free variables of u are all non-linear.

We denote by $\ell\Lambda$ the set of all terms.

We say that a term t *matches* a pattern p, and write $t \lozenge$ p, when: $t \lozenge a$ for all t; $t \lozenge x$ just if $t = u$; and if $t \lozenge$ p and $u \lozenge$ q, then $t \otimes u \lozenge$ p $\otimes$ q. In case $u \lozenge$ p, we define the *substitution* $t[u/\text{p}]$ as follows: $t[u/a]$ is defined as usual; $t[u/x]$ is obtained by substituting $u(i)$ to the unique free occurrence x_i in t, for all $i \in \mathbb{N}$; and $t[u_1 \otimes u_2/\text{p}_1 \otimes \text{p}_2] := t[u_1/\text{p}_1][u_2/\text{p}_2]$.

We define $\rightarrow_s$ (*shallow reduction*) and $\rightsquigarrow$ (*unboxing reduction*) as the smallest binary relations $\ell\Lambda$ such that:

- $(\lambda\text{p}.t)u \rightarrow_s t[u/\text{p}]$ whenever $u \lozenge$ p;
- if $t \rightarrow_s t'$, then $S[t] \rightarrow_s S[t']$ for every shallow context S;
- $u \rightsquigarrow u(0)$ for every box u.

Note that the unboxing step *is not closed under any context*: it applies only to a term which is itself a box, "extracting" its first subterm. Usually, one allows reduction inside boxes. The non-standard definition adopted here is technically simpler for our purposes.

A *redex* is a term of the form $(\lambda\text{p}.t)u$ and such that $u \lozenge$ p. Each $\rightarrow_s$ step is obviously associated with a redex, which we say is *fired* by the reduction step.

Reduction is defined by $\rightarrow := \rightarrow_s \cup \rightsquigarrow$. If $\dashrightarrow$ is any reduction relation, we denote by $\dashrightarrow^*$ its reflexive-transitive closure and by $\dashrightarrow^l$ the composition of exactly $l \in \mathbb{N}$ steps of $\dashrightarrow$. Reduction is obviously confluent and strongly normalizing. Both points are a consequence of the affinity conditions: no redex is duplicated and the size of terms (*i.e.*, the number of symbols) strictly decreases with reduction.

Uniform Structure. As noted above, $\ell\Lambda$ may be seen as a subset of the set of functions $\mathbb{N}^* \longrightarrow \Sigma$. If we equip Σ with the discrete uniformity,[1] then we may endow terms with the *uniformity of uniform convergence on finitely branching trees* (as subsets of $\mathbb{N}^*$). More explicitly, let $(\mathbb{N}^\mathbb{N}, \leq, \vee)$ be the join-semilattice of infinite sequences of natural numbers, ranged over by ξ, with the pointwise ordering. We denote by ξ_i the ith element of the sequence ξ and by $n \cdot \xi$ a sequence whose first element is n. We also write $\alpha \prec \xi$ if $\alpha \in \mathbb{N}^*$ is a prefix of ξ. The uniformity of uniform convergence on finitely branching trees is generated by the following basis of entourages, for $\xi \in \mathbb{N}^\mathbb{N}$:

$$\mathcal{U}_\xi := \{(t, t') \in \ell\Lambda \times \ell\Lambda \mid \forall \alpha \prec \xi' \leq \xi, t[\alpha] = t'[\alpha]\}.$$

---

[1] The word "uniformity" takes here its standard topological sense [4], which is essentially a generalization of the concept of metric still allowing one to speak of Cauchy sequences. This, unfortunately, is completely unrelated to the equally standard meaning more common in computer science (and employed, in particular, in the title of this paper).

The intuition is the following: with each $\xi \in \mathbb{N}^{\mathbb{N}}$ we associate an infinite but finitely branching tree τ_ξ, such that every node at depth i of τ_ξ has exactly $\xi_i + 1$ siblings; then, two terms are ξ-close (*i.e.*, belong to $\mathcal{U}_\xi$) if they coincide on τ_ξ. A basis of open neighborhoods of t for the induced topology is $\mathcal{U}_\xi(t) := \{t' \in \ell\Lambda \mid \forall \alpha \prec \xi' \leq \xi, t'[\alpha] = t[\alpha]\}$, for $\xi \in \mathbb{N}^{\mathbb{N}}$, *i.e.*, the terms coinciding with t on τ_ξ. Observe that the local basis is uncountable. In fact, one can show that no countable local basis exists for this topology, so the space is not metrizable.

The fundamental result concerning the uniform structure on $\ell\Lambda$ is the Cauchy-continuity of reduction. We remind that a function between uniform spaces is Cauchy-continuous when it preserves Cauchy nets. In particular, it is continuous. Given $\alpha \in \mathbb{N}^*$, we define $R_\alpha : \ell\Lambda \longrightarrow \ell\Lambda$ by $R_\alpha(t) := t'$ if $t \to_s t'$ by reducing a redex at position α, or $R_\alpha(t) := t$ if no such reduction applies. Similarly, we define $U(t) := t'$ if $t \rightsquigarrow t'$ and $U(t) := t$ otherwise.

Proposition 1 (Cauchy-continuity of reduction). *For all $\alpha \in \mathbb{N}^*$, R_α is Cauchy-continuous, and so is U.*

Infinitary Terms and Approximations. Intuitively, Cauchy sequences in $\ell\Lambda$ are made of terms that coincide on wider and wider trees, such as $\Delta_n := \lambda x.x_0 \langle x_1, \ldots, x_n \rangle$. Note, however, that $(\Delta_n)_{n \in \mathbb{N}}$ has no limit in $\ell\Lambda$, showing that the space is not complete. We denote by $\ell\Lambda_\infty$ the completion of $\ell\Lambda$. From now on, the word "term" will refer to an element of $\ell\Lambda_\infty$, whereas the elements of $\ell\Lambda$ will be called *finite terms*.

Indeed, the elements of $\ell\Lambda_\infty$ may be seen as infinitary terms. They still verify the affinity constraints and we apply to them the same terminology and notations as for finite terms (free and bound variable, subterm relation $\sqsubseteq$, etc.). A typical example of infinitary term is $\Delta := \lambda x.x_0 \langle x_1, x_2, x_3, \ldots \rangle$, which is the limit of $(\Delta_n)_{n \in \mathbb{N}}$. Apart from being infinitely wide, terms of $\ell\Lambda_\infty$ may also have infinite height, such as $\langle \lambda x.x_0, \lambda x.x_0 x_1, \lambda x.x_0(x_1 x_2), \ldots \rangle$. Nevertheless, one may show that they are always well-founded.

In fact, we will mostly be interested in infinitary terms of finite height, like Δ above, but knowing that all terms are well-founded is quite useful because it allows reasoning by induction. For example, given $t \in \ell\Lambda_\infty$, we may define a default sequence of finite terms converging to t, its *n-th approximations* $\lfloor t \rfloor_n$, as follows: $\lfloor \bot \rfloor_n := \bot$; $\lfloor a \rfloor_n := a$; $\lfloor x_i \rfloor_n := x_i$; $\lfloor \lambda p.t \rfloor_n := \lambda p.\lfloor t \rfloor_n$; $\lfloor tu \rfloor_n = \lfloor t \rfloor_n \lfloor u \rfloor_n$; $\lfloor t \otimes u \rfloor_n := \lfloor t \rfloor_n \otimes \lfloor u \rfloor_n$; $\lfloor \mathbf{u} \rfloor_n := \langle \lfloor \mathbf{u}(0) \rfloor_n, \ldots, \lfloor \mathbf{u}(n) \rfloor_n \rangle$. The definition makes sense because of well-foundedness: technically, what we are saying is that $\lfloor \cdot \rfloor_n$ is a function satisfying the above equalities. One may prove by well-founded induction that such a function is well defined and unique.

Reduction may be defined for terms of $\ell\Lambda_\infty$ in the obvious way, using substitution (which may now require infinitely many substitutions in the case $t[\mathbf{u}/x]$). Nevertheless, we stress that, from a strictly technical point of view, by Proposition 1 we do not need an explicit definition: indeed, Cauchy-continuity is exactly the property guaranteeing that a function on a uniform space uniquely extends to its completion.

It is worthwhile noting that reduction in $\ell\Lambda_\infty$, although still strongly confluent (a topological proof is given in [12]), is no longer normalizing. In fact, if we set $\Omega := \Delta \langle \Delta, \Delta, \Delta, \ldots \rangle$, with Δ as above, we have $\Omega \to_s \Omega$.

Correspondence with a Non-linear λ-calculus. For programming purposes, it will be convenient to consider a more standard, non-linear λ-calculus. We use the same sets of linear and non-linear variables as $\ell\Lambda$ (ranged over by a and x, respectively) but for each non-linear variable we also consider a corresponding §-*variable* denoted by $x^\S$. Patterns p are defined as in $\ell\Lambda$, with the addition of §-variables. Terms and *reduction contexts* are defined as follows:

$$M, N ::= a \mid x^\S \mid x \mid \lambda p.M \mid MN \mid M \otimes N \mid \S M \mid {!}M,$$
$$R ::= \bullet \mid \lambda p.R \mid RM \mid MR \mid R \otimes M \mid M \otimes R.$$

The affinity constraint is on linear variables and §-variables, which must occur at most once. Non-linear variables may occur arbitrarily many times. Also, the simultaneous presence of $x^\S$ and x in a term is excluded. In ${!}M$ (resp. $\S M$), called !-*box* (resp. §-*box*), we require all free variables to be non-linear (resp. to be non-linear or §-variables). The set of terms thus defined is denoted by Λ.

Matching between terms and patterns is defined as in $\ell\Lambda$, with both $\S M \, \between \, x^\S$ and ${!}M \between x^\S$, and ${!}M \between x$. Substitution is also extended in the obvious way: in $M[\S N/x^\S]$, $M[{!}N/x^\S]$ and $M[{!}N/x]$, N is substituted to all free occurrences of x or $x^\S$ in M (so several copies of N may be needed). Reduction, denoted by $\to_\beta$, is the union of $\to_{\beta 0}$ and $\leadsto_\beta$, which are the smallest binary relations on Λ defined as follows:

- $(\lambda p.M)N \to_{\beta 0} M[N/p]$ whenever $N \between p$;
- if $M \to_{\beta 0} M'$, then $R[M] \to_{\beta 0} R[M']$;
- $\S M \leadsto_\beta M$.

We may represent terms of Λ in $\ell\Lambda_\infty$, as follows. Let $\iota : \mathbb{N}^* \longrightarrow \mathbb{N} \setminus \{0\}$ be an injection. If p is a pattern of Λ, we denote by p^- the pattern of $\ell\Lambda_\infty$ obtained by replacing every $x^\S$ with x. Given $\alpha \in \mathbb{N}^*$, we define $[\![M]\!]_\alpha^\iota$ by induction on M: $[\![a]\!]_\alpha^\iota := a$; $[\![x^\S]\!]_\alpha^\iota := x_0$; $[\![x]\!]_\alpha^\iota := x_{\iota(\alpha)}$; $[\![\lambda p.M]\!]_\alpha^\iota := \lambda p^-.[\![M]\!]_\alpha^\iota$; $[\![MN]\!]_\alpha^\iota := [\![M]\!]_{0\cdot\alpha}^\iota [\![N]\!]_{1\cdot\alpha}^\iota$; $[\![M \otimes N]\!]_\alpha^\iota := [\![M]\!]_{0\cdot\alpha}^\iota \otimes [\![N]\!]_{1\cdot\alpha}^\iota$; $[\![\S M]\!]_\alpha^\iota := \langle [\![M]\!]_\alpha^\iota \rangle$; $[\![{!}M]\!]_\alpha^\iota := \langle [\![M]\!]_{0\cdot\alpha}^\iota, [\![M]\!]_{1\cdot\alpha}^\iota, [\![M]\!]_{2\cdot\alpha}^\iota, \ldots \rangle$.

We say that $t \in \ell\Lambda_\infty$ *represents* $M \in \Lambda$, and we write $t \blacktriangleleft M$, if there exist α and ι as above such that $[\![M]\!]_\alpha^\iota = t$.

Proposition 2. *Let* $t \blacktriangleleft M$ *and* $M \to_\beta M'$, *then* $t \to t' \blacktriangleleft M'$.

Proof. Essentially, this is one direction of the isomorphism of [12]. □

The converse of Proposition 2 fails: let $i > 0$, $t := (\lambda x.x_i)\langle I \rangle$ and $M := (\lambda x.x)\S I$; we have $t \blacktriangleleft M$, yet M is a normal form whereas $t \to_s \bot$. The perfect correspondence of [12] could be recovered by modifying the syntax of $\ell\Lambda$ (and $\ell\Lambda_\infty$) but this is inessential for our purposes.

3 The Parsimonious Stratified Calculus

Stratification. The *box-depth* of a specific occurrence of subterm u in $t \in \ell\Lambda_\infty$, denoted by $d_u(t)$, is the number of nested boxes of t in which u is contained. For instance,

$d_u(t) = 0$ iff $t = S[u]$ for some shallow context S. The *box-depth* of t, denoted by $d(t)$, is the supremum of the box-depths of its subterms. It is always finite if the height of t is finite, which will be the case of interest to us.

The *binder-relative box-depth* of an occurrence x_i appearing (free or bound) in t, denoted by $\mathrm{rd}_{x_i}(t)$, is: $d_{x_i}(t)$ if x is free in t; if x is bound, then there exists $\lambda p.u \sqsubseteq t$ such that $x \in p$ and x_i appears in u, in which case $\mathrm{rd}_{x_i}(t) := d_{x_i}(t) - d_{\lambda p.u}(t)$ (which is easily seen to be equal to $d_{x_i}(u)$).

Definition 1 (Stratified term). *A term t is* stratified *if, for all $x_i \sqsubseteq t$, $\mathrm{rd}_{x_i}(t) = 1$. We denote by $\ell\Lambda^s_\infty$ the set of all stratified terms and by $\ell\Lambda^{s0}_\infty$ the set of stratified terms having no free non-linear variable.*

Proposition 3 (Reduction and stratification). *1) If $t \in \ell\Lambda^s_\infty$ and $t \to_s t'$, then $t' \in \ell\Lambda^s_\infty$ and $d(t') \le d(t)$; 2) moreover, if $t \in \ell\Lambda^{s0}_\infty$ and $t \rightsquigarrow t'$, then $t' \in \ell\Lambda^{s0}_\infty$ and $d(t') = d(t) - 1$; 3) every $t \in \ell\Lambda^{s0}_\infty$ of finite height is strongly normalizing.*

Parsimony. Given $m, n \in \mathbb{N}$ and $t, t' \in \ell\Lambda_\infty$, we write:

- $t \sim_n t'$ if t and t' differ only in the indices of the *bound occurrences* of their non-linear variables, and the indices vary by at most n, i.e., if x_i in t corresponds to x_j in t', then $|i - j| \le n$.
- $t :<^m t'$ if t' is obtained from t by replacing *every free occurrence* of non-linear variable x_i with x_{i+m}.
- $t :\lesssim^m_n t'$ iff there is u s.t. $t \sim_n u :<^m t'$ iff there is u' s.t. $t :<^m u' \sim_n t'$ (the latter two conditions are equivalent because the relations act on disjoint occurrences).

Definition 2 (Parsimonious term). *A box $\mathbf{u}$ is* parsimonious *if there exist $c, k \in \mathbb{N}$ such that, for all $i \ge j \ge k$, $\mathbf{u}(i) :\lesssim^{i-j}_c \mathbf{u}(j)$. The smallest $k \in \mathbb{N}$ realizing the above definition is called the* non-uniformity factor *of $\mathbf{u}$, or* n.u. *factor for short. A term $t \in \ell\Lambda_\infty$ is* parsimonious *if all of its boxes are. We denote by $\ell\Lambda^p_\infty$ the set of all parsimonious terms.*

Note that, unlike stratification, parsimony is inherited by subterms. Another difference is that every finite term is parsimonious. In fact, intuitively, the structure of a parsimonious term admits a finite description: in every box $\mathbf{u}$, all $\mathbf{u}(i)$ ultimately have the same "shape". Nonetheless, the term itself may not be finitely describable at all. For instance, if $I_i := \lambda x.x_i$ with $i \in \{0, 1\}$, the term $\langle I_{i_0}, I_{i_1}, I_{i_2}, \ldots \rangle$ is parsimonious (with n.u. factor 0) regardless of the sequence i_n (this will be a key ingredient for encoding infinite binary words). Moreover, we have:

Lemma 1. *1. Every parsimonious term has finite height;*
  2. let $\mathbf{u}$ be parsimonious of n.u. factor k and let x have infinitely many free occurrences in $\mathbf{u}$. Then, for all $h \in \mathbb{N}$, there is exactly one free occurrence x_{j_h} in $\mathbf{u}(k+h)$ and $j_h = j_0 + h$.

Proposition 4 (Reduction and parsimony). *If $t \in \ell\Lambda^p_\infty$ and $t \to t'$, then $t \in \ell\Lambda^p_\infty$.*

Bounds on Parsimonious Stratified Terms. By Propositions 3 and 4, parsimonious stratified terms form a well defined calculus with respect to the reduction relation $\to$.

Definition 3 (The calculus $\ell\Lambda^{ps0}_\infty$). *We define $\ell\Lambda^{ps0}_\infty := \ell\Lambda^{s0}_\infty \cap \ell\Lambda^p_\infty$.*

In what follows, δ_d is the Kronecker symbol, equal to 1 if $d = 0$ and to 0 otherwise. Let $t \in \ell\Lambda^p_\infty$. The *size of t at box-depth d* is defined as follows. First, we define the size of a pattern by setting $|a| := |x| := 1$, and $|p \otimes q| := 1 + |p| + |q|$. Then, we set $|\perp|_d := |a|_d := \delta_d$; $|x_i|_d := (1 + i)\delta_d$; $|\lambda p.t|_d := \delta_d|p| + |t|_d$; $|tu|_d := |t \otimes u|_d := \delta_d + |t|_d + |u|_d$; $|\mathbf{u}|_0 := 1$ and $|\mathbf{u}|_{d+1} := \sum_{i=0}^{k}|\mathbf{u}(i)|_d$, where k is the n.u. factor of $\mathbf{u}$. Finally, we define the *size* by $|t| = \sum_{j=0}^{d(t)}|t|_j$.

Lemma 2. *Let $t \in \ell\Lambda^{ps0}_\infty$ and let $t \to_s^* t'$. Then, for all $j \geq 1$, $|t'|_j \leq |t|_1|t|_j$.*

In what follows, we will make use of the n-th approximations of a term, defined at page 309. We also introduce the following notation: given two terms t, t' and $n \in \mathbb{N}$, $t \backsimeq_n t'$ just if $\lfloor t \rfloor_n = \lfloor t' \rfloor_n$. It is obviously a family of equivalence relations.

Definition 4. *Let $t \in \ell\Lambda_\infty$ and let x appear in t. Given $n \in \mathbb{N}$, we define $v_{x,t}(n) := \sup\{i \in \mathbb{N} \mid x_i \text{ appears in } \lfloor t \rfloor_n\}$.*

Let now $t \to t'$ and $n \in \mathbb{N}$. We define $m_{t \to t'}(n) \in \mathbb{N}$ as follows. If $t \rightsquigarrow t'$, $m_{t \to t'}(n) := n$. Otherwise, $t \to_s t'$ by firing a redex $(\lambda p.u)v$ such that $x^1, \ldots, x^p$ are the non-linear variables appearing in p. Then, we set $m_{t \to t'}(n) := \max(n, \sup\{v_{x^1,t}(n), \ldots, v_{x^p,t}(n)\})$.

Lemma 3. *Let $t \in \ell\Lambda_\infty$ and let $t \to t'$. Then, for all $n \in \mathbb{N}$ and for all $u \in \ell\Lambda_\infty$, $u \backsimeq_{m_{t \to t'}(n)} t$ implies $u \to u'$ such that $u' \backsimeq_n t'$.*

Lemma 4. *Let $t \in \ell\Lambda^p_\infty$ and let x be a bound variable of t. Then, $v_{x,t}(n) \leq |t| + n$, for all $n \in \mathbb{N}$.*

Lemma 5. *Let $t \in \ell\Lambda^{ps0}_\infty$ and let $t \to^l t'$. Then:*
1. *$|t'| \leq |t|^{2^{d(t)}}$ and $l \leq (d(t) + 1)|t|^{2^{d(t)}} + d(t)$;*
2. *for all $n \in \mathbb{N}$, there is $m \leq n + l|t|^{2^{d(t)}}$ such that $\lfloor t \rfloor_m \to^l t'' \backsimeq_n t'$.*

Proof. Point 1 is proved by by induction on the number of $\rightsquigarrow$ steps in the reduction, in a similar way as [5,2]. In synthesis, Lemma 2 and Proposition 3 give $d(t) + 1$ "rounds" each squaring the size. The result follows.

For point 2, we reason by induction on l. The case $l = 0$ is trivial, so let $t \to^{l'} u \to t'$. The induction hypothesis gives us, for all $m_1 \in \mathbb{N}$, $m(m_1) \leq m_1 + l'|t|^{2^{d(t)}}$ such that $\lfloor t \rfloor_{m(m_1)} \to^{l'} u' \backsimeq_{m_1} u$. We then apply Lemma 3 to obtain $\lfloor t \rfloor_{m(m_{u \to t'}(n))} \to^{l'} u' \to t'' \backsimeq_n t'$. To conclude, we need to bound $m(m_{u \to t'}(n)) \leq m_{u \to t'}(n) + l'|t|^{2^{d(t)}}$. By definition, $m_{u \to t'}(n)$ is either n, in which case we are done, or of the form $v_{x,u}(n)$ for some x appearing in u. By Lemma 4, $v_{x,u}(n) \leq |u| + n$. Now, using the size bound of point 1 (which does not depend on l'), we have $|u| \leq |t|^{2^{d(t)}}$, which allows us to conclude. $\square$

Parsimony and Stratification in the Non-linear Calculus. In Λ, the concepts of box-depth and binder-relative box-depth are defined just as in $\ell\Lambda$, with !- and §-boxes both counting as boxes.

Definition 5 (The uniform calculus Λ^{ps0}). *We denote by Λ^{ps0} the subset of terms of Λ satisfying the following requirements, which correspond to those of $\ell\Lambda^{\mathrm{ps0}}_\infty$: 1) occurrences of non-linear variables have binder-relative box-depth 1; 2) every non-linear variable appears in at most one subterm of the form $!M$, in which case it occurs exactly once in M; 3) no §-variable or non-linear variable appears free.*

Proposition 5. *1. If $t \blacktriangleleft M$, then $M \in \Lambda^{\mathrm{ps0}}$ iff $t \in \ell\Lambda^{\mathrm{ps0}}_\infty$;*
2. if $M \in \Lambda^{\mathrm{ps0}}$ and $M \to_\beta M'$, then $M' \in \Lambda^{\mathrm{ps0}}$.

Proof. Point 1 is an immediate consequence of the definitions. Point 2 easily follows from point 1, modulo Propositions 3 and 4. □

4   A Characterization of P/poly

Representing Basic Data and Languages. We consider the usual Church encodings of Booleans and binary strings (the members of $\mathbb{W} := \{0,1\}^*$), adapted to $\ell\Lambda^{\mathrm{ps0}}_\infty$. For the Booleans, we set $\mathsf{tt} := \lambda a.\lambda b.a$ and $\mathsf{ff} := \lambda a.\lambda b.b$. Given $w = w_1 \cdots w_n \in \mathbb{W}$, we say that a term t is a *Church encoding* of w if $t = \lambda s^0.\lambda s^1.\langle \lambda a.s^{w_1}_{i_1}(\ldots s^{w_n}_{i_n}a\ldots)\rangle$, with $i_1,\ldots,i_n \in \mathbb{N}$ arbitrary as long as affinity is assured. For example, the encodings of 010 are all of the form $\lambda s^0.\lambda s^1.\langle \lambda a.s^0_{i_1}(s^1_j(s^0_{i_2}a))\rangle$, with $i_1 \neq i_2$. We denote by $\underline{w}$ a generic Church encoding of w. Observe that, by choosing the indices as small as possible, every $w \in \mathbb{W}$ admits a Church encoding such that $|\underline{w}| = O(|w|^2)$ (where $|w|$ is the length of the string w). On the other hand, $\mathrm{d}(\underline{w}) = 1$ independently of w and of the Church encoding.

Definition 6 (The class C_∞). *We say that a language $L \subseteq \mathbb{W}$ is decidable in $\ell\Lambda^{\mathrm{ps0}}_\infty$ ($L \in \mathsf{C}_\infty$) if there exists $t \in \ell\Lambda^{\mathrm{ps0}}_\infty$ such that, for all $w \in \mathbb{W}$ and for any one of its Church encodings $\underline{w}$, $t\underline{w} \to^* \mathsf{tt}$ if $w \in L$, and $t\underline{w} \to^* \mathsf{ff}$ otherwise.*

Uniform Programming. A good deal of the expressive power of $\ell\Lambda^{\mathrm{ps0}}_\infty$ may be shown using the more standard calculus Λ^{ps0}. This is especially convenient because Λ^{ps0} may be provided with a typing discipline which greatly facilitates programming.

The types are second order intuitionistic linear logic formulas, generated by $A,B ::= X \mid A \multimap B \mid A \otimes B \mid §A \mid !A \mid \forall X.A$, with X ranging over propositional variables. The usual conventions for parentheses are applied ($\multimap$ associates to the right). The typing rules are a decoration of the sequent calculus for a subsystem of intuitionistic linear logic. Typing judgments are of the form $\Gamma \vdash M : A$, where Γ is a finite list of variable assignments of the form $\mathsf{p} : A$. In case $\mathsf{p} = x^§$ (resp. $\mathsf{p} = x$), we require that $A = §B$ (resp. $A = !B$). The rules are as follows:

$$\frac{}{a : A \vdash a : A}\ \mathrm{ax}$$

$$\frac{\Gamma \vdash N : A \quad \Delta, \mathsf{p} : A \vdash M : C}{\Gamma, \Delta \vdash \mathsf{let}\,\mathsf{p} = N\,\mathsf{in}\,M : C}\ \mathrm{cut}$$

$$\frac{\Gamma \vdash M : C}{\Gamma, \mathsf{p} : A \vdash M : C}\ \mathrm{weak}$$

$$\frac{\Gamma, x : !A, y^§ : §A \vdash M : C}{\Gamma, x : !A \vdash M[x/y^§] : C}\ \mathrm{asym\ cntr}$$

$$\frac{\Gamma, \mathsf{p} : A \vdash M : B}{\Gamma \vdash \lambda \mathsf{p}.M : A \multimap B}\ \multimap\!\mathrm{R}$$

$$\frac{\Gamma \vdash N : A \quad \Delta, \mathsf{p} : B \vdash M : C}{\Gamma, \Delta, a : A \multimap B \vdash \mathsf{let}\,\mathsf{p} = aN\,\mathsf{in}\,M : C}\ \multimap\!\mathrm{L}$$

$$\frac{\Gamma \vdash M : A \quad \Delta \vdash N : B}{\Gamma, \Delta \vdash M \otimes N : A \otimes B} \otimes R \qquad \frac{\Gamma, \mathsf{p} : A, \mathsf{q} : B \vdash M : C}{\Gamma, \mathsf{p} \otimes \mathsf{q} : A \otimes B \vdash M : C} \otimes L$$

$$\frac{\overrightarrow{\mathsf{p}} : \overrightarrow{B} \vdash M : A}{\overrightarrow{x}^{\S} : \S\overrightarrow{B} \vdash \S\mathsf{let}\ \overrightarrow{\mathsf{p}} = \overrightarrow{x}^{\S}\ \mathsf{in}\ M : \S A} \S \qquad \frac{\overrightarrow{\mathsf{p}} : \overrightarrow{B} \vdash M : A}{\overrightarrow{x} : !\overrightarrow{B} \vdash !\mathsf{let}\ \overrightarrow{\mathsf{p}} = \overrightarrow{x}\ \mathsf{in}\ M : !A} !$$

$$\frac{\Gamma \vdash M : A}{\Gamma \vdash M : \forall X.A} \forall R\ (X \notin \mathrm{free}\ \Gamma) \qquad \frac{\Gamma, \mathsf{p} : A \vdash M : C}{\Gamma, \mathsf{p} : \forall X.A \vdash M : C} \forall L$$

We used the following notational conventions: in the rules cut and $\multimap$ L, the notation
let $\mathsf{p} = N$ in M stands for $M[N/a]$ in case $\mathsf{p} = a$ or $(\lambda \mathsf{p}.M)N$ otherwise (the obvious n-ary generalization of this notation is used in the $\S$ and ! rules); in rule asym cntr, the substitution $M[x/y^{\S}]$ simply means that the unique occurrence of $y^{\S}$ in M is replaced by x (of which there may already be occurrences); in the $\S$ and ! rules, $\overrightarrow{\mathsf{p}} : \overrightarrow{B}$ means that the context is of the form $\mathsf{p}_1 : B_1, \ldots, \mathsf{p}_n : B_n$ and, in the conclusion, $\overrightarrow{x}^{\S} : \S\overrightarrow{B}$ (resp. $\overrightarrow{x} : !\overrightarrow{B}$) means that every $\mathsf{p}_i : B_i$ is replaced by $x^{i\S} : \S B_i$ (resp. $x^i : !B_i$).

The reader may check that if $\mathsf{p}_1 : A_1, \ldots, \mathsf{p}_n : A_n \vdash M : C$ is derivable, then $\lambda \mathsf{p}_1 \ldots \lambda \mathsf{p}_n.M \in \Lambda^{\mathrm{ps0}}$. The system enjoys subject reduction with respect to $\to_{\beta 0}$ but not $\leadsto_\beta$. This failure is to be expected and is actually rather mild: if $\vdash M : \S A$ and $M \leadsto_\beta N$, then $\vdash N : A$. This is enough for our purposes; subject reduction in itself is not essential for us, because we never use typing as a means of ensuring properties.

The types of Booleans and Church strings are $\mathsf{Bool} := \forall X.X \multimap X \multimap X$ and $\mathsf{Str} := \forall X.!(X \multimap X) \multimap !(X \multimap X) \multimap \S(X \multimap X)$, which are adaptations of the corresponding standard System F types. Booleans are the same as in $\ell \Lambda_\infty^{\mathrm{ps0}}$ and Church strings are obtained by erasing indices. In particular, each string has a unique encoding, e.g. $\underline{010} = \lambda s^0.\lambda s^1.\S(\lambda a.s^0(s^1(s^0 a)))$.

Definition 7 (The class C). *A language L is decidable in Λ^{ps0} ($L \in$ C) if there is a derivation $\vdash M : \mathsf{Str} \multimap \S^k \mathsf{Bool}$, with $\S^k A = \S \cdots \S A$ for some $k \in \mathbb{N}$, s.t. $M\underline{w} \to_\beta^*$ tt if $w \in L$ and $M\underline{w} \to_\beta^*$ ff if $w \notin L$.*

By Propositions 2 and 5 we have $\mathsf{C} \subseteq \mathsf{C}_\infty$.

A slight variant of the type of binary strings gives us the Church numerals, *i.e.*, unary integers, of type $\mathsf{Nat} := \forall X.!(X \multimap X) \multimap \S(X \multimap X)$. These are of the form $\underline{n} := \lambda s.\S(\lambda a.s(\ldots sa \ldots))$, with n occurrences of s. If $a : \Gamma, c : A \vdash F : A$ and $b : \Delta \vdash Z : A$, we define $\mathrm{it}(F, Z) := (\lambda z^{\S}.\S(z^{\S} Z[y^{\S}/b]))(n!(\lambda c.F[x/a]))$. It is readily verified that $x : !\Gamma, y^{\S} : \S\Delta, n : \mathsf{Nat} \vdash \mathrm{it}(F, Z) : \S A$ and that $\mathrm{it}(F, Z)\underline{n} \to_\beta^* (\lambda c.F)(\ldots(\lambda c.F)Z \ldots)$, the n-fold iteration of F on Z. Using iteration, we may define the basic arithmetic functions, including any polynomial, by adapting the usual definitions, much as in [5,2]. Furthermore, following [2], we may define a type Tur of Turing machine configurations and, for any deterministic transition function, a term of type $\mathsf{Tur} \vdash \mathsf{Tur}$ implementing it. One may also easily implement the function building an initial configuration from a string (of type $\mathsf{Str} \vdash \mathsf{Tur}$), the function telling whether a configuration is accepting (of type $\mathsf{Tur} \vdash \S\mathsf{Bool}$) and the function returning the length of a string (of type $\mathsf{Str} \vdash \mathsf{Nat}$). Composing all these, with the help of iteration and the numerical functions shown above, every deterministic Turing machine with a polynomial clock may be implemented in Λ^{ps0}, showing that $\mathsf{P} \subseteq \mathsf{C}$.

The Characterization. We remind that P/poly is the class of languages decided by poly-time Turing machines with polynomial advice, or by polynomial-size Boolean circuits (see for instance [1]).

Theorem 1. $C_\infty = P/poly$ *and* $C = P$.

Proof. The inclusion $C_\infty \subseteq P/poly$ is obtained from Lemma 5, as delineated in Sect. 1: if $t\underline{w} \to^l u$, with u the encoding of a Boolean, we have $\lfloor u \rfloor_0 = u$. By Lemma 5, there exists $m \leq (d(t\underline{w}) + 1)|t\underline{w}|^{2^{d(t\underline{w})+1}} + d(t\underline{w})|t\underline{w}|^{2^{d(t\underline{w})}}$ such that $\lfloor t\underline{w} \rfloor_m = \lfloor t \rfloor_m \underline{w} \to^* u$. But $|t\underline{w}| = 1 + |t| + |\underline{w}| = O(|w|^2)$ (by choosing the suitable Church encoding) and $d(t\underline{w}) = \max(d(t), 1) = O(1)$, so m is polynomial in $|w|$. From this, to prove $C \subseteq P$ it is enough to observe that, in case $t = [\![M]\!]$ with M in Λ^{ps0}, the $\lfloor t \rfloor_m$ are in fact polytime computable (they are actually logspace computable, see Sect. 5).

For the converse, we need to encode Turing machines with advice as terms of $\ell\Lambda^{ps0}_\infty$. We will make the simplifying assumption that the advice strings a_n (where $n \in \mathbb{N}$ is the length of the input) are "cumulative", *i.e.*, for all n, a_n is a polynomially-long prefix of an infinite binary word A. Every polynomial advice may be transformed into a cumulative polynomial advice, so there is no loss of generality. We will show how to encode the infinite string A in $\ell\Lambda^{ps0}_\infty$ and how a prefix of a given length may be extracted. This is enough to conclude, because the rest is all uniform computation which we already know is representable in $\ell\Lambda^{ps0}_\infty$ (via Λ^{ps0}): if w is the input string, the prefix of A to be extracted is of length $q(|w|)$ with q a polynomial, and we know that both q and $|\cdot|$ are representable; the resulting advice string is then fed to the encoding of the suitable polynomially-clocked Turing machine, together with a copy of w.

Let A_j be the j-th bit of A, and let

$$Z_A := \langle\langle\epsilon\rangle\rangle \otimes \langle I_{A_0}, I_{A_1}, I_{A_2}, \ldots\rangle,$$
$$F := \lambda w \otimes x.\langle(\lambda f.\lambda y.\langle f_0 y_0\rangle)(x_0\langle S_0, S_1\rangle)w_0\rangle \otimes \langle x_1, x_2, x_3, \ldots\rangle,$$
$$\mathrm{extr}_A := \lambda n.(\lambda z.\langle(\lambda a \otimes b.a)(z_0 Z_A)\rangle)(n\langle F, F, F, \ldots\rangle),$$

where, for $i \in \{0, 1\}$, $I_i := \lambda x.\langle x_i\rangle$ and S_i represent the two constructors on Church strings (s.t. $S_i\underline{w} \to^* \underline{iw}$). The reader may check that the above terms are all in $\ell\Lambda^{ps0}_\infty$. The term extr_A takes a Church numeral $\underline{n}$ of $\ell\Lambda^{ps0}_\infty$ (which is of the form $\lambda s.\langle\lambda a.s_{i_1}(\ldots s_{i_n} a \ldots)\rangle$, with $i_1, \ldots, i_n$ pairwise distinct but otherwise arbitrary) and iterates n times F on Z_A. The result is a pair, of which the first component is taken as the final result. The term Z_A is where we fully exploit non-uniformity, representing A. Finally, if we disregard boxes, F takes a pair (w, iW), composed of a finite and an infinite string, and returns (iw, W). Therefore, $\mathrm{extr}_A \underline{q(n)} \to^* \underline{a_n}$ for all $n \in \mathbb{N}$. $\square$

5   Affine Lambda-Terms and Boolean Circuits

Let $L \in P$. By Theorem 1, we know that L is decided by $M \in \Lambda^{ps0}$, so deciding whether $w \in L$ amounts to normalizing $M\underline{w}$, which, by Proposition 5, amounts to normalizing $[\![M]\!]\underline{w}$ (with $\underline{w}$ any encoding of w in $\ell\Lambda^{ps0}_\infty$). But, for this, we know that it is enough to normalize $\lfloor[\![M]\!]\rfloor_m \underline{w}$ with m polynomial in $|w|$. By inspecting the definition of $[\![\cdot]\!]$ and $\lfloor\cdot\rfloor_n$ one may see that building $\lfloor[\![M]\!]\rfloor_m$ from M may be done in logarithmic space (in $|w|$), much like building the circuit representing the computation of a polytime Turing machine from the trace of its execution on w.

We have therefore given an alternative proof of the P-completeness of the normalization problem for the affine λ-calculus (given an affine λ-term, decide whether its normal form is the Boolean tt). Mairson [11] showed this by encoding Boolean circuits in affine λ-terms. The interest of the above proof is that it is virtually identical to the usual P-completeness proof of CIRCUIT VALUE [8,14], which is essentially the Cook-Levin theorem and does not rest on the P-completeness of another problem. It is also noteworthy that the "locality of computation" is reflected in the continuity of normalization.

The results of this paper seem to suggest the following "equation":

$$\frac{\text{affine } \lambda\text{-terms}}{\text{(infinitary affine) } \lambda\text{-terms}} = \frac{\text{Boolean circuits}}{\text{Turing machines (with advice)}}$$

The relationship between Boolean circuits and affine calculi was of course already known [11,15]. However, we are seeing a connection here which is deeper than what was shown by any previous result. An interesting perspective given by the above "equation" is to study the notion of uniformity of families of Boolean circuits via the uniformity of the infinitary affine λ-calculus. This may be defined in a purely algebraic way: the terms t such that $t \blacktriangleleft M$ may be characterized by means of a partial equivalence relation, as in [12]. This might be turned into a notion of uniform family of Boolean circuits which is purely intrinsic, *i.e.*, it depends only on the "shape" of the circuits in the family and does not invoke external algorithms producing the circuits themselves. Investigating such a notion is definitely a topic worth further investigation.

Acknowledgments. We wish to thank Kazushige Terui for discussions which greatly contributed to the development of this work. We acknowledge partial support of ANR projects LO-GOI ANR-2010-BLAN-0213-02 and COQUAS ANR-12-JS02-006-01.

References

1. Arora, S., Barak, B.: Computational Complexity – A Modern Approach. Cambridge University Press (2009)
2. Asperti, A., Roversi, L.: Intuitionistic light affine logic. ACM Trans. Comput. Log. 3(1), 137–175 (2002)
3. Bellantoni, S., Cook, S.A.: A new recursion-theoretic characterization of the polytime functions. Computational Complexity 2, 97–110 (1992)
4. Bourbaki, N.: General Topology, ch. 1-4. Springer (1998)
5. Girard, J.Y.: Light linear logic. Inf. Comput. 143(2), 175–204 (1998)
6. Jones, N.D.: Logspace and ptime characterized by programming languages. Theor. Comput. Sci. 228(1-2), 151–174 (1999)
7. Kfoury, A.J.: A linearization of the lambda-calculus and consequences. J. Log. Comput. 10(3), 411–436 (2000)
8. Ladner, R.E.: The circuit value problem is log-space complete for P. SIGACT News 6(2), 18–20 (1975)
9. Lafont, Y.: Soft linear logic and polynomial time. Theor. Comput. Sci. 318(1-2), 163–180 (2004)

10. Leivant, D., Marion, J.Y.: Lambda calculus characterizations of poly-time. Fundam. Inform. 19(1/2) (1993)
11. Mairson, H.G.: Linear lambda calculus and ptime-completeness. J. Funct. Program. 14(6), 623–633 (2004)
12. Mazza, D.: An infinitary affine lambda-calculus isomorphic to the full lambda-calculus. In: Proceedings of LICS, pp. 471–480 (2012)
13. Melliès, P.A.: Asynchronous games 2: The true concurrency of innocence. Theor. Comput. Sci. 358(2-3), 200–228 (2006)
14. Papadimitriou, C.H.: Computational Complexity. Addison-Wesley (1994)
15. Terui, K.: Proof nets and boolean circuits. In: Proceedings of LICS, pp. 182–191 (2004)

On the Positivity Problem
for Simple Linear Recurrence Sequences*,**

Joël Ouaknine and James Worrell

Department of Computer Science, Oxford University, UK

Abstract. Given a linear recurrence sequence (LRS) over the integers, the *Positivity Problem* asks whether all terms of the sequence are positive. We show that, for simple LRS (those whose characteristic polynomial has no repeated roots) of order 9 or less, Positivity is decidable, with complexity in the Counting Hierarchy.

1   Introduction

A (real) ***linear recurrence sequence (LRS)*** is an infinite sequence $\mathbf{u} = \langle u_0, u_1, u_2, \ldots \rangle$ of real numbers having the following property: there exist constants $b_1, b_2, \ldots, b_k$ (with $b_k \neq 0$) such that, for all $n \geq 0$,

$$u_{n+k} = b_1 u_{n+k-1} + b_2 u_{n+k-2} + \ldots + b_k u_n. \tag{1}$$

If the initial values $u_0, \ldots, u_{k-1}$ of the sequence are provided, the recurrence relation defines the rest of the sequence uniquely. Such a sequence is said to have ***order*** k.[1]

The best-known example of an LRS was given by Leonardo of Pisa in the 12th century: the Fibonacci sequence $\langle 0, 1, 1, 2, 3, 5, 8, 13, \ldots \rangle$, which satisfies the recurrence relation $u_{n+2} = u_{n+1} + u_n$. Leonardo of Pisa introduced this sequence as a means to model the growth of an idealised population of rabbits. Not only has the Fibonacci sequence been extensively studied since, but LRS now form a vast subject in their own right, with numerous applications in mathematics and other sciences. A deep and extensive treatise on the mathematical aspects of recurrence sequences is the monograph of Everest *et al.* [9].

Given an LRS $\mathbf{u}$ satisfying the recurrence relation (1), the ***characteristic polynomial*** of $\mathbf{u}$ is

$$p(x) = x^k - b_1 x^{k-1} - \ldots - b_{k-1} x - b_k. \tag{2}$$

An LRS is said to be ***simple*** if its characteristic polynomial has no repeated roots. Simple LRS, such as the Fibonacci sequence, possess a number of desirable

---

* This research was partially supported by EPSRC. We are also grateful to Matt Daws for considerable assistance in the initial stages of this work.
** The full version of this paper is available as [21].
[1] Some authors define the order of an LRS as the *least* k such that the LRS obeys such a recurrence relation. The definition we have chosen allows for a simpler presentation of our results and is algorithmically more convenient.

J. Esparza et al. (Eds.): ICALP 2014, Part II, LNCS 8573, pp. 318–329, 2014.
© Springer-Verlag Berlin Heidelberg 2014

properties which considerably simplify their analysis—see, e.g., [9, 10, 2, 3, 23]. They constitute a large[2] and well-studied class of sequences, and correspond to *diagonalisable* matrices in the matricial formulation of LRS—see Sec. 2.

In this paper, we focus on the **Positivity Problem** for simple LRS over the integers (or equivalently, for our purposes, the rationals): given a simple LRS, are all of its terms positive?[3]

As detailed in [22], the Positivity Problem (and assorted variants) has applications in a wide array of scientific areas, including theoretical biology, economics, software verification, probabilistic model checking, quantum computing, discrete linear dynamical systems, combinatorics, formal languages, statistical physics, generating functions, etc. Positivity also bears an important relationship to the well-known *Skolem Problem*: does a given LRS have a zero? The decidability of the Skolem Problem is generally considered to have been open since the 1930s (notwithstanding the fact that algorithmic decision issues had not at the time acquired the importance that they have today—see [13] for a discussion on this subject; see also [28, p. 258] and [16], in which this state of affairs—the enduring openness of decidability for the Skolem Problem—is described as "faintly outrageous" by Tao and a "mathematical embarrassment" by Lipton). A breakthrough occurred in the mid-1980s, when Mignotte *et al.* [18] and Vereshchagin [30] independently showed decidability for LRS of order 4 or less. These deep results make essential use of Baker's theorem on linear forms in logarithms (which earned Baker the Fields Medal in 1970), as well as a p-adic analogue of Baker's theorem due to van der Poorten. Unfortunately, little progress on that front has since been recorded.[4]

It is considered folklore that the decidability of Positivity (for arbitrary LRS) would entail that of the Skolem Problem [22], noting however that the reduction increases the order of LRS quadratically.[5] Nevertheless, the earliest explicit references in the literature to the Positivity Problem that we have found are from the 1970s (see, e.g., [26, 25, 5]). In [26], the Skolem and Positivity Problems are described as "very difficult", whereas in [24], the authors assert that the Skolem and Positivity Problems are "generally conjectured [to be] decidable". Positivity is again stated as an open problem in [12, 4, 15, 17, 29, 22], among others.

Unsurprisingly, progress on the Positivity Problem over the last few decades has been fairly slow. In the early 1980s, Burke and Webb [8] showed that the closely related problem of *Ultimate Positivity* (are all but finitely many terms

---

[2] In the measure-theoretic sense, almost all LRS are *simple* LRS.

[3] In keeping with established terminology, 'positive' here is taken to mean 'non-negative'.

[4] A proof of decidability of the Skolem Problem for LRS of order 5 was announced in [13]. However, as pointed out in [20], the proof seems to have a serious gap.

[5] It is worth noting that, under this reduction, the decidability of the Positivity Problem for simple LRS of order at most 14 would entail the decidability of the Skolem Problem for simple LRS of order 5, which is open and from which the general case of the Skolem Problem at order 5 would follow, based on the work carried out in [13]; see also [20], which identifies the last unresolved critical case for the Skolem Problem at order 5, involving simple LRS.

of a given LRS positive?) is decidable for LRS of order 2, and nine years later
Nagasaka and Shiue [19] showed the same for LRS of order 3 that have repeated
characteristic roots. Much more recently, Halava *et al.* [12] showed that Positivity is decidable for LRS of order 2, and subsequently Laohakosol and Tangsupphathawat [15] proved that Positivity is decidable for LRS of order 3. In 2012,
an article purporting to show decidability of Positivity for LRS of order 4 was
published [27], with the authors noting they were unable to tackle the case of
order 5. Unfortunately, as pointed out in [22] and acknowledged by the authors
themselves [14], that paper contains a major error, invalidating the order-4 claim.
Very recently, Positivity was nevertheless shown decidable for arbitrary integer
LRS of order 5 or less [22], with complexity in the Counting Hierarchy; moreover,
the same paper shows by way of hardness that the decidability of Positivity for
integer LRS of order 6 would entail major breakthroughs in analytic number
theory (certain longstanding Diophantine-approximation open problems would
become solvable). Finally, in [23], the authors show that Ultimate Positivity for
simple integer LRS of unrestricted order is decidable within PSPACE, and in
polynomial time if the order is fixed.

Main Result. The main result of this paper is that the Positivity Problem for
simple integer LRS of order 9 or less is decidable. An analysis of the decision
procedure shows that its complexity lies in $\text{coNP}^{\text{PP}^{\text{PP}^{\text{PP}}}}$, i.e., within the fourth
level of the Counting Hierarchy (itself contained in PSPACE).[6]

Comparison with Related Work. It is important to note the fundamental
difference between the above result and those of [23]: in the latter, Ultimate
Positivity is shown to be decidable for simple LRS of all orders, but in a *non-
constructive* sense: a given LRS may be certified ultimately positive, yet no
index threshold is provided beyond which all terms of the LRS are positive. At
the time of writing, this appears to be a fundamental difficulty: for simple LRS of
any given order, the ability to compute such index thresholds would immediately
enable one to decide Positivity. Yet as noted earlier, the decidability of Positivity
for simple LRS of order at most 14 would in turn entail the decidability of the
Skolem Problem for arbitrary LRS of order 5, a longstanding and major open
problem.

Our overall approach is similar to that followed in [22], attacking the problem
via the exponential polynomial solution of LRS using sophisticated tools from
analytic and algebraic number theory, Diophantine geometry and approximation,
and real algebraic geometry. However the present paper makes vastly greater and
deeper use of real algebraic geometry, as can be seen from the full version [21].

The present paper also markedly differs from [23]. In fact, aside from sharing
standard material on LRS, the non-constructive approach of [23] eschews most of
the real algebraic geometry of the present paper, as well as Baker's theorem, and is

---

[6] The complexity is as a function of the bit length of the standard representation of
integer LRS; for an LRS of order k as defined by Eq. (1), this representation consists
of the $2k$-tuple $(b_1, \ldots, b_k, u_0, \ldots, u_{k-1})$ of integers.

underpinned instead by non-constructive lower bounds on sums of S-units, which in turn follow from deep results in Diophantine approximation (Schlickewei's p-adic generalisation of Schmidt's Subspace theorem).

We present a high-level overview of our proof strategy—split in two parts—within Sec. 3, and also briefly discuss why the present approach does not seem extendable beyond order 9. As noted earlier, establishing the decidability of Positivity for simple LRS of order 14 would entail a major advance, namely the decidability of the Skolem Problem for arbitrary LRS of order 5. It is an open problem whether similar 'hardness' results can be established for simple LRS of orders 10–13.

In terms of complexity, it is shown in [23] that the Positivity Problem for simple integer LRS of arbitrary order is hard for co∃ℝ, the class of problems whose complements are solvable in the existential theory of the reals, and which is known to contain coNP. However, no lower bounds are known when the order is fixed or bounded, as is the case in the present paper. Either establishing non-trivial lower bounds or improving the Counting-Hierarchy complexity of the present procedure also appear to be challenging open problems.

2   Linear Recurrence Sequences

We recall some fundamental properties of (simple) linear recurrence sequences. Results are stated without proof, and we refer the reader to [9, 13] for details.

Let $\mathbf{u} = \langle u_n \rangle_{n=0}^{\infty}$ be an LRS of order k over the reals satisfying the recurrence relation $u_{n+k} = b_1 u_{n+k-1} + \ldots + b_k u_n$, where $b_k \neq 0$. We denote by $\|\mathbf{u}\|$ the bit length of its representation as a $2k$-tuple of integers, as discussed in the previous section. The **characteristic roots** of $\mathbf{u}$ are the roots of its characteristic polynomial (cf. Eq. (2)), and the **dominant roots** are the roots of maximum modulus. The characteristic roots can be computed in time polynomial in $\|\mathbf{u}\|$—see [21] for further details on algebraic-number manipulations.

The characteristic roots divide naturally into real and non-real ones. Since the characteristic polynomial has real coefficients, non-real roots always arise in conjugate pairs. Thus we may write $\{\rho_1, \ldots, \rho_\ell, \gamma_1, \overline{\gamma_1}, \ldots, \gamma_m, \overline{\gamma_m}\}$ to represent the set of characteristic roots of $\mathbf{u}$, where each $\rho_i \in \mathbb{R}$ and each $\gamma_j \in \mathbb{C} \setminus \mathbb{R}$. If $\mathbf{u}$ is a simple LRS, there are algebraic constants $a_1, \ldots, a_\ell \in \mathbb{R}$ and $c_1, \ldots, c_m$ such that, for all $n \geq 0$,

$$u_n = \sum_{i=1}^{\ell} a_i \rho_i^n + \sum_{j=1}^{m} \left(c_j \gamma_j^n + \overline{c_j} \overline{\gamma_j}^n \right). \tag{3}$$

This expression is referred to as the **exponential polynomial** solution of $\mathbf{u}$. For fixed k, all constants a_i and c_j can be computed in time polynomial in $\|\mathbf{u}\|$, since they can be obtained by solving a system of linear equations involving the first k instances of Eq. (3).

An LRS is said to be **non-degenerate** if it does not have two distinct characteristic roots whose quotient is a root of unity. As pointed out in [9], the study of arbitrary LRS can effectively be reduced to that of non-degenerate LRS, by partitioning the original LRS into finitely many subsequences, each of which is non-degenerate. In general, such a reduction will require exponential time. However, when restricting to LRS of bounded order (in our case, of order at most 9), the reduction can be carried out in polynomial time. In particular, any LRS of order 9 or less can be partitioned in polynomial time into at most $3.9 \cdot 10^7$ non-degenerate LRS of the same order or less.[7] Note that if the original LRS is simple, this process will yield a collection of simple non-degenerate subsequences. In the rest of this paper, we shall therefore assume that all LRS we are given are non-degenerate.

Any LRS **u** of order k can alternately be given in matrix form, in the sense that there is a square matrix M of dimension $k \times k$, together with k-dimensional column vectors $\boldsymbol{v}$ and $\boldsymbol{w}$, such that, for all $n \geq 0$, $u_n = \boldsymbol{v}^T M^n \boldsymbol{w}$. It suffices to take M to be the transpose of the companion matrix of the characteristic polynomial of **u**, let $\boldsymbol{v}$ be the vector $(u_{k-1}, \ldots, u_0)$ of initial terms of **u** in reverse order, and take $\boldsymbol{w}$ to be the vector whose first $k - 1$ entries are 0 and whose kth entry is 1. It is worth noting that the characteristic roots of **u** correspond precisely to the eigenvalues of M, and that if **u** is simple then M is diagonalisable. This translation is instrumental in Sec. 3 to place the Positivity Problem for simple LRS of order at most 9 within the Counting Hierarchy.

Conversely, given any square matrix M of dimension $k \times k$, and any k-dimensional vectors $\boldsymbol{v}$ and $\boldsymbol{w}$, let $u_n = \boldsymbol{v}^T M^n \boldsymbol{w}$. Then $\langle \boldsymbol{v}^T M^n \boldsymbol{w} \rangle_{n=k}^{\infty}$ is an LRS of order at most k whose characteristic polynomial divides that of M, as can be seen by applying the Cayley-Hamilton Theorem.[8] When M is diagonalisable, the resulting LRS is simple.

3   Decidability and Complexity

Let $\mathbf{u} = \langle u_n \rangle_{n=0}^{\infty}$ be an integer LRS of order k. As discussed in the Introduction, we assume that u is presented as a $2k$-tuple of integers $(b_1, \ldots, b_k, u_0, \ldots, u_{k-1}) \in \mathbb{Z}^{2k}$, such that for all $n \geq 0$,

$$u_{n+k} = b_1 u_{n+k-1} + \ldots + b_k u_n. \tag{4}$$

The **Positivity Problem** asks, given such an LRS **u**, whether for all $n \geq 0$, it is the case that $u_n \geq 0$. When this holds, we say that **u** is **positive**.

In this section, we establish the following:

Theorem 1. *The Positivity Problem for simple integer LRS of order 9 or less is decidable in* $\mathrm{coNP}^{\mathrm{PP}^{\mathrm{PP}^{\mathrm{PP}}}}$.

---

[7] We obtained this value using a bespoke enumeration procedure for order 9. A bound of $e^{2\sqrt{6\cdot 9 \log 9}} \leq 2.9 \cdot 10^9$ can be obtained from Cor. 3.3 of [31].

[8] In fact, if none of the eigenvalues of M are zero, it is easy to see that the full sequence $\langle \boldsymbol{v}^T M^n \boldsymbol{w} \rangle_{n=0}^{\infty}$ is an LRS (of order at most k).

Note that deciding whether the characteristic roots are simple can easily be done in polynomial time; cf. [21].

Observe also that Thm. 1 immediately carries over to rational LRS. To see this, consider a rational LRS $\mathbf{u}$ obeying the recurrence relation (4). Let ℓ be the least common multiple of the denominators of the rational numbers $b_1, \ldots, b_k$, $u_0, \ldots, u_{k-1}$, and define an integer sequence $\mathbf{v} = \langle v_n \rangle_{n=0}^{\infty}$ by setting $v_n = \ell^{n+1} u_n$ for all $n \geq 0$. It is easily seen that $\mathbf{v}$ is an integer LRS of the same order as $\mathbf{u}$, and that for all n, $v_n \geq 0$ iff $u_n \geq 0$. Moreover, $\mathbf{v}$ is simple iff $\mathbf{u}$ is simple.

High-Level Synopsis (I). At a high level, the algorithm upon which Thm. 1 rests proceeds as follows. Given an LRS $\mathbf{u}$, we first decide whether or not $\mathbf{u}$ is ultimately positive[9] by studying its exponential polynomial solution—further details on this task are provided shortly. As we prove in this paper, whenever $\mathbf{u}$ is an ultimately positive simple LRS of order 9 or less, there is a polynomial-time computable threshold N of at most exponential magnitude such that all terms of $\mathbf{u}$ beyond N are positive. Clearly $\mathbf{u}$ cannot be positive unless it is ultimately positive. Now in order to assert that an ultimately positive LRS $\mathbf{u}$ is *not* positive, we use a *guess-and-check* procedure: find $n \leq N$ such that $u_n < 0$. By writing $u_n = \mathbf{v}^T M^n \mathbf{w}$, for some square integer matrix M and vectors $\mathbf{v}$ and $\mathbf{w}$ (cf. Sec. 2), we can decide whether $u_n < 0$ in PosSLP[10] via iterative squaring, which yields an NP^{PosSLP} procedure for non-Positivity. Thanks to the work of Allender *et al.* [1], which asserts that PosSLP $\subseteq$ P$^{\text{PP}^{\text{PP}^{\text{PP}}}}$, we obtain the required coNP$^{\text{PP}^{\text{PP}^{\text{PP}}}}$ algorithm for deciding Positivity.

The following is an old result concerning LRS; proofs can be found in [11, Thm. 7.1.1] and [4, Thm. 2]. It also follows easily and directly from either Pringsheim's theorem or from [7, Lem. 4]. It plays an important role in our approach by enabling us to significantly cut down on the number of subcases that must be considered, avoiding the sort of quagmire alluded to in [19].

Proposition 2. *Let* $\langle u_n \rangle_{n=0}^{\infty}$ *be an LRS with no real positive dominant characteristic root. Then there are infinitely many n such that $u_n < 0$ and infinitely many n such that $u_n > 0$.*

By Prop. 2, it suffices to restrict our attention to LRS whose dominant characteristic roots include one real positive value. Given an integer LRS $\mathbf{u}$, note that determining whether the latter holds is easily done in time polynomial in $\|\mathbf{u}\|$ (cf. [21]).

Thus let $\mathbf{u}$ be a non-degenerate simple integer LRS of order $k \leq 9$ having a real positive dominant characteristic root $\rho > 0$. Note that $\mathbf{u}$ cannot have a real negative dominant characteristic root (which would be $-\rho$), since otherwise the quotient $-\rho/\rho = -1$ would be a root of unity, contradicting non-degeneracy. Let

---

[9] A sequence is ultimately positive if all but finitely many of its terms are positive.

[10] Recall that PosSLP is the problem of determining whether an arithmetic circuit, with addition, multiplication, and subtraction gates, evaluates to a positive integer.

us write the characteristic roots as $\{\rho, \gamma_1, \overline{\gamma_1}, \ldots, \gamma_m, \overline{\gamma_m}\} \cup \{\gamma_{m+1}, \gamma_{m+2}, \ldots, \gamma_\ell\}$, where we assume that the roots in the first set all have common modulus ρ, whereas the roots in the second set all have modulus strictly smaller than ρ.

Let $\lambda_i = \gamma_i / \rho$ for $1 \le i \le \ell$. We can then write

$$\frac{u_n}{\rho^n} = a + \sum_{j=1}^{m} \left(c_j \lambda_j^n + \overline{c_j} \overline{\lambda_j}^n \right) + r(n), \tag{5}$$

for some real algebraic constant a and complex algebraic constants $c_1, \ldots, c_m$, where $r(n)$ is a term tending to zero exponentially fast.

Note that none of $\lambda_1, \ldots, \lambda_m$, all of which have modulus 1, can be a root of unity, as each λ_i is a quotient of characteristic roots and $\mathbf{u}$ is assumed to be non-degenerate. Likewise, for $i \ne j$, λ_i / λ_j and $\overline{\lambda_i} / \lambda_j$ cannot be roots of unity.

For $i \in \{1, \ldots, \ell\}$, observe also that as each λ_i is a quotient of two roots of the same polynomial of degree k, it has degree at most $k(k-1)$. In fact, it is easily seen that $\|\lambda_i\| = \|\mathbf{u}\|^{\mathcal{O}(1)}$, $\|a\| = \|\mathbf{u}\|^{\mathcal{O}(1)}$, and $\|c_i\| = \|\mathbf{u}\|^{\mathcal{O}(1)}$ (cf. [21]).

Finally, we place bounds on the rate of convergence of $r(n)$. We have

$$r(n) = c_{m+1} \lambda_{m+1}^n + \ldots + c_\ell \lambda_\ell^n .$$

Combining our estimates on the height and degree of each λ_i together with root-separation bounds (cf. [21]), we get $\frac{1}{1 - |\lambda_i|} = 2^{\|\mathbf{u}\|^{\mathcal{O}(1)}}$, for $m + 1 \le i \le \ell$. Thanks also to the bounds on the height and degree of the constants c_i, it follows that we can find $\varepsilon \in (0, 1)$ and $N \in \mathbb{N}$ such that:

$$1/\varepsilon = 2^{\|\mathbf{u}\|^{\mathcal{O}(1)}} \tag{6}$$

$$N = 2^{\|\mathbf{u}\|^{\mathcal{O}(1)}} \tag{7}$$

$$\text{For all } n > N, \; |r(n)| < (1 - \varepsilon)^n . \tag{8}$$

We can compute such ε and N in time polynomial in $\|\mathbf{u}\|$, since all relevant calculations on algebraic numbers only require polynomial time (cf. [21]).

We now seek to answer positivity and ultimate positivity questions for the LRS $\mathbf{u} = \langle u_n \rangle_{n=0}^{\infty}$ by studying the same for $\langle u_n / \rho^n \rangle_{n=0}^{\infty}$.

In what follows, we assume that $\mathbf{u}$ is as above, i.e., $\mathbf{u}$ is a non-degenerate simple integer LRS having a real positive dominant characteristic root $\rho > 0$.

High-Level Synopsis (II). Before launching into technical details, let us provide a high-level overview of our proof strategy for deciding whether $\mathbf{u}$ is ultimately positive, and when that is the case, for computing an index threshold N beyond which all of its terms are positive. Let us rewrite Eq. (5) as

$$\frac{u_n}{\rho^n} = a + h(\lambda_1^n, \ldots, \lambda_m^n) + r(n), \tag{9}$$

where $h : \mathbb{C}^m \to \mathbb{R}$ is a continuous function. In general, there will be integer multiplicative relationships among the $\lambda_1, \ldots, \lambda_m$, forming a free abelian group

L for which we can compute a basis (cf. [21]). These multiplicative relationships define a torus $T \subseteq \mathbb{C}^m$ on which the joint iterates $\{(\lambda_1^n, \ldots, \lambda_m^n) : n \in \mathbb{N}\}$ are dense, as per Kronecker's theorem (cf. [21]).

Now the critical case arises when $a + \min h\!\restriction_T = 0$, where $h\!\restriction_T$ denotes the function h restricted to the torus T. Provided that $h\!\restriction_T$ achieves its minimum $-a$ at only finitely many points, we can use Baker's theorem (cf. [21]) to bound the iterates $(\lambda_1^n, \ldots, \lambda_m^n)$ away from these points by an inverse polynomial in n. By combining Renegar's results (cf. [21]) with techniques from real algebraic geometry, we then argue that $h(\lambda_1^n, \ldots, \lambda_m^n)$ is itself eventually bounded away from the minimum $-a$ by a (different) inverse polynomial in n, and since $r(n)$ decays to zero exponentially fast, we are able to conclude that u_n/ρ^n is ultimately positive, and can compute a threshold N after which all terms u_n (for $n > N$) are positive.

Note in the above that a key component is the requirement that $h\!\restriction_T$ achieve its minimum at finitely many points. In the full version of this paper [21], we show that this is the case provided that L, the free abelian group of multiplicative relationships among the $\lambda_1, \ldots, \lambda_m$, has rank 0, 1, $m-1$, or m. In fact, simple counterexamples can be manufactured in the other instances, which seems to preclude the use of Baker's theorem. Since non-real characteristic roots always arise in conjugate pairs, the earliest appearance of this vexing state of affairs is at order 10: one real dominant root, $m = 4$ pairs of complex dominant roots, one non-dominant root ensuring that the term $r(n)$ is not identically 0, and a free abelian group L of rank 2. The difficulty encountered there is highly reminiscent of that of the critical unresolved case for the Skolem Problem at order 5, as described in [20].

We now proceed with the formalisation of the above. Recall that $\mathbf{u}$ is assumed to be a non-degenerate simple LRS of order at most 9, with a real positive dominant characteristic root $\rho > 0$ and complex dominant roots $\gamma_1, \overline{\gamma_1}, \ldots, \gamma_m, \overline{\gamma_m} \in \mathbb{C} \setminus \mathbb{R}$. We write $\lambda_j = \gamma_j/\rho$ for $1 \leq j \leq m$.

Note that the number of dominant roots is odd and at most 9. Because of space constraints, we focus solely on the most difficult case of there being exactly 7 dominant roots. (Other cases are handled in the full version [21].) We therefore have $m = 3$ in Eq. (5).

Let $L = \{(v_1, v_2, v_3) \in \mathbb{Z}^3 : \lambda_1^{v_1} \lambda_2^{v_2} \lambda_3^{v_3} = 1\}$ have rank p (as a free abelian group), and let $\{\boldsymbol{\ell_1}, \ldots, \boldsymbol{\ell_p}\}$ be a basis for L. Write $\boldsymbol{\ell_q} = (\ell_{q,1}, \ell_{q,2}, \ell_{q,3})$ for $1 \leq q \leq p$. Such a basis may be computed in polynomial time, and moreover each $\ell_{q,j}$ may be assumed to have magnitude polynomial in $\|\mathbf{u}\|$ (cf. [21]).

Let $\mathbb{T} = \{z \in \mathbb{C} : |z| = 1\}$ and write

$$T = \left\{(z_1, z_2, z_3) \in \mathbb{T}^3 : \text{for each } q \in \{1, \ldots, p\},\ z_1^{\ell_{q,1}} z_2^{\ell_{q,2}} z_3^{\ell_{q,3}} = 1\right\}.$$

Define $h : T \to \mathbb{R}$ by $h(z_1, z_2, z_3) = \sum_{j=1}^3 (c_j z_j + \overline{c_j} \overline{z_j})$, so that for all n,

$$\frac{u^n}{\rho^n} = a + h(\lambda_1^n, \lambda_2^n, \lambda_3^n) + r(n). \tag{10}$$

The set $\{(\lambda_1^n, \lambda_2^n, \lambda_3^n) : n \in \mathbb{N}\}$ is a dense subset of T (cf. [21]). Since h is continuous, we have $\inf\{h(\lambda_1^n, \lambda_2^n, \lambda_3^n) : n \in \mathbb{N}\} = \min h\restriction_T = \mu$, for some $\mu \in \mathbb{R}$.

We can represent μ via the following formula $\tau(y)$:

$$\exists(\zeta_1, \zeta_2, \zeta_3) \in T : (h(\zeta_1, \zeta_2, \zeta_3) = y \land \forall(z_1, z_2, z_3) \in T, \, y \leq h(z_1, z_2, z_3)).$$

We can construct an equivalent formula $\tau'(y)$ in the first-order theory of the reals, over a bounded number of real variables, with $||\tau'(y)|| = ||\mathbf{u}||^{\mathcal{O}(1)}$. As detailed in [21], we can then compute in polynomial time an equivalent quantifier-free formula

$$\chi(y) = \bigvee_{i=1}^{I} \bigwedge_{j=1}^{J_i} h_{i,j}(y) \sim_{i,j} 0.$$

In the above, each $\sim_{i,j}$ is either $>$ or $=$. Now $\chi(y)$ must have a satisfiable disjunct, and since the satisfying assignment to y is unique (namely $y = \mu$), this disjunct must comprise at least one equality predicate. Moreover, the degree and height of each $h_{i,j}$ are bounded by $||\mathbf{u}||^{\mathcal{O}(1)}$ and $2^{||\mathbf{u}||^{\mathcal{O}(1)}}$ respectively, hence we immediately conclude that μ is an algebraic number and moreover that $||\mu|| = ||\mathbf{u}||^{\mathcal{O}(1)}$ (cf. [21] for details).

Returning to Eq. (10), we see that if $a + \mu < 0$, then $\mathbf{u}$ is neither positive nor ultimately positive, whereas if $a + \mu > 0$ then $\mathbf{u}$ is ultimately positive. In the latter case, thanks to our bounds on $||\mu||$, we have $\frac{1}{a+\mu} = 2^{||\mathbf{u}||^{\mathcal{O}(1)}}$ (cf. [21]). The latter, together with Eqs. (6)–(8), implies an exponential upper bound on the index of possible violations of positivity. The actual positivity of $\mathbf{u}$ can then be decided via a coNP procedure that invokes a PosSLP oracle as outlined earlier.

It remains to analyse the case in which $\mu = -a$. To this end, let $\lambda_j = e^{i\theta_j}$ for $1 \leq j \leq 3$. From Eq. (5), we have:

$$\frac{u_n}{\rho^n} = a + \sum_{j=1}^{3} 2|c_j| \cos(n\theta_j + \varphi_j) + r(n).$$

In the above, $c_j = |c_j|e^{i\varphi_j}$ for $1 \leq j \leq 3$. We make the further assumption that each c_j is non-zero; note that if this did not hold, we could simply recast our analysis in a lower dimension.

Let $Z = \{(\zeta_1, \zeta_2, \zeta_3) \in T : h(\zeta_1, \zeta_2, \zeta_3) = \mu\}$ be the set of points of T at which h achieves its minimum μ. One of our key results, the *Zero-Dimensionality Lemma*, proved in the full version of this paper [21], asserts that Z is finite. We concentrate on the set Z_1 of first coordinates of Z. Write

$$\tau_1(x) = \exists z_1 : (\mathrm{Re}(z_1) = x \land z_1 \in Z_1)$$
$$\tau_2(y) = \exists z_1 : (\mathrm{Im}(z_1) = y \land z_1 \in Z_1).$$

Similarly to our earlier construction, $\tau_1(x)$ is equivalent to a formula $\tau_1'(x)$ in the first-order theory of the reals, over a bounded number of real variables, with

$||\tau_1'(x)|| = ||\mathbf{u}||^{\mathcal{O}(1)}$. As shown in [21], we then obtain an equivalent quantifier-free formula

$$\chi_1(x) = \bigvee_{i=1}^{I} \bigwedge_{j=1}^{J_i} h_{i,j}(x) \sim_{i,j} 0.$$

Note that since there can only be finitely many $\hat{x} \in \mathbb{R}$ such that $\chi_1(\hat{x})$ holds, each disjunct of $\chi_1(x)$ must comprise at least one equality predicate, or can otherwise be entirely discarded as having no solution.

A similar exercise can be carried out with $\tau_2(y)$, yielding $\chi_2(y)$. The bounds on the degree and height of each $h_{i,j}$ in $\chi_1(x)$ and $\chi_2(y)$ then enable us to conclude that any $\zeta = \hat{x} + i\hat{y} \in Z_1$ is algebraic, and moreover satisfies $||\zeta|| = ||\mathbf{u}||^{\mathcal{O}(1)}$. In addition, bounds on I and J_i guarantee that the cardinality of Z_1 is at most polynomial in $||\mathbf{u}||$.

Since λ_1 is not a root of unity, for each $\zeta \in Z_1$ there is at most one value of n such that $\lambda_1^n = \zeta$. This value (if it exists) is at most $M = ||\mathbf{u}||^{\mathcal{O}(1)}$, which we can take to be uniform across all $\zeta \in Z_1$ (cf. [21]). We can now invoke Baker's theorem (cf. [21]) to conclude that, for $n > M$, and for all $\zeta \in Z_1$, we have

$$|\lambda_1^n - \zeta| > \frac{1}{n^{||\mathbf{u}||^D}}, \tag{11}$$

where $D \in \mathbb{N}$ is some absolute constant.

Let $b > 0$ be minimal such that the set

$$\{z_1 \in \mathbb{C} : |z_1| = 1 \text{ and, for all } \zeta \in Z_1, |z_1 - \zeta| \geq \frac{1}{b}\}$$

is non-empty. Thanks to our bounds on the cardinality of Z_1, we can use the first-order theory of the reals, together with suitable size bounds (cf. [21]), to conclude that b is algebraic and $||b|| = ||\mathbf{u}||^{\mathcal{O}(1)}$.

Define the function $g : [b, \infty) \to \mathbb{R}$ as follows:

$$g(x) = \min\{h(z_1, z_2, z_3) - \mu : (z_1, z_2, z_3) \in T \text{ and, for all } \zeta \in Z_1, |z_1 - \zeta| \geq \frac{1}{x}\}.$$

It is clear that g is continuous and $g(x) > 0$ for all $x \in [b, \infty)$. Moreover, as before, g can be rewritten as a function in the first-order theory of the reals over a bounded number of variables, with $||g|| = \mathbf{u}^{\mathcal{O}(1)}$. It follows from Prop. 2.6.2 of [6] (invoked with the function $1/g$) that there is a polynomial $P \in \mathbb{Z}[x]$ such that, for all $x \in [b, \infty)$,

$$g(x) \geq \frac{1}{P(x)}. \tag{12}$$

Moreover, an examination of the proof of [6, Prop. 2.6.2] reveals that P is obtained through a process which hinges on quantifier elimination. We are therefore able to conclude that $||P|| = ||\mathbf{u}||^{\mathcal{O}(1)}$, a fact which relies among others on our upper bounds for $||b||$ (cf. [21]).

By Eqs. (6)–(8), we can find $\varepsilon \in (0,1)$ and $N = 2^{||\mathbf{u}||^{\mathcal{O}(1)}}$ such that for all $n > N$, we have $|r(n)| < (1 - \varepsilon)^n$, and moreover $1/\varepsilon = 2^{||\mathbf{u}||^{\mathcal{O}(1)}}$. Moreover, it is shown in [21] that there is $N' = 2^{||\mathbf{u}||^{\mathcal{O}(1)}}$ such that, for all $n \geq N'$,

$$\frac{1}{P(n^{||\mathbf{u}||^D})} > (1 - \varepsilon)^n. \tag{13}$$

Combining Eqs. (10)–(13), we get

$$\begin{aligned}
\frac{u^n}{\rho^n} &= a + h(\lambda_1^n, \lambda_2^n, \lambda_3^n) + r(n) \\
&\geq -\mu + h(\lambda_1^n, \lambda_2^n, \lambda_3^n) - (1 - \varepsilon)^n \\
&\geq g(n^{||\mathbf{u}||^D}) - (1 - \varepsilon)^n \\
&\geq \frac{1}{P(n^{||\mathbf{u}||^D})} - (1 - \varepsilon)^n \\
&\geq 0,
\end{aligned}$$

provided $n > \max\{M, N, N'\}$, which establishes ultimate positivity of $\mathbf{u}$ and provides an exponential upper bound on the index of possible violations of positivity, as required. We can then decide the actual positivity of $\mathbf{u}$ via a coNP^{PosSLP} procedure as detailed earlier.

This completes the proof of Thm. 1.

References

[1] Allender, E., Bürgisser, P., Kjeldgaard-Pedersen, J., Miltersen, P.B.: On the complexity of numerical analysis. SIAM J. Comput. 38(5) (2009)

[2] Amoroso, F., Viada, E.: Small points on subvarieties of a torus. Duke Mathematical Journal 150(3) (2009)

[3] Amoroso, F., Viada, E.: On the zeros of linear recurrence sequences. Acta Arithmetica 147(4) (2011)

[4] Bell, J.P., Gerhold, S.: On the positivity set of a linear recurrence. Israel Jour. Math. 57 (2007)

[5] Berstel, J., Mignotte, M.: Deux propriétés décidables des suites récurrentes linéaires. Bull. Soc. Math. France 104 (1976)

[6] Bochnak, J., Coste, M., Roy, M.-F.: Real Algebraic Geometry. Springer (1998)

[7] Braverman, M.: Termination of integer linear programs. In: Ball, T., Jones, R.B. (eds.) CAV 2006. LNCS, vol. 4144, pp. 372–385. Springer, Heidelberg (2006)

[8] Burke, J.R., Webb, W.A.: Asymptotic behavior of linear recurrences. Fib. Quart. 19(4) (1981)

[9] Everest, G., van der Poorten, A., Shparlinski, I., Ward, T.: Recurrence Sequences. American Mathematical Society (2003)

[10] Evertse, J.H., Schlickewei, H.P., Schmidt, W.M.: Linear equations in variables which lie in a multiplicative group. Ann. Math. 155(3) (2002)

[11] Gyori, I., Ladas, G.: Oscillation Theory of Delay Differential Equations. Oxford Mathematical Monographs. Oxford University Press (1991)

[12] Halava, V., Harju, T., Hirvensalo, M.: Positivity of second order linear recurrent sequences. Discrete Appl. Math. 154(3) (2006)
[13] Halava, V., Harju, T., Hirvensalo, M., Karhumäki, J.: Skolem's problem — on the border between decidability and undecidability. Technical Report 683, Turku Centre for Computer Science (2005)
[14] Laohakosol, V.: Personal communication (July 2013)
[15] Laohakosol, V., Tangsupphathawat, P.: Positivity of third order linear recurrence sequences. Discrete Appl. Math. 157(15) (2009)
[16] Lipton, R.J.: Mathematical embarrassments. Blog entry (December 2009), http://rjlipton.wordpress.com/2009/12/26/mathematical-embarrassments/
[17] Liu, L.L.: Positivity of three-term recurrence sequences. Electr. J. Comb. 17(1) (2010)
[18] Mignotte, M., Shorey, T.N., Tijdeman, R.: The distance between terms of an algebraic recurrence sequence. Journal für die reine und angewandte Mathematik 349 (1984)
[19] Nagasaka, K., Shiue, J.-S.: Asymptotic positiveness of linear recurrence sequences. Fib. Quart. 28(4) (1990)
[20] Ouaknine, J., Worrell, J.: Decision problems for linear recurrence sequences. In: Finkel, A., Leroux, J., Potapov, I. (eds.) RP 2012. LNCS, vol. 7550, pp. 21–28. Springer, Heidelberg (2012)
[21] Ouaknine, J., Worrell, J.: On the Positivity Problem for simple linear recurrence sequences (full version). arXiv:1309.1550 (2013)
[22] Ouaknine, J., Worrell, J.: Positivity problems for low-order linear recurrence sequences. In: Proc. Symp. on Discrete Algorithms (SODA). ACM-SIAM (2014)
[23] Ouaknine, J., Worrell, J.: Ultimate Positivity is decidable for simple linear recurrence sequences. In: Esparza, J., Fraigniaud, P., Husfeldt, T., Koutsoupias, E. (eds.) ICALP 2014, Part II. LNCS, vol. 8573, pp. 330–341. Springer, Heidelberg (2014); Full version as arXiv:1309.1914
[24] Rozenberg, G., Salomaa, A.: Cornerstones of Undecidability. Prentice Hall (1994)
[25] Salomaa, A.: Growth functions of Lindenmayer systems: Some new approaches. In: Lindenmayer, A., Rozenberg, G. (eds.) Automata, Languages, Development. North-Holland (1976)
[26] Soittola, M.: On D0L synthesis problem. In: Lindenmayer, A., Rozenberg, G. (eds.) Automata, Languages, Development. North-Holland (1976)
[27] Tangsupphathawat, P., Punnim, N., Laohakosol, V.: The positivity problem for fourth order linear recurrence sequences is decidable. Colloq. Math. 128(1) (2012)
[28] Tao, T.: Structure and Randomness. American Mathematical Society (2008)
[29] Tarasov, S., Vyalyi, M.: Orbits of linear maps and regular languages. In: Kulikov, A., Vereshchagin, N. (eds.) CSR 2011. LNCS, vol. 6651, pp. 305–316. Springer, Heidelberg (2011)
[30] Vereshchagin, N.K.: The problem of appearance of a zero in a linear recurrence sequence. Mat. Zametki 38(2) (1985) (in Russian)
[31] Yokoyama, K., Li, Z., Nemes, I.: Finding roots of unity among quotients of the roots of an integral polynomial. In: Proc. Intern. Symp. on Symb. and Algebraic Comp. (1995)

Ultimate Positivity is Decidable for Simple Linear Recurrence Sequences*

Joël Ouaknine and James Worrell

Department of Computer Science, University of Oxford, UK

Abstract. We consider the decidability and complexity of the Ultimate Positivity Problem, which asks whether all but finitely many terms of a given rational linear recurrence sequence (LRS) are positive. Using lower bounds in Diophantine approximation concerning sums of S-units, we show that for simple LRS (those whose characteristic polynomial has no repeated roots) the Ultimate Positivity Problem is decidable in polynomial space. If we restrict to simple LRS of a fixed order then we obtain a polynomial-time decision procedure. As a complexity lower bound we show that Ultimate Positivity for simple LRS is at least as hard as the decision problem for the universal theory of the reals: a problem that is known to lie between **coNP** and **PSPACE**.

1   Introduction

A **linear recurrence sequence (LRS)** is an infinite sequence $\boldsymbol{u} = \langle u_0, u_1, \ldots \rangle$ of rational numbers satisfying a recurrence relation

$$u_{n+k} = a_1 u_{n+k-1} + a_2 u_{n+k-2} + \ldots + a_k u_n \tag{1}$$

for all $n \geq 0$, where $a_1, a_2, \ldots, a_k$ are fixed rational numbers with $a_k \neq 0$. Such a sequence is determined by its initial values $u_0, \ldots, u_{k-1}$ and the recurrence relation. We say that the recurrence has **characteristic polynomial**

$$f(x) = x^k - a_1 x^{k-1} - \ldots - a_{k-1}x - a_k.$$

The least k such that $\boldsymbol{u}$ satisfies a recurrence of the form (1) is called the **order** of $\boldsymbol{u}$. If the characteristic polynomial of this (unique) recurrence has no repeated roots then we say that $\boldsymbol{u}$ is **simple**.

Given an LRS $\boldsymbol{u}$ there are polynomials $p_1, \ldots, p_k \in \mathbb{C}[x]$ such that

$$u_n = p_1(n)\gamma_1^n + \ldots + p_k(n)\gamma_k^n,$$

where $\gamma_1, \ldots, \gamma_k$ are the roots of the characteristic polynomial. Moreover $\boldsymbol{u}$ is simple if and only if it admits such a representation in which each polynomial p_i is a constant. Simple LRS are a natural and widely studied subclass of LRS whose analysis nevertheless remains extremely challenging [1,10,12,24].

---

* The full version of this paper is available as [23].

J. Esparza et al. (Eds.): ICALP 2014, Part II, LNCS 8573, pp. 330–341, 2014.

Motivated by questions in language theory and formal power series, Rozenberg, Salomaa, and Soittola [27,29] highlight the following four decision problems concerning LRS. Given an LRS $\langle u_n \rangle_{n=0}^{\infty}$ (represented by a linear recurrence and sequence of initial values):

1. Does $u_n = 0$ for some n?
2. Does $u_n = 0$ for infinitely many n?
3. Is $u_n \geq 0$ for all n?
4. Is $u_n \geq 0$ for all but finitely many n?

Linear recurrence sequences are ubiquitous in mathematics and computer science, and the above four problems (and assorted variants) arise in a variety of settings; see [25] for references. For example, an LRS modelling population size is biologically meaningful only if it never becomes negative.

Problem 1 is known as **Skolem's Problem**, after the Skolem-Mahler-Lech Theorem [18,19,28], which characterises the set $\{n \in \mathbb{N} : u_n = 0\}$ of zeros of an LRS u as an ultimately periodic set. The proof of the Skolem-Mahler-Lech Theorem is non-effective, and the decidability of Skolem's Problem is open. Blondel and Tsitsiklis [6] remark that "the present consensus among number theorists is that an algorithm [for Skolem's Problem] should exist". However, so far decidability is known only for LRS of order at most 4: a result due independently to Vereschagin [32] and Mignotte, Shorey, and Tijdeman [21]. At order 5 decidability is not known, even for simple LRS [22]. Decidability of Skolem's Problem is also listed as an open problem and discussed at length by Tao [30, Section 3.9]. The problem can furthermore be seen as a generalisation of the Orbit Problem, studied by Kannan and Lipton [16, Section 5].

In contrast to the situation with Skolem's Problem, Problem 2—hitting zero infinitely often—was shown to be decidable for arbitrary LRS by Berstel and Mignotte [4].

Problems 3 and 4 are respectively known as the **Positivity** and **Ultimate Positivity** Problems. The problems are stated as open in [2,14,17], among others, while in [27] the authors assert that the problems are "generally conjectured [to be] decidable". Decidability of Positivity entails decidability of Skolem's Problem via a straightforward algebraic transformation of LRS (which however does not preserve the order) [14].

Hitherto, all decidability results for Positivity and Ultimate Positivity have been for low-order sequences. The paper [25] gives a detailed account of these results, obtained over a period of time stretching back some 30 years, and proves decidability of both problems for sequences of order at most 5. It is moreover shown in [25] that obtaining decidability for either Positivity or Ultimate Positivity at order 6 would necessarily entail major breakthroughs in Diophantine approximation.

The main result of this paper is that the Ultimate Positivity Problem for simple LRS of arbitrary order is decidable. The restriction to simple LRS allows us to circumvent the strong "mathematical hardness" result for sequences of order 6 alluded to above. However, our decision procedure is non-constructive: given an ultimately positive LRS $\langle u_n \rangle_{n=0}^{\infty}$, the procedure does not compute a threshold N such that $u_n \geq 0$ for all $n \geq N$. Indeed the ability to compute such a threshold N would immediately yield an algorithm for the Positivity Problem for simple LRS since the signs of $u_0, \ldots, u_{N-1}$ can be checked directly. In turn this would yield decidability of Skolem's Problem for simple LRS. But Skolem's Problem is open for simple LRS of order 5, while (as discussed below) Positivity for simple LRS is only known to be decidable up to order 9.

The non-constructive aspect of our results arises from our use of lower bounds in Diophantine approximation concerning sums of S-units. These bounds were proven in [11,31] using Schlickewei's p-adic generalisation of Schmidt's Subspace Theorem (itself a far-reaching generalisation of the Thue-Siegel-Roth Theorem), and therein applied to study the asymptotic growth of LRS in absolute value. By contrast, in [24] we use Baker's Theorem on linear forms in logarithms to show decidability of Positivity for simple LRS of order at most 9. Unfortunately, while Baker's Theorem yields effective Diophantine-approximation lower bounds, it appears only to be applicable to low-order LRS. In particular, the analytic and geometric arguments that are used in [24] to bring Baker's Theorem to bear (and which give that work a substantially different flavour to the present paper) do not apply beyond order 9.

Relying on complexity bounds for the decision problem for first-order formulas over the field of real numbers, we show that our procedure for deciding Ultimate Positivity requires polynomial space in general and polynomial time for LRS of each fixed order. As a complexity lower bound, we give a polynomial-time reduction of the decision problem for the universal theory of the reals to both the Positivity and Ultimate Positivity Problems for simple LRS. The decision problem for the universal theory of the reals is easily seen to be **coNP**-hard and, from the work of Canny [8], is contained in **PSPACE**. Thus the complexity of the Ultimate Positivity problem for simple LRS lies between **coNP** and **PSPACE**. Hitherto the best lower bound known for either Positivity or Ultimate Positivity was **coNP**-hardness [3].

Full proofs of all results can be found in the long version of this paper [23].

2   Background

Number Theory. A complex number α is **algebraic** if it is a root of a univariate polynomial with integer coefficients. The **defining polynomial** of α, denoted p_α, is the unique integer polynomial of least degree, whose coefficients have no common factor, that has α as a root. The **degree** of α is the degree of p_α, and the **height** of α is the maximum absolute value of the coefficients of p_α. If p_α is monic then we say that α is an **algebraic integer**.

For computational purposes an algebraic number α can be represented by a polynomial f that has α as a root, together with an approximation of α with rational real and imaginary parts of sufficient accuracy to distinguish α from the other roots of f [15]. We denote by $\|\alpha\|$ the length of this representation.[1] It

---

[1] In general we denote by $\|X\|$ the length of the binary representation of a given object X.

can be shown that $||\alpha||$ is polynomial in the degree and logarithm of the height of α. Given a univariate polynomial f, it is moreover known how to obtain representations of each of its roots in time polynomial in $||f||$.

A **number field** K is a finite-dimensional extension of $\mathbb{Q}$. The set of algebraic integers in K forms a ring, denoted $\mathcal{O}$. Given two ideals I, J in $\mathcal{O}$, the product IJ is the ideal generated by the elements ab, where $a \in I$ and $b \in J$. An ideal P of $\mathcal{O}$ is **prime** if $ab \in P$ implies $a \in P$ or $b \in P$. The fundamental theorem of ideal theory states that any non-zero ideal in $\mathcal{O}$ can be written as the product of prime ideals, and the representation is unique if the order of the prime ideals is ignored.

We will need the following classical result of Dirichlet [13].

Theorem 2.1 (Dirichlet). *Let P be the set of primes and $P_{a,b}$ the set of primes congruent to a mod b, where $\gcd(a, b) = 1$. Then*

$$\lim_{n \to \infty} \frac{|P_{a,b} \cap \{1, \ldots, n\}|}{|P \cap \{1, \ldots, n\}|} = \frac{1}{\varphi(b)},$$

where φ denotes Euler's totient function.

Linear Recurrence Sequences. Let $\boldsymbol{u} = \langle u_n \rangle_{n=0}^{\infty}$ be a sequence of rational numbers satisfying the recurrence relation $u_{n+k} = a_1 u_{n+k-1} + \ldots + a_k u_n$. We represent such an LRS as a $2k$-tuple $(a_1, \ldots, a_k, u_0, \ldots, u_{k-1})$ of rational numbers (encoded in binary). Given an arbitrary representation of $\boldsymbol{u}$, we can compute the coefficients of the unique minimal-order recurrence satisfied by $\boldsymbol{u}$ in polynomial time by straightforward linear algebra. Henceforth we will always assume that an LRS is presented in terms of its minimal-order recurrence. By the characteristic polynomial of an LRS we mean the characteristic polynomial of the minimal-order recurrence. The roots of this polynomial are called the **characteristic roots**. The characteristic roots of maximum modulus are said to be **dominant**.

It is well-known (see, e.g., [2, Thm. 2]) that if an LRS $\boldsymbol{u}$ has no real positive dominant characteristic root then there are infinitely many n such that $u_n < 0$ and infinitely many n such that $u_n > 0$. Clearly such an LRS cannot be ultimately positive.

Since the characteristic polynomial of $\boldsymbol{u}$ has real coefficients, its set of roots can be written in the form $\{\rho_1, \ldots, \rho_\ell, \gamma_1, \overline{\gamma_1}, \ldots, \gamma_m, \overline{\gamma_m}\}$, where each $\rho_i \in \mathbb{R}$. If $\boldsymbol{u}$ is simple then there are non-zero real algebraic constants $b_1, \ldots, b_\ell$ and complex algebraic constants $c_1, \ldots, c_m$ such that, for all $n \geq 0$,

$$u_n = \sum_{i=1}^{\ell} b_i \rho_i^n + \sum_{j=1}^{m} \left(c_j \gamma_j^n + \overline{c_j \gamma_j}^n \right). \tag{2}$$

Conversely, a sequence $\boldsymbol{u}$ that admits the representation (2) is a simple LRS over $\mathbb{R}$, with characteristic roots among $\rho_1, \ldots, \rho_\ell, \gamma_1, \overline{\gamma_1}, \ldots, \gamma_m, \overline{\gamma_m}$. Arbitrary LRS admit a more general "exponential-polynomial" representation in which the coefficients b_i and c_j are replaced by polynomials in n.

An LRS is said to be **non-degenerate** if it does not have two distinct characteristic roots whose quotient is a root of unity. A non-degenerate LRS is either identically zero or only has finitely many zeros. The study of arbitrary LRS can effectively be reduced to that of non-degenerate LRS using the following result from [10].

Proposition 2.2. *Let* $\langle u_n \rangle_{n=0}^{\infty}$ *be an LRS of order k over* $\mathbb{Q}$. *There is a constant* $M = 2^{O(k\sqrt{\log k})}$ *such that each subsequence* $\langle u_{Mn+l} \rangle_{n=0}^{\infty}$ *is non-degenerate for* $0 \le l < M$.

The constant M in Proposition 2.2 is the least common multiple of the orders of all roots of unity appearing as quotients of characteristic roots of u. This number can be computed in time polynomial in $||u||$ since determining whether an algebraic number α is a root of unity (and computing the order of the root) can be done in polynomial time in $||\alpha||$ [15]. From the representation (2) we see that if the original LRS is simple with characteristic roots $\lambda_1, \ldots, \lambda_k$, then each subsequence $\langle u_{Mn+l} \rangle_{n=0}^{\infty}$ is also simple, with characteristic roots among $\lambda_1^M, \ldots, \lambda_k^M$.

The following is a celebrated result on LRS [18,19,28].

Theorem 2.3 (Skolem-Mahler-Lech). *The set $\{n : u_n = 0\}$ of zeros of an LRS u comprises a finite set together with a finite number of arithmetic progressions. If u is non-degenerate and not identically zero, then its set of zeros is finite.*

Suppose that u and v are LRS of orders k and l respectively, then the pointwise sum $\langle u_n + v_n \rangle_{n=0}^{\infty}$ is an LRS of order at most $k + l$, and the pointwise product $\langle u_n v_n \rangle_{n=0}^{\infty}$ is an LRS of order at most kl. Given representations of u and v we can compute representations of the sum and product in polynomial time by straightforward linear algebra.

First-Order Theory of the Reals. Let $x = x_1, \ldots, x_m$ be a list of m real-valued variables, and let $\sigma(x)$ be a Boolean combination of atomic predicates of the form $g(x) \sim 0$, where each $g(x)$ is a polynomial with integer coefficients in the variables x, and $\sim$ is either $>$ or $=$. We consider the problem of deciding the truth over the field $\mathbb{R}$ of sentences φ in the form

$$Q_1 x_1 \ldots Q_m x_m \, \sigma(x), \tag{3}$$

where each Q_i is one of the quantifiers $\exists$ or $\forall$. We write $||\varphi||$ for the length of the syntactic representation of φ.

The collection of true sentences of the form (3) is called the **first-order theory of the reals**. Tarski famously showed that this theory admits quantifier elimination and is therefore decidable. In this paper we rely on decision procedures for two fragments of this theory. We use the result of Canny [8] that if each Q_i is a universal quantifier, then the truth of φ can be decided in space polynomial in $||\varphi||$. We also use the result of Renegar [26] that for each fixed $M \in \mathbb{N}$, if the number of variables in φ is at most M, then the truth of φ can be determined in time polynomial in $||\varphi||$.

Given a representation of an algebraic number α, as described in Section 2, both the real and imaginary parts of α are straightforwardly definable by quantifier-free formulas $\varphi(x)$ of size polynomial in $||\alpha||$.

3 Multiplicative Relations

Throughout this section let $\boldsymbol{\lambda} = (\lambda_1, \ldots, \lambda_s)$ be a tuple of algebraic numbers, each of height at most H and degree at most d. Assume that each λ_i is represented in the manner described in Section 2.

We define the group of multiplicative relations holding among the λ_i to be the subgroup $L(\boldsymbol{\lambda})$ of $\mathbb{Z}^s$ defined by

$$L(\boldsymbol{\lambda}) = \left\{ (v_1, \ldots, v_s) \in \mathbb{Z}^s : \lambda_1^{v_1} \ldots \lambda_s^{v_s} = 1 \right\}.$$

Bounds on the complexity of computing a basis of $L(\boldsymbol{\lambda})$, considered as a free abelian group, can be obtained from the following result of Masser [20] which gives an upper bound on the magnitude of the entries of the vectors in such a basis.

Theorem 3.1 (Masser). *The free abelian group $L(\boldsymbol{\lambda})$ has a basis $\boldsymbol{v}_1, \ldots, \boldsymbol{v}_l \in \mathbb{Z}^s$ for which*

$$\max_{1 \leq i \leq l, \, 1 \leq j \leq s} |\boldsymbol{v}_{i,j}| = (d \log H)^{O(s^2)}.$$

Corollary 3.2. *A basis of $L(\boldsymbol{\lambda})$ can be computed in space polynomial in $||\boldsymbol{\lambda}||$. If s and d are fixed, such a basis can be computed in time polynomial in $||\boldsymbol{\lambda}||$.*

Proof. Masser's bound entails that there is a basis $\boldsymbol{v}_1, \ldots, \boldsymbol{v}_l$ whose total bit length is polynomial in s, $\log d$ and $\log \log H$, all of which are polynomial in $||\boldsymbol{\lambda}||$. Moreover the membership problem "$\lambda_1^{v_1} \ldots \lambda_s^{v_s} = 1$?" for a potential basis vector $\boldsymbol{v} \in \mathbb{Z}^s$ is decidable in space polynomial in $||\boldsymbol{\lambda}||$ by reduction to the decision problem for existential sentences over the reals.

A set of vectors $\boldsymbol{v}_1, \ldots, \boldsymbol{v}_l$ in $L(\boldsymbol{\lambda})$ is a basis if every vector $\boldsymbol{v} \in L(\boldsymbol{\lambda})$ whose entries satisfy the bound in Theorem 3.1 lies in the integer span of $\boldsymbol{v}_1, \ldots, \boldsymbol{v}_l$. For each such vector $\boldsymbol{v}$ this can be checked by solving a system of linear equations over the integers. Thus we can compute a basis of $L(\boldsymbol{\lambda})$ in space polynomial in $||\boldsymbol{\lambda}||$ by brute-force search.

If s and d are fixed then the same brute-force search can be done in time polynomial in $||\boldsymbol{\lambda}||$, noting that the number of possible bases is polynomial in $||\boldsymbol{\lambda}||$ and the membership problem "$\lambda_1^{v_1} \ldots \lambda_s^{v_s} = 1$?" is decidable in time polynomial in $||\boldsymbol{\lambda}||$ by reduction to the decision problem for existential sentences over the reals with a fixed number of variables. $\qquad \square$

The following is an easy consequence of Corollary 3.2.

Corollary 3.3. *Given $M \in \mathbb{N}$, a basis of $L(\lambda_1^M, \ldots, \lambda_s^M)$ can be computed in space polynomial in $||M||$ and $||\boldsymbol{\lambda}||$.*

Next we relate the group $L(\boldsymbol{\lambda})$ to the **orbit** $\{(\lambda_1^n, \ldots, \lambda_s^n) \mid n \in \mathbb{N}\}$ of $\boldsymbol{\lambda}$. Recall from [9] the following classical theorem of Kronecker on inhomogeneous Diophantine approximation.

Theorem 3.4 (Kronecker). *Let $\theta_1, \ldots, \theta_s$ and $\psi_1, \ldots, \psi_s$ be real numbers. Suppose moreover that for all integers $u_1, \ldots, u_s$, if $u_1\theta_1 + \ldots + u_s\theta_s \in \mathbb{Z}$ then also $u_1\psi_1 + \ldots + u_s\psi_s \in \mathbb{Z}$, i.e., all integer relations among the θ_i also hold among the ψ_i (modulo $\mathbb{Z}$). Then for each $\varepsilon > 0$, there exist integers $p_1, \ldots, p_s$ and a non-negative integer n such that $|n\theta_i - p_i - \psi_i| \le \varepsilon$.*

Write $\mathbb{T} = \{z \in \mathbb{C} : |z| = 1\}$ and consider the s-dimensional torus $\mathbb{T}^s$ as a group under coordinatewise multiplication. The following can be seen as a multiplicative formulation of Kronecker's Theorem.

Proposition 3.5. *Let $\boldsymbol{\lambda} = (\lambda_1, \ldots, \lambda_s) \in \mathbb{T}^s$ and consider the group $L(\boldsymbol{\lambda})$ of multiplicative relations among the λ_i. Define a subgroup $T(\boldsymbol{\lambda})$ of the torus $\mathbb{T}^s$ by*

$$T(\boldsymbol{\lambda}) = \{(\mu_1, \ldots, \mu_s) \in \mathbb{T}^s \mid \mu_1^{v_1} \ldots \mu_s^{v_s} = 1 \text{ for all } \boldsymbol{v} \in L(\boldsymbol{\lambda})\}.$$

Then the orbit $S = \{(\lambda_1^n, \ldots, \lambda_s^n) \mid n \in \mathbb{N}\}$ is a dense subset of $T(\boldsymbol{\lambda})$.

Proof. For $j = 1, \ldots, s$, let $\theta_j \in \mathbb{R}$ be such that $\lambda_j = e^{2\pi i \theta_j}$. Notice that multiplicative relations $\lambda_1^{v_1} \ldots \lambda_s^{v_s} = 1$ are in one-to-one correspondence with additive relations $\theta_1 v_1 + \ldots + \theta_s v_s \in \mathbb{Z}$. Let $(\mu_1, \ldots, \mu_s)$ be an arbitrary element of $T(\boldsymbol{\lambda})$, with $\mu_j = e^{2\pi i \psi_j}$ for some $\psi_j \in \mathbb{R}$. Then the hypotheses of Theorem 3.4 apply to $\theta_1, \ldots, \theta_s$ and $\psi_1, \ldots, \psi_s$. Thus given $\varepsilon > 0$, there exist $n \ge 0$ and $p_1, \ldots, p_s \in \mathbb{Z}$ such that $|n\theta_j - p_j - \psi_j| \le \varepsilon$ for $j = 1, \ldots, s$. Whence for $j = 1, \ldots, s$,

$$|\lambda_j^n - \mu_j| = |e^{2\pi i(n\theta_j - p_j)} - e^{2\pi i \psi_j}| \le |2\pi(n\theta_j - p_j - \psi_j)| \le 2\pi\varepsilon.$$

It follows that $(\mu_1, \ldots, \mu_s)$ lies in the closure of S. $\qquad\square$

4   Algorithm for Ultimate Positivity

Let K be a number field of degree d over $\mathbb{Q}$. Recall that there are d distinct field monomorphisms $\sigma_1, \ldots, \sigma_d : K \to \mathbb{C}$ (see, e.g., [13]). Given a finite set S of prime ideals in the ring of integers $\mathcal{O}$ of K, we say that $\alpha \in \mathcal{O}$ is an S-**unit** if the principal ideal (α) is a product of prime ideals in S. The following lower bound on the magnitude of sums of S-units, whose key ingredient is Schlickewei's p-adic generalisation of Schmidt's Subspace Theorem, was established in [11,31] to analyse the growth of LRS.

Theorem 4.1 (Evertse, van der Poorten, Schlickewei). *Let m be a positive integer and S a finite set of prime ideals in $\mathcal{O}$. Then for every $\varepsilon > 0$ there exists a constant C, depending only on m, K, S, and ε with the following property: for any set of S-units $x_1, \ldots, x_m \in \mathcal{O}$ such that $\sum_{i \in I} x_i \ne 0$ for all non-empty $I \subseteq \{1, \ldots, m\}$, it holds that*

$$|x_1 + \ldots + x_m| \ge CXY^{-\varepsilon}, \tag{4}$$

where $X = \max\{|x_i| : 1 \le i \le m\}$, $Y = \max\{|\sigma_j(x_i)| : 1 \le i \le m, 1 \le j \le d\}$.

We first consider how to decide Ultimate Positivity in the case of a non-degenerate simple LRS $\boldsymbol{u}$. As explained in Section 2, we can assume without loss of generality that $\boldsymbol{u}$ has a positive real dominant root. Furthermore, by considering the LRS $\langle k^{n+1} u_n \rangle_{n=0}^{\infty}$ for a suitable integer $k \geq 1$, we may assume that the characteristic roots and coefficients in the closed-form solution (2) are all algebraic integers.

Suppose that $\boldsymbol{u}$ has dominant characteristic roots $\rho, \gamma_1, \overline{\gamma_1}, \ldots, \gamma_s, \overline{\gamma_s}$, where ρ is real and positive. Then we can write $\boldsymbol{u}$ in the form

$$u_n = b\rho^n + c_1\gamma_1^n + \overline{c_1\gamma_1}^n + \ldots + c_s\gamma_s^n + \overline{c_s\gamma_s}^n + r(n), \tag{5}$$

where $r(n) = o(\rho^{n(1-\varepsilon)})$ for some $\varepsilon > 0$. Now let $\lambda_i = \gamma_i/\rho$ for $i = 1, \ldots, s$. Then we can write

$$u_n = \rho^n f(\lambda_1^n, \ldots, \lambda_s^n) + r(n), \tag{6}$$

where $f : \mathbb{T}^s \to \mathbb{R}$ is defined by $f(z_1, \ldots, z_s) = b + c_1 z_1 + \overline{c_1 z_1} + \ldots + c_s z_s + \overline{c_s z_s}$.

Proposition 4.2. *The LRS $\langle u_n \rangle_{n=0}^{\infty}$ is ultimately positive if and only if $f(z) \geq 0$ for all $z \in T(\lambda)$.*

Proof. Consider the expression (5). Let K be the number field generated over $\mathbb{Q}$ by the characteristic roots of $\boldsymbol{u}$ and let S be the set of prime ideal divisors of the dominant characteristic roots $\rho, \gamma_1, \overline{\gamma_1}, \ldots, \gamma_s, \overline{\gamma_s}$ and the associated coefficients $b, c_1, \overline{c_1}, \ldots, c_s, \overline{c_s}$. (These coefficients lie in K by straightforward linear algebra.) Then the term

$$b\rho^n + c_1\gamma_1^n + \overline{c_1\gamma_1}^n + \ldots + c_s\gamma_s^n + \overline{c_s\gamma_s}^n \tag{7}$$

is a sum of S-units.

Applying Theorem 4.1 to the sum of S-units in (7), we have $X = C_1\rho^n$ for some constant $C_1 > 0$ and $Y = C_2\rho^n$ for some constant $C_2 > 0$ (since an embedding of K into $\mathbb{C}$ maps characteristic roots to characteristic roots). The theorem tells us that for each $\varepsilon > 0$ there is a constant $C > 0$ such that

$$|b\rho^n + c_1\gamma_1^n + \overline{c_1\gamma_1}^n + \ldots + c_s\gamma_s^n + \overline{c_s\gamma_s}^n| \geq C\rho^{n(1-\varepsilon)}$$

for all but finitely many values of n. (Since $\boldsymbol{u}$ is non-degenerate, it follows from the Skolem-Mahler-Lech Theorem that each non-empty sub-sum of the left-hand side vanishes for finitely many n.)

Now choose $\varepsilon > 0$ such that $r(n) = o(\rho^{n(1-\varepsilon)})$ in (5). Then for all sufficiently large n, $u_n \geq 0$ if and only if $b\rho^n + c_1\gamma_1^n + \overline{c_1\gamma_1}^n + \ldots + c_s\gamma_s^n + \overline{c_s\gamma_s}^n > 0$. Equivalently, looking at (6), for all sufficiently large n we have $u_n \geq 0$ if and only if $f(\lambda_1^n, \ldots, \lambda_s^n) \geq 0$. But the orbit $\{(\lambda_1^n, \ldots, \lambda_s^n) : n \in \mathbb{N}\}$ is a dense subset of $T(\lambda)$ by Proposition 3.5. Thus u_n is ultimately positive if and only if $f(z) \geq 0$ for all $z \in T(\lambda)$. $\square$

We can now state and prove our main result.

Theorem 4.3. *The Ultimate Positivity Problem for simple LRS is decidable in polynomial space in general, and in polynomial time for LRS of fixed order.*

Proof. A decision procedure is given in the table below. Correctness follows from the fact that u is ultimately positive if and only if each of the non-degenerate subsequences v considered in Step 2 is ultimately positive. But ultimate positivity of these subsequences is determined in Step 2.4 using Proposition 4.2. It remains to account for the complexity of each step.

As noted in Section 2, Step 1 requires time polynomial in $||u||$.

For LRS of fixed order, there is an absolute bound on M in Step 2, while for LRS of arbitrary order, M is exponentially bounded in $||u||$ by Proposition 2.2. We show that for each subsequence v, Steps 2.1–2.4 require polynomial time for fixed-order LRS and polynomial space in general.

Using iterated squaring, the coefficients b_i and c_j in the expression (8) for v are definable in terms of the characteristic roots of u and the corresponding coefficients in the closed-form expression for u by a polynomial-size first-order formula that uses only universal quantifiers. This accomplishes Step 2.1.

Combining Corollaries 3.2 and 3.3, Step 2.3 can be done in polynomial space for arbitrary LRS and polynomial time for LRS of fixed order.

Step 2.4 uses a decision procedure for universal sentences over the reals, having already noted that the coefficients b_i and c_j are first-order definable. By the results described in Section 2 this can be done in polynomial space for arbitrary LRS and polynomial time for LRS of fixed order.   □

Decision procedure for ultimate positivity of a simple LRS u

---

1. Compute the characteristic roots $\{\rho_1, \ldots, \rho_\ell, \gamma_1, \overline{\gamma_1}, \ldots, \gamma_m, \overline{\gamma_m}\}$ of u.
   Writing $\alpha \sim \beta$ if α/β is a root of unity, let $M = \mathrm{lcm}\{\mathrm{ord}(\alpha/\beta) : \alpha \sim \beta$ are characteristic roots$\}$. Moreover let $\{\rho_i : i \in I\} \cup \{\gamma_j, \overline{\gamma_j} : j \in J\}$ contain a unique representative from each equivalence class.
2. For $l = 0, \ldots, M - 1$, check ultimate positivity of the non-degenerate subsequence $v_n = u_{Mn+l}$ as follows:
   2.1. Compute the coefficients b_i and c_j in the closed-form solution

   $$v_n = \sum_{i \in I} b_i \rho_i^{Mn} + \sum_{j \in J} \left(c_j \gamma_j^{Mn} + \overline{c_j}\,\overline{\gamma_j}^{Mn} \right). \tag{8}$$

   2.2. If $v \not\equiv 0$ and there is no dominant real characteristic root in (8) then v is not ultimately positive.
   2.3. Let $\rho_1, \gamma_1, \overline{\gamma_1}, \ldots, \gamma_s, \overline{\gamma_s}$ be dominant among the characteristic roots appearing in (8). Define $\lambda_1 = \gamma_1/\rho_1, \ldots, \lambda_s = \gamma_s/\rho_1$ and compute a basis of $L(\lambda_1^M, \ldots, \lambda_s^M)$.
   2.4. Define $f : \mathbb{T}^s \to \mathbb{R}$ by $f(z_1, \ldots, z_s) = b_1 + c_1 z_1 + \overline{c_1 z_1} + \ldots + c_s z_s + \overline{c_s z_s}$. Then v is ultimately positive if and only if $f(z) \geq 0$ for all $z \in T(\lambda^M)$.

We note that a related proof strategy (passing from a finitely generated group to its closure and appealing to the theory of the reals) was used in [5] in the context of threshold problems for quantum automata.

5 Complexity Lower Bound

In this section we give reductions of the decision problem for universal sentences over the field of real numbers to the Positivity and Ultimate Positivity Problems respectively. The former problem is easily seen to be **coNP**-hard and, through the work of Canny [8], is known to be in **PSPACE**. Typically this **PSPACE** upper bound is stated for the complement problem: the decision problem for existential sentences over the field of reals.

It is known that the problem 4-FEAS of whether a degree-4 polynomial has a real root is polynomial-time equivalent to the decision problem for the existential theory of the reals [7]. Here we consider a related problem, 4-POS, which asks whether a degree-4 polynomial $f(x_1, \ldots, x_n)$ with rational coefficients satisfies $f(\boldsymbol{x}) \geq 0$ for all $\boldsymbol{x} \in [0, 1]^n$. Using the above-mentioned result on 4-FEAS in tandem with bounds on magnitude of definable numbers in the existential theory of the reals (see [23] for details) we can show:

Theorem 5.1. *There is a polynomial-time reduction of the decision problem for the universal theory of the reals to the problem 4-POS.*

We now reduce 4-POS to the Positivity and Ultimate Positivity Problems. The first step of the reduction is to compute a collection of s multiplicatively independent algebraic numbers of absolute value 1.

By a classical result of Lagrange, a prime number is congruent to 1 modulo 4 if and only if it can be written as the sum of two squares [13]. By Theorem 2.1, the class of such primes has asymptotic density $1/2$ in the set of all primes, and therefore, by the Prime Number Theorem, asymptotic density $1/(2 \log n)$ in the set of natural numbers. It follows that one can compute the first s such primes $p_1, \ldots, p_s$ and their decomposition as sums of squares in time polynomial in s. Writing $p_j = a_j^2 + b_j^2$, where $a_j, b_j \in \mathbb{Z}$, define $\lambda_j = \frac{a_j + ib_j}{a_j - ib_j}$ for $j = 1, \ldots, s$. Then each λ_j is an algebraic number of degree 2 and absolute value 1.

Proposition 5.2. $\lambda_1, \ldots, \lambda_s$ *are multiplicatively independent.*

Proof. Recall that the ring of Guassian integers $\mathbb{Z}(i)$ is a unique factorisation domain and that $a + ib \in \mathbb{Z}(i)$ is prime iff $a^2 + b^2$ is a rational prime [13]. Now $\lambda_1^{n_1} \ldots \lambda_s^{n_s} = 1$ if and only if

$$(a_1 + ib_1)^{n_1} \ldots (a_s + ib_s)^{n_s} = (a_1 - ib_1)^{n_1} \ldots (a_s - ib_s)^{n_s}$$

But each factor $a_j + ib_j$ and $a_j - ib_j$ is prime by construction. Thus by unique factorisation we must have $n_1 = 0, \ldots, n_s = 0$. $\qquad \square$

Theorem 5.3. *There are polynomial-time reductions from 4-POS to the Positivity and Ultimate Positivity Problems for LRS.*

Proof. Suppose we are given an instance of 4-POS, consisting of a polynomial $f(x_1, \ldots, x_s)$. Let $\lambda_1, \ldots, \lambda_s$ be multiplicatively independent algebraic numbers, constructed as in Proposition 5.2. For $j = 1, \ldots, s$, the sequence $\langle y_{j,n} : n \in \mathbb{N} \rangle$

defined by $y_{j,n} = \frac{1}{2}(\lambda_j^n + \overline{\lambda_j}^n)$ satisfies a second-order linear recurrence $y_{j,n+2} = (2a_j/p_j)y_{j,n+1} - y_{j,n}$ with rational coefficients.

Recall, moreover, that given two simple LRS of respective orders l and m, their sum is a simple LRS of order at most $l + m$, their product is a simple LRS of order at most lm, and representations of both can be computed in polynomial time in the size of the input LRS. Thus the sequence $\boldsymbol{u} = \langle u_n : n \in \mathbb{N} \rangle$ given by $u_n = f(y_{1,n}^2, \ldots, y_{s,n}^2)$ is a simple LRS over the rationals. Since f has degree at most 4, the order of $\boldsymbol{u}$ is at most 4^4 times the number of monomials in f and the recurrence satisfied by $\boldsymbol{u}$ can be computed in time polynomial in $\|f\|$. (Observe that if the degree of f were not fixed, then the above reasoning would yield an upper bound on the order of $\boldsymbol{u}$ that is exponential in the degree of f.)

From Propositions 3.5 and 5.2 it follows that the orbit $\{(\lambda_1^n, \ldots, \lambda_s^n) : n \in \mathbb{N}\}$ is dense in the torus $\mathbb{T}^s$. Thus the set $\{(y_{1,n}^2, \ldots, y_{s,n}^2) : n \in \mathbb{N}\}$ is dense in $[0,1]^s$ and f assumes a strictly negative value on $[0,1]^s$ if and only if $u_n < 0$ for some (equivalently infinitely many) n. This completes the reduction. $\qquad\square$

6   Conclusion

We have shown that the Ultimate Positivity Problem for simple LRS is decidable in polynomial space and as hard as the decision problem for universal sentences over the field of real numbers. A more careful accounting of the complexity of our decision procedure places it in **coNP** with an oracle for the universal theory of the reals. Thus a **PSPACE**-hardness result for Ultimate Positivity would have non-trivial consequences for the complexity of decision problems for first-order logic over the reals. On the other hand, the obstacle to improving the polynomial-space upper bound is the complexity of computing a basis of the group of multiplicative relations among the characteristic roots of the recurrence.

References

1. Amoroso, F., Viada, E.: Small points on subvarieties of a torus. Duke Mathematical Journal 150(3) (2009)
2. Bell, J.P., Gerhold, S.: On the positivity set of a linear recurrence. Israel Jour. Math. 57 (2007)
3. Bell, P., Delvenne, J.-C., Jungers, R., Blondel, V.: The continuous Skolem-Pisot problem. Theor. Comput. Sci. 411(40-42), 3625–3634 (2010)
4. Berstel, J., Mignotte, M.: Deux propriétés décidables des suites récurrentes linéaires. Bull. Soc. Math. France 104 (1976)
5. Blondel, V., Jeandel, E., Koiran, P., Portier, N.: Decidable and undecidable problems about quantum automata. SIAM J. Comput. 34(6), 1464–1473 (2005)
6. Blondel, V., Tsitsiklis, J.: A survey of computational complexity results in systems and control. Automatica 36(9), 1249–1274 (2000)
7. Blum, L., Cucker, F., Shub, M., Smale, S.: Complexity and real computation. Springer (1997)
8. Canny, J.: Some algebraic and geometric computations in PSPACE. In: Proceedings of STOC 1988, pp. 460–467. ACM (1988)

9. Cassels, J.: An introduction to Diophantine approximation. Camb. Univ. Pr. (1965)
10. Everest, G., van der Poorten, A., Shparlinski, I., Ward, T.: Recurrence Sequences. American Mathematical Society (2003)
11. Evertse, J.-H.: On sums of S-units and linear recurrences. Compositio Mathematica 53(2), 225–244 (1984)
12. Evertse, J.-H., Schlickewei, H.P., Schmidt, W.M.: Linear equations in variables which lie in a multiplicative group. Ann. Math. 155(3) (2002)
13. Fröhlich, A., Taylor, M.: Algebraic Number Theory. Camb. Univ. Press (1993)
14. Halava, V., Harju, T., Hirvensalo, M.: Positivity of second order linear recurrent sequences. Discrete Applied Mathematics 154(3) (2006)
15. Halava, V., Harju, T., Hirvensalo, M., Karhumäki, J.: Skolem's problem – on the border between decidability and undecidability. Technical Report 683, Turku Centre for Computer Science (2005)
16. Kannan, R., Lipton, R.J.: Polynomial-time algorithm for the orbit problem. JACM 33(4) (1986)
17. Laohakosol, V., Tangsupphathawat, P.: Positivity of third order linear recurrence sequences. Discrete Applied Mathematics 157(15) (2009)
18. Lech, C.: A note on recurring series. Ark. Mat. 2 (1953)
19. Mahler, K.: Eine arithmetische Eigenschaft der Taylor Koeffizienten rationaler Funktionen. Proc. Akad. Wet. Amsterdam 38 (1935)
20. Masser, D.W.: Linear relations on algebraic groups. In: New Advances in Transcendence Theory. Camb. Univ. Press (1988)
21. Mignotte, M., Shorey, T., Tijdeman, R.: The distance between terms of an algebraic recurrence sequence. J. für die reine und angewandte Math. 349 (1984)
22. Ouaknine, J., Worrell, J.: Decision problems for linear recurrence sequences. In: Finkel, A., Leroux, J., Potapov, I. (eds.) RP 2012. LNCS, vol. 7550, pp. 21–28. Springer, Heidelberg (2012)
23. Ouaknine, J., Worrell, J.: Ultimate Positivity is decidable for simple linear recurrence sequences. CoRR, abs/1309.1914 (2013)
24. Ouaknine, J., Worrell, J.: On the Positivity Problem for simple linear recurrence sequences. In: Esparza, J., Fraigniaud, P., Husfeldt, T. (eds.) ICALP 2014, Part II. LNCS, vol. 8573, Springer, Heidelberg (2014); CoRR, abs/1309.1550
25. Ouaknine, J., Worrell, J.: Positivity problems for low-order linear recurrence sequences. In: Proceedings of SODA 2014. ACM-SIAM (2014)
26. Renegar, J.: On the computational complexity and geometry of the first-order theory of the reals. J. Symb. Comp. (1992)
27. Rozenberg, G., Salomaa, A.: Cornerstones of Undecidability. Prentice Hall (1994)
28. Skolem, T.: Ein Verfahren zur Behandlung gewisser exponentialer Gleichungen. In: Comptes rendus du congrès des mathématiciens scandinaves (1934)
29. Soittola, M.: On D0L synthesis problem. In: Lindenmayer, A., Rozenberg, G. (eds.) Automata, Languages, Development. North-Holland (1976)
30. Tao, T.: Structure and randomness: pages from year one of a mathematical blog. American Mathematical Society (2008)
31. van der Poorten, A., Schlickewei, H.: The growth conditions for recurrence sequences. Macquarie Math. Reports (82-0041) (1982)
32. Vereshchagin, N.K.: The problem of appearance of a zero in a linear recurrence sequence. Mat. Zametki 38(2) (1985) (in Russian)

Going Higher in the First-Order Quantifier Alternation Hierarchy on Words[*]

Thomas Place and Marc Zeitoun

LaBRI, Université de Bordeaux, France

Abstract. We investigate the quantifier alternation hierarchy in first-order logic on finite words. Levels in this hierarchy are defined by counting the number of quantifier alternations in formulas. We prove that one can decide membership of a regular language to the levels $\mathcal{B}\Sigma_2$ (boolean combination of formulas having only 1 alternation) and Σ_3 (formulas having only 2 alternations beginning with an existential block). Our proof works by considering a deeper problem, called separation, which, once solved for lower levels, allows us to solve membership for higher levels.

The connection between logic and automata theory is well known and has a fruitful history in computer science. It was first observed when Büchi, Elgot and Trakhtenbrot proved independently that the regular languages are exactly those that can be defined using a monadic second-order logic (MSO) formula. Since then, many efforts have been made to investigate and understand the expressive power of relevant fragments of MSO. In this field, the yardstick result is often to prove *decidable characterizations*, *i.e.*, to design an algorithm which, given as input a regular language, decides whether it can be defined in the fragment under investigation. More than the algorithm itself, the main motivation is the insight given by its proof. Indeed, in order to prove a decidable characterization, one has to consider and understand *all* properties that can be expressed in the fragment.

The most prominent fragment of MSO is first-order logic (FO) equipped with a predicate "<" for the linear-order. The expressive power of FO is now well-understood over words and a decidable characterization has been obtained. The result, Schützenberger's Theorem [19,9], states that a regular language is definable in FO if and only if its syntactic monoid is aperiodic. The syntactic monoid is a finite algebraic structure that can effectively be computed from any representation of the language. Moreover, aperiodicity can be rephrased as an equation that needs to be satisfied by all elements of the monoid. Therefore, Schützenberger's Theorem can indeed be used to decide definability in FO.

In this paper, we investigate an important hierarchy inside FO, obtained by classifying formulas according to the number of quantifier alternations in their prenex normal form. More precisely, an FO formula is Σ_i if its prenex normal form has at most $(i - 1)$ quantifier alternations and starts with a block of existential quantifiers. The hierarchy also involves the classes $\mathcal{B}\Sigma_i$ of boolean combinations of Σ_i formulas, and the classes Δ_i of languages that can be defined by both a Σ_i and

---

[*] Supported by ANR 2010 BLAN 0202 01 FREC.

J. Esparza et al. (Eds.): ICALP 2014, Part II, LNCS 8573, pp. 342–353, 2014.
© Springer-Verlag Berlin Heidelberg 2014

the negation of a Σ_i formula. The quantifier alternation hierarchy was proved to be strict [6,29]: $\Delta_i \subsetneq \Sigma_i \subsetneq \mathcal{B}\Sigma_i \subsetneq \Delta_{i+1}$. In the literature, many efforts have been made to find decidable characterizations of levels of this well-known hierarchy.

Despite these efforts, only the lower levels are known to be decidable. The class $\mathcal{B}\Sigma_1$ consists exactly of all piecewise testable languages, *i.e.*, such that membership of a word only depends on its subwords up to a fixed size. These languages were characterized by Simon [20] as those whose syntactic monoid is $\mathcal{J}$-trivial. A decidable characterization of Σ_2 (and hence of Δ_2 as well) was proven in [3]. For Δ_2, the literature is very rich [25]. For example, these are exactly the languages definable by the two variable restriction of FO [27]. These are also those whose syntactic monoid is in the class DA [13]. For higher levels in the hierarchy, getting decidable characterizations remained an important open problem. In particular, the case of $\mathcal{B}\Sigma_2$ has a very rich history and a series of combinatorial, logical, and algebraic conjectures have been proposed over the years. We refer to [11,2,10,12] for an exhaustive bibliography. So far, the only known effective result was partial, working only when the alphabet is of size 2 [24]. One of the main motivations for investigating this class in formal language theory is its ties with two other famous hierarchies defined in terms of regular expressions. In the first one, the *Straubing-Thérien hierarchy* [22,26], level i corresponds exactly to the class $\mathcal{B}\Sigma_i$ [28]. In the second one, the *dot-depth hierarchy* [7], level i corresponds to adding a predicate for the successor relation in $\mathcal{B}\Sigma_i$ [28]. Proving decidability for $\mathcal{B}\Sigma_2$ immediately proves decidability of level 2 in the Straubing-Thérien hierarchy, but also in the dot-depth hierarchy using a reduction by Straubing [23].

In this paper, we prove decidability for $\mathcal{B}\Sigma_2$, Δ_3 and Σ_3. These new results are based on a deeper decision problem than decidable characterizations: the separation problem. Fix a class Sep of languages. The Sep-separation problem amounts to decide whether, given two input regular languages, there exists a third language in Sep containing the first language while being disjoint from the second one. This problem generalizes decidable characterizations. Indeed, since regular languages are closed under complement, testing membership in Sep can be achieved by testing whether the input is Sep-separable from its complement. Historically, the separation problem was first investigated as a special case of a deep problem in semigroup theory, see [1]. This line of research gave solutions to the problem for several classes. However, the motivations are disconnected from our own, and the proofs rely on deep, purely algebraic arguments. Recently, a research effort has been made to investigate this problem from a different perspective, with the aim of finding new and self-contained proofs relying on elementary ideas and notions from language theory only [8,15,18,16]. This paper is a continuation of this effort: we solve the separation problem for Σ_2, and use our solution as a basis to obtain decidable characterizations for $\mathcal{B}\Sigma_2$, Δ_3 and Σ_3.

Our solution works as follows: given two regular languages, one can easily construct a monoid morphism $\alpha : A^* \to M$ that recognizes both of them. We then design an algorithm that computes, inside the monoid M, enough Σ_2-related information to answer the Σ_2-separation question for *any* pair of languages that are recognized by α. It turns out that it is also possible (though much more difficult) to use this information to obtain decidability of $\mathcal{B}\Sigma_2$, Δ_3 and Σ_3. This information amounts to the notion of Σ_2-chain, our main tool in the paper. A Σ_2-chain is an *ordered sequence* $s_1, \ldots, s_n \in M$ that witnesses a property of α wrt. Σ_2. Let us give some intuition in the case $n = 2$ – which is enough to make the link with Σ_2-separation. A sequence s_1, s_2 is a Σ_2-chain if any Σ_2 language containing all words in $\alpha^{-1}(s_1)$ intersects $\alpha^{-1}(s_2)$. In terms of separation, this means that $\alpha^{-1}(s_1)$ is *not* separable from $\alpha^{-1}(s_2)$ by a Σ_2 definable language.

This paper contains three main separate and difficult new results: (1) an algorithm to compute Σ_2-chains – hence Σ_2-separability is decidable (2) decidability of Σ_3 (decidability of Δ_3 is an immediate consequence), and (3) decidability of $\mathcal{B}\Sigma_2$. Computing Σ_2-chains is achieved using a fixpoint algorithm that starts with trivial Σ_2-chains such as $s, s, \ldots, s$, and iteratively computes more Σ_2-chains until a fixpoint is reached. Note that its completeness proof relies on the Factorization Forest Theorem of Simon [21]. This is not surprising, as the link between this theorem and the quantifier alternation hierarchy was already observed in [13,4].

For Σ_3, we prove a decidable characterization via an equation on the syntactic monoid of the language. This equation is parametrized by the set of Σ_2-chains of length 2. In other words, we use Σ_2-chains to abstract an infinite set of equations into a single one. The proof relies again on the Factorization Forest Theorem of Simon [21] and is actually generic to all levels in the hierarchy. This means that for any i, we define a notion of Σ_i-chain and characterize Σ_{i+1} using an equation parametrized by Σ_i-chains of length 2. However, decidability of Σ_{i+1} depends on our ability to compute the Σ_i-chains of length 2, which we can only do for $i = 2$.

Our decidable characterization of $\mathcal{B}\Sigma_2$ is the most difficult result of the paper. As for Σ_3, it is presented by two equations parametrized by Σ_2-chains (of length 2 and 3). However, the characterization is this time specific to the case $i = 2$. This is because most of our proof relies on a deep analysis of our algorithm that computes Σ_2-chains, which only works for $i = 2$. The equations share surprising similarities with the ones used in [5] to characterize a totally different formalism: boolean combination of open sets of infinite trees. In [5] also, the authors present their characterization as a set of equations parametrized by a notion of "chain" for open sets of infinite trees (although their "chains" are not explicitly identified as a separation relation). Since the formalisms are of different nature, the way these chains and our Σ_2-chains are constructed are completely independent, which means that the proofs are also mostly independent. However, once the construction analysis of chains has been done, several combinatorial arguments used to make the link with equations are analogous. In particular, we reuse and adapt definitions from [5] to present these combinatorial arguments in our proof. One could say that the proofs are both (very different) setups to apply similar combinatorial arguments in the end.

Organization. We present definitions on languages and logic in Sections 1 and 2 respectively. Section 3 is devoted to the presentation of our main tool: Σ_i-chains. In Section 4, we give our algorithm computing Σ_2-chains. The two remaining sections present our decidable characterizations, for Σ_3 and Δ_3 in Section 5 and for $\mathcal{B}\Sigma_2$ in Section 6. Due to lack of space, proofs can be found in [17].

1   Words and Algebra

Words and Languages. We fix a finite alphabet A and we denote by A^* the set of all words over A. If u, v are words, we denote by $u \cdot v$ or uv the word obtained by concatenation of u and v. If $u \in A^*$ we denote by $\mathsf{alph}(u)$ its alphabet, *i.e.*, the smallest subset B of A such that $u \in B^*$. A *language* is a subset of A^*. In this paper we consider regular languages: these are languages definable by *nondeterministic finite automata*, or equivalently by *finite monoids*. In the paper, we only work with the monoid representation of regular languages.

Monoids. A *semigroup* is a set S equipped with an associative multiplication denoted by '$\cdot$'. A *monoid* M is a semigroup in which there exists a neutral element denoted 1_M. In the paper, we investigate classes of languages, such as Σ_i, that are not closed under complement. For such classes, it is known that one needs to use *ordered monoids*. An ordered monoid is a monoid endowed with a partial order '$\leqslant$' which is compatible with multiplication: $s \leqslant t$ and $s' \leqslant t'$ imply $ss' \leqslant tt'$. Given any finite semigroup S, it is well known that there is a number $\omega(S)$ (denoted by ω when S is understood from the context) such that for each element s of S, s^ω is an idempotent: $s^\omega = s^\omega \cdot s^\omega$.

Let L be a language and M be a monoid. We say that L *is recognized by* M if there exists a monoid morphism $\alpha : A^* \to M$ and an *accepting set* $F \subseteq M$ such that $L = \alpha^{-1}(F)$. It is well known that a language is regular if and only if it can be recognized by a *finite monoid*.

Syntactic Ordered Monoid of a Language. The *syntactic preorder* $\leqslant_L$ of a language L is defined as follows on pairs of words in A^*: $w \leqslant_L w'$ if for all $u, v \in A^*$, $uwv \in L \Rightarrow uw'v \in L$. Similarly, we define $\equiv_L$, the *syntactic equivalence* of L as follows: $w \equiv_L w'$ if $w \leqslant_L w'$ and $w' \leqslant_L w$. One can verify that $\leqslant_L$ and $\equiv_L$ are compatible with multiplication. Therefore, the quotient M_L of A^* by $\equiv_L$ is an ordered monoid for the partial order induced by the preorder $\leqslant_L$. It is well known that M_L can be effectively computed from L. Moreover, M_L recognizes L. We call M_L the *syntactic ordered monoid of* L and the associated morphism the *syntactic morphism*.

Separation. Given three languages L, L_0, L_1, we say that L *separates* L_0 from L_1 if $L_0 \subseteq L$ and $L_1 \cap L = \emptyset$. Set X as a class of languages, we say that L_0 is X-*separable* from L_1 if some language in X separates L_0 from L_1. Observe that when X is not closed under complement, the definition is not symmetrical: L_0 could be X-separable from L_1 while L_1 is not X-separable from L_0.

When working on separation, we consider as input two regular languages L_0, L_1. It will be convenient to have a *single* monoid recognizing both of them, rather than having to deal with two objects. Let M_0, M_1 be monoids recognizing L_0, L_1 together with the morphisms α_0, α_1, respectively. Then, $M_0 \times M_1$ equipped with the componentwise multiplication $(s_0, s_1) \cdot (t_0, t_1) = (s_0 t_0, s_1 t_1)$ is a monoid that recognizes both L_0 and L_1 with the morphism $\alpha : w \mapsto (\alpha_0(w), \alpha_1(w))$. From now on, we work with such a single monoid recognizing both languages.

Chains and Sets of Chains. Set M as a finite monoid. A *chain* for M is a word over the alphabet M, *i.e.*, an element of M^*. A remark about notation is in order here. A word is usually denoted as the concatenation of its letters. Since M is a monoid, this would be ambiguous here since st could either mean a word with 2 letters s and t, or the product of s and t in M. To avoid confusion, we will write $(s_1, \ldots, s_n)$ a chain of length n on the alphabet M.

In the paper, we will consider both sets of chains (denoted by $\mathcal{T}, \mathcal{S}, \ldots$) and sets of sets of chains (denoted by $\mathfrak{T}, \mathfrak{S}, \ldots$). In particular, if $\mathfrak{T}$ is a set of sets of chains, we define $\downarrow\mathfrak{T}$, the *downset* of $\mathfrak{T}$, as the set:

$$\downarrow\mathfrak{T} = \{\mathcal{T} \mid \exists \mathcal{S} \in \mathfrak{T}, \ \mathcal{T} \subseteq \mathcal{S}\}.$$

We will often restrict ourselves to considering only chains of a given fixed length. For $n \in \mathbb{N}$, observe that M^n, the set of chains of length n, is a monoid when equipped with the componentwise multiplication. Similarly the set 2^{M^n} of sets of chains of length n is a monoid for the operation: $\mathcal{S} \cdot \mathcal{T} = \{\bar{s}\bar{t} \in M^n \mid \bar{s} \in \mathcal{S} \ \bar{t} \in \mathcal{T}\}$.

2   First-Order Logic and Quantifier Alternation Hierarchy

We view words as logical structures made of a sequence of positions labeled over A. We denote by $<$ the linear order over the positions. We work with first-order logic FO using unary predicates P_a for all $a \in A$ that select positions labeled with an a, as well as a binary predicate for the linear order $<$. The *quantifier rank* of an FO formula is the length of its longest sequence of nested quantifiers.

One can classify first-order formulas by counting the number of alternations between $\exists$ and $\forall$ quantifiers in the prenex normal form of the formula. Set $i \in \mathbb{N}$, a formula is said to be Σ_i (resp. Π_i) if its prenex normal form has $i - 1$ quantifier alternations (*i.e.*, i blocks of quantifiers) and starts with an $\exists$ (resp. $\forall$) quantification. For example, a formula whose prenex normal form is

$$\forall x_1 \forall x_2 \exists x_3 \forall x_4 \ \varphi(x_1, x_2, x_3, x_4) \quad \text{(with } \varphi \text{ quantifier-free)}$$

is Π_3. Observe that a Π_i formula is by definition the negation of a Σ_i formula. Finally, a $\mathcal{B}\Sigma_i$ formula is a boolean combination of Σ_i formulas. For $X = \text{FO}, \Sigma_i, \Pi_i$ or $\mathcal{B}\Sigma_i$, we say that a language L is X-definable if it can be defined by an X-formula. Finally, we say that a language is Δ_i-definable if it can be defined by *both* a Σ_i and a Π_i formula. It is known that this gives a strict infinite hierarchy of classes of languages as represented in Figure 1.

Preorder for Σ_i. Let $w, w' \in A^*$ and $k, i \in \mathbb{N}$. We write $w \lesssim_i^k w'$ if any Σ_i formula of quantifier rank k satisfied by w is also satisfied by w'. Observe that since a Π_i formula is the negation of a Σ_i formula, we have $w \lesssim_i^k w'$ iff any Π_i formula of quantifier rank k satisfied by w' is also satisfied by w. One can verify that $\lesssim_i^k$ is a preorder for all k, i. Moreover, by definition, a language L can be defined by a Σ_i formula of rank k iff L is saturated by $\lesssim_i^k$, *i.e.*, for all $w \in L$ and all w' such that $w \lesssim_i^k w'$, we have $w' \in L$.

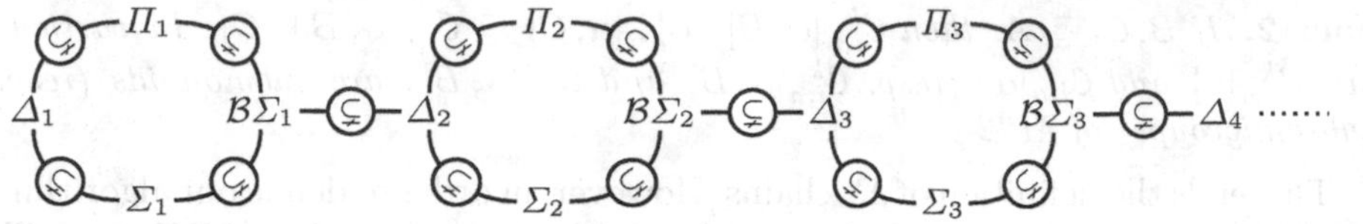

$$\Pi_1 \qquad \Pi_2 \qquad \Pi_3$$

$$\Delta_1 \qquad B\Sigma_1 - \Delta_2 \qquad B\Sigma_2 - \Delta_3 \qquad B\Sigma_3 - \Delta_4 \ \cdots$$

$$\Sigma_1 \qquad \Sigma_2 \qquad \Sigma_3$$

Fig. 1. Quantifier Alternation Hierarchy

3 Σ_i-Chains

We now introduce the main tool of this paper: Σ_i-*chains*. Fix a level i in the quantifier alternation hierarchy and $\alpha : A^* \to M$ a monoid morphism. A Σ_i-*chain* for α is a chain $(s_1, \ldots, s_n) \in M^*$ such that for arbitrarily large $k \in \mathbb{N}$, there exist words $w_1 \lesssim_i^k \cdots \lesssim_i^k w_n$ mapped respectively to $s_1, \ldots, s_n$ by α. Intuitively, this contains information about the limits of the expressive power of the logic Σ_i with respect to α. For example, if (s_1, s_2) is a Σ_i-chain, then any Σ_i language that contains all words of image s_1 must also contain at least one word of image s_2.

In this section, we first give all definitions related to Σ_i-chains. We then present an immediate application of this notion: solving the separation problem for Σ_i can be reduced to computing the Σ_i-chains of length 2.

3.1 Definitions

Σ_i-**Chains.** Fix i a level in the hierarchy, $k \in \mathbb{N}$ and $B \subseteq A$. We define $\mathcal{C}_i^k[\alpha]$ (resp. $\mathcal{C}_i^k[\alpha, B]$) as the *set of $\Sigma_i[k]$-chains for α* (resp. for (α, B)) and $\mathcal{C}_i[\alpha]$ (resp. $\mathcal{C}_i[\alpha, B]$) as the *set of Σ_i-chains for α* (resp. for (α, B)). For $i = 0$, we set $\mathcal{C}_i[\alpha] = \mathcal{C}_i^k[\alpha] = M^*$. Otherwise, let $\bar{s} = (s_1, \ldots, s_n) \in M^*$. We let

- $\bar{s} \in \mathcal{C}_i^k[\alpha]$ if there exist $w_1, \ldots, w_n \in A^*$ verifying $w_1 \lesssim_i^k w_2 \lesssim_i^k \cdots \lesssim_i^k w_n$ and for all j, we have $\alpha(w_j) = s_j$. Moreover, $\bar{s} \in \mathcal{C}_i^k[\alpha, B]$ if the words w_j can be chosen so that they satisfy additionally $\mathsf{alph}(w_j) = B$ for all j.
- $\bar{s} \in \mathcal{C}_i[\alpha]$ if for all k, we have $\bar{s} \in \mathcal{C}_i^k[\alpha]$. That is, $\mathcal{C}_i[\alpha] = \bigcap_k \mathcal{C}_i^k[\alpha]$. In the same way, $\mathcal{C}_i[\alpha, B] = \bigcap_k \mathcal{C}_i^k[\alpha, B]$.

One can check that if $i \geqslant 2$, then $\mathcal{C}_i^k[\alpha] = \bigcup_{B \subseteq A} \mathcal{C}_i^k[\alpha, B]$, since the fragment Σ_i can detect the alphabet (i.e., for $i \geqslant 2$, $w \lesssim_i^k w'$ implies $\mathsf{alph}(w) = \mathsf{alph}(w')$). Similarly for $i \geqslant 2$, the set of Σ_i-chains for α is $\mathcal{C}_i[\alpha] = \bigcup_{B \subseteq A} \mathcal{C}_i[\alpha, B]$. Observe that all these sets are closed under subwords. Therefore, by Higman's lemma, we get the following fact.

Fact 1. *For all $i, k \in \mathbb{N}$ and $B \subseteq A$, $\mathcal{C}_i[\alpha, B]$ and $\mathcal{C}_i^k[\alpha, B]$ are regular languages.*

Fact 1 is interesting but essentially useless in our argument, as Higman's lemma provides no way for actually computing a recognizing device for $\mathcal{C}_i[\alpha, B]$.

For any fixed $n \in \mathbb{N}$, we let $\mathcal{C}_{i,n}^k[\alpha, B]$ be the set of $\Sigma_i[k]$-chains of length n for α, B, i.e., $\mathcal{C}_{i,n}^k[\alpha, B] = \mathcal{C}_i^k[\alpha, B] \cap M^n$. We define $\mathcal{C}_{i,n}[\alpha, B], \mathcal{C}_{i,n}^k[\alpha]$ and $\mathcal{C}_{i,n}[\alpha]$ similarly. The following fact is immediate.

Fact 2. *If* $B, C \subseteq A$, *then* $\mathcal{C}_{i,n}^k[\alpha, B] \cdot \mathcal{C}_{i,n}^k[\alpha, C] \subseteq \mathcal{C}_{i,n}^k[\alpha, B \cup C]$. *In particular*, $\mathcal{C}_{i,n}^k[\alpha]$ *and* $\mathcal{C}_{i,n}[\alpha]$ *(resp.* $\mathcal{C}_{i,n}^k[\alpha, B]$ *and* $\mathcal{C}_{i,n}[\alpha, B]$*) are submonoids (resp. subsemigroups) of* M^n.

This ends the definition of Σ_i-chains. However, in order to define our algorithm for computing Σ_2-chains and state our decidable characterization of $\mathcal{B}\Sigma_2$, we will need a slightly refined notion: *compatible sets of chains* .

Compatible Sets of Σ_i-Chains. In some cases, it will be useful to know that several Σ_i-chains with the same first element can be 'synchronized'. For example take two Σ_i-chains (s, t_1) and (s, t_2) of length 2. By definition, for all k there exist words w_1, w_1', w_2, w_2' whose images under α are s, t_1, s, t_2 respectively, and such that $w_1 \lesssim_i^k w_1'$ and $w_2 \lesssim_i^k w_2'$. In some cases (but not all), it will be possible to choose $w_1 = w_2$ for all k. The goal of the notion of compatible sets of chains is to record the cases in which this is true.

Fix i a level in the hierarchy, $k \in \mathbb{N}$ and $B \subseteq A$. We define two sets of sets of chains: $\mathfrak{C}_i^k[\alpha, B]$, the *set of compatible sets of $\Sigma_i[k]$-chains for (α, B)*, and $\mathfrak{C}_i[\alpha, B]$, the *set of compatible sets of Σ_i-chains for (α, B)*. Let $\mathcal{T}$ be a set of chains, all having the same length n and the same first element s_1.

- $\mathcal{T} \in \mathfrak{C}_i^k[\alpha, B]$ if there exists $w \in A^*$ such that $\mathsf{alph}(w) = B$, $\alpha(w) = s_1$, and for all chains $(s_1, \ldots, s_n) \in \mathcal{T}$, there exist $w_2, \ldots, w_n \in A^*$ verifying $w \lesssim_i^k w_2 \lesssim_i^k \cdots \lesssim_i^k w_n$, and for all $j = 2, \ldots, n$, $\alpha(w_j) = s_j$, and $\mathsf{alph}(w_j) = B$.
- $\mathcal{T} \in \mathfrak{C}_i[\alpha, B]$ if $\mathcal{T} \in \mathfrak{C}_i^k[\alpha, B]$ for all k.

As before we set $\mathfrak{C}_i^k[\alpha]$ and $\mathfrak{C}_i[\alpha]$ as the union of these sets for all $B \subseteq A$. Moreover, we denote by $\mathfrak{C}_{i,n}^k[\alpha, B], \mathfrak{C}_{i,n}[\alpha, B], \mathfrak{C}_{i,n}^k[\alpha]$ and $\mathfrak{C}_{i,n}[\alpha]$ the restriction of these sets to sets of chains of length n (*i.e.*, subsets of 2^{M^n}).

Fact 3. *If* $B, C \subseteq A$, *then* $\mathfrak{C}_{i,n}^k[\alpha, B] \cdot \mathfrak{C}_{i,n}^k[\alpha, C] \subseteq \mathfrak{C}_{i,n}^k[\alpha, B \cup C]$. *In particular*, $\mathfrak{C}_{i,n}^k[\alpha]$ *and* $\mathfrak{C}_{i,n}[\alpha]$ *(resp.* $\mathfrak{C}_{i,n}^k[\alpha, B]$ *and* $\mathfrak{C}_{i,n}[\alpha, B]$*) are submonoids (resp. subsemigroups) of* 2^{M^n}.

3.2 Σ_i-Chains and Separation

We now state a reduction from the separation problem by Σ_i and by Π_i-definable languages to the computation of Σ_i-chains of length 2.

Theorem 4. *Let* L_1, L_2 *be regular languages and* $\alpha : A^* \to M$ *be a morphism into a finite monoid recognizing both languages with accepting sets* $F_1, F_2 \subseteq M$. *Set* $i \in \mathbb{N}$. *Then the following properties hold:*

1. L_1 *is* Σ_i-*separable from* L_2 *iff for all* $s_1, s_2 \in F_1, F_2$, $(s_1, s_2) \notin \mathcal{C}_i[\alpha]$.
2. L_1 *is* Π_i-*separable from* L_2 *iff for all* $s_1, s_2 \in F_1, F_2$, $(s_2, s_1) \notin \mathcal{C}_i[\alpha]$.

The proof of Theorem 4, which is parametrized by Σ_i-chains, is standard and identical to the corresponding theorems in previous separation papers, see *e.g.*, [18]. In Section 4, we present an algorithm computing Σ_i-chains of length 2 at level $i = 2$ of the alternation hierarchy (in fact, our algorithm needs to compute the more general notion of sets of compatible Σ_2-chains). This makes Theorem 4 effective for Σ_2 and Π_2.

4 Computing Σ_2-Chains

In this section, we give an algorithm for computing all Σ_2-chains and sets of compatible Σ_2-chains of a given fixed length. We already know by Theorem 4 that achieving this for length 2 suffices to solve the separation problem for Σ_2 and Π_2. Moreover, we will see in Sections 5 and 6 that this algorithm can be used to obtain decidable characterizations for Σ_3, Π_3, Δ_3 and $\mathcal{B}\Sigma_2$. Note that in this section, we only provide the algorithm and intuition on its correctness.

For the remainder of this section, we fix a morphism $\alpha : A^* \to M$ into a finite monoid M. For any fixed $n \in \mathbb{N}$ and $B \subseteq A$, we need to compute the following:

1. the sets $\mathcal{C}_{2,n}[\alpha, B]$ of Σ_2-chains of length n for α.

2. the sets $\mathfrak{C}_{2,n}[\alpha, B]$ of compatible subsets of $\mathcal{C}_{2,n}[\alpha, B]$.

Our algorithm directly computes the second item, *i.e.*, $\mathfrak{C}_{2,n}[\alpha, B]$. More precisely, we compute the map $B \mapsto \mathfrak{C}_{2,n}[\alpha, B]$. Observe that this is enough to obtain the first item since by definition, $\bar{s} \in \mathcal{C}_{2,n}[\alpha, B]$ iff $\{\bar{s}\} \in \mathfrak{C}_{2,n}[\alpha, B]$. Note that going through compatible subsets is necessary for the technique to work, even if we are only interested in computing the map $B \mapsto \mathcal{C}_{2,n}[\alpha, B]$.

Outline. We begin by explaining what our algorithm does. For this outline, assume $n = 2$. Observe that for all $w \in A^*$ such that $\mathsf{alph}(w) = B$, we have $\{(\alpha(w), \alpha(w))\} \in \mathfrak{C}_{2,n}[\alpha, B]$. The algorithm starts from these trivially compatible sets, and then saturates them with two operations that preserve membership in $\mathfrak{C}_{2,n}[\alpha, B]$. Let us describe these two operations. The first one is multiplication: if $S \in \mathfrak{C}_{2,n}[\alpha, B]$ and $T \in \mathfrak{C}_{2,n}[\alpha, C]$ then $S \cdot T \in \mathfrak{C}_{2,n}[\alpha, B \cup C]$ by Fact 3. The main idea behind the second operation is to exploit the following property of Σ_2:

$$\forall k \ \exists \ell \quad w \lesssim_2^k u, w \lesssim_2^k u' \text{ and } \mathsf{alph}(w') = \mathsf{alph}(w) \quad \Longrightarrow \quad w^{2\ell} \lesssim_2^k u^\ell w' u'^\ell.$$

This is why compatible sets are needed: in order to use this property, we need to have a single word w such that $w \lesssim_2^k u$ and $w \lesssim_2^k u'$, which is information that is not provided by Σ_2-chains. This yields an operation that states that whenever S belongs to $\mathfrak{C}_{2,n}[\alpha, B]$, then so does $S^\omega \cdot T \cdot S^\omega$, where T is the set of chains $(1_M, \alpha(w'))$ with $\mathsf{alph}(w') = B$. Let us now formalize this procedure and generalize it to arbitrary length.

Algorithm. As we explained, our algorithm works by fixpoint, starting from trivial compatible sets. For all $n \in \mathbb{N}$ and $B \subseteq A$, we let $\mathfrak{I}_n[B]$ be the set $\mathfrak{I}_n[B] = \{\{(\alpha(w), \ldots, \alpha(w))\} \mid \mathsf{alph}(w) = B\} \subseteq 2^{M^n}$. Our algorithm will start from the function $f_0 : 2^A \to 2^{2^{M^n}}$ that maps any $C \subseteq A$ to $\mathfrak{I}_n[C]$.

Our algorithm is defined for any fixed length $n \geqslant 1$. We use a procedure Sat_n taking as input a mapping $f : 2^A \to 2^{2^{M^n}}$ and producing another such mapping. The algorithm starts from f_0 and iterates Sat_n until a fixpoint is reached.

When $n \geqslant 2$, the procedure Sat_n is parametrized by $\mathcal{C}_{2,n-1}[\alpha, B]$, the sets of Σ_2-chains of length $n - 1$, for $B \subseteq A$. This means that in order to use Sat_n, one needs to have previously computed the Σ_2-chains of length $n - 1$ with Sat_{n-1}.

We now define the procedure Sat_n. If $\mathcal{S}$ is a set of chains of length $n - 1$ and $s \in M$, we write $(s, \mathcal{S})$ for the set $\{(s, s_1, \ldots, s_{n-1}) \mid (s_1, \ldots, s_{n-1}) \in \mathcal{S}\}$, which consists of chains of length n. Let $f : 2^A \to 2^{2^{M^n}}$ be a mapping, written $f = (C \mapsto \mathfrak{T}_C)$. For all $B \subseteq A$, we define a set $Sat_n[B](f)$ in 2^{M^n}. That is, $B \mapsto Sat_n[B](f)$ is again a mapping from 2^A to $2^{2^{M^n}}$. Observe that when $n = 1$, there is no computation to do since for all B, $\mathfrak{C}_{2,1}[\alpha, B] = \mathfrak{I}_1[B]$ by definition. Therefore, we simply set $Sat_1[B](C \mapsto \mathfrak{T}_C) = \mathfrak{T}_B$. When $n \geqslant 2$, we define $Sat_n[B](C \mapsto \mathfrak{T}_C)$ as the set $\mathfrak{T}_B \cup \mathfrak{M}_B \cup \mathfrak{O}_B$ with

$$\mathfrak{M}_B = \bigcup_{C \cup D = B} (\mathfrak{T}_C \cdot \mathfrak{T}_D) \tag{1}$$

$$\mathfrak{O}_B = \{\mathcal{T}^\omega \cdot (1_M, \mathcal{C}_{2,n-1}[\alpha, B]) \cdot \mathcal{T}^\omega \mid \mathcal{T} \in \mathfrak{T}_B\} \tag{2}$$

This ends the description of the procedure Sat_n. We now formalize how to iterate it. For any mapping $f : 2^A \to 2^{M^n}$ and any $B \subseteq A$, we set $Sat_n^0[B](f) = f(B)$. For all $j \geqslant 1$, we set $Sat_n^j[B](f) = Sat_n[B](C \mapsto Sat_n^{j-1}[C](f))$. By definition of Sat_n, for all $j \geqslant 0$ and $B \subseteq A$, we have $Sat_n^j(f)[B] \subseteq Sat_n^{j+1}(f)[B] \subseteq 2^{M^n}$. Therefore, there exists j such that $Sat_n^j[B](f) = Sat_n^{j+1}[B](f)$. We denote by $Sat_n^*[B](f)$ this set. This finishes the definition of the algorithm. Its correctness and completeness are stated in the following proposition.

Proposition 5. Let $n \geqslant 1$, $B \subseteq A$ and $\ell \geqslant 3|M| \cdot 2^{|A|} \cdot n \cdot 2^{2^{2|M|^n}}$. Then

$$\mathfrak{C}_{2,n}[\alpha, B] = \mathfrak{C}_{2,n}^\ell[\alpha, B] = \downarrow Sat_n^*[B](C \mapsto \mathfrak{I}_n[C]).$$

Proposition 5 states correctness of the algorithm (the set $\downarrow Sat_n^*[B](C \mapsto \mathfrak{I}_n[C])$ *only* consists of compatible sets of Σ_2-chains) and completeness (this set contains *all* such sets). It also establishes a bound ℓ. This bound is a byproduct of the proof of the algorithm. It is of particular interest for separation and Theorem 4. Indeed, one can prove that for any two languages that are Σ_2-separable and recognized by α, the separator can be chosen with quantifier rank ℓ (for $n = 2$).

We will see in Sections 5 and 6 how to use Proposition 5 to get decidable characterizations of Σ_3, Π_3, Δ_3 and $\mathcal{B}\Sigma_2$. We already state the following corollary as a consequence of Theorem 4.

Corollary 6. *Given as input two regular languages L_1, L_2 it is decidable to test whether L_1 can be Σ_2-separated (resp. Π_2-separated) from L_2.*

5   Decidable Characterizations of Σ_3, Π_3, Δ_3

In this section we present our decidable characterizations for Δ_3, Σ_3 and Π_3. We actually give characterizations for all classes Δ_i, Σ_i and Π_i in the quantifier alternation hierarchy. The characterizations are all stated in terms of equations on the syntactic monoid of the language. However, these equations are parametrized by the Σ_{i-1}-chains of length 2. Therefore, getting *decidable* characterizations depends on our ability to compute the set of Σ_{i-1}-chains of length 2, which we are only able to do for $i \leqslant 3$. We begin by stating our characterization for Σ_i, and the characterizations for Π_i and Δ_i will then be simple corollaries.

Theorem 7. *Let L be a regular language and $\alpha : A^* \to M$ be its syntactic morphism. For all $i \geqslant 1$, L is definable in Σ_i iff M satisfies the following property:*

$$s^\omega \leqslant s^\omega t s^\omega \quad \text{for all } (t,s) \in \mathcal{C}_{i-1}[\alpha]. \tag{3}$$

It follows from Theorem 7 that it suffices to compute the Σ_{i-1}-chains of length 2 in order to decide whether a language is definable in Σ_i. Also observe that when $i = 1$, by definition we have $(t, 1_M) \in \mathcal{C}_0[\alpha]$ for all $t \in M$. Therefore, (3) can be rephrased as $1_M \leqslant t$ for all $t \in M$, which is the already known equation for Σ_1, see [13]. Similarly, when $i = 2$, (3) can be rephrased as $s^\omega \leqslant s^\omega t s^\omega$ whenever t is a 'subword' of s, which is the previously known equation for Σ_2 (see [13,4]).

The proof of Theorem 7 is done using Simon's Factorization Forest Theorem and is actually a generalization of a proof of [4] for the special case of Σ_2. Here, we state characterizations of Π_i and Δ_i as immediate corollaries. Recall that a language is Π_i-definable if its complement is Σ_i-definable, and that it is Δ_i-definable if it is both Σ_i-definable and Π_i-definable.

Corollary 8. *Let L be a regular language and let $\alpha : A^* \to M$ be its syntactic morphism. For all $i \geqslant 1$, the following properties hold:*

- *L is definable in Π_i iff M satisfies $s^\omega \geqslant s^\omega t s^\omega$ for all $(t,s) \in \mathcal{C}_{i-1}[\alpha]$.*
- *L is definable in Δ_i iff M satisfies $s^\omega = s^\omega t s^\omega$ for all $(t,s) \in \mathcal{C}_{i-1}[\alpha]$.*

We finish the section by stating decidability for the case $i = 3$. Indeed by Proposition 5, one can compute the Σ_2-chains of length 2 for any morphism. Therefore, we get the following corollary.

Corollary 9. *Definability of a regular language in Δ_3, Σ_3 or Π_3 is decidable.*

6   Decidable Characterization of $\mathcal{B}\Sigma_2$

In this section we present our decidable characterization for $\mathcal{B}\Sigma_2$. In this case, unlike Theorem 7, the characterization is specific to the case $i = 2$ and does not generalize as a non-effective characterization for all levels. The main reason is that both the intuition and the proof of the characterization rests on a deep analysis of our algorithm for computing Σ_2-chains, which is specific to level $i = 2$. The characterization is stated as two equations that must be satisfied by the syntactic morphism of the language. The first one is parametrized by Σ_2-chains of length 3, and the second one by sets of compatible Σ_2-chains of length 2 through a more involved relation that we define below.

Alternation Schema. Let $\alpha : A^* \to M$ be a monoid morphism and let $B \subseteq A$. A B-schema for α is a triple $(s_1, s_2, s_2') \in M^3$ such that there exist $\mathcal{T} \in \mathfrak{C}_2[\alpha, B]$ and $r_1, r_1' \in M$ verifying $s_1 = r_1 r_1'$, $(r_1, s_2) \in \mathcal{C}_2[\alpha, B] \cdot \mathcal{T}^\omega$ and $(r_1', s_2') \in \mathcal{T}^\omega \cdot \mathcal{C}_2[\alpha, B]$. Intuitively, the purpose of B-schemas is to abstract a well-known property of Σ_2 on elements of M: one can prove that if (s_1, s_2, s_2') is a B-schema, then for all $k \in \mathbb{N}$, there exist $w_1, w_2, w_2' \in A^*$, mapped respectively to s_1, s_2, s_2' under α, and such that for all $u \in B^*$, $w_1 \leqslant_2^k w_2 u w_2'$.

Theorem 10. *Let L be a regular language and $\alpha : A^* \to M$ be its syntactic morphism. Then L is definable in $\mathcal{B}\Sigma_2$ iff M satisfies the following properties:*

$$\begin{aligned} s_1^\omega s_3^\omega = s_1^\omega s_2 s_3^\omega \\ s_3^\omega s_1^\omega = s_3^\omega s_2 s_1^\omega \end{aligned} \quad \text{for } (s_1, s_2, s_3) \in \mathcal{C}_2[\alpha] \tag{4}$$

$$(s_2 t_2)^\omega s_1 (t_2' s_2')^\omega = (s_2 t_2)^\omega s_2 t_1 s_2' (t_2' s_2')^\omega$$
$$\text{for } (s_1, s_2, s_2') \text{ and } (t_1, t_2, t_2') \text{ } B\text{-schemas for some } B \subseteq A \tag{5}$$

The proof of Theorem 10 is far more involved than that of Theorem 7. However, a simple consequence is decidability of definability in $\mathcal{B}\Sigma_2$. Indeed, it suffices to compute Σ_2-chains of length 3 and the B-schemas for all $B \subseteq A$ to check validity of both equations. Computing this information is possible by Proposition 5, and therefore, we get the following corollary.

Corollary 11. *Definability of a regular language in $\mathcal{B}\Sigma_2$ is decidable.*

7 Conclusion

We solved the separation problem for Σ_2 using the new notion of Σ_2-chains, and we used our solution to prove decidable characterizations for $\mathcal{B}\Sigma_2$, Δ_3, Σ_3 and Π_3. The main open problem in this field remains to lift up these results to higher levels in the hierarchy. In particular, we proved that for any natural i, generalizing our separation solution to Σ_i (*i.e.*, being able to compute the Σ_i-chains of length 2) would yield a decidable characterization for Σ_{i+1}, Π_{i+1} and Δ_{i+1}.

Our algorithm for computing Σ_2-chains cannot be directly generalized for higher levels. An obvious reason for this is the fact that it considers Σ_2-chains parametrized by sub-alphabets. This parameter is designed to take care of the alternation between levels 1 and 2, but is not adequate for higher levels. However, this is unlikely to be the only problem. In particular, we do have an algorithm that avoids using the alphabet, but it remains difficult to generalize. We leave the presentation of this alternate algorithm for further work.

Another open question is to generalize our results to logical formulas that can use a binary predicate $+1$ for the successor relation. In formal languages, this corresponds to the well-known *dot-depth hierarchy* [7]. It was proved in [23] and [14] that decidability of $\mathcal{B}\Sigma_2(<, +1)$ and $\Sigma_3(<, +1)$ is a consequence of our results for $\mathcal{B}\Sigma_2(<)$ and $\Sigma_3(<)$. However, while the reduction itself is simple, its proof rely on deep algebraic arguments. We believe that our techniques can be generalized to obtain direct proofs of the decidability of $\mathcal{B}\Sigma_2(<, +1)$ and $\Sigma_3(<, +1)$.

References

1. Almeida, J.: Some algorithmic problems for pseudovarieties. Publ. Math. Debrecen 54, 531–552 (1999); Proc. of Automata and Formal Languages, VIII
2. Almeida, J., Klíma, O.: New decidable upper bound of the 2nd level in the Straubing-Thérien concatenation hierarchy of star-free languages. DMTCS (2010)
3. Arfi, M.: Polynomial operations on rational languages. In: Brandenburg, F.J., Wirsing, M., Vidal-Naquet, G. (eds.) STACS 1987. LNCS, vol. 247, pp. 198–206. Springer, Heidelberg (1987)
4. Bojańczyk, M.: Factorization forests. In: Diekert, V., Nowotka, D. (eds.) DLT 2009. LNCS, vol. 5583, pp. 1–17. Springer, Heidelberg (2009)

5. Bojańczyk, M., Place, T.: Regular languages of infinite trees that are boolean combinations of open sets. In: Czumaj, A., Mehlhorn, K., Pitts, A., Wattenhofer, R. (eds.) ICALP 2012, Part II. LNCS, vol. 7392, pp. 104–115. Springer, Heidelberg (2012)
6. Brzozowski, J., Knast, R.: The dot-depth hierarchy of star-free languages is infinite. J. Comp. Syst. Sci. 16(1), 37–55 (1978)
7. Cohen, R.S., Brzozowski, J.: Dot-depth of star-free events. J. Comp. Syst. Sci. 5, 1–16 (1971)
8. Czerwiński, W., Martens, W., Masopust, T.: Efficient separability of regular languages by subsequences and suffixes. In: Fomin, F.V., Freivalds, R., Kwiatkowska, M., Peleg, D. (eds.) ICALP 2013, Part II. LNCS, vol. 7966, pp. 150–161. Springer, Heidelberg (2013)
9. McNaughton, R., Papert, S.: Counter-Free Automata. MIT Press (1971)
10. Pin, J.-É.: Bridges for concatenation hierarchies. In: Larsen, K.G., Skyum, S., Winskel, G. (eds.) ICALP 1998. LNCS, vol. 1443, pp. 431–442. Springer, Heidelberg (1998)
11. Pin, J.-É.: Theme and variations on the concatenation product. In: Winkler, F. (ed.) CAI 2011. LNCS, vol. 6742, pp. 44–64. Springer, Heidelberg (2011)
12. Pin, J.-E., Straubing, H.: Monoids of upper triangular boolean matrices. In: Semigroups. Structure and Universal Algebraic Problems. Colloquia Mathematica Societatis Janos Bolyai, vol. 39, pp. 259–272. North-Holland (1985)
13. Pin, J.-E., Weil, P.: Polynomial closure and unambiguous product. Theory of Computing Systems 30(4), 383–422 (1997)
14. Pin, J.-E., Weil, P.: The wreath product principle for ordered semigroups. Communications in Algebra 30, 5677–5713 (2002)
15. Place, T., van Rooijen, L., Zeitoun, M.: Separating regular languages by piecewise testable and unambiguous languages. In: Chatterjee, K., Sgall, J. (eds.) MFCS 2013. LNCS, vol. 8087, pp. 729–740. Springer, Heidelberg (2013)
16. Place, T., van Rooijen, L., Zeitoun, M.: Separating regular languages by locally testable and locally threshold testable languages. In: FSTTCS 2013. LIPIcs (2013)
17. Place, T., Zeitoun, M.: Going higher in the first-order quantifier alternation hierarchy on words. Arxiv (2014), http://arxiv.org/abs/1404.6832
18. Place, T., Zeitoun, M.: Separating regular languages with first-order logic. In: CSL-LICS 2014 (2014)
19. Schützenberger, M.P.: On finite monoids having only trivial subgroups. Information and Control 8, 190–194 (1965)
20. Simon, I.: Piecewise testable events. In: Brakhage, H. (ed.) GI-Fachtagung 1975. LNCS, vol. 33, pp. 214–222. Springer, Heidelberg (1975)
21. Simon, I.: Factorization forests of finite height. TCS 72(1), 65–94 (1990)
22. Straubing, H.: A generalization of the Schützenberger product of finite monoids. TCS (1981)
23. Straubing, H.: Finite semigroup varieties of the form V * D. J. Pure App. Algebra 36, 53–94 (1985)
24. Straubing, H.: Semigroups and languages of dot-depth two. TCS (1988)
25. Tesson, P., Therien, D.: Diamonds are forever: The variety DA. In: Semigroups, Algorithms, Automata and Languages, pp. 475–500. World Scientific (2002)
26. Thérien, D.: Classification of finite monoids: the language approach. TCS (1981)
27. Thérien, D., Wilke, T.: Over words, two variables are as powerful as one quantifier alternation. In: STOC 1998, pp. 234–240. ACM (1998)
28. Thomas, W.: Classifying regular events in symbolic logic. J. Comp. Syst. Sci. (1982)
29. Thomas, W.: A concatenation game and the dot-depth hierarchy. In: Börger, E. (ed.) Computation Theory and Logic. LNCS, vol. 270, pp. 415–426. Springer, Heidelberg (1987)

Hardness Results for Intersection Non-Emptiness

Michael Wehar

Department of Computer Science and Engineering
University at Buffalo, Buffalo, USA
mwehar@buffalo.edu

Abstract. We carefully reexamine a construction of Karakostas, Lipton, and Viglas (2003) to show that the intersection non-emptiness problem for DFA's (deterministic finite automata) characterizes the complexity class NL. In particular, if restricted to a binary work tape alphabet, then there exist constants c_1 and c_2 such that for every k intersection non-emptiness for k DFA's is solvable in $c_1 k \log(n)$ space, but is not solvable in $c_2 k \log(n)$ space. We optimize the construction to show that for an arbitrary number of DFA's intersection non-emptiness is not solvable in $o(\frac{n}{\log(n)\log(\log(n))})$ space. Furthermore, if there exists a function $f(k) = o(k)$ such that for every k intersection non-emptiness for k DFA's is solvable in $n^{f(k)}$ time, then P $\neq$ NL. If there does not exist a constant c such that for every k intersection non-emptiness for k DFA's is solvable in n^c time, then P does not contain any space complexity class larger than NL.

1 Introduction

Let $\mathcal{A}$ denote a class of machines. The intersection non-emptiness problem for $\mathcal{A}$, denoted by $IE_{\mathcal{A}}$, consists of all finite lists of machines in $\mathcal{A}$ whose underlying languages have a non-empty intersection. By fixing the number of machines in the input to k, one obtains intersection non-emptiness for k machines which we denote by k-$IE_{\mathcal{A}}$. Intersection non-emptiness problems can be motivated by the following scenario. Consider that you are trying to construct an object x for a particular application. You propose a finite list of conditions for x to satisfy such that each condition can be decided by a machine in $\mathcal{A}$. An algorithm that solves intersection non-emptiness for $\mathcal{A}$ provides a method for checking if there exists an object x satisfying the proposed conditions.

Let $IE_{\mathcal{D}}$ denote the intersection non-emptiness problem for DFA's. One can solve $IE_{\mathcal{D}}$ by checking reachability in a product machine. Given an input consisting of k machines each of size at most m, the product machine has size at most m^k. Therefore, checking reachability takes at most m^{ck} time for some constant c. $IE_{\mathcal{D}}$ is a well known PSPACE-complete problem [5]. In [6], it was shown that one can pad strings in $IE_{\mathcal{D}}$ to obtain problems hard for smaller complexity classes such as $NSPACE(g(n) \log(n))$ where g is a slow growing log-space-constructible

J. Esparza et al. (Eds.): ICALP 2014, Part II, LNCS 8573, pp. 354–362, 2014.
© Springer-Verlag Berlin Heidelberg 2014

function such as $\log^*(n)$. In [4], it was shown that improvements to the standard algorithm imply separation results. In particular, if there exists a function $f(k) = o(k)$ such that $\text{IE}_\mathcal{D}$ is solvable in $m_1 \cdot m_2^{f(k)}$ time where m_1 is the size of a designated largest machine and all other machines have size at most m_2, then $\text{NL} \neq \text{P}$.

In this paper, we carefully reexamine and optimize the construction from [4] in order to prove new results. We show that if restricted to a binary work tape alphabet, then there exist constants c_1 and c_2 such that for every k, $k\text{-IE}_\mathcal{D} \in \text{NSPACE}(c_1 k \log(n))$ and $k\text{-IE}_\mathcal{D} \notin \text{NSPACE}(c_2 k \log(n))$. Then, we introduce an optimized construction to show that $\text{IE}_\mathcal{D} \notin \text{NSPACE}(o(\frac{n}{\log(n)\log(\log(n))}))$. Finally, we combine these results with a diagonalization argument to show that if there exists a function $f(k) = o(k)$ such that for every k, $k\text{-IE}_\mathcal{D} \in \text{DTIME}(n^{f(k)})$, then $\text{P} \neq \text{NL}$. If there does not exist a constant c such that for every k, $k\text{-IE}_\mathcal{D} \in \text{DTIME}(n^c)$, then $\text{NSPACE}(f(n)) \nsubseteq \text{P}$ for all $f(n) = \omega(\log(n))$ such that f is space-constructible.

2 Notation and Conventions

The input for $\text{IE}_\mathcal{D}$ is an encoding of a finite list of DFA's. For each encoding, n will denote the length and k will denote the number of machines that are represented. For each natural number k, $k\text{-IE}_\mathcal{D}$ denotes a restriction of the $\text{IE}_\mathcal{D}$ problem such that we only accept inputs that encode at most k machines.

Whenever we use the term Turing machine, we refer to a deterministic or non-deterministic machine with a two-way read only input tape and a two-way read/write work tape. For our purposes, we will only consider Turing machines where the work tape alphabet is binary. A work tape over a binary alphabet will be referred to as a binary work tape. A cell on a binary work tape will be referred to as a bit cell.

For each k, there are acceptance problems for space and time bounded Turing machines denoted by $N_{k \log}^S$ and $D_{n^k}^T$, respectively. $N_{k \log}^S$ refers to the problem where we are given an encoding of a non-deterministic Turing machine M with a binary work tape and an input s. We accept (M, s) if and only if M accepts s using at most $k \log(n)$ work tape bit cells where n denotes the length of s. $D_{n^k}^T$ is defined similarly for n^k deterministic time. We denote by $\text{NSPACE}^2(h(n))$ the set of problems solvable by a non-deterministic Turing machine using at most $h(n)$ work tape bit cells. Such classes are used to measure the binary space complexity of problems [2]. We associate $N_{k \log}^S$ with $\text{NSPACE}^2(k \log(n))$ and $D_{n^k}^T$ with $\text{DTIME}(n^k)$.

3 Binary Space Complexity

We introduce a function $S_{\text{NL}}(k)$ that measures the actual space complexities of the $N_{k \log}^S$ problems. In particular, $S_{\text{NL}}(k)$ is defined as follows:

$$S_{\text{NL}}(k) := \min\{\, d \in \mathbb{N} \mid N_{k \log}^S \in \text{NSPACE}^2(d \log(n)) \,\}. \tag{1}$$

In this section, we sketch how one could apply standard techniques from the space hierarchy theorem to prove that there exist constants c_1 and c_2 such that for every k sufficiently large, $N_{k \log}^S \in \text{NSPACE}^2(c_1 k \log(n))$ and $N_{k \log}^S \notin \text{NSPACE}^2(c_2 k \log(n))$. Using the function $\text{S}_{\text{NL}}(k)$, we express this result as $\text{S}_{\text{NL}}(k) = \Theta(k)$.

Proposition 1. $\text{S}_{\text{NL}}(k) = O(k)$.

Sketch of proof. Using the simulation found in any common proof of the space hierarchy theorem, one shows that $N_{\log}^S \in \text{NL}$. Further, one shows $\text{S}_{\text{NL}}(k) = O(k)$ by using padding to reduce $N_{k \log}^S$ to $N_{\log}^S$ for every k.   □

Proposition 2. $\text{S}_{\text{NL}}(k) = \Omega(k)$.

Sketch of proof. Using the standard diagonalization argument found in any common proof of the non-deterministic space hierarchy theorem, one shows $\text{S}_{\text{NL}}(k) = \Omega(k)$. Notice that in order to carry out the diagonalization one needs to show there exists c such that for all k,

$$\text{NSPACE}^2(k \log(n)) \subseteq co\text{-NSPACE}^2(ck \log(n)). \tag{2}$$

First, one applies the result $\text{NL} = co\text{-NL}$ to show that there exists c such that $N_{\log}^S \in co\text{-NSPACE}^2(c \log(n))$. Further, one shows (2) by using padding to reduce $N_{k \log}^S$ to $N_{\log}^S$ for every k.   □

Corollary 3. $\text{S}_{\text{NL}}(k) = \Theta(k)$.

4   Reductions

We introduce a function $\text{S}_{\text{IE}}(k)$ that measures the actual space complexities of the k-$\text{IE}_{\mathcal{D}}$ problems. In particular, $\text{S}_{\text{IE}}(k)$ is defined as follows:

$$\text{S}_{\text{IE}}(k) := \min\{\, d \in \mathbb{N} \mid k\text{-IE}_{\mathcal{D}} \in \text{NSPACE}^2(d \log(n)) \,\}. \tag{3}$$

In this section, we carefully reexamine the construction from [4] to show that there exist constants c_1 and c_2 such that for every k sufficiently large, k-$\text{IE}_{\mathcal{D}} \in \text{NSPACE}^2(c_1 k \log(n))$ and k-$\text{IE}_{\mathcal{D}} \notin \text{NSPACE}^2(c_2 k \log(n))$. Using the function $\text{S}_{\text{IE}}(k)$, we can express this result as $\text{S}_{\text{IE}}(k) = \Theta(\text{S}_{\text{NL}}(k)) = \Theta(k)$.

Proposition 4. $\text{S}_{\text{IE}}(k) = O(k)$.

Sketch of proof. As was previously discussed, one can solve $\text{IE}_{\mathcal{D}}$ by checking reachability in a product machine. A state of the product machine can be stored as a string of $k \log(n)$ bits. Given such a state, we can non-deterministically guess which state comes next. There exists a path from an initial state to a final state if and only if there exists a path from an initial state to a final state of length at most n^k. Therefore, k-$\text{IE}_{\mathcal{D}}$ is solvable using at most $ck \log(n)$ bits for some constant c.   □

Theorem 5. $S_{IE}(k) = \Omega(S_{NL}(k))$.

Proof. We will describe a reduction from $N_{k\log}^S$ to k-$IE_{\mathcal{D}}$. Then, we will discuss encoding details to show that this is a log-space reduction.

Let a $k\log(n)$ space bounded non-deterministic Turing machine M and an input string s of length n be given. Our first task is to construct k DFA's, denoted by $< D_i >_{i \in [k]}$, each of size at most $p(n)$ for some fixed polynomial p such that M accepts s if and only if $\bigcap_{i \in [k]} L(D_i)$ is non-empty. The DFA's will read in a string that represents a computation of M on s and verify that the computation is valid and accepting. The work tape of M will be split into k sections each consisting of $\log(n)$ sequential bits of memory. The ith DFA, D_i, will keep track of the ith section and verify that it is managed correctly. In addition, all of the DFA's will keep track of the input and work tape head positions. We will achieve a better simulation in Theorem 7 where we split up the management of the tape head positions to separate DFA's. The following two concepts are essential to our construction.

A *section i configuration* of M is a tuple of the form

(state, input position, work position, ith section of work tape).

A *forgetful configuration* of M is a tuple of the form

(state, input position, work position).

We say that a section i configuration r extends a forgetful configuration a if r agrees with a on state, input position, and work position. We say that a section i configuration r_1 transitions to a section i configuration r_2 on input s if either the work position for r_1 is in the ith section and r_2 correctly represents how the tape positions and the ith section could change in one step of the computation on s, or r_1 is not in the ith section and r_1 and r_2 agree on the ith section of the work tape.

The states of D_i are identified with section i configurations. The alphabet characters are identified with forgetful configurations. For D_i, each alphabet character a transitions from a state r_1 to a state r_2 if and only if r_2 extends a and r_1 transitions to r_2 on input s.

We assert without proof that for every string x, x represents a valid accepting computation of M on s if and only if $x \in \bigcap_{i \in [k]} L(D_i)$. Therefore, M accepts s if and only if $\bigcap_{i \in [k]} L(D_i)$ is non-empty.

We show that the D_i's have size at most $p(n)$ for some fixed polynomial p. Each D_i consists of a start state, a list of final states, and a list of transitions where each transition consists of two states and an alphabet character. Each state is represented by a section i configuration and each alphabet character is represented by a forgetful configuration. Let m denote the number of states in M. Therefore, in total there are $m \cdot n \cdot k \log(n) \cdot 2^{\log(n)}$ section i configurations and $m \cdot n \cdot k \log(n)$ forgetful configurations. Hence, there exists a fixed two variable polynomial q such that each D_i has at most $q(n, k)$ states. Since k is fixed, one can blow up the degree of q to get a polynomial p such that p doesn't depend on k and each D_i has size at most $p(n)$.

It should be clear from the preceding that there is a fixed polynomial $t(n)$ such that for every k, $N^S_{k\log}$ is $t(n)$-time reducible to k-IE$_\mathcal{D}$. However, we want to show that there is a constant c such that for every k, $N^S_{k\log}$ is $c\log(n)$-space reducible to k-IE$_\mathcal{D}$. We accomplish this by describing how to print the string encoding of the D_i's to an auxiliary write only output tape using at most $c\log(n)$ space for some constant c.

We will describe how to print the transitions for each D_i and leave the remaining encoding details to the reader. We use a bit string i to represent the current DFA and two bit strings j_1 and j_2 to represent section i configurations. We iterate through every combination of i, j_1, and j_2. If D_i has a transition from j_1 to j_2, then we print (i, j_1, a, j_2) where a is the forgetful configuration such that j_2 extends a. We assert that checking whether to print (i, j_1, a, j_2) requires no more than $d\log(k) + d\log(n)$ bits for some constant d. Therefore, in printing the encoding of the D_i's, we use no more than $c\log(k) + c\log(n)$ bits for some constant c. For each k, when n is sufficiently large, the $\log(k)$ term goes away. It follows that for every k, $N^S_{k\log}$ is $c\log(n)$-space reducible to k-IE$_\mathcal{D}$. $\square$

Corollary 6. $S_{IE}(k) = \Theta(S_{NL}(k)) = \Theta(k)$.

Proof. By Corollary 3, we have $S_{NL}(k) = \Theta(k)$. Applying Proposition 4 and Theorem 5, we get that $S_{IE}(k) = \Theta(S_{NL}(k)) = \Theta(k)$. $\square$

Theorem 7. IE$_\mathcal{D} \notin$ NSPACE$(o(\frac{n}{\log(n)\log(\log(n))}))$.

Proof. By the non-deterministic space hierarchy theorem, we may choose a problem Q such that $Q \in$ NSPACE(n), but $Q \notin$ NSPACE$(o(n))$. Choose $c \in \mathbb{N}$ and a non-deterministic Turing machine M that solves Q using at most cn bit cells. We optimize the construction from the proof of Theorem 5 to show that if IE$_\mathcal{D}$ $\in$ NSPACE$(o(\frac{n}{\log(n)\log(\log(n))}))$, then $Q \in$ NSPACE$(o(n))$. Since we know that $Q \notin$ NSPACE$(o(n))$, it follows that IE$_\mathcal{D} \notin$ NSPACE$(o(\frac{n}{\log(n)\log(\log(n))}))$.

Let an input string s for M of length n be given. Our task is to construct $(c+1) \cdot n$ DFA's each with at most $d\log(n)$ states for some constant d such that M accepts s if and only if the DFA's have a non-empty intersection. The DFA's will read in a bit string that represents a computation of M on s and verify that the computation is valid and accepting. In this construction, we split up the management of the tape head positions to separate DFA's. There are n DFA's, denoted by $< I_i >_{i \in [n]}$, that manage the input tape and there are cn DFA's, denoted by $< W_i >_{i \in [cn]}$, that manage the work tape. The following concept is essential to our construction.

An *informative configuration* of M is a tuple of the form

(state, input position, current input bit, work position, current work bit).

The DFA's will read in a sequence of informative configurations that are encoded as bit strings. In contrast to the previous construction, the DFA's will have a binary input alphabet.

Each DFA is assigned to manage a bit position of either the input tape or work tape. Each I_i stores the ith input tape bit and operates as follows. It reads each informative configuration and checks if it represents the input position i. If it does not, then it ignores the informative configuration and moves on to the next one. However, if it does represent the input position i, then it checks that the stored bit matches the current input bit and uses the current work bit to check that the input position and state validly transition to the next informative configuration. Each W_i stores the ith work tape bit and operates as follows. It reads each informative configuration and checks if it represents the work position i. If it does not, then it ignores the informative configuration and moves on to the next one. However, if it does represent position i, then it checks that the stored bit matches the current work bit and uses the current input bit to modify the stored bit and check that the work position and state validly transition to the next informative configuration. It's important to remark that DFA's for boundary positions such as I_1, I_n, W_1, and W_{cn} cannot allow the input position or work position to go outside $[n]$ or $[cn]$, respectively.

We assert without proof that for every bit string x, x represents a valid accepting computation of M on s if and only if $x \in \bigcap_{i \in [n]} L(I_i)$ and $x \in \bigcap_{i \in [cn]} L(W_i)$. Therefore, M accepts s if and only if there exists a string x such that $x \in \bigcap_{i \in [n]} L(I_i)$ and $x \in \bigcap_{i \in [cn]} L(W_i)$.

A DFA with $\log(cn)$ states can be constructed to recognize a fixed binary number $i \in [cn]$. Since a tape position i could only transition to $i-1$, i, or $i+1$ in one step, it follows that a DFA with $d\log(n)$ states for some constant d can be constructed to check the validity of transitioning to the next informative configuration. Therefore, we can construct each DFA with at most $d\log(n)$ states for some constant d.

We described how to construct $(c+1) \cdot n$ DFA's each with at most $d\log(n)$ states for some constant d whose intersection is non-empty if and only if M accepts s. Since the total length of the string encoding of $< I_i >_{i \in [n]}$ combined with $< W_i >_{i \in [cn]}$ is at most $n\log(n)\log(\log(n))$, it follows that $\mathrm{IE}_{\mathcal{D}} \in \mathrm{NSPACE}(o(\frac{n}{\log(n)\log(\log(n))}))$ implies $Q \in \mathrm{NSPACE}(o(n))$. We obtain the desired result because $Q \notin \mathrm{NSPACE}(o(n))$. $\square$

5   Space vs Time

We introduce functions $\mathrm{R}_{\mathrm{NL}}(k)$ and $\mathrm{R}_{\mathrm{IE}}(k)$ that measure the actual time complexities of $N^S_{k\log}$ and $k\text{-}\mathrm{IE}_{\mathcal{D}}$, respectively. In particular, $\mathrm{R}_{\mathrm{NL}}(k)$ and $\mathrm{R}_{\mathrm{IE}}(k)$ are defined as follows:

$$\mathrm{R}_{\mathrm{NL}}(k) := \min\{\, d \in \mathbb{N} \mid N^S_{k\log} \in \mathrm{DTIME}(n^d)\,\} \tag{4}$$

$$\mathrm{R}_{\mathrm{IE}}(k) := \min\{\, d \in \mathbb{N} \mid k\text{-}\mathrm{IE}_{\mathcal{D}} \in \mathrm{DTIME}(n^d)\,\}. \tag{5}$$

In this section, we show that if there exists a function $f(k) = o(k)$ such that for every k, $N^S_{k\log} \in \mathrm{DTIME}(n^{f(k)})$, then $\mathrm{P} \neq \mathrm{NL}$. Using the function $\mathrm{R}_{\mathrm{NL}}(k)$ we can express this result as if $\mathrm{R}_{\mathrm{NL}}(k) = o(k)$, then $\mathrm{P} \neq \mathrm{NL}$. Notice that by using the reduction from Theorem 5, we also have $R_{IE}(k) = \Theta(R_{NL}(k))$. It follows that if $R_{IE}(k) = o(k)$, then $P \neq NL$.

Proposition 8. $R_{IE}(k) = \Theta(R_{NL}(k))$.

Theorem 9. *If $R_{NL}(k) = o(k)$, then $NL \neq P$.*

Proof. Suppose that $NL = P$. Since $D_n^T \in P$, we have $D_n^T \in NL$. Choose $d \in \mathbb{N}$ such that $D_n^T \in NSPACE^2(d \log(n))$. Further, by using padding to reduce $D_{n^k}^T$ to D_n^T for every k, one can show that there exists d' such that for all k, $D_{n^k}^T \in NSPACE^2(d'k \log(n))$. Choose such a constant d' satisfying for all k, $D_{n^k}^T \in NSPACE^2(d'k \log(n))$.

Suppose for sake of contradiction that $R_{NL}(k) = o(k)$. By Proposition 2, we may choose c such that for all k sufficiently large

$$N_{k \log}^S \notin NSPACE^2(\left\lfloor \frac{k}{c} \right\rfloor \log(n)). \tag{6}$$

Since $R_{NL}(k) = o(k)$, for all k sufficiently large

$$R_{NL}(k) < \left\lfloor \frac{k}{cd'} \right\rfloor. \tag{7}$$

Choose m satisfying $N_{m \log}^S \notin NSPACE^2(\left\lfloor \frac{m}{c} \right\rfloor \log(n))$ and $R_{NL}(m) < \left\lfloor \frac{m}{cd'} \right\rfloor$. Therefore,

$$N_{m \log}^S \in DTIME(o(n^{\left\lfloor \frac{m}{cd'} \right\rfloor})). \tag{8}$$

Since $D_{n^k}^T \in NSPACE^2(d'k \log(n))$ for all k,

$$D_{n^{\left\lfloor \frac{m}{cd'} \right\rfloor}}^T \in NSPACE^2(d' \left\lfloor \frac{m}{cd'} \right\rfloor \log(n)) \subseteq NSPACE^2(\left\lfloor \frac{m}{c} \right\rfloor \log(n)). \tag{9}$$

Since we can trivially reduce every problem in $DTIME(o(n^{\left\lfloor \frac{m}{cd'} \right\rfloor}))$ to $D_{n^{\left\lfloor \frac{m}{cd'} \right\rfloor}}^T$,

$$N_{m \log}^S \in DTIME(o(n^{\left\lfloor \frac{m}{cd'} \right\rfloor})) \subseteq NSPACE^2(\left\lfloor \frac{m}{c} \right\rfloor \log(n)) \tag{10}$$

which is a contradiction because $N_{m \log}^S \notin NSPACE^2(\left\lfloor \frac{m}{c} \right\rfloor \log(n))$. $\square$

Corollary 10. *If $R_{IE}(k) = o(k)$, then $NL \neq P$.*

Next, we show that if $R_{NL}(k)$ is unbounded, then P does not contain any space complexity class larger than NL. Since $R_{IE}(k) = \Theta(R_{NL}(k))$, it follows that if $R_{IE}(k)$ is unbounded, then P does not contain any space complexity class larger than NL.

For every function f, let N_f^S denote the acceptance problem for $f(n)$-space bounded non-deterministic Turing machines. N_f^S is of particular interest to us if it is non-deterministically solvable in $f(n)$ space.

Theorem 11. *If* $R_{NL}(k)$ *is unbounded, then* $N_f^S \notin P$ *for all functions* $f(n) = \omega(\log(n))$.

Proof. We will prove the contrapositive. Suppose that $N_f^S \in P$ for some function $f(n) = \omega(\log(n))$. By assumption, we may choose $c \in \mathbb{N}$ and a deterministic Turing machine T such that T solves N_f^S in at most $O(n^c)$ time. Let $k \in \mathbb{N}$ be given. Choose a non-deterministic Turing machine M that solves $N_{k\log}^S$ using at most $O(\log(n))$ bit cells. We can deterministically solve $N_{k\log}^S$ in at most $O(n^c)$ time by feeding T an encoding of M and the input string. Since k is arbitrary, $N_{k\log}^S$ is solvable in $O(n^c)$ time for every k. It follows that $R_{NL}(k)$ is bounded.
□

Corollary 12. *If* $R_{NL}(k)$ *is unbounded, then* $\mathrm{NSPACE}(f(n)) \nsubseteq P$ *for all* $f(n) = \omega(\log(n))$ *such that* f *is space-constructible.*

Proof. Suppose $R_{NL}(k)$ is unbounded. Let a function $f(n) = \omega(\log(n))$ such that f is space-constructible be given. Apply the preceding theorem to get that $N_f^S \notin P$. Since f is space-constructible, one can use the simulation found in any common proof of the space hierarchy theorem to show that $N_f^S \in \mathrm{NSPACE}(f(n))$. Since $N_f^S \notin P$ and $N_f^S \in \mathrm{NSPACE}(f(n))$, it follows that $\mathrm{NSPACE}(f(n)) \nsubseteq P$.
□

Corollary 13. *If* $R_{IE}(k)$ *is unbounded, then* $\mathrm{NSPACE}(f(n)) \nsubseteq P$ *for all* $f(n) = \omega(\log(n))$ *such that* f *is space-constructible.*

6   Conclusion

In Section 4, we showed that $S_{NL}(k) = S_{IE}(k) = \Theta(k)$. Therefore, we think of intersection non-emptiness for DFA's as characterizing the complexity class NL. Further, we showed that $IE_{\mathcal{D}} \notin \mathrm{NSPACE}(o(\frac{n}{\log(n)\log(\log(n))}))$. In Section 5, we showed that if $R_{IE}(k) = o(k)$, then $\mathrm{NL} \neq P$ and if $R_{IE}(k)$ is unbounded, then $\mathrm{NSPACE}(f(n)) \nsubseteq P$ for all $f(n) = \omega(\log(n))$ such that f is space-constructible. Therefore, the asymptotic complexity of $R_{IE}(k)$ determines the relationship between space and time complexity classes.

There are several related problems that appear to be harder than k-$IE_{\mathcal{D}}$, but easier than $N_{k\log}^S$. For example, consider intersection non-emptiness for k NFA's, non-emptiness for k-turn 2DFA's, and intersection non-emptiness for k DFA's and a one-counter automaton. We can use $S_{NL}(k) = S_{IE}(k)$ and $R_{NL}(k) = R_{IE}(k)$ as squeeze theorems to show that all of these problems are of "equivalent" difficulty. Also, one could define a function that maps the k-$IE_{\mathcal{D}}$ problems to their actual circuit complexities. The asymptotic complexity of such a function could determine the relationship between NL vs NP and P/poly vs space complexity classes [4].

Several related intersection non-emptiness problems have been studied. There are two such problems that we would like to mention. In [10], intersection non-emptiness for acyclic DFA's, which are DFA's without directed cycles, was shown to be NP-complete. We assert that one could modify the construction from the proof of Theorem 5 to reduce the acceptance problem for n-time and $k \log(n)$-space bounded non-deterministic Turing machines to intersection non-emptiness for k acyclic DFA's. Also, in [11], intersection non-emptiness for tree automata was shown to be EXPTIME-complete. In an upcoming paper, the author and Joseph Swernofsky introduce time complexity lower bounds for intersection non-emptiness for tree automata.

Acknowledgments. I greatly appreciate all of the help and suggestions that I received. In particular, I would like to thank Christos Kapoutsis for suggestions related to the constructions, Joseph Swernofsky for proof reading and many discussions, Richard Lipton and Kenneth Regan for calling attention to my results in an article on their blog [8], and the many anonymous referees. I would especially like to thank all those at Carnegie Mellon University who offered their help and support for my honors thesis on the same topic. In particular, I would like to thank my thesis advisor, Klaus Sutner, and my thesis committee members, Manuel Blum and Richard Statman.

References

1. Blondin, M., Krebs, A., McKenzie, P.: The complexity of intersecting finite automata having few final states. In: Computational Complexity, CC (to appear, 2014)
2. Goldreich, O.: Computational Complexity: A Conceptual Perspective. Cambridge University Press, New York (2008)
3. Jones, N.D., Lien, Y.E., Laaser, W.T.: New problems complete for nondeterministic log space. Mathematical Systems Theory 10 (1976)
4. Karakostas, G., Lipton, R.J., Viglas, A.: On the complexity of intersecting finite state automata and NL versus NP. Theoretical Computer Science 302, 257–274 (2003)
5. Kozen, D.: Lower bounds for natural proof systems. In: Proc. 18th Symp. on the Foundations of Computer Science, pp. 254–266 (1977)
6. Lange, K.-J., Rossmanith, P.: The emptiness problem for intersections of regular languages. In: Havel, I.M., Koubek, V. (eds.) MFCS 1992. LNCS, vol. 629, pp. 346–354. Springer, Heidelberg (1992)
7. Lipton, R.J.: On the intersection of finite automata. Gödel's Lost Letter and P=NP (August 2009)
8. Lipton, R.J., Regan, K.W.: The power of guessing. Gödel's Lost Letter and P=NP (November 2012)
9. Rabin, M.O., Scott, D.: Finite automata and their decision problems. IBM Journal (1959)
10. Rampersad, N., Shallit, J.: Detecting patterns in finite regular and context-free languages. Information Processing Letters 110 (2010)
11. Veanes, M.: On computational complexity of basic decision problems of finite tree automata. UPMAIL Technical Report 133 (1997)
12. Wehar, M.: Intersection emptiness for finite automata. Honors thesis, Carnegie Mellon University (2012)

Branching Bisimilarity Checking for PRS

Qiang Yin, Yuxi Fu, Chaodong He, Mingzhang Huang, and Xiuting Tao

BASICS, Department of Computer Science, Shanghai Jiao Tong University

Abstract. Recent studies reveal that branching bisimilarity is decidable for both nBPP (normed Basic Parallel Processes) and nBPA (normed Basic Process Algebras). These results lead to the question if there are any other models in the hierarchy of PRS (Process Rewrite Systems) whose branching bisimilarity is decidable. It is shown in this paper that the branching bisimilarity for both nOCN (normed One Counter Nets) and nPA (normed Process Algebras) is undecidable. These results essentially imply that the question has a negative answer.

1 Introduction

Verification on infinite-state systems has been intensively studied for the past two decades [2,12]. One major concern in these studies is equivalence checking. Given a specification S of an intended behaviour and a claimed implementation I of S, one is supposed to demonstrate that I is correct with respect to S. A standard interpretation of correctness is that an implementation should be behaviourally equivalent to its specification. Among all the behavioural equalities studied so far, bisimilarity stands out as the most abstract and the most tractable one. Two well known bisimilarities are the strong bisimilarity and the weak bisimilarity due to Park [16] and Milner [15]. Considerable amount of effort has been made to investigate the decidability and the algorithmic aspect of the two bisimilarities on various models of infinite state system [18]. These models include pushdown automata, process algebras, Petri nets and their restricted and extended variations. An instructive classification of the models in terms of PRS (Process Rewrite Systems) is given by Mayr [13].

The strong bisimilarity checking problem has been well studied for PRS hierarchy. Influential decidability results include for example [1,4,3,21,8]. On the negative side, Jančar attained in [9] the undecidable result of strong bisimilarity on nPN (normed Petri Nets). The proof makes use of a powerful technique now known as Defender's Forcing [11], which remains a predominant tool to establish negative results about equivalence checking.

In the weak case the picture is less clear. It is widely believed that weak bisimilarity is decidable for both nBPA (normed Basic Process Algebras) and nBPP (normed Basic Parallel Processes). The problem has been open for a long time. Srba [17] showed that weak bisimilarity on nPDA (normed Pushdown Automata) is undecidable by a reduction from the halting problem of Minsky Machine. The undecidability was soon extended to nOCN (normed One Counter Nets), a submodel of both nPDA and nPN, by Mayr [14]. Srba also showed that

J. Esparza et al. (Eds.): ICALP 2014, Part II, LNCS 8573, pp. 363–374, 2014.

	nBPA	nBPP	nPDA	nPA	nPN
Strong Bisimilarity	✓[1]	✓[3]	✓[21]	✓[8]	×[9]
Branching Bisimilarity	✓[7]	✓[5]	×[this paper]	×[this paper]	×[9]
Weak Bisimilarity	?	?	×[14]	×[this paper]	×[9]

Fig. 1. Decidability of Branching Bisimilarity for Normed PRS

the weak bisimilarity on PA (Process Algebras) is undecidable [19]. Later several highly undecidable results were established by Jančar and Srba [20,10,11] for the weak bisimilarity checking problem on PN, PDA and PA.

The decidability of the weak bisimilarity on nBPA and nBPP has been open for well over twenty years. Encouraging progress has been made recently. Czerwiński, Hofman and Lasota proved that branching bisimilarity, a standard refinement of the weak bisimilarity, is decidable on nBPP [5]. The novelty of their approach is the discovery of some kind of normal form for nBPP. Using a quite different technique Fu showed that the branching bisimilarity is also decidable on nBPA [7]. In retrospect one cannot help thinking that more attention should have been paid to the branching bisimilarity. Going back to the original motivation to equivalence checking, one would agree that a specification S normally contains no silent actions because silent actions are about how-to-do. Consequently all the silent actions introduced in an implementation must be bisimulated vacuously by the specification. It follows that S is weakly bisimilar to an implementation $\mathcal{I}$ if and only if S is branching bisimilar to $\mathcal{I}$. What this observation tells us is that as far as verification is concerned the branching bisimilarity ought to play a role no less than the weak bisimilarity.

The above discussion suggests to address the following question: Is there any other model in the PRS hierarchy whose branching bisimilarity is decidable? The purpose of this paper is to resolve this issue. Our contributions are as follows:

- We establish the fact that on both nOCN and nPA every relation between the branching bisimilarity and the weak bisimilarity is undecidable. These are improvement of Mayr's result about the undecidability of the weak bisimilarity on nOCN [14] and Srba's result [19] about the undecidability of the weak bisimilarity on PA. These new results together with the previous (un)decidability results about the *normed* models in PRS are summarized in Fig. 1, where a tick is for 'decidable' and a cross for 'undecidable'.
- We showcase the subtlety of Defender's Forcing technique usable in branching bisimulation game. It is pointed out that the technique must be of a semantic nature for it to be applicable to the branching bisimilarity.

The two negative results imply that in the PRS hierarchy the branching bisimilarity on every normed model above either nBPA or nBPP is undecidable.

The rest of the paper is organized as follows. Section 2 introduces the necessary preliminaries. Section 3 establishes the undecidability result for nOCN and demonstrates Defender's Forcing technique for branching bisimulation game. Section 4 proves the undecidability result about nPA. Section 5 concludes.

2   Preliminaries

A *process algebra* $\mathcal{P}$ is a triple $(\mathcal{C}, \mathcal{A}, \Delta)$, where $\mathcal{C}$ is a finite set of process constants, $\mathcal{A}$ is a finite set of actions ranged over by ℓ, and Δ is a finite set of transition rules. The *processes* defined by $\mathcal{P}$ are generated by the following grammar:

$$P ::= \epsilon \mid X \mid PP' \mid P \| P'.$$

The grammar equality is denoted by $=$. We assume that the sequential composition PP' is associative up to $=$ and the parallel composition $P \| P'$ is associative and commutative up to $=$. We also assume that $\epsilon P = P \epsilon = \epsilon \| P = P \| \epsilon = P$. There is a special symbol τ in $\mathcal{A}$ for silent transition. The set $\mathcal{A} \setminus \{\tau\}$ is ranged over by a, b, c, d. The transition rules in Δ are of the form $X \xrightarrow{\ell} P$. The following labeled transition rules define the operational semantics of the processes.

$$\frac{X \xrightarrow{\ell} P \in \Delta}{X \xrightarrow{\ell} P} \qquad \frac{P \xrightarrow{\ell} P'}{PQ \xrightarrow{\ell} P'Q} \qquad \frac{P \xrightarrow{\ell} P'}{P \| Q \xrightarrow{\ell} P' \| Q} \qquad \frac{Q \xrightarrow{\ell} Q'}{P \| Q \xrightarrow{\ell} P \| Q'}$$

The operational semantics is structural, meaning that $PQ \xrightarrow{\ell} P'Q$, $P \| Q \xrightarrow{\ell} P' \| Q$ and $Q \| P \xrightarrow{\ell} Q \| P'$ whenever $P \xrightarrow{\ell} P'$. We write $\Longrightarrow$ for the reflexive transitive closure of $\xrightarrow{\tau}$, and $\xRightarrow{\hat{\ell}}$ for $\Longrightarrow \xrightarrow{\ell} \Longrightarrow$ if $\ell \neq \tau$ and for $\Longrightarrow$ otherwise.

A *one counter net* $\mathcal{M}$ is a 4-tuple $(\mathcal{Q}, X, \mathcal{A}, \Delta)$, where $\mathcal{Q}$ is a finite set of states ranged over by p, q, r, s, X represents a place, $\mathcal{A}$ is a finite set of actions as in a process algebra, and Δ is a finite set of transition rules. A *process* defined by $\mathcal{M}$ is of the form pX^n, where n indicates the number of tokens in X. A transition rule in Δ is of the form $pX^i \xrightarrow{\ell} qX^j$ with $i < 2$. The semantics is structural in the sense that $pX^{i+k} \xrightarrow{\ell} qX^{j+k}$ whenever $pX^i \xrightarrow{\ell} qX^j$. A process P defined in $\mathcal{P}$, respectively $\mathcal{M}$, is *normed* if $\exists \ell_1, \ldots, \ell_n.P \xrightarrow{\ell_1} \ldots \xrightarrow{\ell_n} \epsilon$, respectively $\exists \ell_1, \ldots, \ell_n, p.(P \xrightarrow{\ell_1} \ldots \xrightarrow{\ell_n} p) \wedge \forall \ell, Q.\neg(p \xrightarrow{\ell} Q)$. We say that $\mathcal{P}/\mathcal{M}$ is normed if only normed processes are definable in it. We write (n)PA for the (normed) Process algebras and (n)OCN for the (normed) One Counter Nets.

In the presence of silent actions two well known process equalities are the weak bisimilarity [15] and the branching bisimilarity [24].

Definition 1. *A relation $\mathcal{R}$ is a* weak bisimulation *if the following are valid:*

1. *Whenever $P\mathcal{R}Q$ and $P \xrightarrow{\ell} P'$, then $Q \xRightarrow{\hat{\ell}} Q'$ and $P'\mathcal{R}Q'$ for some Q'.*

2. *Whenever $P\mathcal{R}Q$ and $Q \xrightarrow{\ell} Q'$, then $P \xRightarrow{\hat{\ell}} P'$ and $P'\mathcal{R}Q'$ for some P'.*
The weak bisimilarity $\approx$ *is the largest weak bisimulation.*

Definition 2. *A relation $\mathcal{R}$ is a* branching bisimulation *if the following hold:*

1. *Whenever $P\mathcal{R}Q$ and $P \xrightarrow{\ell} P'$, then either (i) $Q \Longrightarrow Q'' \xrightarrow{\ell} Q'$ and $P'\mathcal{R}Q'$ and $P\mathcal{R}Q''$ for some Q', Q'' or (ii) $\ell = \tau$ and $P'\mathcal{R}Q$.*

2. *Whenever $P\mathcal{R}Q$ and $Q \xrightarrow{\ell} Q'$, then either (i) $P \Longrightarrow P'' \xrightarrow{\ell} P'$ and $P'\mathcal{R}Q'$ and $P''\mathcal{R}Q$ for some P', P'' or (ii) $\ell = \tau$ and $P\mathcal{R}Q'$.*
The branching bisimilarity $\simeq$ *is the largest branching bisimulation.*

The following lemma, first noticed by van Glabbeek and Weijland [24], plays a fundamental role in the study of branching bisimilarity.

Lemma 1. *If* $P \Longrightarrow P' \Longrightarrow P'' \simeq P$ *then* $P' \simeq P$.

Let $\approx$ be a process equivalence. A silent action $P \xrightarrow{\tau} P'$ is *state preserving* with regards to $\approx$, notation $P \to P'$, if $P' \approx P$; it is *change-of-state* with regards to $\approx$, notation $P \xrightarrow{\iota} P'$, if $P' \not\approx P$. The reflexive and transitive closure of $\to$ is denoted by $\to^*$. Branching bisimilarity strictly refines weak bisimilarity in the sense that only state preserving silent actions can be ignored; a change-of-state must be explicitly bisimulated. Suppose that $P \simeq Q$ and $P \xrightarrow{\ell} P'$ is matched by the transition sequence $Q \xrightarrow{\tau} \cdots \xrightarrow{\tau} Q_i \xrightarrow{\tau} \cdots \xrightarrow{\tau} Q'' \xrightarrow{\ell} Q'$. By definition one has $P \simeq Q''$. It follows from Lemma 1 that $P \simeq Q_i$, meaning that all silent actions in $Q \Longrightarrow Q''$ are necessarily state preserving. This property fails for the weak bisimilarity as the following example demonstrates.

Example 1. Consider the transition system $\{P \xrightarrow{b} \epsilon, \; P \xrightarrow{\tau} P' \xrightarrow{a} \epsilon, \; P \xrightarrow{a} \epsilon; \; Q \xrightarrow{b} \epsilon, \; Q \xrightarrow{\tau} Q' \xrightarrow{a} \epsilon\}$. One has $P \approx Q$. However $P \not\simeq Q$ since $Q \not\simeq Q'$.

A game theoretic characterization of bisimilarity is by *bisimulation game* [22]. Suppose that a pair of processes P, Q, called a *configuration*, are defined in say a process algebra $(\mathcal{C}, \mathcal{A}, \Delta)$. A *branching bisimulation game* for the configuration (P, Q) is played between *Attacker* and *Defender*. The game is played in *rounds*. A new configuration is chosen after each round. Every round consists of three steps defined as follows, assuming (P_0, P_1) is the current configuration:

1. Attacker chooses $i \in \{0, 1\}$, $\ell \in \mathcal{A}$ and some process P_i' such that $P_i \xrightarrow{\ell} P_i'$.
2. Defender may respond in either of the following manner:
   - Choose some P_{1-i}', P_{1-i}'' such that $P_{1-i} \Longrightarrow P_{1-i}'' \xrightarrow{\ell} P_{1-i}'$.
   - Do nothing in the case that $\ell = \tau$.
3. Attacker decides which of (P_i, P_{1-i}''), (P_i', P_{1-i}') is the new configuration if Defender has played. Otherwise the new configuration must be (P_i', P_{1-i}).

In a *weak bisimulation game* a round consists of two steps. The first step is the same as above. In the second step Defender chooses some P_{1-i}' and some transition sequence $P_{1-i} \xRightarrow{\hat{\ell}} P_{1-i}'$. The game then continues with (P_i', P_{1-i}').

Defender wins a game if it never gets stuck; otherwise Attacker wins. We say that Defender/Attacker has a *winning strategy* if it can always win no matter how the opponent plays. The following lemma is well known, a clever use of which often simplifies bisimulation argument considerably.

Lemma 2. *Defender has a winning strategy in the branching, respectively weak, bisimulation game starting from the configuration (P, Q) if and only if $P \simeq Q$, respectively $P \approx Q$.*

Attacker has a winning strategy for the branching bisimulation game of the pair P, Q defined in Example 1. It simply chooses $P \xrightarrow{a} \epsilon$. If Defender chooses $Q \xrightarrow{\tau} Q' \xrightarrow{a} \epsilon$, Attacker chooses the configuration (P, Q') and wins. Defender can win the weak bisimulation game of (P, Q) though.

3   Defender's Forcing with Delayed Justification

A powerful technique for proving lower bounds for bisimilarity checking problem is Defender's Forcing described by Jančar and Srba in [11]. The basic idea is to force Attacker to make a particular choice in a bisimulation game by introducing enough copycat rules. An application of the technique to weak bisimulation game should be careful since both Attacker and Defender can take advantage of silent transitions. The design of a branching bisimulation game is even more subtle. In such a game a sequence of silent transitions used by Defender, except possibly the last one, must all be state preserving. A useful technique, motivated by Lemma 1, is to make use of generating processes. The process G defined by the rules $G \xrightarrow{\tau} GX$ and $GX \xrightarrow{\tau} G$ is $generating$ due to the fact that every process that G may evolve into, say GX^n, is branching bisimilar to G. The presence of other transition rules for G and X would not change the fact that $G \simeq GX^n$ for all n. This technique has already been used in the design of weak bisimulation games [11,14]. The relations these games give rise to are not branching bisimulation because a state-preserving transition may be simulated by a change-of-state silent transition. In what follows we use a small example to expose the subtlety of branching bisimulation game and the technique to apply Defender's Forcing in such a game.

Mayr proved in [14] a general result that the weak bisimilarity is undecidable for any model that subsumes nOCN. The lower bound is achieved by reducing from the halting problem of Minsky machine. A Minsky machine $\mathcal{M}$ with two counters c_1, c_2 is a program of the form $1 : I_1; 2 : I_2; \ldots; m-1 : I_{m-1}; m :$ halt, where for each $i \in \{1, \ldots, m-1\}$ the instruction I_i is in either of the following forms, assuming $j, k \in \{1, \ldots, m-1\}$ and $e \in \{1, 2\}$,

- $c_e := c_e + 1$ and then goto j.
- if $c_e = 0$ then goto j; otherwise $c_e := c_e - 1$ and then goto k.

By encoding a pair of numbers (n_1, n_2) by Gödel number of the form $2^{n_1}3^{n_2}$, Mayr implemented the increment and decrement operations on the counters by multiplying and dividing by 2 and 3 respectively. The central part of Mayr's proof is to show that it is possible to encode these operations and test for divisibility by constant into weak bisimulation games on nOCN. We shall show that Mayr's reduction can be strengthened to produce reductions to branching bisimulation games on nOCN. For every instruction "$i : I_i$" of a Minsky machine $\mathcal{M}$ a pair of states p_i, p'_i are introduced. Suppose "$i : c_2 := c_2 + 1$; goto j" is the i-th instruction of $\mathcal{M}$. The instruction is translated to the rules given in Fig. 2. The model defined in Fig. 2 is open-ended. Transition rules associated to p_j and p'_j are not given. We have however the following interesting property.

Lemma 3. *Let* $n = 2^{n_1}3^{n_2}$ *for some* n_1, n_2. *Defender of the branching bisimulation game of* $(p_j X^{3n}, p'_j X^{3n})$ *has a winning strategy if and only if Defender of the branching bisimulation game of* $(p_i X^n, p'_i X^n)$ *has a winning strategy.*

Proof. The crucial point here is that the copycat rules $p_i \xrightarrow{\tau} G'$ and $p'_i \xrightarrow{\tau} G'$, which syntactically identify what $p_i X^n$ and $p'_i X^n$ may reach in one silent step,

$p_i \xrightarrow{\tau} G'$	$p_i' \xrightarrow{\tau} G'$
$p_i \xrightarrow{a} q_1$	$G' \xrightarrow{a} q_1'$, $G' \xrightarrow{\tau} G'X$, $G'X \xrightarrow{\tau} G'$
$q_1 \xrightarrow{a} q_2$	$q_1' \xrightarrow{a} q_2'$
$q_1 \xrightarrow{t} t_3$	$q_1' \xrightarrow{t} t_1$
$q_2 \xrightarrow{\tau} G$	$q_2' \xrightarrow{\tau} G$
$G \xrightarrow{\tau} GX$, $GX \xrightarrow{\tau} G$, $G \xrightarrow{a} q_3$	$q_2' \xrightarrow{a} q_3'$
$q_3 \xrightarrow{a} p_j$	$q_3' \xrightarrow{a} p_j'$
$q_3 \xrightarrow{t} t_1$	$q_3' \xrightarrow{t} t_1$
$t_3X \xrightarrow{c} t''X$, $t''X \xrightarrow{c} t'X$, $t'X \xrightarrow{c} t_3$	$t_1X \xrightarrow{c} t_1$

Fig. 2. Multiplication Operation on Counter in OCN

do not automatically create a Defender's Forcing situation. The reason is that although $p_i'X^n \to G'X^n$, since $p_i'X^n \xrightarrow{\tau} G'X^n$ is the only action of $p_i'X^n$, it might well be that $p_iX^n \xrightarrow{\iota} G'X^n$. For branching bisimulation syntactical Defender's Forcing is insufficient. One needs Defender's Forcing that works at semantic level. Let's take a look at the development of the game in some detail.

1. If Attacker plays $p_iX^n \xrightarrow{\tau} G'X^n$, Defender plays $p_i'X^n \xrightarrow{\tau} G'X^n$. By Lemma 1 this response is equivalent to any other response from Defender.
2. If Attacker chooses the action $p_iX^n \xrightarrow{a} q_1X^n$, Defender responds with $p_i'X^n \to G'X^n \to^* G'X^{3n} \xrightarrow{a} q_1'X^{3n}$, making use of Lemma 1. Attacker's optimal move is to choose $(q_1X^n, q_1'X^{3n})$ to be the next configuration.
3. Now Attacker would not do a t action since $t_3X^n \simeq t_1X^{3n}$. It chooses the action a and the new configuration $(q_2X^n, q_2'X^{3n})$.
4. Then we come to another semantic Defender's Forcing. If Attacker plays $q_2X^n \xrightarrow{\tau} GX^n$, Defender plays $q_2'X^n \xrightarrow{\tau} GX^{3n}$; and vice versa.
5. If Attacker chooses the transition $q_2'X^{3n} \xrightarrow{a} q_3'X^{3n}$, Defender's response is $q_2X^n \xrightarrow{\tau} GX^n \Longrightarrow GX^{3n} \xrightarrow{a} q_3X^{3n}$, exploiting again Lemma 1. Attacker's nontrivial choice of the new configuration is $(q_3X^{3n}, q_3'X^{3n})$.
6. Finally Attacker would not choose a t_1 action since $t_1X^{3n} \simeq t_1X^{3n}$. So after an a action, the configuration becomes $(q_jX^{3n}, q_j'X^{3n})$.

It is easy to see that the configuration $(q_jX^{3n}, q_j'X^{3n})$ is optimal for both Attacker and Defender. If $q_jX^{3n} \simeq q_j'X^{3n}$ then Defender's Forcing described above is justified. If $q_jX^{3n} \not\simeq q_j'X^{3n}$ the forcing is ineffective since Attacker can choose to play $p_iX^n \xrightarrow{\tau} G'X^n$ and wins.   □

The main result of the section follows easily from Lemma 3 and its proof.

Theorem 1. *On nOCN every relation $\mathcal{R}$ satisfying $\simeq \subseteq \mathcal{R} \subseteq \approx$ is undecidable.*

Proof. Dividing a number by a constant can be encoded in similar fashion. The rest of Mayr's reduction does not refer to any silent transitions. It follows that we can construct a reduction witnessing that "$\mathcal{M}$ halts iff $p_1X \not\simeq p_1'X$". As a matter of fact the reduction supports the stronger correspondence stated as follows: "$\mathcal{M}$ halts iff $p_1X \not\approx p_1'X$".   □

4   Undecidability of nPA

Following [19], our main undecidability result is proved by reducing PCP (Post's Correspondence Problem) to the branching bisimilarity checking problem on nPA. Suppose Σ is a finite set of symbols and Σ^+ is the set of nonempty finite strings over Σ. The size of Σ is at least two. PCP is defined as follows.

POST'S CORRESPONDENCE PROBLEM

Input: $\{(u_1, v_1), (u_2, v_2) \ldots (u_n, v_n) \mid u_i, v_i \in \Sigma^+\}$.
Problem: Are there $i_1, i_2, \ldots i_m \in \{1, 2, \ldots, n\}$ with $m \geq 1$
such that $u_{i_1} u_{i_2} \ldots u_{i_m} = v_{i_1} v_{i_2} \ldots v_{i_m}$?

We will fix a PCP instance INST=$\{(u_1, v_1), (u_2, v_2) \ldots (u_n, v_n) \mid u_i, v_i \in \Sigma^+\}$ in this section. Our task is to construct a normed process algebra $\mathcal{G}=(\mathcal{C}, \mathcal{A}, \Delta)$ containing two process constants X, Y that render true the following equivalence.

$$\text{"INST has a solution" iff } X \simeq Y \text{ iff } X \approx Y. \tag{1}$$

We will prove (1) by validating the following statements:

- "If INST has a solution then $X \simeq Y$". This is Lemma 6 of Section 4.4.
- "If INST has no solution then $X \not\approx Y$". This is Lemma 7 of Section 4.4.

As $X \simeq Y$ implies $X \approx Y$, the main theorem of the paper follows from (1).

Theorem 2. *On* nPA *every relation* $\mathcal{R}$ *satisfying* $\simeq \subseteq \mathcal{R} \subseteq \approx$ *is undecidable.*

In the rest of the section, we firstly define $\mathcal{G}$, and then argue in several steps how the game based on $\mathcal{G}$ works in Defender's favour if INST has a solution.

4.1   The nPA Game

The construction of $\mathcal{G} = (\mathcal{C}, \mathcal{A}, \Delta)$ from INST is based on Srba's reduction [19]. Substantial amount of redesigning effort is necessary to make it work for the *branching* bisimilarity on the *normed* PA. The set $\mathcal{A}$ of actions is defined by

$$\mathcal{A} = \Lambda \cup \mathcal{N} \cup \Sigma \cup \{\tau\},$$

where $\Lambda = \{\lambda_U, \lambda_V, \lambda_D, \lambda_I, \lambda_S, \lambda_Z\}$, $\mathcal{N} = \{1, \ldots, n\}$ and Σ, n are from INST. The set $\mathcal{C}$ of process constants is defined by

$$\mathcal{C} = \{X, Y, Z, I, S, C, C', D, G, G', G_u, G_v, G'_v\} \cup \mathcal{U} \cup \mathcal{V} \cup \mathcal{W},$$
$$\mathcal{U} = \{U_i \mid i \in \mathcal{N}\},$$
$$\mathcal{V} = \{V_i \mid i \in \mathcal{N}\},$$
$$\mathcal{W} = \{W(\omega, i), W(\omega, 0) \mid \omega \in (\mathcal{SF}(u_i) \cup \mathcal{SF}(v_i)) \text{ and } i \in \mathcal{N}\},$$

where for each $\omega \in \Sigma^*$, the notation $\mathcal{SF}(\omega)$ stands for the set of suffixes of ω. The set of transition rules is given in Fig. 3. It is clear from these rules that $\mathcal{G}$ is indeed normed. In particular $P \Longrightarrow \epsilon$ for all $P \in \mathcal{U} \cup \mathcal{V} \cup \mathcal{W}$.

We write $\mathbb{P}_u$, respectively $\mathbb{P}_v$, for a sequential composition of members of $\mathcal{U}$, respectively $\mathcal{V}$. Similarly we write $\mathbb{P}$, respectively $\mathbb{Q}$, for a sequential composition of members of $\mathcal{U} \cup \mathcal{V}$, respectively $\mathcal{U} \cup \mathcal{V} \cup \mathcal{W}$. If for example the sequence u is empty, $\mathbb{P}_u$ is understood to denote ϵ.

$$X \xrightarrow{\lambda_U} D \,\|\, G_v, \ X \xrightarrow{\tau} D; \quad Y \xrightarrow{\tau} D; \quad D \xrightarrow{\tau} D \,\|\, G_u, \ D \xrightarrow{\lambda_D} C;$$

$$G_u \xrightarrow{\tau} G_u U_i, \ G_u \xrightarrow{\lambda_U} G_v U_i; \quad G_u \xrightarrow{\tau} G_v', \ G_v' \xrightarrow{\tau} G_v' V_i, \ G_v' \xrightarrow{\tau} Z;$$

$$G_v \xrightarrow{\tau} G_v V_i, \ G_v \xrightarrow{\tau} \epsilon, \ G_v \xrightarrow{\lambda_V} Z; \quad Z \xrightarrow{\tau} \epsilon, \ Z \xrightarrow{\lambda_Z} \epsilon;$$

$$C \xrightarrow{\lambda_I} I, \ C \xrightarrow{\lambda_S} S, \ C \xrightarrow{\tau} C \,\|\, G, \ C \xrightarrow{\tau} C \,\|\, G_v;$$
$$G \xrightarrow{\tau} G U_i, \ G \xrightarrow{\tau} G V_i, \ G \xrightarrow{\tau} \epsilon;$$

$$I \xrightarrow{\lambda_I} C', \ I \xrightarrow{i} I; \quad S \xrightarrow{\lambda_S} C', \ S \xrightarrow{a} S; \quad C' \xrightarrow{\tau} C' \,\|\, G', \ C' \xrightarrow{\tau} \epsilon;$$
$$G' \xrightarrow{\tau} G' U_i, \ G' \xrightarrow{\tau} G' V_i, \ G' \xrightarrow{\tau} G' W, \ G' \xrightarrow{\tau} G_v, \ G' \xrightarrow{\tau} Z;$$

$$U_i \xrightarrow{\tau} W(u_i, i), \ V_i \xrightarrow{\tau} W(v_i, i);$$
$$W(a\omega, i) \xrightarrow{a} W(\omega, i), \ W(a\omega, 0) \xrightarrow{a} W(\omega, 0), \ W(\omega, i) \xrightarrow{i} W(\omega, 0),$$
$$W(a\omega, i) \xrightarrow{\tau} W(\omega, i), \ W(a\omega, 0) \xrightarrow{\tau} W(\omega, 0), \ W(\omega, i) \xrightarrow{\tau} W(\omega, 0), \ W(\epsilon, 0) \xrightarrow{\tau} \epsilon.$$

In the above rules, i ranges over $\{1, \ldots, n\}$, a ranges over Σ, and W ranges over $\mathcal{W}$.

Fig. 3. Transition Rules for the nPA Game

4.2 Defender's Generator

To explain how the reduction works we start with the generators introduced by the process algebra. A generator should be able to not only produce what is necessary but also do away with what has been produced. The process D for instance can induce circular silent transition sequence of the form

$$D \xrightarrow{\tau} D \,\|\, G_u \Longrightarrow D \,\|\, G_u \mathbb{P}_u \xrightarrow{\tau} D \,\|\, G_v' \mathbb{P}_u \Longrightarrow D \,\|\, G_v' \mathbb{P}_v \mathbb{P}_u \Longrightarrow D.$$

By Lemma 1 all the processes appearing in the above sequence are branching bisimilar. Notice that the only reason the process constant G_v' is introduced is to make available the above circular sequence. The constant G_v' is necessary because G_u cannot reach G_v via silent moves. Similar circular silent transition sequences are also available for C and C'.

Lemma 4. *Suppose $P \in \{D, C, C'\}$ and $P \Longrightarrow P \,\|\, Q$. Then $P \,\|\, Q \longrightarrow P$.*

Corollary 1. *The following equalities are valid for all $\mathbb{P}_u, \mathbb{P}_v, \mathbb{P}, \mathbb{Q}$.*

1. $D \simeq D \,\|\, G_u \mathbb{P}_u \simeq D \,\|\, G_v' \mathbb{P}_v \mathbb{P}_u \simeq D \,\|\, Z \mathbb{P}_v \mathbb{P}_u \simeq D \,\|\, \mathbb{P}_v \mathbb{P}_u \simeq D \,\|\, W \mathbb{P}_v \mathbb{P}_u;$
2. $C \simeq C \,\|\, G\mathbb{P} \simeq C \,\|\, \mathbb{P} \simeq C \,\|\, W\mathbb{P} \simeq C \,\|\, G_v \mathbb{P}_v;$
3. $C' \simeq C' \,\|\, G'\mathbb{Q} \simeq C' \,\|\, G_v \mathbb{Q} \simeq C' \,\|\, Z\mathbb{Q} \simeq C' \,\|\, \mathbb{Q}.$

It has been observed that generating transitions are the most tricky ones in decidability proofs [23,5,7]. Here they are used to Defender's advantage. A generator can start everything all over again from scratch. This gives Defender the ability to copy Attacker if the latter does not make a particular move.

The bisimulation game of (X, Y) is played in two phases. The generating phase comes first. During this phase Defender tries to produce a pair $\mathbb{P}_u, \mathbb{P}_v$, via Defender's Forcing using the generators, that encode a solution to INST. Next comes the checking phase in which Attacker tries to reject the pair $\mathbb{P}_u, \mathbb{P}_v$. In the light of the delayed effect of Defender's Forcing in branching bisimulation games, we will look at the two phases in reverse order.

4.3   Checking Phase

The processes U_i, V_i play two roles. One is to announce u_i, respectively v_i; the other is to reveal the index i. The first role can be suppressed by composing U_i, respectively V_i, with S while the second can be discharged by composing with I [19]. Since I, S are normed, Attacker can choose to remove I, respectively S. In our game the removal can be done by playing $I \xrightarrow{\lambda_I} C'$, respectively $S \xrightarrow{\lambda_S} C'$. According to (3) of Corollary 1 however Attacker would lose immediately if it plays $I \xrightarrow{\lambda_I} C'$, respectively $S \xrightarrow{\lambda_S} C'$, in a branching bisimulation game starting from $(I \parallel \mathbb{Q}, I \parallel \mathbb{Q}')$, respectively $(S \parallel \mathbb{Q}, S \parallel \mathbb{Q}')$. Notice that it is important for a process constant W to ignore the string/index information by doing silent transitions. Otherwise the interleaving between actions in Σ and actions in $\mathcal{N}$ would defeat Defender's attempt to prove string/index equality.

Lemma 5. *Suppose* $\mathbb{U} = U_{i_1} U_{i_2} \dots U_{i_l}$, $\mathbb{V} = V_{j_1} V_{j_2} \dots V_{j_r}$ *and* $B \in \{\epsilon, Z, G_v\}$. *The following statements are valid, where* $\approxeq \, \in \{\simeq, \approx\}$.

1. $I \parallel B\mathbb{P}\mathbb{U} \approxeq I \parallel B\mathbb{P}\mathbb{V}$ *if and only if* $u_{i_1} u_{i_2} \dots u_{i_l} = v_{j_1} v_{j_2} \dots v_{j_r}$.
2. $S \parallel B\mathbb{P}\mathbb{U} \approxeq S \parallel B\mathbb{P}\mathbb{V}$ *if and only if* $i_1 i_2 \dots i_l = j_1 j_2 \dots j_r$.

Proof. Suppose $I \parallel B\mathbb{P}\mathbb{U} \simeq I \parallel B\mathbb{P}\mathbb{V}$ and w.l.o.g. $|u_{i_1} u_{i_2} \dots u_{i_l}| \geq |v_{j_1} v_{j_2} \dots v_{j_r}|$. An action sequence from $I \parallel B\mathbb{P}\mathbb{U}$ to $I \parallel \mathbb{U}$ must be simulated essentially by an action sequence from $I \parallel B\mathbb{P}\mathbb{V}$ to $I \parallel \mathbb{V}$. But then $u_{i_1} u_{i_2} \dots u_{i_l} = v_{j_1} v_{j_2} \dots v_{j_r}$ can be derived from $I \parallel \mathbb{U} \simeq I \parallel \mathbb{V}$. The converse implication follows from the discussion in the above. The second equivalence can be proved similarly.   □

The following proposition, in which $\approxeq \, \in \{\simeq, \approx\}$, says that the constant C can be used to check both string equality and index equality by Attacker's forcing.

Proposition 1. *If* $\mathbb{U} = U_{i_1} U_{i_2} \dots U_{i_l}$ *and* $\mathbb{V} = V_{j_1} V_{j_2} \dots V_{j_r}$, *then for all* $\mathbb{P}$, $C \parallel Z\mathbb{P}\mathbb{U} \approxeq C \parallel Z\mathbb{P}\mathbb{V}$ *iff* $i_1 i_2 \dots i_l = j_1 j_2 \dots j_r$ *and* $u_{i_1} u_{i_2} \dots u_{i_l} = v_{j_1} v_{j_2} \dots v_{j_r}$.

Proof. In one direction we prove that $C \parallel Z\mathbb{P}\mathbb{U} \approx C \parallel Z\mathbb{P}\mathbb{V}$ implies $i_1 i_2 \dots i_l = j_1 j_2 \dots j_r$ and $u_{i_1} u_{i_2} \dots u_{i_l} = v_{j_1} v_{j_2} \dots v_{j_r}$. If $i_1 i_2 \dots i_l \neq j_1 j_2 \dots j_r$, then Attacker chooses $C \parallel Z\mathbb{P}\mathbb{U} \xrightarrow{\lambda_S} S \parallel Z\mathbb{P}\mathbb{U}$. Defender cannot invoke the action $Z \xrightarrow{\tau} \epsilon$ for otherwise an λ_Z action cannot be performed before an λ_V action. The process constant Z is introduced precisely for this blocking effect. Defender's play must be of the form $C \parallel Z\mathbb{P}\mathbb{V} \Longrightarrow C \parallel Q \parallel Z\mathbb{P}\mathbb{V} \xrightarrow{\lambda_S} S \parallel Q \parallel Z\mathbb{P}\mathbb{V} \Longrightarrow S \parallel Q' \parallel Z\mathbb{P}\mathbb{V}$. If Q' can perform any one of $\{\lambda_V, \lambda_Z\} \cup \mathcal{N}$, Attacker wins since S can do none of those. If Q' can do none of those actions, then $S \simeq S \parallel Q'$. By Lemma 5 Attacker has a winning strategy for the weak bisimulation game $(S \parallel Z\mathbb{P}\mathbb{U}, S \parallel Q' \parallel Z\mathbb{P}\mathbb{V})$. If $u_{i_1} u_{i_2} \dots u_{i_l} \neq v_{j_1} v_{j_2} \dots v_{j_r}$, the argument is similar.

Conversely we prove that $i_1 i_2 \dots i_l = j_1 j_2 \dots j_r \wedge u_{i_1} u_{i_2} \dots u_{i_l} = v_{j_1} v_{j_2} \dots v_{j_r}$ implies $C \parallel Z\mathbb{P}\mathbb{U} \simeq C \parallel Z\mathbb{P}\mathbb{V}$. This is done by showing that the relation

$$\left\{ (C \parallel Q \parallel Z\mathbb{P}\mathbb{U}, C \parallel Q \parallel Z\mathbb{P}\mathbb{V}) \, \middle| \, \begin{matrix} i_1 i_2 \dots i_l = j_1 j_2 \dots j_r \\ u_{i_1} u_{i_2} \dots u_{i_l} = v_{j_1} v_{j_2} \dots v_{j_r}. \end{matrix} \right\} \cup \simeq$$

is a branching bisimulation.   □

4.4   Generating Phase

Suppose that INST has a solution $i_1, i_2, \ldots, i_k$. Fix the following abbreviations: $\mathbb{U}^- = U_{i_2} \ldots U_{i_k}$, $\mathbb{U} = U_{i_1} \mathbb{U}^-$ and $\mathbb{V} = V_{i_1} V_{i_2} \ldots V_{i_k}$. We will argue that Defender has a winning strategy in the branching bisimulation game of (X, Y). Defender's basic idea is to produce the pair $\mathbb{U}, \mathbb{V}$ by forcing. Its strategy and Attacker's counter strategy are described below.

(i) By Defender's Forcing Attacker plays $X \xrightarrow{\lambda_U} D \parallel G_v$. Defender proposes $\mathbb{U}$ via the transitions $Y \xrightarrow{\tau} D \xrightarrow{\tau} D \parallel G_u \Longrightarrow D \parallel G_u \mathbb{U}^- \xrightarrow{\lambda_U} D \parallel G_v \mathbb{U}$. The use of an explicit action λ_U guarantees that $\mathbb{U}$ is *nonempty*. Now Attacker has a number of configurations to choose from. But by (1) of Corollary 1, it all boils down to choosing $(D \parallel G_v, D \parallel G_v \mathbb{U})$.

(ii) Due to (1) of Corollary 1 Attacker would not remove G_v using either $G_v \xrightarrow{\tau} \epsilon$ or $G_v \xrightarrow{\lambda_V} Z$. It can generate an element of $\mathcal{V}$ using G_v. It can do an action induced by D or a descendant of D. Defender simply copycats Attacker's actions. The configuration stays in the form $(D \parallel Q \parallel G_v \mathbb{P}_v, D \parallel Q \parallel G_v \mathbb{P}_v \mathbb{U})$.

(iii) To have any chance to win, Attacker must try the action λ_D. Defender does the same action. The configuration becomes $(C \parallel Q \parallel G_v \mathbb{P}_v, C \parallel Q \parallel G_v \mathbb{P}_v \mathbb{U})$. At this point if Attacker plays a harmless action, Defender can copycat the action; and the configuration stays in the same shape.

(iv) An important observation is that if Attacker plays $C \parallel Q \parallel G_v \mathbb{P}_v \xrightarrow{\ell} P_1$, Defender can play $C \parallel Q \parallel G_v \mathbb{P}_v \mathbb{U} \Longrightarrow C \parallel Q \Longrightarrow C \parallel Q \parallel G_v \mathbb{P}_v \xrightarrow{\ell} P_1$ and wins. Here $C \parallel Q \simeq C \parallel Q \parallel G_v \mathbb{P}_v$ by (2) of Corollary 1. To see that the assumptions $i_1 i_2 \ldots i_l = j_1 j_2 \ldots j_r$ and $u_{i_1} u_{i_2} \ldots u_{i_l} = v_{j_1} v_{j_2} \ldots v_{j_r}$ imply $C \parallel Q \parallel G_v \mathbb{P}_v \mathbb{U} \simeq C \parallel Q$, notice that $C \parallel Q \parallel G_v \mathbb{P}_v \mathbb{U} \Longrightarrow C \parallel Q \Longrightarrow C \parallel Q \parallel G_v \mathbb{P}_v \mathbb{V}$ and that $C \parallel Q \parallel G_v \mathbb{P}_v \mathbb{U} \simeq C \parallel Q \parallel G_v \mathbb{P}_v \mathbb{V}$ is a corollary of Proposition 1. Thus Attacker would choose $C \parallel Q \parallel G_v \mathbb{P}_v \mathbb{U}$ to continue.

(v) Attacker would not play $C \parallel Q \parallel G_v \mathbb{P}_v \mathbb{U} \xrightarrow{\tau} C \parallel Q \parallel \mathbb{P}_v \mathbb{U}$ because it would lose right away according to (2) of Corollary 1.

(vi) By Lemma 5 Attacker would not do a λ_I action or a λ_S action. It stands the best chance to play $C \parallel Q \parallel G_v \mathbb{P}_v \mathbb{U} \xrightarrow{\lambda_V} C \parallel Q \parallel Z \mathbb{P}_v \mathbb{U}$. The counter play from Defender is $C \parallel Q \parallel G_v \mathbb{P}_v \Longrightarrow C \parallel Q \parallel G_v \mathbb{P}_v \mathbb{V} \xrightarrow{\lambda_V} C \parallel Q \parallel Z \mathbb{P}_v \mathbb{V}$.

The last configuration $(C \parallel Q \parallel Z \mathbb{P}_v \mathbb{V}, C \parallel Q \parallel Z \mathbb{P}_v \mathbb{U})$ is optimal for Attacker. By Proposition 1 Defender has a winning strategy for the branching bisimulation game of $(C \parallel Q \parallel Z \mathbb{P}_v \mathbb{V}, C \parallel Q \parallel Z \mathbb{P}_v \mathbb{U})$. Hence the following lemma.

Lemma 6. *If INST has a solution then* $X \simeq Y$.

The converse of Lemma 6 also holds. In fact a stronger result is obtainable. In the weak bisimulation game of (X, Y), Attacker has a strategy to force the game to reach a configuration that is essentially of the form $(C \parallel Z \mathbb{P}'_v, C \parallel Z \mathbb{P}_v \mathbb{P}_u)$, where $\mathbb{P}_u \neq \epsilon$. If there is no solution to INST, Proposition 1 implies $C \parallel Z \mathbb{P}'_v \not\simeq C \parallel Z \mathbb{P}_v \mathbb{P}_u$. It follows that Attacker has a winning strategy for the weak bisimulation game of (X, Y).

Lemma 7. *If INST has no solution then* $X \not\simeq Y$.

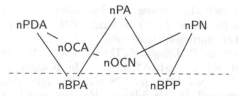

Fig. 4. Decidability Border for Branching Bisimilarity on Normed PRS

5 Conclusion

Putting together the results derived in this paper, we see that there is a decidability border in the normed PRS hierarchy, see Fig. 4. The branching bisimilarity

1. is undecidable on all normed models above either nBPA or nBPP, and
2. is decidable for both nBPP and nBPA [5,7].

We have confirmed that the first statement is valid for the weak bisimilarity, which slightly strengthens the results obtained in [12]. In fact the statement is valid for every relation between the branching bisimilarity and the weak bisimilarity. It has been conjectured that the second statement is also true for the weak bisimilarity. The answers however have remained a secret for us up to now.

Tighter complexity bounds, or even completeness characterizations, would be very welcome. Another avenue for further study is based on the observation that although the undecidability results of both the present paper and the paper of Jančar and Srba [11] are about the same models, the degrees of undecidability are most likely to be different. In [11] it is pointed out that by constraining the silent actions of nPDA, say to ϵ-popping or ϵ-pushing silent moves, the degree of undecidability of the weak bisimilarity goes from the analytic hierarchy down to the arithmetic hierarchy. It is therefore a reasonable hope that the same restriction may lead to decidable results for the branching bisimilarity on some PRS models. Further studies are called for.

Complete proofs of the results stated in this extended abstract can be found in the full paper [25].

Acknowledgement. We gratefully acknowledge the support of the National Science Foundation of China (61033002, ANR 61261130589, 91318301). We thank the anonymous referees and Patrick Totzke for their constructive suggestions.

References

1. Baeten, J.C.M., Bergstra, J.A., Klop, J.W.: Decidability of Bisimulation Equivalence for Processes Generating Context-free Languages. In: de Bakker, J.W., Nijman, A.J., Treleaven, P.C. (eds.) PARLE 1987. LNCS, vol. 259, pp. 94–111. Springer, Heidelberg (1987)
2. Burkart, O., Caucal, D., Moller, F., Steffen, B.: Verification on Infinite Structures. In: Handbook of Process Algebra. Elsevier Science (2001)
3. Christensen, S., Hirshfeld, Y., Moller, F.: Bisimulation Equivalence is Decidable for Basic Parallel Processes. In: Best, E. (ed.) CONCUR 1993. LNCS, vol. 715, pp. 143–157. Springer, Heidelberg (1993)

4. Christensen, S., Hüttel, H., Stirling, C.: Bisimulation Equivalence is Decidable for all Context-free Processes. In: Cleaveland, W.R. (ed.) CONCUR 1992. LNCS, vol. 630, pp. 138–147. Springer, Heidelberg (1992)
5. Czerwiński, W., Hofman, P., Lasota, S.: Decidability of Branching Bisimulation on Normed Commutative Context-free Processes. In: Katoen, J.-P., König, B. (eds.) CONCUR 2011. LNCS, vol. 6901, pp. 528–542. Springer, Heidelberg (2011)
6. De Nicola, R., Montanari, U., Vaandrager, F.: Back and Forth Bisimulations. In: Baeten, J.C.M., Klop, J.W. (eds.) CONCUR 1990. LNCS, vol. 458, pp. 152–165. Springer, Heidelberg (1990)
7. Fu, Y.: Checking Equality and Regularity for Normed BPA with Silent Moves. In: Fomin, F.V., Freivalds, R., Kwiatkowska, M., Peleg, D. (eds.) ICALP 2013, Part II. LNCS, vol. 7966, pp. 238–249. Springer, Heidelberg (2013)
8. Hirshfeld, Y., Jerrum, M.: Bisimulation Equivanlence is Decidable for Normed Process Algebra. In: Wiedermann, J., Van Emde Boas, P., Nielsen, M. (eds.) ICALP 1999. LNCS, vol. 1644, pp. 412–421. Springer, Heidelberg (1999)
9. Jančar, P.: Undecidability of Bisimilarity for Petri Nets and Some Related Problems. Theoretical Computer Science 148, 281–301 (1995)
10. Jančar, P., Brics, J.S.: Highly Undecidable Questions for Process Algebras. In: Levy, J.-J., Mayr, E.W., Mitchell, J.C. (eds.) TCS 2004. IFIP, vol. 155, pp. 507–520. Springer, Boston (2004)
11. Jančar, P., Srba, J.: Undecidability of Bisimilarity by Defender's Forcing. Journal of the ACM 55, 1–26 (2008)
12. Kučera, A., Jančar, P.: Equivalence-Checking on Infinite-State Systems: Techniques and Results. Theory and Practice of Logic Programming 6, 227–264 (2006)
13. Mayr, R.: Process Rewrite Systems. Information and Computation 156, 264–286 (2000)
14. Mayr, R.: Undecidability of Weak Bisimulation Equivalence for 1-Counter Processes. In: Baeten, J.C.M., Lenstra, J.K., Parrow, J., Woeginger, G.J. (eds.) ICALP 2003. LNCS, vol. 2719, pp. 570–583. Springer, Heidelberg (2003)
15. Milner, R.: Communication and Concurrency. Prentice Hall (1989)
16. Park, D.: Concurrency and Automata on Infinite Sequences. In: Deussen, P. (ed.) GI-TCS 1981. LNCS, vol. 104, pp. 167–183. Springer, Heidelberg (1981)
17. Srba, J.: Undecidability of Weak Bisimilarity for Pushdown Processes. In: Brim, L., Jančar, P., Křetínský, M., Kučera, A. (eds.) CONCUR 2002. LNCS, vol. 2421, pp. 579–594. Springer, Heidelberg (2002)
18. Srba, J.: Roadmap of Infinite Results. EATCS 78, 163–175 (2002)
19. Srba, J.: Undecidability of Weak Bisimilarity for PA-Processes. In: Ito, M., Toyama, M. (eds.) DLT 2002. LNCS, vol. 2450, pp. 197–208. Springer, Heidelberg (2003)
20. Srba, J.: Completeness Results for Undecidable Bisimilarity Problems. Electronic Notes in Computer Science 98, 5–19 (2004)
21. Stirling, C.: Decidability of Bisimulation Equivalence for Normed Pushdown Processes. Theoretical Computer Science 195, 113–131 (1998)
22. Stirling, C.: The Joys of Bisimulation. In: Brim, L., Gruska, J., Zlatuška, J. (eds.) MFCS 1998. LNCS, vol. 1450, pp. 142–151. Springer, Heidelberg (1998)
23. Stirling, C.: Decidability of Weak Bisimilarity for a Subset of Basic Parallel Processes. In: Honsell, F., Miculan, M. (eds.) FOSSACS 2001. LNCS, vol. 2030, pp. 379–393. Springer, Heidelberg (2001)
24. van Glabbeek, R., Weijland, W.: Branching Time and Abstraction in Bisimulation Semantics. Journal of ACM 43, 555–600 (1996)
25. Yin, Q., Fu, Y., He, C., Huang, M., Tao, X.: Branching Bisimilarity Checking for PRS (2014), http://arxiv.org/abs/1402.0050

Labeling Schemes for Bounded Degree Graphs

David Adjiashvili[1] and Noy Rotbart[2]

[1] Institute for Operations Research, ETH Rämistrasse 101,8092 Zürich, Switzerland
addavid@ethz.ch
[2] Department of Computer Science, University of Copenhagen
Universitetsparken 5, 2100 Copenhagen, Denmark
noyro@diku.dk

Abstract. We investigate adjacency labeling schemes for graphs of bounded degree $\Delta = O(1)$. In particular, we present an optimal (up to an additive constant) $\log n + O(1)$ adjacency labeling scheme for bounded degree trees. The latter scheme is derived from a labeling scheme for bounded degree outerplanar graphs. Our results complement a similar bound recently obtained for bounded depth trees [Fraigniaud and Korman, SODA 2010], and may provide new insights for closing the long standing gap for adjacency in trees [Alstrup and Rauhe, FOCS 2002]. We also provide improved labeling schemes for bounded degree planar graphs. Finally, we use combinatorial number systems and present an improved adjacency labeling schemes for graphs of bounded degree Δ with $(e+1)\sqrt{n} < \Delta \leq n/5$.

1 Introduction

A labeling scheme is a method of distributing the information about the structure of a graph among its vertices by assigning short *labels*, such that a selected function on pairs of vertices can be computed using only their labels. The quality of a labeling scheme is mostly measured by its *size*: that is, the maximum number of bits used in a label. Additional important attributes of labeling schemes are the running times of the label generation algorithm (*encoder*), and the decoding algorithm (*decoder*), which replies to a query given a pair of labels.

Among all labeling schemes, that of *adjacency* is perhaps the most fundamental, as it directly comprises an implicit representation of the graph. For a graph G and any two of its vertices u, v, the decoder of an adjacency labeling scheme is required to deduce whether u and v are adjacent in G directly from their labels. Adjacency queries for bounded degree graphs appear naturally in networks of small dilation [4], peer-to-peer (P2P) [18] and wireless ad-hoc networks [22].

Our main contribution is an optimal (up to an additive constant) $\log n + O(1)$ size adjacency labeling scheme for bounded-degree outerplanar graphs. As a special case thereof, we obtain an optimal labeling scheme for bounded degree trees. We summarize this result in the following theorem.

J. Esparza et al. (Eds.): ICALP 2014, Part II, LNCS 8573, pp. 375–386, 2014.
© Springer-Verlag Berlin Heidelberg 2014

Theorem 1. *For every fixed $\Delta \geq 1$, the class $\mathcal{O}(n, \Delta)$ of bounded-degree outerplanar graphs admits an adjacency labeling scheme of size $\log n + O(1)$, with encoding complexity $O(n \log n)$ and decoding complexity $O(\log \log n)$.*

Our labeling scheme utilizes a novel technique based on *edge*-universal graphs[1] for bounded degree outerplanar graphs. Unlike other results in the field which rely on a tight connection to induced-universal graphs[2] [1,3,14,17], our technique embeds the input graph into a small edge-universal graph. Moreover, to the best of our knowledge, our labeling scheme is the first to use the total label size to separate the different components of the label. In contrast, other labeling schemes, such as [21,2], introduce an extra overhead to support such separation.

Kannan, Naor and Rudich [15] showed that if a graph class $\mathcal{R}$ admits an adjacency labeling scheme with maximum label length $g(n)$, then there exists an induced-universal graph with $2^{g(n)}$ vertices for $\mathcal{R}$, efficiently constructible from the labeling scheme. The opposite relation holds in a weaker sense. The existence of an induced-universal graph with $2^{g(n)}$ vertices for a family $\mathcal{R}$ of graphs implies the existence of labeling scheme with size $g(n)$. This transformation is however not efficient, namely the resulting scheme has exponential running time. In light of the existing linear size induced-universal graphs for bounded degree trees [9], our contribution is in devising an *efficient* labeling scheme of optimal size.

As a corollary of Theorem 1, we also obtain an efficient $(\lfloor \Delta/2 \rfloor + 1) \log n$ size labeling scheme for graphs with maximum degree Δ. For the case of bounded degree planar graphs we construct a $\frac{\lceil \Delta/2 \rceil + 1}{2} \log n$ size labeling scheme with average label size of $(1 + o(1)) \log n + O(\log \log n)$, improving the best known construction for $\Delta \leq 4$. Finally, we observe that a simple application of combinatorial number systems [16] gives an adjacency labeling scheme for all graphs with maximum degree $\Delta(n)$, improving the known bounds for $\Delta(n) \in [(e + 1)\sqrt{n}, n/5]$.

We summarize all known results for adjacency labeling schemes in Table 1, and our contributions in Table 2. Our results for outerplanar graphs and general graphs are presented in Section 2, and Section 3, respectively. *Our planar graph result and all technical proofs are deferred to the full version of the paper.*

Table 1. Best known adjacency labeling schemes for graphs with at most n vertices

Family	Upper bound	Lower bound	Encoding	Decoding	Ref.
Trees	$\log n + O(\log^* n)$	$\log n + \Omega(1)$	$O(n \log^* n)$	$O(\log^* n)$	[3]
Binary trees	$\log n + O(1)$	$\log n + \Omega(1)$	$O(n)$	$O(1)$	[5]
Bd. depth δ trees	$\log n + O(1)$	$\log n + \Omega(1)$	$O(n)$	$O(1)$	[11]
Bd. deg. Δ trees	$\log n + O(\log^* n)$	$\log n + \Omega(1)$	$O(n \log^* n)$	$O(\log^* n)$	[3]
Planar graphs	$2 \log n + O(\log \log n)$	$\log n + \Omega(1)$	$O(n)$	$O(1)$	[13]
Outerplanar graphs	$\log n + O(\log \log n)$	$\log n + \Omega(1)$	$O(n)$	$O(1)$	[13]
Bd. deg. $\Delta(n)$ graphs	$(\lfloor \frac{\Delta(n)}{2} \rfloor + 1) \log n + O(\log^* n)$	$\frac{\Delta(n)}{2} \log n$	$O(n)$	$O(\log^* n)$	[3]
Graphs	$\lfloor \frac{1}{2} n \rfloor + \lceil \log n \rceil$	$\lfloor \frac{1}{2} n \rfloor - 1$	$O(n)$	$O(1)$	[19]

---

[1] A graph G is *edge-universal* for a class $\mathcal{R}$ of graphs, if every graph in $\mathcal{R}$ appears as a subgraph in G (not necessarily induced).

[2] A graph G is *induced-universal* for a class $\mathcal{R}$ of graphs, if every graph in $\mathcal{R}$ appears as an induced subgraph in G.

Table 2. Our contribution for families of graphs with bounded degree Δ

Family	Upper bound	Tight	Encoding	Decoding	Ref.
Trees	$\log n + O(1)$	yes	$O(n \log n)$	$O(\log \log n)$	Cor. 1
Outerplanar	$\log n + O(1)$	yes	$O(n \log n)$	$O(\log \log n)$	Thm. 1
Planar ($\Delta \le 4$)	$\frac{3}{2} \log n + O(\log \log n)$	no	$O(n \log n)$	$O(\log \log n)$	
Graphs, $\Delta(n)$	$\log \binom{n}{\lceil \Delta(n)/2 \rceil} + 2\log n$	no	$O(n^2)$	$O(\Delta(n) \log n)$	Thm. 3
Graphs (unbounded)	$(\lfloor \frac{\Delta}{2} \rfloor + 1) \log n + O(1)$	no	$O(n \log n)$	$O(\log \log n)$	Cor. 2

1.1   Previous Work

Alstrup and Rauhe [3] proved that forests (and trees) in $\mathcal{G}(n)$ have an adjacency labeling scheme of $\log n + O(\log^* n)$.[3] Their technique uses a recursive decomposition of the tree which yields the same $\log n + O(\log^* n)$ label size for bounded degree trees. Fraigniaud and Korman [11] showed that bounded depth trees have a labeling scheme of size $\log n + O(1)$. Bonichon et al. [5] proved that caterpillars and binary trees enjoy a labeling scheme of size $\log(n) + O(1)$ using a method called "Traversal and Jumping". In a follow up paper, Bonichon et al. [6] claimed without proof that the aforementioned methods can be used to achieve the same bound for bounded degree trees. Chung [9] showed the existence of an induced-universal graph with $O(n)$ vertices for bounded degree trees.

Graphs with maximum degree Δ have *arboricity*[4] $k(\Delta) = \lfloor \Delta/2 \rfloor + 1$ [15] thus, by the theorem of Nash-Williams [20], they can be decomposed into $\lfloor \Delta/2 \rfloor + 1$ forests. Alstrup and Rauhe [3] combined this result with their labeling scheme for forests to obtain a labeling scheme of size $(\lfloor \Delta/2 \rfloor + 1) \log n + O(\log^* n)$ for bounded degree graphs.

Butler [7] constructed an induced-universal graph for graphs with maximum degree Δ with $O(n^{\lceil \frac{\Delta+1}{2} \rceil})$ vertices. The author notes that any induced-universal graph must have at least $cn^{\lfloor \Delta/2 \rfloor + 1}$ vertices for some c depending only on Δ, which implies that the bounds are optimal when Δ is even. For odd Δ, Esperet et al. [10] showed a smaller induced-universal graph with $O(n^{\lceil \Delta/2 \rceil - 1/\Delta} \log^{2+2/\Delta} n)$ vertices. It follows that there exists a labeling scheme for $\mathcal{G}(n, \Delta)$ of size $\frac{\Delta}{2} \log n$ bits for even Δ, and $(\lceil \frac{\Delta}{2} \rceil - 1/\Delta) \log n + \log(\log^{2+2/\Delta} n)$ bits for an odd Δ (but that is not necessary efficient). We summarize the best known bounds for adjacency labeling schemes in Table 1.

1.2   Preliminaries

For two integers $0 \le k_1 \le k_2$ we denote $[k_1] = \{1, \cdots, k_1\}$ and $[k_1, k_2] = \{k_1, \cdots, k_2\}$. A binary string x is a member of the set $\{0, 1\}^*$, and we denote its length by $|x|$. We denote the concatenation of two binary strings x, y by $x \circ y$.

For a graph G we denote its set of vertices and edges by $V(G)$ and $E(G)$, respectively. The family of all graphs is denoted $\mathcal{G}$. For any graph family $\mathcal{R}$, let

---

[3] $\log^* n$ denotes the iterated logarithm of n.

[4] The *arboricity* of a graph G is the minimum number of edge-disjoint acyclic subgraphs whose union is G.

$\mathcal{R}(n) \subseteq \mathcal{R}$ denote the subfamily containing the graphs of at most n vertices. The collection of graphs with bounded degree Δ in $\mathcal{R}(n)$ is denoted $\mathcal{R}(n, \Delta)$. The collection of planar graphs, outerplanar graphs, and trees, in $\mathcal{G}(n)$ is denoted $\mathcal{P}(n), \mathcal{O}(n)$ and $\mathcal{T}(n)$, resp. Unless otherwise stated, we assume hereafter Δ to be constant. To simplify the presentation we suppress all dependencies on Δ in all our bounds and running time estimations. All these dependencies can be computed and shown to be at most a multiplicative factor of $O(\Delta \log \Delta)$ times the claimed bounds. We defer the exact details to the journal version of the paper. Non constant bounds on the degree are denoted by $\Delta(n)$. We note that all results work for disconnected graphs. We assume trees to be rooted, and denote $\log_2 n$ as $\log n$. For a set of vertices $S \subset V(G)$ we define $G - S$ to be the graph obtained from G by removing the vertices in S and all incident edges. The set of edges in $G = (V, E)$ incident to a vertex $v \in V$ is denoted E_v.

Let $G = (V, E) \in \mathcal{G}(n)$, and let $u, v \in V$. The boolean function $adjacency(v, u)$, define over vertices in $G \in \mathcal{G}$, returns **true** if and only if u and v are adjacent in G. A *label assignment* for $G \in \mathcal{G}$ is a mapping of each $v \in V(G)$ to a bit string $\mathcal{L}(v)$, called the *label* of v. An *adjacency labeling scheme* for $\mathcal{G}$ consists of an encoder and decoder. The *encoder* is an algorithm that receives $G \in \mathcal{G}$ as input and computes the label assignment e_G. The *decoder* is an algorithm that receives any two labels $\mathcal{L}(v), \mathcal{L}(u)$ and computes the query $d(\mathcal{L}(v), \mathcal{L}(u))$, such that $d(\mathcal{L}(v), \mathcal{L}(u)) = adjacency(v, u)$. The *size* of the labeling scheme is the maximum label length. Hereafter, we refer to adjacency labeling schemes simply as labeling schemes. For the encoding and decoding algorithms, we assume a $\Omega(\log n)$ word size RAM model (see [1] for additional details).

2 $\log n + O(1)$ Labeling Scheme for Bounded-Degree Outerplanar Graphs

In this section we describe a labeling scheme for outerplanar. A graph is *outerplanar* if it admits a planar embedding with the property that all vertices lie on the unbounded face. graphs with bounded degree Δ. Our method relies on an embedding technique of Bhatt, Chung, Leighton and Rosenberg [4] for bounded degree outerplanar graphs. In their paper, the authors were concerned with *edge-universal graphs* for various families of bounded degree graphs. In particular, they show that for every $n \in \mathbb{N}$ there exists a graph H_n with $O(n)$ vertices and $O(n)$ edges that contains every bounded degree outerplanar graph $G \in \mathcal{O}(n, \Delta)$ as a subgraph (not necessarily induced).

2.1   Our Methods

Our main tool is an embedding technique due to Bhatt et al. [4] of outerplanar graphs into H_n. On the one hand, the embedding is simple to compute. This fact will lead to an efficient $O(n \log n)$ time encoder. On the other hand, the embedding satisfies a useful locality property. This property allows our labels to contain both unique vertex identifiers of the graph H_n and edge identifiers, without exceeding the desired label size $\log n + O(1)$.

To obtain the latter label size via an embedding into H_n we need to overcome several difficulties. Although H_n has a linear number of edges, its maximum degree is $\Omega(\log n)$, thus, unique edge identifiers require $\Omega(\log \log n)$ bits, in general. Since also $|V(H_n)| = \Omega(n)$, it follows that a label cannot contain an arbitrary combination of vertex identifiers in V_n and edge identifiers at the same time, as it would lead to labels with size $\log n + \Omega(\log \log n)$. This difficulty is overcome by exploiting the structure of H_n further and constructing unique vertex identifiers in a particular way that allows reducing the encoding length. This solution creates an additional difficulty of separating the different parts of the label in the decoding phase. This difficulty is overcome by designing careful encoding lengths that minimize the ambiguity, and storing an additional constant amount of information to eliminate it altogether.

2.2   A Compact Edge-Universal Graph for Bounded-Degree Outerplanar Graphs

We describe next the edge-universal graph $H_n = (V_n, E_n)$ constructed by Bhatt et al. [4] for $\mathcal{O}(n, \Delta)$. We let $k = \min\{s \in \mathbb{N} : 2^s - 1 \geq n\}$ and set $N = 2^k - 1$. The construction uses two constants $c = c(\Delta), g = g(\Delta)$, that depend only on Δ.

The graph H_n is constructed from the complete binary tree T on N vertices as follows. To obtain the vertex set V_n, split every vertex $v \in V(T)$ at level t in T into $\gamma_t = c \log(N/2^t)$ vertices $w_1(v), \cdots, w_{\gamma_t}(v)$. The latter set of vertices is called the *cluster* of v. For $w \in V_n$ we denote by $t(w)$ the *level* of w, that is the level in the binary tree T of the cluster to which w belongs.

The edge set E_n is defined as follows. Two vertices $w_i(v), w_j(u) \in V_n$ are adjacent if and only if the clusters they belong to in T are at distance at most g in T. Note that $w_1 w_2 \in E_n$ implies $|t(w_1) - t(w_2)| \leq g$. This completes the construction of the graph. One can easily check that $|V_n| = O(n)$. $|E_n| = O(n)$ also holds, but we do not use this fact directly. The graph H_n is illustrated in Figure 1. Our labeling scheme relies on the following result of Bhatt et al. [4].

Theorem 2 (Bhatt et al. [4]). *H_n is edge-universal for the class of bounded degree outerplanar graphs $\mathcal{O}(n, \Delta)$.*

2.3   Warm-Up: A $\log n + O(\log \log n)$ Labeling Scheme

We briefly describe a simple $\log n + O(\log \log n)$ labeling scheme. First, assign unique identifiers Id to the vertices in H_n. Since $|V_n| = O(n)$ we can assume that $|Id(v)| = \log n + O(1)$ for every $v \in V_n$. Next, for every $v \in V_n$ assign unique identifiers to the edges incident to v. Since every vertex in H_n has $O(\log n)$ neighbours, each edge can be encoded using $\log \log n + O(1)$ bits.

To obtain labels for a given outerplanar graph $G = (V, E)$, compute first an embedding $\phi : V \to V_n$ of G into H_n. Next, define the label of vertex $v \in V$ to

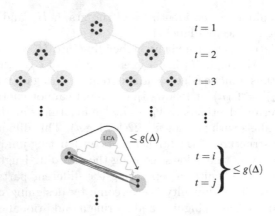

Fig. 1. An illustration of the graph H_n

be the concatenation of $Id(\phi(v))$ with the identifiers of all edges in E_n leading to images under the embedding ϕ of neighbouring vertices, namely all $\phi(u)\phi(v)$ such that $uv \in E$. Since the maximum degree in G is bounded by the constant Δ, this results in a $\log n + O(\log \log n)$ label size. It is not difficult to see that encoding and decoding can be performed efficiently (we elaborate on it later on).

2.4   The Encoder

To reduce the size of the labels to $\log n + O(1)$ we need to refine the latter scheme significantly. As a first step we employ *differential sizing*, a technique first used in the context of labeling schemes in [21]. In differential sizing some parts of the label do not have a fixed number of allocated bits across all labels. Concretely, we use differential sizing for both vertex and edge identifiers.

The resulting labels will have the desired length, but will also contain an undesired ambiguity, that will prohibit correct decoding. We will then append a short prefix to the label that will resolve this ambiguity.

Differential Sizing - The Suffix of a Label. Let us first formally define our naming scheme for vertex and edge identifiers in H_n.

Definition 1. *A naming of H_n is an injective function $Id : V_n \to \mathbb{N}$ and a collection of injective functions $EId_v : E_v \to \mathbb{N}$ for every $v \in V_n$. A naming is coherent if for every $v, v_1, v_2 \in V_n$*

1. *$Id(v_1) > Id(v_2)$ implies that $t(v_1) > t(v_2)$, or $t(v_1) = t(v_2)$ and the cluster of v_2 appears to the left of the cluster of v_1 in T.*
2. *$EId_v(vv_1) > EId_v(vv_2)$ implies that $Id(v_1) > Id(v_2)$.*

We compute a coherent naming by assigning the identifiers 1 through $|V_n|$ to V_n level by level, traversing the clusters in any single level in T from left to right, and then naming the edges incident to $v \in V_n$ from 1 to $|E_v|$ in a way that is consistent with the vertex naming. For $v, v' \in V$ define $\alpha(v) := \lceil \log Id(v) \rceil$ and $\beta_v(vv') := \lceil \log EId_v(vv') \rceil$ and let

$$\alpha_t = \max_{v \,:\, t(v) \leq t} \alpha(v) \quad \text{and} \quad \beta_t = \max_{vv' \,:\, t(v) \leq t} \beta_v(vv')$$

be the maximal number of bits required to encode a vertex name and an edge name for vertices with level at most t. The simple labels described in the beginning of this section store $\log n + O(1)$ and $\log \log n + O(1)$ bits for *every* vertex name, and every edge name, respectively. In contrast, our label for a vertex in level t stores α_t bits for a vertex name, and β_t for edge names. In the following lemma we prove that the new labels have the desired size.

Lemma 1. *For every $t \leq \log N$ it holds that $\alpha_t \leq t + \lceil \log(\log N - t) \rceil + O(1)$, $\beta_t \leq \lceil \log(\log N - t) \rceil + O(1)$ and $\alpha_t + \Delta\beta_t = \log n + O(1)$.*

We henceforth denote the part of the label containing the vertex name and all its edge identifiers as the *suffix*.

Resolving Ambiguity. Since the vertex name does not occupy a fixed number of bits across all labels, it is a priori unclear which part of the label contains it. To resolve this ambiguity we analyze the following function, which represents the final length of our labels for vertices in level t (up to a fixed constant). Let $D = [\lceil \log N \rceil]$ and $f : D \to \mathbb{N}$ be defined as

$$f(t) = \alpha_t + \Delta\beta_t = t + (\Delta + 1)\lceil \log(\log N - t)) \rceil.$$

The following lemma states that all but a constant number (depending on Δ) of values in D have at most two pre-images under f. This observation is useful, since it implies that the knowledge of the level $t(v)$ of the vertex v can resolve all remaining ambiguities in its label, as the vertex name occupies exactly $\alpha_{t(v)}$ bits.

Lemma 2. *Let $r(\Delta) = \lceil 8(\Delta + 1) \log(\Delta + 1) \rceil$. For every $t \in [\lceil \log N \rceil - r(\Delta)]$ the number of integers $t' \in D \setminus \{t\}$ that satisfy $f(t) = f(t')$, is at most one.*

Remark 1. It is natural to ask if having equal label lengths for vertices in different levels can be avoided altogether (thus making Lemma 2 unnecessary). This seems not to be the case for the following reason. The number of vertices in every level is at least $\Omega(\log N)$, thus, with label size $\ell = o(\log \log N)$ one can not uniquely represent all vertices in *any* level. Furthermore, the label length is also restricted to $\log n + O(1)$, and the number of levels is $\lceil \log N \rceil$. Thus, a function assigning levels to label lengths would need to have domain $[\lceil \log N \rceil]$ and range $[g(N), \lceil \log N \rceil]$ for $g(N) = \Omega(\log \log N)$, implying that it cannot be one-to-one.

Recall that the length of the suffix of vertex $v \in V$ is exactly $\alpha_{t(v)} + \Delta\beta_{t(v)}$. We next show how the structural property proved here allows to construct a constant size *prefix*, that will eliminate the ambiguity caused by differential sizing.

Constructing the Prefix. For a formal description of the prefix we need the following definition. We let $r(\Delta) = \lceil 8(\Delta + 1) \log(\Delta + 1) \rceil$, as in Lemma 2.

Definition 2. *A vertex $v \in V$ is called* shallow *if its level $t(v)$ is at most $\lceil \log N \rceil - r(\Delta)$. We call a shallow vertex* early *if $t(v)$ is the smallest pre-image of $f(t(v))$. A shallow vertex that is not early is called* late.

A vertex $v \in V$ that is not shallow is called deep. *A deep vertex is of type τ, if its level satisfies $t(v) = \lceil \log N \rceil - \tau$.*

It is easy to verify the following properties. Lemma 2 guarantees that if v is shallow, then $f(t(v))$ has at most two pre-images under f. If v is shallow and there is only one pre-image for $f(t(v))$, then v is early. Finally, observe that the type of deep vertices ranges in the interval $[1, r(\Delta)]$.

We are now ready to define the prefix of a label $\mathcal{L}(v)$ for a vertex $v \in V$. Every prefix starts with a single bit $D(v)$ that is set to 0 if v is shallow, and to 1 if v is deep. The second bit $R(v)$ in every prefix indicates whether a shallow vertex is early, in which case it is set to 0, or late, in which case it is set to 1. The bit $R(v)$ is always set to 0 in labels of deep vertices. The next part $Type(v)$ of the prefix contains $\lceil \log r(\Delta) \rceil$ bits representing the type of the vertex v, in case v is deep. If v is shallow this field is set to zero. This concludes the definition of the prefix. Observe that the prefix contains $O(\log \Delta) = O(1)$ bits. We stress that length s_p of the prefix is identical across all labels.

It is evident that the prefix of a label eliminates any remaining ambiguity. This follows from the fact that the level $t(v)$ of a vertex $v \in V$ can be computed from the length of the suffix and the additional information stored in the prefix. The level, in turn, allows to decompose the suffix into the vertex name and the incident edge names, which can then be used for decoding. We elaborate on the decoding algorithm later on.

The Final Labels. The final label is obtained by concatenating the suffix to the prefix, namely $\mathcal{L}(v)$ is defined as follows.

$$\mathcal{L}(v) = \underbrace{D(v) \circ R(v) \circ Type(v)}_{\text{prefix}} \circ \underbrace{Id(\phi(v)) \circ EId_{\phi(v)}(e_1) \circ \cdots \circ EId_{\phi(v)}(e_\Delta)}_{\text{suffix}}.$$

Figure 2.4 illustrates the label structure as a function of the level of the vertex. Note that $|\mathcal{L}(v)| = s_p + f(t(v))$, thus the level $t(v)$ of v determines $|\mathcal{L}(v)|$. Note that Lemma 1 and the fact that the prefix has constant size guarantees that $|\mathcal{L}(v)| = \log n + O(1)$, as desired. We also pad each label with sufficiently many 0's and a single '1', to arrive at a uniform length. The latter simple modification allows the decoder to work without knowing n in advance (see [12] for details).

Although it is not necessary for the correctness of our labeling scheme, we prove here uniqueness of the labels. In other words, we show that two different vertices in G necessarily get different labels.

Lemma 3. *For every two distinct $u, v \in V(G)$ it holds that $\mathcal{L}(u) \neq \mathcal{L}(v)$.*

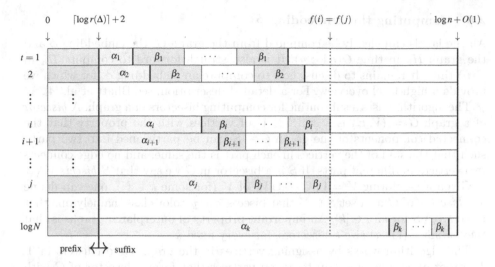

Fig. 2. The composition of labels in our labeling scheme for vertices in different levels $t = 1, \cdots, \log N$. The size of the prefix is seen to be constant in every level, while the fields of lengths α_t and β_t, used to store vertex and edge identifiers, respectively, have variable sizes. The levels i and j comprise a collision with respect to the function f, thus labels of vertices in these levels have the same length.

2.5 Decoding

Consider two labels $\mathcal{L}(u)$ and $\mathcal{L}(v)$ for vertices $u, v \in V$. The decoder first extracts the levels $t(u)$ and $t(v)$ of u and v respectively, using the following simple procedure, which we describe for v. If $D(v) = 0$, v is shallow. To this end, the decoder computes all pre-images of the length of the suffix, $|\mathcal{L}(v)| - s_p$, under f. Recall, that by Lemma 2, the number of pre-images is at most two. Let $t_1 \le t_2$ be the computed pre-images. Next, the decoder inspects the bit $R(v)$. According to the definition of the labels, $t(v) = t_1$ if $R(v) = 0$, and $t(v) = t_2$, otherwise. Consider next the case $D(v) = 1$, namely that v is deep. In this case, the decoder inspects $Type(v)$. The level of v is $t(v) = \lceil \log N \rceil - Type(v)$, by definition of a type of a deep vertex. It is obvious by the definition of the labels that the decoder extracts $t(v)$ correctly. Assume next that the decoder extracted $t(u)$ and $t(v)$. The decoder can now extract $Id(\phi(u))$ and $Id(\phi(v))$, by inspecting the first $\alpha_{t(u)}$ and $\alpha_{t(v)}$ bits of the suffix of $\mathcal{L}(u)$ and $\mathcal{L}(v)$, respectively. Next, the decoder checks if $\phi(u)\phi(v) \notin E_n$, in which case it reports **false**. Finally, if $\phi(u)\phi(v) \in E_n$ the decoder scans all Δ blocks of $\beta_{t(u)}$ bits each, succeeding $Id(\phi(u))$ in the suffix of $\mathcal{L}(u)$, checking if one of them contains the edge-identifier $EId_{\phi(u)}(\phi(u)\phi(v))$. If this identifier is found the decoder reports **true**. Otherwise, it reports **false**. The correctness of the decoding is clear from the label definition and Lemma 2.

Lemma 4. *The decoding of the labels can be performed in time $O(\log \log n)$.*

2.6   Computing the Embedding ϕ

All the labels can clearly be computed from the graph G, the embedding ϕ and the graph H_n in time $O(n \log n)$. It is also straightforward to compute H_n in $O(n)$ time. It remains to discuss how to compute an embedding ϕ, for which we provide a high-level overview. For a detailed description, see Bhatt et al. [4].

The algorithm uses a subroutine for computing bisectors of a graph. A *bisector* of a graph $G = (V, E)$ is a set $S \subset V$ of vertices with the property that the connected components of the graph $G - S$ can be partitioned into two parts, such that the sum of the vertices in each part is the same, and no edge connects two vertices in different parts. If S is a bisector in G we say that S *bisects* $V \setminus S$.

Given a k-coloring $V = V_1 \cup \cdots \cup V_k$ of V (for some $k \in \mathbb{N}$), one can define a k-*bisector* of G as a set $S \subset V$ that bisects *every* color class, namely one that bisects $V_i \setminus S$ for all $i \in [k]$. An important property of outerplanar graphs is that they admit $O(\log n)$ size k-bisectors, for every fixed k.

The algorithm works by assigning vertices in the graph G to clusters in T. The root of T is assigned up to $c \log n$ vertices that form a bisector of G with parts G_1, G_2. In the next iteration, vertices adjacent to vertices in the bisector are given a new color. Next, two new 2-bisectors are found, one in each part G_1, G_2, and they are assigned to the corresponding decedents of the root of T.

Let $k(\Delta) = \log \Delta + 1$. In general, the vertices stored at a vertex of T at level t correspond to a $k(\Delta)$-bisector. The colors of this bisector correspond to the neighbors of vertex-sets stored at $k(\Delta) - 1$ nearest ancestors of the current vertex in T. The last color is reserved to the remaining vertices. Also stored in this vertex are all neighbors of the vertex-set stored in the ancestor of the current vertex at distance exactly $k(\Delta)$, that were not yet assigned to some other cluster. We refer the reader to [4] for an analysis of the sizes of clusters.

Let $T(n)$ be the running time of the latter algorithm in a graph with n vertices. $T(n)$ clearly satisfies $T(n) \leq 2T(n/2) + O(h(n))$, where $h(n)$ is the complexity of finding an $O(1)$-bisector of $O(\log n)$ size in an n-vertex graph. For outerplanar graphs the latter can be done in linear time [8,4], thus the labels of our labeling scheme can be computed in $O(n \log n)$ time.

2.7   Improvements and Special Cases

Several well-known techniques can be easily applied on top of our construction to reduce the additive constant in the label size. First, since graphs of maximum degree Δ have arboricity $\lfloor \frac{\Delta}{2} \rfloor + 1$, one can reduce the number of edge identifiers stored in each label to the latter number (see Kannan et al. [15]). We later show a simpler procedure that works for bounded degree graphs.

Finally, for bounded-degree trees $\mathcal{T}(n, \Delta)$, it suffices to store a single edge identifier (corresponding to the edge connecting a vertex to its parent in G). We summarize this result in the following corollary of Theorem 1.

Corollary 1. *For every fixed $\Delta \geq 1$, the class $\mathcal{T}(n, \Delta)$ of bounded-degree trees admits a labeling scheme of size $\log n + O(1)$, with encoding complexity $O(n \log n)$ and decoding complexity $O(\log \log n)$.*

3   Labeling Schemes for $\mathcal{G}(n, \Delta)$ and $\mathcal{G}(n, \Delta(n))$

First we note that Theorem 1 implies almost directly a $(\lfloor \Delta/2 \rfloor + 1) \log n$ label-ing scheme for graphs of fixed bounded degree Δ. The result follows from the technique of Alstrup and Rauhe [3], Lemma 3 and the fact that any subtree of a bounded degree graph has bounded degree.

Corollary 2. *For every $\Delta \geq 1$, the class $\mathcal{G}(n, \Delta)$ of bounded-degree graphs ad-mits labeling schemes of size $(\lfloor \Delta/2 \rfloor + 1) \log n + O(1)$, with encoding complexity $O(n \log n)$ and decoding complexity $O(\log \log n)$.*

From here on, we discuss labeling schemes for graphs of non-constant bounded degree $k = \Delta(n)$. Adjacency relation between any two vertices may be reported in only one of the labels representing them. For bounded degree graphs, the method of Kannan et al. [15] of decomposition into forests can be replaced with a simpler procedure, using Eulerian circuits, as we prove in the following.

Lemma 5. *Let $G = (V, E)$ be a graph with degree bounded by k. It is possible to partition E into sets S_v, $v \in V$, with the properties that all edges in S_v are incident to v and $|S_v| \leq \lceil \frac{k}{2} \rceil$ for all $v \in V$. This partition implies a labeling scheme with size $(\lceil \frac{k}{2} \rceil + 1) \log n$ for graphs with degree bounded by k.*

The current best labeling schemes for graphs works in two modes, according to the range of k. If $k \leq n/\log n$, a $k/2 \log n$ labeling scheme can be achieved [15], essentially by encoding an adjacency list. For larger k, labels defined through the adjacency matrix of the graph have size $n/2 + \log n$ [19]. Our improved labels use the well-known *combinatorial number system* (see e.g. [16]).

Lemma 6. *Let $L = \binom{n}{k}$. There is a bijective mapping $\sigma : S_k \to [0, L-1]$ between the set of strictly increasing sequences S_k of the form $0 \leq t_1 < t_2 \cdots < t_k < n$ and $[0, L-1]$ given by*

$$\sigma(t_1, \cdots, t_k) = \sum_{i=1}^{k} \binom{t_i}{i}.$$

We use Lemma 6 to prove the following theorem.

Theorem 3. *For $1 \leq k \leq n$, there exist an adjacency labeling scheme for $\mathcal{G}(n, k)$ with size:* $\log \binom{n}{\lceil k/2 \rceil} + \lceil \log n \rceil + \lceil \log k \rceil$.

The labeling scheme suggested implies a label size of approximately $(k+2) \log n$ bits, when k is small and $\Theta(n)$ when $k = \Theta(n)$. The following lemma identifies the range of k for which our labeling scheme improves on the best known bounds.

Lemma 7. *For $(e+1)\sqrt{n} \leq k \leq \frac{n}{5}$, and $f(n, k) = \binom{n}{\lceil k/2 \rceil} + \log k + \log n$ it holds that a) $f(n, k) < \frac{n}{2}$; and b) $f(n, k) < \lceil k/2 \rceil + 2 \log(n)$.*

We conclude from Lemma 7 that our labeling scheme is preferable to [19] for graphs of bounded degree k for $(e+1)\sqrt{n} \leq k \leq \frac{n}{5}$.

Acknowledgements. The authors thank Prof. Julia Lawall for her useful comments.

References

1. Alstrup, S., Bille, P., Rauhe, T.: Labeling schemes for small distances in trees. SIAM J. Disc. Math. 19(2), 448–462 (2005)
2. Alstrup, S., Gavoille, C., Kaplan, H., Rauhe, T.: Nearest common ancestors: A survey and a new distributed algorithm. In: Proc. of the Fourteenth Annual ACM Symposium on Parallel Algorithms and Architectures, SPAA 2002, pp. 258–264 (2002)
3. Alstrup, S., Rauhe, T.: Small induced-universal graphs and compact implicit graph representations. In: Proc. the 43rd Annual IEEE Symposium on Foundations of Computer Science, pp. 53–62. IEEE (2002)
4. Bhatt, S., Chung, F.R.K., Leighton, T., Rosenberg, A.: Universal Graphs for Bounded-Degree Trees and Planar Graphs. SIAM J. Disc. Math. 2(2), 145–155 (1989)
5. Bonichon, N., Gavoille, C., Labourel, A.: Short labels by traversal and jumping. In: Flocchini, P., Gąsieniec, L. (eds.) SIROCCO 2006. LNCS, vol. 4056, pp. 143–156. Springer, Heidelberg (2006)
6. Bonichon, N., Gavoille, C., Labourel, A.: Short labels by traversal and jumping. Electr. Notes in Discrete Math. 28, 153–160 (2007)
7. Butler, S.: Induced-universal graphs for graphs with bounded maximum degree. Graphs Combinator. 25(4), 461–468 (2009)
8. Chung, F.R.K.: Separator theorems and their applications. Forschungsinst. Für Diskrete Mathematik (1989)
9. Chung, F.R.K.: Universal graphs and induced-universal graphs. J. Graph Theor. 14(4), 443–454 (1990)
10. Esperet, L., Labourel, A., Ochem, P.: On induced-universal graphs for the class of bounded-degree graphs. Inform. Process. Lett. 108(5), 255–260 (2008)
11. Fraigniaud, P., Korman, A.: Compact ancestry labeling schemes for xml trees. In: Proc. of the 21st Annual ACM-SIAM Symposium on Discrete Algorithms, SODA 2010, pp. 458–466 (2010)
12. Fraigniaud, P., Korman, A.: Compact ancestry labeling schemes for xml trees. In: Proc. 21st ACM-SIAM Symp. on Discrete Algorithms, SODA 2010 (2010)
13. Gavoille, C., Labourel, A.: Shorter implicit representation for planar graphs and bounded treewidth graphs. In: Arge, L., Hoffmann, M., Welzl, E. (eds.) ESA 2007. LNCS, vol. 4698, pp. 582–593. Springer, Heidelberg (2007)
14. Gavoille, C., Paul, C.: Distance labeling scheme and split decomposition. Discrete Math. 273, 115–130 (2003)
15. Kannan, S., Naor, M., Rudich, S.: Implicit representation of graphs. In: Proc. of the 20th ACM Symposium on Theory of Computing, STOC 1988, pp. 334–343 (1988)
16. Knuth, D.: Combinatorial algorithms. The Art of Computer Programming, vol. 4a (2011)
17. Korman, A., Peleg, D., Rodeh, Y.: Constructing labeling schemes through universal matrices. In: Asano, T. (ed.) ISAAC 2006. LNCS, vol. 4288, pp. 409–418. Springer, Heidelberg (2006)
18. Laoutaris, N., Rajaraman, R., Sundaram, R., Teng, S.H.: A bounded-degree network formation game. arXiv preprint cs/0701071 (2007)
19. Moon, J.W.: On minimal n-universal graphs. In: Proc. of the Glasgow Math. Assoc. vol. 7, pp. 32–33 (1965)
20. Nash-Williams, C.: Edge-disjoint spanning trees of finite graphs. J. London Math. Soc. 1(1), 445–450 (1961)
21. Thorup, M., Zwick, U.: Compact routing schemes. In: SPAA 2001, New York, pp. 1–10 (2001)
22. Wang, Y., Li, X.Y.: Localized construction of bounded degree and planar spanner for wireless ad hoc networks. Mobile Networks and Applications 11(2), 161–175 (2006)

Bounded-Angle Spanning Tree: Modeling Networks with Angular Constraints*

Rom Aschner and Matthew J. Katz

Department of Computer Science, Ben-Gurion University, Israel
{romas,matya}@cs.bgu.ac.il

Abstract. We introduce a new structure for a set of points in the plane and an angle α, which is similar in flavor to a bounded-degree MST. We name this structure α-MST. Let P be a set of points in the plane and let $0 < \alpha \leq 2\pi$ be an angle. An α-ST of P is a spanning tree of the complete Euclidean graph induced by P, with the additional property that for each point $p \in P$, the smallest angle around p containing all the edges adjacent to p is at most α. An α-MST of P is then an α-ST of P of minimum weight. For $\alpha < \pi/3$, an α-ST does not always exist, and, for $\alpha \geq \pi/3$, it always exists [1,2,9]. In this paper, we study the problem of computing an α-MST for several common values of α.

Motivated by wireless networks, we formulate the problem in terms of directional antennas. With each point $p \in P$, we associate a wedge W_p of angle α and apex p. The goal is to assign an orientation and a radius r_p to each wedge W_p, such that the resulting graph is connected and its MST is an α-MST. (We draw an edge between p and q if $p \in \mathrm{W}_q$, $q \in \mathrm{W}_p$, and $|pq| \leq r_p, r_q$.) Unsurprisingly, the problem of computing an α-MST is NP-hard, at least for $\alpha = \pi$ and $\alpha = 2\pi/3$. We present constant-factor approximation algorithms for $\alpha = \pi/2, 2\pi/3, \pi$.

One of our major results is a surprising theorem for $\alpha = 2\pi/3$, which, besides being interesting from a geometric point of view, has important applications. For example, the theorem guarantees that given *any* set P of $3n$ points in the plane and *any* partitioning of the points into n triplets, one can orient the wedges of each triplet *independently*, such that the graph induced by P is connected. We apply the theorem to the *antenna conversion* problem.

1 Introduction

Let P be a set of points in the plane and let $0 < \alpha \leq 2\pi$ be an angle. An α-ST of P is a spanning tree of the complete Euclidean graph induced by P, with the additional property that for each point $p \in P$, the smallest angle around p

---

* A version including the missing proofs can be found at http://arxiv.org/abs/ 1402.6096.

Work by R. Aschner was partially supported by the Lynn and William Frankel Center for Computer Sciences. Work by M. Katz was partially supported by grant 1045/10 from the Israel Science Foundation. Work by M. Katz and R. Aschner was partially supported by grant 2010074 from the United States – Israel Binational Science Foundation.

J. Esparza et al. (Eds.): ICALP 2014, Part II, LNCS 8573, pp. 387–398, 2014.
© Springer-Verlag Berlin Heidelberg 2014

containing all the edges adjacent to p is at most α. An α-MST of P is then an α-ST of P of minimum weight.

In this paper, we study the problem of computing an α-MST for several common values of α. For $\alpha < \pi/3$, an α-ST does not always exist (consider, e.g., an equilateral triangle). Moreover, it is well known that there always exists a Euclidean MST of degree at most 5. Therefore, it is interesting to focus on the range $\pi/3 \leq \alpha < 8\pi/5$.

Carmi et al. [9] showed that, for $\alpha = \pi/3$, an α-ST always exists. A somewhat simpler construction was subsequently proposed by Ackerman et al. [1]. Aichholzer et al. [2] have also obtained this result (together with additional related results), independently. However, in all these papers, the goal is to construct an α-ST (for $\alpha = \pi/3$) and not an α-MST.

The problem of computing an α-MST is similar in flavor to the problem of computing a Euclidean minimum weight degree-k spanning tree, which has been studied extensively (see, e.g., [4, 10, 16, 17, 19]). A minimum weight degree-k spanning tree is a minimum weight spanning tree, such that the degree of each point is at most k, where the interesting values of k are 2,3, and 4. Notice that for $k = 2$ we get the Euclidean traveling salesman path problem.

The problem of computing an α-ST is closely related to problems in which one needs to compute a Hamiltonian path or cycle, with some restrictions on the angles. Fekete and Woeginger [14] showed that every set of points has a Hamiltonian *path*, such that all its angles are bounded by $\pi/2$. An alternative construction was given later in [9]. Fekete and Woeginger also conjectured that for every set of $2k \geq 8$ points there exists a Hamiltonian *cycle*, such that all its angles are bounded by $\pi/2$. Recently, Dumitrescu et al. [12] showed how to construct a Hamiltonian cycle whose angles are bounded by $2\pi/3$. As for lower bound, in [9] and, independently, in [12] it is shown that, for any $\varepsilon > 0$, there exists a set of points, for which any Hamiltonian path has an angle greater than $\pi/2 - \varepsilon$. The problem of finding Hamiltonian paths with large angles was also considered in [14], where it is conjectured that every point set admits a Hamiltonian path, whose angles are at least $\pi/6$; Bárány et al. [6] showed how to construct a path, whose angles are at least $\pi/9$.

Unsurprisingly, the problem of computing an α-MST is NP-hard, at least for $\alpha = \pi$ and $\alpha = 2\pi/3$. For $\alpha = \pi$, one can show this by a reduction from the problem of finding a Hamiltonian path in grid graphs of degree at most 3, which is known to be NP-hard [15]. The reduction is similar to the one described for the problem of computing a minimum weight degree-3 spanning tree [20], with a few simple adaptations. For $\alpha = 2\pi/3$, one can show this by a straightforward reduction from Hamiltonian path in hexagonal grid graphs. Arkin et al. [3] showed that the problem of finding a Hamiltonian cycle in hexagonal grid graphs is NP-hard. However, with not too much effort, one can prove that finding a Hamiltonian path in hexagonal grid graphs is NP-hard as well. The NP-hardness proofs can be found in the full version.

Motivated by wireless networks, we formulate the problem of computing an α-MST in terms of directional antennas. In the last few years, directional antennas have received considerable attention (see, e.g., [7, 8, 18]), as they have some noticeable advantages over omni-directional antennas. In particular, they require less energy to reach a receiver at a given distance, and when broadcasting to this receiver the affected region is much smaller, reducing the probability of causing interference at friendly receivers or being subject to eves dropping by hostile receivers. With each point $p \in P$, we associate a wedge W_p of angle α and apex p. The goal now is to assign an orientation and a radius r_p to each wedge W_p, such that the resulting graph is connected and its MST is an α-MST. (We draw an edge between p and q if $p \in W_q$, $q \in W_p$, and $|pq| \le r_p, r_q$.)

An interesting related problem is the *antenna conversion* problem. The *unit disk graph* of P, denoted UDG(P), is the graph in which there is an edge between p and q if $|pq| \le 1$. This is the communication graph induced by P, where each point in P represents a transceiver equipped with an omni-directional antenna of radius 1. We assume that UDG(P) is connected. Suppose that one wishes to replace the omni-directional antennas with directional antennas of angle α. The goal now is to assign an orientation to each of the wedges W_p and to fix a common range $\delta = \delta(\alpha)$, such that the resulting (symmetric) communication graph is a c-hop-spanner of UDG(P), where $c = c(\alpha)$. Moreover, δ and c should be small constants. Aschner et al. [5] considered this problem for $\alpha = \pi/2$. Here we solve it for $\alpha = 2\pi/3$, using significantly smaller constants. Recently, it has been brought to our attention that Dobrev et al. [11] have also been considering the antenna conversion problem.

Our results. In Section 2 we focus on the case $\alpha = 2\pi/3$. We begin by describing a simple gadget: Given any set S of three points in the plane, we show how to orient the wedges associated with the points of S, such that G_S, the graph induced by S, is connected, and, moreover, the union of the wedges of S covers the plane. We then prove a surprising theorem, which, besides being interesting from a geometric point of view, has far-reaching applications, such as the one mentioned in the abstract. Informally, the theorem states that any two such gadgets are connected. That is, let S_1 and S_2 be two triplets of points in the plane, and assume that the wedges (associated with the points) of S_1 and, independently, of S_2 are oriented according to the gadget construction instructions, then the graph induced by $S_1 \cup S_2$ is connected. Proving this theorem turned out to be a very challenging task, due to the huge number of possible configurations that must be considered, and only after arriving at the current three-stage proof structure (see Section 2.2), were we able to complete the proof.

In Section 3, we present constant-factor approximation algorithms for computing an α-MST. In particular, we compute a 2-approximation for a π-MST, a 6-approximation for a $2\pi/3$-MST, and a 16-approximation for a $\pi/2$-MST. These approximations are actually with respect to a Euclidean MST, which is a lower bound for an α-MST, for any α. In Section 4, we present a solution to the antenna conversion problem for $\alpha = 2\pi/3$, based on the theorem above. Specifically, we construct, in $O(n \log n)$ time, a 6-hop-spanner of UDG(P), in which each edge is of length at most 7. Finally, NP-hardness proofs for the problem of computing an α-MST, for $\alpha = \pi$ and $\alpha = 2\pi/3$, can be found in the full version.

2 $\alpha = \frac{2\pi}{3}$

Notation. Let p be a point and let α be an angle. We denote the wedge of angle α and apex p by w_p. The left ray bounding w_p (when looking from p into w_p) is denoted by $\overleftarrow{w}_p$ and the right ray by $\overrightarrow{w}_p$. The bisector of w_p is denoted by $bis(w_p)$. The orientations of $\overleftarrow{w}_p$, $\overrightarrow{w}_p$, and $bis(w_p)$ are denoted by $\theta(\overleftarrow{w}_p)$, $\theta(\overrightarrow{w}_p)$, and $\theta(bis(w_p))$, respectively. The *orientation* of w_p is the orientation of its bisector and is denoted by $\theta(w_p)$. We denote the ray emanating from p of orientation $\theta(bis(w_p)) + 180$ by $\widetilde{w}_p$; its orientation is denoted by $\theta(\widetilde{w}_p)$.

Let S be a set of points, where each point $p \in S$ is associated with a wedge w_p of some orientation. The graph induced by S, denoted G_S, is the graph in which there is an edge between $p, q \in S$ if and only if $p \in w_q$ and $q \in w_p$. If there is an edge between p and q, we say that p and q are *connected* and denote this by $\{p\} \leftrightarrow \{q\}$. Similarly, if S_1 and S_2 are two such sets of points, and there exist a point p in S_1 and a point q in S_2 such that p and q are connected, then we say that S_1 and S_2 are *connected* and denote this by $\{S_1\} \leftrightarrow \{S_2\}$. The notation $\{p\} \nleftrightarrow \{q\}$ means that p and q are not connected, and, similarly, $\{S_1\} \nleftrightarrow \{S_2\}$ means that there does not exist a point in S_1 and a point in S_2 such that these points are connected.

2.1   The Basic Gadget

Claim. Let $S = \{a, b, c\}$ be a set of three points in the plane, and set $\alpha = 2\pi/3$. Then, one can orient the wedges of S, such that G_S, the induced graph of S, contains a $2\pi/3$-ST of S, and the wedges of S cover the plane.

Proof. Consider $\triangle abc$, and assume w.l.o.g. that $\angle b \leq \angle c \leq \angle a$. Then, $\angle b \leq 60$ and $\angle c < 90$. Draw $\triangle abc$, such that $\overline{bc}$ is horizontal (with b to the left of c) and a is not below the line containing $\overline{bc}$. Orient the wedges of S as follows (see Figure 1(a)): $\theta(w_a) = 240$, $\theta(w_b) = 0$, $\theta(w_c) = 120$.

It is easy to see that the non-directed edges (a, b) and (b, c) are in the induced graph G_S. Thus, G_S contains a $2\pi/3$-ST. As for the second requirement, notice that w_a contains the wedge $w_a{}'$ of orientation $\theta(w_a)$ and apex b, and w_c contains the wedge $w_c{}'$ of orientation $\theta(w_c)$ and apex b. But, clearly, $w_a{}' \cup w_b \cup w_c{}' = \mathbb{R}$.

The gadget of Claim 2.1 has some noticeable properties:

Property 1. For any $x \in S$, the orientations of the wedges of S are $\theta(w_x)$ and $\theta(w_x) \pm 120$.

Property 2. For any $x \in S$, the orientations of the rays bounding the wedges of S are $\theta(w_x) \pm 60$ and $\theta(w_x) + 180$. Moreover, each of these three orientations appears exactly twice, once as the orientation of a left ray bounding some wedge and once as the orientation of a right ray bounding some other wedge (see Figure 1(b)).

Property 3. Consider any two wedges w_x and w_y and the four rays defining them. Then, by Property 2, exactly two of these rays, ρ_1 from w_x and ρ_2 from w_y, have the same orientation. Let l be a line intersecting both ρ_1 and ρ_2 and perpendicular to ρ_1 (and to ρ_2). Then, $w_x \cup w_y$ covers the halfplane defined by l that does not include the points x and y.

Finally, let R_i denote the range $((i-1)60, i60)$, for $1 \leq i \leq 6$ (see Figure 1(c)).

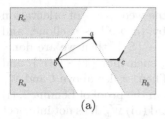 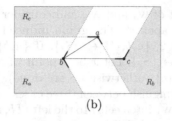 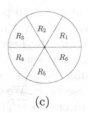

(a)   (b)   (c)

Fig. 1. (a) The basic gadget of Claim 2.1. $\theta(w_a) = 240$, $\theta(w_b) = 0$, and $\theta(w_c) = 120$. A point p is in region R_a if and only if $p \in w_a$ and $p \notin w_b, w_c$, i.e., $R_a = w_a \setminus (w_b \cup w_c)$. Regions R_b and R_c are defined analogously. (b) $\theta(\overleftarrow{w}_a) = \theta(\overleftarrow{w}_b) = \theta(\widetilde{w}_c) = 300$, $\theta(\overrightarrow{w}_a) = \theta(\widetilde{w}_b) = \theta(\overleftarrow{w}_c) = 180$, and $\theta(\overrightarrow{w}_a) = \theta(\overleftarrow{w}_b) = \theta(\overrightarrow{w}_c) = 60$. (c) The six ranges $R_1, \ldots, R_6$.

2.2 The Induced Graph of $S_1 \cup S_2$ is Connected

In this section, we prove the following surprising theorem (Theorem 1), which, as mentioned, has far-reaching applications. Let $S_1 = \{a, b, c\}$ and S_2 be two triplets of points in the plane, and assume that the wedges (associated with the points) of S_1 and, independently, of S_2 are oriented according to the proof of Claim 2.1. Then, the induced graph of $S_1 \cup S_2$ is connected.

In order to cope with the huge number of cases, we prove Theorem 1 in three stages. In the first stage (Lemma 1), we prove the statement assuming that both induced graphs of S_1 and of S_2 are cliques. In the second stage (Lemma 2), we prove the statement assuming only one of the induced graphs is a clique, using, of course, Lemma 1. Finally, in the third stage (Theorem 1), we prove the statement without any additional assumptions, using Lemma 2.

Throughout this section, we assume (as in the proof of Claim 2.1) that, in $\triangle abc$, $\angle b \leq \angle c \leq \angle a$, $\overline{bc}$ is horizontal, with b to the left of c, and a is not below the line l containing $\overline{bc}$ (see Figure 1(a)).

The proofs of Lemma 1 and Lemma 2 can be found in the full version.

Lemma 1 (Two cliques). *Let $S_1 = \{a, b, c\}$ and S_2 be two sets of points and let $\alpha = 2\pi/3$. Assume that the wedges (associated with the points) of S_1 and, independently, of S_2 are oriented according to the proof of Claim 2.1, and that both induced graphs, G_{S_1} and G_{S_2}, are cliques. Then, the induced graph $G_{S_1 \cup S_2}$ is connected.*

Lemma 2 (One clique). *Let $S_1 = \{a, b, c\}$ and S_2 be two sets of points and let $\alpha = 2\pi/3$. Assume that the wedges of S_1 and, independently, of S_2 are oriented according to the proof of Claim 2.1, and that the induced graph G_{S_2} is a clique. Then, the induced graph $G_{S_1 \cup S_2}$ is connected.*

Theorem 1. *Let $S_1 = \{a, b, c\}$ and S_2 be two sets of points and let $\alpha = 2\pi/3$. Assume that the wedges of S_1 and, independently, of S_2 are oriented according to the proof of Claim 2.1. Then, the induced graph $G_{S_1 \cup S_2}$ is connected.*

Proof. If one (or both) of the induced graphs G_{S_1}, G_{S_2} is a clique, then, by Lemma 2, we are done. Assume therefore that none of them is a clique. Let c' be the intersection point of $\overleftarrow{W}_a$ and $\overleftarrow{W}_c$, and consider the wedge $W_{c'}$ of orientation $\theta(W_{c'}) = \theta(W_c)$ and apex c'. The graph induced by $\{a, b, c'\}$ is a clique, and therefore, by Lemma 2, $\{a, b, c'\} \leftrightarrow \{S_2\}$. If $\{a, b\} \leftrightarrow \{S_2\}$, then we are done, so assume that $\{c'\} \leftrightarrow \{S_2\}$. Let x be a point of S_2 such that $\{x\} \leftrightarrow \{c'\}$, and assume that $\{x\} \not\leftrightarrow \{c\}$ (otherwise we are done). Then, x lies above l and $\overleftarrow{W}_x$ intersects $\overline{cc'}$. We distinguish between three cases, as in the proof of Lemma 2: (i) $\overrightarrow{W}_x$ intersects $\overline{bc'}$, (ii) $\overrightarrow{W}_x$ intersects l to the left of b, and (iii) $\overrightarrow{W}_x$ does not intersect l. As mentioned in the proof of Lemma 2, our arguments there for Case (i) and Cases (ii)(1) and (ii)(2) do not use the extra assumption that G_{S_2} is a clique. Therefore, we can reuse them here. It remains to show that $\{S_1\} \leftrightarrow \{S_2\}$ in Cases (ii)(3) and (iii).

Case (ii)(3): W_y covers exactly one point of S_2, namely, c. We know that either $\theta(W_y) \in R_1$ or $\theta(W_y) \in R_3$. In the latter case, W_y must also cover b, which is impossible. In the former case, if y is above l, then $\{y\} \leftrightarrow \{c\}$, so y is necessarily below l. Let z be the remaining point. Then, $\theta(W_z) \in R_3$. At this point, we would like to show, as in the proof of Lemma 2, that $\{z\} \leftrightarrow \{a, b\}$. However, we cannot assume now that $\{y\} \leftrightarrow \{z\}$. So, we first prove that $\{y\} \leftrightarrow \{z\}$, by proving that $\{x\} \not\leftrightarrow \{y\}$, and then we proceed as in the proof of Lemma 2.

Thus, our goal now is to prove that $\{x\} \not\leftrightarrow \{y\}$. Let p be the midpoint of $\overline{bc}$, and let a' be the projection of a onto l. According to the construction in the proof of Claim 2.1, a' lies somewhere between p and c (not including c). Let o be the intersection point of $\overleftarrow{W}_x$ and l. We know that o is somewhere between c' and c (not including c). Finally, let t be the intersection point of $bis(W_x)$ and l (see Figure 2(a)). We show that t lies to the left of p and therefore also to the left of a'. If t is to the left of b (or $t = b$), then this is clear. Assume therefore that t is to the right of b, and consider the two triangles $\triangle xto$ and $\triangle xbt$. Recall first that x is above $\overleftarrow{W}_b$ and notice that it is below $bis(W_c)$ (since, if x were above $bis(W_c)$, then $\{x\} \leftrightarrow \{c\}$). Therefore $\angle xbt > 60$ and the projection of x onto l lies to the left of p. Now, in $\triangle xto$, $\angle xot \le 60$ and $\angle txo = 60$, and therefore $|xt| \le |to|$. And, in $\triangle xbt$, $\angle bxt < 60$ and $\angle xbt > 60$, and therefore $|bt| < |xt|$. Together, we get that $|bt| < |to| < |tc|$, so t lies to the left of p and therefore to the left of a'.

Since the projection of x onto l lies to the left of p and so does t, we have that a lies to the left of $bis(W_x)$. Now, if $\{x\} \leftrightarrow \{y\}$, then y must lie to the right of $bis(W_x)$ and therefore cover a, which is impossible. We conclude that $\{x\} \not\leftrightarrow \{y\}$, and therefore $\{y\} \leftrightarrow \{z\}$ (and $\{x\} \leftrightarrow \{z\}$).

From this point, we continue as in the proof of Lemma 2. Notice that $\overleftarrow{W}_y$ separates between a and c and between b and c, since $\theta(W_y) \in R_1$ and W_y covers only c. Since $\{y\} \leftrightarrow \{z\}$, we know that z lies to the right of $bis(W_y)$. Clearly, a and b lie to the left of $\overrightarrow{W}_z$ (whose orientation is in R_2), and to the right of $\overleftarrow{W}_z$ (whose orientation is in R_4). In other words, W_z covers both a and b. Notice also that $z \notin R_c$, since $bis(W_y)$ (whose orientation is in R_1) intersects l to the right of b, and z lies to the right of $bis(W_y)$. Therefore, either W_a or W_b (or both) covers z. We conclude that $\{z\} \leftrightarrow \{a, b\}$.

Case (iii): $\overrightarrow{W}_x$ does not intersect l, implying that $\theta(\overrightarrow{W}_x) < 180$. Notice that in this case $b \in W_x$, so we assume that $x \notin W_b$, implying that $\theta(\overleftarrow{W}_x) > 240$. It follows that $\theta(\mathbf{w}_x) \in R_4$, $\theta(\overleftarrow{\mathbf{w}}_x) \in R_5$, $\theta(\overrightarrow{\mathbf{w}}_x) \in R_3$, and $\theta(\widetilde{\mathbf{w}}_x) \in R_1$. Notice also that $bis(\mathbf{w}_x)$, whose orientation is in R_4, intersects l to the left of b.

Let y be the point of S_2 such that $\theta(\overleftarrow{\mathbf{w}}_y) \in R_3$ and $\theta(\overrightarrow{\mathbf{w}}_y) \in R_1$, and let z be the point of S_2 such that $\theta(\overleftarrow{\mathbf{w}}_z) \in R_1$ and $\theta(\overrightarrow{\mathbf{w}}_z) \in R_5$. Notice that for $\overleftarrow{\mathbf{w}}_x$ to intersect l to the right of c', x must lie above $l(\overleftarrow{\mathbf{w}}_a)$, and, therefore, $\mathbf{w}_x$ covers a.

We first show that if $\{x\} \leftrightarrow \{z\}$, then $\{S_1\} \leftrightarrow \{S_2\}$. Indeed, if $\{x\} \leftrightarrow \{z\}$, then z must lie to the right of $bis(\mathbf{w}_x)$. If z is above l, then $\{z\} \leftrightarrow \{c\}$. Assume, therefore, that z is below l. Notice that $\overleftarrow{\mathbf{w}}_z$ intersects $\overleftarrow{\mathbf{w}}_b$ at a point above x, implying that $\overleftarrow{\mathbf{w}}_z$ passes above a. Moreover, $\overrightarrow{\mathbf{w}}_z$ passes below a, since it is directed downwards. It follows that $\mathbf{w}_z$ covers a. But, $z \in \mathbf{w}_a$, since z lies to the right of $bis(\mathbf{w}_x)$, which intersects l to the left of b. We conclude that $\{z\} \leftrightarrow \{a\}$.

Next, we address the most difficult case, in which $\{x\} \not\leftrightarrow \{z\}$. If $\{x\} \not\leftrightarrow \{z\}$, then necessarily y is connected to both x and z. Notice that z must lie below $\overleftarrow{\mathbf{w}}_x$. Also, if it is above l, then $\{z\} \leftrightarrow \{c\}$. Assume, therefore, that z is below l. Since $\mathbf{w}_y$'s rays are directed upwards and $\{y\} \leftrightarrow \{z\}$, we know that y is below z and therefore also below l. According to the construction in the proof of Claim 2.1, either x or z lies on $bis(\mathbf{w}_y)$, and the angle at this point in $\triangle xyz$ does not exceed the angle at the other point. It follows that the point that lies on $bis(\mathbf{w}_y)$ is necessarily x. Since, if it were z, then $\angle yzx \geq 120$, as it contains $\mathbf{w}_z$.

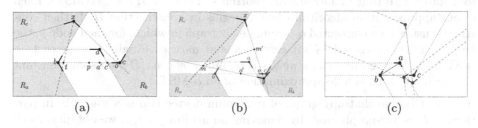

Fig. 2. (a) t lies to the left of a'. (b) x lies on $bis(\mathbf{w}_y)$. (c) Each of the triplets induces a connected graph and covers the plane, but the graph of their union is not connected.

The case where x lies on $bis(\mathbf{w}_y)$ is also impossible, as we show below (see Figure 2(b)). If $y \notin \mathbf{w}_b$, then $\{y\} \leftrightarrow \{a\}$, since $\overleftarrow{\mathbf{w}}_y$ is below $l(\overrightarrow{\mathbf{w}}_b)$ and $\overrightarrow{\mathbf{w}}_y$ is above $l(\overleftarrow{\mathbf{w}}_x)$. Assume, therefore, that $y \in \mathbf{w}_b$ but $\{y\} \not\leftrightarrow \{b\}$. Let m be the intersection point of $\overleftarrow{\mathbf{w}}_y$ and $bis(\mathbf{w}_x)$. Then, m is above l (since otherwise $\{y\} \leftrightarrow \{b\}$). Notice that $\triangle xmy$ is equilateral, and consider the bisector of $\angle xmy$. Let m' be the intersection point of this bisector and side $\overline{xy}$. Then, $\overline{mm'}$ is the perpendicular bisector of $\overline{xy}$.

Next, we show that m' lies above l. Let o be the intersection point of $\overline{xy}$ and l, and let o' the intersection point of $\overline{my}$ and l. We show that $|\overline{yo}| < |\overline{xo}|$, implying that m' is somewhere between o and x and thus above l. Consider $\triangle yoo'$. Since $\theta(\overleftarrow{\mathbf{w}}_y) \in R_3$, we know that $\angle yo'o < 60$. But $\angle oyo' = 60$, so we get that $|\overline{oy}| < |\overline{oo'}|$. Now, consider $\triangle xbo$. $\angle xbo > 60$ and $\angle bxo < 60$, and therefore $|\overline{ox}| > |\overline{ob}|$. It follows that $|\overline{oy}| < |\overline{oo'}| < |\overline{ob}| < |\overline{ox}|$.

Since all its corners lie above l, $\triangle mm'x$ is above l. Since $\{y\} \leftrightarrow \{z\}$ and z is below l, we have that $z \in \triangle yo'o \subseteq \triangle ymm'$, and therefore z is closer to y than to x – contradiction the construction of Claim 2.1.

Remark. Theorem 1 above proves that when the wedges of each of the triplets are oriented, independently, according to the construction of Claim 2.1, then there is always an edge between the two triplets. This is not necessarily true for other constructions with similar properties. For example, the wedges of each of the triplets in Figure 2(c) form a connected graph and cover the plane, but there is no edge between the triplets.

3   Approximating the α-MST

Let P be a set of n points in the plane. In this section we consider the problem of computing an α-MST of P, for $\alpha = \pi, 2\pi/3, \pi/2$. For each of these angles, we devise a constant-factor approximation algorithm. The approximation ratios are actually with respect to the weight of a Euclidean MST, which is a lower bound for the weight of an α-MST, for any α.

Consider the TSP tour $\Pi = e_0, e_1, \ldots, e_{n-1}$ obtained by applying the standard 2-approximation algorithm for metric TSP. This algorithm first duplicates the edges of a MST to obtain an Eulerian tour, and then transforms the Eulerian tour into a TSP tour by introducing shortcuts. Thus, $wt(\Pi) \leq 2wt(\text{MST})$. Each of our approximation algorithms below begins by constructing Π. It then constructs, using Π, a connected α-graph, i.e., a graph in which, for each node p, the angle spanned by the edges adjacent to p is at most α. Finally, it construct an α-ST from the α-graph, whose weight is bounded by $c \cdot wt(\Pi)$, for some constant $c = c(\alpha)$, and thus is a $2c$-approximation of an α-MST.

$\boldsymbol{\alpha = \pi}$. Observe that any graph of maximum degree two is a π-graph. In particular, Π is a π-graph, and, by removing an arbitrary edge, we obtain a π-ST of weight at most $2wt(\pi\text{-MST})$.

$\boldsymbol{\alpha = 2\pi/3}$. Assume, for convenience, that $n = 3m$, for some integer m. We partition P into m triplets, by traversing Π from an arbitrary point $p \in P$. That is, each of the triplets consists of three consecutive points along Π. Orient the wedges of each triplet, independently, according to Claim 2.1. By Theorem 1, the graph induced by P, denoted here G_α (instead of G_P), is connected. In particular, for any two consecutive triplets t, t' along Π, there exists an edge of the graph between a point of t and a point of t'.

Next, we construct a $2\pi/3$-ST, T, and show that $wt(T) \leq 6 \cdot wt(2\pi/3\text{-MST})$. Initially, T has no edges. For each of the m triplets t, add to T any two edges (of the at least two edges) of G_α connecting between pairs of points of t. We call these edges *inner-edges*. Next, for each of the m pairs of consecutive triplets t, t' along Π (except for the pair consisting of the 'last' triplet and the 'first' triplet), add to T any edge (of the at least one edge) of G_α connecting between a point of t and a point of t'. We call these edges *connecting-edges*. T is connected and has $2n/3$ inner-edges and $n/3 - 1$ connecting-edges, thus the total number of edges is $n - 1$, and T is a $2\pi/3$-ST.

We now bound the weight of T. By the triangle inequality, the weight of an edge (u, v) of T does not exceed the weight of the shorter path (in terms of number of edges) in Π between u and v. We charge the weight of this path for the edge (u, v). Each edge of Π between two points of the same triplet t is charged at most four times. Twice for the two inner-edges chosen for t, and twice for the two connecting-edges that connect t to its two adjacent triplets along Π. Each edge of Π between two consecutive triplets t, t' (except for the edge between the last and first) is charged only once for the corresponding connecting-edge of T. Thus, each edge of Π is charged at most four times, and $wt(T) = \Sigma_{e \in T}|e| \leq 4\Sigma_{e \in \Pi}|e| = 4 \cdot wt(\Pi) \leq 8 \cdot wt(\text{MST}) \leq 8 \cdot wt(2\pi/3\text{-MST})$.

Next, we improve the approximation ratio. Observe, that there are three possible ways to partition Π into m triplets. In other words, the set of edges of Π connecting between the triplets can be either E_0, E_1, or E_2, where $E_j = \{e_i \in E : i = (j \mod 3)\}$, for $0 \leq j \leq 2$. By the pigeon hole principle, the weight of one of these sets, say E_2, is at least $\frac{1}{3} \cdot wt(\Pi)$. We partition Π into triplets, such that the set of edges connecting between the triples is E_2. Now, each of the edges of E_2 (except e_{n-1}) is charged exactly once, and each of the edges of $E_0 \cup E_1$ is charged at most four times. Thus, $wt(T) \leq wt(E_2) + 4(wt(E_0) + wt(E_1)) = wt(\Pi) + 3(wt(E_0) + wt(E_1)) \leq wt(\Pi) + 3 \cdot \frac{2}{3}wt(\Pi) = 3 \cdot wt(\Pi) \leq 6 \cdot wt(\text{MST}) \leq 6 \cdot wt(2\pi/3\text{-MST})$.

$\boldsymbol{\alpha = \pi/2}$. Assume, for convenience, that $n = 8m$, for some integer m. Our construction for $\alpha = \pi/2$ is similar to the one for $\alpha = 2\pi/3$, but slightly more complicated. It is based on a basic gadget described by Aschner et al. [5] for a set S of four points, indicating the locations of four $\pi/2$-wedges. This gadget is presented as the proof for the claim that one can orient the wedges of S, such that the induced graph is connected, and the wedges of S cover the plane. Unfortunately, we cannot claim that two quadruplets, whose wedges are oriented independently, are connected. However, if they are separable by a line, then they are connected, see [5].

We use this latter claim in our construction. We partition the tour Π into m sections, each consisting of 8 consecutive points along Π. Then, we partition each of the sections into two quadruplets, a left quadruplet consisting of the 4 leftmost points of the section and a right quadruplet consisting of the 4 rightmost points. (Notice that the points of a quadruplet are not necessarily consecutive along Π.) Thus, in each section, the two quadruplets are separable by a (vertical) line. Now, orient the wedges of each quadruplet, independently, such that their induced graph is connected and the wedges cover the plane. Let G_α be the graph induced by P. Observe that G_α is connected, since, for any two consecutive sections, there exists two quadruplets, one from each section, that are separable by a (vertical) line and thus connected.

Next, we construct the tree T from G_α. We distinguish between three types of edges. The first type are the *inner-edges*, which connect between points of the same quadruplet. For each quadruplet, we pick three such edges that make the quadruplet connected. The second type are the *q-connecting-edges*, which connect between quadruplets of the same section. For each section, we pick one such edge. The third type are the *s-connecting-edges*, which connect between consecutive sections along Π. For each pair of consecutive sections along Π (except for the pair consisting of the last and first sections), we pick one such edge. Notice that T is a $\pi/2$-ST, since it is connected and it has $n-1$ edges, i.e., $3n/4$ inner-edges, $n/8$ q-connecting-edges, and $n/8-1$ s-connecting-edges.

We compute the approximation ratio by charging the edges of Π. Each edge of Π either connects between points of the same section, or between points of consecutive sections. An edge of the former kind is charged at most nine times. Since for a section, we have six inner-edges, one q-connecting-edge, and two s-connecting-edges. An edge of the latter kind is charged only once.

As for $\alpha = 2\pi/3$, we can choose the subset of edges of Π that connect between consecutive sections, so that its weight is at least $\frac{1}{8} \cdot wt(\Pi)$. Let E_7 denote this subset. Then, $wt(T) \leq wt(E_7) + 9 \cdot wt(E \setminus E_7) \leq wt(\Pi) + 8 \cdot wt(E \setminus E_7) \leq wt(\Pi) + 8 \cdot \frac{7}{8}wt(\Pi) = 8 \cdot wt(\Pi) \leq 16 \cdot wt(\text{MST}) \leq 16 \cdot wt(\pi/2\text{-MST})$.

The following theorem summarizes the results of this section.

Theorem 2. *Let P be a set of points in the plane. Then, one can construct (i) a π-ST of weight at most $2 \cdot wt(\pi\text{-MST})$, (ii) a $2\pi/3$-ST of weight at most $6 \cdot wt(2\pi/3\text{-MST})$, and (iii) a $\pi/2$-ST of weight at most $16 \cdot wt(\pi/2\text{-MST})$.*

Remark. As mentioned, the approximation ratios above are with respect to $wt(\text{MST})$, which is a lower bound for $wt(\alpha\text{-MST})$. It is possible that by comparing the weight of the constructed α-ST with that of an α-MST, one can get better ratios, but it is not clear how to do so. Moreover, it is easy to see that, for $\alpha \in [60, 180)$, 2 is a lower bound on the ratio with respect to a MST, e.g., consider n points on a line. And, for $\alpha \in [180, 240)$, $\frac{2+\sqrt{3}}{3} \approx 1.244$ is a lower bound on the ratio, e.g., consider 3 points at the corners of an equilateral triangle and a fourth point at the center of the circle passing through them. Finally, for $\alpha \in [60, 90)$, it is easy to give an example where $wt(\alpha\text{-MST})/wt(\text{MST}) \to n-1$. Therefore, any algorithm for an angle α in this range, should be analyzed with respect to $wt(\alpha\text{-MST})$.

4   Constant Range Hop-Spanner for $\alpha = 2\pi/3$

In this section we apply Theorem 1 to obtain a solution to a problem that arises in wireless communication networks. Let P be a set of n points in the plane, where each point in P represents a transceiver equipped with an omnidirectional antenna. The coverage region of p's antenna is modeled by a disk centered at p, and assume that all disks are of radius 1. Then, the resulting communication graph is the *unit disk graph* of P, denoted $\text{UDG}(P)$. (I.e., there is an edge between points p and q if the distance between them is at most 1.) As mentioned in the introduction, directional antennas have some advantages over omni-directional antennas and are gaining popularity. The coverage region of a directional antenna of angle α is modeled by a circular sector of angle α.

Assume that $\text{UDG}(P)$ is connected. Before stating our problem, we need the following definition. A graph $G = (P, E)$ is a *c-hop-spanner* of $\text{UDG}(P)$, for some constant c, if for any two points $p, q \in P$, the minimum number of hops between p and q in G is at most c times this number in $\text{UDG}(P)$. That is, for each edge $e = (p, q)$ in $\text{UDG}(P)$, there exists a path in G between p and q

consisting of at most c edges. Assume now that one replaces each of the omni-directional antennas by a directional antenna of angle $2\pi/3$. We address the following *Antenna Conversion* problem: Orient the directional antennas and fix a range $\delta = O(1)$, such that the resulting (symmetric) communication graph is a c-hop-spanner of $\text{UDG}(P)$, for some constant c. I.e., construct a $2\pi/3$-graph, such that the length of its edges is bounded by δ and it is a c-hop-spanner of $\text{UDG}(P)$.

We show how to construct such a graph with $\delta = 7$ and $c = 6$, in $O(n \log n)$ time. We first partition the points of P into connected components (of $\text{UDG}(P)$) of size at most three. This is done greedily. Set $Q = P$. As long as $Q \neq \emptyset$, perform the following step, which finds the next component C. Pick any point $a \in Q$, add it to C (which is initially empty), and remove it from Q. Now, if $Q \neq \emptyset$ and there exists a point in Q whose distance from a is at most 1, then pick any such point $b \in Q$, add it to C, and remove it from Q. Finally, if $Q \neq \emptyset$ and there exists a point in Q whose distance to either a or b (or both) is at most 1, then pick any such point $c \in Q$, add it to C, and remove it from Q.

Claim. Let C be a connected component of size one or two. Then, each of the neighbors of C in $\text{UDG}(P)$ belongs to a component of size three.

Proof. Assume that one of the neighbors of C belongs to a component C' of size one or two, i.e., there exists an edge of $\text{UDG}(P)$ between a point in C and a point in C'. Moreover, assume, e.g., that C was found before C'. Then, in the iteration in which C was found, we would have found a larger component, i.e., with at least one additional point.

Now, consider the connected components that were found. We first orient the wedges of each connected component of size exactly three, independently, according to the proof of Claim 2.1. Next, for each connected component C of size one or two, let C' be any connected component of size exactly three, such that C has a neighbor in C'. Recall that the wedges of C' cover the plane. We orient each of the wedges of C (alternatively, the single wedge of C) towards the wedge of C' that covers it. Observe that if the length of the edges is not limited, then the $2\pi/3$-graph, G_α, that is induced by the wedges of P is connected. Moreover, it is easy to verify that G_α is a c-hop-spanner, for $c = 5$. However, our goal is to limit the length of the edges without increasing c by much.

Let C be a component of size one or two. Then, the edge of G_α connecting between C and C', where C' is the component of size three to which C was connected, is of length at most 4. Moreover, consider any two components of size three C' and C'', such that C has a neighbor both in C' and in C''. Then, the edge of G_α connecting between C' and C'' is of length at most 7. Finally, the edge of G_α connecting between two neighboring components of size three is of length at most 5. Therefore, one can drop all edges of length greater than 7 from G_α, without disconnecting it.

Finally, it is easy to see that the resulting graph is a 6-hop spanner.

The following theorem summarizes the result of this section. For the missing details, see the full version.

Theorem 3. *Let P be a set of points in the plane and assume that $\text{UDG}(P)$ is connected. Let $\alpha = 2\pi/3$. Then, one can construct, in $O(n \log n)$ time, a 6-hop-spanner of $\text{UDG}(P)$, in which each edge is of length at most 7.*

References

1. Ackerman, E., Gelander, T., Pinchasi, R.: Ice-creams and wedge graphs. Comput. Geom.: Theory & Applications 46(3), 213–218 (2013)
2. Aichholzer, O., Hackl, T., Hoffmann, M., Huemer, C., Pór, A., Santos, F., Speckmann, B., Vogtenhuber, B.: Maximizing maximal angles for plane straight-line graphs. Comput. Geom.: Theory & Applications 46(1), 17–28 (2013)
3. Arkin, E.M., Fekete, S.P., Islam, K., Meijer, H., Mitchell, J.S.B., Rodríguez, Y.N., Polishchuk, V., Rappaport, D., Xiao, H.: Not being (super) thin or solid is hard: A study of grid Hamiltonicity. Comput. Geom.: Theory & Applications 42(6-7), 582–605 (2009)
4. Arora, S.: Polynomial time approximation schemes for Euclidean traveling salesman and other geometric problems. J. ACM 45(5), 753–782 (1998)
5. Aschner, R., Katz, M.J., Morgenstern, G.: Symmetric connectivity with directional antennas. Comput. Geom.: Theory & Applications 46(9), 1017–1026 (2013)
6. Bárány, I., Pór, A., Valtr, P.: Paths with no small angles. SIAM Journal Discrete Mathematics 23(4), 1655–1666 (2009)
7. Bose, P., Carmi, P., Damian, M., Flatland, R., Katz, M.J., Maheshwari, A.: Switching to directional antennas with constant increase in radius and hop distance. In: Dehne, F., Iacono, J., Sack, J.-R. (eds.) WADS 2011. LNCS, vol. 6844, pp. 134–146. Springer, Heidelberg (2011)
8. Caragiannis, I., Kaklamanis, C., Kranakis, E., Krizanc, D., Wiese, A.: Communication in wireless networks with directional antennas. In: 20th ACM Sympos. on Parallelism in Algorithms and Architectures, pp. 344–351 (2008)
9. Carmi, P., Katz, M.J., Lotker, Z., Rosén, A.: Connectivity guarantees for wireless networks with directional antennas. Comput. Geom.: Theory & Applications 44(9), 477–485 (2011)
10. Chan, T.M.: Euclidean bounded-degree spanning tree ratios. Discrete & Computational Geometry 32(2), 177–194 (2004)
11. Dobrev, S., Eftekhari, M., MacQuarrie, F., Manuch, J., Morales-Ponce, O., Narayanan, L., Opatrny, J., Stacho, L.: Connectivity with directional antennas in the symmetric communication model. In: Mexican Conf. on Discrete Mathematics and Computational Geometry (2013)
12. Dumitrescu, A., Pach, J., Tóth, G.: Drawing Hamiltonian cycles with no large angles. Electronic Journal of Combinatorics 19(2), P31 (2012)
13. Efrat, A., Itai, A., Katz, M.J.: Geometry helps in bottleneck matching and related problems. Algorithmica 31(1), 1–28 (2001)
14. Fekete, S.P., Woeginger, G.J.: Angle-restricted tours in the plane. Comput. Geom.: Theory & Applications 8, 195–218 (1997)
15. Itai, A., Papadimitriou, C.H., Szwarcfiter, J.L.: Hamilton paths in grid graphs. SIAM Journal on Computing 11(4), 676–686 (1982)
16. Jothi, R., Raghavachari, B.: Degree-bounded minimum spanning trees. Discrete Applied Mathematics 157(5), 960–970 (2009)
17. Khuller, S., Raghavachari, B., Young, N.E.: Low-degree spanning trees of small weight. SIAM Journal on Computing 25(2), 355–368 (1996)
18. Kranakis, E., Krizanc, D., Morales, O.: Maintaining connectivity in sensor networks using directional antennae. In: Nikoletseas, S., Rolim, J.D.P. (eds.) Theoretical Aspects of Distributed Computing in Sensor Networks, ch. 3, Springer
19. Mitchell, J.S.B.: Guillotine subdivisions approximate polygonal subdivisions: a simple polynomial-time approximation scheme for geometric TSP, k-MST, and related problems. SIAM Journal on Computing 28(4), 1298–1309 (1999)
20. Papadimitriou, C.H., Vazirani, U.V.: On two geometric problems related to the travelling salesman problem. Journal of Algorithms 5(2), 231–246 (1984)

Distributed Computing on Core-Periphery Networks: Axiom-Based Design*

Chen Avin[1,**], Michael Borokhovich[1], Zvi Lotker[1], and David Peleg[2]

[1] Ben-Gurion University of the Negev, Israel
{avin,borokhom,zvilo}@cse.bgu.ac.il
[2] The Weizmann Institute, Israel
david.peleg@weizmann.ac.il

Abstract. Inspired by social networks and complex systems, we propose a *core-periphery* network architecture that supports fast computation for many distributed algorithms and is robust and efficient in number of links. Rather than providing a concrete network model, we take an axiom-based design approach. We provide three intuitive (and independent) algorithmic axioms and prove that any network that satisfies all axioms enjoys an efficient algorithm for a range of tasks (e.g., MST, sparse matrix multiplication, etc.). We also show the *minimality* of our axiom set: for networks that satisfy any subset of the axioms, the same efficiency cannot be guaranteed for *any* deterministic algorithm.

1 Introduction

A fundamental goal in distributed computing is designing a network architecture that allows fast running times for various distributed algorithms, but at the same time is cost-efficient in terms of minimizing the number of communication links between machines and the amount of memory used by each machine.

For illustration, let's consider three basic networks topologies: a star, a clique and a constant degree expander. The *star graph* has only a linear number of links and can compute every computable function after only one round of communication. But clearly, such an architecture has two major disadvantages: the memory requirements of the central node do not scale, and the network is not robust. The *complete graph*, on the other hand, is very robust and can support extremely fast performance for tasks such as information dissemination, distributed sorting and minimum spanning tree, to name a few [1,2,3]. Also, in a complete graph the amount of memory used by a single processor is minimal. But obviously, the main drawback of that architecture is the high number of links it uses. *Constant degree expanders* are a family of graphs that support efficient computation for many tasks. They also have linear number of links and can effectively balance the workload between many machines. But the diameter of these graphs is lower bounded by $\Omega(\log n)$ which implies similar lower bound for most of the interesting tasks one can consider.

---

* Supported in part by the Israel Science Foundation (grant 1549/13).
** Part of this work was done while the author was visiting ICERM, Brown university.

J. Esparza et al. (Eds.): ICALP 2014, Part II, LNCS 8573, pp. 399–410, 2014.
© Springer-Verlag Berlin Heidelberg 2014

A natural question is therefore whether there are other candidate topologies with guaranteed good performance. We are interested in the best compromise solution: a network on which distributed algorithms have small running times, memory requirements at each node are limited, the architecture is robust to link and node failures, and the total number of links is minimized (preferably linear).

To try to answer this question we adopt in this paper an *axiomatic* approach to the design of efficient networks. In contrast to the direct approach to network design, which is based on providing a *concrete* type of networks (by deterministic or random construction) and showing its efficiency, the axiomatic approach attempts to abstract away the algorithmic requirements that are imposed on the concrete model. This allows one to isolate and identify the basic requirements that a network needs for a certain type of tasks. Moreover, while usually the performance of distributed algorithms is dictated by specific *structural* properties of a network (e.g., diameter, conductance, degree, etc.), the axioms proposed in this work are expressed in terms of desired *algorithmic* properties that the network should have.

The axioms[1] proposed in the current work are motivated and inspired by the *core-periphery* structure exhibited by many social networks and complex systems. A core-periphery network is a network structured of two distinct groups of nodes, namely, a large, sparse, and weakly connected group identified as the *periphery*, which is loosely organized around a small, cohesive and densely connected group identified as the *core*. Such a dichotomic structure appears in many areas of our life, and has been observed in many social organizations including modern social networks [4]. It can also be found in urban and even global systems (e.g., in global economy, the wealthiest countries constitute the core which is highly connected by trade and transportation routes) [5,6,7]. There are also peer-to-peer networks that use a similar hierarchical structure, e.g., FastTrack [8] and Skype [9], in which the supernodes can be viewed as the core and the regular users as the periphery.

The main technical contribution of this paper is proposing a minimal set of simple core-periphery-oriented axioms and demonstrating that networks satisfying these axioms achieve efficient running time for various distributed computing tasks while being able to maintain linear number of edges and limited memory use. We identify three basic, abstract and conceptually simple (parameterized) properties, that turn out to be highly relevant to the effective interplay between core and periphery. For each of these three properties, we propose a corresponding axiom, which in our belief captures some intuitive aspect of the desired behavior expected of a network based on a core-periphery structure. Let us briefly describe our three properties, along with their "real life" interpretation, technical formulation, and associated axioms.

The three axioms are: (i) clique-like structure of the core, (ii) fast convergecast from periphery to the core and (iii) balanced boundary between the core and

---

[1] One may ask whether the properties we define qualify as "axioms". Our answer is that the axiomatic lens helps us focus attention on the fundamental issues of minimality, independence and necessity of our properties.

Table 1. Summary of algorithms for core-periphery networks

Task	Running time on $\mathcal{CP}$ networks	Lower bounds All Axioms	Any 2 Axioms
MST *	$O(\log^2 n)$	$\Omega(1)$	$\tilde{\Omega}(\sqrt[4]{n})$
Matrix transposition	$O(k)$	$\Omega(k)$	$\Omega(n)$
Vector by matrix multiplication	$O(k)$	$\Omega(k/\log n)$	$\Omega(n/\log n)$
Matrix multiplication	$O(k^2)$	$\Omega(k^2)$	$\Omega(n/\log n)$
Find my rank	$O(1)$	$\Omega(1)$	$\Omega(n)$
Find median	$O(1)$	$\Omega(1)$	$\Omega(\log n)$
Find mode	$O(1)$	$\Omega(1)$	$\Omega(n/\log n)$
Find number of distinct values	$O(1)$	$\Omega(1)$	$\Omega(n/\log n)$
Top r ranked by areas	$O(r)$	$\Omega(r)$	$\Omega(r\sqrt{n})$

k - maximum number of nonzero entries in a row or column. * - randomized algorithm

periphery. The first property deals with the flow of information within the core. It is guided by the key observation that to be influential, the core must be able to accomplish fast information dissemination internally among its members. The corresponding Axiom $\mathcal{A}_E$ postulates that the core must be a $\Theta(1)$-clique emulator (to be defined formally later). Note that this requirement is stronger than just requiring the core to possess a dense interconnection subgraph, since the latter permits the existence of "bottlenecks", whereas the requirement of the axiom disallows such bottlenecks.

The second property focuses on the flow of information from the periphery to the core and measures its efficiency. The core-periphery structure of the network is said to be a γ-convergecaster if this data collection operation can be performed in time γ. The corresponding Axiom $\mathcal{A}_C$ postulates that information can flow from the periphery nodes to the core efficiently (i.e., in constant time). Note that one implication of this requirement is that the presence of periphery nodes that are far away from the core, or bottleneck edges that bridge between many periphery nodes and the core, is forbidden.

The third and last property concerns the "boundary" between the core and the periphery and claim that core nodes are *"effective ambassadors"*. Ambassadors serve as bidirectional channels through which information flows into the core and influence flows from the core to the periphery. However, to be effective as an ambassador, the core node must maintain a balance between its interactions with the "external" periphery and its interactions with the other core members, serving as its natural "support"; a core node which is significantly more connected to the periphery than to the core becomes ineffective as a channel of influence. In distributed computing terms, a core node that has many connections to the periphery has to be able to distribute all the information it collected from them to other core nodes. The corresponding Axiom $\mathcal{A}_B$ states that the core must have a $\Theta(1)$-balanced boundary (to be defined formally later).

To support and justify our selection of axioms, we examine their usefulness for effective distributed computations on core-periphery networks. We consider a collection of different types of tasks, and show that they can be efficiently solved on core-periphery networks, by providing a distributed algorithm for each task and bounding its running time. Moreover, for each task we argue the necessity of all three axioms, by showing that if at least one of the axioms is not satisfied by the network under consideration, then the same efficiency can not be guaranteed by *any* algorithm for the given task.

Table 1 provides an overview of the main tasks we studied along with the upper and lower bounds on the running time when the network satisfies our axioms and a worst case lower bound on the time required when at least one of the axioms is not satisfied. For each task we provide an algorithm and prove formally its running time and the necessity of the axioms. As it turns out, some of the necessity proofs make use of an interesting connection to known communication complexity results.

The most technically challenging part of the paper is the distributed construction of a *minimum-weight spanning tree* (MST), a significant task in both the distributed systems world (cf.[10,11,12]) and the social networks world [13,14,15]. Thus, the main algorithmic result of the current paper is proving that MST can be computed efficiently (in $O(\log^2 n)$ rounds) on core-periphery networks. To position this result in context we recall that for the complete graph $G = K_n$, an MST can be constructed distributedly in $O(\log \log n)$ time [1]. For the wider class of graphs of diameter at most 2, this task can still be performed in time $O(\log n)$. In contrast, taking the next step, and considering graphs of diameter 3, drastically changes the picture, as there are examples of such graphs for which any distributed MST construction requires $\Omega\left(\sqrt[4]{n}\right)$ time [16].

The rest of the paper is organized as follows. Section 2 formally describes core-periphery networks, the axioms and their basic structural implications. Section 3 provides an overview on the MST algorithm and Section 4 an overview on the rest of the task we study. Due to lack of space we defer many of the technical details and proofs to the report [17].

2   Axiomatic Design for Core-Periphery Networks

Preliminaries. Let $G(V, E)$ denote our (simple, undirected) network, where V is the set of nodes, $|V| = n$, and E is the set of edges, $|E| = m$. The network can be thought of as representing a distributed system. We assume the synchronous CONGEST model (cf. [12]), where communication proceeds in *rounds* and in each round each node can send a message of at most $O(\log n)$ bits to each of its neighbors. Initially each node has a unique ID of $O(\log n)$ bits.

For a node v, let $N(v)$ denote its set of neighbors and $d(v) = |N(v)|$ its degree. For a set $S \subset V$ and a node $v \in S$, let $N_{in}(v, S) = N(v) \cap S$ denote its set of neighbors within S and denote the number of neighbors of v in the set S by $d_{in}(v, S) = |N_{in}(v, S)|$. Analogously, let $N_{out}(v, S) = N(v) \cap V \setminus S$ denote v's set of neighbors outside the set S and let $d_{out}(v) = |N_{out}(v, S)|$. For two subsets $S, T \subseteq V$, let $\partial(S, T)$ be the *edge boundary* (or cut) of S and T, namely the set of edges with exactly one endpoint in S and one in T and $|\partial(S, T)| = \sum_{v \in S} |N_{out}(v, S) \cap T|$. Let $\partial(S)$ denote the special case where $T = V \setminus S$.

Core-Periphery Networks. Given a network $G(V, E)$, a $\langle C, P \rangle$-partition is a partition of the nodes of V into two sets, the *core* C and the *periphery* P. Denote the sizes of the core and the periphery by n_C and n_P respectively. To represent the partition along with the network itself, we denote the *partitioned network* by $G(V, E, C, P)$.

Intuitively, the core C consists of a relatively small group of strong and highly connected machines designed to act as central servers, whereas the periphery P consists of the remaining nodes, typically acting as clients. The periphery machines are expected to be weaker and less well connected than the core machines, and they may perform much of their communication via the dense interconnection network of the core. In particular, a central component in many of our algorithms for various coordination and computational tasks is based on assigning each node v a *representative* core node $r(v)$, essentially a neighbor acting as a "channel" between v and the core. The representative chosen for each periphery node is fixed.

For a partitioned network to be effective, the $\langle C, P \rangle$ partition must possess certain desirable properties. In particular, a partitioned network $G(V, E, C, P)$ is called a *core-periphery network*, or CP-*network* for short, if the $\langle C, P \rangle$-partition satisfies three properties, defined formally later on in the form of three axioms.

Core-periphery Properties and Axioms. We first define certain key parameterized properties of node groups in networks that are of particular relevance to the relationships between core and periphery in our partitioned network architectures. We then state our axioms, which capture the expected behavior of those properties in core-periphery networks, and demonstrate their independence and necessity. Our three basic properties are:

α-**Balanced Boundary.** *A subset of nodes S is said to have an α-balanced boundary iff $\frac{d_{\mathrm{out}}(v,S)}{d_{\mathrm{in}}(v,S)+1} = O(\alpha)$ for every node $v \in S$.*

β-**Clique Emulation.** *The task of* clique emulation *on an n-node graph G involves delivering a distinct message $M_{v,w}$ from v to w for every pair of nodes v, w in $V(G)$. An n-node graph G is a β-clique-emulator if it is possible to perform clique emulation on G within β rounds (in the CONGEST model).*

γ-**convergecast.** *For $S, T \subseteq V$, the task of $\langle S, T \rangle$-convergecast on a graph G involves delivering $|S|$ distinct messages M_v, originated at the nodes $v \in S$, to some nodes in T (i.e., each message must reach at least one node in T). The sets $S, T \subset V$ form a γ-convergecaster if it is possible to perform $\langle S, T \rangle$-convergecast on G in γ rounds (in the CONGEST model).*

Consider a partitioned network $G(V, E, C, P)$. We propose the following set of *axioms* concerning the core C and periphery P.

$\mathcal{A}_B$. **Core Boundary.** The core C has a $\Theta(1)$-balanced boundary.

$\mathcal{A}_E$. **Clique Emulation.** The core C is a $\Theta(1)$-clique emulator.

$\mathcal{A}_C$. **Periphery-Core Convergecast**. The periphery P and the core C form a $\Theta(1)$-convergecaster.

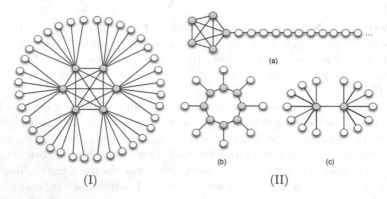

Fig. 1. (I) An example for a 36-node $\mathcal{CP}$-network that satisfies all three axioms. The 6 core nodes (in gray) are connected in clique. In this example every core node is also an ambassadors with equal number of edges to the core and outside from the core. The core and periphery form a convergecaster since the periphery can send all its information to the core in one round. (II) Networks used in proofs: (a) The lollipop partitioned network L_{25}. (b) The sun partitioned network S_{16}. (c) The dumbbell partitioned network D_{16}.

Let us briefly explain the axioms. Axiom $\mathcal{A}_B$ talks about the boundary between the core and periphery. Think of core nodes with a high out-degree (i.e., with many links to the periphery) as "*ambassadors*" of the core to the periphery. Axiom $\mathcal{A}_B$ states that while not all nodes in the core must serve as ambassadors, if a node is indeed an ambassador, then it must also have many links *within* the core. Axiom $\mathcal{A}_E$ talks about the flow of information within the core, and postulates that the core must be dense, and in a sense behave almost like a complete graph: "everyone must know everyone else". The clique-emulation requirement is actually stronger than just being a dense subgraph, since the latter permits the existence of bottlenecks nodes, which a clique-emulator must avoid. Axiom $\mathcal{A}_C$ also concerns the boundary between the core and periphery, but in addition it refers also to the structure of the periphery. It postulates that information can flow efficiently from the periphery to the core. For example, it forbids the presence of periphery nodes that are far away from the core, or bottleneck edges that bridge between many periphery nodes and the core. Fig. 1 (I) provides an example for $\mathcal{CP}$-network satisfying the three axioms.

We next show that the axioms are independent. Later, we prove the necessity of the axioms for the efficient performance of a variety of computational tasks.

Theorem 1. *Axioms $\mathcal{A}_B$, $\mathcal{A}_E$, $\mathcal{A}_C$ are independent, namely, assuming any two of them does not imply the third.*

We prove this theorem by considering three examples of partitioned networks, described next, each of which satisfies two of the axioms but not the third (hence they are not $\mathcal{CP}$-networks), implying independence.

The lollipop partitioned network L_n: (Fig. 1 (II)(a) The lollipop graph consists of a $\sqrt{n}$-node clique and an $n - \sqrt{n}$-node line attached to some node of the clique.

Set the core $\mathcal{C}$ to be the clique and the periphery $\mathcal{P}$ to be the line. Observe that Axioms $\mathcal{A}_E$ and $\mathcal{A}_B$ hold but $\mathcal{A}_C$ is not satisfied since the long line will require linear time for periphery to core convergcast.

The sun partitioned network S_n: (Fig. 1 (II)(b)) The sun graph consists of an $n/2$-node cycle with an additional leaf node attached to each cycle node. Set the core $\mathcal{C}$ to be the cycle and the periphery $\mathcal{P}$ to contain the $n/2$ leaves. Axioms $\mathcal{A}_C$ and $\mathcal{A}_B$ hold but Axiom $\mathcal{A}_E$ does not, since the distance between diametrically opposing nodes in the cycle is $n/4$, preventing fast clique emulation.

The dumbbell partitioned network D_n: (Fig. 1 (II)(c)) The dumbbell graph is composed of two stars, each consisting of a center node connected to $n/2 - 1$ leaves, whose centers are connected by an edge. Set the core $\mathcal{C}$ to be the two centers, and the periphery $\mathcal{P}$ to contain all the leaves. Then Axioms $\mathcal{A}_E$ and $\mathcal{A}_C$ hold while Axiom $\mathcal{A}_B$ does not.

Structural Implications of the Axioms. The axioms imply a number of simple properties of the network structure.

Theorem 2. *If $G(V, E, \mathcal{C}, \mathcal{P})$ is a core-periphery network (i.e., it satisfies Axioms $\mathcal{A}_B$, $\mathcal{A}_E$ and $\mathcal{A}_C$), then the following properties hold:*

1. *The size of the core satisfies $\Omega(\sqrt{n}) \le n_c \le O(\sqrt{m})$.*
2. *Every node v in the core satisfies $d_{\mathrm{out}}(v, \mathcal{C}) = O(n_c)$ and $d_{\mathrm{in}}(v, \mathcal{C}) = \Omega(n_c)$.*
3. *The number of outgoing edges from the core is $|\partial(\mathcal{C})| = \Theta(n_c^2)$.*
4. *The core is dense, i.e., the number of edges in it is $\sum_{v \in \mathcal{C}} d_{\mathrm{in}}(v, \mathcal{C}) = \Theta(n_c^2)$.*

Proof. Axiom $\mathcal{A}_E$ necessitates that the inner degree of each node v is $d_{\mathrm{in}}(v, \mathcal{C}) = \Omega(n_c)$ (or else it would not be possible to complete clique emulation in constant time), implying the second part of claim 2. It follows that the number of edges in the core is $\sum_{v \in \mathcal{C}} d_{\mathrm{in}}(v, \mathcal{C}) = \Theta(n_c^2)$, hence it is dense; claim 4 follows. Since also $\sum_{v \in \mathcal{C}} d_{\mathrm{in}}(v, \mathcal{C}) \le 2m$, we must have the upper bound of claim 1, that is, $n_c = O(\sqrt{m})$. Axiom $\mathcal{A}_B$ yields that for every v, $d_{\mathrm{out}}(v, \mathcal{C}) = O(n_c)$, so the first part of claim 2 follows. Note that $|\partial(\mathcal{C})| = \sum_{v \in \mathcal{C}} d_{\mathrm{out}}(v, \mathcal{C}) = O(n_c^2)$, so the upper bound of claim 3 follows. To give a lower bound on n_c, note that by Axiom $\mathcal{A}_C$ we have $|\partial(\mathcal{C})| = \Omega(n - n_c)$ (otherwise the information from the $n - n_c$ nodes of $\mathcal{P}$ could not flow in $O(1)$ time to $\mathcal{C}$), so $n_c = \Omega(\sqrt{n})$ and the lower bounds of claims 1 and 3 follow. $\qquad\square$

An interesting case for efficient networks is where the number of edges is linear in the number of nodes. In this case we have the following corollary.

Corollary 1. *In a core-periphery network where $m = O(n)$, the following properties hold:*

1. *The size of the core satisfies $n_c = \Theta(\sqrt{n})$*
2. *The number of outgoing edges from the core is $|\partial(\mathcal{C})| = \Theta(n)$.*
3. *The number of edges in the core is $\sum_{v \in \mathcal{C}} d_{\mathrm{in}}(v, \mathcal{C}) = \Theta(n)$.*

Now we show a key property relating our axioms to the network diameter.

Claim 1. *If the partitioned network $G(V, E, \mathcal{C}, \mathcal{P})$ satisfies Axioms $\mathcal{A}_E$ and $\mathcal{A}_C$ then its diameter is $\Theta(1)$.*

The following claim shows that the above conditions are necessary.

Claim 2. *For $X \in \{E, C\}$, there exists a family of n-node partitioned networks $G_X(V, E, \mathcal{C}, \mathcal{P})$ of diameter $\Omega(n)$ that satisfy all axioms except $\mathcal{A}_X$.*

3   MST on a Core-Periphery Network

In this section we present a time-efficient randomized distributed algorithm for computing a *minimum-weight spanning tree* (MST) on a core-periphery network. In particular, we consider an n-node core periphery network $G(V, E, \mathcal{C}, \mathcal{P})$, namely, a partitioned network satisfying all three axioms, and show that an MST can be distributedly computed on such a network in $O(\log^2 n)$ rounds with high probability. Upon termination, each node knows which of its edges belong to the MST. Formally, we state the following theorem.

Theorem 3. *On a $\mathcal{CP}$-network $G(V, E, \mathcal{C}, \mathcal{P})$, Algorithm $\mathcal{CP}$-MST constructs an MST in $O(\log^2 n)$ rounds with high probability.*

We also show that Axioms $\mathcal{A}_B$, $\mathcal{A}_E$, and $\mathcal{A}_C$ are indeed necessary for our distributed MST algorithm to be efficient.

Theorem 4. *For each $X \in \{B, E, C\}$ there exists a family $\mathcal{F}_X = \{G_X(V, E, \mathcal{C}, \mathcal{P})(n)\}$ of partitioned networks that do not satisfy Axiom $\mathcal{A}_X$ but satisfy the other two axioms, and the time complexity of any distributed MST algorithm on $\mathcal{F}_X$ (as a function of the network size n) is $\Omega(n^{\alpha_X})$ for some constant $\alpha_X > 0$.*

The formal proof of Theorem 4 can be found in [17], but the idea of the proof is as following. For each case of Theorem 4 we show a graph in which, for a certain weight assignment, there exist two nodes s and r such that in order to decide which of the edges incident to r belong to the MST, it is required to know the weights of all the edges incident to s. Thus, at least $deg(s)$ (i.e., degree of s) messages have to be delivered from s to r in order to complete the MST task, which implies a lower bound on any distributed MST algorithm.

Now let us give a high level description of the algorithm. Our $\mathcal{CP}$-MST algorithm is based on Boruvka's MST algorithm [18], and runs in $O(\log n)$ phases, each consisting of several steps. The algorithm proceeds by maintaining a forest of tree *fragments* (initially singletons), and merging fragments until the forest converges to a single tree. Throughout the execution, each node has two *officials*, namely, core nodes that represent it. In particular, recall that each node v is assigned a *representative* core neighbor $r(v)$ passing information between v and the core. In addition, v is also managed by the *leader* $l(i)$ of its current fragment i. An important distinction between these two roles is that the representative of each node is fixed, while its fragment leader may change in each phase (as its fragment grows). At the beginning of each phase, every node knows the IDs of its fragment and its leader. Then, every node finds its minimum weight outgoing edge, i.e., the edge with the second endpoint belonging to the other fragment and having the minimum weight. This information is delivered to the core by the means of the representative nodes, which receive the information, aggregate it (as much as possible) and forward it to the leaders of the appropriate fragments. The leaders decide on the fragment merging and inform all the nodes about new fragments IDs.

The correctness of the algorithm follows from emulating Boruvka's algorithm and the correctness of the fragments merging procedure, described in the technical report [17]. The main challenges in obtaining the proof were in bounding the running time, which required careful analysis and observations. There are two major sources of problems that can cause delays in the algorithm. The first involves sending information between officials (representatives to leaders and vice versa). Note that there are only $O(\sqrt{m})$ officials, but they may need to send information about m edges, which can lead to congestion. For example, if more than $\alpha \cdot \sqrt{m}$ messages need to be sent to an officials of degree $\sqrt{m}$, then this will take at least α rounds. We use randomization of leaders and the property of clique emulation to avoid this situation and make sure that officials do not have to send or receive more than $O(\sqrt{m} \log m)$ messages in a phase. The second source for delays is the fragments merging procedure. This further splits into two types of problems. The first is that a chain of fragments that need to be merged could be long, and in the basic distributed Boruvka's algorithm will take long time (up to n) to resolve. This problem is overcome by using a modified pointer jumping technique similar to [16]. The second problem is that the number of fragments that need to be merged could be large, resulting in a large number of *merging* messages that contain, for example, the new fragment ID. This problem is overcome by using randomization and by reducing the number of messages needed for resolving a merge. Full description of the algorithm along with the proofs of correctness and running time can be found in [17].

4   Additional Algorithms in Core-Periphery Networks

In addition to MST, we have considered a number of other distributed problems of different types, and developed algorithms for these problems that can be efficiently executed on core-periphery networks. In particular, we dealt with the following set of tasks related to matrix operations. (M1) Sparse matrix transposition. (M2) Multiplication of a sparse matrix by a vector. (M3) Multiplication of two sparse matrices.

We then considered problems related to calculating aggregate functions of initial values initially stored one at each node in V. In particular, we developed efficient algorithms for the following problems. (A1) Finding the rank of each value, assuming the values are ordered. (As output, each node should know the rank of the element it stores.) (A2) Finding the median of the values. (A3) Finding the (statistical) mode, namely, the most frequent value. (A4) Finding the number of distinct values stored in the network. Each of these problems requires $\Omega(Diam)$ rounds on general networks, whereas on a $\mathcal{CP}$-network it can be performed in $O(1)$ rounds.

An additional interesting task is defined in a setting where the initial values are split into disjoint groups, and requires finding the r largest values of each group. This task can be used, for example, for finding the most popular headlines in each area of news. Here, there is an $O(r)$ round solution on a $\mathcal{CP}$-network, whereas in general networks the diameter is still a lower bound.

In all of these problems, we also establish the necessity of all 3 axioms, by showing that there are network families satisfying 2 of the 3 axioms for which the general lower bound holds. Due to space limitation, we discuss in this section only one of these problems, namely, multiplication of a vector by a sparse matrix. Our results for the other problems can be found in [17].

A few definitions are in place. Let A be a matrix in which each entry $A(i, j)$ can be represented by $O(\log n)$ bits (i.e., it fits in a single message in the CONGEST model). Denote by $A_{i,*}$ (respectively, $A_{*,i}$) the ith row (resp., column) of A. Denote the ith entry of a vector s by $s(i)$. We assume that the nodes in $\mathcal{C}$ have IDs $[1, \ldots, n_c]$ and this is known to all of them. A square $n \times n$ matrix A with $O(k)$ nonzero entries in each row and each column is hereafter referred to as an $O(k)$-*sparse matrix*.

Let s be a vector of size n and A be a square $n \times n$ $O(k)$-sparse matrix. Initially, each node in V holds one entry of s (along with the index of the entry) and one row of A (along with the index of the row). The task is to distributively calculate vector $s' = sA$ and store its entries at the corresponding nodes in V, such that the node that initially stored $s(i)$ will store $s'(i)$. We start with a claim on the lower bound (the proof can be found in [17]).

Claim 3. *The* lower bound *for any algorithm for multiplication of a vector by a sparse matrix on any network is* $\Omega(D)$, *and on a $\mathcal{CP}$-network is* $\Omega(k/\log n)$.

Algorithm 1. The following algorithm solves the task in $O(k)$ rounds on a $\mathcal{CP}$-network $G(V, E, \mathcal{C}, \mathcal{P})$.

1. Each $u \in V$ sends the entry of s it has (along with the index of the entry) to its representative $r(u) \in \mathcal{C}$ (recall that if $u \in \mathcal{C}$ then $r(u) = u$).
2. $\mathcal{C}$ nodes redistribute the s entries among them so that the node with ID i stores indices $[1 + (n/n_c)(i - 1), \ldots, (n/n_c)i]$ (assume n/n_c is integer).
3. Each $u \in V$ sends the index of the row of A it has to $r(u) \in \mathcal{C}$.
4. Each representative requests the $s(i)$ entries corresponding to rows $A_{i,*}$ that it represents from the $\mathcal{C}$ node storing it.
5. Each representative gets the required elements of s and sends them to the nodes in $\mathcal{P}$ it represents.
6. Each $u \in V$ sends the products $\{A(i, j)s(i)\}_{j=1}^{n}$ to its representative.
7. Each representative sends each nonzero value $A(i, j)s(i)$ it has (up to $O(kn_c)$ values) to the representative responsible for $s(j)$, so it can calculate $s'(j)$.
8. Now, each node $u \in V$ that initially stored $s(i)$, requests $s'(i)$ from its representative. The representative gets the entry from the corresponding node in $\mathcal{C}$ and sends it back to u.

We state the following results regarding the running time of Algorithm 1.

Theorem 5. *On a $\mathcal{CP}$-network $G(V, E, \mathcal{C}, \mathcal{P})$, the multiplication of a $O(k)$-sparse matrix by a vector can be completed in $O(k)$ rounds w.h.p.*

Before we start with the proof, we present the following theorem from [2].

Theorem 6 ([2]). *Consider a fully connected system of n_c nodes. Each node is given up to M_s messages to send, and each node is the destination of at most M_r messages. There exists algorithm that delivers all the messages to their destinations in $O\left(\frac{M_s + M_r}{n_c}\right)$ rounds w.h.p.*

This theorem will be extensively used by our algorithms since it gives running time bound on messages delivery in a core that satisfies Axiom $\mathcal{A}_E$. The result of the theorem holds with high probability which implies that it exploits a randomized algorithm. Nevertheless, our algorithms can be considered deterministic in the sense that all the decisions they make are deterministic. The randomness of the information delivery algorithm of Theorem 6 does not affect our algorithms since the decisions when and what message will be sent along with the message source and destination, are deterministically controlled by our algorithms.

Proof of Theorem 5. Consider Algorithm 1 and the $\mathcal{CP}$-network $G(V, E, \mathcal{C}, \mathcal{P})$. At Step 1, due to $\mathcal{A}_B$ and $\mathcal{A}_C$, each representative will obtain $O(n_c)$ entries of s in $O(1)$ rounds. For Step 2, we use Theorem 6 with the parameters: $M_s = O(n_c)$ and $M_r = O(n/n_c)$, and thus such a redistribution will take $O((n_c + n/n_c)/n_c) = O(1)$ rounds. At Step 3, due to $\mathcal{A}_B$ and $\mathcal{A}_C$ each representative will obtain $O(n_c)$ row indices of A in $O(1)$ rounds.

For Step 4, we again use Theorem 6 with the parameters: $M_s = O(n_c)$ (indices of rows each representative has), $M_r = O(n/n_c)$ (number of entries of s stored in each node in $\mathcal{C}$), and obtain the running time for this step: $O((n_c + n/n_c)/n_c) = O(1)$ rounds. At Step 5, each representative gets the required elements of s which takes running time is $O(1)$ due to Theorem 6, and then sends them to the nodes in $\mathcal{P}$ it represents which also takes $O(1)$ due to $\mathcal{A}_C$. Step 6 takes $O(k)$ rounds since A has up to k nonzero entries in each row. Step 7 again uses Theorem 6 with parameters $M_s = O(kn_c)$, $M_r = O(n/n_c)$, and thus the running time is $O(kn/n_c^2) = O(k)$.

At Step 8, a single message is sent by each node to its representative (takes $O(1)$ due to $\mathcal{A}_C$), then the requests are delivered to the appropriate nodes in $\mathcal{C}$ and the replies with the appropriate entries of s' are received back by the representatives. All this takes $O(1)$ rounds due to the Axiom $\mathcal{A}_E$ and Theorem 6. Then the entries of s' are delivered to the nodes that have requested them. Due to $\mathcal{A}_C$ this will also take $O(1)$ rounds. $\square$

The following theorem shows the necessity of the axioms for achieving $O(k)$ running time. The proof of the theorem can be found in [17].

Theorem 7. *For each $X \in \{B, E, C\}$ there exist a family $\mathcal{F}_X = \{G_X(V, E, \mathcal{C}, \mathcal{P})(n)\}$ of partitioned networks that do not satisfy Axiom $\mathcal{A}_X$ but satisfy the other* *two axioms, and input matrices of size $n \times n$ and vectors of size n, for every n, such that the time complexity of any algorithm for multiplying a vector by a matrix on the networks of $\mathcal{F}_X$ with the corresponding-size inputs is $\Omega(n/\log n)$.*

References

1. Lotker, Z., Patt-Shamir, B., Pavlov, E., Peleg, D.: Minimum-weight spanning tree construction in o(log log n) communication rounds. SIAM J. Computing 35(1), 120–131 (2005)
2. Lenzen, C., Wattenhofer, R.: Tight bounds for parallel randomized load balancing. In: STOC, pp. 11–20 (2011)
3. Lenzen, C.: Optimal deterministic routing and sorting on the congested clique. In: PODC, pp. 42–50 (2013)
4. Avin, C., Lotker, Z., Pignolet, Y.A., Turkel, I.: From caesar to twitter: An axiomatic approach to elites of social networks. CoRR abs/1111.3374 (2012)
5. Fujita, M., Krugman, P.R., Venables, A.J.: The spatial economy: Cities, regions, and international trade. MIT Press (2001)
6. Krugman, P.: Increasing Returns and Economic Geography. The Journal of Political Economy 99(3), 483–499 (1991)
7. Holme, P.: Core-periphery organization of complex networks. Physical Review E 72, 46111 (2005)
8. Liang, J., Kumar, R., Ross, K.W.: The fasttrack overlay: A measurement study. Computer Networks 50, 842 (2006)
9. Baset, S., Schulzrinne, H.: An analysis of the skype peer-to-peer internet telephony protocol. In: INFOCOM, pp. 1–11 (2006)
10. Attiya, H., Welch, J.: Distributed Computing: Fundamentals, Simulations and Advanced Topics. McGraw-Hill (1998)
11. Lynch, N.: Distributed Algorithms. Morgan Kaufmann (1995)
12. Peleg, D.: Distributed Computing: A Locality-Sensitive Approach. SIAM (2000)
13. Adamic, L.: The small world web. Research and Advanced Technology for Digital Libraries, 852–852 (1999)
14. Bonanno, G., Caldarelli, G., Lillo, F., Mantegna, R.: Topology of correlation-based minimal spanning trees in real and model markets. Phys. Rev. E 68 (2003)
15. Chen, C., Morris, S.: Visualizing evolving networks: Minimum spanning trees versus pathfinder networks. In: INFOVIS, pp. 67–74 (2003)
16. Lotker, Z., Patt-Shamir, B., Peleg, D.: Distributed MST for constant diameter graphs. Distributed Computing 18(6), 453–460 (2006)
17. Avin, C., Borokhovich, M., Lotker, Z., Peleg, D.: Distributed computing on core-periphery networks: Axiom-based design. CoRR abs/1404.6561 (2014)
18. Nesetril, J., Milkova, E., Nesetrilova, H.: Otakar boruvka on minimum spanning tree problem translation of both the 1926 papers, comments, history. Discrete Mathematics 233(1-3), 3–36 (2001)

Fault-Tolerant Rendezvous in Networks*

Jérémie Chalopin[1],**, Yoann Dieudonné[2], Arnaud Labourel[1],**,
and Andrzej Pelc[3],***

[1] LIF, CNRS & Aix-Marseille University, Marseille, France
[2] MIS, Université de Picardie Jules Verne, France
[3] Département d'informatique, Université du Québec en Outaouais,
Gatineau, Québec, Canada

Abstract. Two mobile agents, starting from different nodes of an
unknown network, have to meet at the same node. Agents move in synchronous rounds using a deterministic algorithm. Each agent has a different label, which it can use in the execution of the algorithm, but it does
not know the label of the other agent. Agents do not know any bound on
the size of the network. In each round an agent decides if it remains idle
or if it wants to move to one of the adjacent nodes. Agents are subject
to *delay faults*: if an agent incurs a fault in a given round, it remains in
the current node, regardless of its decision. If it planned to move and
the fault happened, the agent is aware of it. We consider three scenarios
of fault distribution: random (independently in each round and for each
agent with constant probability $0 < p < 1$), unbounded adversarial (the
adversary can delay an agent for an arbitrary finite number of consecutive rounds) and bounded adversarial (the adversary can delay an agent
for at most c consecutive rounds, where c is unknown to the agents). The
quality measure of a rendezvous algorithm is its cost, which is the total
number of edge traversals.

For random faults, we show an algorithm with cost polynomial in the
size n of the network and *polylogarithmic* in the larger label L, which
achieves rendezvous with very high probability in arbitrary networks.
By contrast, for unbounded adversarial faults we show that rendezvous
is not feasible, even in the class of rings. Under this scenario we give
a rendezvous algorithm with cost $O(n\ell)$, where ℓ is the smaller label,
working in arbitrary trees, and we show that $\Omega(\ell)$ is the lower bound
on rendezvous cost, even for the two-node tree. For bounded adversarial
faults, we give a rendezvous algorithm working for arbitrary networks,
with cost polynomial in n, and *logarithmic* in the bound c and in the
larger label L.

Keywords: rendezvous, deterministic algorithm, mobile agent, delay
fault.

---

* The full version of the paper is available at http://arxiv.org/abs/1402.2760
** Partially supported by the ANR project MACARON (anr-13-js02-0002).
*** Partially supported by NSERC discovery grant and by the Research Chair in Dis-
tributed Computing at the Université du Québec en Outaouais.

J. Esparza et al. (Eds.): ICALP 2014, Part II, LNCS 8573, pp. 411–422, 2014.
© Springer-Verlag Berlin Heidelberg 2014

1   Introduction

The Background. Two mobile entities, called agents, starting from different nodes of a network, have to meet at the same node. This task is known as *rendezvous* and has been extensively studied in the literature. Mobile entities may represent software agents in computer networks, mobile robots, if the network is composed of corridors in a mine, or people who want to meet in an unknown city whose streets form a network. The reason to meet may be to exchange data previously collected by the agents, or to coordinate a future network maintenance task. In this paper we study a fault-tolerant version of the rendezvous problem: agents have to meet in spite of delay faults that they can incur during navigation. Such faults may be due to mechanical reasons in the case of robots and to network congestion in the case of software agents.

The Model and the Problem. The network is modeled as an undirected connected graph. We seek deterministic rendezvous algorithms that do not rely on the knowledge of node identifiers, and can work in anonymous graphs as well (cf. [3]). The importance of designing such algorithms is motivated by the fact that, even when nodes are equipped with distinct identifiers, agents may be unable to perceive them because of limited sensory capabilities (a robot may be unable to read signs at corridor crossings), or nodes may refuse to reveal their identifiers to software agents, e.g., due to security or privacy reasons. Note that, if nodes had distinct identifiers visible to the agents, the agents might explore the graph and meet at the node with smallest identifier, hence rendezvous would reduce to graph exploration. On the other hand, we assume that edges incident to a node v have distinct labels (visible to the agents) in $\{0, \ldots, d-1\}$, where d is the degree of v. Thus every undirected edge $\{u, v\}$ has two labels, which are called its *port numbers* at u and at v. Port numbering is *local*, i.e., there is no relation between port numbers at u and at v. Note that in the absence of port numbers, edges incident to a node would be undistinguishable for agents and thus rendezvous would be often impossible, as the adversary could prevent an agent from taking some edge incident to the current node. Security and privacy reasons for not revealing node identifiers to software agents are irrelevant in the case of port numbers, and port numbers in the case of a mine or labyrinth can be made implicit, e.g., by marking one edge at each crossing (using a simple mark legible by the robot), considering it as corresponding to port 0 and all other port numbers increasing clockwise.

Agents start at different nodes of the graph and traverse its edges in synchronous rounds. They cannot mark visited nodes or traversed edges in any way. The adversary wakes up each of the agents in possibly different rounds. Each agent starts executing the algorithm in the round of its wake-up. It has a clock measuring rounds that starts at its wake-up round. In each round an agent decides if it remains idle or if it chooses a port to move to one of the adjacent nodes. Agents are subject to *delay faults* in rounds in which they decide to move: if an agent incurs a fault in such a round, it remains at the current node and is aware of the fault. We consider three scenarios of fault distribution: random (independently in each round and for each agent with constant probability $0 < p < 1$), unbounded adversarial (the adversary can delay an agent for an arbitrary finite number of consecutive rounds) and bounded adversarial (the adversary can delay an agent for at most c consecutive rounds, where c is unknown to the agents). Agents do not know the topology of the graph or any bound on its size. Each agent has a different positive integer label which it knows and can use in the execution of the rendezvous algorithm, but it does not know the label of the other agent nor its starting round. When an agent enters a node, it learns its degree and the port of entry. When agents cross each other on an edge, traversing it simultaneously in different directions, they do not notice this fact. We assume that the memory of the agents is unlimited: from the computational point of view they are modeled as Turing machines.

The quality measure of a rendezvous algorithm is its *cost*, which is the total number of edge traversals. For each of the considered fault distributions we are interested in deterministic algorithms working at low cost. For both scenarios with adversarial faults we say that a deterministic rendezvous algorithm works at a cost at most C for a given class of graphs if for any initial positions in a graph of this class both agents meet after at most C traversals, regardless of the faults imposed by the adversary obeying the given scenario. In the case of random faults the algorithm is also deterministic, but, due to the stochastic nature of faults, the estimate of its cost is with high probability.

Our Results. For random faults, we show an algorithm which achieves rendezvous in arbitrary networks at cost polynomial in the size n of the network and *polylogarithmic* in the larger label L, with very high probability. More precisely, our algorithm achieves rendezvous with probability 1, and its cost exceeds a polynomial in n and $\log L$ with probability inverse exponential in n and $\log L$. By contrast, for unbounded adversarial faults, we show that rendezvous is not feasible, even in the class of rings. Under this scenario we give a rendezvous algorithm with cost $O(n\ell)$, where ℓ is the smaller label, working in arbitrary trees, and we show that $\Omega(\ell)$ is the lower bound on rendezvous cost, even for the two-node tree. For bounded adversarial faults we give a rendezvous algorithm working for arbitrary networks, with cost polynomial in n, and *logarithmic* in the bound c and in the larger label L.

Due to lack of space, all the proofs are omitted. The full version of the paper with all the proofs is available on-line.

Related Work. The problem of rendezvous has been studied both under the randomized and the deterministic scenarios. An extensive survey of randomized rendezvous in various models can be found in [3], cf. also [1,2,4,8,24]. Deterministic rendezvous in networks has been surveyed in [31]. Several authors considered the geometric scenario (rendezvous in an interval of the real line, see, e.g., [8,9], or in the plane, see, e.g., [5,6]). Gathering more than two agents has been studied, e.g., in [22,24,29,34].

For the deterministic setting many authors studied the feasibility of synchronous rendezvous, and the time required to achieve this task, when feasible. For instance, deterministic rendezvous of agents equipped with tokens used to mark nodes was considered, e.g., in [28]. Deterministic rendezvous of two agents that cannot mark nodes but have unique labels was discussed in [17,26,33]. Since this is our scenario, these papers are the most relevant in our context. All of them are concerned with the time of rendezvous in arbitrary graphs. In [17] the authors show a rendezvous algorithm polynomial in the size of the graph, in the length of the shorter label and in the delay between the starting times of the agents. In [26,33] rendezvous time is polynomial in the first two of these parameters and independent of the delay.

Memory required by the agents to achieve deterministic rendezvous has been studied in [23] for trees and in [12] for general graphs. Memory needed for randomized rendezvous in the ring is discussed, e.g., in [27].

Apart from the synchronous model used in this paper, several authors investigated asynchronous rendezvous in the plane [11,22] and in network environments [7,13,16,20]. In the latter scenario the agent chooses the edge which it decides to traverse but the adversary controls the speed of the agent. Under this assumption rendezvous in a node cannot be guaranteed even in very simple graphs and hence the rendezvous requirement is relaxed to permit meeting inside an edge.

Fault-tolerant aspects of the rendezvous problem have been investigated in [10,14,15,19,21]. Faulty unmovable tokens were considered in the context of the task of gathering many agents at one node. In [14,21] the authors considered gathering in rings, and in [15] gathering was studied in arbitrary graphs, under the assumption that an unmovable token is located in the starting node of each agent. Tokens could disappear during the execution of the algorithm, but they could not reappear again. Byzantine tokens which can appear and disappear arbitrarily have been considered in [18] for the related task of network exploration. A different fault scenario for gathering many agents was investigated in [19]. The authors assumed that some number of agents are Byzantine and they studied the problem of how many good agents are needed to guarantee meeting of all of them despite the actions of Byzantine agents. To the best of our knowledge rendezvous with delay faults considered in the present paper has never been studied before.

2   Preliminaries

Throughout the paper, the number of nodes of a graph is called its size. In this section we recall two procedures known from the literature, that will be used as building blocks in some of our algorithms. The aim of the first procedure is graph exploration, i.e., visiting all nodes and traversing all edges of the graph by a single agent. The procedure, based on universal exploration sequences (UXS) [25], is a corollary of the result of Reingold [32]. Given any positive integer m, it allows the agent to traverse all edges of any graph of size at most m, starting from any node of this graph, using $P(m)$ edge traversals, where P is some polynomial. (The original procedure of Reingold only visits all nodes, but it can be transformed to traverse all edges by visiting all neighbors of each visited node before going to the next node.) After entering a node of degree d by some port p, the agent can compute the port q by which it has to exit; more precisely $q = (p + x_i) \mod d$, where x_i is the corresponding term of the UXS.

A *trajectory* is a sequence of nodes of a graph, in which each node is adjacent to the preceding one. Given any starting node v, we denote by $R(m, v)$ the trajectory obtained by Reingold's procedure followed by its reverse. (Hence the trajectory starts and ends at node v.) The procedure can be applied in any graph starting at any node, giving some trajectory. We say that the agent *follows* a trajectory if it executes the above procedure used to construct it. This trajectory will be called *integral*, if the corresponding route covers all edges of the graph. By definition, the trajectory $R(m, v)$ is integral if it is obtained by Reingold's procedure applied in any graph of size at most m starting at any node v.

The second auxiliary procedure is the Algorithm RV-asynch-poly from [20] that guarantees rendezvous of two agents under the *asynchronous* scenario. Unlike in the synchronous scenario used in the present paper, in the asynchronous scenario each agent chooses consecutive ports that it wants to use but the adversary controls the speed of the agent, changing it arbitrarily during navigation. Rendezvous is guaranteed in the asynchronous scenario, if it occurs for any behavior of the adversary. Under this assumption rendezvous in a node cannot be guaranteed even in very simple graphs and hence the rendezvous requirement is relaxed to permit the agents to meet inside an edge. Recall that in our synchronous scenario, agents crossing each other on an edge traversing it simultaneously in different directions, not only do not meet but do not even notice the fact of crossing.

Algorithm RV-asynch-poly works at cost polynomial in the size n of the graph in which the agents operate and in the length of the smaller label. Let A be a polynomial, such that if two agents with different labels λ_1 and λ_2 execute Algorithm RV-asynch-poly in an n-node graph, then the agents meet in the asynchronous model, after at most $A(n, \min(\log \lambda_1, \log \lambda_2))$ steps.

3   Random Faults

In this section we consider the scenario when agents are subject to random and independent faults. More precisely, for each agent and each round the probability that the agent is delayed in this round is $0 < p < 1$, where p is a constant, and the events of delaying are independent for each round and each agent. Under this scenario we construct a deterministic rendezvous algorithm that achieves rendezvous in any connected graph with probability 1 and its cost exceeds a polynomial in n and $\log L$ with probability inverse exponential in n and $\log L$, where n is the size of the graph and L is the larger label.

The intuition behind the algorithm is the following. Since the occurrence of random faults represents a possible behavior of the asynchronous adversary in Algorithm RV-asynch-poly from [20], an idea to get the guarantee of a meeting with random faults at polynomial cost might be to only use this algorithm. However, this meeting may occur either at a node or inside an edge, according to the model from [20]. In the synchronous model with random faults considered in this section, the second type of meeting is not considered as rendezvous, in fact agents do not even notice it. Hence we must construct a *deterministic* mechanism which guarantees a legitimate meeting at a node, with high probability, soon after an "illegitimate" meeting inside an edge. Constructing this mechanism and proving its correctness is the main challenge of rendezvous with random faults.

Before describing the algorithm we define the following transformation of the label λ of an agent. Let $\Phi(0) = (0011)$ and $\Phi(1) = (1100)$. Let $(c_1 \ldots c_k)$ be the binary representation of the label λ. We define the *modified label* λ^* of the agent as the concatenation of sequences $\Phi(c_1), \ldots, \Phi(c_k)$ and (10). Note that if labels of two agents are different, then their transformed labels are different and none of them is a prefix of the other.

We first describe the procedure Dance (λ, x, y) executed by an agent with label λ located at node y at the start of the procedure. Node x is a node adjacent to y.

Procedure Dance (λ, x, y)

Let $\lambda^* = (b_1, \ldots, b_m)$.

Stage 1.

Stay idle at y for 10 rounds.

Stage 2.

for $i = 1$ **to** m **do**

    if $b_i = 0$

    then stay idle for two rounds

    else go to x and in the next round return to y.

Stage 3.

Traverse the edge $\{x, y\}$ 12 times (i.e., go back and forth 6 times on this edge). ◇

Note that procedure Dance (λ, x, y) has cost $O(\log \lambda)$.

We will also use procedure Asynch(λ) executed by an agent with label λ starting at any node x_0 of a graph. This procedure produces an infinite walk $(x_0, x_1, x_2, \ldots)$ in the graph resulting from applying Algorithm RV-asynch-poly by a single agent with label λ.

Using these procedures we now describe Algorithm RV-RF (for rendezvous with random faults), that works for an agent with label λ starting at an arbitrary node of any connected graph.

Algorithm RV-RF

The algorithm works in two phases interleaved in a way depending on faults occurring in the execution and repeated until rendezvous. The agent starts executing the algorithm in phase Progress.

Phase Progress

This phase proceeds in stages. Let $(x_0, x_1, x_2, \ldots)$ be the infinite walk produced by the agent starting at node x_0 and applying Asynch(λ). The ith stage of phase Progress, for $i \geq 1$, is the traversal of the edge $\{x_{i-1}, x_i\}$ from x_{i-1} to x_i, followed by the execution of Dance (λ, x_{i-1}, x_i). The agent executes consecutive stages of phase Progress until a fault occurs.

If a fault occurs in the first round of the ith stage, then the agent repeats the attempt of this traversal again, until success and then continues with Dance (λ, x_{i-1}, x_i). If a fault occurs in the tth round of the ith stage, for $t > 1$, i.e., during the execution of procedure Dance (λ, x_{i-1}, x_i) in the ith stage, then this execution is interrupted and phase Correction is launched starting at the node where the agent was situated when the fault occurred.

Phase Correction

Let e denote the edge $\{x_{i-1}, x_i\}$ and let w be the node at which the agent was situated when the last fault occurred during the execution of Dance (λ, x_{i-1}, x_i). Hence w is either x_{i-1} or x_i.

Stage 1. Stay idle at w for 20 rounds.

Stage 2. Traverse edge e 20 times.

Stage 3. If the agent is not at w, then go to w.

If a fault occurs during the execution of phase Correction, then the execution of this phase is dropped and a new phase Correction is launched from the beginning, starting at the node where the agent was situated when the fault occurred. Upon completing an execution of the phase Correction without any fault the agent is at node w. It resumes the execution of the tth round of the ith stage of phase Progress.   ◇

The following theorem shows the correctness and estimates the performance of Algorithm RV-RF. More precisely, it shows that Algorithm RV-RF achieves rendezvous at polynomial cost with very high probability, under the random fault model.

Theorem 1. *Consider two agents a and b, with different labels λ_1 and λ_2, respectively, starting at arbitrary different nodes of an n-node graph, where n is unknown to the agents. Suppose that delay faults occur randomly and independently with constant probability $0 < p < 1$ in each round and for each agent. Algorithm RV-RF guarantees rendezvous of the agents with probability 1. Moreover, there exists a polynomial B such that rendezvous at some node occurs at cost $\tau = O(B(n, \max(\log \lambda_1, \log \lambda_2)))$ with probability at least $1 - e^{-O(\tau)}$.*

The proof of this theorem relies on the following property: with very high probability, each agent can execute sufficiently many times a block of at least 42 consecutive rounds without being subject to any fault. This permits the agents to execute sufficiently many steps of phase Progress to achieve rendezvous because an entire execution of phase Correction consists of at most 41 rounds.

4   Unbounded Adversarial Faults

In this section we consider the scenario when the adversary can delay each of the agents for any finite number of consecutive rounds. Under this scenario the time (number of rounds until rendezvous) depends entirely on the adversary, so the only meaningful measure of efficiency of a rendezvous algorithm is its cost. However, it turns out that, under this harsh fault scenario, even feasibility of rendezvous is usually not guaranteed, even for quite simple graphs. Recall that we *do not* assume knowledge of any upper bound on the size of the graph.

The following theorem establishes the impossibility of rendezvous with unbounded adversarial faults.

Theorem 2. *Rendezvous with unbounded adversarial faults is not feasible, even in the class of rings.*

In view of Theorem 2, it is natural to ask if rendezvous with unbounded adversarial faults can be accomplished in the class of connected graphs not containing cycles, i.e., in the class of trees, and if so, at what cost it can be done. The rest of this section is devoted to a partial answer to this problem. Our goal is to present an efficient rendezvous algorithm working for arbitrary trees. We will use the following notion. Consider any tree T. A *basic walk* in T, starting from node v is a traversal of all edges of the tree ending at the starting node v and defined as follows. Node v is left by port 0; when the walk enters a node by port i, it leaves it by port $(i+1) \bmod d$, where d is the degree of the node. Any basic walk consists of $2(n-1)$ edge traversals. An agent completing the basic walk knows that this happened and learns the size n of the tree and the length $2(n-1)$ of the basic walk.

The following Algorithm `Tree-RV-UF` (for rendezvous in trees with unbounded faults) works for an agent with label λ, starting at an arbitrary node of any tree T.

Algorithm Tree-RV-UF

Repeat 2λ basic walks starting from the initial position and stop.   ◇

Theorem 3. *Algorithm* `Tree-RV-UF` *is a correct rendezvous algorithm with unbounded adversarial faults in arbitrary trees, and works at cost $O(n\ell)$, where n is the size of the tree and ℓ is the smaller label.*

We do not know if Algorithm `Tree-RV-UF` has optimal cost, i.e., if a lower bound $\Omega(n\ell)$ can be proved on the cost of any rendezvous algorithm with unbounded adversarial faults, working in arbitrary trees of size n. However, we establish a weaker lower bound. It is clear that no algorithm can beat cost $\Theta(n)$ for rendezvous in n-node trees, even without faults. Our next result shows that, for unbounded adversarial faults, $\Omega(\ell)$ is a lower bound on the cost of any rendezvous algorithm, even for the simplest tree, that of two nodes.

Proposition 1. *Let T be the two-node tree. Every rendezvous algorithm with unbounded adversarial faults, working for the tree T, has cost $\Omega(\ell)$, where ℓ is the smaller label.*

5   Bounded Adversarial Faults

In this section we consider the scenario when the adversary can delay each of the agents for at most c consecutive rounds, where c is a positive integer, called the *fault bound*. First note that if c is known to the agents, then, given any synchronous rendezvous algorithm working without faults for arbitrary networks, it is possible to obtain an algorithm working for bounded adversarial faults and for arbitrary networks, at the same cost. Let $\mathcal{A}$ be a synchronous rendezvous algorithm for the scenario without faults, working for arbitrary networks. Consider the following algorithm $\mathcal{A}(c)$ working for bounded adversarial faults with parameter c. Each agent replaces each round r of algorithm $\mathcal{A}$ by a segment of $2c + 1$ rounds. If in round r of algorithm $\mathcal{A}$ the agent was idle, this round is replaced by $2c + 1$ consecutive rounds in which the agent is idle. If in round r the agent left the current node by port p, this round is replaced by a segment of $2c + 1$ rounds in each of which the agent makes an attempt to leave the current node v by port p until it succeeds, and in the remaining rounds of the segment it stays idle at the node adjacent to v that it has just entered.

We associate the first segment of the later starting agent with the (unique) segment of the earlier agent that it intersects in at least $c + 1$ rounds. Let it be the ith segment of the earlier agent. We then associate the jth segment of the later agent with the $(j + i - 1)$th segment of the earlier agent, for $j > 1$. Hence, regardless of the delay between starting rounds of the agents, corresponding segments intersect in at least $c + 1$ rounds. If the agents met at node x in the jth round of the later agent, according to algorithm $\mathcal{A}$, then, according to algorithm $\mathcal{A}(c)$, in the last $c + 1$ rounds of its jth segment the later agent is at x and in the last $c + 1$ rounds of its $(j + i - 1)$th segment the earlier agent is at x. Since these segments intersect in at least $c + 1$ rounds, there is a round in which both agents are at node x according to algorithm $\mathcal{A}(c)$, regardless of the actions of the adversary, permitted by the bounded adversarial fault scenario. This shows that algorithm $\mathcal{A}(c)$ is correct. Notice that the cost of algorithm $\mathcal{A}(c)$ is the same as that of algorithm $\mathcal{A}$, because in each segment corresponding to an idle round of algorithm $\mathcal{A}$, an agent stays idle in algorithm $\mathcal{A}(c)$ and in each segment corresponding to a round in which an agent traverses an edge in algorithm $\mathcal{A}$, the agent makes exactly one traversal in algorithm $\mathcal{A}(c)$.

In the rest of this section we concentrate on the more difficult situation when the fault bound c is unknown to the agents. The following Algorithm Graph-RV-BF (for rendezvous in graphs with bounded faults) works for an agent with label λ starting at an arbitrary node of any graph.

Algorithm Graph-RV-BF is divided into phases. The i-th phase is composed of 2^i stages, each lasting $s_i = 2^{i+4}$ rounds. Hence the i-th phase lasts $p_i = 2^{2i+4}$ rounds. The λ-th stage of the i-th phase consists of two parts: the *busy* part of $b_i = 3 \cdot 2^i$ rounds and the *waiting* part of $w_i = 13 \cdot 2^i$ rounds. During the busy part of the λ-th stage of phase i, the agent (with label λ) tries to explore the graph three times (each exploration attempt lasts at most $e_i = 2^i$ rounds), using a UXS. We say that the agent is *active* during the busy part of the λ-th stage of each phase $i \geq q = \lceil \log(\lambda + 1) \rceil$. In order to explore the graph, the agent keeps estimates of the values of c and $P(n)$. (Recall that the latter is the length of a UXS that allows to traverse all edges of any graph of size at most n, starting from any node). The values of these estimates in phase i are called c_i and u_i, respectively, and grow depending on the strategy of the adversary. For the first phase q in which the agent is active, we set $u_q = 1$ and $c_q = 2^q$. In phase i the agent uses the UXS of length u_i. Call this sequence S. The agent uses this UXS proceeding by steps. Steps correspond to terms of the sequence S. During phase i, the k-th step consists of s_i rounds during which the agent tries to move, using port $(p + S[k] \mod d)$ (where d is the degree of the current node and p is the port by which the agent entered the current node), until it succeeds or until the s_i rounds of the k-th step are over. If it succeeded to move, it waits until the s_i rounds of the step are over. If the agent succeeds to perform all of its three UXS explorations during a phase i, i.e., if it succeeds to move once in each step, then we set $u_{i+1} = 2u_i$ and $c_{i+1} = c_i$. Otherwise, we set $u_{i+1} = u_i$ and $c_{i+1} = 2c_i$. When the agent is not active, it waits at its current node. The agent executes this algorithm until it meets the other agent.

Theorem 4. *Algorithm* `Graph-RV-BF` *is a correct rendezvous algorithm with bounded adversarial faults in arbitrary graphs, and works at cost polynomial in the size n of the graph, and logarithmic in the fault bound c and in the larger label L.*

Notice that in the bounded fault scenario (as opposed to the unbounded fault scenario) it makes sense to speak about the time of a rendezvous algorithm execution (i.e., the number of rounds from the start of the earlier agent until rendezvous), apart from its cost. Indeed, now the time can be controlled by the algorithm. Our last result gives an estimate on the execution time of Algorithm `Graph-RV-BF`.

Theorem 5. *Algorithm* `Graph-RV-BF` *works in time polynomial in the size n of the graph, in the fault bound c and in the larger label L.*

Notice the difference between the estimates of cost and of time of Algorithm `Graph-RV-BF`: while we showed that cost is polylogarithmic in L and c, for time we were only able to show that it is polynomial in L and c. Indeed, Algorithm `Graph-RV-BF` relies on a technique similar to "coding by silence" in the time-slice algorithm for leader election [30]: "most of the time" both agents stay idle, in order to guarantee that agents rarely move simultaneously. It remains open whether there exists a rendezvous algorithm with bounded adversarial faults, working for arbitrary graphs, whose both cost and time are polynomial in the size n of the graph, and polylogarithmic in the fault bound c and in the smaller label ℓ.

6   Conclusion

We presented algorithms for rendezvous with delay faults under various distributions of faults. Since we assumed no knowledge of any bound on the size of the graph, for unbounded adversarial faults rendezvous is impossible, even for the class of rings. Hence it is natural to ask how the situation changes if a polynomial upper bound m on the size of the graph is known to the agents. In this case, even under the harshest model of unbounded adversarial faults, a simple rendezvous algorithm can be given. In fact this algorithm mimics the asynchronous rendezvous algorithm (without faults) from [16]. An agent with label λ, starting at node v of a graph of size at most m, repeats $(P(m)+1)^\lambda$ times the trajectory $R(m, v)$, which starts and ends at node v, and stops. Indeed, in this case, the number of integral trajectories $R(m, v)$ performed by the agent with larger label is larger than the number of edge traversals by the other agent, and consequently, if they have not met before, the larger agent must meet the smaller one after the smaller agent stops, because the larger agent will still perform at least one entire trajectory afterwards. The drawback of this algorithm is that, while its cost is polynomial in m, it is exponential in the smaller label ℓ. We know from Theorem 1 that the cost of any rendezvous algorithm must be at least linear in ℓ, even for the two-node tree. Hence an interesting open problem is:

Does there exist a deterministic rendezvous algorithm, working in arbitrary graphs for unbounded adversarial faults, with cost polynomial in the size of the graph and in the smaller label, if a polynomial upper bound on the size of the graph is known to the agents?

References

1. Alpern, S.: The rendezvous search problem. SIAM J. on Control and Optimization 33, 673–683 (1995)
2. Alpern, S.: Rendezvous search on labelled networks. Naval Reaserch Logistics 49, 256–274 (2002)
3. Alpern, S., Gal, S.: The theory of search games and rendezvous. Int. Series in Operations research and Management Science. Kluwer Academic Publisher (2002)
4. Anderson, E., Weber, R.: The rendezvous problem on discrete locations. Journal of Applied Probability 28, 839–851 (1990)
5. Anderson, E., Fekete, S.: Asymmetric rendezvous on the plane. In: Proc. 14th Annual ACM Symp. on Computational Geometry, pp. 365–373 (1998)
6. Anderson, E., Fekete, S.: Two-dimensional rendezvous search. Operations Research 49, 107–118 (2001)
7. Bampas, E., Czyzowicz, J., Gąsieniec, L., Ilcinkas, D., Labourel, A.: Almost optimal asynchronous rendezvous in infinite multidimensional grids. In: Lynch, N.A., Shvartsman, A.A. (eds.) DISC 2010. LNCS, vol. 6343, pp. 297–311. Springer, Heidelberg (2010)
8. Baston, V., Gal, S.: Rendezvous on the line when the players' initial distance is given by an unknown probability distribution. SIAM J. on Control and Opt. 36, 1880–1889 (1998)
9. Baston, V., Gal, S.: Rendezvous search when marks are left at the starting points. Naval Reaserch Logistics 48, 722–731 (2001)
10. Chalopin, J., Das, S., Widmayer, P.: Deterministic symmetric rendezvous in arbitrary graphs: Overcoming anonymity, failures and uncertainty. In: Alpern, S., et al. (eds.) Search Theory: A Game Theoretic Perspective, pp. 175–195. Springer (2013)
11. Cieliebak, M., Flocchini, P., Prencipe, G., Santoro, N.: Distributed computing by mobile robots: Gathering. SIAM J. Comput. 41, 829–879 (2012)
12. Czyzowicz, J., Kosowski, A., Pelc, A.: How to meet when you forget: Log-space rendezvous in arbitrary graphs. Distributed Computing 25, 165–178 (2012)
13. Czyzowicz, J., Labourel, A., Pelc, A.: How to meet asynchronously (almost) everywhere. ACM Transactions on Algorithms 8 (2012)
14. Das, S.: Mobile Agent Rendezvous in a Ring Using Faulty Tokens. In: Rao, S., Chatterjee, M., Jayanti, P., Murthy, C.S.R., Saha, S.K. (eds.) ICDCN 2008. LNCS, vol. 4904, pp. 292–297. Springer, Heidelberg (2008)

15. Das, S., Mihalák, M., Šrámek, R., Vicari, E., Widmayer, P.: Rendezvous of Mobile Agents When Tokens Fail Anytime. In: Baker, T.P., Bui, A., Tixeuil, S. (eds.) OPODIS 2008. LNCS, vol. 5401, pp. 463–480. Springer, Heidelberg (2008)

16. De Marco, G., Gargano, L., Kranakis, E., Krizanc, D., Pelc, A., Vaccaro, U.: Asynchronous deterministic rendezvous in graphs. Theoretical Computer Science 355, 315–326 (2006)

17. Dessmark, A., Fraigniaud, P., Kowalski, D., Pelc, A.: Deterministic rendezvous in graphs. Algorithmica 46, 69–96 (2006)

18. Dieudonné, Y., Pelc, A.: Deterministic network exploration by a single agent with Byzantine tokens. Information Processing Letters 112, 467–470 (2012)

19. Dieudonné, Y., Pelc, A., Peleg, D.: Gathering despite mischief. In: Proc. 23rd Annual ACM-SIAM Symposium on Discrete Algorithms (SODA 2012), pp. 527–540 (2012)

20. Dieudonné, Y., Pelc, A., Villain, V.: How to meet asynchronously at polynomial cost. In: Proc. 32nd Annual ACM Symposium on Principles of Distributed Computing (PODC 2013), pp. 92–99 (2013)

21. Flocchini, P., An, H.-C., Krizanc, D., Luccio, F.L., Santoro, N., Sawchuk, C.: Mobile Agents Rendezvous When Tokens Fail. In: Kralovic, R., Sýkora, O. (eds.) SIROCCO 2004. LNCS, vol. 3104, pp. 161–172. Springer, Heidelberg (2004)

22. Flocchini, P., Prencipe, G., Santoro, N., Widmayer, P.: Gathering of asynchronous robots with limited visibility. Theoretical Computer Science 337, 147–168 (2005)

23. Fraigniaud, P., Pelc, A.: Delays induce an exponential memory gap for rendezvous in trees. ACM Transactions on Algorithms 9, article 17 (2013)

24. Israeli, A., Jalfon, M.: Token management schemes and random walks yield self stabilizing mutual exclusion. In: Proc. 9th Annual ACM Symposium on Principles of Distributed Computing (PODC 1990), pp. 119–131 (1990)

25. Koucký, M.: Universal traversal sequences with backtracking. Journal of Computer and System Sciences 65, 717–726 (2002)

26. Kowalski, D.R., Malinowski, A.: How to Meet in Anonymous Network. In: Flocchini, P., Gąsieniec, L. (eds.) SIROCCO 2006. LNCS, vol. 4056, pp. 44–58. Springer, Heidelberg (2006)

27. Kranakis, E., Krizanc, D., Morin, P.: Randomized Rendez-Vous with Limited Memory. In: Laber, E.S., Bornstein, C., Nogueira, L.T., Faria, L. (eds.) LATIN 2008. LNCS, vol. 4957, pp. 605–616. Springer, Heidelberg (2008)

28. Kranakis, E., Krizanc, D., Santoro, N., Sawchuk, C.: Mobile agent rendezvous in a ring. In: Proc. 23rd Int. Conference on Distributed Computing Systems (ICDCS 2003), pp. 592–599 (2003)

29. Lim, W., Alpern, S.: Minimax rendezvous on the line. SIAM J. on Control and Optimization 34, 1650–1665 (1996)

30. Lynch, N.L.: Distributed algorithms, Morgan Kaufmann Publ. Inc., San Francisco (1996)

31. Pelc, A.: Deterministic rendezvous in networks: A comprehensive survey. Networks 59, 331–347 (2012)

32. Reingold, O.: Undirected connectivity in log-space. Journal of the ACM 55 (2008)

33. Ta-Shma, A., Zwick, U.: Deterministic rendezvous, treasure hunts and strongly universal exploration sequences. In: Proc. 18th ACM-SIAM Symposium on Discrete Algorithms (SODA 2007), pp. 599–608 (2007)

34. Thomas, L.: Finding your kids when they are lost. Journal on Operational Res. Soc. 43, 637–639 (1992)

Data Delivery by Energy-Constrained Mobile Agents on a Line

Jérémie Chalopin[1], Riko Jacob[2], Matúš Mihalák[2], and Peter Widmayer[2]

[1] LIF, Aix-Marseille University & CNRS, Marseille, France
[2] Institute of Theoretical Computer Science, ETH Zurich, Zürich, Switzerland

Abstract. We consider n mobile agents of limited energy that are placed on a straight line and that need to collectively deliver a single piece of data from a given source point s to a given target point t on the line. Agents can move only as far as their batteries allow. They can hand over the data when they meet. In this paper we show that deciding whether the agents can deliver the data is (weakly) NP-complete, and for instances where all input values are integers, we present a quasi-, pseudo-polynomial time algorithm that runs in time $O(\Delta^2 \cdot n^{1+4\log\Delta})$, where Δ is the distance between s and t. This answers an open problem stated by Anaya et al. (DISC 2012).

Keywords: Mobile agents and robots; data aggregation and delivery; power-awareness; algorithms; complexity.

1 Introduction

The production of inexpensive, simple-built, mobile robots has led to new research questions in how to employ and operate a *swarm* of such robots to achieve desired goals. One of the fundamental goals of robotics is the delivery of data from given sources to specified targets. An *energy-efficient* operation of mobile robots becomes crucial when batteries of the robots are limited.

In this paper we study how to efficiently operate robots (called *agents* in this paper) of limited batteries that need to collectively deliver one piece of data along a line from a *single* source to a *single* target. Formally, our setting is given by a *source* s and a *target* t placed along a line, and n autonomous mobile agents, where agent i, $i = 1, 2, \ldots, n$, has an initial *position* a_i and an initial *range* R_i, denoting the maximum length of a walk the agent can do. We ask whether the agents can *deliver* a *message* from source s to target t. The message is *picked up* by the first agent that reaches point s. An agent i with the message can *pass on* the message to agent j, if i and j meet at the same point on the line. The message is *delivered* if an agent with the message reaches target t. No agent $i = 1, \ldots, n$ can travel more than its range R_i. We refer to this problem as DataDelivery.

Obviously, it makes no sense for an agent to carry the message more than once. Then, even though the agents can in principle move simultaneously at a

J. Esparza et al. (Eds.): ICALP 2014, Part II, LNCS 8573, pp. 423–434, 2014.

time, it is easy to observe that for the sake of completing the task only,[1] the agents can move in turns: in the first turn, the agent picking up the message at s moves; then, the agent taking over from the first agent moves; then, the agent taking over from the second agent moves, and so on. In this view, a solution to DATADELIVERY can be given in form of a *schedule* that prescribes the subset of the agents that move and the order in which they move. We call a schedule which indeed delivers the message from s to t a *feasible schedule*.

Previously, this problem has been studied in edge-weighted graphs and with multiple sources [5]. Besides other results, it has been shown that the problem is NP-complete (for general graphs), and a $\min\{3, 1 + \max_{i,j} \frac{R_i}{R_j}\}$-*resource augmented* algorithm has been presented; here, a *polynomial-time* algorithm is called a γ-*resource augmented*, $\gamma > 1$, if either the algorithm (correctly) answers that there is no feasible schedule, or it finds a feasible schedule for the modified (augmented) powers $R_i' := \gamma \cdot R_i$. The complexity of the problem for the case when the graph is a line has been left open (it has been raised as an open question by Anaya et al. [3], but not studied).

In this paper, we close the open problem in that we show that DATADELIVERY is *weakly* NP-complete (even if all input values are integers), and at the same time, if all input values are integers, we present a quasi-, pseudo-polynomial time algorithm running in time $O\left(\Delta^2 \cdot n^{1+4\log\Delta}\right)$ and in time $O\left(\Delta^2 \cdot n^{1+4\log(R_{\max}+1)}\right)$, where Δ is the distance between s and t, and $R_{\max} := \max_i R_i$.

Related Work

There are very few papers studying explicitly the algorithmic question of data-aggregation-like problems by mobile agents with limited batteries. Besides the already mentioned paper by Chalopin et al. [5], the work of Anaya et al. [3] comes closest to our problem. Anaya et al. [3] study the *convergecast* problem: given a set of mobile agents in an edge-weighted graph, each agent possessing a certain piece of data, and having a uniform battery power B, the agents need to move, not more than what the battery allows, such that at some point at least one agent knows all data. Obviously, there are no fixed source and target nodes, which constitutes a difference to our problem. However, the main difference is that in [3], it is assumed that all agents have the same range. In this case, the problem is polynomial on a line, but is NP-hard if the graph is a tree.

Our problem has a flavour of the well studied problem of data aggregation in (wireless) sensor networks [8], where the general computational problem is to schedule the communication between (mostly) stationary sensor nodes so that all data collected by the individual sensors eventually arrive in a pre-specified aggregation node. While in data aggregation, the data is being sent over communication channels, in our setting, the agents physically deliver the data.

---

[1] It is an interesting, more general, and thus an even more difficult algorihtmic question to also minimize the time needed for the delivery; for this objective, parallel simultaneous moving of the agents would be crucial.

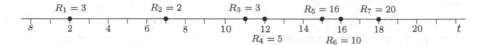

Fig. 1. A solvable instance of DATADELIVERY on a line with agents $a_1, a_2, \ldots, a_7$ depicted by the small full disks, and annotated by their respective ranges $R_1, R_2, \ldots, R_7$

Power-aware computation with mobile agents is a relatively new research area, and, consequently, there is little algorithm-theoretical research. As an exception, Heo and Varshney [7] study self-deployment of agents in this context.

A related and intensively studied research question is that of minimizing the total travelled distance by agents (which have unlimited battery) [1], [2], [4].

Further Terminology, Notation, and Model Refinement

We consider the positions of s, t and all n agents to be given by their distance from s on the line. For simplicity, we identify the source s and the target t with this distance, i.e., we set $s = 0$ and we interpret $t > 0$ as the distance of the target from s. The position of agent i, $i = 1, \ldots, n$, is given by its distance a_i from s. Sometimes, we will refer to agent i by its position, i.e., we say agent a_i to denote agent i. Recall that R_i is the range of agent i, and let $R_{\max} = \max\{R_i \mid 1 \leq i \leq n\}$. Figure 1 gives a (solvable)[2] instance of DATADELIVERY on a line.

We will assume, without loss of generality, that all a_i are between s and t: if there is an agent a_i lying left of s, we can move it to s (and reduce its range correspondingly), and similarly, we can move any agent a_j right of t to t (and reduce its range correspondingly). Obviously, the original instance has a solution if and only if the modified instance has one. In this adjusted problem instance, we may assume that $R_i < 2t$ for any agent i (as otherwise agent i with $R_i \geq 2t$ can deliver the message on its own).

Furthermore, we can assume that there is no agent at s. If there should be such an agent a_i, we use $s' = s + R_i$ as the new starting position. Now, any schedule from s to t can still be used (starting from the point in time when the packet passed position s') to deliver the data from s' to t, the only agent no longer available is a_i who cannot be used beyond s' anyway. Finally, we adjust all agents to the left of s' as described above. If this leads to an agent being positioned at s', we repeat the process.

2   The Quasi-, Pseudo-Polynomial Time Algorithm

In this section we present a dynamic-programming based algorithm for finding a feasible schedule, if the ranges and the positions of the agents are integers. We restrict ourselves to a specific class of feasible schedules: we call a feasible schedule $(a_{i_1}, a_{i_2}, \ldots, a_{i_j}, \ldots)$ *normalized*, if

---

[2] A solution is the schedule $(a_5, a_1, a_7, a_2, a_4, a_3, a_6)$.

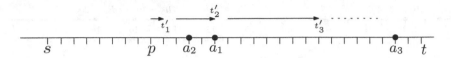

Fig. 2. Agents $a_1, a_2, a_3 \in A_p$ are p-crossing. The intervals (s'_i, t'_i) are lower bounds on the growth of the intervals (s_i, t_i): $s_i \geq s'_i$ and $t_i \geq t'_i$.

1. The positions where agents exchange the message are integers.
2. Every agent a_{i_j} walks with the message as far towards t as its range allows, with the exception when the agent reaches the initial position of the next agent in the feasible schedule, i.e., agent $a_{i_{j+1}}$.
3. The length of the schedule is minimal, i.e., we cannot remove any agent from the schedule and maintain its feasibility. This means, for example, that no agent a_{i_j} can reach (by exhausting its range) the point at which $a_{i_{j+2}}$ picks up the data from $a_{i_{j+1}}$ (making $a_{i_{j+1}}$ obsolete).

It is easy to see that in our "integer setting", if there exists a feasible schedule, there always exists a normalized feasible schedule, and thus our restriction to these schedules is without loss of generality.

As every agent moves once to the left (to pick up the message – this move can be of zero length), and then once to the right, we have that in every normalized schedule every move of an agent to the right equals the advancement of the message done by this agent.

In the following we prove a structural lemma about normalized schedules, which will be a crucial ingredient in designing our algorithm. We will use the following notation. For an integer position p between s and t we say that an agent a (not necessarily from the schedule) is p-*crossing* if a lies at p or to the right of p, and at the same time the range of a allows the agent to walk left of p (to at least position $p - 1$). By X_p we denote the set of all p-crossing agents. We denote by $A_p \subseteq X_p$ the agents of the schedule that are p-crossing, and that in the schedule never move left of p (i.e., they only move on the part of the line that is to the right of p). Thus, A_p are agents that could possibly help advancing the message in the part of the line to the left of p, but they do not (because they are used right of p).

Lemma 1. *Let p be a position (an integer) such that $s < p < t$. Then $|A_p| \leq 2 \cdot \log t$ (for $t \geq 2$), and $|A_p| \leq 2 \cdot \log(R_{max} + 1)$.*

Proof. Let $a_1, a_2, \ldots, a_\ell$ be the agents in A_p sorted in the order as they appear in delivering the message from s to t. Each agent a_i is responsible for advancing the message on a certain interval $I_i = [s_i, t_i]$ between p and t (recall that none of A_p moves left of p in the feasible schedule), where the order of the segments appearing on the line is identical to the order of the agents in which they move.

Recall that in a normalized schedule, agent a_i can stop before using all its range if it reaches the position of the next agent in the schedule. Let $t'_i \geq t_i$ be the point which a_i reaches if it uses all its range (to walk from s_i). It follows that every two intervals $I_i' := [s_i, t_i']$ and $I_{i+2}' := [s_{i+2}, t_{i+2}']$ are disjoint (otherwise we can remove agent a_{i+1} from the schedule; a contradiction that the schedule is minimal). Thus, $s_{i+2} \geq t_i' + 1$. Furthermore, the position of a_{i+2} is (strictly) to the right of t_i' (as again we could remove agent a_{i+1} from the schedule).

Expressing t_i' in the form $p + \Delta_i$, we show that Δ_i grows exponentially with i and thus there can be at most (roughly) $\log t$ many agents in A_p before t_i' reaches t. Figure 2 illustrates the discussion of the proof. We start with t_1': Clearly, agent a_1 can reach $p - 1$ and thus, when reaching p, it can move at least to position $p + 1$: $t_1' \geq p + 1$. We can continue with t_3': Since $s_3 \geq t_1' + 1$, we get $s_3 \geq p + 2$; Furthermore, since agent a_3 is strictly to the right of t_1', i.e., $a_3 \geq t_1' + 1 \geq p + 2$, and a_3 can reach $p - 1$, a_3 has at $t_1' + 1$ enough energy to move to the right to position $(t_1' + 1) + ((t_1' + 1) - (p - 1)) \geq (t_1' + 1) + 3 = t_1' + 4$ (i.e., $t_3' \geq p + 5$). Similarly, $s_5 \geq t_3' + 1$, agent a_5 lies to the right of t_3', and agent a_5 has at $t_3' + 1$ enough energy to walk to $(t_3' + 1) + 7 = t_3' + 8$. In a similar spirit (i.e., by an easy induction), it follows that $t_{2i-1}' \geq t_{2i-3}' + 2^i$, $i = 2, 3, \ldots$. Thus, $t_\ell' \geq t_1' + 2^2 + 2^3 + \ldots + 2^{\lceil \ell/2 \rceil} \geq (p+1) + (2^{\lceil \ell/2 \rceil + 1} - 4) = p + 2^{\lceil \ell/2 \rceil + 1} - 3$. By setting $t_\ell' \leq t$, we get $2^{\lceil \ell/2 \rceil + 1} \leq t + 3 - p \leq t + 2$, which implies $\ell/2 \leq \log(t + 2) - 1$, which implies $\ell/2 \leq \log t$ (for $t \geq 2$), i.e., $\ell \leq 2 \cdot \log t$ (the first claim of the lemma).

At the same time, since a_ℓ is at least $t_{\ell-2}' + 1$, and the agent has energy to reach $p - 1$, it follows that $R_\ell \geq p + 2^{\lceil \ell/2 \rceil} - 2 - (p - 1) = 2^{\lceil \ell/2 \rceil} - 1$. Since the range of any agent is at most $R_{\max}$, we get $\ell \leq 2 \cdot \log(R_{\max} + 1)$. $\square$

We now present our quasi-, pseudo-polynomial time algorithm. For simplicity of exposition, we will use the upper bound $|A_p| \leq 2 \log t$. We will further assume that $t \geq 2$ (to be able to apply Lemma 1): If $t = 1$, finding a solution is trivial, we just try every agent and see whether it can deliver the message on its own.

The main idea of our dynamic-programming based algorithm is to scan the line backwards from t to s and to gradually build a feasible schedule from the last agent delivering the message to t to the first agent picking the message at s and at every intermediate step p to remember the set $A_p \subseteq X_p$ of p-crossing agents that were used so far to the right of p, and thus are not available to be used to the left of p.

Formally, we define a boolean table $T[p, A_p]$ for every (integer) point p between s and t (including s and t) and every set $A_p \subseteq X_p$ of cardinality $|A_p| \leq 2 \cdot \log t$. We interpret the table as $T[p, A_p] = \mathtt{true}$ if and only if there is a feasible schedule which advances the message from p to t, and among all p-crossing agents X_p it uses only the agents in A_p.

We fill the table as follows. We initialize $T[t, \emptyset] = \mathtt{true}$. Then, for every $p = t-1, t-2, t-3 \ldots, s$ we enumerate all sets $A_p \subseteq X_p$ of cardinality $|A_p| \leq 2 \cdot \log t$ and set $T[p, A_p] = \mathtt{true}$, if and only if

$$\exists p' > p, \exists A_{p'} \subseteq X_{p'}, \exists \text{ agent } a_p \in A_p \setminus A_{p'} \text{ such that:}$$

$$a_p \text{ can bring the message from } p \text{ to } p', \tag{1}$$

$$A_p = \{a_p\} \cup (A_{p'} \cap X_p), \tag{2}$$

$$T[p', A_{p'}] = \mathtt{true}. \tag{3}$$

After filling the table, the algorithm checks, whether for some set A_s there exists an entry $T[s, A_s] = \texttt{true}$, and if yes, it outputs a schedule of agents $a_{p_1}, a_{p_2}, a_{p_3}, \ldots, a_{p_\ell}$ that are, according to our dynamic program, recursively responsible for setting $T[p_i, A_{p_i}]$ to be $\texttt{true}$ (this can be done by standard book-keeping techniques); Otherwise, the algorithm decides that the agents cannot deliver the message from s to t.

Theorem 1. *The presented algorithm solves any instance of* DATADELIVERY *in time* $O\left(t^2 \cdot n^{1+4\log t}\right)$.

Proof. We will prove that there exists a solution to a given instance of DATADE-LIVERY if and only if the algorithm finds one.

If there is a solution to a given instance, i.e., a feasible schedule, then, by our observations, there also exists a normalized schedule $(a_{p_1}, a_{p_2}, \ldots, a_{p_\ell})$ of agents indexed with points on the line where they pick up the message, i.e., where agent a_{p_i} picks up the message at p_i and advances it to p_{i+1} (where we set $p_{\ell+1} := t$, and where, naturally, $p_1 = s$). Let S be the agents of this solution, i.e., $S = \{a_{p_1}, a_{p_2}, \ldots, a_{p_\ell}\}$. At any of these points p_i, $i = 1, 2, \ldots, \ell$, let $A_{p_i} \subseteq S \cap X_{p_i}$ be the set of agents from the solution S that are p_i-crossing. Furthermore, set $A_{\ell+1} = \emptyset$. Then, by Lemma 1, $|A_{p_i}| \leq 2 \cdot \log t$. Therefore, for every such set A_{p_i}, $i = 1, 2, \ldots, \ell+1$, there will be an entry $T[p_i, A_{p_i}]$ in our table. Furthermore, it follows that $A_{p_i} = \{a_{p_i}\} \cup (A_{p_{i+1}} \cap X_{p_i})$, $i = 1, \ldots, \ell$. Therefore, according to the rules of our dynamic programming (Eqs. (1), (2), (3)), all entries $T[p_i, A_{p_i}]$, $k = \ell, \ell-1, \ldots, 3, 2, 1$, will be set to $\texttt{true}$ because of the previous entry $T[p_{i+1}, A_{p_{i+1}}]$. Thus, our algorithm finds a solution (e.g., the one just derived from the normalized feasible schedule).

Assume now that the algorithm finds a schedule $a_{p_1}, \ldots, a_{p_\ell}$ of agents, indexed by the points p_i at which the corresponding entry $T[p_i, A_{p_i}]$ was set to true. By the rules of filling the table, it follows that the agents can deliver the message from s to t. What remains is to argue that no two agents $a_{p_i}, a_{p_{i+\Delta}}$ from the returned schedule are the same agent. Assume, for the sake of contradiction, that this is the case, i.e., $a = a_{p_i} = a_{p_{i+\Delta}}$, and that Δ is the smallest such number. By Eq. (2), $a_{p_{i+\Delta}} \in A_{p_{i+\Delta}}$; At the same time, since $a \in X_{p_i}$ and because of Eq. (2), a appears in all sets $A_{p_{i+\Delta-1}}, A_{p_{i+\Delta-2}}, \ldots, A_{p_{i+1}}$, and especially in $A_{p_{i+1}}$. But this contradicts the fact that $a = a_{p_i} \in A_{p_i} \setminus A_{p_{i+1}}$.

The runtime of the algorithm follows from the size of the table T and the way we fill in the table: For every p, we enumerate at most $\sum_{i=1}^{2\log t} \binom{n}{i} \leq O(n^{2\log t})$ many sets A_p. To fill in an entry $T[p, A_p]$ we try all possible values p', $A_{p'}$ and a_p, and check whether conditions in Eqs. (1), (2), and (3) hold, which can be done in time linear in size of A_p, $A_{p'}$, and X_p (if we store the elements of the sets sorted according to their position on the line). This results in total running time of $O\left(t^2 \cdot n^{1+4\log t}\right)$ (with small constants hidden in the big-oh notation). □

Using the upper bound $|A_p| \leq 2 \cdot \log(R_{\max} + 1)$, we can bound the running time of the algorithm by $O\left(t^2 \cdot n^{1+4\log(R_{\max}+1)}\right)$.

3   NP-Completeness

We first create an auxiliary NP-hard problem WEIGHTED-4-PARTITION, which we then reduce to DATADELIVERY. Along the way, we use the NP-hard problem 4-PARTITION-FROM-4-SETS, which has been shown NP-complete in [6] as a step in proving the NP-hardness of 3-PARTITION.

4-PARTITION-FROM-4-SETS

Input: Four sets of positive integers $A' = (a'_i)_{1 \le i \le q}$, $B' = (b'_i)_{1 \le i \le q}$, $C' = (c'_i)_{1 \le i \le q}$, $D' = (d'_i)_{1 \le i \le q}$, and an integer S'.
Question: Does there exist three permutations π_A, π_B, π_C of $[1, q]$ such that for every i, $a'_{\pi_A(i)} + b'_{\pi_B(i)} + c'_{\pi_C(i)} + d'_i = S'$?

WEIGHTED-4-PARTITION

Input: A set of positive integers $E = (e_i)_{1 \le i \le 4q}$ and an integer S such that for every partition of E into 4 sets A, B, C, D, each of size q, $\sum_{a \in A} a + \sum_{b \in B} 2b + \sum_{c \in C} 4c + \sum_{d \in D} 8d \le qS$.
Question: Does there exist a partition of E into q sets $E_1, E_2, \ldots, E_q$, each of size 4, such that for every $1 \le i \le q$, if $E_i = \{a, b, c, d\}$ with $a \le b \le c \le d$, $a + 2b + 4c + 8d = S$? We call such a partition a *weighted partition*.

Theorem 2. WEIGHTED-4-PARTITION *is NP-hard.*

Proof. From an instance of 4-PARTITION-FROM-4-SETS, we construct an instance of WEIGHTED-4-PARTITION as follows. Note that we can assume that $\sum_{x' \in A' \cup B' \cup C' \cup D'} x' = qS'$, and that for every $x' \in A' \cup B' \cup C' \cup D'$, $x' \le S'$.

For each $i \in [1, q]$, let $a_i = 8a'_i$, $b_i = 4b'_i + 32S'$, $c_i = 2c'_i + 128S'$ and $d_i = d'_i + 512S'$. Let $A = \{a_i\}_{1 \le i \le q}$, $B = \{b_i\}_{1 \le i \le q}$, $C = \{c_i\}_{1 \le i \le q}$ and $D = \{d_i\}_{1 \le i \le q}$. Let $E = A \cup B \cup C \cup D$ and let $S = 4680S'$. Note that $\forall a \in A, b \in B, c \in C, d \in D, a < b < c < d$. Consequently, for any partition of E into 4 sets A^*, B^*, C^*, D^* of size q, $\sum_{a \in A^*} a + \sum_{b \in B^*} 2b + \sum_{c \in C^*} 4c + \sum_{d \in D^*} 8d \le \sum_{a \in A} a + \sum_{b \in B} 2b + \sum_{c \in C} 4c + \sum_{d \in D} 8d = \sum_{a' \in A'} 8a' + \sum_{b' \in B'} (8b' + 64S') + \sum_{c' \in C'} (8c' + 512S') + \sum_{d' \in D'} (8d' + 4096S') = 8 \sum_{x' \in A' \cup B' \cup C' \cup D'} x' + 4672qS' = 4680qS' = qS$.

Suppose that we are given a solution π_A, π_B, π_C of 4-PARTITION-FROM-4-SETS such that for every i, $a'_{\pi_A(i)} + b'_{\pi_B(i)} + c'_{\pi_C(i)} + d'_i = S'$. Then for every i, $a_{\pi_A(i)} + 2b_{\pi_B(i)} + 4c_{\pi_C(i)} + 8d_i = 8a'_{\pi_A(i)} + 8b'_{\pi_B(i)} + 64S' + 8c'_{\pi_B(i)} + 512S' + 8d'_i + 4096S' = 4672S' + 8(a'_{\pi_A(i)} + b'_{\pi_B(i)} + c'_{\pi_B(i)} + d'_i) = 4680S' = S$. Since $a_{\pi_A(i)} \le 8S' < 32S' \le b_{\pi_B(i)} \le 36S' < 128S' \le c_{\pi_C(i)} \le 130S' < 512S' \le d_i$, we have found a solution to WEIGHTED-4-PARTITION, i.e., a weighted partition.

Conversely, suppose that there exists a partition $E_1, E_2, \ldots E_q$ that is a solution of the instance (E, S) of WEIGHTED-4-PARTITION. For every $1 \le i \le q$, let $E_i = \{w_i, x_i, y_i, z_i\}$ with $w_i \le x_i \le y_i \le z_i$.

Suppose first that there exists $1 \le i \le q$ such that $E_i \cap D = \emptyset$. Then, $w_i \le x_i \le y_i \le z_i \le 130S'$, and thus $w_i + 2x_i + 4y_i + 8z_i \le 15 \cdot 130S' = 1930S' < 4680S'$, contradicting the fact that $E_1, E_2, \ldots E_q$ is a weighted partition of E. Consequently, for every $1 \le i \le q$, $z_i \in D$ and thus, $w_i, x_i, y_i \in A \cup B \cup C$.

Suppose now that there exists $1 \leq i \leq q$ such that $E_i \cap C = \emptyset$. Then, $w_i \leq x_i \leq y_i \leq 36S'$ and $z_i \leq 513S'$. Thus, $w_i + 2x_i + 4y_i + 8z_i \leq 7 \cdot 36S' + 8 \cdot 513S' = 4356S' < 4680S'$, a contradiction. Consequently, for every $1 \leq i \leq q$, $y_i \in C$ and thus, $w_i, x_i \in A \cup B$.

Suppose now that there exists $1 \leq i \leq q$ such that $E_i \cap B = \emptyset$. Then, $w_i \leq x_i \leq 8S'$, $y_i \leq 130S'$ and $z_i \leq 513S'$. Thus, $w_i + 2x_i + 4y_i + 8z_i \leq 3 \cdot 8S' + 4 \cdot 130S' + 8 \cdot 513S' = 4648S' < 4680S'$, a contradiction. Therefore, for every $1 \leq i \leq q$, $x_i \in B$ and thus, $w_i \in A$.

Consequently, each E_i contains one element of A, one element of B, one element of C and one element of D. Without loss of generality (by reordering the elements in D), assume that for every $1 \leq i \leq q$, $d_i \in E_i$. For every $1 \leq i \leq q$, let $\pi_A(i) = j$ iff $a_j \in E_i$, let $\pi_B(i) = j$ iff $b_j \in E_i$, and let $\pi_C(i) = j$ iff $c_j \in E_i$. Consequently, for every i, $a_{\pi_A(i)} + 2b_{\pi_B(i)} + 4c_{\pi_C(i)} + 8d_i = 4680S'$, i.e., $8a'_{\pi_A(i)} + 8b'_{\pi_B(i)} + 64S' + 8c'_{\pi_B(i)} + 512S' + 8d'_i + 4096S' = 4680S'$. Consequently, $4672S' + 8(a'_{\pi_A(i)} + b'_{\pi_B(i)} + c'_{\pi_B(i)} + d'_i) = 4680S'$ and thus $a'_{\pi_A(i)} + b'_{\pi_B(i)} + c'_{\pi_B(i)} + d'_i = S'$. Consequently, π_A, π_B, π_C is a solution of the instance (A', B', C', D', S') of 4-PARTITION-FROM-4-SETS. □

Given an instance of WEIGHTED-4-PARTITION (E, S) given by a set $E = \{e_1, e_2, \ldots, e_{4q}\}$ and by a target value S, we construct an instance of DATADE-LIVERY as follows. There will be two types of agents: $4q$ "big" agents corresponding to the elements of E, and $q + 1$ "small" agents.

Let $M = \max_i\{e_i\}$, and let $r = \max\{15M - S, \frac{32^q - 1}{31}S\}$. There is a small agent s_i starting at position $x_i = \frac{16^i - 1}{15}(S + 16r)$ for $0 \leq i \leq q$. The range of each small agent is r. Note that $x_0 = 0$ and that $x_{i+1} = 16(x_i + r) + S$. Note also that all the positions and ranges are integers. We set $s = 0$, and $t = x_q + r$. For each $1 \leq j \leq 4q$, there is a big agent b_j starting at position t with a range equal to $t + e_j$.

Note that every big agent b_j starts at t and its range enables it to reach any point between s and t. Moreover, if b_j collects the message at a point $l \in [s, t]$, the furthest point where it can deliver the message is $2l + e_j$.

We claim that there exists a weighted partition of (E, S) if and only if the set of agents $A = \{s_i\}_{0 \leq i \leq q} \cup \{b_j\}_{1 \leq j \leq 4q}$ can deliver the message from s to t. The first direction is straightforward, and shown in Lemma 2. The other direction is more complicated, and is shown after Lemma 2 in a series of claims.

Lemma 2. *If there exists a weighted partition $E_1, E_2, \ldots, E_q$ of E, then there is a feasible schedule for* DATADELIVERY *from s to t.*

Proof. Suppose that we are given 4 integers $a \leq b \leq c \leq d$ such that $a + 2b + 4c + 8d = S$. Consider four agents a', b', c', d' initially located on t with respective ranges $t + a, t + b, t + c, t + d$. We claim that if some message is on $x_i + r$ for $i < q$, then agents a', b', c', d' can move the message to x_{i+1} when they are activated in the following order: d', c', b', a'.

Since $x_i + r < x_q + r = t$, d' can move the message to $2(x_i+r)+d$. Since $2(x_i+r)+d < t$, c' can move the message to $4(x_i+r)+2d+c$. Since $4(x_i+r)+2d+c < t$, b' can move the message to $8(x_i+r)+4d+2c+b$. Since $8(x_i+r)+4d+2c+b < t$, a' can move the message to $16(x_i+r)+8d+4c+2b+a = 16(x_i+r)+S = x_{i+1}$.

Recall that every small agent s_i, $0 \le i \le q$, can move the message from from x_i to $x_i + r$. Thus, we can use alternatively an agent s_i and the agents corresponding to E_{i+1} to move the message from $s = x_0 = 0$ to $t = x_q + r$.   □

We now show the other direction. In the rest of the section, we assume that there exists a feasible schedule for the created instance of DATADELIVERY, and we show that there exists a weighted-partition of (E, S).

Up to rearranging the elements of E, assume that the big agents are activated in the order $b_1, b_2, \ldots, b_{4q}$. For every $1 \le i \le q$, let $B_i = \{b_{4i-3}, b_{4i-2}, b_{4i-1}, b_{4i}\}$ and let $\delta_i = 8e_{4i-3} + 4e_{4i-2} + 2e_{4i-1} + e_{4i} - S$. Note that for every i, $-S \le \delta_i \le 15M - S$.

Note that if we activate all agents from B_i consecutively (without activating a small agent in between) and if b_{4i-3} collects the message in $l \in [s, t]$, b_{4i-3} can deliver the message to $2l + e_{4i-3}$, b_{4i-2} can deliver the message to $4l + 2e_{4i-3} + e_{4i-2}$, b_{4i-1} can deliver the message to $8l + 4e_{4i-3} + 2e_{4i-2} + e_{4i-3}$, and b_{4i} can deliver the message to $16l + 8e_{4i-3} + 4e_{4i-2} + 2e_{4i-1} + e_{4i} = 16l + S + \delta_i$.

We denote by u_i the furthest point where, in the considered feasible schedule, b_{4i} can deliver the message. We denote by y_i the furthest point where s_i can deliver the message.

In the next two lemmas, we show that we can assume that for each $0 \le i \le q$, s_i is activated after b_{4i} and before b_{4i+1}.

Lemma 3. *For every $0 \le i \le q$, $y_i \le x_i + r$. For every $1 \le i \le q$, s_i cannot be activated before b_{4i}, and $u_i \le x_i + r$.*

Proof. The first assertion of the lemma is trivial since s_i starts in x_i and its range is r.

We prove the second assertion by induction on i. Let $i \ge 0$ and assume that $y_i \le x_i + r$ and that $u_i \le x_i + r$ if $i \ge 1$. Suppose that s_{i+1} is activated before b_{4i+4}.

Since $\max\{u_i, y_i\} \le x_i + r$, and since $e_{4i+1}, e_{4i+2}, e_{4i+3} \le M$, b_{4i+1} cannot deliver the message further than $2(x_i+r)+M$, b_{4i+2} cannot deliver the message further than $4(x_i + r) + 3M$, and b_{4i+3} cannot deliver the message further than $8(x_i + r) + 7M$. Since s_{i+1} cannot collect the message before $x_{i+1} - r = 16(x_i + r) + S - r = 16x_i + S + 15r$, it is enough to show that $7r + S > 7M$ in order to prove that s_{i+1} cannot be activated before b_{4i+4}. Since $r \ge 15M - S$ and $S \le 15M$, $7r + S \ge 7 \cdot 15M - 6S \ge 15M > 7M$ and s_{i+1} cannot collect the message before b_{4i+4} has been activated.

Note that b_{4i+4} cannot deliver the message further than $u_{i+1} = 16(x_i + r) + 15M = x_{i+1} + 15M - S \le x_{i+1} + r$.   □

Lemma 4. *There exists a feasible schedule for DATADELIVERY from s to t such that for every $0 \le i \le q$, s_i is activated before b_{4i+1} and $y_i \ge x_i + r - \frac{2S}{31}(32^i - 1)$. For every $1 \le i \le q$, $u_i \ge x_i - \frac{S}{31}(32^i - 1) \ge x_i - r$.*

Proof. We prove the lemma by induction on i.

Note that since $x_0 = s$, $y_0 = x_0 + r$ (no matter when s_0 is activated). Suppose now that s_0 is not activated first. If when s_0 is activated, the message has already reached $x_0 + r$, then it means that there exists a feasible schedule for DATADELIVERY from s to t for $A' = A \setminus \{s_0\}$. Suppose now that when s_0 is activated, the message has not reached $x_0 + r$ and let i_0 be the maximal index i such that b_i has been activated before s_0. In this case, it means that there exists a feasible schedule for DATADELIVERY from $x_0 + r > s$ to t for $A' = A \setminus \{s_0, b_1, b_2, \ldots, b_{i_0}\}$. In both cases, it means that there exists a feasible schedule for DATADELIVERY from $x_0 + r > s$ to t for $A' = A \setminus \{s_0\}$ and thus there exists a feasible schedule for DATADELIVERY from s to t for A where s_0 is activated first.

Suppose now that s_i has been activated before b_{4i+1}, and that $y_i \geq x_i + r - \frac{2S}{31}(32^i - 1)$. Since s_{i+1} cannot be activated before b_{4i+4} (Lemma 3), we can assume that b_{4i+4} delivers the message to $u_{i+1} \geq 16y_i + S + \delta_{i+1} \geq 16(x_i + r) - \frac{32S}{31}(32^i - 1) = x_{i+1} - S - \frac{32S}{31}(32^i - 1) = x_{i+1} - \frac{S}{31}(32^{i+1} - 1)$. Since $r \geq \frac{S}{31}(32^{i+1} - 1)$, $u_{i+1} \geq x_{i+1} - r$.

Consequently, s_{i+1} can always be activated after b_{4i+4}. If s_{i+1} is activated before b_{4i+5}, either $u_{i+1} \geq x_{i+1}$ and $y_{i+1} = x_{i+1} + r \geq x_{i+1} + r - \frac{2S}{31}(32^{i+1} - 1)$, or $u_{i+1} < x_{i+1}$ and $y_{i+1} = 2u_{i+1} + r - x_{i+1} \geq x_{i+1} + r - \frac{2S}{31}(32^{i+1} - 1)$.

Suppose that we activate b_{4i+5} before s_{i+1}, then b_{4i+5} can deliver the message to a point $z \geq 2(x_{i+1} - r) + e_{4i+5} \geq x_{i+1} + (x_{i+1} - 2r) = x_{i+1} + (16x_i + 14r + S) > x_{i+1} + r$. That is, at this moment, s_{i+1} is useless. Consequently, there exists a feasible schedule for DATADELIVERY from z to t for $A' = A \setminus \{s_0, \ldots, s_i, s_{i+1}, b_1, \ldots, b_{4i+4}, b_{4i+5}\}$. This implies that there exists a schedule for DATADELIVERY from $2u_{i+1} + r - x_{i+1} \geq u_{i+1}$ to t for $A' \cup \{b_{4i+5}\}$, and thus, there exists a schedule DATADELIVERY from u_{i+1} to t for $A' \cup \{s_{i+1}, b_{4i+5}\}$ where s_{i+1} is activated first.   □

From Lemmas 3 and 4, there exists a feasible schedule for DATADELIVERY from s to t where we activate alternatively a small agent and four big agents. Consequently, we can assume that for every $1 \leq i \leq q$, $u_i = 16y_{i-1} + S + \delta_i$ and that $y_i = 2u_i + r - x_i$ if $u_i < x_i$ and $y_i = x_i + r$ otherwise.

In the next lemma, we show that $\sum_{i=1}^{q} \delta_i \geq 0$ and that this inequality is strict if at least one small agent has to go back to collect the message.

Lemma 5. *For any two indices $i < j$ such that $y_i = x_i + r$, $u_l < x_l$ for every $i + 1 \leq l \leq j - 1$ and $u_j \geq x_j$, we have $\sum_{l=i+1}^{j} \delta_l \geq 0$. Moreover, this inequality is strict if $j > i + 1$.*

Proof. If $j = i + 1$, $u_{i+1} = 16(x_i + r) + S + \delta_{i+1} = x_{i+1} + \delta_{i+1}$ and consequently, $\delta_{i+1} \geq 0$. In the following, we assume that $j > i + 1$.

For every $i \leq l \leq j$, let $z_l = x_l + r - y_l$. Note that by Lemma 3, $z_l \geq 0$. Moreover, $z_i = z_j = 0$ and for every $i + 1 \leq l \leq j - 1$, $z_l > 0$.

For every integer $i \leq l \leq j - 1$, $u_{l+1} = 16y_l + S + \delta_{l+1}$. For $i \leq l \leq j - 2$, $u_{l+1} < x_{l+1}$ and thus, $y_{l+1} = 2u_{l+1} + r - x_{l+1} = 32y_l + 2S + 2\delta_{l+1} + r - x_{l+1}$.

Consequently, $z_{l+1} = x_{l+1} + r - y_{l+1} = 2x_{l+1} - (32y_l + 2\delta_{l+1} + 2S) = 32(x_l + r) + 2S - (32y_l + 2\delta_{l+1} + 2S) = 32(x_l + r - y_l) - 2\delta_{l+1} = 32z_l - 2\delta_{l+1}$. Thus, for every $i + 1 \le l \le j - 1$, $z_l = -2\sum_{t=i+1}^{l} 32^{l-t}\delta_t$.

Moreover, $u_j = 16y_{j-1} + S + \delta_j = 16(x_{j-1} + r) - 16z_{j-1} + S + \delta_j = x_j - 16z_{j-1} + \delta_j = x_j + \delta_j + 32\sum_{t=i+1}^{j-1} 32^{j-1-t}\delta_t = x_j + \sum_{t=i+1}^{j} 32^{j-t}\delta_t$.

Let $S_1 = u_j - x_j$ and $S_2 = \sum_{l=i+1}^{j-1} -\frac{z_l}{2}$, i.e., $S_1 = \sum_{t=i+1}^{j} 32^{j-t}\delta_t$ and $S_2 = \sum_{l=i+1}^{j-1}(\sum_{t=i+1}^{l} 32^{l-t}\delta_t)$. Since $u_j \ge x_j$, $S_1 \ge 0$, and since for every $i + 1 \le l \le j - 1$, $z_l > 0$, it follows that $S_2 < 0$. Consequently, $S_1 - 31S_2 > 0$. We claim that $\sum_{t=i+1}^{j} \delta_t = S_1 - 31S_2$. We get

$$S_1 - 31S_2 = S_1 - 31\sum_{l=i+1}^{j-1}\sum_{t=i+1}^{l} 32^{l-t}\delta_t = S_1 - 31\sum_{t=i+1}^{j-1}\sum_{l=t}^{j-1} 32^{l-t}\delta_t$$

$$= S_1 - 31\sum_{t=i+1}^{j-1} \frac{32^{j-t} - 1}{31}\delta_t = \sum_{t=i+1}^{j} 32^{j-t}\delta_t - \sum_{t=i+1}^{j-1}(32^{j-t} - 1)\delta_t$$

$$= \sum_{t=i+1}^{j} \delta_t.$$

Consequently, $\sum_{t=i+1}^{j} \delta_t = S_1 - 31S_2 > 0$. □

Proposition 1. $(E_1, \ldots, E_q)$ *is a weighted-partition of* (E, S) *where for each* $1 \le i \le q$, $E_i = \{e_{4i-3}, e_{4i-2}, e_{4i-1}, e_{4i}\}$.

Proof. Since for every partition of E into 4 sets A, B, C, D of size q, $\sum_{a \in A} a + \sum_{b \in B} 2b + \sum_{c \in C} 4c + \sum_{d \in D} 8d \le qS$, $\sum_{i=1}^{q} \delta_i = \sum_{i=1}^{q}(8e_{4i-3} + 4e_{4i-2} + 2e_{4i-1} + e_{4i} - S) \le 0$.

Since we have a feasible schedule for DATADELIVERY from $s = 0$ to $t = x_q + r$ where s_q is activated after b_{4q}, s_q delivers the message to $t = x_q + r$, and thus, $u_q \ge x_q$. By Lemma 5, $\sum_{i=1}^{q} \delta_i \ge 0$, and thus, $\sum_{i=1}^{q} \delta_i = 0$.

Moreover, if there exists $1 \le i < q$ such that $u_i < x_i$, then $y_i < x_i + r$ and from Lemma 5, $\sum_{i=1}^{q} \delta_i > 0$, which is impossible. Consequently, for each $1 \le i \le q$, $u_i \ge x_i$ and $y_i = x_i + r$. Since $u_i = 16y_{i-1} + S + \delta_i = 16(x_{i-1} + r) + S + \delta_i = x_i + \delta_i$, it implies that for each i, $\delta_i \ge 0$.

Since $\sum_{i=1}^{q} \delta_i = 0$, we get that $\delta_i = 0$ for every $1 \le i \le q$, i.e., $8e_{4i-3} + 4e_{4i-2} + 2e_{4i-1} + e_{4i} = S$. Consequently $(E_1, E_2, \ldots, E_q)$ is a solution to the instance (E, S) of the WEIGHTED-4-PARTITION problem. □

This ends the proof of the NP-hardness of DATADELIVERY. Note that one can check quickly whether a given permutation $\sigma = (a_1, \ldots, a_{n'})$ of a subset A' of the agents can solve an instance (A, s, t) of DATADELIVERY. Indeed, setting $t_0 = s$, agent a_{i+1} can reach t_i if and only if $a_{i+1} - R_{i+1} \le t_i$. Moreover, if a_{i+1} can reach t_i, the furthest point agent a_{i+1} can reach with the information is $a_{i+1} + R_{a_{i+1}}$ if $t_i \ge a_{i+1}$ and $2t_i + R_{a_i} - a_i$ otherwise. Thus, one can iteratively compute the t_is until we find an index $t_i \ge t$, or until we find that $t_i < a_{i+1} - R_{i+1}$ or that

$t_{n'} < t$. This can be done by performing $O(n)$ arithmetical operations and the values we handle are smaller than $2t + R_{max}$: this can be done in polynomial time and thus DATADELIVERY is in NP. Consequently, we get the following theorem.

Theorem 3. DATADELIVERY *is NP-complete.*

4   Conclusions and Open Problems

We have shown that DATADELIVERY on a line is NP-hard. This answers the open problem raised by Anaya et al. [3]. It actually is a surprising result, because everyone we talked to about the problem believed it to be polynomial. We accompanied the result with a quasi-, pseudo- polynomial time algorithm. It remains an open problem, whether a pseudo-polynomial time algorithm exists. It also is an interesting problem to provide good γ-resource augmented algorithms.

Acknowledgements. We are grateful for the valuable comments of the anonymous reviewers. Jérémie Chalopin acknowledges a partial support by ANR project MACARON (ANR-13-JS02-0002).

References

1. Albers, S., Henzinger, M.R.: Exploring unknown environments. SIAM Journal on Computing 29(4), 1164–1188 (2000)
2. Alpern, S., Gal, S.: The theory of search games and rendezvous, vol. 55. Kluwer Academic Pub. (2002)
3. Anaya, J., Chalopin, J., Czyzowicz, J., Labourel, A., Pelc, A., Vaxès, Y.: Collecting information by power-aware mobile agents. In: Aguilera, M.K. (ed.) DISC 2012. LNCS, vol. 7611, pp. 46–60. Springer, Heidelberg (2012)
4. Blum, A., Raghavan, P., Schieber, B.: Navigating in unfamiliar geometric terrain. SIAM Journal on Computing 26(1), 110–137 (1997)
5. Chalopin, J., Das, S., Mihalák, M., Penna, P., Widmayer, P.: Data delivery by energy-constrained mobile agents. In: Proc. 9th International Symposium on Algorithms and Experiments for Sensor Systems, Wireless Networks and Distributed Robotics (ALGOSENSORS), pp. 111–122 (2013)
6. Garey, M.R., Johnson, D.S.: Computers and Intractability: A Guide to the Theory of NP-Completeness. W. H. Freeman & Co., New York (1979)
7. Heo, N., Varshney, P.K.: Energy-efficient deployment of intelligent mobile sensor networks. IEEE Transactions on Systems, Man, and CyberNetics (Part A) 35(1), 78–92 (2005)
8. Rajagopalan, R., Varshney, P.K.: Data-aggregation techniques in sensor networks: a survey. IEEE Communications Surveys & Tutorials 8(4), 48–63 (2006)

The Power of Two Choices
in Distributed Voting*

Colin Cooper[1], Robert Elsässer[2], and Tomasz Radzik[1]

[1] Department of Informatics, King's College London, United Kingdom
{colin.cooper,tomasz.radzik}@kcl.ac.uk
[2] Department of Computer Sciences, University of Salzburg, Austria
elsa@cosy.sbg.ac.at

Abstract. Distributed voting is a fundamental topic in distributed computing. In pull voting, in each step every vertex chooses a neighbour uniformly at random, and adopts its opinion. The voting is completed when all vertices hold the same opinion. On many graph classes including regular graphs, pull voting requires $\Omega(n)$ expected steps to complete, even if initially there are only two distinct opinions.

In this paper we consider a related process which we call two-sample voting: every vertex chooses two random neighbours in each step. If the opinions of these neighbours coincide, then the vertex revises its opinion according to the chosen sample. Otherwise, it keeps its own opinion. We consider the performance of this process in the case where two different opinions reside on vertices of some (arbitrary) sets A and B, respectively. Here, $|A| + |B| = n$ is the number of vertices of the graph.

We show that there is a constant K such that if the initial imbalance between the two opinions is $\nu_0 = (|A| - |B|)/n \geq K\sqrt{(1/d) + (d/n)}$, then with high probability two sample voting completes in a random d regular graph in $O(\log n)$ steps and the initial majority opinion wins. We also show the same performance for any regular graph, if $\nu_0 \geq K\lambda_2$, where λ_2 is the second largest eigenvalue of the transition matrix. In the graphs we consider, standard pull voting requires $\Omega(n)$ steps, and the minority can still win with probability $|B|/n$.

1 Introduction

Distributed voting has applications in various fields including consensus and leader election in large networks [5,18], serialisation of read/write in replicated data-bases [17], and the analysis of social behaviour in game theory [11]. Voting algorithms are usually simple, fault-tolerant, and easy to implement [18,20].

---

* The full version of this paper is available at arxiv.org/abs/1404.7479. This work was partially supported by EPSRC grant EP/J006300/1, "Random Walks on Computer Networks", the Austrian Science Fund (FWF) under contract P25214-N23 "Analysis of Epidemic Processes and Algorithms in Large Networks", and the 2012 SAMSUNG Global Research Outreach (GRO) grant "Fast Low Cost Methods to Learn Structure of Large Networks."

J. Esparza et al. (Eds.): ICALP 2014, Part II, LNCS 8573, pp. 435–446, 2014.
© Springer-Verlag Berlin Heidelberg 2014

One straightforward form of distributed voting is *pull voting*. In the beginning each vertex of a connected undirected graph $G = (V, E)$ has an initial opinion. The voting process proceeds synchronously in discrete time steps called rounds. During each round, each vertex independently contacts a random neighbour and adopts the opinion of that neighbour. The completion time T is the number of rounds needed for a single opinion to emerge. We showed in [7] that with high probability (w.h.p.) the completion time is $O(n/(\nu(1 - \lambda_2)))$, where n is the number of vertices, λ_2 is the second largest eigenvalue of the transition matrix, $\nu = \sum_{v \in V} d^2(v)/(d^2 n)$, $d(v)$ is the degree of vertex v and d is the average degree.

In the *two-party voter model*, vertices initially hold one of two opinions A and B. As usual, the pull voting is completed when all vertices have the same opinion. Hassin and Peleg [18] and Nakata *et al.* [22] considered the discrete-time two-party voter model on connected graphs, and discussed its application to consensus problems in distributed systems. Both papers focus on analysing the probability that all vertices will eventually adopt the opinion which is initially held by a given group of vertices.

Let A and B denote also the sets of vertices with opinions A and B, respectively; $A \cup B = V$. Let $d(X)$ be the sum of the degrees of the vertices in a set X. We say that opinion A wins, if all vertices eventually adopt this opinion. The central result of [18] and [22] is that the probability that opinion A wins is $P_A = d(A)/2m$, where m is the number of edges in G. Thus in the case of connected regular graphs, the probability that A wins is proportional to the original size of A, irrespective of the graph structure. Apart from the probability of winning the vote, another quantity of interest is the time T taken for voting to complete. In [18] it is proven that $\mathbf{E}T = O(n^3 \log n)$ for general connected n vertex graphs. In the case of random d-regular graphs, w.h.p. $\mathbf{E}T \sim 2n(d-1)/(d-2)$ [8]. It follows from the proof of this result that w.h.p. two-party voting needs $\Theta(n)$ time to complete on random d-regular graphs.

The performance of the two-party pull-voting seems unsatisfactory in two ways. Firstly, it is reasonable to require that a clear majority opinion wins with high probability. However, even if initially only a single vertex v holds opinion A, this opinion wins with probability $P_A = d(v)/2m$. Secondly, the expected completion time is at least $\Omega(n)$ on many classes of graphs, including regular expanders and complete graphs. This seems a long time to wait to resolve a dispute between two opinions. A more reasonable waiting time would depend on the graph diameter, which is $O(\log n)$ for many important graph classes.

To address these issues, we consider a modified version of pull voting in which each vertex v randomly queries two neighbours at each step. If both neighbours have the same opinion, the calling vertex v adopts this opinion. If the two opinions differ, the calling vertex retains its current opinion in this round. To distinguish this process from the conventional pull voting, we use terms *single-sample voting* and *two-sample voting*. The aim of the two-sample voting is to ensure (w.h.p.) that voting finishes quickly and the initial majority opinion wins.

In [6] we analysed a different two-sample process called min-voting. Here, initially each vertex holds a distinct opinion. In each step every vertex chooses two random neighbours and takes the smaller opinion of the two. For graphs with good expansion properties we proved that w.h.p. min-voting completes in time $O(\log n)$. Thus min-voting is fast, but an adversary with rather limited power could break the system by introducing small numbers into the network.

In this paper we analyse two-sample voting for two classes of d-regular graphs: random graphs and expanders. Our results depend only on the initial imbalance $\nu_0 = (|A| - |B|)/n$. As an example, for random d-regular graphs there is a constant K, independent of d, such that if

$$\nu_0 \geq K\sqrt{\frac{d}{n} + \frac{1}{d}},$$

w.h.p. two-sample voting is completed in $O(\log n)$ steps and the winner is the initial majority opinion A. Thus our two-sample voting achieves both, logarithmic completion time and a high probability that the initial majority opinion wins.

It seems interesting to enquire further how the performance of pull voting systems depends on the range of choices available in the design. We restrict our discussion to two-party voting. The main issues seem to be the number of neighbours k to contact at each step, and the rule used to reach a decision based on the opinions obtained. In the case $k = 1$, this is single-sample pull voting, as discussed above. For $k = 2$, a simple rule is to adopt the opinion if both neighbours agree (the voting protocol analysed in this paper). For $k \geq 3$ odd, a comparable rule is to adopt the majority opinion. Interestingly, the number of neighbours contacted at each step can substantially influence the performance of the process in at least three ways: the time to completion, the nature of the final outcome, and the robustness of the system against adversarial attacks.

We briefly compare the performance of such systems for two-party voting on random d-regular graphs. Surprisingly, a clearly defined complexity hierarchy emerges, which distinguishes between the cases $k = 1$, $k = 2$ and $k \geq 5$ odd.

- $k = 1$. As previously mentioned, the expected time to completion in this case is $\Theta(n)$ w.h.p. Let A be the size of the initial majority opinion. The probability that opinion A wins is A/n. Thus if $A = cn$, opinion A wins with probability $c < 1$, even if A is a clear majority.
- $k = 2$. This is the topic of this paper. We show that if the initial imbalance between the opinions is not too small, then the time to completion is $\Theta(\log n)$ w.h.p., resulting in an exponential speed up over the case $k = 1$, and the majority wins w.h.p. More details are given in the next section.
- $k \geq 5$. It follows from the proof presented by Abdullah and Draief [1], that for k odd and $d \geq k$ constant, if the initial allocation of the opinions is chosen randomly, the initial imbalance is sufficiently large, and the selection of k neighbours is done without replacement, then w.h.p. the majority wins, and the voting completes in $\Theta(\log \log n)$ rounds.

The performance of two-party voting is well studied in the particular case of the complete graph K_n. Becchetti et al. [4] consider $k = 3$ and focus on the completion time as a function of the number of opinions. The result for two opinions is $O(\log n)$

steps, provided that the initial imbalance is not too small. Cruise and Ganesh [10] consider a more general but asynchronous model. Their work includes the case $k = 2$, and gives a $\Theta(\log n)$ time bound. A variant of two-sample voting in K_n has been considered by Doerr et al. [12], for the case of opinions drawn from $\{1, 2, \ldots, n\}$. In their model, whenever a vertex v contacts two vertices u and w, it adopts the median of the opinions of v, u and w. Once the system is left with two opinions, this protocol is equivalent to the two-sample voting considered in this paper. It is shown in [12] that if initially there are s opinions, the median voting converges to a so-called "stable consensus" in $O(\log s \log \log n + \log n)$ steps.

An alternative approach to k-sample voting is to use a *majority dynamic*: in each step, each vertex adopts the most popular opinion among *all* its neighbours. Majority dynamics were studied by Mossel et al. in [21] who gave bounds for different scenarios. They consider a model where initially opinions from $\{1, 2, \ldots, r\}$ are assigned to the vertices independently according to a probability distribution. Then, the following deterministic process is considered, which is fully defined by the initial distribution of opinions. For T time steps, each vertex adopts the opinion held by the majority of its neighbours. After step T a fair and monotone election function is applied to the opinions of the vertices, resulting in a winning opinion. It is shown in [21] that (under certain assumptions) for the two-party model this process results in the correct (initial majority) answer.

Recently, Abdullah and Draief [1] have considered two-party majority voting on fixed degree sequence random graphs. The initial opinions are distributed randomly according to a biassed distribution. They show that if the initial bias toward one opinion is large enough, then w.h.p. this opinion is adopted by all vertices within $\Theta(\log \log n)$ steps.

2   Our Results for Two-Sample Voting

Initially each vertex holds one of two opinions. For convenience, A and B denotes these opinions, the two sets of vertices which have these opinions, and the sizes of these sets, depending on the context. If opinion A is the majority, then the imbalance ν (the relative difference between the votes, or the advantage of the A vote) is given by $A - B = \nu n$. We show that for sufficiently large initial imbalance, w.h.p. two-sample voting on d-regular random graphs and expanders completes in logarithmic time and the initial majority opinion wins.

The results hold w.h.p., that is, with probability which tends to 1 with increasing n, and depends on the selection of a graph (in the case of random graphs) as well as on the voting process, which is itself probabilistic. A random d-regular graph is sampled uniformly at random from the set of all (simple) d-regular graphs.

Theorem 1. *Let G be a random n-vertex d-regular graph with opinions A and B and with initial imbalance $\nu_0 = (A - B)/n$. There is an absolute constant K (independent of d) such that, provided*

$$\nu_0 \geq K\sqrt{\frac{d}{n} + \frac{1}{d}}, \tag{1}$$

with high probability two-sample voting is completed in $O(\log n)$ steps and the winner is the initial majority opinion A.

We give a similar result for expanders, that is, for a d-regular graph G with a small second eigenvalue $\lambda_G = \max\{\lambda_2, |\lambda_n|\}$, where $\lambda_1 \geq \lambda_2 \geq \cdots \geq \lambda_n$ are the eigenvalues of the transition matrix $P = (1/d)A$ of a random walk on G and A is the adjacency matrix of G.

Theorem 2. *Let G be an n-vertex d-regular graph with opinions A and B and with the initial imbalance ν_0. There is an absolute constant K (independent of d and λ_G) such that if $\nu_0 \geq K\lambda_G$, then with high probability two-sample voting is completed in $O(\log n)$ steps and the winner is the opinion with the initial majority.*

Observe that for the above results to be non-trivial, we should consider $K^2 \leq d \leq n/K^2$ in Theorem 1 and $\lambda_G \leq 1/K$ in Theorem 2. These theorems say that to guarantee (w.h.p.) that two-sample voting starting with small imbalance ν_0 completes within $O(\log n)$ steps and the initial majority opinion wins, it suffices to take a random d-regular graph with appropriately large degree d, or an expander graph with appropriately small λ_G. We show that the initial majority wins in $O(\log n)$ steps also in random regular graphs with small degree and in expanders with λ_G not too small, if the initial minority is small enough.

Theorem 3. *Let $d > 10$ and let G be a random d-regular n-vertex graph with votes A and B. There is a constant $c > 0$ (independent of d) such that if the initial minority is at most cn, then w.h.p. two-sample voting is completed in $O(\log n)$ steps and the winner is the initial majority.*

Theorem 4. *Let G be a d-regular n-vertex graph with $\lambda = \lambda_G = 3/5 - \epsilon$. If the initial minority is at most $(\epsilon/5)n$, then w.h.p. two-sample voting is completed in $O(\log n)$ steps and the winner is the initial majority.*

The above theorems hold under following adversarial conditions. The adversary has full knowledge of the graph, decides the initial distribution of the opinions among the vertices, and can arbitrarily redistribute the opinions at the start of each voting step. The adversary cannot change the number of opinions of each type. However, one can trace our analysis to see that if we allow the adversary to change the opinions of at most $f = o(\nu_0 n)$ vertices at each step, then under the conditions considered in Theorems 1-4, after $O(\log n)$ voting steps all but $O(f)$ vertices adopt the majority opinion (cf. [12]).

As described before, there seems to be a clearly defined hierarchy w.r.t. distributed voting in random regular graphs. If every node is only allowed to consult one single neighbour (and adopt its opinion), then as shown in [7], $\Theta(n)$ steps are required to converge to one opinion. If every node can consult two neighbours (selecting them randomly with or without replacement) and adopt the opinion of these two vertices if they are the same, then the running time is $O(\log n)$, so exponentially faster. On the other hand, even if the adversary is not allowed to re-distribute votes, $\Omega(\log_d n)$ is a natural lower bound in any d-regular graph.

This holds since there might be initially $\Theta(n)$ edges such that the end vertices of all these edges have the same minority opinion B. The end vertices of such an edge choose each other with probability $\Theta(1/d^2)$, in which case they both keep their opinion B. Thus, the protocol needs $\Omega(\log_d n)$ steps in order to guarantee that in none of these $\Theta(n)$ edges the end vertices choose each other all the time. This lower bound holds also for k-sample voting for a constant $k \geq 3$, if the selection of k neighbours is done with replacement (if a B vertex v has a B neighbour, then v does not change its opinion in the current step with probability $\Omega(1/d^k)$). If every node may contact at least five different neighbours (selection without replacement) and adopts the majority opinion among them, then on random regular graphs with randomly distributed opinions (biased toward A), $\Theta(\log \log n)$ steps suffice until A wins (this follows from the analysis in [1]).

We should mention that the last result does not hold if the opinions are not randomly distributed. An adversary could assign the minority opinion B to a vertex v as well as all vertices which are at distance at most $Diam/3$ to v, where $Diam$ denotes the diameter of the graph. Clearly, the voting protocol would need at least $Diam/3$ steps. Also, the result w.r.t. 5-sample voting [1] requires graphs such that for each vertex, the neighbourhood of this vertex of depth $O(\log \log n)$ is (almost) a tree. It might therefore be difficult to extend that result beyond the class of random graphs.

Concerning our results, the constant eigenvalue gap $1 - \lambda_G$ (as in Theorem 4) seems to be needed. For example, consider a hypercube, where $d = \log n$ and $1 - \lambda_G = o(1)$. If the adversary is allowed to rearrange the opinions in each step, then we may have for $\Omega(d^2)$ steps configurations in which all vertices of a subcube of dimension $d - c$ have opinion B, where c is a constant. Such a B-vertex converts to A with probability $(c/d)^2$, so $\Omega(d^2) = \Omega(\log^2 n)$ steps are needed for the protocol to finish.

3   Background Material and Outline of Proof

The analysis of two-sample voting is made in the following three phases, where B is the minority vote.

Phase I: $cn \leq B \leq n(1 - \nu_0)/2$.

Phase II: $\omega \leq B \leq cn$.

Phase III: $1 \leq B \leq \omega$.

Let $B(t)$ denote the set of vertices with opinion B and the size of this set in step t. Whenever it is clear from the context, we write B instead of $B(t)$. Phase I reduces $B(t)$ from $B(0) = n(1 - \nu_0)/2$ to $B(T) \leq cn$, for some small constant c, in a sequence of $T = O(\log(1/\nu_0))$ rounds. The reduction in $B(t)$ in Phase II is more dramatic. The ω threshold between phases II and III is a function slowly growing with n. In Phase III things can slow down again and the last few steps can be viewed a biassed random walk.

The following Chernoff–Hoeffding inequalities are used throughout the proofs. Let $Z = Z_1 + Z_2 + \cdots Z_N$ be the sum of the independent random variables $0 \leq Z_i \leq 1$, $i = 1, 2, \ldots, N$, $\mathbf{E}(Z_1 + Z_2 + \cdots + Z_N) = N\mu$, and $0 \leq \epsilon \leq 1$. Then

$$\mathbf{Pr}(Z \leq (1 - \epsilon)N\mu) \leq e^{-\epsilon^2 N\mu/3}, \text{ and } \mathbf{Pr}(Z \geq (1 + \epsilon)N\mu) \leq e^{-\epsilon^2 N\mu/2}. \quad (2)$$

Our proofs for the case of random graphs are made using the configuration model of d-regular n-vertex multigraphs. Let $\mathcal{C}_{n,d}$ be the space of d-regular n-vertex configurations, and let $\mathcal{C}_{n,d}^*$ be the sub-space of $\mathcal{C}_{n,d}$ of the configurations whose underlying graphs are simple. A configuration S is a matching of the nd "configuration points" (each vertex is represented by d points). Every simple graph maps to the same number of configurations, so $\mathcal{C}_{n,d}^*$ maps uniformly onto $\mathcal{G}_{n,d}$, the space of d-regular n-vertex graphs. We use the following result of [15] for the size of $|\mathcal{C}^*|/|\mathcal{C}|$. See e.g. [9] for a proof.

Lemma 1. *Let* $1 \leq d \leq n/8$. *For a random configuration* $S \in \mathcal{C}_{n,d}$,

$$\mathbf{Pr}(S \in \mathcal{C}_{n,d}^*) \geq e^{-20d^2}. \quad (3)$$

This lemma is used in the following way. Let Q be a property of d-regular n vertex multigraphs. Then, denoting by $G(S)$ the underlying multigraph of configuration S,

$$\mathbf{Pr}_{\mathcal{G}}(G \in Q) = \mathbf{Pr}_{\mathcal{C}}(G(S) \in Q \mid S \in \mathcal{C}^*) \leq \mathbf{Pr}_{\mathcal{C}}(G(S) \in Q) \cdot e^{20d^2}. \quad (4)$$

At any step t of the voting process, let $\Delta_{AB} = \Delta_{AB}(t)$ be the number of A vertices converting to B during this step. Similarly, let Δ_{BA} be the number of B vertices converting to A during step t. At each step we obtain a lower bound on $\mathbf{E}\Delta_{BA}$, an upper bound on $\mathbf{E}\Delta_{AB}$, and use the concentration of these two random variables given by (2) to get a w.h.p. value of $\Delta = \Delta_{BA} - \Delta_{AB}$, the increase of the number of A vertices in this step.

For a vertex v and a set of vertices C, let d_v^C be the number of vertices in C which are adjacent to v. For $v \in A$, let $X_v = 1$ if v chooses twice in B at step t, and 0 otherwise. Thus

$$\Delta_{AB} = X_A = \sum_{v \in A} X_v$$

The X_v are independent $\{0, 1\}$ random variables with the expected value depending whether the neighbours are selected with or without replacement:

$$\mathbf{E}X_v(\text{with replacement}) = \left(\frac{d_v^B}{d}\right)^2, \quad \mathbf{E}X_v(\text{no replacement}) = \frac{(d_v^B)(d_v^B - 1)}{d(d - 1)}.$$

We give analysis for sampling with replacement. The case when sampling is without replacement is similar because

$$\mathbf{E}X_v(\text{no replacement}) = \frac{d}{d - 1}\mathbf{E}X_v(\text{with replacement}),$$

so any inequalities for expected values in one model imply similar inequalities in the other model.

We omit most of the proofs; see the full version on `arXiv` for details.

4   Phase I of Analysis: $cn \leq B \leq n(1 - \nu_0)/2$

Lemma 2 below gives a sufficient condition for a fast reduction of the minority B-vote from $(1 - \nu_0)n/2$ to cn. The condition in Lemma 2 says that the number $E(X, Y)$ of edges between any disjoint large subsets of vertices X and Y is close to the value dXY/n expected in the random regular graph. This condition is of the form as in the Expander Mixing Lemma (stated below as Lemma 3), so Lemma 2 can be immediately applied to expanders (see Corollary 1). Lemma 2 can also be applied without a reference to the second eigenvalue (if the second eigenvalue is not known or is not good enough) by directly checking that large subsets of vertices are connected by many edges. We illustrate this by considering random d-regular graphs (see Lemma 4 and Corollary 2).

The parameters c and α in the above lemma can be considered as some small constants, but they can also depend on d (and decrease with increasing d).

Lemma 2. *Let $0 < c \leq 1/2$, $0 < \alpha \leq c^{3/2}/36$, and $\alpha^2 c^2 n = \Omega(n^\epsilon)$, for a constant $\epsilon > 0$. Let G be a d-regular n-vertex connected graph such that*

$$\left| E(X, Y) - \frac{dXY}{n} \right| \leq \alpha d \sqrt{XY}, \tag{5}$$

for each pair X and Y of disjoint subsets of vertices of sizes $Y \geq cn$ and $X \geq (2/3)\alpha c^{3/2}n$. There exist absolute constants K and K' (independent of d, c and α) such that, if the initial advantage of the A-vote in G is

$$\nu_0 \geq K\alpha, \tag{6}$$

then with probability at least $1 - e^{-\Theta(\alpha^2 c^2 n)}$, the advantage of the A-vote increases to $1 - 2c$ (the B-vote decreases to cn) within $K'(\log(1/\nu_0) + \log(1/c))$ voting steps.

Proof (Sketch). Consider one voting step, when the vote imbalance $\nu = (A - B)/n$ is $K\alpha \leq \nu \leq 1 - 2c$, for some large constant K. Using (5), show that

$$\mathbf{E}\Delta_{BA} = \sum_{v \in B} \left(\frac{d_v^A}{d} \right)^2 \geq \frac{1}{Bd^2} \left(\sum_{v \in B} d_v^A \right)^2 \geq \frac{A^2 B}{n^2} (1 - \Theta(\eta)), \tag{7}$$

where $\eta \equiv \eta_{AB} = \alpha n/\sqrt{AB}$. Then show that the expectation $\mathbf{E}\Delta_{AB}$ is

$$\mathbf{E}\Delta_{AB} = \sum_{v \in A} \left(\frac{d_v^B}{d} \right)^2 \leq \frac{AB^2}{n^2} (1 + \Theta(\eta)), \tag{8}$$

by partitioning set A into sets C_i, for $i = 1, 2, \ldots, q = \Theta(\log(n/(\eta B^2)))$:

$$C_i = \left\{ v \in A : (1 + 2^{i-1}\eta)\frac{dB}{n} \leq d_v^B < (1 + 2^i\eta)\frac{dB}{n} \right\}.$$

The bounds (7) and (8), and the concentration bounds (2) imply

$$A(t+1) \geq A(t) + \frac{A(t)B(t)(A(t) - B(t))}{n^2} - \Theta(\eta(t)) \cdot n, \tag{9}$$

and analysing this recurrence leads to the bound on the number of steps.   □

Lemma 3. (Expander Mixing Lemma [3]). *Let $G = (V, E)$ be a d-regular n-vertex graph and denote $\lambda = \lambda_G = \max\{|\lambda_2|, |\lambda_n|\}$. Then for all $S, T \subseteq V$,*

$$\left| E(S, T) - \frac{dST}{n} \right| \leq \lambda d \sqrt{ST}.$$

Lemmas 2 and 3 imply the following corollary.

Corollary 1. *For any constant $0 < c < 1/2$, there exist constants K_1 and K_2 (which depend on c) such that for any regular n-vertex graph G with the initial advantage of the A-vote $\nu_0 \geq K_1 \lambda_G$, the minority vote B decreases to cn within $K_2 \log(1/\nu_0)$ voting steps, with probability at least $1 - e^{-\Theta(\lambda_G^2 n)}$.*

Proof. Let $K_1 = \max\{K, 36/c^{3/2}\}$, where constant K is from Lemma 2. If $\lambda = \lambda_G \geq 1/K_1$, then the corollary is trivial. If $\lambda \leq 1/K_1$, then $\lambda \leq c^{3/2}/36$ and we can apply Lemma 2 with c and $\alpha = \lambda$. Now Lemmas 2 and 3 imply that if $\nu_0 \geq K_1 \lambda$, then with probability at least $1 - e^{-\Theta(\lambda^2 n)}$, the size of the B-vote decreases to cn within $K'(\log(1/\nu_0) + \log(1/c)) = K_2(\log(1/\nu_0))$ voting steps.

Consider now random regular graphs. If $d = O(1)$, then a random d-regular graph has $\lambda_G \leq (2\sqrt{d-1} + \epsilon)/d$, w.h.p., where $\epsilon > 0$ can be any small constant [16]. Thus, for $d = O(1)$ Corollary 1 applies. To apply Lemma 2 to random regular graphs with some degree which may grow with the number of vertices, we need to establish a suitable α for (5) without referring to λ_G. The bound we show in the next lemma is stronger than a similar bound shown in [14], which would lead to a weaker relation between ν_0 and d than in Theorem 1.

Lemma 4. *For given set sizes $X \leq Y$, in a random d-regular n-vertex graph $G = (E, V)$, with probability at least $1 - 2e^{-Y}$, for each pair of disjoint subset of vertices $\mathcal{X}$ and $\mathcal{Y}$ of sizes X and Y,*

$$\left| E(\mathcal{X}, \mathcal{Y}) - \frac{dXY}{n} \right| \leq d\sqrt{XY} \sqrt{\frac{1}{d} 24 \log(ne/Y) + \frac{d}{Y} 160}. \tag{10}$$

Corollary 2. *For any constant $0 < c < 1/2$, there exist constants K_1 and K_2 (which depend on c) such that for a random d-regular n-vertex graph with the initial advantage of the A-vote*

$$\nu_0 \geq K_1 \sqrt{\frac{1}{d} + \frac{d}{n}}, \tag{11}$$

the minority vote B decreases within $K_2 \log(1/\nu_0)$ steps to cn, with probability at least $1 - e^{-\Theta(n^{1/2})}$.

5   Phase II of Analysis: $\omega \leq B \leq cn$

The analysis of this middle phase needs the property that small sets of vertices do not induce many edges, which holds for expanders and random regular graphs. Lemma 5 shows that for a graph with such a property, if the minority vote is still substantial, then one voting step reduces this minority by a constant factor with high probability. This implies that with high probability the minority vote reduces from cn to ω within $O(\log n)$ steps (Corollary 3).

Lemma 5. *Let G be a d-regular n-vertex graph with A and B votes, $A > B$. Let $0 < \alpha \le 3/10$ and $\gamma = \gamma(\alpha) = (1/2)(1 - 2\alpha)(1 - 3\alpha) > 0$. If the set B is such that every superset $S \supseteq B$ of size at most $(1 + 1/\alpha)B$ spans at most αdS edges (that is, $|E(S)| \le \alpha dS$), then one voting step reduces B at least by a factor $1 - \gamma$, with probability at least $1 - e^{-\tilde{\gamma}B}$, for a constant $\tilde{\gamma}$.*

Corollary 3. *Let G be a d-regular n-vertex graph, and let $0 < g < 1$ be such that for each subset of vertices S of size at most gn, $|E(S)| \le (3/10)dS$. Then the minority vote B is reduced from $(3/13)gn$ to at most ω within $O(\log n)$ steps with probability at least $1 - e^{-\Theta(\omega)}$.*

Lemma 6. *Let G be a d-regular n-vertex graph with $\lambda = \lambda_G < 3/5$. Then the minority vote B reduces from $(3/13)(3/5 - \lambda)n$ to ω within $O(\log n)$ steps with probability at least $1 - e^{-\Theta(\omega)}$.*

Proof. This lemma follows from Corollary 3 applied with $c = 1 - (2/5)(1-\lambda)^{-1} > 0$, after checking that $|E(S)| \le (3/10)dS$ whenever $S \le cn$. It is shown in [19] that the conductance of graph $G = (V, E)$ defined as

$$\Phi_G = \min_{\emptyset \ne S \subset V} \frac{nE(S, \bar{S})}{dS\bar{S}},$$

is at least $1 - \lambda$. This implies that $E(S, \bar{S}) \ge (1 - \lambda)dS\bar{S}/n$, so if $S \le cn$, then

$$|E(S)| = \frac{1}{2}\left(dS - E(S, \bar{S})\right) \le \frac{1}{2}dS\left(1 - (1 - \lambda)\bar{S}/n\right) \le \frac{3}{10}dS.$$

As mentioned before, for constant d a random d-regular graph has eigenvalue $\lambda_G \approx 2/\sqrt{d}$, w.h.p. Thus, for constant d, the result which we have obtained for expanders (Lemma 6) applies to random regular graphs as well, provided that d is sufficiently large to guarantee $\lambda_G < 3/5$. To consider random regular graphs with degree which may grow with the number of vertices, we show the following lemma. This is a stronger version of a result from [9] that w.h.p. for $3 \le d \le cn$ no set of vertices of size $|S| \le n/70$ induces more than $d|S|/12$ edges.

Lemma 7. *Let $600 \le d \le n/K$ for some large constant K, and let $G = (V, E)$ be a random d-regular n vertex graph. Let $\alpha = 1/12$ and consider the event*

$$Q = \{\exists S \subseteq V : |S| \le n/15 \text{ and } S \text{ spans at least } \alpha d|S| \text{ edges }\}.$$

Then $\mathbf{Pr}(Q) \le n^{-\delta}$, for some constant $\delta > 0$.

Lemma 8. *Let $d > 10$ and G be a random d-regular n-vertex graph with votes A and B. There is a constant $c > 0$ (independent of d) such that the minority vote B reduces from cn to ω within $O(\log n)$ steps with probability at least $1 - e^{-\Theta(\omega)} - o(1/n)$.*

Proof. For $11 \le d < 600$ use Lemma 6: in this case, a random d-regular graph has $\lambda_G \le (2\sqrt{d-1} + \epsilon)/d < 3/5$.

For $600 \le d \le n/K$, Lemma 7 implies that with probability at least $1 - o(1/n)$, $E(S) \le (1/12)dS < (3/10)dS$, for each subset of vertices S of size at most $n/15$. Thus, applying Corollary 3 with $g = 1/15$, we conclude that the minority vote B is reduced from $(3/13)gn = (1/65)n$ to at most ω within $O(\log n)$ steps with probability at least $(1 - o(1/n))(1 - e^{-\Theta(\omega)})$.

6 Phase III of Analysis: $1 \leq B \leq \omega$

Lemma 9. *Let $\omega = \omega(1)$ grow with n and $\omega = o(n)$. Let G be a d-regular n-vertex graph such that for each subset of vertices S of size at most $(13/3)\omega$, $|E(S)| \leq (3/10)dS$. Then the minority vote B is reduced from ω to 0 within $O(\omega \log \omega)$ steps with probability at least $1 - e^{-\Theta(\omega)}$.*

Corollary 4. *Let $\omega = \omega(1)$ grow with n and $\omega = o(n)$. If $G = (V, E)$ is a d-regular expander with $\lambda_G < 3/5$ or it is a random d-regular graph with $d > 10$, then voting reduces B from at most ω to 0 in $O(\omega \log \omega)$ steps with probability at least $1 - e^{-\Theta(\omega)}$.*

Proof. If G is a d-regular expander with $\lambda_G \leq 3/5$, then the assumptions of Lemma 9 are fulfilled for G, as shown in the proof of Lemma 6. If G is a random d-regular graph with $d = \omega(1)$, then the assumptions of Lemma 9 are also fulfilled for G according to Lemma 7. If $d > 10$ but $d = O(1)$, then G has eigenvalue $\lambda_G < 3/5$, w.h.p. [16].

7 Putting the Phases Together

To conclude the proof of our main Theorems 1 and 2, it remains to check how the three phases fit together. For expanders (Theorem 2), first use Corollary 1 with $c = 1/10$ to get constant $K = K(c)$ such that if the initial imbalance of vote is $\nu_0 \geq K\lambda_G$, then the minority vote reduces to $n/10$ within $O(\log(1/\nu_0))$ steps. Then use Lemma 6 with $\omega = \log n / \log \log n$ and assume that $\lambda_G \leq 1/6$ to show that the minority vote reduces from $n/10$ to ω in $O(\log n)$ steps. Finally, apply Corollary 4 with the same ω to show that the minority vote decreases from ω to 0 in $O(\log n)$ steps.

For random regular graphs, Lemma 8 gives a constant $c < 1/2$ which defines the beginning of phase II. Then Corollary 2 can be used to find the constant $K = K(c)$ for Theorem 1. The transition from phase II to phase III is at the same $\omega = \log n / \log \log n$ as before.

According to our analysis, we can also derive the following corollary.

Corollary 5. *Assume an adversary can change the opinion of at most $f = o(\nu_0 n)$ vertices during the execution of the algorithm. Then, under the assumptions of Theorems 1 and 2, w.h.p. all but $O(f)$ vertices will adopt opinion A within $O(\log n)$ steps.*

References

1. Abdullah, M., Draief, M.: Consensus on the Initial Global Majority by Local Majority Polling for a Class of Sparse Graphs (2013), http://www.arXiv.org
2. Aldous, D., Fill, J.: Reversible Markov Chains and Random Walks on Graphs, http://stat-www.berkeley.edu/pub/users/aldous/RWG/book.html

3. Alon, N., Chung, F.R.K.: Explicit construction of linear sized tolerant networks. Discrete Math. 72, 15–19 (1989)
4. Becchetti, L., Clementi, A., Natale, E., Pasquale, F., Silvestri, R., Trevisan, L.: Simple Dynamics for Majority Consensus (2013), http://www.arXiv.org
5. Brahma, S., Macharla, S., Pal, S.P., Singh, S.K.: Fair Leader Election by Randomized Voting. In: Ghosh, R.K., Mohanty, H. (eds.) ICDCIT 2004. LNCS, vol. 3347, pp. 22–31. Springer, Heidelberg (2004)
6. Cooper, C., Elsässer, R., Ono, H., Radzik, T.: Coalescing Random Walks and Voting on Graphs. In: PODC 2012, pp. 47–56 (2012)
7. Cooper, C., Elsässer, R., Ono, H., Radzik, T.: Coalescing Random Walks and Voting on Connected Graphs. SIAM J. on Discrete Math. 27(4), 1748–1758 (2013)
8. Cooper, C., Frieze, A., Radzik, B.: Multiple Random Walks in Random Regular Graphs. SIAM J. on Discrete Math. 23(4), 1738–1761 (2009)
9. Cooper, C., Frieze, A., Reed, B.: Random regular graphs of non-constant degree: connectivity and Hamilton cycles. Combinatorics Prob. & Comp. 11, 249–262 (2002)
10. Cruise, J., Ganesh, A.: Probabilistic consensus via polling and majority rules. arXiv:1311.4805
11. Deng, X., Papadimitriou, C.: On the Complexity of Cooperative Solution Concepts. Mathematics of Operations Research 19(2), 257–266 (1994)
12. Doerr, B., Goldberg, L.A., Minder, L., Sauerwald, T., Scheideler, C.: Stabilizing Consensus with the Power of Two Choices. In: SPAA 2011, pp. 149–158 (2011)
13. Donnelly, P., Welsh, D.: Finite particle systems and infection models. Math. Proc. Camb. Phil. Soc. 94(1), 167–182 (1983)
14. Fountoulakis, N., Panagiotou, K.: Rumor Spreading on Random Regular Graphs and Expanders. In: Serna, M., Shaltiel, R., Jansen, K., Rolim, J. (eds.) APPROX 2010. LNCS, vol. 6302, pp. 560–573. Springer, Heidelberg (2010)
15. Frieze, A., Łuczak, T.: On the independence and chromatik numbers of random graphs. J. Combinatorial Theory, Ser. B 54, 123–132 (1992)
16. Friedman, J.: A proof of Alon's second eigenvalue conjecture. In: STOC 2003, pp. 720–724 (2003)
17. Gifford, D.: Weighted Voting for Replicated Data. In: SOSP 1979, pp. 150–162 (1979)
18. Hassin, Y., Peleg, D.: Distributed probabilistic polling and applications to proportionate agreement. Information & Computation 171(2), 248–268 (2001)
19. Jerrum, M., Sinclair, A.: Conductance and the rapid mixing property for Markov chains: the approximation of permanent resolved. In: STOC 1988, pp. 235–244 (1988)
20. Johnson, B.: Design and Analysis of Fault Tolerant Digital Systems. Addison-Wesley (1989)
21. Mossel, E., Neeman, J., Tamuz, O.: Majority Dynamics and Aggregation of Information in Social Networks, arXiv:1207.0893 (2012)
22. Nakata, T., Imahayashi, H., Yamashita, M.: Probabilistic local majority voting for the agreement problem on finite graphs. In: Asano, T., Imai, H., Lee, D.T., Nakano, S.-i., Tokuyama, T. (eds.) COCOON 1999. LNCS, vol. 1627, pp. 330–338. Springer, Heidelberg (1999)
23. Oliviera, R.I.: On the Coalescence Time of Reversible Random Walks. Trans. Amer. Math. Soc. 364, 2109–2128 (2012)
24. Wormald, N.C.: Models of random regular graphs. In: Lamb, J.D., Preece, D.A. (eds.) Surveys in Combinatorics, pp. 239–298

Jamming-Resistant Learning
in Wireless Networks*

Johannes Dams[1], Martin Hoefer[2], and Thomas Kesselheim[3]

[1] Dept. of Computer Science, RWTH Aachen University, Germany
dams@cs.rwth-aachen.de
[2] Max-Planck-Institut für Informatik and Saarland University, Germany
mhoefer@mpi-inf.mpg.de
[3] Dept. of Computer Science, Cornell University, USA
kesselheim@cs.cornell.edu

Abstract. We consider capacity maximization in wireless networks under adversarial interference conditions. There are n links, each consisting of a sender and a receiver, which repeatedly try to perform a successful transmission. In each time step, the success of attempted transmissions depends on interference conditions, which are captured by an interference model (e.g. the SINR model). Additionally, an adversarial jammer can render a $(1 - \delta)$-fraction of time steps unsuccessful. For this scenario, we analyze a framework for distributed no-regret learning algorithms. We obtain an $O(1/\delta)$-approximation for the problem of maximizing the number of successful transmissions. Our approach provides even a constant-factor approximation when the jammer exactly blocks a $(1 - \delta)$-fraction of time steps. In addition, we consider the parameters of the jammer being unknown to the algorithm, and we also consider a stochastic jammer, for which we obtain a constant-factor approximation after a polynomial number of time steps. We extend our results to more general settings, in which links arrive and depart dynamically.

1 Introduction

One of the algorithmic challenges in this domain of wireless communication is referred to as *capacity maximization*. The goal is to maximize the number of simultaneous successful transmissions in a given network. More formally, the wireless network is represented by a set of n communication requests (or *links*), each consisting of a pair of sender and receiver. The resulting algorithmic problem is to find a maximum cardinality subset of successful links, where "successful" is defined by the absence of conflicts at receivers in an interference model. Most prominently traditional models like disk graphs or the recently popular SINR model [16] are used in such analyses to capture the impact of simultaneous

---

* A full version of the paper is available online [6]. This work has been supported by DFG through Cluster of Excellence "MMCI" at Saarland University, UMIC Research Centre at RWTH Aachen University, grant Ho 3831/3-1 and by a fellowship within the Postdoc-Programme of the German Academic Exchange Service (DAAD).

J. Esparza et al. (Eds.): ICALP 2014, Part II, LNCS 8573, pp. 447–458, 2014.
© Springer-Verlag Berlin Heidelberg 2014

transmission. For example in the SINR model "success" (or being conflict-free) is defined by the sum of interference from other links being below a certain threshold.

To this date, many algorithms for capacity maximization that provide provable worst-case guarantees are centralized [10–13]. In contrast, wireless networks are inherently decentralized and, hence, there is a need for algorithms with senders making transmission decisions in a distributed way not knowing the behavior of other links. Distributed algorithms often assume perfect, static conditions including all links behaving as given by the algorithm. By contrast, the environment can change rapidly, particularly in the presence of a co-existing network or even maliciously behaving wireless transmitters. A common way to take these effects into consideration is by modeling interference conditions as if they were determined by an adversary, adapting to the algorithm's efforts over time. Thus algorithms need to be very robust.

In this paper, we address this issue and extend capacity maximization to this scenario by studying distributed learning algorithms with adversarial jamming. Links iteratively adapt their behavior to maximize the capacity of the single time steps. We consider a very powerful adversary model of a $(T', 1-\delta)$-bounded jammer [4]. Such an adversary is allowed to make all transmissions unsuccessful during a $(1 - \delta)$-fraction of any time window of T' time steps. Beyond such a worst-case scenario, we also address a stochastic jammer that blocks each time step independently at random with a probability of $(1 - \delta)$.

We assume that links have no prior knowledge about the size or structure of the network. Giving such information to links can be infeasible when considering, e.g., distributed large scale sensor networks or ad-hoc networks. The only feedback they obtain is whether their respective own previous transmissions were successful or not. Links must adjust their behavior over time and decide about transmission attempts given only the previous feedback. Our algorithms are based on no-regret learning techniques to exploit the non-jammed time steps as efficiently as possible. A no-regret learning algorithm is an iterative randomized procedure that repeatedly decides which of multiple possible actions to take. After choosing an action, the algorithm receives a utility as feedback for its choice. Based on this feedback, it adjusts its internal probability distribution over choices, thereby obtaining a "no-regret" property over time. Each link can run such an algorithm independently of other links – even without knowing the number of links or the network structure. Our analysis shows how one can use such algorithms and their no-regret property to obtain provable approximation factors for capacity maximization under adversarial jamming. This can even be achieved without knowing the bound on the jammer (i.e., T' and δ).

We adapt no-regret learning algorithms to obtain a constant-factor approximation of the maximum possible number of successful transmissions if the adversary jams exactly a $(1 - \delta)$-fraction of the time. If the adversary jams less time steps, our algorithms still guarantee an $O(1/\delta)$-approximation. While our algorithms need to know the parameters T' and δ of the adversary, they are oblivious to the number n of links and the exact topology of the network.

More generally, we can even obtain the similar results if T' or δ is unknown. Based on these results, we show that for a stochastic jammer, the same results hold with high probability after a polynomial number of time steps.

Our results are obtained using a proof template based on linear programming. This way, we can significantly generalize previous approaches for online learning in wireless networks. We identify and base our approach on several key parameters of the sequences of transmission attempts resulting from our algorithms. We then show how to adjust no-regret learning algorithms to compute such sequences with suitable values for the key parameters. This approach turns out to be very flexible. Besides adversarial and stochastic jamming, we can successfully address even further generalizations of the scenario with little overhead extending our results to incorporate natural aspects that have not been subject to worst-case analysis in the literature so far, even without adversarial jamming.

For example, we consider a scenario where links can join and leave the network, which introduces additional difficulties for the algorithms to adjust their behavior to the network. In this case, our approximation guarantee increases only by a factor of $O(\log n)$. By applying our analysis directly with the proof template, we can easily combine this with all results on adversarial jamming above if links remain in the network sufficiently long to guarantee the properties necessary for applying our template. The template can also be applied to scenarios where a "link" consists of a single sender and multiple receivers. We obtain the same results as before when a successful transmission means that for a sender (a) at least one or (b) all receivers are conflict-free (i.e., receive the respective transmission successfully). Due to spatial constraints details on this issue and most proofs can only be found in the full version of this paper [6].

1.1 Related Work

Capacity maximization has been a central algorithmic research topic over the last decade. Many papers consider graph-based interference models, mainly restricted to simple models like disk graphs [9, 17, 22]. This neglects some of the main characteristics of wireless networks, and recently the focus has shifted to more realistic settings. Most prominently, Moscibroda and Wattenhofer [16] popularized models based on the signal-to-interference-plus-noise-ratio (SINR).

Our work is closely related to results on learning and capacity maximization in the SINR model with uniform powers (see e.g. [1,8,10,18]). In fact, we consider a more general scenario including a variety of interference models that satisfy a property called C-independence, which is similarly used in [2].

The effect of jammers on wireless networks was studied in [4, 19–21]. These works focus on the simpler graph-based interference models. A recent approach by Ogierman et al. [18] specializes in the SINR model with jammers. They tailor the adversary to the SINR model rather specifically – the adversary has a budget of power to influence ambient noise. In contrast, our work considers a general class of interference models. The network model also differs. It is not link-centered, but it consists of single nodes able to transmit and to receive messages from all other nodes. The number of successful transmissions is defined as the number of receivers successfully decoding no matter from which sender the transmission originates. They analyze the algorithm in terms of competitiveness proving a constant competitive-factor under certain conditions.

While we obtain a similar approximation ratio for a link-centered scenario, we are able to extend it in various directions. The regret-learning techniques allow a very distributed approach with little feedback. We do not assume that a specific algorithm is used but instead rely on the (external) no-regret property of existing algorithms yielding some key properties to apply our proof template. All algorithms that satisfy these conditions are suitable for application within our framework.

In a recent paper [7], we study no-regret learning algorithms for multiple channels. An adversary draws stochastic availabilities that are presented to the links in the beginning of each round and links have to decide on which channel to transmit or not to transmit at all. Having multiple channels and knowing which channels are available before deciding whether to transmit gives the problem quite a different flavor. While there are similarities in the analysis, we apply more intricate no-sleeping-expert regret algorithms.

1.2   Formal Problem Description

Network Model and Adversary. We consider the network consisting of a set V of n wireless links $\ell_v = (s_v, r_v)$ for $v \in V$ composed of sender s_v and receiver r_v. We assume the time steps to be synchronized and all links to use the same channel, i.e., all transmission attempts increase the interference for each other. An adversary is able to jam a restricted number of time slots. The overall goal in *capacity maximization* is to maximize the total number of transmission over time. Whenever some link $v \in V$ transmits successfully in some time step, this counts as one successful transmission. Success is defined using an interference model as specified below. We aim to maximize the sum of successful transmissions over all links and all time steps. With full knowledge of the jammer, an optimum solution is constructed by picking in each time step a set of non-jammed links $V' \subseteq V$ with maximum cardinality such that their transmissions are simultaneously successful. Obviously, this approach requires global knowledge, centralized control, and is known to be NP-hard. Instead, we design distributed learning algorithms that provably approximate the optimum number of successful transmissions.

Similarly as in previous work [4, 19, 20] we assume there is an adversary that can render transmission attempts unsuccessful. The jammer is prevented from blocking all time steps and making communication impossible as follows.

- A *(global)* $(T', 1 - \delta)$-*bounded* adversary can jam at most a $(1 - \delta)$-fraction of the time steps in any time window of length T' or larger.
- We will also consider the special case of an *(global)* $(T', 1-\delta)$-*exact* adversary, which exactly jams an $(1 - \delta)$-fraction of any time window of length T'.
- As a third variant, we treat a *(global)* *stochastic* adversary, where we assume any time step to be independently jammed with a probability $1 - \delta$.

Whereas these adversaries jam the channel globally for all links, an *individual* adversary can block each link individually. This leads to similar definitions of *individual* $(T', 1-\delta)$-*bounded, individual* $(T', 1-\delta)$-*exact* and *individual stochastic* adversaries. They obey the same restrictions on the type and number of jammed time slots for each link, but decide individually for each link if a slot is jammed. Note that the random trials of the individual stochastic adversary can be correlated between links but are assumed to be independent between time steps.

The stochastic adversary is obviously not adaptive. In all other cases, the adversary can jam arbitrarily (subject to the given constraints). Thus, the adversary can be a adaptive online adversary knowing the history of actions played by the algorithm or a reactive one even knowing the actions chosen by the algorithm beforehand.

When the (individual) adversary jams a time slot, every attempted transmission (of the jammed link) in this time slot becomes unsuccessful. Links receive as information only success or failure of their own transmissions, i.e., they cannot distinguish whether a transmission failed due to adversarial jamming or interference from other transmissions. Thus, a protocol has to base the decisions about transmission only on the feedback of success or failure of previous time steps. The optimum differs in different time steps due to jamming and we will consider the average optimum for comparison later.

Interference Model. We use a general framework based on edge-weighted conflict graphs that encompasses a variety of interference models, including the SINR model or models based on bounded-independence graphs like unit-disk graphs [22]. A *conflict graph* is a directed graph $G = (V, E)$ consisting of the links as vertices and weights $b_v(w)$ for any edge $(v, w) \in E$. Given a subset L of links transmitting, we say that $\ell_w \in L$ is *successful* iff $\sum_{v \in L} b_v(w) \leq 1$ (i.e., the sum of incoming edge weights from other transmitting links is bounded by 1). Such a set of links is called *feasible* if all links in this set can successfully transmit simultaneously. We use the notion of C-independence as a key parameter for the connection between interference model and performance of the algorithm.

Definition 1. *A conflict graph is called C-independent if for any feasible set L there exists a subset $L' \subseteq L$ with $|L'| \geq 1/2 \cdot |L|$ and $\sum_{v \in L'} b_u(v) \leq C$ for all $u \in V$, where $|L|$ and $|L'|$ denote the number of transmitting links in these sets.*

C-independence generalizes the bounded-independence property popular in the distributed computing literature. It has been observed, e.g., in [7,15] that successful transmissions in the SINR model can easily be represented by this framework using edge weights based on the notion of affectance [13].

If the gain matrix in the SINR model is based on metric distances and we use uniform power for transmission, this results in a C-independent conflict graph with constant $C = O(1)$ (cf. [2, Lemma 11]).

While we assume such a constant C-independence for simplicity, our results can be generalized straightforward to arbitrary conflict graphs losing a factor of C in the approximation guarantee.

No-Regret Learning. Our algorithms for capacity maximization are based on no-regret learning. Links decide independently in every round (or time slot) whether to transmit or not using an appropriate learning algorithm. The algorithms adjust their behavior based on the outcome of previous decisions. This outcome is either a successful transmission or an unsuccessful one. The quality of an outcome is measured by a suitable utility function $u_i^{(t)}(a_i^{(t)})$ depending on action $a_i^{(t)}$ chosen by player i in round t and depending on actions chosen by other players in t.

In our case, there are only two possible actions in each round – sending or not sending. We use utility functions $u_i^{(t)}$ defined in the subsequent sections that strike a balance between interference minimization and throughput maximization, where we also account for different forms of adversarial jamming. Given this setup with appropriate utility functions, we assume links apply arbitrary no-regret learning algorithms that minimize external regret. The *(external) regret* for an algorithm or a sequence of chosen actions is defined as follows.

Definition 2. *Let* $a_i^{(1)}, \ldots, a_i^{(T)}$ *be a sequence of action vectors. The external regret of this sequence for link* i *is defined by* $\max_{a_i' \in \mathcal{A}} \sum_{t=1}^{T} u_i^{(t)}(a_i') - \sum_{t=1}^{T} u_i^{(t)}(a_i^{(t)})$, *where* $\mathcal{A}$ *denotes the set of actions. An algorithm has the no-external regret property if the external regret of the computed sequence of actions grows in* $o(T)$.

Note that algorithms yielding the no-regret properties do not actually calculate the regret. Here, the regret depicts the difference between the actual gained utilities and the best action in hindsight (either transmitting or not) with fixed actions of others and the jammer.

One algorithm achieving regret low enough to apply our results with high probability after a polynomial number of time steps is given by Auer et al. [3] (see also [2]). This algorithm works by updating a probability distribution in a multiplicative weights fashion. It is applicable in the bandit model, where after each round the weights are only updated for the action chosen by the algorithm.

2   General Approach

In this section, we present a general template to analyze capacity maximization algorithms with adversarial jamming. Our approach here unifies and extends previous analyses of simpler problem variants. We adapt no-regret learning algorithms by defining appropriate utility functions and altering the number of time steps between learning (i.e., updating the probabilities). This way we achieve that certain key properties discussed below, on which our analysis relies, hold. A central idea in our construction is to divide time into phases. Here, a *phase* refers to a consecutive interval of k time steps (where k will be chosen appropriately in the respective settings). Our algorithms are assumed to decide about an action at the beginning of each phase. A link will either transmit in every time step or not at all during a phase. This way, we adapt no-regret learning algorithms such that one round (update step) of the algorithm coincides with a phase and not with a single time step. We denote by $\mathcal{R}_v$ the set of phases for link ℓ_v and in general we do not assume the phases of different links to be synchronized.

We label a phase as either successful or unsuccessful. It is successful if link ℓ_v attempts transmission throughout the phase and at least a fraction $\mu \in (0, 1]$ of time steps within the phase is successful. We adjust μ in specific settings below. For the analysis of the computed sequence of actions, let q_v denote the fraction of phases in which ℓ_v attempted transmission and w_v the fraction of successful phases.

As the first step, we identify a relation between attempted and successful transmissions. This and the property later on are useful for our analysis and capture the intuition of a good approximation algorithm. Being (γ, ϵ)-successful implies that an $(2/\gamma)$-fraction of phases with attempted transmissions in a computed sequence of actions must be successful. Otherwise the algorithm would have rather decided not to transmit. In subsequent sections, we will see that the no-regret property can be used to yield this property. Our proofs rely on parameter ϵ, which denotes the regret averaged over the phases.

Definition 3. *A sequence of action vectors is* (γ, ϵ)-*successful if* $\frac{2w_v + \epsilon}{\gamma} \geq q_v$.

The attempted transmissions allow to obtain a bound on the incoming edge weights from other transmitting links. Mirroring the (γ, ϵ)-successfulness, intuitively an algorithm only decides to send seldomly if there is interference. To model this property, f_v in the following definition is the fraction of unsuccessful phases not restricted to those phases in which ℓ_v transmits.

Definition 4. *A sequence of action vectors is* η-*blocking if for every link with* $q_v \leq \frac{1}{4}\eta$ *we have for the fraction of unsuccessful phases due to other links (independent of whether* ℓ_v *transmits)* $f_v \geq \frac{1}{4}\eta$ *and* $\sum_{u \in V} b_u(v)q_u \geq \frac{1}{8}\eta$.

Given these conditions, we can obtain a bound on the performance of the algorithm for capacity maximization. We consider the approximation factor in terms of the number of successful transmissions summed over all time steps and the optimum also summed over all time steps.

Theorem 1. *Suppose an algorithm computes a sequence of actions which is* η-*blocking and* (γ, ϵ)-*successful with* $\epsilon < \frac{1}{4n}\gamma\eta$. *Against an (individual)* $(T', 1 - \delta)$-*bounded adversary the average throughput of the computed action sequence yields an approximation factor of* $O\left(\frac{C}{\mu \cdot \gamma \cdot \eta}\right)$.

Proof. We will prove the theorem using a primal-dual approach. The following primal linear program corresponds to the optimal scheduling (c.f. [14]).

$$\text{Max. } \sum_{v \in V} x_v$$
$$\text{s.t. } \sum_{v \in V} b_u(v)x_v \leq C \ \forall u \in V$$
$$x_v \leq 1 \ \forall v \in V$$
$$x_v \geq 0 \ \forall v \in V$$

Let OPT' denote the set L' for $L = OPT$ from the definition of C-independence. For a global adversary we can choose x_v to correspond to the single slot optimum without jammer by setting $x_v = 1$ if link ℓ_v is transmitting in OPT' and $x_v = 0$ otherwise. Due to C-independence, this solution is feasible.

Let T be the set of all time steps. For an individual $(T', 1-\delta)$-bounded adversary, different time steps yield different optima denoted by OPT'_t. Therefore, we define $x_v = |\{t \in T \mid \ell_v \in OPT'_t\}|/|T|$ as the fraction of time steps in which ℓ_v is in the optimum of all time steps. As every single OPT'_t is C-independent, this average is also C-independent. This yields a feasible solution for the LP.

By primal-dual arguments we bound the value of the primal optimum.

$$\text{Min. } \sum_{v \in V} C \cdot y_v + \sum_{v \in V} z_v$$
$$\text{s.t. } \sum_{u \in V} b_u(v) y_u + z_v \geq 1 \; \forall v \in V$$
$$y_v, z_v \geq 0 \; \forall v \in V$$

To construct a feasible solution for the dual LP we set $y_v = \frac{1}{\eta} \cdot 8 q_v$ and $z_v = \frac{1}{\eta} \cdot 4 q_v$. If $q_v \geq \frac{1}{4}\eta$, this directly fulfills the constraints due to $z_v \geq 1$. Otherwise, by Definition 4 it holds that the interference from other links over all phases (including phases in which ℓ_v does not send) is at least $\frac{1}{8}\eta$. This yields $\sum_{u \in V} b_u(v) q_u \geq \frac{1}{8}\eta$ and plugging in fulfills the constraints.

For the objective functions by Definition 3 and get $\sum_{v \in V} \frac{|\{t \in T \mid \ell_v \in OPT'_t\}|}{T} \leq \sum_{v \in V} C \cdot \frac{12}{\eta} \cdot \frac{1}{\gamma}(2w_v + \epsilon)$.

Remember that a phase is of length k. As a successful phase has link ℓ_v being successful in at least μk time steps, we can conclude that w_v and the total number of successful steps are related by a factor of μ. This yields an approximation factor of $O(C/(\eta\gamma\mu))$ for $\epsilon < \frac{1}{4n}\eta\gamma$ with respect to the primal optimum. $\qquad \square$

Note that for an $(T', 1-\delta)$-exact adversary for all $T' \leq T$, where T is the length of the sequence of actions, the average optimum is in fact a factor δ worse than the single-slot optimum without adversary. As mentioned in the proof above, the guarantee also holds with respect to the single-slot optimum improving it for global exact jammers by a factor of $1/\delta$.

Corollary 1. *Consider an algorithm with conditions as in Theorem 1. Against any global $(T', 1-\delta)$-exact adversary with $T' \leq T$, the average throughput of the computed action sequence yields an approximation factor of $O\left(C \cdot \delta/(\mu \cdot \gamma \cdot \eta)\right)$.*

3   Bounded Adversary

$(T', 1-\delta)$-**bounded Adversary.** In this section we construct no-regret algorithms that provide constant and $O(1/\delta)$-factors approximation for diminishing regret against $(T', 1-\delta)$-exact and bounded adversaries, respectively. While we will first assume that the parameters T' and δ are known to the links and can be used by the algorithm, we will later relax this assumption. We will describe how to embed any no-regret learning algorithm into our general approach from Section 2. In particular, we define appropriate utility functions for feedback. Based on these, the no-regret property implies suitable bounds for γ, ϵ and η. We can allow different links to use different no-regret algorithms.

Each no-regret algorithm has two actions available (sending and not sending). We set the length of a phase $k = T'$ and thus assume each algorithm sticks to a chosen action for T' time steps before changing its decision. We consider a phase to be successful iff more than $\mu = \frac{1}{2}\delta$ time steps throughout the phase are successful. After a phase the following utility function inspired by [1] is used to give feedback to the no-regret algorithms to adjust the sending probabilities. Let w_u^R denote the fraction of successful transmissions during phase R.

$$u_i^{(R)}(s_i, s_{-i}) = \begin{cases} 1 & \text{if } \ell_i \text{ transmits and } w_u^R \geq \frac{1}{2}\delta \\ -1 & \text{if } \ell_i \text{ transmits and } w_u^R < \frac{1}{2}\delta \\ 0 & \text{otherwise.} \end{cases}$$

A no-regret algorithm embedded as described above will converge to an $O(1/\delta)$-approximation for both $(T', 1 - \delta)$-bounded and individually-$(T', 1-\delta)$-bounded adversaries. We use Theorem 1 and establish the necessary properties in Lemma 1.

Theorem 2. *Every sequence of action vectors with average regret per phase of $\epsilon \leq \frac{1}{4n}$ for all links yields an $O\left(1/\delta\right)$-approximation against individual $(T', 1-\delta)$-bounded adversaries.*

Lemma 1 (cf. [2, 5]). *Every no-regret algorithm with average regret per phase $\epsilon < \frac{1}{4}$ using the utility above computes an action sequence that is $(1, \epsilon)$-successful and 1-blocking.*

Combing these insights with $\mu = \frac{1}{2}\delta$, Theorem 1 implies an approximation factor in $O(C/\delta)$ for (individual) $(T', 1 - \delta)$-bounded jammers. Additionally, the following corollary follows from Corollary 1.

Corollary 2. *Every sequence of action vectors with average regret per phase of $\epsilon \leq \frac{1}{4n}$ for all links yields an $O(1)$-approximation against global $(T', 1-\delta)$-exact adversaries.*

Unknown T'. For the previous results it is necessary to know both T' and δ to design utility function and phase length. Ommitting this assumption, we show how to use regret-learning to reach an $O\left(1/\delta\right)$-approximation if the bound on T' is not known. Thus, we consider only δ to be known to the links. We use the following utility function and learn in every time step by setting the phase length to be $k = 1$.

$$u_i^{(t)}(s_i, s_{-i}) = \begin{cases} 1 & \text{if } \ell_i \text{ transmits successfully} \\ -\frac{\delta_v}{2-\delta_v} & \text{if } \ell_i \text{ transmits unsuccessfully} \\ 0 & \text{otherwise} \end{cases}$$

Theorem 3. *Every sequence of action vectors with average regret $\epsilon \leq \frac{1}{4n} \cdot \frac{\delta^2}{2-\delta}$ for all links yields an $O(1/\delta^2)$-approximation against individual $(T', 1 - \delta)$-bounded adversaries and an $O(1/\delta)$-approximation against $(T', 1 - \delta)$-exact adversaries.*

In this setting, every no-regret algorithm computes sequences of action vectors that is $\left(\frac{\delta}{2}, \epsilon\right)$-successful and δ-blocking. Together with $\mu = 1$ from the utility function, the theorem follows from Theorem 1 and Corollary 1.

Lemma 2. *Every no-regret algorithm with average regret per time step $\epsilon < \frac{1}{4} \cdot \frac{\delta^2}{2-\delta}$ using the given utility computes an action sequence that is $\left(\frac{\delta}{2}, \epsilon\right)$-successful and δ-blocking.*

Unknown δ. For asynchronous regret learning it seems to be necessary to know δ, as guessing a larger δ can have the jammer tripping an algorithm into experiencing much interference and crediting this to other links. As soon as the guessed δ is at least twice the actual one, the no-regret algorithm can be arbitrarily bad. The adversary can force the no-regret algorithm to consider not-sending to be the best strategy in hindsight.

While the learning algorithms for known δ in Section 3 easily adjust to links joining later, we here give a synchronized algorithm for unknown δ, in which all links start the algorithm at the same time. The basic idea is to test different values for δ in a coordinated fashion – half of all phases $\delta = \frac{1}{2}$ is assumed, in a quarter of all phases $\delta = \frac{1}{4}$ and so on. This implies that the correct δ (up to a factor of 2) is considered in a δ-fraction of all phases.

This way, in a δ-fraction of all phases our synchronized algorithm assumes the jammer to be $(T', 1-\delta)$-bounded. In the phases where the correct δ is tried, the algorithm achieves a constant-factor approximation due to Theorem 2 or an $O\left(1/\delta\right)$-approximation due to Theorem 3 when T' is not known. Note that the running time increases by a factor of $1/\delta$ over the asynchronous case, as we need the regret to be sufficiently low in the phases with the correct assumption on δ.

Theorem 4. *Consider any $(T', 1-\delta)$-bounded adversary. Then there exists an algorithm yielding an $O(1/\delta)$-approximation without knowledge of δ, and an algorithm yielding an $O(1/\delta^2)$-approximation without knowledge of δ and T'.*

4   Stochastic Adversary

In this section we extend results for the bounded adversary to the stochastic adversary. After a sufficient number of time steps an algorithm obtains very similar guarantees against a stochastic adversary as against a corresponding $(T', 1-\delta)$-exact adversary considered before. We consider no-regret algorithms with utility functions as discussed before and apply slight modifications as follows. For algorithms where $\mu < \delta$ and $k > 1$ we adjust the length of phases in order to bound the number of phases caused to be unsuccessful by the adversary. This allows to concentrate the behavior of the stochastic jammer to an "expected" exact jammer. It also allows us to show that in the stochastic setting an algorithm loses at most a constant factor in its η-blocking property after a sufficiently long time. We observe that against the non-individual stochastic adversary, the optimum is at most $\frac{9 \cdot \delta}{8}$-th of a single-slot optimum. The respective proofs can be found in the full version.

Let p_z denote the probability that the adversary renders a phase unsuccessful.

Lemma 3. *Let $\mu < \delta$. Then for any $k \geq 1$ it holds $p_z \leq \exp\left(-(\mu/\delta)^2 \delta k/2\right)$.*

Lemma 4. *Consider an algorithm that computes a sequence of actions which is η-blocking against an $(T', 1 - \delta)$-exact adversary. After $T \geq \frac{\max\{p_z, 1-p_z\}}{\eta^2} \cdot 8^2 \cdot 3 \cdot c \cdot \ln(n) + \ln(n)$ phases, the computed sequence is $\frac{\eta}{2}$-blocking against a stochastic adversary with probability at least $1 - \frac{1}{n^c}$.*

Additionally to using Lemmas 3 and 4, we will also bound the number of time steps till the jammer converges to an exact one to yield that the optimum against the stochastic adversary is close to the one against an exact adversary.

Lemma 5. *After $T \geq \frac{8^2}{3\delta} \cdot c \cdot \ln(n)$ time steps it holds with probability $1 - \frac{1}{n^c}$ that the optimum against the stochastic adversary is at most $\frac{9}{8}$ of the optimum against an exact adversary.*

In total, we obtain the corollary below matching the results in Section 3. It follows from the previous lemmas with $\mu = \delta/2$ and $\eta = 1$. Setting $k = \frac{2}{\delta} \cdot \ln(8)$ yields the probability $1 - p_z > 7/8$ by Lemma 3. Together with Lemma 4 and 5 this yields the claim. We require that the jammer is close to expectation, and the algorithms obtain low regret with high probability. For a suitable algorithm that achieves this after a time polynomial in the number of links see, e.g., [3].

Corollary 3. *With high probability, by setting $k = \frac{2}{\delta} \cdot \ln(8)$ the algorithm in Section 3 yields a $O(1)$-approximation after $T \in O(\ln(n))$ phases against an (global) stochastic adversary.*

5   Joining and Leaving Links

Our general approach in Section 2 does not require that links join at the same time. Still, we have to assume all links stay within the network. Here, we relax this assumption and consider links being able to leave the network earlier. However, they are assumed to stay until they obtain an action sequence in which their *own* regret is low. For this we prove convergence to an $O(\log(n)/\delta)$-approximation against an $(T', 1 - \delta)$-bounded adversary.

More formally, each link comes with an interval of phases $\mathcal{R}_v$ in which it is present in the network. In these phases it can transmit and observe the outcome of his actions. Outside of its interval a link cannot transmit or learn. The following theorem adjusts our general approach for this more general case.

Theorem 5. *Consider an algorithm as in Theorem 1. Against an (individual) $(T', 1 - \delta)$-bounded adversary the average throughput of the computed action sequence yields an approximation factor of $O\left(\left(\log n + \log\left(\frac{1}{\eta}\right)\right) \frac{C}{\mu \cdot \gamma \cdot \eta}\right)$.*

For the proof we extend the primal-dual approach from Section 2 to a more complex LP, which increases the approximation guarantee (details in the full version). The theorem allows to transfer all guarantees for all the above settings to the case where links are allowed to join and leave the network. This increases the guarantees by a factor of $O(\log n + \log 1/\eta)$. In particular, Theorem 5 also implies that without adversaries, we can use no-regret learning techniques to obtain an $O(\log n)$-approximation guarantee.

References

1. Andrews, M., Dinitz, M.: Maximizing capacity in arbitrary wireless networks in the SINR model: Complexity and game theory. In: Proc. 28th INFOCOM, pp. 1332–1340 (2009)
2. Asgeirsson, E.I., Mitra, P.: On a game theoretic approach to capacity maximization in wireless networks. In: Proc. 30th INFOCOM, pp. 3029–3037 (2011)
3. Auer, P., Cesa-Bianchi, N., Freund, Y., Schapire, R.E.: The nonstochastic multi-armed bandit problem. SIAM Journal on Computing 32(1), 48–77 (2002)
4. Awerbuch, B., Richa, A.W., Scheideler, C.: A jamming-resistant mac protocol for single-hop wireless networks. In: PODC, pp. 45–54 (2008)
5. Dams, J., Hoefer, M., Kesselheim, T.: Scheduling in wireless networks with rayleigh-fading interference. In: Proc. 24th SPAA (2012)
6. Dams, J., Hoefer, M., Kesselheim, T.: Jamming-resistant learning in wireless networks (full version). CoRR abs/1307.5290 (2013)
7. Dams, J., Hoefer, M., Kesselheim, T.: Sleeping experts in wireless networks. In: Proc. 27th DISC, pp. 344–357 (2013)
8. Dinitz, M.: Distributed algorithms for approximating wireless network capacity. In: Proc. 29th INFOCOM, pp. 1397–1405 (2010)
9. Erlebach, T., Jansen, K., Seidel, E.: Polynomial-time approximation schemes for geometric intersection graphs. SIAM J. Comput. 34(6), 1302–1323 (2005)
10. Goussevskaia, O., Halldórsson, M., Wattenhofer, R., Welzl, E.: Capacity of arbitrary wireless networks. In: Proc. 28th INFOCOM, pp. 1872–1880 (2009)
11. Goussevskaia, O., Oswald, Y.A., Wattenhofer, R.: Complexity in geometric SINR. In: Proc. 8th MobiHoc, pp. 100–109 (2007)
12. Gupta, P., Kumar, P.R.: The capacity of wireless networks. IEEE Trans. Inf. Theory 46(2), 388–404 (2000)
13. Halldórsson, M.M., Wattenhofer, R.: Wireless communication is in APX. In: Albers, S., Marchetti-Spaccamela, A., Matias, Y., Nikoletseas, S., Thomas, W. (eds.) ICALP 2009, Part I. LNCS, vol. 5555, pp. 525–536. Springer, Heidelberg (2009)
14. Halldórsson, M.M., Mitra, P.: Wireless capacity and admission control in cognitive radio. In: Proc. 31st INFOCOM, pp. 855–863 (2012)
15. Hoefer, M., Kesselheim, T., Vöcking, B.: Approximation algorithms for secondary spectrum auctions. In: Proc. 23rd SPAA, pp. 177–186 (2011)
16. Moscibroda, T., Wattenhofer, R.: The complexity of connectivity in wireless networks. In: Proc. 25th INFOCOM, pp. 1–13 (2006)
17. Nieberg, T., Hurink, J., Kern, W.: Approximation schemes for wireless networks. ACM Trans. Algorithms 4(4) (2008)
18. Ogierman, A., Richa, A., Scheideler, C., Schmid, S., Zhang, J.: Competitive medium sharing under adversarial sinr (unpublished paper)
19. Richa, A.W., Scheideler, C., Schmid, S., Zhang, J.: A jamming-resistant mac protocol for multi-hop wireless networks. In: DISC, pp. 179–193 (2010)
20. Richa, A.W., Scheideler, C., Schmid, S., Zhang, J.: Competitive and fair medium access despite reactive jamming. In: ICDCS, pp. 507–516 (2011)
21. Richa, A.W., Scheideler, C., Schmid, S., Zhang, J.: Competitive and fair throughput for co-existing networks under adversarial interference. In: Proc. 31st PODC, pp. 291–300 (2012)
22. Schneider, J., Wattenhofer, R.: An optimal maximal independent set algorithm for bounded-independence graphs. Distributed Computing 22(5-6), 349–361 (2010)

Facility Location in Evolving Metrics*

David Eisenstat[1], Claire Mathieu[2], and Nicolas Schabanel[3,4]

[1] Brown University, USA
http://www.davideisenstat.com/
[2] CNRS, École normale supérieure UMR 8548, France
[3] CNRS, Université Paris Diderot, France
http://www.liafa.univ-paris-diderot.fr/$\sim$nschaban/
[4] IXXI, École normale supérieure de Lyon, France

Abstract. Understanding the dynamics of evolving social or infrastructure networks is a challenge in applied areas such as epidemiology, viral marketing, and urban planning. During the past decade, data has been collected on such networks but has yet to be analyzed fully. We propose to use information on the dynamics of the data to find stable partitions of the network into groups. For that purpose, we introduce a time-dependent, dynamic version of the facility location problem, which includes a switching cost when a client's assignment changes from one facility to another. This might provide a better representation of an evolving network, emphasizing the abrupt change of relationships between subjects rather than the continuous evolution of the underlying network. We show for some realistic examples that this model yields better hypotheses than its counterpart without switching costs, where each snapshot can be optimized independently. For our model, we present an $O(\log nT)$-approximation algorithm and a matching hardness result, where n is the number of clients and T is the number of timesteps. We also give another algorithm with approximation ratio $O(\log nT)$ for a variant model where the decision to open a facility is made independently at each timestep.

1 Introduction

During the past decade, a massive amount of data has been collected on diverse networks such as the web (pages and links), social networks (e.g., Facebook, Twitter, and LinkedIn), and social encounters in hospitals, schools, companies, and conferences [18,21]. These networks evolve over time, and their dynamics have a considerable impact on their structure and effectiveness [19,14]. Understanding the dynamics of evolving networks is a central question in many applied areas such as epidemiology, vaccination planning, anti-virus design, management of human resources, and viral marketing. A relevant clustering of the data often is needed to design informative representations of massive data sets. Algorithmic approaches have yielded useful insights on real networks such as the social interaction networks of zebras [22].

* This work was partially supported by the ANR-2010-BLAN-0204 Magnum and ANR-12-BS02-005 RDAM grants.

J. Esparza et al. (Eds.): ICALP 2014, Part II, LNCS 8573, pp. 459–470, 2014.

The dynamics of real-life evolving networks, however, are not yet well understood, partly because it is difficult to observe and analyze such large, sparsely connected networks over time. Some basic mechanisms such as preferential attachment and copy/paste have been observed, but more specific structures remain to be discovered. In this article, we propose a new formulation of the facility location problem adapted to these evolving networks. We show that, in many realistic situations, solutions that are stable over time match the ground truth more closely than those obtained by independent optimization with respect to each snapshot of the network.

The problem. We focus on a generalized facility location problem where clients are moving in some metric space over time. We look for a set of *open* facilities (also called centers) and a dynamic many-to-one assignment of clients to open facilities that minimizes the sum of three costs, of which the first two are inherited from the classical facility location problem. The *distance cost* is the sum over each (client,timestep) pair of the distance from the client to its assigned facility at that timestep. This cost tends to ensure that assigned facilities are representative with respect to position. The *opening cost* is linear in the number of facilities. This cost tends to ensure that only the most meaningful facilities are open. The new cost, *switching*, is linear in the number of (client,timestep) pairs where the client is assigned to a different facility at the next timestep. This cost tends to ensure that clients switch facilities only in response to significant and lasting changes in position. We argue that, in many realistic situations, the switching cost makes solutions close to the ground truth relatively more attractive (see Section 2.1).

Related work. The facility location problem has been studied extensively in the offline, online, and incremental settings [12]. The offline setting was a case study accompanying the development of approximation techniques: primal-dual and dual fitting methods and local search, for example. A series of papers [20,16,13,2,5,3,15] obtained almost matching upper and lower bounds on the polynomially achievable approximation ratio: $\Theta(\log n)$ in general and $[1.463, 1.488]$ in the metric case, where the specified distances satisfy the triangle inequality.

The online setting, where clients arrive over time and the algorithm gradually opens more and more facilities to serve them, was addressed first by [17], which obtained the asymptotically tight bound $\Theta(\log n/\log \log n)$ on the competitive ratio of the best online algorithm. Subsequent work considered the special case where the clients are drawn from some distribution [1] and other special cases [11]. Since many clustering applications benefit from the flexibility to change the solution over time, incremental settings also have been studied. Such variants may allow better (i.e., constant) competitive ratios, e.g., the metric case with streaming constraints [10] and the Euclidean metric setting where facilities may be moved as new clients arrive [8]. We also mention the related clustering problem in which clusters may be merged but not split [4].

Our setting differs from previous dynamic settings because the distances between clients and facilities may vary over time and because it is desirable to achieve a trade-off between the *stability* of the solution – the assignment should be modified slowly – and its *adaptability* – the assignment should be modified if the distances change significantly. Given the existence of experiments such as [21], we assume access to the whole evolution of the network ahead of time. We show that constructing an independent optimal solution for each snapshot of the network yields results that, in a large variety of realistic situations, are not only unstable (and thus arbitrarily bad according to our objective) but also undesirable with respect to network dynamics analysis.

As far as we know, settings where the distances between locations vary over time are still largely unexplored.

Our results. After defining the problem formally in Section 2.1 and giving examples showing the benefits that one can expect from solving this problem in the context of metrics evolving over time, we give in Section 2.3 an $O(\log nT)$-approximation algorithm for this problem, where n is the number of clients and T is the number of timesteps.

Theorem 1 (Fixed opening cost) *For the dynamic facility location problem with fixed opening cost, there exists a polynomial-time randomized algorithm that, on all inputs, with probability at least $1/4$, outputs a solution satisfying*

$$cost \leqslant 8\log(2nT) \cdot \mathrm{LP} \leqslant 8\log(2nT) \cdot \mathrm{OPT},$$

where OPT *is the cost of an optimal solution and* LP *is the value of LP (1), defined at the end of Section 2.1.*

Through repetition, running the algorithm t times and taking the best of the t solutions constructed, the probability $1/4$ can be improved to $1 - (3/4)^t$. The constant 8 can be improved as well.

We show in Section 2.4 that this approximation ratio is asymptotically optimal, even for a very special case.

Theorem 2 (Hardness for Fixed Opening Cost) *Unless $P = NP$, for the dynamic facility location problem with fixed opening cost, there is no $o(\log T)$-approximation.*

The lower bound holds even for the metric case with one client and two locations. This new problem differs significantly from the classic facility location problem, which admits no $o(\log n)$-approximation for nonmetric distances but can be 1.488-approximated when the distances satisfy the triangle inequality [15]. In Section 3, we show how to extend our approximation algorithm to the setting where facilities can be opened and closed at each timestep. The opening cost in this setting is equal to f times the number of (facility,timestep) pairs such that the facility is open at that timestep.

Theorem 3 (Hourly Opening Cost) *For the dynamic facility location problem with hourly opening cost, there exists a polynomial-time randomized algorithm that, on all inputs, with probability at least $1/4$, outputs a solution satisfying*

$$cost \leqslant 8 \log(2nT) \cdot LP \leqslant 8 \log(2nT) \cdot OPT,$$

where OPT *is the cost of an optimal solution and* LP *is the value of LP* (2), *defined at the end of Section 3.1.*

Again, through repetition, running the algorithm t times and taking the best of the t solutions constructed, the probability $1/4$ can be improved to $1 - (3/4)^t$. The constant 8 can be improved as well. This article concludes with several open questions and possible extensions of this work.

2   Facility Location in Evolving Metrics

2.1   Definition

We denote by $[n] = \{1, \ldots, n\}$ the subset of integers from 1 to n inclusive.

Dynamic facility location problem with fixed opening cost. We are given a set F of m *facilities* and a set C of n *clients* together with a finite sequence of distances $(d_t)_{t \in [T]}$ over $F \times C$, a nonnegative *facility opening cost* f, and a nonnegative *client switching cost* g. The goal is to output a subset $A \subseteq F$ of *open* facilities and, for each timestep $t \in [T]$, an assignment $\phi_t : C \to A$ of clients to open facilities so as to minimize

$$f \cdot |A| + \sum_{t \in [T], j \in C} d_t(\phi_t(j), j) + g \cdot \sum_{t \in [T-1], j \in C} \mathbb{1}\{\phi_t(j) \neq \phi_{t+1}(j)\},$$

namely, the sum of the *opening cost* (f for each open facility), the *distance cost* (the sum over each (client,timestep) pair from the client to its assigned facility at that timestep), and the *switching cost* (g for each (client,timestep) pair where the client is assigned to a different facility at the next timestep).

Examples. The two examples in Figure 1 show how facility location in the dynamic setting is quite different from facility location in the static setting and yields more desirable partitions of the clients. In both examples, a facility can be opened at every client (so that electing a facility consists of electing a representative for every significantly different behavior).

In example 1(a), we see a classroom with students split into five groups and a teacher moving from group to group in cyclic order. When the number of students is large, static facility location isolates the five groups and moves the teacher from one group to the next between snapshots. Dynamic facility location isolates every group of students and puts the teacher in a sixth group.

In example 1(b), we see two groups of people passing through each other, on a street for instance. Static facility location outputs first the two groups, then a

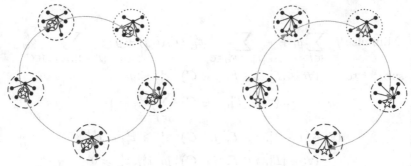

Optimal Dynamic Facility Location   Optimal Static Facility Location

(a) The classroom: one teacher cycling between 5 groups of students.

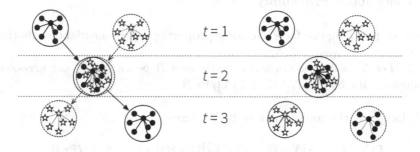

Optimal Dynamic Facility Location   Optimal Static Facility Location

(b) Two groups crossing.

Fig. 1. Dynamic versus static facility location

single group, then two groups again. Dynamic facility location, however, keeps the same groups for the whole time period, with the same representatives.

Assuming in the first example that the distances between individuals are very small and in the second that they are very large, the ratio of the (dynamic) cost between the dynamic solution and the sequence of static solutions can be made arbitrarily large, because the switching cost grows for the sequence of static solutions as $\Omega(T)$ and $\Omega(n)$ respectively.

Fact 4. *The (dynamic) cost of a sequence of optimal static facility location solutions for each snapshot can be larger than the cost of an optimal dynamic facility location solution by a factor $\Omega(T + n)$.*

A linear relaxation. For an integer programming formulation, we define indicator 0-1 variables y_i, x_{ij}^t, z_{ij}^t for $t \in [T]$ and $i \in F$ and $j \in C$. We let $y_i = 1$ if and only if facility i is open; $x_{ij}^t = 1$ if and only if client j is assigned to facility i at timestep t; and $z_{ij}^t = 1$ if and only if client j is assigned to facility i at timestep t but not at timestep $t + 1$. The dynamic facility location problem is equivalent to finding an integer solution to the following linear programming relaxation.

$$
\begin{cases}
\text{Minimize } f \cdot \sum_{i \in F} y_i + \sum_{t \in [T], i \in F, j \in C} x_{ij}^t \cdot d_t(i,j) + g \cdot \sum_{t \in [T-1], i \in F, j \in C} z_{ij}^t \\
\text{subject to} \quad (\forall t \in [T], \ i \in F, \ j \in C) \ x_{ij}^t \leqslant y_i \\
\qquad\qquad (\forall t \in [T], \ j \in C) \ \sum_{i \in F} x_{ij}^t = 1 \\
(\forall t \in [T-1], \ i \in F, \ j \in C) \ z_{ij}^t \geqslant x_{ij}^t - x_{ij}^{t+1} \\
(\forall t \in [T], \ i \in F, \ j \in C) \ y_i, x_{ij}^t, z_{ij}^t \geqslant 0
\end{cases}
\tag{1}
$$

2.2 Facts about Probability

We use the following two facts and some properties of exponential distributions.

Fact 5. *Let $X \geqslant 0$ be a random variable and B be an event, not necessarily independent. We have $E[X \mid B] \leqslant E[X]/\Pr B$.*

Proof. Let $\overline{B}$ be the complement of B. We have

$$
E[X \mid B] \leqslant E[X \mid B] + E[X \mid \overline{B}] \Pr \overline{B}/\Pr B = E[X]/\Pr B. \qquad \square
$$

Fact 6 (Markov's Inequality). *Let $X \geqslant 0$ be a random variable. For every $x > 0$, we have $\Pr\{X > x\} \leqslant E[X]/x$.*

A random variable X is *exponentially distributed* with rate λ if and only if, for every $x \geqslant 0$, it satisfies $\Pr\{X > x\} = e^{-\lambda x}$.

Fact 7. *If X is exponentially distributed with rate λ, then, for every $c > 0$, the distribution of X/c is exponential with rate $c\lambda$.*

Fact 8. *Let $(X_i)_{i \in F}$ be a sequence of independent random variables, where X_i is exponentially distributed with rate λ_i. Then $\min_{i \in F} X_i$ is exponentially distributed with rate $\sum_{i \in F} \lambda_i$, and the argument of the minimum is i with probability $\lambda_i / \sum_{k \in F} \lambda_k$.*

Proof. Indeed, $\Pr\{\min_{i \in F} X_i > x\} = \prod_{i \in F} \Pr\{X_i > x\} = e^{-\sum_{i \in F} \lambda_i x}$. As for the second claim,

$$
\Pr\{\arg \min_{k \in F} X_k = i\} = \int_{x=0}^{\infty} \Pr\{(\forall k \neq i) \ X_k > x\} \cdot \Pr\{X_i \in [x, x+dx]\}
$$

$$
= \int_{x=0}^{\infty} e^{-\sum_{k \neq i} \lambda_k x} \cdot \lambda_i e^{-\lambda_i x} dx = \lambda_i \Big/ \sum_{k \in F} \lambda_k. \qquad \square
$$

2.3 Approximation Algorithm

In order to determine a solution, we need to (1) decide which facilities to open, (2) decide when each client switches from one facility to another, and (3) decide which facility to connect each client to between switches. After computing an optimal (fractional) solution (x, y, z) to LP (1), Algorithm 1 proceeds as follows. Decision (1) is made by sampling the facilities according to $(y_i)_i$ approximately $O(\log nT)$ times. As we will show, this ensures that every client selects a sampled facility with high probability.

Regarding decision (2), since $\sum_i x_{ij}^t = 1$, one can view $(x_{ij}^t)_i$ as the desired distribution for the facility assigned to client j at timestep t. The subroutine Algorithm 2 partitions time, independently for each client j, into intervals during which the distribution $(x_{ij}^t)_i$ remains stable enough, i.e., the distributions $(x_{ij}^t)_i$ *share a large enough common probability mass* during each time interval of the partition. The common probability mass of the distributions $(x_{ij}^t)_i$ during a time interval U is defined as the sum over all facilities i of the minimum probability $\hat{x}_{ij}^U = \min_{t \in U} x_{ij}^t$ of assigning client j to i over U. The rule defining the partition is that each interval (except the last one) is maximal subject to the constraint that the common probability mass is at least $1/2$. This ensures two key properties. First, the distributions $(x_{ij}^t)_i$ for $t \in U$ are close enough to each other to be compatible and also, due to the first LP constraint, close enough to $(y_i)_i$ to match the sampling of the facilities. Second, the distributions are deemed to have changed too much when the x_{ij}^ts have had a combined decrease of at least $1/2$, which implies by the third LP constraint that the corresponding z_{ij}^ts sum to at least $1/2$, covering the cost of switching to another facility.

Decision (3) is made simply by assigning each client to the most likely of its preferred facilities to be open.

We propose two versions of the algorithm. The first assigns clients to open facilities via an optimal dynamic program, while the second uses the intuitive strategy described in Algorithm 2. We analyze the latter, as its approximation ratio is no worse than that of the former.

Theorem 1 states that Algorithm 1 outputs an $O(\log nT)$-approximation with positive constant probability. In the next section, we will show that this is asymptotically optimal (unless $P = NP$).

Proof (Theorem 1). Note that Algorithm 2 may produce an assignment that is not feasible. We bound the expected cost without conditioning on feasibility, bound the probability of feasibility, and finish by applying Fact 5 and Markov's inequality.

Algorithm 1. Fixed opening cost

- Solve the linear program LP (1) to obtain an optimal (fractional) solution (x, y, z).
- Choose the open facilities A randomly as follows. For each facility i, choose Y_i having exponential distribution with rate $2\log(2nT)$. Let $A = \{i \in F : Y_i \leqslant y_i\}$.
- With a dynamic program, determine how to assign optimally clients to facilities in A. Alternatively, for the purposes of analysis, use Algorithm 2.

Algorithm 2. Intuition-driven assignment of clients to facilities

for each client j **do**
- Partition time greedily into ℓ_j intervals $[t_k^j, t_{k+1}^j)$ where ℓ_j and $(t_k^j)_{k \in [\ell_j + 1]}$ are defined as follows: $t_1^j = 1$, and t_{k+1}^j is defined inductively as the greatest $t \in (t_k^j, T+1]$ such that $\sum_{i \in F} \left(\min_{t_k^j \leqslant u < t} x_{ij}^u \right) \geqslant 1/2$. Let $t_{\ell_j+1}^j = T+1$.
- For each time interval $U = [t_k^j, t_{k+1}^j)$, assign client j to argument of $\min_{i \in F}(Y_i / \hat{x}_{ij}^U)$, where $\hat{x}_{ij}^U = \min_{u \in U} x_{ij}^u$.

end for

The unconditional expected facility opening cost is

$$f \cdot \sum_{i \in F}(1 - e^{-2y_i \log(2nT)}) \leqslant (2 \log(2nT))f \cdot \sum_{i \in F} y_i$$

by the well known inequality $1 + x \leqslant e^x$. The right-hand side is $2 \log(2nT)$ times the corresponding term in the LP objective.

To analyze the unconditional expected distance cost, we define, for each client j, all of its time intervals U, and all $t \in U$, a fictitious independent event B_j^t such that $\Pr B_j^t = \sum_{k \in F} \hat{x}_{kj}^U \in [1/2, 1]$ by the LP and the definition of U.[1] We use this fictitious event B_j^t to define a random variable $I_j^t \in F$ by letting $\Pr\{I_j^t = i \mid B_j^t\} = \hat{x}_{ij}^U / \sum_{k \in F} \hat{x}_{kj}^U$ and $\Pr\{I_j^t = i \mid \overline{B}_j^t\} = (x_{ij}^t - \hat{x}_{ij}^U) / \sum_{k \in F}(x_{kj}^t - \hat{x}_{kj}^U)$, where $\overline{B}_j^t$ is the complement of B_j^t. Note that $\Pr \overline{B}_j^t = \sum_{k \in F}(x_{kj}^t - \hat{x}_{kj}^U)$ since $\sum_{k \in F} x_{kj}^t = 1$ by LP (1). The unconditional distribution of I_j^t thus is described by $\Pr\{I_j^t = i\} = x_{ij}^t$, so the expected distance from j to I_j^t is $E[d_t(I_j^t, j)] = \sum_{i \in F} x_{ij}^t \cdot d_t(i, j)$. Since $\arg\min_{i \in F}(Y_i / \hat{x}_{ij}^U)$ is i with probability $\hat{x}_{ij}^U / \sum_{k \in F} \hat{x}_{kj}^U$ by Fact 8, the actual assignment of Algorithm 2 is made according to the conditional distribution of I_j^t given B_j^t, so by applying Fact 5 and summing, the total unconditional expected distance cost is at most

$$2 \cdot \sum_{t \in [T], i \in F, j \in C} x_{ij}^t \cdot d_t(i, j),$$

which is twice the corresponding term in the LP objective.

To bound the switching cost, which is deterministic, we prove that, for each client j and all of its time intervals U except the last one,

$$\sum_{t \in U, i \in F} z_{ij}^t > 1/2.$$

---

[1] More concretely, let B_j^t be an event corresponding to the outcome HEADS of an independent biased coin flip that results a priori in HEADS with probability $\sum_{i \in F} \hat{x}_{ij}^U$. This event represents our ability to sample from the common probability mass of the distributions $(x_{ij}^t)_i$ for $t \in U$.

Each client switches only after its non-last intervals. Since each variable z_{ij}^t appears in exactly one sum, the total switching cost is bounded above by

$$2g \cdot \sum_{t \in [T-1], i \in F, j \in C} z_{ij}^t,$$

which is twice the corresponding term in the LP objective.

The z-variables measure decreases in the corresponding x-variables. Specifically, for every $t_1 \leqslant t_2$, the LP inequalities telescope to yield $x_{ij}^{t_1} - x_{ij}^{t_2} \leqslant \sum_{u \in [t_1, t_2)} z_{ij}^u$. By letting t_1 be the first time in $U = [t_1, t_3)$ and t_2 be the argument of the minimum $\min_{u \in [t_1, t_3]} x_{ij}^u$, whose domain is $U \cup \{t_3\}$, we sum to obtain the inequality

$$1/2 = 1 - 1/2 < 1 - \sum_{i \in F} \min_{u \in [t_1, t_3]} x_{ij}^u = \sum_{i \in F} (x_{ij}^{t_1} - \min_{u \in [t_1, t_3]} x_{ij}^u) \leqslant \sum_{u \in U, i \in F} z_{ij}^u,$$

where the first inequality is a consequence of defining U maximally.

As the next to last step, we bound the probability that every client is assigned to an open facility. Recall that, at each timestep t, Algorithm 2 assigns each client j to the argument i^* of the minimum $\min_{i \in F}(Y_i / \hat{x}_{ij}^U)$, where $U \ni t$ is the corresponding interval for j. This facility is open if and only if $Y_{i^*} \leqslant y_{i^*}$. Since $\hat{x}_{ij}^U \leqslant x_{ij}^t \leqslant y_i$,

$$\Pr\{Y_{i^*} \leqslant y_{i^*}\} \geqslant \Pr\{Y_{i^*} \leqslant \hat{x}_{i^*j}^U\} = \Pr\{\min_{i \in F}(Y_i / \hat{x}_{ij}^U) \leqslant 1\}.$$

The quantity $\min_{i \in F}(Y_i / \hat{x}_{ij}^U)$ is exponentially distributed with rate $2\log(2nT) \cdot \sum_{i \in F} \hat{x}_{ij}^U \geqslant \log(2nT)$ since $\sum_{i \in F} \hat{x}_{ij}^U \geqslant 1/2$, so $\Pr\{Y_{i^*} \leqslant y_{i^*}\} \geqslant 1 - 1/(2nT)$. By a union bound over all clients and timesteps, the probability of a feasible assignment is at least $1/2$.

In conclusion, we observe that the unconditional expected cost is a $2\log(2nT)$-approximation of the LP objective, and the probability of a feasible assignment is at least $1/2$. By Fact 5, the conditional expected cost given feasibility is a $4\log(2nT)$-approximation. By Markov's inequality, with probability at least $1/2 \cdot 1/2 = 1/4$, the output is a feasible $8\log(2nT)$-approximation. $\qquad\square$

2.4   Hardness of Approximation

Proof (Theorem 2). We exhibit an objective-preserving reduction from the set cover problem. Fix an instance of set cover with T elements and m sets. We define the following instance of dynamic facility location. There is one timestep t for each element of the set cover instance, one facility i for each set of the set cover instance, and a single client. We set $g = 0$ (i.e., g is small enough with respect to f and $1/n$ and $1/T$). Assume that the only possible positions for the client and facilities are two locations a and b at distance ∞ (i.e., large enough) from each other (note that this metric satisfies the triangle inequality). At every timestep t, the client's position is location a. For each set i of the set cover instance, the position of the corresponding facility is location a if set i contains element t and location b otherwise.

Since the distance between the two locations is infinite, a solution for our instance of dynamic facility location has finite cost if and only if, at every timestep, some open facility has position a, i.e., the set of open facilities corresponds to a cover. The cost of such a solution is f times the number of open facilities. We conclude that the $\Omega(\ln T)$-inapproximability result for set cover with T elements [7] implies the same inapproximability result for our problem. $\qquad\square$

3   Hourly Opening Cost

3.1   Dynamic Facility Location with Hourly Opening Cost

We now focus on a variant of the problem studied in the previous section, where each facility may be open or closed independently at each timestep and where the opening cost f is paid for each (facility,timestep) pair where the facility is open at that timestep. In other words, the cost of a facility is not its construction cost but its rental cost.

Dynamic facility location problem with hourly *opening cost.* We are given a set F of m facilities and a set C of n clients together with a finite sequence of distances $(d_t)_{t\in[T]}$ over $F \times C$ and two nonnegative values f and g. The goal is to output a sequence of subsets $A_t \subseteq F$ of facilities and, for each timestep $t \in [T]$, an assignment $\phi_t : C \to A_t$ of clients to facilities so as to minimize

$$f \cdot \sum_{t\in[T]} |A_t| + \sum_{t\in[T],j\in C} d_t(\phi_t(j),j) + g \cdot \sum_{t\in[T-1],j\in C} \mathbb{1}\{\phi_t(j) \neq \phi_{t+1}(j)\}.$$

Linear relaxation. LP (1) can easily be adapted to this variant, with new variables y_i^t replacing y_i. The interpretation of y_i^t is that it equals 1 if and only if facility i is open at timestep t.

$$
\begin{cases}
\text{Minimize } f \sum_{t\in[T],i\in A} y_i^t + \sum_{t\in[T],i\in F,j\in C} x_{ij}^t \cdot d_t(i,j) + g \sum_{t\in[T-1],i\in F,j\in C} z_{ij}^t \\
\text{subject to} \quad (\forall t \in [T],\ i \in F,\ j \in C)\ x_{ij}^t \leqslant y_i^t \\
\qquad\qquad\qquad (\forall t \in [T],\ j \in C)\ \sum_{i\in F} x_{ij}^t = 1 \qquad\qquad\qquad (2) \\
(\forall t \in [T-1],\ i \in F,\ j \in C)\ z_{ij}^t \geqslant x_{ij}^t - x_{ij}^{t+1} \\
(\forall t \in [T],\ i \in F,\ j \in C)\ y_i^t, x_{ij}^t, z_{ij}^t \geqslant 0
\end{cases}
$$

3.2   Approximation Algorithm

Our algorithm for hourly costs, Algorithm 3, is very similar to Algorithm 1, for fixed costs. The key idea is to choose the random variables Y_i only once to ensure

---

Algorithm 3. Hourly opening cost

- Solve the linear program LP (2) to obtain an optimal (fractional) solution (x, y, z).
- For each timestep t, choose the open facilities A_t randomly as follows. **Once**, for each facility i, choose Y_i having exponential distribution with rate $2 \log(2nT)$. Let $A_t = \{i \in F : Y_i \leqslant y_i^t\}$.
- With a dynamic program, determine how optimally to assign clients to facilities in A_t. Alternatively, for the purposes of analysis, use Algorithm 2 (as done in Algorithm 1).

---

that the set of open facilities is stable. The statements of correctness, Theorems 1 and 3, are proved by exactly the same arguments. The only difference is that, in order for the facility $i^* = \arg\min_{i \in F} Y_i / \hat{x}_{ij}^U$ to be open to client j throughout its time interval U, we need $Y_{i^*} \leqslant y_{i^*}^t$ for all $t \in U$. For each choice of j and U, this family of inequalities is satisfied with probability at least $1 - 1/(2nT)$, the same bound as before, since the fact that $x_{ij}^U \leqslant x_{ij}^t \leqslant y_i^t$ for all $t \in U$ and all $i \in F$ implies as before that $\Pr\{(\forall t \in U)\ Y_{i^*} \leqslant y_{i^*}^t\} \geqslant \Pr\{Y_{i^*} \leqslant \hat{x}_{i^*j}^U\} = \Pr\{\min_{i \in F} Y_i / \hat{x}_{ij}^U \leqslant 1\} \geqslant 1 - 1/(2nT)$. The rest of the proof requires no change.

4 Conclusion and Open Questions

Algorithm 1 applies even if the distances between clients and facilities do not satisfy the triangle inequality, and it extends directly to nonuniform opening costs as well as arrival and departure dates for clients. It is striking that instances with distances satisfying the triangle inequality are not easier in the dynamic setting as opposed to the classic static setting (the approximation ratio $\Theta(\log nT)$ of Algorithm 1 is tight in both dynamic cases). Algorithm 3 also extends directly to the setting of opening costs that are nonuniform in time. The last section naturally raises the question of whether there exists an $\omega(1)$-hardness result / $O(1)$-approximation algorithm for the general hourly opening cost case.

We believe that our dynamic setting should be helpful in designing better *static* representations of dynamic graphs (e.g., two dimensional flowcharts of clients navigating between facilities over time). Another natural extension of our work is to study other objective functions for the distance cost, such as the sum of the diameters of the reported clusters over all timesteps (i.e., the sum of the distance of the farthest client assigned to each facility, see, e.g., [6] for a static formulation). As it turns out, the optimal dynamic solutions with respect to this objective tend to exhibit very intriguing behaviors, even in the simplest case of clients moving along a fixed line [9].

References

1. Anagnostopoulos, A., Bent, R., Upfal, E., Van Hentenryck, P.: A simple and deterministic competitive algorithm for online facility location. Information and Computation 194, 175–202 (2004)

2. Arya, V., Garg, N., Khandekar, R., Meyerson, A., Munagala, K., Pandit, V.: Local search heuristics for k-median and facility location problems. SIAM J. on Computing 33(3), 544–562 (2004)
3. Byrka, J., Aardal, K.: An optimal bifactor approximation algorithm for the metric uncapacitated facility location problem. SIAM J. on Computing 39(6), 2212–2231 (2010)
4. Charikar, M., Chekuri, C., Feder, T., Motwani, R.: Incremental clustering and dynamic information retrieval. In: STOC, pp. 626–635 (1997)
5. Charikar, M., Guha, S.: Improved combinatorial algorithms for facility location problems. SIAM J. on Computing 34(4), 803–824 (2005)
6. Charikar, M., Panigrahy, R.: Clustering to minimize the sum of cluster diameters. In: STOC, pp. 1–10 (2001)
7. Dinur, I., Steurer, D.: Analytical approach to parallel repetition. In: STOC 2014, arXiv:1305.1979 (2014)
8. Divéki, G., Imreh, C.: Online facility location with facility movements. Central European J. of Operations Research 19(2), 191–200 (2010)
9. Fernandes, C.G., Oshiro, M.I., Schabanel, N.: Dynamic clustering of evolving networks: some results on the line. In: AlgoTel, 4 p. (2013), hal-00818985
10. Fotakis, D.: Incremental algorithms for facility location and k-median. Theoretical Computer Science 361(2-3), 275–313 (2006)
11. Fotakis, D.: On the competitive ratio for online facility location. Algorithmica 50(1), 1–57 (2008)
12. Fotakis, D.: Online and incremental algorithms for facility location. SIGACT News 42(1), 97–131 (2011)
13. Jain, K., Mahdian, M., Markakis, E., Saberi, A., Vazirani, V.: Greedy facility location algorithms analyzed using dual fitting with factor-revealing lp. J. ACM 50(6), 795–824 (2003)
14. Kleinberg, J.M.: The small-world phenomenon and decentralized search. SIAM News 37(3) (2004)
15. Li, S.: A 1.488 approximation algorithm for the uncapacitated facility location problem. In: Aceto, L., Henzinger, M., Sgall, J. (eds.) ICALP 2011, Part II. LNCS, vol. 6756, pp. 77–88. Springer, Heidelberg (2011)
16. Mahdian, M., Ye, Y., Zhang, J.: Improved approximation algorithms for metric facility location problems. In: Jansen, K., Leonardi, S., Vazirani, V.V. (eds.) APPROX 2002. LNCS, vol. 2462, pp. 229–242. Springer, Heidelberg (2002)
17. Meyerson, A.: Online facility location. In: FOCS, vol. 42, pp. 426–431 (2001)
18. Newman, M.E.J.: The structure and function of complex networks. SIAM Review 45(2), 167–256 (2003)
19. Pastor-Satorras, R., Vespignani, A.: Epidemic spreading in scale-free networks. Physical Review Letters 86, 3200–3203 (2001)
20. Shmoys, D.B., Tardos, E., Aardal, K.I.: Approximation algorithms for facility location problems. In: STOC, vol. 29, pp. 265–274 (1997)
21. Stehlé, J., Voirin, N., Barrat, A., Cattuto, C., Isella, L., Pinton, J.-F., Quaggiotto, M., Van den Broeck, W., Régis, C., Lina, B., Vanhems, P.: High-resolution measurements of face-to-face contact patterns in a primary school. PLoS ONE 6(8), 23176 (2011)
22. Tantipathananandh, C., Berger-Wolf, T.Y., Kempe, D.: A framework for community identification in dynamic social networks. In: KDD, pp. 717–726 (2007)

Solving the ANTS Problem with Asynchronous Finite State Machines

Yuval Emek[1], Tobias Langner[2], Jara Uitto[2], and Roger Wattenhofer[2]

[1] Technion, Israel
[2] ETH Zürich, Switzerland

Abstract. Consider the *Ants Nearby Treasure Search (ANTS)* problem introduced by Feinerman, Korman, Lotker, and Sereni (PODC 2012), where n mobile agents, initially placed in a single cell of an infinite grid, collaboratively search for an adversarially hidden treasure. In this paper, the model of Feinerman et al. is adapted such that each agent is controlled by an asynchronous (randomized) finite state machine: they possess a constant-size memory and can locally communicate with each other through constant-size messages. Despite the restriction to constant-size memory, we show that their collaborative performance remains the same by presenting a distributed algorithm that matches a lower bound established by Feinerman et al. on the run-time of any ANTS algorithm.

1 Introduction

"They operate without any central control. Their collective behavior arises from local interactions." The last quote is arguably the mantra of distributed computing, however, in this case, "they" are not nodes in a distributed system; rather, this quote is taken from a biology paper that studies social insect colonies [16]. Understanding the behavior of insect colonies from a distributed computing perspective will hopefully prove to be a big step for both disciplines.

In this paper, we study the process of food finding and gathering by ant colonies from a distributed computing point of view. Inspired by the model of Feinerman et al. [11], we consider a colony of n ants whose nest is located at the origin of an infinite grid that collaboratively search for an adversarially hidden food source. An ant can move between neighboring grid cells and can communicate with the ants that share the same grid cell. However, the ant's navigation and communication capabilities are very limited since its actions are controlled by a randomized *finite state machine* (FSM) operating in an asynchronous environment — refer to the model section for a formal definition. Nevertheless, we design a distributed algorithm ensuring that the ants locate the food source within $\mathcal{O}(D + D^2/n)$ time units w.h.p., where D denotes the distance between the food source and the nest.[1] It is not difficult to show that a matching lower

---

[1] We say that an event occurs *with high probability*, abbreviated by w.h.p., if the event occurs with probability at least $1 - n^{-c}$, where c is an arbitrarily large constant.

J. Esparza et al. (Eds.): ICALP 2014, Part II, LNCS 8573, pp. 471–482, 2014.

bound holds even under the assumptions that the ants have unbounded memory (i.e., are controlled by a Turing machine) and know the parameter n.

Related Work. Feinerman et al. [10,11] introduce the aforementioned problem called *ants nearby treasure search (ANTS)* and study it, assuming that the ants (a.k.a. *agents*) are controlled by a Turing machine (with or without space bounds) and do not communicate with each other at all. They show that if the n agents know a constant approximation of n, then they can find the food source (a.k.a. *treasure*) in time $\mathcal{O}(D + D^2/n)$. Moreover, Feinerman et al. observe a matching lower bound and prove that this lower bound cannot be matched without some knowledge of n. In contrast to the model studied in [10,11], the agents in our model can communicate anywhere on the grid as long as they share the same grid cell. However, due to their weak control unit (a FSM), their communication capabilities are very limited even when they do share the same grid cell. Notice that the stronger computational model assumed by Feinerman et al. enables an individual agent to perform tasks way beyond the capabilities of a (single) agent in our setting, e.g., list the grid cells it has already visited or perform spiral searches (that play a major role in their upper bound).

Distributed computing by finite state machines has been studied in several different contexts including *population protocols* [4,5] and the recent work [9] from which we borrowed the agents communication model. In that regard, the line of work closest to our paper is probably the one studying graph exploration by FSM controlled agents, see, e.g., [12].

Graph exploration in general is a fundamental problem in computer science. In the typical case, the goal is for a single agent to visit all nodes in a given graph [1,7,8,15,17]. It is well-known that random walks allow a single agent to visit all nodes of a finite undirected graph in polynomial time [2]. Notice that in an infinite grid, the expected time it takes for a random walk to reach any designated cell is infinite.

Finding treasures in unknown locations has been previously studied, for example, in the context of the classic *cow-path* problem. In the typical setup, the goal is to locate a treasure on a line as quickly as possible and the performance is measured as a function of the distance to the treasure. It has been shown that there is a deterministic algorithm with a competitive ratio 9 and that a spiral search algorithm is close to optimal in the 2-dimensional case [6]. The study of the cow-path problem was extended to the case of multiple agents by López-Ortiz and Sweet [14]. In their study, the agents are assumed to have unique identifiers, whereas our agents cannot be distinguished from each other (at least not at the beginning of the execution).

Model. We consider a variant of [11]'s ANTS problem, where a set of mobile *agents* search the infinite grid for an adversarially hidden treasure. The agents are controlled by asynchronous randomized finite state machines with a common sense of direction and communicate only with agents sharing the same grid cell.

More formally, consider n mobile agents that explore $\mathbb{Z}^2$. In the beginning of the execution, all agents are positioned in a designated grid cell referred to as the *origin* (say, the cell with coordinates $(0,0) \in \mathbb{Z}^2$). We assume for simplicity that the agents can distinguish between the origin and the other cells. We denote the cells with either x or y-coordinate being 0 as *north/east/south/west-axis*, depending on their location.

The main difference between our variation of the ANTS model and the original one lies in the agents' computation and communication capabilities. In both variants, all agents run the same (randomized) protocol, however, under the model considered in the present paper, the agents are controlled by an asynchronous randomized *finite state machine* (FSM). This means that the individual agent has a constant memory and thus, in general, can store neither coordinates in $\mathbb{Z}^2$ nor the number of agents. On the other hand, in contrast to the model considered in [11], our agents may communicate with each other. Specifically, under our model, an agent a positioned in cell $c \in \mathbb{Z}^2$ can communicate with all other agents positioned in cell c at the same time. This communication is quite limited though: agent a merely senses for each state q of the finite state machine, whether there exists at least one agent $a' \neq a$ in cell c whose current state is q. Notice that this communication scheme is a special case of the one-two-many communication scheme introduced in [9] with bounding parameter $b = 1$.

The *distance* between two grid cells $(x, y), (x', y') \in \mathbb{Z}^2$ is defined with respect to the ℓ_1 norm (a.k.a. Manhattan distance), that is, $|x - x'| + |y - y'|$. Two cells are called *neighbors* if the distance between them is 1. In each step of the execution, agent a positioned in cell $(x, y) \in \mathbb{Z}^2$ can either move to one of the four neighboring cells $(x, y + 1), (x, y - 1), (x + 1, y), (x - 1, y)$, or stay put in cell (x, y). The former four *position transitions* are denoted by the corresponding cardinal directions N, S, E, W, whereas the latter (stationary) position transition is denoted by P (standing for "stay put"). We point out that the agents have a common sense of orientation, i.e., the cardinal directions are aligned with the corresponding grid axes for every agent in every cell.

The agents operate in an asynchronous environment. Each agent's execution progresses in discrete (asynchronous) steps indexed by the non-negative integers and we denote the time at which agent a completed step $i > 0$ by $t_a(i) > 0$. Following the common practice, we assume that the time stamps $t_a(i)$ are determined by the policy ψ of an adversary that knows the protocol but is oblivious to its random bits, whereas the agents do not have any sense of time.

Formally, the agents' protocol is captured by the 3-tuple $\Pi = \langle Q, s_0, \delta \rangle$, where Q is the finite set of *states*; $s_0 \in Q$ is the *initial state*; and

$$\delta : Q \times 2^Q \to 2^{Q \times \{N, S, E, W, P\}}$$

is the *transition function*. At time 0, all agents are in state s_0 and positioned in the origin. Suppose that at time $t_a(i)$, agent a is in state $q \in Q$ and positioned in cell $c \in \mathbb{Z}^2$. Then, the state $q' \in Q$ of a at time $t_a(i + 1)$ and its corresponding position transition $\tau \in \{N, S, E, W, P\}$ are dictated based on the transition function δ by picking the pair $(q', \tau) \in \delta(q, Q_a)$, uniformly at random from $\delta(q, Q_a)$, where $Q_a \subseteq Q$ contains state $p \in Q$ if and only if there exists some (at least one) agent $a' \neq a$ such that a' is in state p and positioned in cell c at time $t_a(i)$. (Step i is deterministic if $|\delta(q, Q_a)| = 1$.) For simplicity, we assume that while the state subset Q_a (input to δ) is determined based on the status of cell c at time $t_a(i)$, the actual application of the transition function δ occurs instantaneously at the end of the step, i.e., agent a is considered to be in state q and positioned in cell c throughout the time interval $[t_a(i), t_a(i+1))$.

The goal of the agents is to locate an adversarially hidden *treasure*, i.e., to bring at least one agent to the cell in which the treasure is positioned. The distance of the treasure from the origin is denoted by D. As in [11], we measure the performance of a protocol in terms of its run-time, where the time is scaled so that $t_a(i+1) - t_a(i) \leq 1$ for every agent a and step $i \geq 0$. Although we express the run-time complexity in terms of the parameters n and D, we point out that neither of these two parameters is known to the agents (who cannot even store them in their very limited memory).

2   Parallel Rectangle Search

In this section, we introduce the collaborative search strategy RS (Rectangle-Search) that depends on an *emission scheme*, which divides all participating agents in the origin into *teams* of size ten and emits these teams continuously from the origin until all search teams have been emitted. We delay the description of our emission scheme until Section 3 and describe for now the general search strategy (without a concrete emission scheme). We assume, for the sake of the following informal explanation, an environment in which the agents operate in synchronous rounds and then explain how we can lift this assumption.

The RS strategy consists of two stages. The first stage works as follows: Whenever a team is emitted, one agent becomes an *explorer* and four agents become *guides*, one for each cardinal direction. The remaining five agents become *scouts*, whose function will be explained later. Now, each guide walks into its respective direction until it hits the first cell that is not occupied by another guide. The explorer follows the north-guide and when they hit the non-occupied cell $(0, d) \in \mathbb{Z}^2$ for some $d > 0$, the explorer starts a *rectangle search* by first walking south-west towards the west-guide. When it hits a guide, the explorer changes direction to south-east, then to north-east, and finally to north-west. This way, it traverses all cells in distance d from the origin, referred to as hereafter as *level d*, (and also almost all cells in distance $d+1$). When the explorer meets a guide on its way, the guide enters a *sleep* state to be awoken again in the second stage. The explorer also enters a sleep state after arriving again at the north-guide, thereby completing the first stage of the rectangle search.

The second stage of RS is started when the last search team is emitted from the origin. At this point in time, $\Theta(n)$ cells are occupied by sleeping guides/explorers in all four cardinal directions. The last search team wakes up the innermost sleeping search team upon which it resumes its job and walks outwards to explore the next unexplored level in the same way as in the first stage. Each team recursively wakes up the search team of the next level until all sleeping teams have been woken up and resumed the search. A search in the second stage has one important difference in comparison to a search in the first stage: When an explorer meets a guide g during a search, instead of entering a sleep state, g moves outwards to the next unexplored level, hopping over all the other stationary guides on its way, and waits there for the next appearance of an explorer. When the explorer has finished its rectangle by reaching the north-guide again, it moves north (with the north-guide) to the first unexplored level and starts another search there. Knowing that all other guides have reached their target positions in the same level as well, a new search can begin.

Note that the (temporary) assumption of a synchronous environment is crucial for the correctness of the algorithm described so far as we assume that whenever an explorer crosses a coordinate axis, the respective cell contains a guide. In an asynchronous setting, the guide might still be on its way to that particular cell and hence, the explorer would continue walking diagonally ad infinitum. We counter this problem by coupling the searches for different levels in such a way that a search in level ℓ can never have progressed further than a search in level $\ell' < \ell$. This implies that a search in level ℓ cannot start/finish earlier than a search in level $\ell' < \ell$ starts/finishes. This coupling is implemented by equipping each explorer with a *scout* that essentially allows the explorer e_ℓ in level ℓ to check whether the explorer $e_{\ell-1}$ of the preceding level has already progressed at least as far as e_ℓ and to move only then. On top of that, explorer e_ℓ only leaves a coordinate axis after ensuring, again by means of its scout, that there is already a guide present in level $\ell + 1$. This additional check (together with a few technicalities described later) suffices to ensure that the searches are "nested" properly and the corresponding guides of each explorer are waiting in the right positions along the coordinate axes when those are hit by the explorer.

As a (much desirable) byproduct of the aforementioned explorers' logic, it is guaranteed that during the execution of the RS strategy, every cell contains at most one explorer of each possible state. To ensure that the same holds for the guides, they are also equipped with scouts whose role is to check that during a guide's journey outwards, it does not move into a cell which is already occupied by a guide, unless the latter is in a stationary state (waiting for its explorer).

2.1 The RS strategy

Emission scheme. Initially, all n agents are located at the origin. Until all agents become involved in the RS strategy, an *emission scheme* is responsible for emitting new teams (each consisting of ten agents) from the origin. The emission of the teams is spaced apart in time in the sense that no two teams are emitted at the exact same time. (Under a synchronous schedule, a spacing of 20 time units is guaranteed.) To formally express (and analyze) the emission rate, we introduce the notion of an *emission function* $f_n : \mathbb{N}_0 \to \mathbb{N}_0$, where, until all teams are emitted, $f_n(t)$ bounds from below the number of teams emitted up to time t. For simplicity, we assume that there are enough agents to execute our algorithms, i.e., that $n \geq 30$.

Let $k + 2$, $k \in \Theta(n)$, be the total number of emitted teams where we assume $k \geq 2$. The first and last emitted teams have a special role as *signal teams* in our protocol. The k remaining teams $s_1, \ldots, s_k$ will be referred to as *search teams*.

Whenever a search team becomes ready, four of the ten agents become MGuides — one for each cardinal direction — and walk outwards in their corresponding directions, while the fifth one becomes a MExp and follows the north-MGuide (see below for a detailed description of the agent types). Each MGuide and MExp is accompanied by a Scout that will stick to this particular agent for the rest of the execution.

Agent Types. In the remainder of the paper, we will refer to several different types of agents. Since there is only a constant number of different types, these can be modeled by having individual finite automata for the various types. We essentially use six different types and explain their specific behavior in the following: Scout, Guide, MGuide (for moving guide), MExp (for moving explorer), WExp (for waiting explorer), and Exp (for explorer). We will use the terms "outwards" and "inwards" in the context of agents of the two Guide-types (recall that they are associated with a cardinal direction) to indicate the respective direction away from or towards the origin. We subsume the types Exp/MExp/WExp and Guide/MGuide under the name *explorers* and *guides*, resp. During the process of the algorithm, each non-Scout agent will be accompanied by a Scout, whose type is specific to the type of the agent it is accompanying — its *owner*. Since all different Scout-types have very similar tasks, we first give a general description of a Scout's function and then explain its type-specific behavior together with the owner's behavior.

Scout. The function of each Scout-type is to control when its owner is allowed to move further. It does so by moving to one of the four neighbor cells of the owner – the *scout position* – and waiting for a certain condition (the presence/absence of a certain type of agent) to become true in that cell. When the condition is met, the Scout moves back to the cell containing its owner and notifies the owner. When the owner moves to a new position, the Scout moves along. As Scouts only play an auxiliary role in our protocol, we may refer to a cell as *empty* even if it contains Scouts.

Guide. A Guide waits until a Exp performing a search (this can be encoded in the state of the Exp) has entered its cell. When (i) the cell one coordinate inwards is empty and (ii) its cell contains neither MGuide nor MExp, it becomes a MGuide.

MGuide. A MGuide moves outwards (at least one cell) until it hits a cell c that contains no Guide. The north-MGuide moves north together with the MExp of the same search team. It does so by verifying before each move (after the first) that the MExp has caught up and is in the same cell. Otherwise, it waits for the MExp to catch up. Upon arriving in c, the MGuide becomes a Guide, and waits for an Exp to visit. A MGuide uses its Scout to prevent moving in a cell that contains a MGuide, MExp or Exp.

MExp. A MExp repeatedly moves north together with the north-MGuide of the same search team. More precisely, it only moves north when there is no MGuide in its cell, implying that the MGuide is already one cell further and waits there for the MExp to catch up. The MExp moves until it hits the first cell c that contains neither a Guide that already has an Exp searching (this can be encoded in the state of the Guide) nor an Exp. As soon as cell c contains a Guide (the north-Guide of this explorer's team), it becomes an Exp. A MExp uses its Scout to prevent moving into a cell that contains another MExp while walking outwards.

WExp. A WExp waits until its cell is empty and then becomes a MExp.

Exp. An Exp does the bulk of the actual search process by moving along the sides of a rectangle using Guides on its way to change direction. In the process, it moves south-west, then south-east, north-east, and north-west, in this order. Initially, an Exp performs one move west and then alternatingly south and west.

During a diagonal walk, an Exp uses its Scout to prevent it from overtaking Exps closer to the origin during their search as follows. Consider an Exp e in the north-west quarter-plane (walking south-west). The Scout is sent to the south-neighbor cell, referred to as the *scouting position*, and notifies e, when no Exp present there (which might immediately be the case). Only then, the Exp and the Scout move one cell further where the Scout again enters the scouting position.

When the Exp meets a west/south/east-Guide in an axis cell c, it changes its moving direction. Before leaving the axis, it waits until c does contain neither Guide nor MGuide (thereby ensuring that there is a Guide one cell outwards). Upon arrival back at the north Guide after the rectangle search is completed, it becomes a WExp.

The Exp of the search team exploring level 1 counts its steps (the exploration journey at this level contains exactly 8 cells) and uses the Scout to make sure that the cells on the coordinate axes contain a Guide before entering them.

The signal teams. The first and last emitted teams, s_0 and s_{k+1}, resp., have a special role and they do not actively participate in the exploration of the grid (which is handled by $s_1, \ldots, s_k$). Their job is solely to signal to the other teams when the second stage of the protocol begins.

The first team s_0 enters a special *signal* state and stays at the origin until the last team s_{k+1} has been emitted. (Due to the design of our emission scheme in Section 3, the agents in team s_{k+1} know that they belong to the last emitted team and are able to notify the agents of s_0 accordingly.) The aforementioned logic of the agents in RS ensures that as long as there is an agent present in the origin, the Guides and Exps of the innermost search team (and recursively all other search teams) cannot move outwards. When s_0 is notified by s_{k+1}, the agents in both teams switch to a designated idle state, ignored by all other agents. As now the origin appears to be empty, the Guides and Exps of the innermost (and eventually the other search teams) can move outwards to continue searching — the second stage has begun.

2.2 Correctness

In this section we establish the correctness of the RS strategy by proving that each cell is eventually explored and no agent is lost in the process. We say that a cell in level ℓ is *explored* after it has been visited by an Exp exploring level ℓ, where we recall that level $\ell \in \mathbb{N}_0$ consists of all cells in distance ℓ from the origin. An Exp is said to *start* a (rectangle) search in level ℓ at time t if it moves west from the cell $(0, \ell)$ (containing the north Guide) at time t and it *finishes* a (rectangle) search in level ℓ at time t if it enters the cell $(0, \ell)$ from the east at time t. The *start time* t_ℓ^S, *finish time* t_ℓ^F, and *move time* t_ℓ^M are given by the times at which an Exp starts a search in level ℓ, finishes a search in level ℓ, and when the WExp in level ℓ becomes a MExp, resp. An Exp *explores* level ℓ at time t, if $t_\ell^S < t < t_\ell^F$. The design of RS ensures that regardless of the emission scheme used, the Guides in every cardinal direction occupy a contiguous segment of cells. It also implies the following observation and lemma.

Observation 1. *(Proof deferred to full version) For two levels $\ell' > \ell$, we have $t_{\ell'}^S > t_\ell^S$, $t_{\ell'}^F > t_\ell^F$, and $t_{\ell'}^M > t_\ell^M$.*

Lemma 2. *(Proof deferred to full version) Outside the origin, no two agents of the same type occupy the same cell at the same time.*

Each Exp relies on Guides to indicate when it has to change the search direction in order to search a specific level. The next lemma gives a guarantee for this.

Lemma 3. *Whenever an Exp enters a cell c on an axis, cell c contains a Guide.*

The Canonical Paths. In what follows, we use paths in the infinite grid in their usual graph-theoretic sense, viewing a path p as a (finite or infinite) sequence of cells, where $p(i)$ and $p(i + 1)$ are grid neighbors for every $i \geq 1$. Notice that unless stated otherwise, the paths mentioned are not necessarily simple.

Let $s_1, \ldots, s_k$ be the search teams emitted from the origin (ignoring the two signal teams s_0 and s_{k+1}) ordered by ascending emission time and consider some agent a participating in one of the search teams $s_1, \ldots, s_k$. Given some adversarial policy ψ, let p_a^ψ be the path traversed by a during the execution of the algorithm under ψ starting at the time at which a is emitted from the origin. We extend the sequence defined by p_a^ψ, fixing $p_a^\psi(0) = (0, 0)$. We shall refer to p_a^ψ as the *execution path* of a (under ψ).

The logic of the guides directly implies that if agent a is a north/south/east/west guide, then its execution path satisfies $p_a^\psi(i) = (0, i)/(0, -i)/(i, 0)/(-i, 0)$ for every adversarial policy ψ. In other words, the path traversed by a guide does not depend on the adversarial policy. We argue that this is in fact the case for all agent types, introducing the notion of a *canonical path*.

Lemma 4. *(Proof deferred to full version) For every $1 \leq i \leq k$ and for each agent role ρ (among the 10 different roles in a search team), there exists a canonical path $p_{i,\rho}^*$ such that if agent a is the ρ-agent in search team s_i, then $p_a^\psi = p_{i,\rho}^*$, regardless of the adversarial policy ψ.*

It will sometimes be convenient to use the notation p_a^* for the canonical path $p_{i,\rho}^*$ when agent a is the ρ-agent of search team s_i. The key to Lemma 4's proof is the observation that since MExps do not overtake each other, the explorers maintain a *cyclic order* between them in terms of the levels they explore. The exact same argument can be applied to the guides, concluding that the agents of a search team "stick together" throughout the execution.

Corollary 5. *The agents that were emitted from the origin as guides of search team s_i serve as Guides in levels $\ell = z \cdot k + i$ for $z = 0, 1, \ldots$*

Preventing Dead/live-locks. We now turn to prove that RS does not run into deadlocks. Recall that during the execution of RS, agents often wait for other agents to complete some task before they can proceed. In particular, we say that agent a is *delayed* by agent a' at time t, denoted $a \to_t a'$, if at time t, a is positioned in some cell c and resides in some state q and the RS strategy dictates that a can neither leave cell c nor move to any state other than q until a' performs some action in cell c that may take the form of entering cell c, leaving cell c, or moving to some state within cell c. For example, a guide in an axis cell c is delayed by its corresponding explorer until the latter reaches c. Another example is an explorer which is delayed by its scout in some north-west quarter-plane cell (x, y), while the latter is delayed until the explorer exploring the previous level leaves cell $(x, y - 1)$. To avoid the necessity to account for the scouts, we extend the definition of delays in the context of the correctness proof, allowing for agent a in cell c to be delayed by agent a' in a neighboring cell c' if a is actually delayed by its scout in c who is delayed by a' in c'.

Let $\mathcal{D}_t$ be the directed graph that corresponds to the binary relation $\to_t$ over the set of agents. We prove that RS does not run into deadlocks by establishing the following lemma.

Lemma 6. *The directed graph $\mathcal{D}_t$ does not admit any (directed) cycle at all times t.*

Proof. Consider a snapshot of the agents' states and positions at time t. Examining the RS strategy, one realizes that the outermost MExp and MGuides are not delayed by any other agent and that the ith outermost MExp and MGuides can only be delayed by the $(i - 1)^{\text{th}}$ outermost MExp and MGuides. The innermost Exp e is not delayed by any agent as long as it is not in an axis cell. In an axis cell, e can only be delayed by the corresponding guide. An innermost guide in cell c is delayed by its corresponding explorer until the latter reaches cell c and since then, it can only be delayed by the corresponding innermost MGuide. Non-innermost Exp and Guides in level ℓ can only be delayed by the Exp and Guides in level $\ell - 1$ or by the MExps and MGuides. The assertion follows.   □

The following corollary is derived due to Lemma 6 since there is a constant number of state transitions an agent positioned in cell c can perform before it leaves cell c.

Corollary 7. *Agent a reaches cell $p_a^*(i)$ within finite time for every $i \geq 1$.*

Since the canonical path p_a^* contains infinitely many different nodes for every agent a, we can deduce from Corollary 7 that RS does not run into livelocks, thus establishing the following theorem.

Theorem 8. *The cell containing the treasure is explored in finite time.*

2.3   Runtime Analysis

For the sake of a clearer run-time analysis, we analyze RS employing an ideal emission scheme with emission function $f_n(t) = \Omega(t)$, i.e., a new search team is emitted from the origin every constant number of time units. We do not know how to implement such a scheme, but in Section 3, we will describe an emission scheme with an almost ideal emission function of $f_n(t) = \Omega(t - \log n)$ and in Section 4, we will show how to compensate for the gap.

Our proof consists of two parts. First, we analyze the run-time of RS assuming a "synchronous" adversarial policy ψ^s, where $t_a(i) = i$ for all a and i. Then, we lift this assumption by showing that ψ^s is actually the worst case policy. We start with the following lemmas.

Lemma 9. *(Proof deferred to full version) Under ψ^s, we have $t^M_{\ell+1} - t^M_\ell \geq 4$ and $t^S_{\ell+1} - t^S_\ell \geq 4$.*

Lemma 10. *(Proof deferred to full version) Under ψ^s, the explorer of search team s_i is not delayed after time t^M_i.*

Lemma 11. *(Proof deferred to full version) Under ψ^s, we have $t^F_\ell \in \mathcal{O}(\ell + \ell^2/n)$ for any level $\ell > 0$.*

We now turn to show that the run-time of RS under any adversarial policy ψ is at most the run-time under ψ^s. By definition, policy ψ^s maximizes the length of the time between consecutive completion times of the agents' steps. Informally, we have to prove that by speeding up some agents, the adversary cannot cause larger delays later on.

To that end, consider two agents a and a' and recall that Lemma 4 guarantees that they follow the canonical paths p^*_a and $p^*_{a'}$, resp., regardless of the adversarial policy. The agents can delay each other only when they are in the same cell, so suppose that there exist two indices i and i' such that $p^*_a(i) = p^*_{a'}(i') = c$.

Given some adversarial policy ψ, let $t^\psi_{in}(a)$ (resp., $t^\psi_{in}(a')$) be the time at which agent a (resp., a') enters c in the step corresponding to $p^*_a(i)$ (resp., $p^*_{a'}(i')$) under ψ and let $t^\psi_{out}(a)$ (resp., $t^\psi_{out}(a')$) be the time at which agent a (resp., a') exits c for the first time following $t^\psi_{in}(a)$ (resp., $t^\psi_{in}(a')$) under ψ. The key observation now is that the adversarial policy does not affect the order in which a and a' enter/exit cell c.

Observation 12. *For every two adversarial policies ψ_1, ψ_2, we have $t^{\psi_1}_{in}(a) < t^{\psi_1}_{in}(a')$ if and only if $t^{\psi_2}_{in}(a) < t^{\psi_2}_{in}(a')$ and $t^{\psi_1}_{out}(a) < t^{\psi_1}_{out}(a')$ if and only if $t^{\psi_2}_{out}(a) < t^{\psi_2}_{out}(a')$.*

Therefore, the adversary may decide to modify its policy relatively to ψ^s by speeding up some steps of some agents, but this modification cannot delay the progression of the agents along their canonical paths. Corollary 13 now follows from Lemma 11.

Corollary 13. *Under any adversarial policy, $t^F_\ell \in \mathcal{O}(\ell + \ell^2/n)$ for any level $\ell > 0$.*

3   An Almost Optimal Emission Scheme

We introduce the emission scheme PTA (ParallelTeamAssignment) that w.h.p. guarantees an emission function of $f_n(t) = \Omega(t - \log n)$. In Section 4, we describe the search strategy GS (GeometricSearch), that yields an optimal run-time of $\mathcal{O}(D + D^2/n)$ when combined with RS. The main goal of this section is to establish the following theorem.

Theorem 14. *Employing the PTA emission scheme, RS locates the treasure in time* $\mathcal{O}(D + D^2/n + \log n)$ *w.h.p.*

Our first goal is to describe the process FS (FastSpread), where n agents spread out along the east ray R consisting of the cells $(x, 0)$ for $x \in \mathbb{N}_{>0}$ such that each cell in some prefix of R is eventually assigned to a unique agent. The main idea behind the implementation of FS is that on every step, agent a throws a fair coin and moves outwards (towards east) if the coin shows heads and stays put otherwise. If a senses that it is the only agent occupying cell c, then it marks itself as *ready* and stops moving; cell c is also said to be *ready* following this event. Furthermore, when a walks onto a ready cell, it moves outwards deterministically.

To prevent any cell from becoming empty, the agents employ a mechanism that ensures that at least one agent stays put in each cell. To implement this mechanism, the agents decide in advance, i.e., in step i, if they want to move in step $i + 1$ and report their decision to the other agents. In other words, an agent a throws a coin in step i and enters a state H or T that correspond to throwing heads or tails, resp. Then, a moves outwards in step $i + 1$ if and only if it entered state H in step i and if it senses at least one other agent in state T. Informally, a only moves if at least one other agent has promised to stay put next time it acts.

Next, we show that the protocol works correctly, i.e., no cell in the prefix of R will become empty before getting ready. Suppose for contradiction that there is a cell c, such that c becomes empty at time t. Let a be an agent and i a step of a such that for all agents a' in cell c and all steps j, it holds that $t_{a'}(j) \le t_a(i) < t$. In other words, no agent in c changes its state during time $t_a(i) < t' < t$. According to the design of our protocol, a must sense some other agent a' in state T precisely at time $t_a(i)$. Since a' does not wake up after $t_a(i)$ and before t, it follows that a' resides in state T at time t, which is a contradiction.

Lemma 15. *(Proof deferred to full version) For every positive integer* $s \le 16n$, *the first* $s/16$ *cells of the ray* R *are ready after* $s + \mathcal{O}(\log n)$ *time units w.h.p.*

The last step of the protocol is to gather the teams and move to the origin. This is performed by dedicating the agent of every tenth cell to a specific role. More precisely, the cell in distance $i \cdot 10 + j$ for $i \in \{0 \le i \le \lfloor n/10 \rfloor - 1\}$ and $j \in \{1, \ldots, 10\}$ is dedicated to j-th member of search team s_i where the different values of j correspond to the ten different roles. After an Exp (corresponding to

$j = 1$) becomes ready, it walks outwards to gather the other ready agents of its team, after which they all walk to the origin.

4   Optimal Rectangle Search

The goal of this section is to establish Theorem 16 by presenting the search strategy HybridSearch that locates the treasure with optimal run-time of $\mathcal{O}(D + D^2/n)$. This is achieved by combining RS employing the PTA emission scheme with the randomized search strategy GS that is optimal if the treasure is close to the origin, or more precisely, if $D \leq \log n/2$. GS assigns an agent randomly to one of the four quarter-planes and then lets it walk to a random cell in that quarter-plane in a geometrically distributed distance from the origin, hence the name. HybridSearch initially splits the set of agents into half by tossing a fair coin and then assigns the two halves to perform either GS or RS, thereby combining the virtues of both strategies. A thorough treatment of HybridSearch is deferred to the full version.

Theorem 16. *HybridSearch locates the treasure in time* $\mathcal{O}(D + D^2/n)$ *w.h.p.*

References

1. Albers, S., Henzinger, M.: Exploring Unknown Environments. In: SICOMP (2000)
2. Aleliunas, R., Karp, R.M., Lipton, R.J., Lovasz, L., Rackoff, C.: Random Walks, Universal Traversal Sequences, and the Complexity of Maze Problems. In: SFCS (1979)
3. Alon, N., Avin, C., Koucky, M., Kozma, G., Lotker, Z., Tuttle, M.R.: Many Random Walks are Faster Than One. In: SPAA (2008)
4. Angluin, D., Aspnes, J., Diamadi, Z., Fischer, M.J., Peralta, R.: Computation in Networks of Passively Mobile Finite-State Sensors. Distributed Computing (2006)
5. Aspnes, J., Ruppert, E.: An Introduction to Population Protocols. In: Middleware for Network Eccentric and Mobile Applications
6. Baeza-Yates, R.A., Culberson, J.C., Rawlins, G.J.E.: Searching in the Plane. Information and Computation (1993)
7. Deng, X., Papadimitriou, C.: Exploring an Unknown Graph. JGT (1999)
8. Diks, K., Fraigniaud, P., Kranakis, E., Pelc, A.: Tree Exploration with Little Memory. Journal of Algorithms (2004)
9. Emek, Y., Wattenhofer, R.: Stone Age Distributed Computing. In: PODC (2013)
10. Feinerman, O., Korman, A.: Memory Lower Bounds for Randomized Collaborative Search and Implications for Biology. In: DISC (2012)
11. Feinerman, O., Korman, A., Lotker, Z., Sereni, J.S.: Collaborative Search on the Plane Without Communication. In: PODC (2012)
12. Fraigniaud, P., Ilcinkas, D., Peer, G., Pelc, A., Peleg, D.: Graph Exploration by a Finite Automaton. In: TCS (2005)
13. Förster, K.T., Wattenhofer, R.: Directed Graph Exploration. In: OPODIS (2012)
14. López-Ortiz, A., Sweet, G.: Parallel Searching on a Lattice. In: CCCG (2001)
15. Panaite, P., Pelc, A.: Exploring Unknown Undirected Graphs. In: SODA (1998)
16. Prabhakar, B., Dektar, K.N., Gordon, D.M.: The Regulation of Ant Colony Foraging Activity Without Spatial Information. PLoS Computational Biology (2012)
17. Reingold, O.: Undirected Connectivity in Log-Space. JACM (2008)

Near-Optimal Distributed Approximation
of Minimum-Weight Connected Dominating Set

Mohsen Ghaffari

MIT, USA
ghaffari@mit.edu

Abstract. This paper[1] presents a near-optimal distributed approximation algorithm for the minimum-weight connected dominating set (MCDS) problem. We use the standard distributed message passing model called the CONGEST model in which in each round each node can send $\mathcal{O}(\log n)$ bits to each neighbor. The presented algorithm finds an $\mathcal{O}(\log n)$ approximation in $\tilde{\mathcal{O}}(D + \sqrt{n})$ rounds, where D is the network diameter and n is the number of nodes. MCDS is a classical NP-hard problem and the achieved approximation factor $\mathcal{O}(\log n)$ is known to be optimal up to a constant factor, unless P = NP. Furthermore, the $\tilde{\mathcal{O}}(D+\sqrt{n})$ round complexity is known to be optimal modulo logarithmic factors (for any approximation), following [Das Sarma et al.—STOC'11].

1 Introduction and Related Work

Connected dominating set (CDS) is one of the classical structures studied in graph optimization problems which also has deep roots in networked computation. For instance, CDSs have been used rather extensively in distributed algorithms for wireless networks (see e.g. [2, 3, 5–10, 31, 39, 40]), typically as a global-connectivity backbone.

This paper investigates distributed algorithms for approximating *minimum-weight connected dominating set* (MCDS) while taking *congestion* into account. We first take a closer look at what each of these terms means.

1.1 A Closeup of MCDS, in Contrast with MST

Given a graph $G = (V, E)$, a set $S \subseteq V$ is called a *dominating set* if each node $v \notin S$ has a neighbor in S, and it is called a *connected dominating set* (CDS) if the subgraph induced by S is connected. Figure 1 shows an example. In the *minim-weight CDS* (MCDS) problem, each node has a weight and the objective is to find a CDS with the minimum total weight.

The MCDS problem is often viewed as the node-weighted analogue of the *minimum-weight spanning tree* (MST) problem. Here, we recap this connection. The natural interpretation of the definition of CDS is that a CDS is a selection of *nodes* that provides *global-connectivity*—that is, any two nodes of the graph are connected via a path that its internal nodes are in the CDS. On the counterpart, a *spanning tree* is a (minimal) selection of *edges* that provides global-connectivity. In both cases, the problem of interest is to minimize the total weight needed for global-connectivity. In one case, each edge has a weight and the problem becomes MST; in the other, each node has a weight and the problem becomes MCDS.

---

[1] A full version can be found in [20].

J. Esparza et al. (Eds.): ICALP 2014, Part II, LNCS 8573, pp. 483–494, 2014.
© Springer-Verlag Berlin Heidelberg 2014

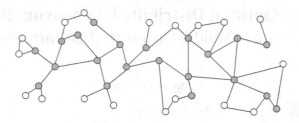

Fig. 1. The green nodes represent a connected dominating set (CDS) of the graph

Despite the seemingly analogous nature of the two problems, MCDS turns out to be a significantly harder problem: The MST problem can be computed sequentially in (almost) $\mathcal{O}(m)$ time, where m is the number of edges. On the other hand, MCDS is NP-hard [19], and in fact, unless P = NP, no polynomial time algorithm can find any approximation better than $\Theta(\log n)$-factor for it (see [1, 16, 37]). Furthermore, the known sequential algorithms for $\mathcal{O}(\log n)$ approximation of MCDS (see [22, 23]) have unspecified polynomial time complexity, which are at least $\Theta(n^3)$.

1.2   Congestion in Distributed Algorithms

Two central issues in distributed computing are *locality* and *congestion* [34]. Classically, locality has received more attention and most graph problems were studied in the LOCAL model, where congestion is ignored and messages can have unbounded size. The recent years have seen a surge in focus on understanding the effect of congestion in graph problems (see e.g., [11–13, 17, 21, 24, 30, 32, 33]). The standard distributed model that takes congestion into account is called CONGEST [34], where in each round, each node can send B bits to each of its neighbors, and normally one assumes $B = \mathcal{O}(\log n)$. The pioneering problem in the study of the CONGEST model was MST: A beautiful line of work shows that MST can be solved in $\mathcal{O}(D + \sqrt{n}\log^* n)$ rounds [18, 29] and that this is (existentially) optimal modulo logarithmic factors [11, 15, 35], and a similar lower bound also applies to many other distributed graph problems [11]. Since then, achieving an $\tilde{\mathcal{O}}(D + \sqrt{n})$ round complexity is viewed as sort of a golden standard for (non-local) problems in the CONGEST model. The area is quite active and in the last couple of years, a few classical graph optimization problems (which are in P) are shown to have approximation matching this standard or getting close to it: some distance-related problems such as shortest-path approximations [30, 32] or diameter and girth approximations [24], and minimum-cut approximation [21].

1.3   Result

The contribution of this paper is to show that in the CONGEST model, MCDS can be solved—that is, approximated optimally—in a time close to that of MST.

Theorem 1. *There is a randomized distributed algorithm in the* CONGEST *model that, with high probability, finds an* $\mathcal{O}(\log n)$ *approximation of the minimum-weight connected dominating set, using* $\tilde{\mathcal{O}}(D + \sqrt{n})$ *rounds.*

This algorithm is (near) optimal in both round complexity and approximation factor: Using techniques of [11], one can reduce the *two-party set-disjointness communication complexity* problem on $\Theta(\sqrt{n})$-bit inputs to MCDS, proving that the round complexity is optimal, up to logarithmic factors, for any approximation (see Appendix B in the full version). As mentioned above, the $\mathcal{O}(\log n)$ approximation factor is known to be optimal up to a constant factor, unless P = NP, assuming that nodes can only perform polynomial-time computations. Note that this assumption is usual, see e.g. [14, 25, 28].

1.4   Other Related Work

To the best of our knowledge, no efficient algorithm was known before for MCDS in the CONGEST model. Notice that in the LOCAL model, MCDS boils down to a triviality and is thus never addressed in the literature: it is folklore[2] that in this model, D rounds is both necessary and sufficient for any approximation of MCDS. However, a special case of MCDS is interesting in the LOCAL model; the so-called "unweighted case" where all nodes have equal weight. Although, the unweighted-case has a significantly different nature as it makes the problem "*local*": Dubhashi et al. [14] present a nice and simple $\mathcal{O}(\log n)$ approximation for the unweighted-case algorithm which uses $\mathcal{O}(\log^2 n)$ rounds of the LOCAL model. To our knowledge, the unweighted case has not been addressed in the CONGEST model, but we briefly comment in Appendix A of the full version that one can solve it in $\mathcal{O}(\log^2 n)$ rounds of the CONGEST model as well, by combining the dominating set approximation of Jia et al. [25] with the *linear skeleton* of Pettie [36] and a simple trick for handling congestion. Another problem which has a name resembling MCDS is the *minimum-weight dominating set* (MDS) problem. However, MDS is also quite different from MCDS as the former is "*local*", even in the weighted case and the CONGEST model: an $\mathcal{O}(\log n)$ factor approximation can be found in $\mathcal{O}(\log^2 n)$ rounds [25, 28] (see also [27]).

2   Preliminaries

Distributed Model: As stated above, we use the CONGEST model: communication between nodes happens in lock-step rounds where in each round, one B-bits message can be sent on each direction of each edge, and we particularly focus on the standard case of $B = \mathcal{O}(\log n)$. The only global knowledge assumed is that nodes know an upper bound $N = \text{poly}(n)$ on n. We use the phrase *with high probability* (w.h.p.) to indicate a probability being at least $1 - \frac{1}{n^\beta}$, for a constant $\beta \geq 2$.

Notations and Basic Definitions: We work with an undirected graph $G = (V, E)$, $n = |V|$, and for each vertex $v \in V$, $c(v)$ denotes the weight (i.e., cost) of node v. Throughout the paper, we will use the words *cost* and *weight* interchangeably. For each subset $T \subseteq V$, we define $\text{cost}(T) = \sum_{v \in T} c(v)$. We assume the weights are at most polynomial in n, so each weight can fit in one message (such assumptions are usual, e.g. [18]). We use notation OPT to denote the CDS with the minimum cost. Also, for

---

[2] In a cycle with $2D$ nodes, nodes need to learn the weight of the node at the opposite side, which is D hops away and requires D rounds.

convenience and when it does not lead to any ambiguity, we sometimes use OPT to refer to the cost of the optimal CDS.

Problem Statement: Initially, each node v knows only its own weight $c(v)$. The objective is to find a set S in a distributed fashion—that is, each node v will need to output whether $v \in S$ or not—such that $\text{cost}(S) = \mathcal{O}(\text{OPT} \cdot \log n)$.

A Basic Tool (Thurimella's Algorithm): A basic tool that we frequently use is a *connected component identification algorithm* presented by Thurimella [38], which itself is a simple application of the MST algorithm of Kutten and Peleg [29]. Given a subgraph $H = (V, E')$ of the main network graph $G = (V, E)$, this algorithm identifies the connected components of H by giving a label $\ell(v)$ to each v such that $\ell(v) = \ell(u)$ if an only if v and u are in the same connected component of H. This algorithm uses $\mathcal{O}(D + \sqrt{n} \log^* n)$ rounds of the CONGEST model. It is easy to see that the same strategy can be adapted to solve the following problems also in $\mathcal{O}(D + \sqrt{n} \log^* n)$ rounds. Suppose each node v has an input $x(v)$. For each node v, which is in a component $\mathcal{C}$ of H, we can make $\ell(v)$ be equal to: (A) the *maximum* value $x(u)$ for nodes $u \in \mathcal{C}$ in the connected component of v, or (B) the *list of $k = \mathcal{O}(1)$ largest* values $x(u)$ for nodes $u \in \mathcal{C}$, or (C) the *summation* of values $x(u)$ for nodes $u \in \mathcal{C}$.

3 The Algorithm for MCDS

3.1 The Outline

The top-level view of the approach is as follows: We start by using the $\mathcal{O}(\log^2 n)$ rounds algorithm of [25] to find a dominating set S with cost $\mathcal{O}(\log n \cdot \text{OPT})$. The challenge is in adding enough nodes to connect the dominating set, while spending extra cost of $\mathcal{O}(\log n \cdot \text{OPT})$. We achieve connectivity in $\mathcal{O}(\log n)$ phases. In each phase, we add some nodes to set S so that we reduce the number of connected components of S by a constant factor, while spending a cost of $\mathcal{O}(\text{OPT})$. After $\mathcal{O}(\log n)$ phases, the number of connected components goes down to 1, meaning that we have achieved connectivity. Each phase uses $\tilde{\mathcal{O}}(D + \sqrt{n})$ rounds of the CONGEST model. What remains is to explain how a phase works.

The reader might recall that such "component-growing" approaches are typical in the MST algorithms, e.g., [18, 29]. While in MST, the choice of the edge to be added to each component is clear (*the lightest outgoing edge*), the choice of the nodes to be added in MCDS is not clear (and in fact can be shown to be an NP-hard problem, itself).

The problem addressed in one phase can be formally recapped as follows (the reader might find the illustration in Figure 2 helpful here): We are given a dominating subset $S \subseteq V$ and the objective is to find a subset $S' \subseteq V \setminus S$ with $\text{cost}(S') = \mathcal{O}(\text{OPT})$ such that the following condition is satisfied. Let $\mathcal{F}$ be the set of subsets of S such that each $\mathcal{C} \in \mathcal{F}$ is a connected component of $G[S]$. Call a connected component $\mathcal{C} \in \mathcal{F}$ *satisfied* if in $G[S \cup S']$, $\mathcal{C}$ is connected to at least one other component $\mathcal{C}' \in \mathcal{F}$. We want S' to be such that at least half of the connected components of $G[S]$ are satisfied. Note that if this happens, then the number of connected components goes down by a $3/4$ factor. To refer to the nodes easier, we assume that all nodes that are in S at the start of the phase are colored *green* and all the other nodes are *white*, initially. During the phase, some white nodes will become gray meaning that they joined S'.

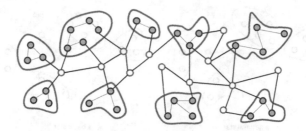

Fig. 2. An example scenario at the start of a phase. Green nodes indicate those in S and white nodes are $V \setminus S$. Unrelated nodes and edges are discarded from the picture.

Before moving on to the algorithm, we emphasize two key points:

(1) It is critical to seek satisfying only a constant fraction of the components of $G[S]$. Using a simple reduction from the set cover problem, it can be shown that satisfying all components might require a cost $\mathcal{O}(\text{OPT} \log n)$ for a phase. Then, at least in the straightforward analysis, the overall approximation factor would become $\mathcal{O}(\log^2 n)$.

(2) In each phase, we *freeze* the set of components $\mathcal{F}$ of $G[S]$. That is, although we continuously add nodes to the CDS and thus the components grow, we will not try to satisfy the newly formed components. We keep track of whether a component $\mathcal{C} \in \mathcal{F}$ is satisfied and the satisfied ones become "inactive" for the rest of the phase, meaning that we will not try to satisfy them again. However, satisfied components will be used in satisfying the others.

3.2 A High-Level View of the Algorithm for One Phase

Note that since S is a dominating set, $\mathcal{C} \in \mathcal{F}$ is satisfied iff there exist one or two nodes that connect $\mathcal{C}$ to another component $\mathcal{C}' \in \mathcal{F}$. That is, either there is a node v such that path $\mathcal{C}$-v-$\mathcal{C}'$ connects component $\mathcal{C}$ to component $\mathcal{C}'$ or there are two adjacent nodes v and w such that path $\mathcal{C}$-v-w-$\mathcal{C}'$ does that. Having this in mind, and motivated by the solution for the unweighted case [14], a naive approach would be that, for each component $\mathcal{C}$, we pick one or two nodes—with smallest total weight—that connect $\mathcal{C}$ to an-

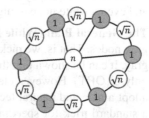

Fig. 3. The naive approach

other component, and we do this for each component $\mathcal{C}$ independently. However, in the weighted case, this naive idea would perform terribly. To see why, let us consider a simple example (see Figure 3): take a cycle with $n-1$ nodes where every other node has weight 1 and the others have weight $\sqrt{n}$, and then add one additional node at the center with weight n, which is connected to all weight-1 nodes. Clearly, the set of weight-1 nodes gives us an optimal dominating set. However, naively connecting this dominating set following the above approach would make us include at least half of the $\sqrt{n}$-weight nodes, leading to overall weight of $\Theta(n\sqrt{n})$. On the other hand, simply adding the center node s to the dominating set would provide us with a CDS of weight $\mathcal{O}(n)$.

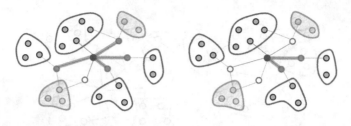

a general star   a basic star

Fig. 4. A basic-star. The opaque components indicate those that are already satisfied and thus deactivated. Two legs of the general star (colored red, on the left) are discarded in the basic-star (colored red, on the right), as each of them forms a useful star, meaning that the leg itself can satisfy at least one active component.

Inspired by this simple example, we view *stars* as the key elements of optimization (instead of 2 or 3 hop paths). We next define what we mean by a star and outline how we use it. We note that the concept is also similar to the notion of *spiders* used in [26] for the node-weighted Steiner trees problem.

Definition 1. *(Stars) A star X is simply a set of white nodes with a center $s \in X$ such that each non-center node in the star is connected to the center s. Naturally, we say a star X satisfies an active component $C \in \mathcal{F}$ if adding this star to S'—that is, coloring its nodes gray—would connect C to some other component and thus make it satisfied. Let $\Phi(X)$ be the set of unsatisfied components in $\mathcal{F}$ that would be satisfied by X. We say a star is useless if $\Phi(X) = \emptyset$. The cost of a star X is $\text{cost}(X) = \sum_{w \in X} c(w)$ and its efficiency is $\rho(X) = \frac{|\Phi(X)|}{\text{cost}(X)}$. We say X is ρ'-efficient if $\rho(X) \geq \rho'$.*

In Figure 3, each white node is one star, the center has efficiency $\Theta(1)$ and every other star has efficiency $\Theta(1/\sqrt{n})$. Notice that in general, different stars might intersect and even a white node v might be the center of up to $2^{\Theta(n)}$ different stars.

The General Plan (While Ignoring Some Difficulties): We greedily add stars to the gray nodes. That is, we pick a star that has the maximum efficiency and color its nodes gray. It can be shown that this greedy idea would satisfy half of components using cost only $\mathcal{O}(\text{OPT})$. However, clearly adding stars one by one would be too slow. Instead we adopt a nice and natural technique due to Berger et al. [4] which by now has become a standard trick for speeding up greedy approaches via parallelizing their steps. The key point is, stars that have efficiency within a constant factor of the max-efficiency are essentially as good as the max-efficient star and hence, we can add those as well. The only catch is, one needs to make sure that adding many stars simultaneously does not lead to (too much) *double counting* in the efficiency calculations. In other words, if there are many stars that try to satisfy the same small set of components, even if each of these stars is very efficient, adding all of them is not a good idea. The remedy is to probabilistically add stars while the probabilities are chosen such that not too many selected stars try to satisfy one component.

While this general outline roughly explains what we will do, the plan faces a number of critical issues. We next briefly hint at two of these challenges and present the definitions that we use in handling them.

Challenge 1: The first step in the above outline is to compute (or approximate) the efficiency of the max-efficient star. Doing this for the general class of stars turns out the be a hard problem in the CONGEST model. Note that for a white node v to find (or approximate) the most-efficient star centered on it, v would need to know which components are adjacent to each of its white neighbors. As each white node might be adjacent to many components, this is like learning the 2-neighborhood of v and appears to be intrinsically slow in the CONGEST model. Instead, we will focus on a special form of stars, which we call *basic-stars* and explain next. Figure 4 shows an example.

Definition 2. *(Basic-Stars) Call a white node u self-sufficient if u is adjacent to two or more components, at least one of which is not satisfied. A star X is called* basic *if for each non-center node $w \in X$, w is not self-sufficient. That is, the star $X' = \{w\}$ is useless.*

We argue later that, considering only the basic-stars will be sufficient for our purposes (sacrificing only a constant factor in the approximation quality) and that we can indeed evaluate the max-efficiency of the basic-stars.

Challenge 2: The other issue, which is a bit more subtle but in fact significantly more problematic, is as follows: as we color some white nodes gray, some components grow and thus, the efficiencies of the stars change. For instances, a useless star $X = \{v\}$ might now become useful–e.g., it gets connected to a satisfied component C' via a node u that just got colored gray, and X can now satisfy an adjacent unsatisfied component C by connecting it to C'. Another example, which is rooted also in the congestion related issues, is as follows: During our algorithm, to be able to cope with communication issues, each white node v will work actively on only one max-efficient basic-star centered on v. But, v might be the center of many such stars and even if one of them looses the efficiency after this iteration, another max-efficient star which existed before might be now considered actively by v.

We note that, if there were no such *"new-stars"* issues, we could use here standard methods such as (a modification of) the LP relaxation based technique of Kuhn and Wattenhofer [28]. However, these changes break that approach and it is not even clear how to formulate the problem as an LP (or even a convex optimization problem, for that matter).

If not controlled, these changes in the stars can slow down our plan significantly. For example, if for a given almost-maximum efficiency $\tilde{\rho}$, in each iteration a small number of $\tilde{\rho}$-efficient new basic-stars are considered actively, we will have to spend some time on these stars but as the result, we would satisfy only very few components, which would become prohibitively slow. To remedy this, when coloring stars gray, we will do it for certain types of $\tilde{\rho}$-efficient basic-stars, which we define next, and after that, we do some clean up work to remove the new $\tilde{\rho}$-efficient basic-stars that would be considered actively later on.

Definition 3. *(ρ^*-Augmented Basic-Stars) A ρ^*-efficient basic-star X centered on node $v \in X$ is called ρ^*-minimal if for any other star $X' \subset X$ centered on v, we have, $\rho(X') < \rho^*$. For a ρ^*-minimal basic-star X centered on v, a good auxiliary-leg is a white node $u \notin X$ that is adjacent to v and furthermore, the following conditions are* *satisfied: u is adjacent to only one component $\mathcal{C} \in \mathcal{F}$, component $\mathcal{C}$ is not satisfied and it is not adjacent to X, and we have $\text{cost}(u) \leq 2/\rho^*$. A ρ^*-Augmented Basic-Star X' is one that can be derived by (one-by-one) adding to ρ^*-minimal basic-star X all good auxiliary-legs adjacent to its center.*

An example is shown in Figure 5. The actual reasoning for why this definition is good is somewhat subtle to be explained intuitively. A very rough version is as follows: after coloring some ρ^*-augmented basic-stars gray, by just handling the nodes which each have cost at most $1/\rho^*$ (in a step we call clean up), we will be able to remove any new ρ^*-augmented basic-star. The point should become clear after seeing the algorithm .

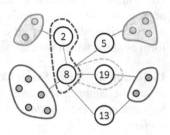

Observation 2. *Each ρ^*-Augmented Basic-Star X has efficiency $\rho(X) \geq \frac{\rho^*}{2}$. Furthermore, if a ρ^*-Augmented Basic-Stars X contains a white node w, then all unsatisfied components adjacent to w get satisfied by X.*

Fig. 5. A 0.1-augmented basic-star is indicated with the dashed lines; the red part is a minimal 0.1-efficient basic-star and the orange part is a good auxiliary leg

3.3　The Algorithm For One Phase

The objective of the algorithm is to satisfy at least half of the components, using a cost $\mathcal{O}(\text{OPT})$, and in $\mathcal{O}((D + \sqrt{n}\log^* n)\log^3 n)$ rounds. Throughout the phase, each non-white node will keep track of whether its component in $\mathcal{F}$ is satisfied or not. Let $N = |\mathcal{F}|$ and also, make all nodes know N by running Thurimella's connected component identification at the start of the phase and then globally gathering the number of components.

While at least $\lfloor N/2 \rfloor$ components in $\mathcal{F}$ remain unsatisfied, we repeat the following iteration, which has 8 steps—$\mathcal{S}1$ to $\mathcal{S}8$—and each step uses $\mathcal{O}(D + \sqrt{n}\log^* n)$ rounds:

($\mathcal{S}1$) We first use Thurimella's algorithm (see Section 2) to identify the connected components of non-white nodes and also to find out whether each component is satisfied (i.e. if it contains a gray node). These take $\mathcal{O}(D + \sqrt{n}\log^* n)$ rounds. Each non-white node broadcasts its component id and whether its component is satisfied to all neighbors. We also find the total number of unsatisfied connected components and if it is less than $N/2$, we call this phase finished and start the next phase.

($\mathcal{S}2$) We now find the globally-maximum efficiency ρ^* of the basic-stars.

They key part is to compute the efficiency of the most-efficient basic-star centered on each white node. After that, the global-maximum can be found in $\mathcal{O}(D)$ rounds easily. We first use one round of message exchanges between the white nodes so that each white node knows all the basic-stars it centers.

Each white node v does as follows: if v is adjacent to only one component (satisfied or unsatisfied), it sends the id of this component, its satisfied/unsatisfied status and v_{id} to its neighbors. If v is adjacent to two or more components, but all of them are satisfied, then v sends a message to its neighbors containing v_{id} and an indicator message "*all-satisfied*". If v is adjacent to two or more components, at least one of which is unsatisfied, then v does not send any message.

This is because, by Definition 2, node v is *self-sufficient* and it thus can be only in basic-stars centered on v. At the end of this round, each white node v has received some messages from its white neighbors. These messages contain all the information needed for forming all the basic-stars centered on v and calculating their efficiency. Node v finds the most-efficient of these basic-stars. It is easy to see that this can indeed be done in polynomial-time local computation. We emphasize that the basic-stars found in this step are not important and the only thing that we want is to find the globally-maximum efficiency ρ^*.

($S3$) Let $\tilde{\rho} = 2^{\lfloor \log_2 \rho^* \rfloor}$, i.e., $\tilde{\rho}$ is equal to ρ^* rounded down to the closest power of 2. We pick at most one $\tilde{\rho}$-augmented basic-star X_v^i (see Definition 3) centered on each white node v, where i is the iteration number.

We reuse the messages exchanged in the previous step. First, each white node v finds a minimal $\tilde{\rho}$-efficient basic-star centered on v, if there is one. Call this the *core-star of* v. Then, v adds to this core-star any good auxiliary-legs available (one by one), to find its ρ^*-augmented basic-star X_v^i. This is the only star centered on v that will be considered for the rest of this iteration. Thus, at most one star X_v^i centered on each white node v remains active for the rest of iteration i. Note that all active remaining stars are $\tilde{\rho}/2$-efficient.

For each active-remaining star X_v^i and each unsatisfied component $\mathcal{C}$ it satisfies, the center v elects one of the white nodes of the star to be *responsible for communicating* with $\mathcal{C}$. If $\mathcal{C}$ has at least one non-center neighbor in X_v^i, then one such non-center node u (selected arbitrarily) is called *responsible for communicating with $\mathcal{C}$*. Otherwise, the center v is responsible[3] for communicating with $\mathcal{C}$.

($S4$) For each unsatisfied component $\mathcal{C} \in \mathcal{F}$, we find the number of active stars that satisfy $\mathcal{C}$. The objective is to find the maximum such number $\Delta_{\tilde{\rho}}^*$, over all unsatisfied components. First, each white node v that centers an active star X_v^i reports this star to each non-center node u of it, by just sending v_{id}, special message *active-star*, and the id of the component $\mathcal{C}$ for which u is responsible for communicating with (if there is one). Then, for each white node w and each unsatisfied component $\mathcal{C}$ that w is responsible for communicating with it in any star, node u sends to one of the nodes of $\mathcal{C}$ the number of stars in which u is responsible for communicating with $\mathcal{C}$. These counts are summed up in each component $\mathcal{C}$ via Thurimella's algorithm, and it is called the *active-degree* of $\mathcal{C}$. The maximum active-degree is found globally and called $\Delta_{\tilde{\rho}}^*$.

($S5$) Next, some active stars propose to their adjacent unsatisfied components. We mark each active star with probability $\frac{1}{5\Delta_{\tilde{\rho}}^*}$, where the decision is made randomly by the center of the star and sent to the other nodes of the star (if there is any). Then, these marks are sent to the components that get satisfied by the marked stars, as proposals, via the white nodes that are responsible for communicating with the components. If v is self-sufficient, it would need to send at most one

---

[3] Since any white node u that is not self-sufficient is adjacent to at most one unsatisfied component, in any basic-star that contains u, node u can be responsible only for this one unsatisfied adjacent component. On the other hand, if v is self-sufficient, it will be only in one star X_v^i.

proposal to each adjacent component (it would be to those components for which v is responsible for communicating with them in X_v^i). However, if v is not self-sufficient, then v might want to send many proposals to an unsatisfied component adjacent to it (there is at most one such component). This is not feasible in the CONGEST model. Instead, v selects at most 3 of these proposals (arbitrarily) and just submits these 3 proposals.

($S6$) Each component grants at most 3 of the proposals it receives. This is done via Thurimella's algorithm, where 3 proposals with largest center ids are granted. Finally components report the granted proposals to the adjacent white nodes.

($S7$) Each marked star collects how many of its proposals are granted. If at least $1/3$ of the proposals of this star were granted, then all nodes of this marked star become gray. After that, we use Thurimella's algorithm again to identify the green nodes which their component (in $\mathcal{F}$) is satisfied (by checking if their component has a gray node).

($S8$) Finally, we have a *clean up* step, which removes the newly-formed $\tilde{\rho}$-augmented basic-stars that if not removed now, might be active in the next iterations. Temporarily (just for this clean up step) color each white node *blue* if its cost is at most $1/\tilde{\rho}$. For each unsatisfied component $\mathcal{C} \in \mathcal{F}$ that can be satisfied using only blue nodes, we find one or two blue nodes that connect $\mathcal{C}$ to some other component in $\mathcal{F}$ and we color these blue nodes gray, thus making $\mathcal{C}$ satisfied. In the first round, for each blue node v, if v is adjacent to only one component, it sends the id of this component and its own id v_{id}. If v is adjacent to two or more components, it just sends its own id with an indicator symbol "*two-or-more*". In the second round, for each blue node u, if u is adjacent to an unsatisfied component $\mathcal{C}$, node u creates a proposal for $\mathcal{C}$ as follows: if u is adjacent to at least one other component $\mathcal{C}' \in \mathcal{F}$, then the proposal is simply the id of u. If u is not adjacent to any other component $\mathcal{C}'$ but there is a blue neighbor w of u such that in the first round, w sent the id of a component $\mathcal{C}'' \neq \mathcal{C}$ or w sent the "*two-or-more*" indicator symbol, then the proposal contains the ids of u and v. Otherwise, the proposal is empty. Each unsatisfied component picks one (nonempty) proposal, if it receives any, and grants it. The granted proposal is reported to all nodes adjacent to the component and if the proposal of u is granted, it becomes gray and if this granted proposal contained a blue neighbor w, then u informs w about the granted proposal which means that w also becomes gray.

Due to the space limitations, the analysis are deferred to the full version.

Acknowledgment. We thank Fabian Kuhn for valuable discussions. We also thank Stephan Holzer and Christoph Lenzen for helpful comments about the presentation.

This work was supported by Simons award for graduate students in theoretical Computer Science (number 318723), AFOSR contract number FA9550-13-1-0042, NSF award 0939370-CCF, NSF award CCF-1217506, and NSF award CCF-AF-0937274.

References

1. Alon, N., Moshkovitz, D., Safra, S.: Algorithmic construction of sets for k-restrictions. ACM Trans. Algorithms 2(2), 153–177 (2006)
2. Alzoubi, K.M., Wan, P.-J., Frieder, O.: Message-optimal connected dominating sets in mobile ad hoc networks. In: the Proceedings of the Int'l Symp. on Mobile Ad Hoc Net. and Comput, pp. 157–164 (2002)
3. Alzoubi, K.M., Wan, P.-J., Frieder, O.: New distributed algorithm for connected dominating set in wireless ad hoc networks. In: Proceedings of the 35th Annual Hawaii International Conference on System Sciences (HICSS), pp. 3849–3855. IEEE (2002)
4. Berger, B., Rompel, J., Shor, P.W.: Efficient NC algorithms for set cover with applications to learning and geometry. In: Proc. of the Symp. on Found. of Comp. Sci. (FOCS), pp. 454–477 (1994)
5. Blum, J., Ding, M., Thaeler, A., Cheng, X.: Connected dominating set in sensor networks and manets. In: Handbook of Combinatorial Optimization, pp. 329–369. Springer (2005)
6. Chen, Y.P., Liestman, A.L.: Approximating minimum size weakly-connected dominating sets for clustering mobile ad hoc networks. In: Proceedings of the 3rd ACM International Symposium on Mobile ad Hoc Networking & Computing, pp. 165–172. ACM (2002)
7. Cheng, X., Huang, X., Li, D., Wu, W., Du, D.-Z.: A polynomial-time approximation scheme for the minimum-connected dominating set in ad hoc wireless networks. Networks 42(4), 202–208 (2003)
8. Cheng, X., Wang, F., Du., D.-Z.: Connected dominating set. In: Encyclopedia of Algorithms, pp. 1–99. Springer (2008)
9. Dai, F., Wu, J.: An extended localized algorithm for connected dominating set formation in ad hoc wireless networks. IEEE Transactions on Parallel and Distributed Systems 15(10), 908–920 (2004)
10. Das, B., Bharghavan, V.: Routing in ad-hoc networks using minimum connected dominating sets. In: Proc. of the IEEE Int'l Conf. on Communications (ICC), vol. 1, pp. 376–380. IEEE (1997)
11. Das Sarma, A., Holzer, S., Kor, L., Korman, A., Nanongkai, D., Pandurangan, G., Peleg, D., Wattenhofer, R.: Distributed verification and hardness of distributed approximation. In: Proc. of the Symp. on Theory of Comp. (STOC), pp. 363–372 (2011)
12. Das Sarma, A., Nanongkai, D., Pandurangan, G.: Fast distributed random walks. In: The Proc. of the Int'l Symp. on Princ. of Dist. Comp. (PODC), pp. 161–170 (2009)
13. Das Sarma, A., Nanongkai, D., Pandurangan, G., Tetali, P.: Efficient distributed random walks with applications. In: The Proc. of the Int'l Symp. on Princ. of Dist. Comp. (PODC), pp. 201–210 (2010)
14. Dubhashi, D., Mei, A., Panconesi, A., Radhakrishnan, J., Srinivasan, A.: Fast distributed algorithms for (weakly) connected dominating sets and linear-size skeletons. In: Pro. of ACM-SIAM Symp. on Disc. Alg. (SODA), pp. 717–724 (2003)
15. Elkin, M.: Unconditional lower bounds on the time-approximation tradeoffs for the distributed minimum spanning tree problem. In: Proc. of the Symp. on Theory of Comp. (STOC), pp. 331–340 (2004)
16. Feige, U.: A threshold of $\ln n$ for approximating set cover (preliminary version). In: Proc. of the Symp. on Theory of Comp. (STOC), pp. 314–318 (1996)
17. Frischknecht, S., Holzer, S., Wattenhofer, R.: Networks cannot compute their diameter in sublinear time. In: Pro. of ACM-SIAM Symp. on Disc. Alg. (SODA), pp. 1150–1162 (2012)
18. Garay, J., Kutten, S., Peleg, D.: A sub-linear time distributed algorithm for minimum-weight spanning trees. In: Proc. of the Symp. on Found. of Comp. Sci. FOCS (1993)

19. Garey, M.R., Johnson, D.S.: Computers and Intractability; A Guide to the Theory of NP-Completeness. W. H. Freeman & Co., New York (1990)
20. Ghaffari, M.: Near-optimal distributed approximation of minimum-weight connected dominating set, http://people.csail.mit.edu/ghaffari/papers/CDS.pdf
21. Ghaffari, M., Kuhn, F.: Distributed minimum cut approximation. In: Proc. of the Int'l Symp. on Dist. Comp. (DISC), pp. 1–15 (2013)
22. Guha, S., Khuller, S.: Approximation algorithms for connected dominating sets. Algorithmica 20(4), 374–387 (1998)
23. Guha, S., Khuller, S.: Improved methods for approximating node weighted steiner trees and connected dominating sets. Information and computation 150(1), 57–74 (1999)
24. Holzer, S., Wattenhofer, R.: Optimal distributed all pairs shortest paths and applications. In: The Proc. of the Int'l Symp. on Princ. of Dist. Comp. (PODC), pp. 355–364 (2012)
25. Jia, L., Rajaraman, R., Suel, T.: An efficient distributed algorithm for constructing small dominating sets. In: The Proc. of the Int'l Symp. on Princ. of Dist. Comp. (PODC), pp. 32–42 (2001)
26. Klein, P., Ravi, R.: A nearly best-possible approximation algorithm for node-weighted steiner trees. Journal of Algorithms 19(1), 104–115 (1995)
27. Kuhn, F., Moscibroda, T., Wattenhofer, R.: What cannot be computed locally? In: The Proc. of the Int'l Symp. on Princ. of Dist. Comp. (PODC), pp. 300–309 (2004)
28. Kuhn, F., Wattenhofer, R.: Constant-time distributed dominating set approximation. In: The Proc. of the Int'l Symp. on Princ. of Dist. Comp. (PODC), pp. 25–32 (2003)
29. Kutten, S., Peleg, D.: Fast distributed construction of k-dominating sets and applications. In: The Proc. of the Int'l Symp. on Princ. of Dist. Comp. (PODC), pp. 238–251 (1995)
30. Lenzen, C., Patt-Shamir, B.: Fast routing table construction using small messages: Extended abstract. In: Proc. of the Symp. on Theory of Comp. (STOC), pp. 381–390 (2013)
31. Min, M., Du, H., Jia, X., Huang, C.X., Huang, S.C.-H., Wu, W.: Improving construction for connected dominating set with steiner tree in wireless sensor networks. Journal of Global Optimization 35(1), 111–119 (2006)
32. Nanongkai, D.: Distributed approximation algorithms for weighted shortest paths. In: Proc. of the Symp. on Theory of Comp. (STOC) (to appear, 2014)
33. Nanongkai, D., Das Sarma, A., Pandurangan, G.: A tight unconditional lower bound on distributed randomwalk computation. In: The Proc. of the Int'l Symp. on Princ. of Dist. Comp. (PODC), pp. 257–266 (2011)
34. Peleg, D.: Distributed Computing: A Locality-sensitive Approach. In: Society for Industrial and Applied Mathematics, Philadelphia, PA, USA (2000)
35. Peleg, D., Rubinovich, V.: A near-tight lower bound on the time complexity of distributed MST construction. In: Proc. of the Symp. on Found. of Comp. Sci. (FOCS), p. 253 (1999)
36. Pettie, S.: Distributed algorithms for ultrasparse spanners and linear size skeletons. In: The Proc. of the Int'l Symp. on Princ. of Dist. Comp. (PODC), pp. 253–262 (2008)
37. Raz, R., Safra, S.: A sub-constant error-probability low-degree test, and a sub-constant error-probability PCP characterization of NP. In: Proc. of the Symp. on Theory of Comp. (STOC), pp. 475–484 (1997)
38. Thurimella, R.: Sub-linear distributed algorithms for sparse certificates and biconnected components. In: The Proc. of the Int'l Symp. on Princ. of Dist. Comp. (PODC), pp. 28–37 (1995)
39. Wan, P.-J., Alzoubi, K.M., Frieder, O.: Distributed construction of connected dominating set in wireless ad hoc networks. In: The Proc. of IEEE Int'l Conf. on Computer Communications (INFOCOM), vol. 3, pp. 1597–1604 (2002)
40. Wu, J., Gao, M., Stojmenovic, I.: On calculating power-aware connected dominating sets for efficient routing in ad hoc wireless networks. In: IEEE's International Conference on Parallel Processing (ICPP), pp. 346–354 (2001)

Randomized Rumor Spreading
in Dynamic Graphs

George Giakkoupis[1], Thomas Sauerwald[2], and Alexandre Stauffer[3]

[1] INRIA Rennes, France
[2] Computer Laboratory, University of Cambridge, UK
[3] Department of Mathematical Sciences, University of Bath, UK

Abstract. We consider the well-studied rumor spreading model in which nodes contact a random neighbor in each round in order to push or pull the rumor. Unlike most previous works which focus on static topologies, we look at a dynamic graph model where an adversary is allowed to rewire the connections between vertices before each round, giving rise to a sequence of graphs, $G_1, G_2, \ldots$ Our first result is a bound on the rumor spreading time in terms of the conductance of those graphs. We show that if the degree of each node does not change much during the protocol (that is, by at most a constant factor), then the spread completes within t rounds for some t such that the sum of conductances of the graphs G_1 up to G_t is $O(\log n)$. This result holds even against an *adaptive* adversary whose decisions in a round may depend on the set of informed vertices before the round, and implies the known tight bound with conductance for static graphs. Next we show that for the alternative expansion measure of vertex expansion, the situation is different. An adaptive adversary can delay the spread of rumor significantly even if graphs are regular and have high expansion, unlike in the static graph case where high expansion is known to guarantee fast rumor spreading. However, if the adversary is *oblivious*, i.e., the graph sequence is decided before the protocol begins, then we show that a bound close to the one for the static case holds for any sequence of regular graphs.

1 Introduction

Randomized rumor spreading is a popular epidemic protocol for disseminating information in large distributed networks. The protocol proceeds in a sequence of synchronous rounds. Initially, in round 0, an arbitrary node has a piece of information, the *rumor*. This rumor is then spread iteratively to the other nodes: In each round, every informed node (i.e., every node that learned the rumor in a previous round) chooses a random neighbor to which it sends the rumor. This is the so-called PUSH protocol. The PULL protocol is symmetric: In each round, every *un*informed node chooses a random neighbor, and if this neighbor knows the rumor it transmits it to the uninformed node. Finally, the PUSH-PULL protocol is the combination of both strategies: In each round, every node chooses a random neighbor to send the rumor to, if the node is informed, or to request the rumor from, otherwise.

J. Esparza et al. (Eds.): ICALP 2014, Part II, LNCS 8573, pp. 495–507, 2014.
© Springer-Verlag Berlin Heidelberg 2014

Several aspects of rumor spreading have been analyzed, including its running time (i.e., the number of rounds until all nodes get informed), the corresponding number of messages, and the amount of randomness needed. The running time is arguably the most fundamental and well-studied of those aspects. In particular, it has been shown that just a logarithmic number of rounds suffice to spread a rumor with high probability (*w.h.p.*) on several topologies, from basic communication networks, such as complete graphs, hypercubes and random graphs [27,11,19], to more complex structures, such as preferential attachment graphs or power-law random graphs modeling social networks [9,12]. Recently, a number of studies have extended this line of work by establishing bounds on the running time of rumor spreading in terms of expansion parameters of the underlying graph, namely conductance [24,4,14,3] and vertex expansion [29,16,15]. This connection between rumor spreading and graph theory is also relevant for understanding social and other real networks, as studies have indicated that such networks have good expansion properties [23,5].

A limitation of the above results is that they require the graph to be fixed throughout the execution of the rumor spreading protocol, whereas many of the prevalent topologies, such as peer-to-peer or wireless networks, are inherently dynamic. In particular, the structure of these networks may change more quickly than a rumor spreads.

In this paper we analyze the running time of randomized rumor spreading in a dynamic setting, given by a sequence of graphs $G_1, G_2, \ldots$ with the same vertex set of size n, but possibly distinct edge sets. In this setting, at each round t, a vertex contacts a random neighbor *in graph* G_t in order to push or pull the rumor. We assume that the edge set for each round is determined by an adversary. The adversary can be either *adaptive*, i.e., it decides the edge set for each round at the beginning of the round, knowing the set of informed vertices at the time, or *oblivious*, i.e., it fixes the complete sequence of graphs before rumor spreading starts, knowing just the source of the rumor.

Our first result is an upper bound on the running time of PUSH–PULL in terms of the conductances of graphs $G_1, G_2, \ldots$ Suppose that during the execution of the protocol, we have for each node that the ratio of its maximum over minimum degree is bounded by some $\rho \geq 1$. We show that rumor spreading then completes within t rounds for some t such that the sum of conductances of the graphs G_1 up to G_t is $O(\rho \log n)$. Moreover, this bound holds even if the adversary is adaptive.

Theorem 1. *Let $G_1, G_2, \ldots$ be a sequence of graphs determined by an adaptive adversary such that for each vertex u, its degree on each graph G_t is at least $\delta_u > 0$ and at most Δ_u. The degree bounds δ_u, Δ_u may be different for each u, and are fixed (in advance) by the adversary. Let $\rho = \max_u(\Delta_u/\delta_u)$. Also for each $t \geq 1$, let ϕ_t be the conductance of graph G_t.*

For any constant $\beta > 0$, there exists a constant $b > 0$ so that the following bound holds for PUSH–PULL. Let τ be the first round for which $\sum_{t=1}^{\tau} \phi_t \geq b\rho \log n$; $\tau = \infty$ if not such round exists. Then with probability $1 - n^{-\beta}$, either $\tau = \infty$ or all nodes have been informed by the end of round τ.

If all graphs $G_1, G_2, \ldots$ are the same and have conductance ϕ, then $\rho = 1$ and the bound of Theorem 1 implies the optimal bound for static graphs established in [14], stating that $O(\log(n)/\phi)$ rounds suffice w.h.p. to spread a rumor on any graph with conductance ϕ.

The dependence of the bound in Theorem 1 on ρ is not an artifact of our analysis. For example, even an oblivious adversary can construct a sequence of graphs in which every graph has constant conductance (but the degrees of nodes change widely and thus ρ is large) so that PUSH-PULL needs a linear number of rounds to inform all nodes (see Proposition 1, in Sect. 5).

We point out that Theorem 1 is shown for a more general setting, where multiple edges and self-loops are allowed. Moreover, it holds even if we define ϕ_t to be the conductance in G_t of the *set of informed vertices* before round t. This can be much larger than the conductance of G_t, which is the minimum conductance of any set of vertices in G_t.

For static graphs, conductance and vertex expansion yield very similar types of bounds on the running time of rumor spreading. For dynamic graphs, however, the situation is different. In particular, we reveal a separation between the adaptive and oblivious adversary models for the case of vertex expansion, which is not observed for conductance: An adaptive adversary can construct a sequence of regular graphs with constant vertex expansion so that PUSH-PULL takes a polynomial number of rounds (see Proposition 2)—by Theorem 1, this is not possible if conductance is considered in place of vertex expansion. For an oblivious adversary, on the other hand, we show that a bound similar to the one in Theorem 1 holds with vertex expansion, for any sequence of regular graphs.

Theorem 2. *Let $G_1, G_2, \ldots$ be a sequence of d-regular graphs determined by an oblivious adversary. For each $t \geq 1$, let α_t be the vertex expansion of G_t. Then, for any constant $\beta > 0$, there exists a constant $c > 0$ so that, if there exists a round t with $\sum_{s=1}^{t} \alpha_s \geq c \cdot \log^4 n \log^2 d$, then PUSH-PULL informs all nodes within t rounds with probability at least $1 - n^{-\beta}$.*

If all graphs $G_1, G_2, \ldots$ are the same and have vertex expansion α, then the bound of Theorem 2 matches within a polylogarithmic factor the optimal bound for static graphs from [15], which states that $O(\log(n) \log(d)/\alpha)$ rounds suffice w.h.p. to spread a rumor on any graph (even non-regular) with vertex expansion α and maximum degree d. Whether Theorem 2 extends also to non-regular graph sequences is an interesting open problem.

Our proofs are non-trivial extensions of previous analyses for static graphs, in particular from [14] and [29]. The dynamic setting, and also the adaptivity of the adversary for the case of conductance, add new challenges to the problem. The proof of Theorem 1 is based on a new martingale argument which exposes the outcome of each round gradually, one vertex at a time. The proof of Theorem 2 has to overcome the problem that the standard symmetry argument relating push and pull no longer holds, and this breaks key arguments used in existing proofs of bounds with vertex expansion. Moreover, tighter proofs from [16,15] employ potential functions based on the boundary of the informed nodes, which may largely fluctuate in the dynamic setting.

Related Work. There have been several studies on information spreading processes in dynamic graphs. Perhaps the closest one to our work is a recent work by Clementi et al. [6] about PUSH in a random edge-Markovian model, yielding a dynamic variant of the Erdős-Rényi random graph. Motivated by the increasing importance of wireless networks, works [20,26,22,25] analyzed the dynamics of information dissemination among moving objects in d-dimensional grids. There are also several analyses on the *flooding* process—a variant of rumor spreading where *every* neighbor of an informed node becomes informed in a round [8,7,2]. We note that in all these works, the graph dynamics are governed by a random process, whereas in our model the dynamics are controlled by an adversary.

Avin et al. [1] analyzed the cover time of random walks on dynamic graphs specified by an oblivious adversary. They constructed graphs in which a simple random walk has an exponential cover time, but also proved that a lazy random walk has a polynomial cover time for any sequence of connected graphs.

Kuhn et al. [21] introduced the so-called *k-token dissemination problem* in a synchronous setting with a dynamically changing network. They considered a worst-case scenario, where the communication links are chosen by an adversary, and nodes do not know who their neighbors are for the current round before they send their messages. In contrast to our model, the connectivity (expansion) assumptions are weaker and correspondingly the time complexity bounds are much larger, i.e., at least polynomial in n [21,17,10,18,28]. Georgiou et al. [13] considered the complexity of asynchronous gossip in a fault-prone distributed setting. While their model is quite different from ours, they also exhibited a separation between an adaptive and oblivious adversary.

2 Model

We consider the standard PUSH, PULL, and PUSH–PULL rumor spreading protocols. We will denote by I_t the set of informed vertices after the first t rounds of the protocol, and by U_t the set $V \setminus I_t$ of uninformed vertices. In particular, I_0 is the singleton set containing just the source, which is an arbitrarily chosen node.

The dynamic graph model we consider is an infinite sequence of graphs $G_1, G_2, \ldots$ on the same set of n vertices, but possibly with different edge sets. In each round t of rumor spreading, if a vertex must choose a neighbor to push the rumor to or pull the rumor from, then it chooses a random one among its neighbors *in graph G_t* (if it has neighbors in G_t).

For each $t \geq 1$, $G_t = (V, E_t)$, where $V = \{1, 2, \ldots, n\}$, and the edge set E_t is determined by an *adversary*. We distinguish between two adversarial models, *adaptive* and *oblivious*. An *adaptive adversary* decides the edge set E_t knowing the outcome of all rounds before round t; precisely, E_t is a function of $I_0, \ldots, I_{t-1}$. An *oblivious adversary*, on the other hand, has to specify the entire graph sequence in advance; precisely, E_t is just a function of I_0. An adversary can be either deterministic or randomized, where in the latter case the sequence of E_t is also a function of a random bit string.

We recall now the definitions of two standard graph expansion parameters we use. For a graph $G = (V, E)$, the *conductance* of a non-empty vertex set $S \subset V$ and the conductance of graph G are defined respectively as

$$\phi(S) = \frac{|E(S, V \setminus S)|}{\min\{\text{vol}(S), \text{vol}(V \setminus S)\}} \quad \text{and} \quad \phi(G) = \min_{S \subset V, \, S \neq \emptyset} \phi(S),$$

where $E(S, V \setminus S)$ is the set of edges with one endpoint in each of the sets S and $V \setminus S$; and $\text{vol}(S) = \sum_{u \in S} \deg(u)$ is the *volume* of S, with $\deg(u)$ denoting the degree of u. The *vertex expansion* of S and G are respectively

$$\alpha(S) = \frac{|\partial S|}{\min\{|S|, |V \setminus S|\}} \quad \text{and} \quad \alpha(G) = \min_{S \subset V, \, S \neq \emptyset} \alpha(S),$$

where $\partial S := N(S) \setminus S$ is the set of vertices outside S that are adjacent to some vertex in S. For any graph G, both $\phi(G)$ and $\alpha(G)$ are between 0 and 1, with high values indicating that the graph is well connected. If G is disconnected then $\phi(G) = \alpha(G) = 0$.

In the following, when we write $\text{vol}(I_t)$, ∂I_t, $\phi(I_t)$, etc. (and similarly for U_t), we will assume that these quantities refer to graph G_{t+1} (and not to G_t), unless mentioned otherwise. This is convenient since I_t is the set of informed vertices at the beginning of round $t + 1$.

Theorem 1 holds even if the graph sequence consists of *multigraphs*, with parallel edges and self-loops (as long as the degree of each vertex is at most polynomially large in n). To compute the conductance of a multigraph, parallel edges are counted with respect to their multiplicity and every self-loop counts as a single edge. Further, when a vertex must pick a random neighbor, this is done proportional to the multiplicity of a (parallel) edge or self-loop. If a self-loop is chosen then no communication takes place.

3   Proof of the Bound with Conductance (Theorem 1)

We observe that it suffices to consider just deterministic adversaries, since in case of a randomized adversary, we can just expose all its randomness (i.e., its random bit string) before the protocol starts, and then proceed deterministically.

Not all vertices are guaranteed to get informed eventually, as the adversary may permanently disconnect the network. However, it is not difficult to show that one can always modify a (deterministic) adversary, in such a way that this does not happen and it suffices to consider the modified adversary in the analysis. Thus, we will assume that the expected number of rounds until all nodes are informed is finite, i.e.,

$$\mathbf{E}[\min\{t \colon I_t = V\}] < \infty. \tag{1}$$

For each round $t \geq 1$, let $\Phi_t = \phi_1 + \cdots + \phi_t$ be the sum of the conductances of the graphs in the first t rounds. We must show that for $\tau = \inf\{t\colon \Phi_t \geq b\rho \log n\}$, we have w.h.p. that all nodes have been informed within τ rounds.

For each set $S \subseteq V$ of vertices, we define the *min-volume* of S as $\mathrm{vol}^*(S) = \sum_{i \in S} \delta_i$. Thus $\mathrm{vol}^*(S)$ is a lower bound for the volume of S on any of the graphs G_t. Our proof is based on an analysis of the growth of $\mathrm{vol}^*(I_t)$.

We will show the following lemma stating (a) if the min-volume of informed vertices is smaller than a constant fraction of the total min-volume, then the sum of conductances ϕ_t until the min-volume doubles is bounded in expectation by $O(\rho)$; and (b) if the min-volume of informed vertices is larger, then the sum of ϕ_t until the min-volume of uninformed vertices halves is bounded by $O(\rho)$.

Lemma 1. *There is a fixed constant $c > 0$ such that for any round $t \geq 1$,*

(a) *If* $\mathrm{vol}^*(I_t) \leq \mathrm{vol}^*(V)/3$ *and* $\tau_t = \min\{k\colon \mathrm{vol}^*(I_k) \geq 2\,\mathrm{vol}^*(I_t)\}$, *then* $\mathbf{E}[\Phi_{\tau_t} - \Phi_t \mid I_t] \leq c\rho$.

(b) *If* $\mathrm{vol}^*(I_t) > \mathrm{vol}^*(V)/3$ *and* $\tau_t = \min\{k\colon \mathrm{vol}^*(U_k) \leq \mathrm{vol}^*(U_t)/2\}$, *then* $\mathbf{E}[\Phi_{\tau_t} - \Phi_t \mid I_t] \leq c\rho$.

From this result, the bound of Theorem 1 follows easily: From Lemma 1(a) and Markov's Inequality it follows that when $\mathrm{vol}^*(I_t) \leq \mathrm{vol}^*(V)/3$, we have $\mathbf{Pr}[\Phi_{\tau_t} - \Phi_t \leq 2c\rho \mid I_t] \geq 1/2$, i.e., with probability at least $1/2$, $\mathrm{vol}^*(I_t)$ doubles after Φ_t has increased by at most $2c\rho$. It follows then by Chernoff bounds that $1/3$ of the total min-volume gets informed with probability at least $1 - n^{-\beta}/2$, for any fixed β, after a number t_1 of rounds such that $\Phi_{t_1} = 2c\rho \cdot O(\log(\mathrm{vol}^*(V)/3)) = O(\rho \log n)$, as $\log(\mathrm{vol}^*(V)) = O(\log n)$. A similar argument using Lemma 1(b) shows that if $1/3$ of the total min-volume has been informed by some round t, then an additional t_2 rounds, such that $\Phi_{t+t_2} - \Phi_t = O(\rho \log n)$, suffice to inform the remaining vertices with probability at least $1 - n^{-\beta}/2$. From these two results and the union bound, Theorem 1 follows.

3.1   Proof of Lemma 1

Recall that during the spread, the degree of each vertex $i \in V$ is lower bounded by $\delta_i > 0$ and upper bounded by Δ_i, and $\Delta_i/\delta_i \leq \rho$. Let $\delta = \max_i \delta_i$.

The proof distinguishes three cases, depending on the min-volume of informed vertices initially: (i) $\mathrm{vol}^*(I_t) < \delta$, (ii) $\delta \leq \mathrm{vol}^*(I_t) \leq \mathrm{vol}^*(V)/3$, and (iii) $\mathrm{vol}^*(I_t) > \mathrm{vol}^*(V)/3$. For cases (ii) and (iii) it suffices that we consider only pull operations, while for case (i) we must consider both push and pull. Due to space limitations we only give the proof for case (ii). In this case, $\tau_t = \min\{k \geq t\colon \mathrm{vol}^*(I_k) \geq 2\,\mathrm{vol}^*(I_t)\}$.

Claim 1. *If* $\delta \leq \mathrm{vol}^*(I_t) \leq \mathrm{vol}^*(V)/3$ *then* $\mathbf{E}[\Phi_{\tau_t} - \Phi_t \mid I_t] < 2\rho + 1$.

Proof. We use a martingale argument that relates the min-volume of vertices informed by pull transmissions to the number of edges between informed and uninformed vertices. In this argument, the outcome of each round is exposed gradually, one vertex at a time.[1]

Assume I_t is fixed. We divide each round $k > t$ into $|\partial I_{k-1}|$ *steps* (∂I_{k-1} is the set of uniformed vertices at the beginning of round k that have some informed neighbor). Each of those $|\partial I_{k-1}|$ steps reveals the push and pull transmissions of the rumor in round k to a single vertex $i \in \partial I_{k-1}$. The order in which vertices $i \in \partial I_{k-1}$ are considered can be arbitrary. We look at the sequence of all those steps, from round $t + 1$ until round τ_t. For each step $s = 1, 2, \ldots$, let i_s be the vertex considered in step s, let k_s be the round in which step s takes place, d_s be the degree of i_s during round k_s (i.e., i_s's degree in G_{k_s}), and γ_s be the number of informed neighbors of i_s in G_{k_s} at the beginning of round k_s.

Below we first show that the sum of all γ_s until the step when the min-volume of informed vertices has doubled is bounded in expectation by $\rho(\mathrm{vol}^*(I_t) + \delta)$. Then we bound the corresponding increase in Φ_t in terms of the sum of γ_s, and combine the two results to obtain the claim.

Let X_s be the indicator variable that is 1 if i_s pulls the rumor in step s, and $X_s = 0$ otherwise. Further, let Z_s be the indicator variable that is 1 if i_s gets informed in step s, and 0 otherwise. Note that $Z_s \geq X_s$, since i_s may get informed by a push transmission. Note also that the sequence $Z_1, \ldots, Z_{s-1}$ completely determines the evolution of the set of informed vertices in the first $s - 1$ steps, and thus determines i_s, k_s, d_s, and γ_s. In order for i_s to pull the rumor in step s, it must choose one of its γ_s informed neighbors, among its d_s neighbors in total. It follows that

$$\mathbf{E}[X_s \mid Z_1 \ldots Z_{s-1}] = \gamma_s/d_s. \tag{2}$$

For each $s \geq 0$, we define $Y_s = \sum_{1 \leq j \leq s} (X_j d_j - \gamma_j)$. The sequence $Y_0, Y_1, \ldots$ is a martingale with respect to $Z_1, Z_2, \ldots$, because

$$\mathbf{E}[Y_s \mid Z_1 \ldots Z_{s-1}] = Y_{s-1} + \mathbf{E}[X_s \mid Z_1 \ldots Z_{s-1}] \cdot d_s - \gamma_s \overset{(2)}{=} Y_{s-1}.$$

Let T be the number of steps after round t until the min-volume of informed vertices doubles, i.e., $T = \min\{s : \sum_{1 \leq j \leq s} Z_j \delta_{i_j} \geq \mathrm{vol}^*(I_t)\}$. (Observe, $k_T = \tau_t$.) Since T is a stopping time for $Z_1, Z_2, \ldots$, and we have that $\mathbf{E}[T] < \infty$ [2] and the differences $Y_s - Y_{s-1}$ are bounded, it follows from the Optional Stopping Theorem that $\mathbf{E}[Y_T] = \mathbf{E}[Y_0] = 0$. Substituting the definition of Y_T and rearranging gives

$$\mathbf{E}\left[\sum_{1 \leq j \leq T} \gamma_j\right] = \mathbf{E}\left[\sum_{1 \leq j \leq T} X_j d_j\right]. \tag{3}$$

---

[1] The reason we expose one vertex at a time (rather than all at once), is that by stopping this process right after the min-volume of informed vertices has doubled, we have the guarantee that the min-volume at that time is by at most $\delta - 1 < \mathrm{vol}^*(I_t)$ larger than $2\,\mathrm{vol}^*(I_t)$. (This is used in the line right above Eq.(4), in order to obtain (4).) On the other hand, the min-volume right after the round during which the min-volume doubles may be much larger than $2\,\mathrm{vol}^*(I_t)$.

[2] This is immediate from Eq.(1).

We have $\sum_{1 \leq j \leq T} X_j d_j \leq \rho \sum_{1 \leq j \leq T} Z_j \delta_{i_j}$, because $X_j \leq Z_j$ and $d_j \leq \Delta_{i_j} \leq \rho \delta_{i_j}$; and from $\bar{T}$'s definition, $\sum_{1 \leq j \leq T} Z_j \delta_{i_j} < \mathrm{vol}^*(I_t) + \delta$. It follows

$$\mathbf{E}\left[\sum_{1 \leq j \leq T} \gamma_j\right] < \rho(\mathrm{vol}^*(I_t) + \delta). \tag{4}$$

Next we bound $\Phi_{\tau_t} - \Phi_t$ in terms of $\sum_{1 \leq j \leq T} \gamma_j$, and apply the above inequality to bound $\mathbf{E}[\Phi_{\tau_t} - \Phi_t]$. For each round k with $t < k \leq \tau_t$, the conductance ϕ_k of G_k is bounded by the conductance of I_{k-1} in G_k, and thus

$$\phi_k \leq \frac{\sum_{s:k_s=k} \gamma_s}{\min\{\mathrm{vol}(I_{k-1}), \mathrm{vol}(U_{k-1})\}} \leq \frac{\sum_{s:k_s=k} \gamma_s}{\mathrm{vol}^*(I_t)},$$

where the second inequality holds because $\mathrm{vol}(I_{k-1}) \geq \mathrm{vol}^*(I_{k-1}) \geq \mathrm{vol}^*(I_t)$ and

$$\mathrm{vol}(U_{k-1}) \geq \mathrm{vol}^*(U_{k-1}) \geq \mathrm{vol}^*(U_{\tau_t-1}) = \mathrm{vol}^*(V) - \mathrm{vol}^*(I_{\tau_t-1})$$
$$> \mathrm{vol}^*(V) - 2\,\mathrm{vol}^*(I_t) \geq \mathrm{vol}^*(I_t),$$

as $\mathrm{vol}^*(I_t) \leq \mathrm{vol}^*(V)/3$. From the above bound on ϕ_k applied for $t < k < \tau_t$, we obtain

$$\Phi_{\tau_t} - \Phi_t = (\phi_{t+1} + \cdots + \phi_{\tau_t-1}) + \phi_{\tau_t} \leq \frac{\sum_{s:k_s<\tau_t} \gamma_s}{\mathrm{vol}^*(I_t)} + 1.$$

From T's definition, $k_T = \tau_t$, and thus the sum above is $\sum_{s:k_s<\tau_t} \gamma_s < \sum_{s \leq T} \gamma_s$. Thus,

$$\mathbf{E}[\Phi_{\tau_t} - \Phi_t] < \frac{\mathbf{E}\left[\sum_{s \leq T} \gamma_s\right]}{\mathrm{vol}^*(I_t)} + 1 \overset{(4)}{<} \frac{\rho(\mathrm{vol}^*(I_t) + \delta)}{\mathrm{vol}^*(I_t)} + 1 \leq 2\rho + 1, \tag{5}$$

as $\delta \leq \mathrm{vol}^*(I_t)$. This completes the proof of Claim 1. $\square$

The proof for the case of $\mathrm{vol}^*(I_t) > \mathrm{vol}^*(V)/3$ is similar, but considers the set of uninformed vertices instead of the set of informed ones.

Claim 2. If $\mathrm{vol}^*(I_t) > \mathrm{vol}^*(V)/3$ then $\mathbf{E}[\Phi_{\tau_t} - \Phi_t \mid I_t] < 2\rho + 1$.

The analysis in the proof of Claim 1 does not carry over to the case of $\mathrm{vol}^*(I_t) < \delta$: The final inequality in (5) does not hold, as the ratio $\delta/\mathrm{vol}^*(I_t)$ may be very large. In fact any analysis that relies only on pull transmissions is bound to fail, for otherwise Theorem 1 would hold even if only PULL were used, which is easily seen to be wrong because of the star graph counter-example. To show the next claim, we extend the approach of Claim 1 by taking into account also push transmissions.

Claim 3. If $\mathrm{vol}^*(I_t) < \min\{\delta, \mathrm{vol}^*(V)/3\}$ then $\mathbf{E}[\Phi_{\tau_t} - \Phi_t \mid I_t] < 3\rho + 1$.

4   Proof of the Bound with Vertex Expansion (Theorem 2)

Theorem 2 can be deduced easily from the following result, in which we assume a uniform lower bound on the vertex expansion. (If $d = O(\log^3 n)$ then Theorem 2 follows directly from Theorem 1 since $\alpha \leq d \cdot \Phi$.)

Theorem 3. *Let t be any integer, and $G_1, G_2, \ldots, G_t$ be a sequence of d-regular graphs with $d = \Omega(\log^3 n)$ determined by an oblivious adversary, so that for each $1 \le s \le t$, G_s has vertex expansion at least $\alpha > 0$. Then, for any constant $\beta > 0$, there exists a constant $\widetilde{c} > 0$ such that, if t satisfies $t \cdot \alpha \ge \widetilde{c} \cdot \log^3 n \log^2 d$, then* PUSH-PULL *informs all nodes within t rounds with probability at least $1 - n^{-\beta}$.*

4.1   Proof of Theorem 3

The analysis of Theorem 3 is divided into three phases, according to the number of informed nodes. Before we analyze the different phases, we provide some tools we will use in the analysis. The proofs of all statements in this section are omitted due to space limitations.

The next lemma establishes a (nearly) exponential growth of the number of informed nodes until that number reaches $d/\log n$.

Lemma 2. *Let $u \in V$ be arbitrary with $I_0 = \{u\}$ and let $\kappa := \log_3(d/(2\log n)) + 440$. Then the following statements hold.*

1. $\mathbf{Pr}\left[I_\kappa \ge \frac{d}{6\log n}\right] \ge 1/2$.
2. $\mathbf{Pr}\left[I_\kappa \le 3^{220} \cdot 10^6 \cdot d\right] \ge 1 - n^{-4}$, *and more generally, for any round $1 \le s \le \kappa$,* $\mathbf{Pr}\left[I_s \le 2 \cdot 10^6 \cdot \log n \cdot 3^s\right] \ge 1 - n^{-4}$.

The first statement of Lemma 2 motivates the following definition.

Definition 1 (Friend). *Let $U \subseteq V$ be any subset. Then a node u is a friend of U in round t if, for κ as in Lemma 2, $\mathbf{Pr}\left[|I_{t+\kappa} \cap U| \ge \frac{d}{12\log n} \;\middle|\; I_t = \{u\}\right] \ge 1/4$.*

A similar notion of a friend was defined in [29, Definition 3.1] for a static graph; our definition depends on the sequence of graphs $G_{t+1}, G_{t+2}, \ldots, G_{t+\kappa}$, and so in particular, on the choice of t. Applying the first statement of Lemma 2, we have that for every subset $U \subseteq V$, every node $u \in V$, and every round t, node u is a friend of either U or $V \setminus U$ in round t.

Next we consider the situation where I_t is of any size, and half of the nodes in ∂I_t are friends of I_t. We shall prove that an almost constant fraction of nodes in I_t gets informed after κ rounds. It should be noted that this is relatively straightforward in the case of static graphs, as it follows from a standard symmetry argument relating PUSH and PULL (cf. [29, Lemma 3.1]). Here, however, the analysis is considerably more involved, as we are dealing with dynamic graphs.

Lemma 3 (Key Lemma). *Consider a round t with a fixed set of informed nodes I_t, where $1 \le |I_t| \le n/2$. Let $S \subseteq V \setminus I_t$ be a set of vertices which are friends of I_t in round $t+1$. Then, there is a constant $0 < C < 1$ so that, with probability at least $1/16$, at least $C \cdot \frac{|S|}{\log^2 n \log d}$ nodes get informed after κ additional rounds, for κ defined as in Lemma 2.*

Next we analyze the growth of informed nodes in 3 phases: $|I_t| \in [1, \frac{d}{6\log n}]$, $|I_t| \in [\frac{d}{6\log n}, \frac{d}{\alpha}]$, and $|I_t| \in [\frac{d}{\alpha}, \frac{n}{2}]$. In the following, κ is defined as in Lemma 2.

Lemma 4 (Phase 1). *Assume that* $|I_0| = 1$. *Then after* $t_1 := \kappa$ *rounds, we have* $|I_{t_1}| \geq \frac{d}{6 \log n}$ *with probability at least* $1/2$.

The above result for the first phase follows immediately from Lemma 2.

Lemma 5 (Phase 2). *Let* t_1 *be the first round for which* $|I_{t_1}| \geq \frac{d}{6 \log n}$. *Then, for any constant* $\beta > 0$, *there exists a constant* $c > 0$ *so that, for* $t_2 := t_1 + \frac{c \log^3 n \log d}{\alpha} \cdot \kappa$, *we have* $|I_{t_2}| \geq \frac{d}{\alpha}$ *with probability at least* $1 - n^{-\beta}$.

The analysis of the second phase is more involved. We only consider every $(\kappa + 1)$-th round and distinguish between two cases. If half of the nodes in ∂I_t are friends of I_t, then Lemma 3 implies that a large fraction of these nodes becomes informed after $\kappa + 1$ rounds. If half of the nodes in ∂I_t are friends of $V \setminus I_t$, then, if such a node in ∂I_t pulls the rumor in round $t + 1$, then after κ additional rounds $d/(12 \log n)$ nodes get informed in $V \setminus I_t$. Expressing the progress of both cases via a submartingale and applying the Optional Stopping Theorem completes the proof of Lemma 5.

Lemma 6 (Phase 3). *Let* t_2 *be the first round for which* $|I_{t_2}| \geq \frac{d}{\alpha}$. *Then, for any constant* $\beta > 0$, *there exists a constant* $c > 0$ *so that for* $t_3 := t_2 + c \frac{\log^3 n \log d}{\alpha} \cdot \kappa$, *we have* $|I_{t_3}| \geq \frac{n}{2}$ *with probability at least* $1 - n^{-\beta}$.

The analysis of the third phase is somewhat similar to the analysis of the second phase. However, the case where half of the nodes in ∂I_t are friends of $V \setminus I_t$ requires a more careful analysis, since we have to analyze the propagation of the rumor within $V \setminus I_t$ from *several* nodes in ∂I_t in parallel.

From Lemmas 4–6 it follows that, with probability at least $1 - n^{-\beta}$, Phase 1 is completed after $O(\kappa \log n) = O(\log n \log d)$ rounds, and Phases 2 and 3 are completed after $O(\frac{\log^3 n \log^2 d}{\alpha})$ rounds, thus proving the bound of Theorem 3.

5   Counter-Examples

High conductance is not sufficient to guarantee fast rumor spreading in our dynamic graph model. Even an oblivious adversary can construct a sequence of high-conductance graphs (in which the degrees of nodes change widely), so that PUSH-PULL needs a linear number of rounds to inform all nodes.

Proposition 1. *An oblivious adversary can construct a sequence of graphs* $G_1, G_2, \ldots$, *each of which has conductance* 1, *so that* PUSH-PULL *needs* $n - 1$ *rounds to complete.*

Proof. Instead of analyzing the PUSH-PULL protocol we analyze a different process in which at each round all neighbors of the set of informed nodes become informed (i.e., $I_{t+1} = I_t \cup \partial I_t$). The strategy of the adversary is as follows. In every round t, G_t is a star graph with n vertices. The source of the rumor is a vertex of degree 1, and in each round except for the last one, the informed nodes are vertices of degree 1. At first it may seem that the adversary has to be adaptive to employ this strategy. But, given that the source of the rumor is known, then I_0 is a deterministic set. Given G_1 and I_0, then I_1 is again a deterministic set, and so on. Therefore, the construction above for the sequence $G_1, G_2, \ldots$ can be done in an oblivious manner. With this, our process satisfies $|I_t| = t + 1$, for every $0 \leq t \leq n - 1$. Since the runtime of this process is at most the runtime of the PUSH-PULL protocol, the claim of the proposition follows. □

An adaptive adversary can significantly delay the spread of the rumor, even if all graphs in the sequence are regular and have high vertex expansion. Thus Theorem 2 does not hold when the adversary is adaptive rather than oblivious.

Proposition 2. *An adaptive adversary can construct a sequence of regular isomorphic graphs $G_1, G_2, \ldots$, each of which has constant vertex expansion, so that PUSH-PULL needs $\Omega(\sqrt{n})$ rounds with probability at least $1/2$.*

Proof. Let G be the Cartesian product of a clique of size $\sqrt{n}$ with a 3-regular expander graph of size $\sqrt{n}$. This graph has $\sqrt{n} \cdot \sqrt{n}$ vertices and is regular, with all vertices having degree $\sqrt{n} - 1 + 3 = \sqrt{n} + 2$. Observe that G can be seen as a collection of $\sqrt{n}$ cliques of size $\sqrt{n}$, with every vertex in each clique connected to 3 vertices in other cliques. By [29, Lemma 4.2], G has constant vertex expansion. Every graph in the sequence $G_1, G_2, \ldots$ will be isomorphic to G.

In each round $t \geq 1$, the adversary permutes the vertices in a way so that there is at most one clique that is not fully informed or fully uninformed (i.e., whose number of informed nodes is in the interval $[1, \sqrt{n} - 1]$). All other cliques contain either $\sqrt{n}$ informed nodes or none. Consider now a clique in which all $\sqrt{n}$ nodes are informed. The expected number of push transmissions that reach a node outside the clique is bounded by $\sqrt{n} \cdot \frac{3}{\sqrt{n}+2} \leq 3$. Similarly, the expected number of pull transmissions coming from outside is bounded by $3 \cdot \sqrt{n} \cdot \frac{1}{\sqrt{n}+2} \leq 3$. Hence every clique which is completely informed contributes at most 6 to the expected number of newly informed node.

For the single clique which is not fully informed nor fully uninformed, its contribution in expectation is at most $3\sqrt{n}$ newly informed nodes within the same clique and at most 6 newly informed nodes outside the clique. Hence,

$$\mathbf{E}\left[|I_{t+1}|\right] \leq |I_t| + (|I_t|/\sqrt{n}) \cdot 6 + 3 \cdot \sqrt{n} + 6 \leq |I_t| + 7 \cdot \sqrt{n},$$

as long as $|I_t| \leq n/2$. Therefore, $\mathbf{E}\left[|I_{t+\sqrt{n}/14-1}|\right] \leq 1 + (\sqrt{n}/14 - 1) \cdot 7\sqrt{n} < n/2$. Hence by Markov's inequality, $\mathbf{Pr}\left[|I_{t+\sqrt{n}/14-1}| \geq 2\mathbf{E}\left[|I_{t+\sqrt{n}/14-1}|\right]\right] \leq 1/2$. □

References

1. Avin, C., Koucký, M., Lotker, Z.: How to explore a fast-changing world (Cover time of a simple random walk on evolving graphs). In: Aceto, L., Damgård, I., Goldberg, L.A., Halldórsson, M.M., Ingólfsdóttir, A., Walukiewicz, I. (eds.) ICALP 2008, Part I. LNCS, vol. 5125, pp. 121–132. Springer, Heidelberg (2008)
2. Baumann, H., Crescenzi, P., Fraigniaud, P.: Parsimonious flooding in dynamic graphs. Distributed Computing 24(1), 31–44 (2011)

3. Censor-Hillel, K., Shachnai, H.: Fast information spreading in graphs with large weak conductance. SIAM J. Comput. 41(6), 1451–1465 (2012)
4. Chierichetti, F., Lattanzi, S., Panconesi, A.: Almost tight bounds for rumour spreading with conductance. In: Proc. 42nd STOC, pp. 399–408 (2010)
5. Chierichetti, F., Lattanzi, S., Panconesi, A.: Rumor spreading in social networks. Theor. Comput. Sci. 412(24), 2602–2610 (2011)
6. Clementi, A., Crescenzi, P., Doerr, C., Fraigniaud, P., Isopi, M., Panconesi, A., Pasquale, F., Silvestri, R.: Rumor spreading in random evolving graphs. In: Bodlaender, H.L., Italiano, G.F. (eds.) ESA 2013. LNCS, vol. 8125, pp. 325–336. Springer, Heidelberg (2013)
7. Clementi, A.E.F., Monti, A., Pasquale, F., Silvestri, R.: Information spreading in stationary markovian evolving graphs. IEEE Trans. Parallel Distrib. Syst. 22(9), 1425–1432 (2011)
8. Clementi, A.E.F., Silvestri, R., Trevisan, L.: Information spreading in dynamic graphs. In: Proc. 31st PODC, pp. 37–46 (2012)
9. Doerr, B., Fouz, M., Friedrich, T.: Social networks spread rumors in sublogarithmic time. In: Proc. 43rd STOC, pp. 21–30 (2011)
10. Dutta, C., Pandurangan, G., Rajaraman, R., Sun, Z., Viola, E.: On the complexity of information spreading in dynamic networks. In: Proc. 24th SODA, pp. 717–736 (2013)
11. Feige, U., Peleg, D., Raghavan, P., Upfal, E.: Randomized broadcast in networks. Random Struct. Algorithms 1(4), 447–460 (1990)
12. Fountoulakis, N., Panagiotou, K., Sauerwald, T.: Ultra-fast rumor spreading in social networks. In: Proc. 23rd SODA, pp. 1642–1660 (2012)
13. Georgiou, C., Gilbert, S., Guerraoui, R., Kowalski, D.R.: Asynchronous gossip. J. ACM 60(2), 11 (2013)
14. Giakkoupis, G.: Tight bounds for rumor spreading in graphs of a given conductance. In: Proc. 28th STACS, pp. 57–68 (2011)
15. Giakkoupis, G.: Tight bounds for rumor spreading with vertex expansion. In: Proc. 25th SODA, pp. 801–815 (2014)
16. Giakkoupis, G., Sauerwald, T.: Rumor spreading and vertex expansion. In: Proc. 23rd SODA, pp. 1623–1641 (2012)
17. Haeupler, B., Karger, D.R.: Faster information dissemination in dynamic networks via network coding. In: Proc. 30th PODC, pp. 381–390 (2011)
18. Haeupler, B., Kuhn, F.: Lower bounds on information dissemination in dynamic networks. In: Proc. 26th DISC, pp. 166–180 (2012)
19. Karp, R., Schindelhauer, C., Shenker, S., Vöcking, B.: Randomized rumor spreading. In: Proc. 41st FOCS, pp. 565–557 (2000)
20. Kesten, H., Sidoravicius, V.: The spread of a rumor or infection in a moving population. Annals of Probability 33, 2402–2462 (2005)
21. Kuhn, F., Lynch, N., Oshman, R.: Distributed computation in dynamic networks. In: Proc. 42nd STOC, pp. 513–522 (2010)
22. Lam, H., Liu, Z., Mitzenmacher, M., Sun, X., Wang, Y.: Information dissemination via random walks in d-dimensional space. In: Proc. 23rd SODA, pp. 1612–1622 (2012)
23. Leskovec, J., Lang, K.J., Dasgupta, A., Mahoney, M.W.: Statistical properties of community structure in large social and information networks. In: Proc. 17th WWW, pp. 695–704 (2008)
24. Mosk-Aoyama, D., Shah, D.: Fast distributed algorithms for computing separable functions. IEEEToIT 54(7), 2997–3007 (2008)

25. Peres, Y., Sinclair, A., Sousi, P., Stauffer, A.: Mobile geometric graphs: Detection, coverage and percolation. In: Proc. 22nd SODA, pp. 412–428 (2011)
26. Pettarin, A., Pietracaprina, A., Pucci, G., Upfal, E.: Tight bounds on information dissemination in sparse mobile networks. In: Proc. 30th PODC, pp. 355–362 (2011)
27. Pittel, B.: On spreading a rumor. SIAM J. Applied Math. 47(1), 213–223 (1987)
28. Sarma, A.D., Molla, A.R., Pandurangan, G.: Fast distributed computation in dynamic networks via random walks. In: Proc. 26th DISC, pp. 136–150 (2012)
29. Sauerwald, T., Stauffer, A.: Rumor spreading and vertex expansion on regular graphs. In: Proc. 22nd SODA, pp. 462–475 (2011)

Online Independent Set Beyond the Worst-Case: Secretaries, Prophets, and Periods*

Oliver Göbel[1], Martin Hoefer[2], Thomas Kesselheim[3],
Thomas Schleiden[1], and Berthold Vöcking[1]

[1] Dept. of Computer Science, RWTH Aachen University, Germany
{goebel,voecking}@cs.rwth-aachen.de
[2] Max-Planck-Institut für Informatik and Saarland University, Germany
mhoefer@mpi-inf.mpg.de
[3] Dept. of Computer Science, Cornell University, Ithaca, NY, USA
kesselheim@cs.cornell.edu

Abstract. We investigate online algorithms for maximum (weight) independent set on graph classes with bounded inductive independence number ρ like interval and disk graphs with applications to, e.g., task scheduling, spectrum allocation and admission control. In the online setting, nodes of an unknown graph arrive one by one over time. An online algorithm has to decide whether an arriving node should be included into the independent set.

Traditional (worst-case) competitive analysis yields only devastating results. Hence, we conduct a stochastic analysis of the problem and introduce a generic sampling approach that allows to devise online algorithms for a variety of input models. It bridges between models of quite different nature – it covers the secretary model, in which an adversarial graph is presented in random order, and the prophet-inequality model, in which a randomly generated graph is presented in adversarial order.

Our first result is an online algorithm for maximum independent set with a competitive ratio of $O(\rho^2)$ in all considered models. It can be extended to maximum-weight independent set by losing only a factor of $O(\log n)$, with n denoting the (expected) number of nodes. This upper bound is complemented by a lower bound of $\Omega(\log n / \log^2 \log n)$ showing that our sampling approach achieves nearly the optimal competitive ratio in all considered models. In addition, we present various extensions, e.g., towards admission control in wireless networks under SINR constraints.

1 Introduction

Various scheduling and resource allocation problems can be formulated as independent set problems for different graph classes, where nodes represent tasks or

---

* Full version appeared as [11]. Supported by DFG through Research Training Group AlgoSyn and UMIC Research Center at RWTH Aachen University, Cluster of Excellence M2CI at Saarland University, grant Ho 3831/3-1, and by a fellowship within the Postdoc-Programme of the German Academic Exchange Service (DAAD).

J. Esparza et al. (Eds.): ICALP 2014, Part II, LNCS 8573, pp. 508–519, 2014.
© Springer-Verlag Berlin Heidelberg 2014

requests that are connected by an edge if they are in mutual conflict. An independent set corresponds to a conflict-free subset of tasks or requests that can be executed or served simultaneously. In the Max-IS problem, the objective is to find an independent set of maximum cardinality. In the Max-Weight-IS problem, the nodes come with weights and the objective is to find an independent set of maximum total weight. Previous work on independent set problems is mostly concerned with offline optimization where the complete input is known in advance. However, in many application contexts, such as online admission control, requests arrive over time. An online algorithm has to make irrevocable decisions about acceptance or rejection of arriving requests without knowing future requests. This corresponds to online variants of independent set where nodes arrive over time. Each node comes with information about its incident edges to previously arrived nodes. The online algorithm has to decide which of the nodes should be included into the independent set and which should be rejected.

Unfortunately, even for rather restrictive (but in the context of scheduling and admission control highly relevant) graph classes like interval and disk graphs, the classical worst-case competitive analysis of online algorithms for independent set problems does not make much sense – we show a strong lower bound of $\Omega(n)$ in the full version. The alternative is a stochastic analysis, but it is already challenging to choose the *right* stochastic input model. On the one hand, it should allow online algorithms with meaningful performance guarantees and, on the other hand, should be reasonable from a practical point of view. We approach this challenge by studying not only one but a variety of stochastic input models. In particular, we study two stochastic input models – one inspired by the classic secretary problem with arrivals in random order, and one with arrivals in adversarial order and stochastic predictions based on so-called prophet-inequalities. Our study is complemented by a third model motivated by a practical admission control problem. In each of these models, an input sequence is generated by a different mix of stochastic and adversarial processes.

In our analysis we focus on graph classes of bounded inductive independence. The inductive independence number ρ of a graph is the smallest number for which there is an order $\prec$ such that for any independent set $S \subseteq V$ and any $v \in V$, we have $|\{u \in S \mid u \succ v$ and $\{u, v\} \in E\}| \leq \rho$. The inductive independence number is a useful concept and bounded in many prominent graph classes (for a further discussion see full version). We casually refer to interval and disk graphs, which have bounded inductive independence number 1 and 5. In disk graphs, this bound is achieved by ordering the disks according to their geometric diameter, beginning with the smallest disk. For any fixed disk, there can be at most five disks of equal or larger size intersecting the fixed disk without being in mutual conflict with each other. In the above mentioned online scenario, a node's position in the ordering with respect to its neighbors emerges as soon as the node arrives. The reason for referring to intervals and disks is that interval graphs are an established model for scheduling problems, where nodes are tasks with start and finishing times. Disk graphs generalize interval graphs from one to two dimensions. They are frequently used to describe spectrum allocation problems in wireless networks. In addition, we also study the independent set problem with respect to more advanced interference models for wireless networks [17], including models based on SINR (signal-to-interference-plus-noise ratio) that yield small bounds for inductive independence.

1.1   Description of the Models

We study the following stochastic input models:

- *Secretary Model:* The adversary defines a node-weighted graph $G = (V, E, w)$ with n nodes. (For simplicity, we assume integer weights. In case of Max-IS, all nodes have weight 1.) A priori, the algorithm knows n but neither G nor the weights. The nodes of G are presented in random order where each permutation of the nodes is assumed to occur equally likely.
- *Prophet-Inequality Model:* The adversary defines a graph $G = (V, E)$, and for each node a separate probability distribution on its weight. A priori, the algorithm knows G and the probability distributions but not their outcomes. The nodes of G are presented in adversarial order to the online algorithm where the actual weight is revealed only when the node arrives.
- *Period Model:* Let $G = (V, E, w)$ be an arbitrary node-weighted graph. Time is partitioned into periods. For each period $t = 1, 2, \ldots$ and each node $i \in V$, the adversary defines a probability $p_i^t \in [0, 1]$ such that, for $t \geq 2$, $p_i^t \in [p_i^{t-1}/c, p_i^{t-1} \cdot c]$, where $c \geq 1$ is assumed to be constant. Let X_i^t denote independent binary variables with $\mathbf{Pr}\left[X_i^t = 1\right] = p_i^t$ and $\mathbf{Pr}\left[X_i^t = 0\right] = 1 - p_i^t$. Let $V_t = \{i \in V \mid X_i^t = 1\}$. In every period $t \geq 2$, the nodes in V_t are presented in adversarial order to the online algorithm which aims at finding an independent set among the nodes in V_t. The probabilities p_i^t, the graph G and the nodes' arrival order in V_t are not assumed to be known a priori.

The first of these stochastic input models is inspired by the classical secretary problem where n secretaries are presented in random order. The second model is in spirit of problems with prophet-inequality, where candidates come in adversarial order but each candidate has a publicly known distribution of his weight. In the base case of either setting, one has to select one of n entities that are presented online and have to be accepted or rejected immediately at arrival. Each entity comes with a weight that is revealed upon arrival. The objective is to maximize the weight of the one accepted entity. In the secretary problem, weights are determined adversarily, but the adversary cannot fix the order of arrival. In the prophet-inequality model, weights are drawn at random from publicly known distributions, but the adversary can fix distributions and arrival order.

The third model is motivated by admission control protocols that have to decide about requests using stochastic knowledge from previous "corresponding" periods. For example, to make decisions in the time period on this week Friday from 9am to 10am, an admission control algorithm might want to learn from events in the same time window(s) of previous Friday(s). The graph G describes a potentially very large universe of possible requests. It might represent disks of various sizes at different positions which are requested with certain probabilities. An adversary fixes a distribution which generates requests by picking a set of nodes from G at random. Distributions might change over time but the deviation from period to period is bounded as specified by the global constant $c \geq 1$. The order in which the requests are presented in the period model is adversarial, which assumes implicitly that the order in which requests arrive within a period is unpredictable. We assume that the online algorithm has access to the distribution in the prophet-inequality model. In the period model it can only observe samples obtained from similar distributions.

We evaluate online algorithms in terms of the *competitive ratio* which is defined as $\mathbf{E}\,[\text{OPT}]\,/\mathbf{E}\,[\text{ALG}]$. OPT denotes the maximum weight of an independent set for the given instance, and ALG denotes the weight of the independent set selected by the online algorithm. The expectation is with respect to the stochastic input model and random coin flips of the algorithm. In case of the secretary model, the weight of OPT is fixed so that the competitive ratio simplifies to $\text{OPT}/\mathbf{E}\,[\text{ALG}]$. In case of the period model, we study the competitive ratio with respect to any fixed period $t \geq 2$.

1.2   Our Contribution

The models described above are conceptually quite different. To cope with these differences, we present a unifying *graph sampling model*, where the online algorithm is initially equipped with a sample graph generated from a distribution that is stochastically similar to the distribution of the input graph. In Section 2, we introduce this model formally and show that it can be simulated by each of the other input models. This approach enables us to devise online algorithms achieving, up to small constant factors, the same competitive ratio for all models.

Based on the graph sampling model, we present an online algorithm for Max-IS with competitive ratio $O(\rho^2)$ for graphs with inductive independence number ρ in Section 3. As a consequence, we achieve competitive ratio $O(1)$ for independent set on interval and disk graphs in all considered input models. Our analytic approach shows that we do not require specific stochastic assumptions in order to break through the $\Omega(n)$ worst-case lower bound. Indeed, the same kind of online algorithm performs well under a variety of stochastic assumptions.

In Section 4, we present upper and lower bounds for Max-Weight-IS. We adapt the algorithm for Max-IS to Max-Weight-IS and obtain a competitive ratio of $O\left(\rho^2 \log n\right)$ in the graph sampling model (and, hence, all of the models), where n denotes an upper bound on the (expected) number of nodes that are presented to the algorithm (in the considered period). As a technical highlight, we show that this bound is almost best possible for interval and disk graphs. In particular, we prove a lower bound for Max-Weight-IS on interval graphs of order $\Omega\left(\log n/\log^2 \log n\right)$ in the secretary and the prophet-inequality models. The same bound applies to the period and the unifying model, and it holds even for randomized algorithms.

Motivated by admission control and scheduling applications, in Section 5 we additionally study a variant in which nodes have different arrival and departure times. The adversary is allowed to fix in advance the conflict graph and for each node a time interval in which the node is present. Only the nodes being active at the same time have to be independent. We show how to solve variants of Max-(Weight-)IS with arrival and departure times by using the previous algorithms as subroutines, losing only a factor $O(\log n)$ in the competitive ratio.

Finally, in Section 6 we show how to transfer our results to edge-weighted conflict graphs. This way, more sophisticated wireless interference models can be analyzed, for example, the commonly studied ones based on SINR constraints. We present an $O\left(\rho^2 \log^2 n\right)$-competitive algorithm for Max-IS in this case.

1.3   Related Work

In the 1970s, Frank [10] gave an algorithm for Max-IS on chordal graphs and, thus, also on interval graphs. For disk graphs, the problem is NP-hard but admits a PTAS [7]. These graph classes are extended to graphs with bounded inductive independence number in [1,27]. Irani's work [18] enables coloring these graphs online with $O(\rho\chi(G)\log n)$ colors.

The use of disk graphs was often motivated by interference in wireless networks. Also a number of approximation algorithms using the more realistic SINR model exist for different variants of maximizing the number of successful simultaneous transmissions [12,16,19]. Interestingly, Hoefer et al. [17] showed that any of these problems can also be described as a maximum independent set problem in an edge-weighted conflict graph. Moreover, the inductive independence number of these graphs turns out to be bounded by a constant or $O(\log n)$ (see also [15]). Besides, also the graphs arising from a number of further simple interference models have a constant inductive independence number as well.

To bypass the trivial lower bound of $\Omega(n)$ in online worst-case optimization, often restricted instances are considered. For example, for interval graphs of value density bounded by k, Koren and Shasha [23] give an optimal $(1 + \sqrt{k})^2$-competitive algorithm. In a similar spirit, Fanghänel et al. [8] use geometric parameters to get tight bounds in SINR.

For the problem of selecting the highest ranked entity, which is pointless in a worst-case setting, optimal algorithms have already been presented in [6] for the secretary model and in [25] for the prophet-inequality settings. Both models have strong connections to online auctions, where bidders arrive one by one, have to be served, and incentive compatibility is desired, see [13,14]. Recently, they have been analyzed with respect to non-trivial combinatorial optimization problems. The matroid independent set problem was considered in secretary [4] and prophet-inequality models [22], similarly variants of matching and set packing problems were studied in both models [24,20,2]. Our algorithm for unweighted independent set is inspired by [24], which uses a greedy algorithm to guide the online-computation of a weighted matching.

More general packing problems have been studied in the secretary model as well, e.g. the knapsack problem by Babaioff et al. [3]. Allowing multiple constraints, the problem becomes solving linear packing problems online. Though in [21,9,5,26] almost-optimal solutions to this kind of online problems are shown to exist when given capacities are large enough, even under this restricting assumption the described algorithms are not applicable for independent set problems.

2   Graph Sampling Model

In this section, we present a technically motivated but rather intuitive stochastic input model that bridges between the three input models from the introduction. In our *graph sampling model*, the online algorithm is initially equipped with a sample graph that is stochastically similar to an input graph presented subsequently in online fashion. In the following, we first describe the properties of this model formally and then explain how it can be simulated by each of the other three models. Thereby, competitive ratios achieved for the graph sampling model hold for those models, too.

Let $G = (V, E)$ be an arbitrary graph from the considered class. We derive two induced subgraphs, the *input graph* $G[V^I]$ with weights w^I and the *sample graph* $G[V^S]$ with weights w^S, where $V^I, V^S \subseteq V$. The sets V^I and V^S are generated implicitly by drawing non-negative weights $w^I(v)$ and $w^S(v)$ at random, for each node $v \in V$. To ease notation, we assume that node weights are integral.

We set $V^I = \{v \in V | w^I(v) > 0\}$ and $V^S = \{v \in V | w^S(v) > 0\}$. The weights w^I and w^S need not to be drawn according to exactly identical distributions, but they must satisfy the following assumptions.[1]

- *Stochastic similarity:* For every $v \in V$ and every integer $b > 0$, it holds $\mathbf{Pr}\left[w^I(v) = b\right] \leq c\mathbf{Pr}\left[w^S(v) = b\right]$ and $\mathbf{Pr}\left[w^S(v) = b\right] \leq c\mathbf{Pr}\left[w^I(v) = b\right]$ with $c \geq 1$ denoting a fixed, constant term.
- *Stochastic independence:* For every $v \in V$, the weights $w^I(v)$ and $w^S(v)$ do not depend on the weights w^I and w^S of other nodes.

We explicitly point out that for any node $v \in V$ the weights $w^I(v)$ and $w^S(v)$ might be correlated. These possible dependencies are crucial for the simulation of the graph sampling model by the secretary model.

Let us now describe how the input is presented to the online algorithm. A priori, the algorithm does not know G, the weights w^I, w^S, or even the probability distributions for the weights. As initial input, it receives the sample graph $G[V^S]$ together with weights w^S for the nodes in V^S and the order $\prec$ among these. Nodes in V^I arrive one by one in adversarial order. When a node $v \in V^I$ arrives, the algorithm gets to know the weight $w^I(v)$ as well as the edges from v to nodes in V^S and to the nodes in V^I that arrived before v and their relative order with respect to $\prec$. If v is also contained in V^S, it is revealed that these

---

[1] We show competitive ratios that do not depend on the size of the graph G, but only on the expected size of the graph $G[V^I]$ presented to the online algorithm. For this reason, the model can be extended to infinite graphs representing, e.g., all possible disks in Euclidean space. In such an extension, probability distributions might be continuous rather than discrete. Only for notational simplicity, we focus on finite graphs, integer weights, and discrete probability distributions.

are identical. Based on this information, the online algorithm has to irrevocably decide whether v should be included into the independent set or rejected.

The *competitive ratio* of an algorithm in the graph sampling model is defined as $\mathbf{E}\left[\text{OPT}(w^I)\right]/\mathbf{E}\left[\text{ALG}\right]$, where $\text{OPT}(w^I)$ is the maximum weight of an independent set with respect to the weights w^I. The next proposition shows that an upper bound on the competitive ratio for the graph sampling model implies upper bounds on the competitive ratios for the other stochastic input models.

Proposition 1. *If there is an α-competitive algorithm for Max-IS (Max-Weight-IS) in the graph sampling model, then there are $O(\alpha)$-competitive algorithms for Max-IS (Max-Weight-IS) in the prophet-inequality model, the period model, and the secretary model.*

Due to space limitations, parts of the proof of this proposition as well as the other missing proofs are presented in the full version.

Proof (Sketch for Secretary Model). Let $\alpha = \alpha(c)$ denote the competitive ratio in the graph sampling model with c denoting the constant term from the similarity condition. We draw a random number k from the Binomial distribution $B(n, \frac{1}{2})$ with $n = |V|$ and set $w^I(v) = 0$, $w^S(v) = w(v)$, for the first k nodes, and $w^I(v) = w(v)$, $w^S(v) = 0$, for the remaining nodes, where $w(v)$ denotes the adversarial weights from the secretary model. As nodes arrive in random order and k is determined by the binomial distribution, this is stochastically equivalent to choosing weight tuples $(w^I(v), w^S(v))$ independently, uniformly at random from $\{(0, w(v)), (w(v), 0)\}$, for all nodes $v \in V$. Thus, stochastic independence and stochastic similarity (with $c = 1$) are satisfied. The online algorithm is $\alpha(1)$-competitive with respect to $G[V^I]$, that is, $\mathbf{E}\left[\text{ALG}\right] \geq \mathbf{E}\left[\text{OPT}(w^I)\right]/\alpha(1)$. Furthermore, by symmetry, $\mathbf{E}\left[\text{OPT}(w^I)\right] = \mathbf{E}\left[\text{OPT}(w^S)\right]$, which implies $\text{OPT}(w) = \mathbf{E}\left[\text{OPT}(w^I + w^S)\right] \leq \mathbf{E}\left[\text{OPT}(w^I)\right] + \mathbf{E}\left[\text{OPT}(w^S)\right] = 2\mathbf{E}\left[\text{OPT}(w^I)\right]$. Consequently, $\mathbf{E}\left[\text{ALG}\right] \geq \mathbf{E}\left[\text{OPT}(w)\right]/2\alpha(1)$ so that the competitive ratio for the secretary model is upper-bounded by $2\alpha(1)$. □

Proposition 1 allows to focus on the graph sampling model when proving upper bounds on the competitive ratio. The following lemma shows that it is indeed sufficient to compare the independent set computed by the algorithm to the maximum-weight independent set with respect to w^S instead of w^I. For the purpose of upper bounding the competitive ratio within constant factors, it suffices to upper-bound $\mathbf{E}\left[\text{OPT}(w^S)\right]/\mathbf{E}\left[\text{ALG}\right]$ instead of $\mathbf{E}\left[\text{OPT}(w^I)\right]/\mathbf{E}\left[\text{ALG}\right]$.

Lemma 1. $\mathbf{E}\left[\text{OPT}(w^S)\right] \geq \frac{1}{c}\mathbf{E}\left[\text{OPT}(w^I)\right]$.

3   Unweighted Independent Set

We study Max-IS on graphs with bounded inductive independence number in the graph sampling model. We consider the input model from Section 2 restricted to $\{0, 1\}$-weights and assume that the underlying graph $G = (V, E)$ has bounded

---

Algorithm 1: Unweighted Online-Max-IS

Input: $G[V^S]$

$M_1, M_2, M_3, M_4 \leftarrow \emptyset$;

forall the $v \in V^S$ *in order according to* $\prec$ **do**
    if $M_1 \cup \{v\}$ *is independent* **then** $M_1 \leftarrow M_1 \cup \{v\}$

forall the $v \in V^I$ *in order of arrival* **do**
    if $\nexists u \in M_1, u \prec v$ with $\{u, v\} \in E$ **then** $M_2 \leftarrow M_2 \cup \{v\}$;
    if $v \in M_2$ **then** w/prob $q := \frac{1}{2pc}$: $M_3 \leftarrow M_3 \cup \{v\}$;
    if $v \in M_3$ and $\nexists u \in M_4$ s.t. $\{v, u\} \in E$ **then** $M_4 \leftarrow M_4 \cup \{v\}$

return M_4;

---

inductive independence number $\rho \geq 1$. The restriction to $\{0, 1\}$-weights simplifies the graph sampling model as follows. One picks two subsets V^I and V^S from V at random. The induced graphs $G[V^I]$ and $G[V^S]$ are the input and the sample graph, respectively. For a node $v \in V$, the events $v \in V^I$ and $v \in V^S$ might be correlated. By stochastic independence, however, these events do not depend on events for other nodes. Stochastic similarity yields for $\{0, 1\}$-weights

$$\frac{1}{c} \mathbf{Pr} \left[v \in V^S \right] \leq \mathbf{Pr} \left[v \in V^I \right] \leq c \, \mathbf{Pr} \left[v \in V^S \right] \ . \tag{1}$$

Our online algorithm applies a greedy algorithm for independent set to the sample graph $G[V^S]$ and employs the output of this algorithm to guide the online computation on the input graph $G[V^I]$. This technique is similar in spirit to the one used for online matching in [24]. However, in our case a single node can introduce arbitrarily many conflicts. Fortunately, applying a more global perspective allows us to show that the overall number of nodes that have to be removed is bounded nonetheless. In the offline setting, a greedy algorithm for independent set on graphs with bounded inductive independence number starts with $I = \emptyset$ and considers all nodes of V iteratively according to $\prec$. It adds a node to I when it is not in conflict with other nodes already in I. This yields a ρ-approximation due to the bound on the inductive independence number: Selecting a node not in the optimal solution prevents at most ρ many neighbors from being selected to I, cf., e.g., [1,27].

In more detail, Algorithm 1 computes two sets $M_1 \subseteq V^S$ and $M_2 \subseteq V^I$. M_1 is the output of the greedy algorithm applied to $G[V^S]$. M_2 is obtained by going through the nodes in V^I in adversarial order and checking for each $v \in V^I$ whether it would have been taken by the greedy algorithm on $G[V^S \cup \{v\}]$. In our analysis, we show that the expected value of M_2 is of the same order as the expected value of M_1 and, hence, an $O(\rho)$-approximation of $OPT(w^S)$. By Lemma 1, this implies that M_2 is an $O(\rho)$-approximation of $\mathbf{E} \left[OPT(w^I) \right]$. Unfortunately, however, M_2 is not an independent set. Feasibility is achieved by two further steps: We first obtain a set M_3 by randomly sparsifying M_2, which loses another factor of $O(\rho)$ in the competitive ratio. The remaining conflicts are resolved by moving to a set M_4 (the output of the algorithm) only those nodes that are not adjacent to nodes previously inserted into M_4. A stochastic analysis of the conflicts in M_3 shows that this final resolution step loses only a constant factor in the competitive ratio.

Theorem 1. *Algorithm 1 is $4c^3\rho^2$-competitive.*

Proof (Sketch). The set M_1 is determined by applying the greedy algorithm to $G[V^S]$ and, hence, it is a ρ-approximation of $\mathrm{OPT}(w^S)$. Combining this with Lemma 1 gives $\mathbf{E}\left[\|M_1\|\right] \geq \frac{1}{c\rho}\mathbf{E}\left[\mathrm{OPT}(w^I)\right]$. By applying *stochastic similarity*, we obtain $\mathbf{E}\left[\|M_2\|\right] \geq \frac{1}{c}\mathbf{E}\left[\|M_1\|\right]$. Furthermore, sparsifying from M_2 to M_3 causes losing the factor $q = \frac{1}{2\rho c}$, i.e., $\mathbf{E}\left[\|M_3\|\right] = q\mathbf{E}\left[\|M_2\|\right]$. Thus, we obtain $\mathbf{E}\left[\|M_3\|\right] \geq \frac{q}{c^2\rho}\mathbf{E}\left[\mathrm{OPT}(w^I)\right]$.

It remains to analyze the final conflict resolution, where only nodes without any conflict are selected in the final output set. The consequence of this approach is that for each conflict which would appear in the offline setting, exactly the node arriving first in the online setting is chosen by our algorithm. We define $C = \{\{u, v\} \in E \mid u, v \in M_3\}$. Note that the size of C is an upper bound to the overall number of nodes that are lost in the conflict resolution.

We show that $\mathbf{E}\left[\|C\|\right] \leq \mathbf{E}\left[\|M_3\|\right]q\rho c$. To this end, we define $C_v = \{u \in V \mid \{u, v\} \in E \text{ and } v \prec u\}$. Using *stochastic similarity* one more time, we get $\mathbf{E}\left[\|C_v \cap M_3\| \mid v \in M_3\right] \leq q\rho c$ and derive $\mathbf{E}\left[\|C\|\right] \leq \mathbf{E}\left[\|M_3\|\right]q\rho c$.

Since the output is the set M_3 lowered by at most one node per existing conflict, $\mathbf{E}\left[\|M_4\|\right] \geq \mathbf{E}\left[\|M_3\|\right] - \mathbf{E}\left[\|C\|\right]$ holds. Combining this with the bound on $\mathbf{E}\left[\|C\|\right]$ gives $\mathbf{E}\left[\|M_4\|\right] \geq (1 - q\rho c)\frac{q}{c^2\rho}\mathbf{E}\left[\mathrm{OPT}(w^I)\right] = \frac{1}{4c^3\rho^2}\mathbf{E}\left[\mathrm{OPT}(w^I)\right]$.   □

4   Weighted Independent Set

In this section, we turn to the Max-Weight-IS problem. We construct an algorithm by dividing nodes into roughly $\log n$ weight classes and running the algorithm for the unweighted problem on a randomly selected class. While this is a common approach in online maximization, we have to deal with technical difficulties here. Neither $|V^I|$ nor the maximum weight are known a priori, and the sample is generated by a stochastically similar rather than by the same distribution. For details see the full version.

Theorem 2. *There is an $O(\alpha \cdot \log(n))$-competitive algorithm for Max-Weight-IS, where $\alpha = O(\rho^2)$ is the competitive ratio of Algorithm 1 and $n = \mathbf{E}\left[|V^I|\right]$.*

The approach of constructing weight classes is fairly generic. However, the competitive ratio turns out to be almost optimal not only for the general problem but even for all special cases mentioned in the introduction.

Theorem 3. *For any algorithm for online maximum-weight independent set, we have $\mathbf{E}\left[\mathrm{ALG}\right] = \Omega\left(\frac{\log^2\log n}{\log n}\right)\mathbf{E}\left[\mathrm{OPT}\right]$, even in interval graphs, and even in the secretary and prophet-inequality model.*

Proof (Sketch). We use a fixed graph known to the algorithm in advance. The node weights are drawn independently from probability distributions known in advance. The precise outcome, however, is only revealed at time of arrival. The times of arrival are in uniform random order. This way, we restrict the adversary to become weaker than in both the secretary and the prophet-inequality models.

We set $d = \Theta(\log^2 n/(\log^2 \log n))$ and construct the graph by nesting intervals into each other, starting with an interval of length 1 and continuing by always putting d intervals of length d^{-i} next to each other into an interval of length d^{-i+1}. The resulting graph is a complete d-ary tree with additional "shortcuts" on the paths from the root to the leaves skipping over some levels. The total number of levels is $h = \Theta\left(\frac{\log n}{\log \log n}\right)$. For a node v on level i, we set the weight $w(v)$ at random to d^{h-i} with probability $p = \frac{1}{2h}$ and 0 otherwise.

It is helpful to consider the paths from the root node to the leaves, which include h nodes each. There are in total d^h such paths. In every independent set, there can be at most one node on any path. In case a node has non-zero weight, its weight directly corresponds to the number of paths it lies on. Therefore, we can equivalently express the weight of an independent set by the number of paths that are covered, i.e., on which a non-zero node is selected.

The optimal online algorithm on this graph is the following HIGHSTAKES policy. This algorithm accepts a node if and only if it has non-zero weight and there is no more ancestor to come that could cover this node. In other words, we reject a node of non-zero weight if there is a chance that an ancestor could still be selected (because it has not arrived and none of its other descendants have been selected so far). No online algorithm can be better than HIGHSTAKES. We show this by proving inductively that at any point in time, no matter which nodes have been selected so far, the conditional expectation is maximized by continuing following the HIGHSTAKES policy.

HIGHSTAKES accepts the jth node on a path only if it has non-zero weight and if the $j-1$ nodes on higher levels occur before this node in the random order. The combined probability of this event is $\frac{p}{j}$. Therefore, the overall probability that any node on a path of length h is accepted is at most $\sum_{j=1}^{h} \frac{p}{j} = O(p \log h)$. The expected value of the solution computed by HIGHSTAKES is exactly the expected number of paths that are covered. By the above considerations, we get
$$\mathbf{E}[\text{ALG}] \leq d^h O(p \log h) = O\left(d^h \frac{\log^2 \log n}{\log n}\right).$$

On the other hand, we get a feasible offline solution by greedily accepting vertices going down the tree. In this procedure a path of length h is only left uncovered if all nodes on it have zero weight. This happens with probability $(1-p)^h = (1 - \frac{1}{2h})^h \leq \frac{1}{\sqrt{e}} = \Omega(1)$. Therefore, this solution has value $\mathbf{E}[\text{OPT}] = \Omega(d^h)$. In total, $\mathbf{E}[\text{ALG}] = \Omega\left(\frac{\log^2 \log n}{\log n}\right) \mathbf{E}[\text{OPT}]$, showing the claim. $\qquad\square$

5   Arrivals and Departures

Interval graphs are often motivated by problems in which two tasks cannot be processed at the same time. Disk graphs in turn capture the requirement of spatial separation. In this section, we introduce an approach to combine both temporal and spatial separation. Again, we assume that requests are nodes in a graph $G = (V, E)$, which models the geometric properties. Furthermore, each node $v \in V$ in this graph has an arrival time $\text{arrival}(v) \in \mathbb{R}$ and a departure time $\text{departure}(v)$. We say that $u \in V$ and $v \in V$ are conflicting if $\{u, v\} \in E$

and $[\text{arrival}(u), \text{departure}(u)] \cap [\text{arrival}(v), \text{departure}(v)] \neq \emptyset$. Still, we make no assumption on the order in which requests in a period are presented to the online algorithm. In particular, this includes the most natural case, in which requests are ordered by arrival times.

Given an algorithm $\mathcal{A}$ that approximately solves the online (weighted) independent set problem on the graph G with competitive ratio γ, Algorithm SPLIT (see full version) achieves $O(\gamma \log n)$ as the overall competitive ratio, where $n = |V^I|$. In the previous sections, we devised such algorithms for the unweighted problem with $\gamma = O(\rho^2)$ and the weighted variant with $\gamma = O(\rho^2 \log n)$.

Theorem 4. *The algorithm* SPLIT *is* $O(\gamma \log n)$-*competitive.*

6   Edge-Weighted Conflict Graphs

To capture more realistic wireless interference models, such as, e.g., the ones based on SINR, we extend our approach to edge-weighted conflict graphs, following [17]. We assume that between any pair of nodes $u, v \in V$, there exists a (directed) weight $w(u, v) \in [0, 1]$. We define $S \subseteq V$ as independent set if $\sum_{u \in S} w(u, v) < 1$ for all $v \in S$. Now the inductive independence number is the smallest number ρ for which there is an ordering $\prec$ such that for all independent sets S, we have $\sum_{u \in S, u \succ v} w(u, v) + w(v, u) \leq \rho$ for all $v \in V$.

The major challenge compared to the case of unweighted conflict graphs is that conflicts become asymmetric. In unweighted graphs, node u has a conflict with node v if and only if node v has a conflict with node u. In edge-weighted conflict graphs, there might be many nodes $u_1, u_2, \ldots$ that can feasibly be placed into the independent set when considering previously added nodes, but this might violate feasibility of some other node v added before. Our solution to this issue is as follows (see full version for details): After computing a set M_3 in a similar fashion as in Algorithm 1, we apply an additional randomized selection step to build the final set M_4, losing only a polylogarithmic factor in the competitive ratio. The output set M_4 of the algorithm is guaranteed to be feasible with high probability. Therefore, one can guarantee feasibility without loss by adding an arbitrary conflict-resolution filter at the end.

Theorem 5. *There is an* $O(\rho^2 \log^2 n)$-*competitive competitive algorithm for Online Max-IS in edge-weighted graphs.*

References

1. Akcoglu, K., Aspnes, J., DasGupta, B., Kao, M.-Y.: Opportunity cost algorithms for combinatorial auctions. CoRR, cs.CE/0010031 (2000)
2. Alaei, S., Hajiaghayi, M., Liaghat, V.: Online prophet-inequality matching with applications to ad allocation. In: Proc. 13th EC, pp. 18–35 (2012)
3. Babaioff, M., Immorlica, N., Kempe, D., Kleinberg, R.D.: A knapsack secretary problem with applications. In: Charikar, M., Jansen, K., Reingold, O., Rolim, J.D.P. (eds.) RANDOM 2007 and APPROX 2007. LNCS, vol. 4627, pp. 16–28. Springer, Heidelberg (2007)
4. Babaioff, M., Immorlica, N., Kleinberg, R.: Matroids, secretary problems, and online mechanisms. In: Proc. 18th SODA, pp. 434–443 (2007)

5. Devanur, N.R., Jain, K., Sivan, B., Wilkens, C.A.: Near optimal online algorithms and fast approximation algorithms for resource allocation problems. In: Proc. 12th EC, pp. 29–38 (2011)
6. Dynkin, E.B.: The optimum choice of the instant for stopping a markov process. Sov. Math. Dokl 4, 627–629 (1963)
7. Erlebach, T., Jansen, K., Seidel, E.: Polynomial-time approximation schemes for geometric intersection graphs. SIAM J. Comput. 34(6), 1302–1323 (2005)
8. Fanghänel, A., Geulen, S., Hoefer, M., Vöcking, B.: Online capacity maximization in wireless networks. In: Proc. 22nd SPAA, pp. 92–99 (2010)
9. Feldman, J., Henzinger, M., Korula, N., Mirrokni, V.S., Stein, C.: Online stochastic packing applied to display ad allocation. In: Proc. 18th ESA, pp. 182–194 (2010)
10. Frank, A.: Some polynomial algorithms for certain graphs and hypergraphs. In: Proc. 5th British Combinatorial Conference, pp. 211–226 (1975)
11. Göbel, O., Hoefer, M., Kesselheim, T., Schleiden, T., Vöcking, B.: Online independent set beyond the worst-case: Secretaries, prophets, and periods. CoRR, abs/1307.3192 (2013)
12. Goussevskaia, O., Wattenhofer, R., Halldórsson, M.M., Welzl, E.: Capacity of arbitrary wireless networks. In: Proc. 28th INFOCOM, pp. 1872–1880 (2009)
13. Hajiaghayi, M., Kleinberg, R., Parkes, D.C.: Adaptive limited-supply online auctions. In: Proc. 5th EC, pp. 71–80 (2004)
14. Hajiaghayi, M., Kleinberg, R.D., Sandholm, T.: Automated online mechanism design and prophet inequalities. In: Proc. 22nd AAAI, pp. 58–65 (2007)
15. Halldórsson, M.M., Holzer, S., Mitra, P., Wattenhofer, R.: The power of non-uniform wireless power. In: Proc. 24th SODA, pp. 1595–1606 (2013)
16. Halldórsson, M.M., Mitra, P.: Wireless capacity with oblivious power in general metrics. In: Proc. 22nd SODA, pp. 1538–1548 (2011)
17. Hoefer, M., Kesselheim, T., Vöcking, B.: Approximation algorithms for secondary spectrum auctions. In: Proc. 23rd SPAA, pp. 177–186 (2011)
18. Irani, S.: Coloring inductive graphs on-line. Algorithmica 11(1), 53–72 (1994)
19. Kesselheim, T.: A constant-factor approximation for wireless capacity maximization with power control in the SINR model. In: Proc. SODA, pp. 1549–1559 (2011)
20. Kesselheim, T., Radke, K., Tönnis, A., Vöcking, B.: An optimal online algorithm for weighted bipartite matching and extensions to combinatorial auctions. In: Bodlaender, H.L., Italiano, G.F. (eds.) ESA 2013. LNCS, vol. 8125, pp. 589–600. Springer, Heidelberg (2013)
21. Kesselheim, T., Radke, K., Tönnis, A., Vöcking, B.: Primal beats dual on online packing LPs in the random-order model. In: Proc. 46th STOC (2014)
22. Kleinberg, R., Weinberg, S.M.: Matroid prophet inequalities. In: Proc. 44th STOC, pp. 123–136 (2012)
23. Koren, G., Shasha, D.: D^{over}: An optimal on-line scheduling algorithm for overloaded uniprocessor real-time systems. SIAM J. Comput. 24(2), 318–339 (1995)
24. Korula, N., Pál, M.: Algorithms for secretary problems on graphs and hypergraphs. In: Albers, S., Marchetti-Spaccamela, A., Matias, Y., Nikoletseas, S., Thomas, W. (eds.) ICALP 2009, Part II. LNCS, vol. 5556, pp. 508–520. Springer, Heidelberg (2009)
25. Krengel, U., Sucheston, L.: Semiamarts and finite values. Bull. Amer. Math. Soc. 83, 745–747 (1977)
26. Molinaro, M., Ravi, R.: Geometry of online packing linear programs. In: Czumaj, A., Mehlhorn, K., Pitts, A., Wattenhofer, R. (eds.) ICALP 2012, Part I. LNCS, vol. 7391, pp. 701–713. Springer, Heidelberg (2012)
27. Ye, Y., Borodin, A.: Elimination graphs. In: Albers, S., Marchetti-Spaccamela, A., Matias, Y., Nikoletseas, S., Thomas, W. (eds.) ICALP 2009, Part I. LNCS, vol. 5555, pp. 774–785. Springer, Heidelberg (2009)

Optimal Competitiveness for Symmetric Rectilinear Steiner Arborescence and Related Problems

Erez Kantor[1],[*] and Shay Kutten[2],[**]

[1] MIT CSAIL, Cambridge, MA
erezk@csail.mit.edu
[2] Technion, Haifa 32000, Israel
kutten@ie.technion.ac.il

Abstract. We present optimal competitive algorithms for two interrelated known problems involving Steiner Arborescence. One is the continuous problem of the Symmetric Rectilinear Steiner Arborescence ($SRSA$), whose online version was studied by Berman and Coulston as a symmetric version of the known Rectilinear Steiner Arborescence (RSA) problem. A very related, but discrete problem (studied separately in the past) is the online Multimedia Content Delivery (MCD) problem on line networks, presented originally by Papadimitriou, Ramanathan, and Rangan. An efficient content delivery was modeled as a low cost Steiner arborescence in a grid of network×time they defined. We study here the version studied by Charikar, Halperin, and Motwani (who used the same problem definitions, but removed some constraints on the inputs). The bounds on the competitive ratios introduced separately in the above papers were similar for the two problems: $O(\log N)$ for the continuous problem and $O(\log n)$ for the network problem, where N was the number of terminals to serve, and n was the size of the network. The lower bounds were $\Omega(\sqrt{\log N})$ and $\Omega(\sqrt{\log n})$ correspondingly.

Berman and Coulston conjectured that both the upper bound and the lower bound could be improved. We disprove this conjecture and close these quadratic gaps for both problems. We present deterministic algorithms that are competitive optimal: $O(\sqrt{\log N})$ for $SRSA$ and $O(\min\{\sqrt{\log n}, \sqrt{\log N}\})$ for MCD, matching the lower bounds for these two online problems. We also present a $\Omega(\sqrt[3]{\log n})$ lower bound on the competitiveness of any randomized algorithm that solves the online MCD problem.

1 Introduction

We present optimal online algorithms for two known interrelated problems involving Steiner Arborescences. The continuous one is the *Symmetric Rectilinear Steiner Arborescence (SRSA)* problem [3,5]. The online Steiner *arborescence* problems are

[*] Supported in a part by AFOSR FA9550-13-1-0042 and by NSF grants Nos. CCF-1217506, CCF-0939370 and CCF-AF-0937274.
[**] Supported in part by the ISF and by the Technion Gordon Center.

J. Esparza et al. (Eds.): ICALP 2014, Part II, LNCS 8573, pp. 520–531, 2014.
© Springer-Verlag Berlin Heidelberg 2014

useful in modeling the time dimension in a process. Intuitively (see, e.g. Papadim-itriou at al., [11]), directed edges represent the passing of time. Since there is no way to go back in time in such processes, all the directed edges are directed away from the initial state of the problem, resulting in an arborescence. Additional examples given in the literature included processes in constructing a Very Large Scale In-tegrated electronic circuits (VLSI), optimization problems computed in iterations (where it was not feasible to return to results of earlier iterations), dynamic pro-gramming, and problems involving DNA, see, e.g. [3,5,8,2].

The $SRSA$ Problem: A rectilinear line segment in the plane is either horizontal or vertical. A rectilinear path contains only rectilinear line segments. This path is also *y-monotone* if during the traversal, the y coordinates of the successive points are never decreasing. The input is a set of *requests* $\mathcal{R} = \{(x_1, y_1), ..., (x_N, y_N)\}$ called Steiner terminals (or points) in the positive quadrant of the plane. A feasible solution to the problem is a set of rectilinear segments connecting all the N terminals to the origin, where the path from the origin to each terminal is a rectilinear y-monotone path. The goal is to find a feasible solution in which the sum of lengths of all the segments is the minimum possible. If we also had require the path connecting the origin to any point to be some shortest path (both x-monotone and y-monotone), then the problem would have been referred to as the *Rectilinear Steiner Arborescence (RSA)* problem [9,14,3,10,6].

Online Model: In the *online* version of $SRSA$ [3], the given points are pre-sented to the algorithm with nondecreasing y-coordinates. After receiving a new given point (terminal), the on-line $SRSA$ algorithm must extend the existing arborescence solution to incorporate the new point. There are two limitations: (1) a line, once drawn, cannot be deleted, and (2) lines can only be drawn in the region between the previous given point y-coordinates and upwards.

A very related, but discrete problem is the online *Multimedia Content Deliv-ery (MCD)* problem on line networks, presented originally by Papadimitriou, Ramanathan, and Rangan [11]. (The formal definitions appear in Section 2). The MCD problem considered a movie residing initially at some *origin* node and a set of requests, each arriving at some node at some time. *Serving* a request at a node v at time t meant delivering a movie copy to the requesting node v from some node u which has a copy at time t; or delivering a copy to v at some time $t' < t$ from u (which has a copy at time t') and then storing the copy at v from time t' until time t.

There are two types of costs, the *delivery cost* associated with the cost of sending a movie copy over the network edges and the *storage cost* associated with the cost of storing a copies at the nodes. MCD captured the tradeoff between the storage and the delivery costs. The goal is to serve all the requests with minimal costs. An example of an algorithm would be to store, always, a movie copy at the origin and serve every request by delivering a copy from the origin at the time of the request. Such an algorithm would incur a high delivery cost. Alternatively, a copy already delivered to some nodes, could be stored there, and delivered later from there. This could reduce delivery costs, but incur storage costs. Papadimitriou et al. defined a grid of network×time (detailed in Section

2), were a request at a node u at time t was translated into a grid point (u, t). A copy stored at a node u they modeled as an edge along the "time dimension" in the above grid (from grid point (u, t) to grid point $(u, t + 1)$), while the delivery they modeled as edges along the "network dimension". A solution (an efficient content delivery plan), was modeled as a low cost Steiner arborescence leading from the origin (node 0 at time 0) to all the requests (the Steiner points). Since time is irreversible, their Steiner tree was (semi) directed away from the origin.

Papadimitriou et al. assumed some constraints on the input. Those constraints were lifted in the paper of Charikar, Halperin and Motwani [4]. The upper bound (in Charikar et al.) on the competitive ratio was $O(\log n)$ for the network problem (where n was the size of the network) and the lower bound was $\Omega(\sqrt{\log n})$. The bounds of Berman and Coulston for $SRSA$ were very similar. The upper bound was $O(\log N)$, where N was the number of terminals[1]. The lower bound was $\Omega(\sqrt{\log N})$. Clearly, these upper bounds were quadratic in the lower bounds. Berman and Coulston conjectured that both the upper bound and the lower bound could be improved.

Our Results. In this paper, we disprove the above conjecture and close these quadratic gaps for both problems. We first present an $O(\sqrt{\log n})$ deterministic competitive algorithm for MCD on the line. We then translate the online algorithm to become a competitive optimal algorithm SRSA^{on} for $SRSA$. The competitive ratio is $O(\sqrt{\log N})$. Finally, we translate SRSA^{on} back to solve the MCD problem. This reverse translation improves the upper bound to $O(\min\{\sqrt{\log n}, \sqrt{\log N}\})$. That is, this final algorithm is competitive optimal for MCD even in the case that the number of requests is small. Intuitively, the "reverse translation" gets rid of the dependance on the network size, using the fact that in the definition of $SRSA$, there is no network. (This last trick may be a useful twist on the common idea of a translation between continuous and discrete problems). We also present an $\Omega(\sqrt[3]{\log n})$ lower bound on the competitiveness of any randomized algorithm that solves the online MCD problem.

Some Additional Related Work. As pointed out in [4], MCD also motivated as a variant of a problem that is useful for data structures for the maintenance of kinematic structures, with numerous applications. Of course, Steiner trees, in general, have many applications, see e.g. [7] for a rather early survey that already included hundreds of items. $SRSA$ is a variant of the Rectilinear Steiner Arborescence (continuous) problem RSA. The offline version of RSA was studied e.g. by Rao, Sadayappan, Hwang, and Shor [14]. RSA was attributed to [10,6] who gave two different exponential time algorithms. PTAS for RSA and $SRSA$ were presented by [9] and [5], respectively. A generalization of the logarithmic upper bound of online MCD to general networks appears in [1].

Paper Structure. Section 3 contains an optimal upper bound on the competitive ratio for MCD as a function of the network size. In Section 4, the above is

---

[1] In fact, the parameter they used was p, the *normalized* size of the network. For simplicity, we present results for n, the size of the network. However, the same results for p follow easily from Sections 4 and 5.

translated to a tight upper bound for $SRSA$. In Section 5, we use the solution of $SRSA$ in order to improve the solution of MCD (to be optimal also as a function of the number of terminals). Finally, Section 6 includes the lower bound for randomized algorithms. Because of space considerations, most of the details in the last three sections were moved and will appear in the full version.

2   Preliminaries

The $SRSA$ problem and its online version was given in the introduction. This section contains formal definitions and notations for the network×time grid, as well as the MCD problem and its online version on that grid. Finally, it contains the offline algorithm of [4] for MCD, which we use later as a tool.

The Network×Time Grid. A *line network* $L(n) = (V_n, E_n)$ is a network whose vertex set is $V_n = \{1, ..., n\}$ and its edge set is $E_n = \{(i, i + 1) \mid i = 1, ..., n - 1\}$. Given a line network $L(n) = (V_n, E_n)$, construct "time-line" graph $\mathcal{L}(n) = (\mathcal{V}_n, \mathcal{E}_n)$, intuitively, by "layering" multiple replicas of $L(n)$, one per time unit, where in addition, each node in each replica is connected to the same node in the next replica . Formally, the node set $\mathcal{V}_n$ contains a *node replica* (sometimes called just a *replica*) (v, t) of every $v \in V_n$, for every time step $t \in \mathbb{N}$. That is, $\mathcal{V}_n = \{(v, t) \mid v \in V_n, t \in \mathbb{N}\}$. The set of edges $\mathcal{E}_n = \mathcal{H}_n \cup \mathcal{A}_n$ contains *horizontal edges* $\mathcal{H}_n = \{((u, t), (v, t)) \mid (u, v) \in E_n, t \in \mathbb{N}\}$, connecting network edges in every time step (round), and directed *vertical edges*, called *arcs*, $\mathcal{A}_n = \{((v, t), (v, t + 1)) \mid v \in V_n, t \in \mathbb{N}\}$, connecting different copies of V_n. When it is clear from the context, we may omit n from X_n and write just X, for every $X \in \{V, E, \mathcal{V}, \mathcal{H}, \mathcal{A}\}$. Notice that $\mathcal{L}(n)$ can be viewed geometrically as a grid of n by ∞ whose grid points are the replicas. Let $d((u, s), (v, t))$ be the distance from (u, s) to (v, t). Formally, $d((u, s), (v, t)) = t - s + |v - u|$ (if $s \leq t$, otherwise, ∞).

MCD: We are given a line network $L(n)$, an *origin* node $v_0 \in V$, and a set of *requests* $\mathcal{R} \subseteq \mathcal{V}$. A feasible solution is a subset of edges $\mathcal{F} \subseteq \mathcal{E}$ such that for every request $r \in \mathcal{R}$, there exists a path in $\mathcal{F}$ from the origin $(v_0, 0)$ to r. A horizontal edge $((v, t), (v + 1, t)) \in \mathcal{F} \cap \mathcal{H}$ stands for sending a copy of the movie (or *copy*, for short) from node v to node $v + 1$, or from node $v + 1$ to node v at time t, while a vertical (directed) edge $((v, t), (v, t + 1)) \in \mathcal{F} \cap \mathcal{A}$ stands for keeping the movie in v's cache at time step t for time $t + 1$. For convenience, the endpoints $\mathcal{V}_{\mathcal{F}}$ of edges in $\mathcal{F}$ are also considered parts of the solution. For a given algorithm A, let $\mathcal{F}_A$ be the solution of A, and let $cost(A, \mathcal{R})$, (the cost of algorithm A), be $|\mathcal{F}_A|$. The goal is to find a minimum cost feasible solution. In our analysis, OPT is the set of edges in some optimal solution whose cost is $|\text{OPT}|$.

Online Model. In the online versions of the problem, the algorithm receives as input a sequence of events. One type of events is a request in the (ordered) set $\mathcal{R}$ of requests $\mathcal{R} = \{r_1, r_2, ..., r_N\}$ (like in $SRSA$). A second type of events is a time event (this event does not exists in $SRSA$), where we assume a clock that tells the algorithm that no additional requests for time t are about to arrive (or that there are no requests for some time t at all). The algorithm then still has the opportunity to complete its calculation for time t (e.g., add arcs from some replica (v,t) to $(v, t+1)$). Then time $t+1$ arrives.

When handling an event ev, the algorithm only knows the following: (a) all the previous requests $r_1, ..., r_i$; (b) time t; and (c) the solution arborescence $\mathcal{F}_{ev}$ it constructed so far (originally containing only the origin). In each event, the algorithm may need to make decisions of two types, before seeing future events:

(1.MCD) If the event is the arrival of a request $r_i = (v_i, t_i)$, then from which *current* (time t_i) cache (a point already in the solution arborescence $\mathcal{F}_{ev}$ when r_i arrives) to serve r_i by adding horizontal edges to $\mathcal{F}_{ev}$.

(2.MCD) If this is the time event for time t, then at which nodes to store a copy for time $t+1$, for future use: select some replica (or replicas) (v,t) already in the solution $\mathcal{F}_{ev}$ and add to $\mathcal{F}_{ev}$ an edge directed from (v,t) to $(v, t+1)$.

Note that, at time t, the online algorithm cannot add nor delete any edge with an endpoint that corresponds to previous times. Similarly to e.g. [1,11,13,12,4], we assume that at least one copy must remain in the network at all times[2].

A Tool: The Offline Algorithm TRIANGLE of Charikar et al. Consider a request set $\mathcal{R} = \{r_0 = (v_0, 0), r_1 = (v_1, t_1), ..., r_N = (v_N, t_N)\}$ such that $0 \le t_1 \le t_2 \le ... \le t_N$. When Algorithm TRIANGLE starts, the solution includes just $r_0 = (v_0, 0)$ (intuitively, a "pseudo request"). Then, TRIANGLE handles, first, request r_1, then request r_2, etc... In handling a request r_i, the algorithm may add some (possibly "past") edges to the solution. (It never deletes any edge from the solution.) After handling r_i, the solution is an arborescence rooted at r_0 that spans the request replicas $r_1, ..., r_i$. For each such request $r_i \in \mathcal{R}$, TRIANGLE performs the following.

(T1) Choose a replica $q_i^T = (u_i^T, s_i^T)$ s.t. q_i^T is already in the solution and the distance from q_i^T to r_i is minimum (over the replicas already in the solution). Call q_i^T the *serving replica* of r_i.

(T2) Define the *radius* ρ_i^T of r_i as $\rho_i^T = d(q_i^T, r_i)$. Also define the *base*[3] BASE(i) of r_i as the set of replicas at time t_i of distance at most ρ_i^T from r_i. That is, BASE(i) = $\{q = (v, t_i) \in \mathcal{V} \mid d(r_i, q) \le \rho_i^T\}$. Similarly, the *edge base* of r_i is BASE$_\mathcal{H}(i)$ = $\{(r, q) \in \mathcal{H} \mid r, q \in \text{BASE}(i)\}$.

(T3) Deliver a copy to each replica in BASE(i). That is, node u_i^T stores a copy from time s_i^T to time t_i. More formally, add the arcs of $\mathcal{P}_\mathcal{A}[(u_i^T, s_i^T),$ $(u_i^T, t_i)] = \{((u_i^T, z), (u_i^T, z+1)) \mid s_i^T \le z < t_i\}$ to the solution.

(T4) Deliver a copy to all replicas in BASE(i). That is, add all the edges of BASE$_\mathcal{H}(i)$ to the solution, except the ones that close circle[4] (if such exists).

It is easy to verify [4] that the cost of TRIANGLE for serving r_i is at most $3\rho_i^T$. Denote by $\mathcal{F}^T = \mathcal{H}^T \cup \mathcal{A}^T$ the feasible solution of TRIANGLE, where

---

[2] Alternatively, the system (not the algorithm) can have the option to delete the movie altogether, this decision must then be made known to the algorithm. At least one of these natural assumptions is also necessary for having a competitive algorithm.

[3] The word "base" comes from the notation used in [4] for Algorithm TRIANGLE. There, BASE(i) is a base of the triangle.

[4] For convenience of the analysis we want the solution to be a tree.

$\mathcal{H}^{\mathrm{T}} \subseteq \cup_{i=1}^{N}\mathrm{BASE}_{\mathcal{H}}(i)$ and $\mathcal{A}^{\mathrm{T}} = \cup_{i=1}^{N}\mathcal{P}_{\mathcal{A}}[(u_i^{\mathrm{T}}, s_i^{\mathrm{T}}),(u_i^{\mathrm{T}}, t_i)]$. Note that $\mathcal{F}^{\mathrm{T}}$ is an arborescence rooted at $(v_0, 0)$ spanning the base replicas of $\mathrm{BASE} = \cup_{i=1}^{N}\mathrm{BASE}(i)$. Rewording the theorem of [4], somewhat,

Theorem 21. *[4]* $\mathcal{F}^{\mathrm{T}}$ *is a 3-approximate solution. Also,* $\sum_{i=1}^{N}\rho_i^{\mathrm{T}} \leq |\mathrm{OPT}|$.

3 Optimal Online Algorithm for MCD

Algorithm LINE^{on}. Like Algorithm TRIANGLE, Algorithm LINE^{on} handles requests one by one, according to the order of arrival. However, in step (T3), TRIANGLE may perform an operation that no online algorithm can perform (if $s_i^{\mathrm{T}} < t_i$). Serving a request r_i must be performed from some replica $q_i^{\mathrm{on}} = (u_i^{\mathrm{on}}, t_i) \in \mathcal{V}[t_i]$ that holds a copy at time t_i in the execution of the online algorithm on $\mathcal{R}$. Thus (in addition to selecting from which nodes to deliver copies), algorithm LINE^{on} at time $t_i - 1$ had to also select the nodes that store copies for the consecutive time t_i (so that q_i^{on} mentioned above would be one of them). Let us start with some definitions.

General Definitions and Notations. Consider an interval $J = \{v, v + 1, ..., v+\rho\} \subseteq V$ and two integers $s, t \in \mathbb{N}$, s.t. $s \leq t$. Let $J[s, t]$ be the *"rectangle subgraph"* of $\mathcal{L}(n)$ corresponding to vertex set J and time interval $[s, t]$. This rectangle consists of the replicas and edges of the nodes of J corresponding to time interval $[s, t]$. For a given subsets $\mathcal{V}' \subseteq \mathcal{V}$, $\mathcal{H}' \subseteq \mathcal{H}$ and $\mathcal{A}' \subseteq \mathcal{A}$, denote by (1) $\mathcal{V}'[s, t]$ replicas of $\mathcal{V}'$ corresponding to times $s, ..., t$. Define similarly (2) $\mathcal{H}'[s, t]$ for horizontal edges of $\mathcal{H}'$; and (3) $\mathcal{A}'[s, t]$ arcs of $\mathcal{A}'$. (When $s = t$, we may write $\mathcal{X}[t] = \mathcal{X}[s, t]$, for $\mathcal{X} \in \{J, \mathcal{V}', \mathcal{H}'\}$.) Consider also two nodes $v, u \in V$. Let $\mathcal{P}_{\mathcal{H}}[(v, t), (u, t)] = \mathcal{P}_{\mathcal{H}}[(u, t), (v, t)]$ be the set of horizontal edges of the shortest path from (v, t) to (u, t).

Partitions of $[1, n]$ into Intervals. Define $m = n/\Delta$ for some positive integer Δ to be chosen later. For convenience, we assume that $m = n/\Delta$ is a power of 2. (It is trivial to generalize it). Define $\log m + 1$ *levels* of partitions of the interval $[1, n]$. In level l, partition $[1, n]$ into $m/2^l = n/\Delta 2^l$ intervals, $I_1^l, I_2^l, ..., I_{m/2^l}^l$, each of size $\Delta 2^l$. $I_j^l = \{\Delta(j-1) \cdot 2^l + k \mid k = 1, ..., \Delta 2^l\}$, for every $1 \leq j \leq m/2^l$ and every $0 \leq l \leq \log m$. Let $\mathcal{I}$ be the set of all such intervals. Let $\ell(I)$ be the *level* of an interval $I \in \mathcal{I}$, i.e., $\ell(I_j^l) = l$. We say that I is a level $\ell(I)$ interval. Denote by $I^l(v)$ (for every node $v \in V$ and every level $l = 0, ..., \log m$) the interval in level l that contains v. That is, $I^l(v) = I_k^l$, where $k = \lfloor \frac{v}{\Delta 2^l} \rfloor + 1$.

For a given interval $I_j^l \in \mathcal{I}$, denote by $N^R(I_j^l)$, for $1 \leq j < m/2^l$ (respectively, $N^L(I_j^l)$, for $1 < j \leq m/2^l$) the *neighbor* interval of level l that is on the right (resp., left) of I_j^l. That is, $N^L(I_j^l) = I_{j-1}^l$ and $N^R(I_j^l) = I_{j+1}^l$. Define that $N^L(I_1^i) = \emptyset$ and $N^R(I_{m/2^l}^i) = \emptyset$. Let $N(I) = N^L(I) \cup I \cup N^R(I)$. We say that $N(I)$ is the *neighborhood* of I.

Active intervals. An interval $I \in \mathcal{I}$ is called *active* at time t, if $\mathrm{BASE} \cap I[t - 2^{\ell(I)}, t] \neq \emptyset$. Intuitively, TRIANGLE kept a movie copy in, at least, one of the nodes of I, at least once, and "not too long" before time t. We say that I

stays-active, intuitively, if I is **not** "just about to stop being active", that is, if $\text{BASE} \cap I[t - 2^{\ell(I)} + 1, t] \neq \emptyset$.

Denote by $\mathcal{C}_{t+1}$ the set of replicas corresponding to the nodes that store copies from time t to time $t + 1$ in a LINE^{on} execution. Also, $\mathcal{C}_0 = \{r_0 = (v_0, 0)\}$ (we choose to store a copy in v_0 always). To help us later in the analysis, we also added an auxiliary set $\text{COMMIT} \subseteq \{\langle I, t \rangle \mid I \in \mathcal{I}, t \in \mathbb{N}\}$. Initially, $\text{COMMIT} \leftarrow \emptyset$. For each time $t = 0, 1, 2, ...$, consider first the case that there exists at least one request corresponding to time t, i.e., $\mathcal{R}[t] = \{r_j, ..., r_k\} \neq \emptyset$. Then, for each request $r_i \in \mathcal{R}[t]$, LINE^{on} simulates TRIANGLE to find the radius ρ_i^{T} and the set of base replicas $\text{BASE}(i)$ of r_i. Next, LINE^{on} delivers a copy to every such base replica $r \in \text{BASE}(i)$ (this is called the *"delivery phase"*). That is, for each $i = j, ..., k$ do:

(D1) choose a closest (to r_i) replica $q_i^{\text{on}} = (u_i^{\text{on}}, t)$ of time t already in the solution;
(D2) add the path $\mathcal{H}^{\text{on}}(i) = \mathcal{P}_\mathcal{H}[q_i^{\text{on}}, r_i] \cup \text{BASE}_\mathcal{H}(i)$ to the solution.

Let $\mathcal{V}^{\text{on}}(i) = \{r \mid (r, q) \in \mathcal{H}^{\text{on}}(i)\}$. (Note that r_j is served from $\mathcal{C}_t$, after that, r_{j+1} is served from $\mathcal{C}_t \cup \mathcal{V}^{\text{on}}(j)$, etc.) Clearly, the delivery phase of time t ensures that (at least) the nodes of $\mathcal{C}_t \cup \text{BASE}[t]$ have copies at the end of that phase. It is left to decide which of the above copies to store for time $t + 1$. That is (the *"storage phase"*), LINE^{on} chooses the set $\mathcal{C}_{t+1} \subseteq \mathcal{C}_t \cup \text{BASE}[t]$. Initially, $\mathcal{C}_{t+1} \leftarrow \{(v_0, t + 1)\}$ (as we choose to store a copy at v_0). Then, for each level $l = 0, ..., \log m$, in an *increasing* order, select as follows.

(S1) While there exists a level l interval $I \in \mathcal{I}$ that is (i) stays-active at t; but (ii) no replica has been selected in I's neighborhood (i.e., $\mathcal{C}_{t+1} \cap N(I)[t+1] = \emptyset$), then perform steps (S1.1-S1.3) below.
(S1.1) Add the tuple $\langle I, t \rangle$ to the set COMMIT (we say that I *commits* at time t).
(S1.2) Select some replica $(v, t) \in \text{BASE}[t] \cup \mathcal{C}_t$ such that $v \in N(I)$ (by Observation 1 below, such a replica does exist).
(S1.3) Add $(v, t + 1)$ to $\mathcal{C}_{t+1}$ and add the arc $((v, t), (v, t + 1))$ to the solution.

The pseudo code of LINE^{on} and an example for an execution of LINE^{on} are given in the full version. The solution constructed by LINE^{on} is denoted $\mathcal{F}^{\text{on}} = \mathcal{H}^{\text{on}} \cup \mathcal{A}^{\text{on}}$, where $\mathcal{H}^{\text{on}} = \cup_{i=1}^N \mathcal{H}^{\text{on}}(i)$ represents the horizontal edges added in the delivery phases and $\mathcal{A}^{\text{on}} = \{((v, t), (v, t+1)) \mid (v, t+1) \in \mathcal{C}_{t+1} \text{ and } t = 0, ..., t_N\}$ represents the arcs added in the storage phase. Before the main analysis, we make some easy-to-prove but crucial observations. For completeness, their proofs appear in full version (however, they are pretty clear from step S1). Recall that the notation of active (including stays-active) refer to the fact that the nodes of some base replicas belong to some interval I in the ("recent") past. Observations 1 and 2 state, intuitively, that LINE^{on} leaves a copy in the *neighborhood* $N(I)$ of I as long as I is active.

Observation 1. (*"WELL DEFINED"*). *If an interval $I \in \mathcal{I}$ is stays-active at time t, then there exists a replica $(v, t) \in \mathcal{C}_t \cup \text{BASE}[t]$ such that $v \in N(I)$.*

Observation 2. (*"AN ACTIVE INTERVAL HAS A NEARBY COPY"*). *If an interval I is active at time t, then, either (i) there is some base replica in I's neighborhood* at t $(\text{BASE} \cap N(I)[t] \neq \emptyset)$, or (ii) at least one of the nodes of $N(I)$ stores a copy for time t $(N(I)[t] \cap \mathcal{C}_t \neq \emptyset)$.

Observation 3. ("BOUND FROM ABOVE ON $|\mathcal{A}^{\text{on}}|$"). $|\mathcal{A}^{\text{on}}| \leq |\text{COMMIT}| + t_N$.

3.1 Analysis of LINE^{on}

We, actually, prove that $\frac{cost(\text{LINE}^{\text{on}}, \mathcal{R})}{cost(\text{TRIANGLE}, \mathcal{R})} = O(\sqrt{\log n})$. This implies the desired competitive ratio of $O(\sqrt{\log n})$ by Theorem 21. We first show, that the number of horizontal edges in $\mathcal{H}^{\text{on}}$ ("delivery cost") is $O(\Delta \cdot cost(\text{TRIANGLE}, \mathcal{R}))$. Then, we show, that the the number of arcs in $\mathcal{A}^{\text{on}}$ ("storage cost") is $O(\frac{\log n}{\Delta} \cdot cost(\text{TRIANGLE}, \mathcal{R}))$. Optimizing Δ, we get a competitiveness of $O(\sqrt{\log n})$.

Delivery Cost Analysis. For each request $r_i \in \mathcal{R}$, the delivery phase (step (D2)) adds $\mathcal{H}^{\text{on}}(i) = \mathcal{P}_{\mathcal{H}}[q_i^{\text{on}}, r_i] \cup \text{BASE}_{\mathcal{H}}(i)$ to the solution. Define the *online radius* of r_i as $\rho_i^{\text{on}} = d(q_i^{\text{on}}, r_i)$. Since $|\text{BASE}_{\mathcal{H}}(i)| \leq 2\rho_i^{\text{T}}$, it follows that,

$$|\mathcal{H}^{\text{on}}| \leq \sum_{i=1}^{N} \left(\rho_i^{\text{on}} + 2\rho_i^{\text{T}} \right). \tag{1}$$

It remains to bound ρ_i^{on} as a function of ρ_i^{T} from above. Intuitively, ρ_i^{T} includes the distance from some base replica $q_i = (u_i, s_i) \in \text{BASE}$ to $r_i = (v_i, t_i)$. That is, ρ_i^{T} includes the distance from v_i to u_i and the time difference between s_i and t_i. Restating Observation 2 somewhat differently (Claim 4 below), we can use the distance $|v_i - u_i| \leq \rho_i^{\text{T}}$ and the time difference $t_i - s_i \leq \rho_i^{\text{T}}$ for bounding ρ_i^{on}. That is, we show that LINE^{on} has a copy at time t_i (of r_i) at a distance at most $4\Delta\rho_i^{\text{T}}$ from u_i (of q_i). Since, $|v_i, u_i| \leq \rho_i^{\text{T}}$, LINE^{on} has a copy at distance at most $(4\Delta + 1)\rho_i^{\text{T}}$ from v_i (of r_i). Throughout, since lack of space some of the proofs are omitted.

Lemma 4. *Consider some base replica $(v, t) \in \text{BASE}$ and some $\rho > 0$, such that, $t + \rho \leq t_N$. Then, there exists a replica $(w, t + \rho) \in \mathcal{C}_{t+\rho}$ such that $|v - w| \leq 4\Delta\rho$.*

Lemma 5. $\rho_i^{\text{on}} \leq (4\Delta + 1) \cdot \rho_i^{\text{T}}$.

The following corollary follows from the above lemma, Ineq. (1) and Theorem 21.

Corollary 1. $|\mathcal{H}^{\text{on}}| \leq (4\Delta + 3) \cdot |\text{OPT}|$.

Storage Cost Analysis. By Observation 3, it remains to bound the size of $|\text{COMMIT}|$ from above. Let $commit(I, t) = 1$ if $\langle I, t \rangle \in \text{COMMIT}$ (otherwise 0). Hence, $|\text{COMMIT}| = \sum_{I \in \mathcal{I}} \sum_{t=0}^{\infty} commit(I, t)$. We begin by bounding the number of commitments in LINE^{on} made by level $l = 0$ intervals.

Observation 6. $\sum_{I \in \{J \in \mathcal{I} | \ell(J) = 0\}} commit(I, t) \leq |\text{BASE}|$.

The following is our main lemma;

Lemma 7. $|\text{COMMIT}| \leq 3|\mathcal{A}^{\text{T}}| + \frac{6 \log n}{\Delta}|\mathcal{H}^{\text{T}}| + |\text{BASE}|$.

Proof Sketch. The |BASE| term in the statement of the lemma follows from Observation 6 for level $l = 0$ intervals. The rest of the proof deals with commitments in intervals $I \in \mathcal{I}$ whose level $\ell(I) > 0$. We now group the commitments of each such an interval into *"bins"*. Later, we shall "charge" the commitments in each bin on certain costs of the offline algorithm TRIANGLE.

Consider an input $\mathcal{R}$ and some interval $I \in \mathcal{I}$ of level $\ell(I) > 0$. We say that I is a *committed-interval* if I commits at least once in the execution of LINE^{on} on $\mathcal{R}$. For each committed-interval I (of level $\ell(I) > 0$), we define (almost) non-overlapping *"sessions"* (one session may end at the same time the next session starts; hence, two consecutive sessions may overlap on their boundaries). The first session of I does *not* contain any commitments (and is termed an *uncommitted-session*); it begins at time 0 and ends at the first time that I contains some base replica. Every other session (of I) contains at least one commitment (and is termed a *committed-session*).

Each commitment (in LINE^{on}) of I belongs to some committed session. Given a commitment $\langle I, t \rangle \in$ COMMIT that I makes at time t, let us identify $\langle I, t \rangle$'s session. Let $t^- < t$ be the last time (before t) there was a base replica in I. Similarly, let $t^+ > t$ be the next time (after t) there will be a base replica in I (if such a time does exist; otherwise, $t^+ = \infty$). The session of commitment $\langle I, t \rangle$ starts at t^- and ends at t^+. Similarly, when talking about the i's session of interval I, we say that the session starts at $t_i^-(I)$ and ends at $t_i^+(I)$. When I is clear from the context, we may omit (I) and write t_i^-, t_i^+. A bin is a couple (I, i) of a commitment-interval and the ith commitment-session of I. Clearly, we assigned all the commitments (of level $l > 0$ intervals) into bins.

Observation 8. *The bins do not overlap (except, perhaps, on their boundaries).*

Let us now point at costs of algorithm TRIANGLE on which we shall "charge" the set of commitments COMMIT(I, i) in bin (I, i). We now consider only a bin (I, i) whose committed session is not the last. Note that the bin corresponds to a rectangle of $|I|$ by $t_i^+ - t_i^-$ replicas. Expand the bin by $|I|$ replicas left and $|I|$ replicas right, if such exist (to I's neighborhood $N(I)$). This yields the *payer* of bin (I, i); that is the payer is a rectangle subgraph of $|N(I)|$ by $t_i^+ - t_i^-$ replicas. We point at specific costs TRIANGLE had in this payer.

Recall that every session of I, except may the last, must ends with a base replica in I. Let $(v, t_i^+) \in$ BASE$\cap I[t_i^+]$ be some base replica in I at the ending time of the session. The solution of TRIANGLE must contain a route (TRIANGLE route) that starts at the root and reaches (v, t_i^+) by the definition of a base replica. For the charging, we use *some* (detailed below) of the edges in the intersection of the TRIANGLE route and the payer rectangle.

The easiest case (**EB**, for Entrance from Below) is that the TRIANGLE route enters the payer at the payer's bottom (t_i^-) and stays in the payer until t_i^+. Then, each time $(t_i^- < t < t_i^+)$ there is a commitment in the bin, there is also an arc a_t in the TRIANGLE route (from time t to time $t+1$). We charge that commitment on that arc a_t. Intuitively, the same arc a_t may be charged also for one bin on the left of (I, i) and one bin on its right, since the payer rectangles are 3 times wider than the bins. Note that arc a_t may also belong to additional $O(\log n)$ payers (of bins of intervals that contain I or are contained in I). The crucial point is that a_t is *not* charged for those additional bins. That is, we claim that there are no commitments for those other bins. Intuitively, LINE^{on} was designed such that if I commits at time t, LINE^{on} also stores a copy in I's neighborhood for time $t + 1$. Hence, an interval J whose neighborhood contains the neighborhood of I, does not need to commit (and the test fails in (S1) in LINE^{on}). Thus, an arc of the TRIANGLE route is charged only by 3 commitments at most.

In the remaining case (**SE**, for Side Entrance), the TRIANGLE route enters the payer from either the left or the right side of the payer. (That is, TRIANGLE delivers a copy from some other node u outside I's neighborhood, rather than stores copies at I's neighborhood from some earlier time. Therefore, the route must "cross" either the left neighbor interval of I or the right neighbor interval in that payer. Thus, there exists at least $|I| = \Delta 2^{\ell(I)}$ horizontal edges in the intersection between the payer $(payer(I, i))$, of (I, i) and the TRIANGLE route. On the other hand, the number of commitments in bin (I, i) is $2^{\ell(I)}$ at most. (To commit, an interval must be active; to be active, it needs a base replica in the last $2^{\ell(i)}$ times; a new base replica would end the session.) That is, we charged the payer Δ times more horizontal edges than there are commitments in the bin. On the other hand, each horizontal edge participates in $O(\log n)$ payers (payers of 3 intervals at most in each level; and payers of 2 bins of each interval at most, since two consecutive sessions may intersect only at their boundaries). This leads to the term $\frac{6 \log n}{\Delta}$ before the $|\mathcal{H}^{\text{T}}|$ in the statement of the lemma.

For each interval I, it is left to account for commitments in I's last session. That is, we now handle the bin (I, i') where I has i' commitment-sessions. This session may not end with a base replica in I, so we cannot apply the argument above that TRIANGLE must have a route reaching a replica in I at $t_{i'}^+$. On the other hand, the first session of I (the uncommitted-session) does end with a base replica in I, but has no commitments. Intuitively, we use the payer of the first session of I to pay for the commitments of the last session of I. Specifically, in the first session, the TRIANGLE route must enter the neighborhood of I from the side; (Note that the TRIANGLE route still starts outside I; this because the origin v_0 who holds a copy, is not in I's neighborhood; otherwise, I would not have been a committed interval.) Hence, we apply the argument of case SE above. *(End of Proof sketch.)* ∎

We now optimize a tradeoff between the storage coast and the delivery cost of LINE^{on}. On the one hand, Lemma 7 shows that a large Δ reduces the number of commitments. By Observation 3, this means a large Δ reduces the storage cost of LINE^{on}. On the other hand, corollary 1 shows that a *small* Δ reduces the delivery cost. To balance this tradeoff, we need to "manipulate" Lemma 7 somewhat, since it uses variables that are different from those used in corollary 1. We use the following observation (1) $t_N \le |\text{OPT}| \le cost(\text{TRIANGLE}, \mathcal{R})$; (2) $|\mathcal{A}^{\text{T}}| + |\mathcal{H}^{\text{T}}| = cost(\text{TRIANGLE}, \mathcal{R})$; and (3) $|\text{BASE}| \le cost(\text{TRIANGLE}, \mathcal{R})$. Substituting the above (1)–(3) in Observation 3 and Lemma 7,

$$|\mathcal{A}^{\text{on}}| \le \left(5 + \frac{3 \log n}{\Delta}\right) \cdot cost(\text{TRIANGLE}, \mathcal{R}). \tag{2}$$

To optimize the tradeoff, fix $\Delta = \sqrt{10 \log n}$. Corollary 1, and inequality (2) imply that $cost(\text{LINE}^{\text{on}}, \mathcal{R}) = |\mathcal{A}^{\text{on}}| + |\mathcal{H}^{\text{on}}| \leq (8 + \sqrt{10 \log n}) \cdot cost(\text{TRIANGLE}, \mathcal{R})$. Thus, by Theorem 21, the following holds.

Theorem 31. LINE^{on} *is* $O(\sqrt{\log n})$*-competitive for* MCD *on the line network.*

4   Optimal Online Algorithm for *SRSA*

Note that our solution for MCD (Section 3) does not yet solve $SRSA$. In MCD, the X coordinate of every request (in the set $\mathcal{R}$) is taken from a known set of size n (the network nodes $\{1, 2, ..., n\}$). On the other hand, in $SRSA$, the X coordinate of a *point* is arbitrary. Let us now transform, in three conceptual stages LINE^{on} into an optimal algorithm for the online problem of $SRSA$:

1. Given an instance of $SRSA$, assume temporarily (and remove the assumption later) that the number N of points is known, as well as M, the maximum X coordinate any request may have. Then, simulate a network where $n \geq N$ and $n = O(\sqrt{\log N})$, and the n nodes are spaced evenly on the interval between 0 and M. Transform each $SRSA$ request to the nearest grid point. Solve the resulting MCD problem.
2. Translate these results to results of the original $SRSA$ instance.
3. Get rid of the assumptions.

The first stage is easy. It turns out that "getting rid of the assumptions" is also relatively easy. To simulate the assumption that M is known, guess that M is some M_j. Whenever a guess fails, (a request $r_i = (x_i, t_i)$ arrives, where $x_i > M_j$), continue with an increased guess M_{j+1}. A similar trick is used for guessing N. In implementing this idea, our algorithm turned out paying a cost of ΣM_j (M_j for a failed guess), while an algorithm that knew M could pay M only once. IF M_{j+1} is "sufficiently" larger than M_j, then $\Sigma M_j = O(M)$. The "sufficiently larger" part turned out somewhat trickier for guessing N than for guessing M.

The second stage above (translate the results) proved to be more difficult, even in the case that N and M are known (and even equal). Intuitively, following the first stage, each request $r_i = (x_i, t_i)$ is in some grid square, where the corners of the square are points of the simulated MCD problem. If we normalize M to be N, then the left bottom left corner of that square is $(\lfloor x_i \rfloor, \lfloor t_i \rfloor))$. Had we wanted an *offline* algorithm, we could have solved an instance of MCD, where the points are $(\lfloor x_1 \rfloor, \lfloor t_1 \rfloor), (\lfloor x_2 \rfloor, \lfloor t_2 \rfloor), (\lfloor x_3 \rfloor, \lfloor t_3 \rfloor), ...$ Then, translating the results of MCD would have meant just augmenting with segments connecting each $(\lfloor x_i \rfloor, \lfloor t_i \rfloor)$ to (x_i, t_i). Unfortunately, this is not possible in an *online* algorithm, since (x_i, t_i) is not yet known at $(\lfloor t_i \rfloor)$. Similarly, we cannot use the upper left corner of the square (for example) that way, since at time $\lceil t_i \rceil$, the algorithm may no longer be allowed to add segments reaching the earlier time t_i. Because of the lack of space, we moved the rest of this proof.

Theorem 41. *Algorithm* SRSA^{on} *is optimal and is* $O(\sqrt{\log N})$*-competitive.*

5   Optimizing MCD for a Small Number of Requests

Algorithm LINE^{On} was optimal only as the function of the network size (Theorem 31). Recall that our solution for $SRSA$ was optimal as a function of the number of requests. We transform that algorithm back to solve MCD, and obtain the promised competitiveness, $O(\min\{\sqrt{\log n}, \sqrt{\log N}\})$.

6   Randomized Lower Bound for Line Networks

Our lower bound on the competitive ratio of randomized algorithms then follows from Yaos min-max principle [15] appears in the full version.

Theorem 61. *The competitive ratio of any randomized online algorithm for MCD on line networks $\Omega(\sqrt[3]{\log n})$.*

Acknowledgment. We would like to thank to Reuven Bar-Yehuda and Dror Rawitz for insights and helpful dissections.

References

1. Bar-Yehuda, R., Kantor, E., Kutten, S., Rawitz, D.: Growing half-balls: Minimizing storage and communication costs in cDNs. In: Czumaj, A., Mehlhorn, K., Pitts, A., Wattenhofer, R. (eds.) ICALP 2012, Part II. LNCS, vol. 7392, pp. 416–427. Springer, Heidelberg (2012)
2. Bein, W., Golin, M., Larmore, L., Zhang, Y.: The knuth-yao quadrangle-inequality speedup is a consequence of total monotonicity. TOPLAS 6(1) (2009)
3. Berman, P., Coulston, C.: On-line algorrithms for steiner tree problems. In: STOC 1997, pp. 344–353 (1997)
4. Charikar, M., Halperin, D., Motwani, R.: The dynamic servers problem. In: SODA 1998, pp. 410–419 (1998)
5. Cheng, X., Dasgupta, B., Lu, B.: Polynomial time approximation scheme for symmetric rectilinear steiner arborescence problem. J. Global Optim. 21(4), 385–396 (2001)
6. Ladeira de Matos, R.R.: A rectilinear arborescence problem. Dissertation, University of Alabama (1979)
7. Richards, D.S., Hwang, F.K.: Steiner tree problems. Networks 22(1), 55–897 (1992)
8. Kahng, A., Robins, G.: On optimal interconnects for vlsi. Kluwer (1995)
9. Lu, B., Ruan, L.: Polynomial time approximation scheme for rectilinear steiner arborescence problem. Combinatorial Optimization 4(3), 357–363 (2000)
10. Nastansky, L., Selkow, S.M., Stewart, N.F.: Cost minimum trees in directed acyclic graphs. Z. Oper. Res. 18, 59–67 (1974)
11. Papadimitriou, C.H., Ramanathan, S., Rangan, P.V.: Information caching for delivery of personalized video programs for home entertainment channels. In: IEEE International Conf. on Multimedia Computing and Systems, pp. 214–223 (1994)
12. Papadimitriou, C.H., Ramanathan, S., Rangan, P.V.: Optimal information delivery. In: 6th ISAAC, pp. 181–187 (1995)
13. Papadimitriou, C.H., Ramanathan, S., Rangan, P.V., Sampathkumar, S.: Multimedia information caching for personalized video-on demand. Computer Communications 18(3), 204–216 (1995)
14. Rao, S., Sadayappan, P., Hwang, F., Shor, P.: The rectilinear steiner arborescence problem. Algorithmica, 277–288 (1992)
15. Yao, A.C.-C.: Probabilistic computations: Toward a unified measure of complexity. In: FOCS 1977, pp. 222–227 (1977)

Orienting Fully Dynamic Graphs
with Worst-Case Time Bounds*

Tsvi Kopelowitz[1],[**], Robert Krauthgamer[2], [***], Ely Porat[3],
and Shay Solomon[4],[†]

[1] University of Michigan, USA
kopelot@gmail.com
[2] Weizmann Institute of Science, Israel
robert.krauthgamer@weizmann.ac.il
[3] Bar-Ilan University, Israel
porately@cs.biu.ac.il
[4] Weizmann Institute of Science, Israel
shay.solomon@weizmann.ac.il

Abstract. In edge orientations, the goal is usually to orient (direct) the
edges of an undirected network (modeled by a graph) such that all out-
degrees are bounded. When the network is fully dynamic, i.e., admits
edge insertions and deletions, we wish to maintain such an orientation
while keeping a tab on the update time. Low out-degree orientations
turned out to be a surprisingly useful tool for managing networks.

Brodal and Fagerberg (1999) initiated the study of the edge orientation problem in terms of the graph's arboricity, which is very natural in
this context. Their solution achieves a constant out-degree and a logarithmic *amortized* update time for all graphs with constant arboricity,
which include all planar and excluded-minor graphs. It remained an open
question – first proposed by Brodal and Fagerberg, later by Erickson and
others – to obtain similar bounds with *worst-case* update time.

We address this 15 year old question by providing a simple algorithm with worst-case bounds that nearly match the previous amortized
bounds. Our algorithm is based on a new approach of maintaining a
combinatorial invariant, and achieves a logarithmic out-degree with logarithmic worst-case update times. This result has applications to various
dynamic network problems such as maintaining a maximal matching,
where we obtain logarithmic worst-case update time compared to a similar amortized update time of Neiman and Solomon (2013).

1 Introduction

A very useful algorithmic tool for managing networks is to *orient* (direct) the
edges while providing a guaranteed upper bound on the out-degree of every vertex.

* A full version appears at http://arxiv.org/abs/1312.1382
** This work is supported by NSF grants CCF-1217338 and CNS-1318294.
*** Work supported in part by a US-Israel BSF grant #2010418, an Israel Science
Foundation grant #897/13, and by the Citi Foundation.
† This work is supported by the Koshland Center for basic Research.

J. Esparza et al. (Eds.): ICALP 2014, Part II, LNCS 8573, pp. 532–543, 2014.
© Springer-Verlag Berlin Heidelberg 2014

Formally, an orientation of an undirected graph $G = (V, E)$ is called a *c-orientation* if every vertex has out-degree at most $c \geq 1$.

There are many examples where orientations are used in the design, maintenance and manipulation of networks, including both static and dynamic networks, and both centralized algorithms and distributed ones. One exciting example of the power of graph orientations can be seen in the seminal paper introducing "color-coding" [1], where orientations are used to develop more efficient algorithms for finding simple cycles and paths. Another fundamental example is in data structures for quickly answering adjacency queries [2,3,4], where a c-orientation of a (dynamic) graph G is used to answer adjacency queries in $O(c)$ time using only linear space. These techniques [2,3,4] were further generalized to answer short-path queries [5]. Additional examples for the algorithmic use of low-degree orientations include load balancing [6], maximal matchings [7], counting subgraphs in sparse graphs [8], prize-collecting TSPs and Steiner Trees [9], reporting all maximal independent sets [10], answering dominance queries [10], subgraph listing problems (listing triangles and 4-cliques) in planar graphs [2], and computing the girth [5].

Efficient Data Communication. The efficiency of network communication can often be improved significantly by assigning one endpoint of every edge as "responsible" for all data transfers occurring on that edge. Such a responsibility assignment can be naturally obtained by orienting the graph's edges and letting each vertex be responsible only for its outgoing edges. Consider, for example, the task of computing some aggregate function of dynamic data that resides locally at a vertex and its neighbors – can this task be carried out without scanning all the neighbors of that vertex? Given a c-orientation, whenever the local data in a vertex u changes, u updates all its outgoing neighbors (neighbors of u through edges oriented out of u). In contrast, u need not update any of its (possibly many) incoming neighbors (neighbors of u through edges oriented into u) about this change. When u wishes to compute the function, it only needs to scan its outgoing neighbors in order to gather the full up-to-date data. Such responsibility assignment is particularly useful in dynamic networks, see [7] for an example, and could be very effective also in many standard tasks in distributed, self-stabilizing, peer-to-peer, or ad-hoc networks, such as reducing the message complexity, or reducing local memory constraints (e.g., a router would only store information about its c outgoing neighbors).

Dynamic Graphs. Our focus here is on maintaining low out-degree orientations of fully dynamic graphs on n fixed vertices, where edge updates (insertions and deletions) take place over time. The goal is to develop efficient and simple algorithms that guarantee that the maximum out-degree in the (dynamic) orientation of the graph is small. In particular, we are interested in obtaining non-trivial update times that hold (1) in the *worst-case*, and (2) *deterministically*. Notice that in order for an update algorithm to be efficient, the number of *edge re-orientations* (done when performing an edge update) must be small, as this number is clearly a lower bound for the algorithm's update time.

The out-degree bound achieved by our algorithms will be expressed in terms of the sparsity of the graph, as measured by the *arboricity* of G (defined below), which is a natural lower bound for the maximum out-degree of any orientation.

Arboricity. The *arboricity* of an undirected graph $G = (V, E)$ is defined as $\alpha(G) = \max_{U \subseteq V} \left\lceil \frac{|E(U)|}{|U|-1} \right\rceil$, where $E(U)$ is the set of edges induced by U (which we assume has size $|U| \geq 2$). This is a concrete formalism for the notion of everywhere-sparse graphs — every subgraph of G has arboricity at most $\alpha(G)$ as well. Arboricity and its related sparseness measures of *thickness, degeneracy* or *density*, which are all equal up to a constant factor, were studied extensively. Most notable in this context is the family of graphs with constant arboricity, which includes all excluded-minor graphs, and in particular planar graphs and bounded-treewidth graphs.

A key property of bounded arboricity graphs that has been exploited in various algorithmic applications is the following Nash-Williams Theorem.

Theorem 1 (Nash-Williams [11,12]). *A graph $G = (V, E)$ has arboricity $\alpha(G)$ if and only if $\alpha(G) > 0$ is the smallest number of sets $E_1, \ldots, E_{\alpha(G)}$ that E can be partitioned into, such that each subgraph (V, E_i) is a forest.*

This theorem implies that the edges of an undirected graph $G = (V, E)$ can be oriented such that the out-degree of each vertex is at most $\alpha(G)$. To see this, consider the guaranteed partition $E_1, \ldots, E_{\alpha(G)}$. For each forest (V, E_i) and each tree in that forest, designate one arbitrary vertex as the root of that tree, and orient all edges towards that root. In each oriented forest the out-degree of every vertex is at most 1, hence the union of the oriented forests has out-degree bounded by $\alpha(G)$. There exists a polynomial-time algorithm that computes for a (static) graph G the exact arboricity $\alpha(G)$ [13], and a linear-time algorithm that computes a $(2\alpha(G) - 1)$-orientation [14].

For every (static) graph G, the minimum-possible maximum out-degree is closely related to $\alpha(G)$: the argument above provides an orientation with maximum out-degree at most $\alpha(G)$, but the maximum out-degree is also easily seen to be at least $\alpha(G) - 1$ (for every orientation).[1] In other words, the arboricity measure of sparsity is a natural baseline for low out-degree orientations.

1.1   Main Result

We obtain efficient algorithms for maintaining a low out-degree orientation of a fully dynamic graph G such that the out-degree of each vertex is small and the running time of all update operations is bounded in the worst-case. Specifically, we present two algorithms. The first algorithm achieves (at any point in time)

---

[1] To see this, let $U \subset V$ be such that $\left\lceil \frac{|E(U)|}{|U|-1} \right\rceil = \alpha(G)$, hence $\frac{|E(U)|}{|U|-1} > \alpha(G) - 1$. For every orientation, the maximum out-degree in G is at least the average out-degree of vertices in U, which in turn is at least $\frac{|E(U)|}{|U|} > \frac{|U|-1}{|U|}(\alpha(G) - 1)$. The bound now follows from both $\alpha(G)$ and the maximum out-degree being integers.

- a maximum out-degree $\Delta \leq \inf_{\beta>1}\{\beta \cdot \alpha(G) + \lceil \log_\beta n \rceil\}$, and
- insertion and deletion update times $O(\Delta^2)$ and $O(\Delta)$, respectively.

The second algorithm works with two parameters $\hat{\alpha}$ and $\hat{\beta} > 1$ both known by the algorithm. The parameter $\hat{\alpha}$ is a set upper bound on $\alpha(G)$ while $\hat{\beta}$ can be chosen arbitrarily and only affects the complexities of the algorithm. This algorithm achieves (at any point in time)

- a maximum out-degree $\Delta \leq \hat{\beta} \cdot \hat{\alpha} + \lceil \log_{\hat{\beta}} n \rceil$, and
- insertion and deletion update times $O(\hat{\beta} \cdot \hat{\alpha} \cdot \Delta)$ and $O(\Delta)$, respectively.

Notice that the first algorithm does not need to know $\alpha(G)$ (hence its bounds change with time together with the graph G), while the second algorithm assumes knowledge of an upper bound on $\alpha(G)$. On the other hand, the second algorithm has faster insertion time, because in the worst-case $\hat{\beta} \cdot \hat{\alpha} \cdot \Delta < \Delta^2$.

All our algorithms are deterministic, and they change the orientation of at most $\Delta + 1$ edges per edge update. Perhaps most importantly, they are relatively simple (especially the first one) to describe and to analyze, which is a great virtue for potential implementation, and also for further extensions and refinements. We should nevertheless point out that the apparent simplicity relies heavily on a fine selection of an effective combinatorial invariant; finding such invariants can be very tricky, and it constitutes the main technical challenge in this work.

Notice that in our second algorithm if $\hat{\alpha}$ is constant we can set $\hat{\beta} = 2$ and all of our bounds translate to $O(\log n)$. In other words, for fully dynamic graphs with a constant upper bound on the arboricity we can maintain an $O(\log n)$-orientation with $O(\log n)$ worst-case update time. Previous work, which is discussed next, only obtained efficient *amortized* update time bounds, in contrast to our bounds which are all in the worst-case. Our results address an open question raised by Brodal and Fagerberg [3] and restated by Erickson [15], of obtaining good worst-case bounds (although the ultimate goal is obviously worst-case time $O(1)$ for all updates, if that is at all possible).

1.2   Comparison with Previous Work

The dynamic setting in our context was pioneered by Brodal and Fagerberg [3], who showed that it is possible to maintain a $4\hat{\alpha}$-orientation of a fully dynamic graph G whose arboricity is always at most $\hat{\alpha}$. They proved that their algorithm is $O(1)$-competitive against the number of re-orientations made by any algorithm, regardless of that algorithm's actual running time. They then provided a specific strategy for re-orienting edges which shows that, for $\hat{\alpha} = O(1)$, their algorithm's insertion time is *amortized* $O(1)$ while the deletion time is *amortized* $O(\log n)$. Kowalik [4] showed that a different analysis of Brodal and Fagerberg's algorithm achieves insertion update time that is *amortized* $O(\log n)$ and the deletion time that is *worst-case* $O(1)$. Kowalik further showed it is possible to support insertions in *amortized* $O(1)$ time and deletions in worst-case $O(1)$ time by using an $O(\log n)$-orientation. These algorithms have been used as black-box components in several applications of dynamic graphs. Recently Gupta et.al. [16] showed that if only insertions are allowed then an amortized 2 edge reorientations suffice for maintaining a maximum out-degree of $O(\alpha(G))$.

Algorithms with amortized runtime bounds may be insufficient for many real-time applications where infrequent costly operations might cause congestion in the system at critical times. Exploring the boundaries between amortized and worst-case bounds is also important from a theoretical point of view, and has received a lot of research attention. The algorithms of Brodal and Fagerberg [3] and Kowalik [4] both incur a *linear worst-case update time*, on which we show an exponential improvement. As mentioned above, our results address an open question raised by Brodal and Fagerberg [3] and restated by Erickson in [15].

1.3   Our Techniques

The algorithm of Brodal and Fagerberg [3] is very elegant, but it is not clear if it can be deamortized as it is inherently amortized. The key technical idea we introduce is to maintain a combinatorial invariant, which is very simple in its basic form: for every vertex $u \in V$, at least (roughly) $\hat{\alpha}$ outgoing edges are directed towards vertices with almost as large out-degree, namely at least $d_{out}(u) - 1$ (where $d_{out}(u)$ is the out-degree of u). Such edges are called *valid* edges. We prove in Section 2 that this combinatorial invariant immediately implies the claimed upper bound on Δ.

An overview of the algorithms that we use for, say, insertion, is as follows. When a new edge (u, v) is added, we first orient it, say, from u to v guaranteeing that the edge is valid. We now check if the invariant holds, but the only culprit is u, whose out-degree has increased. If we know which of the edges leaving u are the "special" valid edges needed to maintain the invariant, we scan them to see if any of them are no longer valid (as a result of the insertion), and if there is such an edge we *flip* its orientation, and continue recursively with the other endpoint of the flipped edge. This process indeed works, but it causes difficulty during an edge deletion — when one of the $\hat{\alpha}$ special valid edges leaving u is deleted, a replacement may not even exist.

Here, our expedition splits into two different parts. We first show an extremely simple (but less efficient) algorithm that maintains a stronger invariant in which for every vertex $u \in V$, *all* of its out-going edges are valid. This approach immediately gives the claimed upper bound on Δ, with update time roughly $O((\frac{\log n}{\log \log n})^2)$ for graphs with constant arboricity.

In the second part we refine the invariant using another idea of *spectrum-validity*, which roughly speaking uses the following invariant: for every vertex $u \in V$ and for every $1 \leq i \leq \frac{\deg(u)}{\hat{\alpha}}$, at least $i \cdot \hat{\alpha}$ of its outgoing edges are directed towards vertices with degree at least $d_{out}(u) - i$. This invariant is stronger than the first invariant (which seemed algorithmically challenging) and weaker than the second invariant (whose bounds were less efficient than desired as it needed to guarantee validness for all edges). Furthermore, maintaining this invariant is more involved algorithmically, and one interesting aspect of our algorithm is that during an insertion process, it does not scan the roughly $\hat{\alpha}$ neighbors with degree at least $d_{out}(u) - 1$, as one would expect, but rather some other neighbors picked in a careful manner. Ultimately, this methodology yields the improved time bounds claim in Section 1.1.

1.4   Selected Applications

We only mention two applications here by stating their theorems for graphs with arboricity bounded by a constant. We discuss these applications and some other ones with more detail in the full version.

Theorem 2 (Maximal matching in fully dynamic graphs). *Let* $G = (V, E)$ *be an undirected fully dynamic graph with arboricity bounded by a constant. Then one can deterministically maintain a maximal matching of* G *such that the worst-case time per edge update is* $O(\log n)$.

Theorem 3 (Adjacency queries in fully dynamic graphs). *Let* $G = (V, E)$ *be an undirected fully dynamic graph with arboricity bounded by a constant. Then one can deterministically answer adjacency queries on* G *in* $O(\log \log \log n)$ *worst-case time where the deterministic worst-case time per edge update is* $O(\log n \cdot \log \log \log n)$.

1.5   Preliminaries

An *orientation* of the undirected edges of G assigns a direction to every edge $e \in E$, thereby turning G into a digraph. We will use the notation $u \to v$ to indicate that the edge $e = (u, v)$ is oriented from u to v. Given such an orientation, let $N^+(u) := \{v \in V : u \to v\}$ denote the set of *outgoing neighbors* of u, i.e., the vertices connected to u via an edge leaving it, and let $d_{out}(u) := |N^+(u)|$ denote the number of *outgoing edges* of u in this orientation, i.e., the *out-degree* of u. Similarly, let $N^-(u) := \{v \in V : v \to u\}$ denote the set of *incoming neighbors* of u, and let $d_{in}(u) := |N^-(u)|$. Finally, we denote by $\Delta := \max_{v \in V} d_{out}(v)$ the maximum out-degree of a vertex in the graph (under the given orientation).

Our algorithms will make use of the following heap-like data structure (the proof is left for the full version).

Lemma 1. *Let* X *be a dynamic set, where each element* $x_i \in X$ *has a key* $k_i \in \mathbb{N}$ *that may change with time, and designate a fixed element* $x_0 \in X$ *to be the* center *of* X *(although its key* k_0 *may change with time). Then there is a data structure that maintains* X *using* $O(|X| + k_0)$ *words of space, and supports the following operations with* $O(1)$ *worst-case time bound (unless specified otherwise):*

- ReportMax(X): *return a pointer to an element from* X *with maximum key.*
- Increment(X, x): *given a pointer to* $x \in X \setminus \{x_0\}$, *increment the key of* x.
- Decrement(X, x): *given a pointer to* $x \in X \setminus \{x_0\}$, *decrement the key of* x.
- Insert(X, x_i, k_i): *insert a new element* x_i *with key* $k_i \le k_0 + 1$ *into* X.
- Delete(X, x): *given a pointer to an element* $x \in X \setminus \{x_0\}$, *remove* x *from* X.
- IncrementCenter(X): *increment* k_0 *in* $O(k_0)$ *worst-case time.*
- DecrementCenter(X): *decrement* k_0 *(unless* $k_0 = 1$*) in* $O(k_0)$ *worst-case time.*

For each vertex $w \in V$, consider the (dynamic) set X_w that contains w and all its incoming neighbors, where the key of each element in X is given by its out-degree. The center element of X_w will be w itself. Each vertex w will have its own data structure (using Lemma 1) for maintaining X_w. In what follows, we denote this data structure by H_w, and use it to find an incoming neighbor of w with out-degree at least $d_{out}(w) + 2$ (if one exists) in $O(1)$ time.

Lemma 2. *The total space used to store the data structures H_w for all $w \in V$ is $O(n + m)$ words, where m stands for the number of edges in the (current) graph.*

Proof. By Lemma 1, for each $w \in V$ the space usage is at most $O(1 + d_{in}(w) + d_{out}(w))$. Summing over all vertices $w \in V$, the total space is $\sum_{w \in V} O(1 + d_{in}(w) + d_{out}(w)) = O(n + m)$.   □

2   Invariants for Bounding the Largest Out-Degree

We assume throughout that the dynamic graph G has, at all times, arboricity $\alpha(G)$ bounded by some parameter $\hat{\alpha}$, i.e., $\alpha(G) \leq \hat{\alpha}$. Let $\hat{\beta} > 1$ be a parameter that may possibly depend on n and $\hat{\alpha}$ (it will be chosen later to optimize our bounds), and define $\gamma := \hat{\beta} \cdot \hat{\alpha}$.

An edge $(u, v) \in E$ oriented such that $u \to v$ is called *valid* if $d_{out}(u) \leq d_{out}(v) + 1$, and is called *violated* otherwise. The following condition provides control (upper bound) on Δ, as proved in Theorem 4. We refer to it as an *invariant*, because we shall maintain the orientation so that the condition is satisfied at all times.

Invariant 3. *For each vertex w, at least $\min\{d_{out}(w), \gamma\}$ outgoing edges of w are valid.*

Theorem 4. *If Invariant 3 holds, then $\Delta \leq \hat{\beta} \cdot \hat{\alpha} + \lceil \log_{\hat{\beta}} n \rceil$.*

Proof. Assume Invariant 3 holds, and suppose for contradiction there is a "source" vertex $s \in V$ satisfying $d_{out}(s) > \gamma + \lceil \log_{\hat{\beta}} n \rceil$. Now consider the set V_i of vertices reachable from s by directed paths of length at most i that use only valid edges. Observe that for every $1 \leq i \leq \lceil \log_{\hat{\beta}} n \rceil$ and every vertex $w \in V_i$,

$$d_{out}(w) \geq d_{out}(s) - i > \gamma + \lceil \log_{\hat{\beta}} n \rceil - i \geq \gamma,$$

implying that at least γ outgoing edges of w are valid.

We next prove by induction on i that $|V_i| > \hat{\beta}^i$ for all $1 \leq i \leq \lceil \log_{\hat{\beta}} n \rceil$. For the base case $i = 1$, notice that s has at least γ valid outgoing edges and all of the corresponding outgoing neighbors of s belong to V_1. Furthermore, s belongs to V_1 as well. Thus $|V_1| \geq \gamma + 1 > \gamma \geq \hat{\beta}$. For the inductive step, suppose $|V_{i-1}| > \hat{\beta}^{i-1}$; observe that the total number of valid outgoing edges from vertices in V_{i-1} is at least $\gamma|V_{i-1}|$, and furthermore all these edges are incident only to vertices in V_i. Since the graph's arboricity is $\alpha(G) \leq \hat{\alpha}$, we can bound $|V_i| - 1 \geq \gamma|V_{i-1}|/\alpha(G) \geq \hat{\beta}|V_{i-1}| > \hat{\beta}^i$, as claimed.

We conclude that $|V_{\lceil \log_{\hat{\beta}} n \rceil}| > \hat{\beta}^{\lceil \log_{\hat{\beta}} n \rceil} \geq n$, yielding a contradiction.   □

Invariant 3 provides a relatively weak guarantee as if $d_{out}(w) > \gamma$, then we know only that γ outgoing edges of w are valid, and have no guarantee on the out-degree of the other $d_{out}(w) - \gamma$ outgoing neighbors of w. Consequently, it is nontrivial to maintain Invariant 3 efficiently, and in particular, if one of the γ valid edges (outgoing from w) is deleted, the invariant might become violated, and it is unclear how to restore it efficiently. We thus need another invariant, namely, a stronger condition (so that a similar theorem still applies) that is also easy to maintain. The next invariant is a natural candidate, as it is simple to maintain (with reasonable efficiency).

Invariant 4. *All edges in G are valid.*

Theorem 5. *If Invariant 4 holds, then $\Delta \leq \inf_{\beta > 1} \beta \cdot \alpha(G) + \lceil \log_\beta n \rceil$.*

The proof of Theorem 5 is similar to the proof of Theorem 4 and is left for the full version.

We first present in Section 3 a very simple algorithm that maintains Invariant 4 with update times $O(\Delta^2)$ and $O(\Delta)$ for insertion and deletion (of an edge), respectively. This algorithm provides a strong basis for a more sophisticated algorithm, developed in Section 4, which maintains an intermediate invariant (stronger than Invariant 3 but weaker than Invariant 4) with update times $O(\gamma \cdot \Delta)$ and $O(\Delta)$ for insertion and deletion, respectively.

3   Worst-Case Algorithm

We consider an infinite sequence of graphs $G_0, G_1, \ldots$ on a fixed vertex set V, where each graph $G_i = (V, E_i)$ is obtained from the previous graph G_{i-1} by either adding or deleting a single edge. For simplicity, we assume that G_0 has no edges. Denote by $\alpha_i = \alpha(G_i)$ the arboricity of G_i. We will maintain Invariant 4 while edges are inserted and deleted into and from the graph, which by Theorem 5 implies that the maximum out-degree Δ_i in the orientation of G_i is bounded by $O(\inf_{\beta > 1}\{\beta \cdot \alpha_i + \log_\beta n\})$.

For the rest of this section we fix i and consider a graph G_i obtained from a graph G_{i-1} satisfying Invariant 4 by either adding or deleting edge $e = (u, v)$.

3.1   Insertions

Suppose that edge (u, v) is added to G_{i-1} thereby obtaining G_i. We begin by orienting the edge from the endpoint with lower out-degree to the endpoint with larger out-degree (breaking a tie in an arbitrary manner). So without loss of generality we now have $u \rightarrow v$. Notice that the only edges that may be violated now are edges outgoing from u, as $d_{out}(u)$ is the only out-degree that has been incremented. Furthermore, if some edge $u \rightarrow v'$ is violated now, then removing this edge will guarantee that there are no violated edges. However, the resulting graph would be missing the edge (u, v') just removed. So we recursively insert the edge (u, v'), but orient it in the opposite direction (i.e., $v' \rightarrow u$). This means that we have actually *flipped* the orientation of (u, v'), reverting $d_{out}(u)$ to its value before the entire insertion process took place. This recursive process continues until all edges of the graph are valid. Moreover, at any given time there is at most one "missing" edge, and the graph obtained at the end of the process has no missing edges. Our choice to remove a violated edge outgoing from u (if such an edge exists) guarantees that the number of recursive steps is at most Δ, as we will show later. This insertion process is described in Algorithm 1.

Algorithm 1. Recursive-Insertion$(G, (u, v))$

/* Assume without loss of generality $d_{out}(u) \leq d_{out}(v)$ */
1. add (u, v) to G with orientation $u \to v$
2. Insert$(X_v, u, d_{out}(u) - 1)$ /* this key will be incremented in line 10 if needed */
3. **for** $v' \in N^+(u)$ **do**
4.   **if** $d_{out}(u) > d_{out}(v') + 1$ **then**
5.   remove (u, v') from G /* now edge (u, v') is missing */
6.   Delete$(X_{v'}, u)$
7.   Recursive-Insertion$(G, (v', u))$ /* recursively insert (u, v'), but oriented $v' \to u$
        */
8.   return
9. **for** $v' \in N^+(u)$ **do**
10.   Increment$(X_{v'}, u)$
11. IncrementCenter(X_u)

We remark that although in line 1 the out-degree of u is incremented by 1, we do not update the new key of u in the appropriate structures (i.e., H_u and $H_{v'}$ for all $v' \in N^+(u)$), because if the condition in line 4 succeeds for some $v' \in N^+(u)$, the out-degree of u will return to its original value, and we want to save the cost of incrementing and then decrementing the key for u in all structures. However, if that condition fails for all v', we will perform the update in lines 9–11.

Correctness and Runtime Analysis. The following lemmas, provide the correctness and runtime analysis of the insertion process. Due to space constraints, the proofs are omitted here and appear in the full version. Notice that the proofs mostly follow from the discussion above.

Lemma 5. *At the end of the execution of* Recursive-Insertion *on an input graph which has an orientation satisfying Invariant 4, Invariant 4 holds for the resulting graph and orientation.*

Lemma 6. *The total number of recursive calls (and hence re-orientations) of* Recursive-Insertion *due to an insertion into G is at most $\Delta + 1$, and the total runtime is bounded by $O(\Delta^2)$.*

3.2 Deletions

Suppose that edge (u, v) is deleted from G_{i-1} thereby obtaining G_i. Assume without loss of generality that in the orientation of G_{i-1} we had $u \to v$.

We begin by removing (u, v) from our data structure. Notice that the only edges that may be violated now are edges incoming into u. Furthermore, if there is an edge $v' \to u$ that is violated now, then adding to the graph another copy of (u, v') (producing a multi-graph) that is oriented in the opposite direction (i.e., $u \to v'$) will guarantee that there are no violated edges. However, the resulting multi-graph has an extra edge that should be deleted. So we now recursively delete the original copy of edge (u, v') (not the copy that was just added, oriented $u \to v'$, which we keep). This means that we have actually *flipped* the orientation of (u, v'), reverting $d_{out}(u)$ to its value before the entire deletion process took place. This recursive process will continue until all edges of the graph are valid. Moreover, there is at most one duplicated edge at any given time, and the graph obtained at the end of the process has no duplicated edges. Our choice to add a copy of a violated edge incoming to u (if such an edge exists) guarantees that the number of recursive steps is at most Δ.

Due to space limitations, more details and correctness of the deletion process are described in the full version. Overall, we prove the following theorem.

Theorem 6. *There exists a deterministic algorithm for maintaining an orientation of a fully dynamic graph on n vertices while supporting the following:*
- *The maximum out-degree is $\Delta \leq \inf_{\beta > 1} \{\beta \cdot \alpha(G) + \log_\beta n\}$,*
- *The worst-case time to execute an edge insertion is $O(\Delta^2)$,*
- *The worst-case time to execute an edge deletion is $O(\Delta)$, and*
- *The worst-case number of orientations performed per update is $\Delta + 1$.*

4   A More Efficient Algorithm

In this section we present a more efficient, though more involved, algorithm that improves the insertion update time from $O(\Delta^2)$ to $O(\gamma \cdot \Delta)$, without increasing any of the other measures, at the cost of setting $\hat{\alpha}$ and $\hat{\beta}$ in advance.

An Intermediate Invariant: So far we have introduced two invariants. On one extreme, the stronger Invariant 4 guarantees that all edges are valid, and this led to our simple algorithm in Section 3. On the other extreme, the weaker Invariant 3 only guarantees that γ outgoing edges of each vertex are valid. On an intuitive level, the benefit of having the weaker Invariant 3 being maintained comes into play during the insertion process of edge (u, v) that is oriented as $u \to v$, where instead of scanning all of the outgoing edges of u looking for a violated edge, it is enough to scan only γ edges. If such a guarantee could be made to work, the insertion update time would be reduced to $O(\gamma \cdot \Delta)$. However, it is unclear how to efficiently maintain Invariant 3 as deletions take place. Specifically, when one of the γ outgoing valid edges of a vertex is deleted, it is possible that there is no other valid outgoing edge to replace it.

Our strategy is not to maintain Invariant 3 directly, but rather to define and maintain an intermediate invariant (see Invariant 7), which is stronger than Invariant 3 but still weak enough so that we only need to scan γ outgoing edges of u during the insertion process. The additional strength of the intermediate invariant will assist us in efficiently supporting deletions. Before stating the invariant, we define the following. For any $i \geq 1$, an edge (u, v) oriented as $u \to v$ is called i-*valid* if $d_{out}(v) \geq d_{out}(u) - i$; if it is not i-valid then it is i-*violated*. We also say that a vertex w is *spectrum-valid* if the set E_w of its outgoing edges can be partitioned into $q = q_w = \lceil \frac{|E_w|}{\gamma} \rceil$ sets $E_w^1, \cdots, E_w^q$ such that for each $1 \leq i \leq q$, the following holds: (1) $|E_w^i| = \gamma$ (except for the residue set E_w^q which contains the remaining $|E_w| - (q - 1) \cdot \gamma$ edges, i.e., $|E_w^q| = |E_w| - (q - 1) \cdot \gamma$), and (2) all edges in E_w^i are i-valid. If a vertex is not spectrum-valid then it is *spectrum-violated*.

Invariant 7. *Each vertex w is spectrum-valid.*

We will call E_w^1 (E_w^q) the first (last) set of edges for w. To give some intuition as to why Invariant 7 helps us support deletions efficiently, notice that once an edge (u, v) that is oriented as $u \to v$ is deleted and needs to be replaced, it will either be replaced by a flip of some violated incoming edge (which will become valid after the flip), or it can be replaced by one of the edges from E_u^2, as these edges were previously 2-valid, and after the deletion they are all 1-valid. We emphasize already here that during the insertion process we do not scan the γ edges of the first set (i.e., those that are guaranteed to be 1-valid prior to the insertion), but rather scan the γ (in fact, $\gamma - 1$) edges of the last set (and possibly of the set before last) that are only guaranteed to be q-valid.

In order to facilitate the use of Invariant 7, each vertex w will maintain its outgoing edges in a doubly linked list $\mathcal{L}_w$. We say that $\mathcal{L}_w$ is *valid* if for every $1 \leq i \leq q$, the edges between location $\gamma \cdot (i - 1) + 1$ and location $(\gamma \cdot i)$ in the list are all i-valid. These locations for a given i are called the i-*block* of $\mathcal{L}_w$. So, in a valid $\mathcal{L}_w$ the first location must be 1-valid and belongs to the 1-block, the last location must be q-valid and belongs to the q-block, etc. Note that for $i = q$ the number of locations (i.e., $|E_w| - (q - 1) \cdot \gamma$) may be smaller than γ. If $\mathcal{L}_w$ is not valid then it is *violated*.

We now provide an overview of the more efficient algorithms for insertion and deletion. Due to space limitations, the full details are given in the full version.

Insertions: Suppose that edge (u, v) is added to G_{i-1} thereby obtaining G_i. The process of inserting the new edge is performed as in Section 3 with the following modifications. Instead of scanning all outgoing edges of u in order to find a violated edge, we only scan the *last* $\gamma - 1$ edges in $\mathcal{L}_u$; if there are less than $\gamma - 1$ edges then we scan them all. If one of these edges, say (u, v'), is violated then we remove (u, v') from the graph, replace (u, v') with (u, v) in $\mathcal{L}_u$, and recursively insert (u, v') with the flipped orientation (just like in Section 3). If all of these edges are valid, we move them together with the new edge (u, v) to front of $\mathcal{L}_u$.

Deletions: Suppose that edge (u, v) is deleted from G_{i-1} thereby obtaining G_i. The process of deleting the edge is performed as in Section 3 with the following modifications. If an edge incoming into u, say (u, v'), is violated and is flipped (just like in Section 3), then we replace (u,v) with (u,v') in $\mathcal{L}_u$ and continue recursively to delete the original copy of (u,v'). If all incoming edges of u are valid, we remove (u,v) from $\mathcal{L}_u$.

Theorem 7. *There exists a deterministic algorithm for maintaining an orientation of a fully dynamic graph on n vertices that has arboricity at most $\hat{\alpha}$ (at all times), while supporting the following:*
- *The maximum out-degree is $\Delta \leq \hat{\beta} \cdot \hat{\alpha} + \log_{\hat{\beta}} n$,*
- *The worst-case time to execute an edge insertion is $O(\hat{\beta} \cdot \hat{\alpha} \cdot \Delta)$,*
- *The worst-case time to execute an edge deletion is $O(\Delta)$, and*
- *The worst-case number of orientations performed per update is $\Delta + 1$.*

Acknowledgments. The fourth-named author is grateful to Ofer Neiman for helpful discussions.

References

1. Alon, N., Yuster, R., Zwick, U.: Color-coding. J. ACM 42, 844–856 (1995)
2. Chrobak, M., Eppstein, D.: Planar orientations with low out-degree and compaction of adjacency matrices. Theor. Comput. Sci. 86, 243–266 (1991)
3. Brodal, G.S., Fagerberg, R.: Dynamic representation of sparse graphs. In: 6th International Workshop on Algorithms and Data Structures, WADS, pp. 342–351 (1999)
4. Kowalik, L.: Adjacency queries in dynamic sparse graphs. Inf. Process. Lett. 102, 191–195 (2007)
5. Kowalik, L., Kurowski, M.: Oracles for bounded-length shortest paths in planar graphs. ACM Transactions on Algorithms 2, 335–363 (2006)
6. Cain, J.A., Sanders, P., Wormald, N.: The random graph threshold for k-orientiability and a fast algorithm for optimal multiple-choice allocation. In: 18th Annual ACM-SIAM Symposium on Discrete Algorithms, pp. 469–476. SIAM (2007)
7. Neiman, O., Solomon, S.: Simple deterministic algorithms for fully dynamic maximal matching. In: Proc. of 45th STOC, pp. 745–754 (2013)
8. Dvořák, Z., Tůma, V.: A dynamic data structure for counting subgraphs in sparse graphs. In: Dehne, F., Solis-Oba, R., Sack, J.-R. (eds.) WADS 2013. LNCS, vol. 8037, pp. 304–315. Springer, Heidelberg (2013)
9. Eisenstat, D., Klein, P.N., Mathieu, C.: An efficient polynomial-time approximation scheme for steiner forest in planar graphs. In: Proc. of SODA, pp. 626–638 (2012)
10. Eppstein, D.: All maximal independent sets and dynamic dominance for sparse graphs. ACM Transactions on Algorithms 5 (2009)
11. Nash-Williams, C.S.J.A.: Edge-disjoint spanning trees in finite graphs. Journal of the London Mathematical Society 36(1), 445–450 (1961)
12. Nash-Williams, C.S.J.A.: Decomposition of finite graphs into forests. Journal of the London Mathematical Society 39(1), 12 (1964)
13. Gabow, H.N., Westermann, H.H.: Forests, frames, and games: Algorithms for matroid sums and applications. Algorithmica 7, 465–497 (1992)
14. Arikati, S.R., Maheshwari, A., Zaroliagis, C.D.: Efficient computation of implicit representations of sparse graphs. Discrete Appl. Math. 78, 1–16 (1997)
15. Erickson, J.: (2006), http://www.cs.uiuc.edu/ jeffe/teaching/datastructures/ 2006/problems/Bill-arboricity.pdf (retrieved November 2013)
16. Gupta, A., Kumar, A., Stein, C.: Maintaining assignments online: Matching, scheduling, and flows. In: Chekuri, C. (ed.) SODA, pp. 468–479. SIAM (2014)

Does Adding More Agents Make a Difference? A Case Study of Cover Time for the Rotor-Router[*]

Adrian Kosowski[1] and Dominik Pająk[2]

[1] Inria Paris-Rocquencourt, France
adrian.kosowski@inria.fr
[2] LaBRI, Inria Bordeaux Sud-Ouest, France
dominik.pajak@inria.fr

Abstract. We consider the problem of graph exploration by a team of k agents, which follow the so-called rotor router mechanism. Agents move in synchronous rounds, and each node successively propagates agents which visit it along its outgoing arcs in round-robin fashion. It has recently been established by Dereniowski et al. (STACS 2014) that the rotor-router cover time of a graph G, i.e., the number of steps required by the team of agents to visit all of the nodes of G, satisfies a lower bound of $\Omega(mD/k)$ and an upper bound of $O(mD/\log k)$ for any graph with m edges and diameter D. In this paper, we consider the question of how the cover time of the rotor-router depends on k for many important graph classes. We determine the precise asymptotic value of the rotor-router cover time for all values of k for degree-restricted expanders, random graphs, and constant-dimensional tori. For hypercubes, we also resolve the question precisely, except for values of k much larger than n. Our results can be compared to those obtained by Elsässer and Sauerwald (ICALP 2009) in an analogous study of the cover time of k independent parallel random walks in a graph; for the rotor-router, we obtain tight bounds in a slightly broader spectrum of cases. Our proofs take advantage of a relation which we develop, linking the cover time of the rotor-router to the mixing time of the random walk and the local divergence of a discrete diffusion process on the considered graph.

1 Introduction

Graph exploration is a task in which a team of agents is initially placed on a subset of nodes of the graph, and the agents are required to move around the graph so that each node is visited by at least one agent. Exploration with multiple walks is usually studied in a scenario where k agents are placed on some set of starting nodes and deployed in parallel, in synchronous steps. The principal parameter of interest is the *cover time* of the process, i.e., the number of steps until each node of the graph has been visited by at least one agent, for a worst case initial placement of agents in the graph. The agents may be endowed with different capabilities, ranging from a

---

[*] Research partially supported by ANR project DISPLEXITY and by NCN under contract DEC-2011/02/A/ST6/00201.

J. Esparza et al. (Eds.): ICALP 2014, Part II, LNCS 8573, pp. 544–555, 2014.
© Springer-Verlag Berlin Heidelberg 2014

priory complete knowledge of the graph topology, through online scenarios in which they need to discover a map of the graph, to the most restrictive, where agents are in some sense passive (oblivious), and their movement is governed by simple local rules within the system.

In this context, a fundamental problem concerns the cover time of k independent parallel random walks in a graph. Alon *et al.* [3], Efremenko and Reingold [9], and Elsässer and Sauerwald [11] have considered the notion of the *speedup* of the random walk for an undirected graph G, defined as the ratio between the cover time of a k-agent walk in G for worst-case initial positions of agents and that of a single-agent walk in G starting from a worst-case initial position, as a function of k. A characterization of the speedup has been achieved for many graph classes, although the question of the minimum and maximal values of speedup attainable in general is still open. The smallest known value of speedup for the random walk is $\Theta(\log k)$, attained e.g. for the cycle, while the largest known value is $\Theta(k)$, attained in a bounded range of values of k for many graph classes, such as expanders, cliques, and stars.

Our focus in this paper is on the deterministic model of walks in graphs known as the *rotor-router*. In the rotor-router model, introduced by Priezzhev *et al.* [16], the behaviour of the agent is fully controlled by the the undirected graph in which it operates. The edges outgoing from each node v are arranged in a fixed cyclic order known as a *port ordering*, which does not change during the exploration. Each node v maintains a *pointer* which indicates the edge to be traversed by the agent during its next visit to v. If the agent has not visited node v yet, then the pointer points to an arbitrary edge adjacent to v. The next time when the agent enters node v, it is directed along the edge indicated by the pointer, which is then advanced to the next edge in the cyclic order of the edges adjacent to v. Each agent propagated by the rotor-router is a memoryless entity, but due to the existence of pointers, the rotor-router system as a whole is not Markovian. On the other hand, the system requires no special initialization, and its state at any moment of time is a valid starting state for the process.

State-of-the-Art for the Rotor-Router. For the case of a single agent, it is known that for any n-node graph of m edges and diameter D, the cover time of the rotor-router in a worst-case initialization in the graph is precisely $\Theta(mD)$ [19,4]. After $\Theta(mD)$ time, the trajectory of the agent stabilizes to a periodic Eulerian traversal of the set of directed edges of the graph.

For $k > 1$ agents, no similar structural properties are observed, and in particular the rotor-router system may stabilize to a limit cycle of configurations of length exponential in n [18]. Recently, Klasing *et al.* [14] have provided the first evidence of speedup, showing that for the special case when G is a cycle, a k-agent system explores an n-node cycle $\Theta(\log k)$ times more quickly than a single agent system for $k < n^{1/11}$. A result for general graphs has been obtained by Dereniowski *et al.* [7], who show that the cover time of a k-agent system is always between $\Theta(mD/k)$ and $\Theta(mD/\log k)$, for any graph.

There exist interesting similarities between the behavior of the rotor-router and the random walk. For a single agent, the (deterministic) cover time of the rotor-router and the (expected) cover time of the random walk prove to be surprisingly convergent for many special graph classes, such as cycles and constant-degree expanders (same order of cover time for a single agent), cliques and stars (where the rotor-router is faster by a factor of $\Theta(\log n)$), or hypercubes (where the rotor-router is slower by a factor of $\Theta(\log n)$). A larger difference in cover time is observed, e.g., for the 2-dimensional grid, where the cover time of the rotor-router for a single agent is $\Theta(n^{3/2})$, as compared to the $\Theta(n \log n \log \log n)$ cover time for the random walk. For general graphs, the $\Theta(mD)$ bound on the cover time of the rotor-router can be compared to the upper bound of $O(mD \log n)$ on the cover time of the random walk, although the latter bound is far from tight for many graph classes.

Our Results. In this paper, we ask about how the cover time of the k-agent rotor router depends on the number of agents k for specific graph classes, and investigate whether the similarities in terms of cover time of the rotor-router and the random-walk extend beyond the single agent case. We determine the precise asymptotic value of the cover time for degree-restricted expanders, random graphs, constant-dimensional tori, and hypercubes. Our results can be seen as complementary to those of Elsässer and Sauerwald [11], who studied the cover time of k multiple random walks in the same graph classes. We show that for all of the considered graph classes (except cycles), the cover time of both the rotor-router and the random walk admits a speedup of precisely $\Theta(k)$ for relatively small values of k, but above a certain threshold, the speedup of the rotor-router and random walk become divergent. (For cycles, both processes admit a speedup of $\Theta(\log k)$.) Our results are succinctly presented in Table 1.

We recall that for $k = 1$, the cover time of the rotor-router is $\Theta(mD)$, and note that for sufficiently large k ($k > n\Delta^D$ for a graph of maximum degree Δ), the cover time of the rotor-router is equal to precisely D, since the graph can be flooded with agents starting from a fixed node initially having Δ^D agents. Above this threshold ($k > n\Delta^D$), adding new agents to the system does not speed up exploration. The results we obtain show that for complete graphs, random graphs, and expanders, a cover time of $\Theta(D)$ is attained already for much smaller teams of agents. These graphs also display dichotomous behaviour: up to a certain threshold value of $k_1 = \Theta(m)$, the cover time decreases linearly with the number of agents, and above this threshold, the cover time remains fixed at $\Theta(D)$. We show that the cycle also admits this type of single-threshold behaviour, but with logarithmic speedup in the range of small k, with a cover time of $\Theta(n^2/\log k)$ for $k < k_1 = 2^n$, and a cover time of $\Theta(n)$ for $k \geq k_1$.

Interestingly, we prove that the d-dimensional torus for constant d (with $D = n^{1/d}$) admits precisely two threshold values of k (cf. Table 1). For $k < k_1 = n^{1-1/d}$, the speedup is linear with k; for $k_1 \leq k < k_2 = 2^{n^{1/d}}$, the cover time further decreases with $\log(k/k_1)$, and above k_2, the cover time is asymptotically fixed at $\Theta(n^{1/d})$. We remark that the for parallel random walks, the situation appears to be similar, however the question of obtaining a complete characterization remains open. For the hypercube, we also prove threshold

Table 1. Cover time of the k-agent rotor-router system for different values of k in a n-node graph with m edges and diameter D. The results for d-dimensional tori are presented for d constant. The result for expanders concerns the case when the ratio of the maximum degree and the minimum degree of the graph is $O(1)$. The result for random graphs holds in the Erdős-Renyi model with edge probability $p > (1 + \varepsilon)\frac{\log n}{n}$, $\varepsilon > 0$, a.s.

Graph	k	Cover time	Reference
General graph	$\leq poly(n)$	$O\left(\frac{mD}{\log k}\right)$	[7]
		$\Omega\left(\frac{mD}{k}\right)$	[7]
Cycle	$< 2^n$	$\Theta\left(\frac{n^2}{\log k}\right)$	[14] (for $k < n^{1/11}$); Thm. 4
	$\geq 2^n$	$\Theta(n)$	
d-dim. torus	$< n^{1-1/d}$	$\Theta\left(\frac{n^{1+1/d}}{k}\right)$	Thm. 3
	$\in [n^{1-1/d}, 2^{n^{1/d}}]$	$\Theta\left(\frac{n^{2/d}}{\log(k/n^{1-1/d})}\right)$	Thm. 3
	$> 2^{n^{1/d}}$	$\Theta(n^{1/d})$	
Hypercube	$< n\frac{\log n}{\log\log n}$	$\Theta\left(\frac{n\log^2 n}{k}\right)$	Cor. 1
	$\in \left[n\frac{\log n}{\log\log n}, n2^{\log^{1-\varepsilon} n}\right]$ (for any $\varepsilon > 0$)	$\Theta(\log n \log\log n)$	Thm. 5
	$> n2^{\log^{1-\varepsilon} n}$	$O(\log n \log\log n)$	Thm. 5
	$> n^{\log_2 n}$	$\Theta(\log n)$	
Complete	$< n^2$	$\Theta\left(\frac{n^2}{k}\right)$	Thm. 2
	$\geq n^2$	$\Theta(1)$	
Expander	$< n\log n$	$\Theta\left(\frac{n\log^2 n}{k}\right)$	Thm. 2
	$\geq n\log n$	$\Theta(\log n)$	
Random graph	$< n\log n$	$\Theta\left(\frac{n\log^2 n}{k}\right)$	Thm. 2
	$\geq n\log n$	$\Theta(\log n)$	

behaviour for the speedup of the k-agent rotor-router, showing that there exist at least three threshold values of k (linear speedup for small k, a flat period with no speed-up for k slightly larger than n, a further period of slow growth, and finally a flat period for extremely large k). We also completely characterize the cover time of the hypercube for k up to a point beyond the first threshold.

Intuition of Approach: Exploration vs. Diffusion. In contrast to the case of parallel random walks, in the rotor-router system multiple agents interact with the same set of pointers at nodes, and the agents cannot be considered independent. However, the link between the multi-agent rotor-router and the parallel random walk processes becomes more apparent when the number of agents is extremely large ($k \gg n$), so that multiple agents are located at each node of the graph. Then, a fixed node v of degree d in the graph, which contains

$a_t(v)$ agents at a given moment of time t, will send them out along outgoing links in the next step of the rotor-router process, propagating the pointer at each step, so that each of its neighbours receives either $\lfloor a_t(v)/d \rfloor$ or $\lceil a_t(v)/d \rceil$ agents. In an analogous parallel random walk process, the *expected* number of agents following each of the outgoing links of a node v containing $a_t(v)$ agents will be $a_t(v)/d$. In fact, both the random walk and the rotor-router can be seen as different forms of discretization of the *continuous diffusion process*, in which a node having real-valued load $a_t(v)$ sends out precisely $a_t(v)/d$ load to each of its neighbours in the given time step. Discrete diffusion processes appear in research areas including statistical physics and distributed load balancing problems, and some studies of rotor-router-type systems have also been devoted to their diffusive properties. It is known, in particular, that, at any moment time, the difference of the number of agents located at a node between the rotor-router system and that in continuous diffusion is bounded by $\Theta(d \log n \mu^{-1})$ for d-regular graphs with eigenvalue gap μ [17], given identical initialization. This difference can even be bounded by constant for the case of lines [6] and grids [8]. Some other results in the area can also be found in [13,1]. In this paper, we observe that, somewhat counter-intuitively, the link between continuous diffusion and the rotor-router can also be exploited for small values of k ($k \ll n$), for which agents as a rule occupy distinct nodes ($a_t(v) = 1$), and rounding $a_t(v)/d$ up or down to the nearest integer makes a major difference.

Organization of Results. In Section 2, we provide a formalization of the rotor-router model and some notation. In Section 3, we outline the technique which we subsequently use to bound the cover time in different graph classes. The main theorem of Section 4 captures the link between the cover time of the k-rotor-router system, the mixing time $\mathsf{MIX}_{1/4}$ of the random walk process in the graph, and a graph parameter known as its discrepancy Ψ [17,5], in its simplest form. This result directly provides tight bounds on the cover time for most of the considered graph classes, admitting small mixing time. The remaining cases of tori, cycles, and hypercubes are considered in Sections 5, 6, and 7, respectively.

2 Preliminaries

We consider an undirected connected graph $G = (V, E)$ with n nodes, m edges and diameter D. We denote the neighborhood of a node $v \in V$ by $\Gamma(v)$. The degree of a node v is denoted by $\deg(v)$, the maximum degree of the graph is denoted by Δ, and the minimum degree by δ. The directed graph $\overrightarrow{G} = (V, \overrightarrow{E})$ is the directed symmetric version of G, where the set of arcs $\overrightarrow{E} = \{(v, u) : \{v, u\} \in E\}$. We will denote arc (v, u) by $v \rightarrow u$.

Model Definition. We study the rotor-router model (on graph G) with $k \geq 1$ indistinguishable agents, which run in steps, synchronized by a global clock. In each step, each agent moves in discrete steps from node to node along the arcs of graph $\vec{G}$. A *configuration* at the current step is defined as a triple $((\rho_v)_{v\in V}, (\pi_v)_{v\in V}, \{r_1, \ldots, r_k\})$, where ρ_v is a cyclic order of the arcs (in graph $\vec{G}$) outgoing from node v, π_v is an arc outgoing from node v, which is referred to as *the (current) port pointer at node v*, and $\{r_1, \ldots, r_k\}$ is the (multi-)set of nodes currently containing an agent. For each node $v \in V$, the cyclic order ρ_v of the arcs outgoing from v is fixed at the beginning of exploration and does not change in any way from step to step.

For an arc $v \to u$, let $next(v \to u)$ denote the arc next after arc $(v \to u)$ in the cyclic order ρ_v. The exploration starts from some initial configuration and then keeps running in all future rounds, without ever terminating. During the current step, first each agent i is moved from node r_i traversing the arc π_{r_i}, and then the port pointer π_{r_i} at node r_i is advanced to the next arc outgoing from r_i (that is, π_{r_i} becomes (π_{r_i})). This is performed sequentially for all k agents. Note that the order in which agents are released within the same step is irrelevant from the perspective of the system, since agents are indistinguishable. For example, if a node v contained two agents at the start of a step, then it will send one of the agents along the arc π_v, and the other along the arc $(v, next(\pi_v))$. For simplicity of notation only, we will assume that the ports outgoing from a node od degree d are numbered with consecutive integers $\{0, \ldots, d-1\}$, that the function $next$ advances the pointer from port i to port $(i+1) \bmod d$, $0 \le i < d$, and that the pointers at all nodes of the graph all initially point towards port 0.

Notation. We will say that a node is *visited* by an agent in round t if the agent is located at this node at the start of round $t+1$. By $\mathbf{n}_t(v)$ we will denote the total number of visits of agents to node v, counting from the initialization of the system to the end of round t of the considered rotor-router process. In particular, $\mathbf{n}_0(v)$ refers to the number of agents at a node directly after initialization (at the start of round 1). Henceforth, we will treat $\mathbf{n}_t$ as a non-negative integer-valued vector of dimension n. The worst-case cover time of the rotor-router for an initialization on graph G with k agents (i.e., for an initialization satisfying $\|\mathbf{n}_0\|_1 = k$), will be denoted by $C_{rr}^k(G)$.

We also introduce some auxiliary notation related to random walks and diffusion on the graph. We will denote by $P_t(v, u)$ the probability that a simple random walk, starting at node v of the graph, is located at u after exactly t steps of the walk, $t \ge 0$. The transition matrix of the random walk will be denoted by $\mathbf{M}$. For a node $u \in V$, $\mathbf{u}$ will denote a vector of length n with $\mathbf{u}(u) = 1$ and all other entries 0. We recall that the cells of the t-th power of this matrix satisfy the following relation: $\mathbf{u}^\mathsf{T}\mathbf{M}^t\mathbf{v} = P_t(v, u)$ [2]. The mixing time after which the random walk on the graph G reaches a total variation distance of at most $1/4$ from its stationary distribution will be denoted by $\mathsf{MIX}_{1/4}(G) = \max_{v \in V} \min\{t : \|P_t(v, \cdot) - \pi\|_{TV}\}$, where π denotes the vector of the stationary distribution of the random walk. By $P_t(v, \cdot)$ we denote the vector of probability distribution of the t-step random walk starting from v. For vector $P_t(v, \cdot) - \pi$, the value $\|P_t(v, \cdot) - \pi\|_{TV}$ is the total variation distance defined as follows: $\|P_t(v, \cdot) - \pi\|_{TV} = \frac{1}{2}\sum_{u \in V}|P_t(v, u) - \pi_u|$. This definition of mixing time can be compared with the following definition of the mixing time used in [11].

We will denote the mixing time defined according to this second definition by $\mathsf{MIX}^*_{1/2}(G) = \max_{v \in V} \min \left\{ t : \forall_{u \in V} \frac{3\pi_u}{2} \geq P_t(v, u) \geq \frac{\pi_u}{2} \right\}$. In our considerations we will use a similar value $t_{1/2}(G)$, which satisfies slightly relaxed constraints: $t_{1/2}(G) = \max_{v \in V} \min \left\{ t : \forall_{u \in V} P_t(v, u) \geq \frac{\pi_u}{2} \right\}$, which denotes time after which probability of being at any node is at least half of the stationary probability regardless of the starting node of the random walk.

Clearly $t_{1/2}(G) \leq \mathsf{MIX}^*_{1/2}(G)$, and thus to upper bound $t_{1/2}(G)$, we can use results from [11], where authors present upper bounds on the value of $\mathsf{MIX}^*_{1/2}(G)$ for some graph classes.

3   The Main Technique

To bound the cover time of the rotor-router, for any moment of time t, we will estimate the difference between the number of visits of the rotor-router to a node $x \in V$ up to time t, and the corresponding *expected* number of visits of parallel random walks, starting from the same initial placement of agents in the graph, to the same node x. (The latter notion can be equivalently interpreted as the total amount of load arriving in rounds 1 to t in a similarly initialized continuous diffusion process in load balancing.) It turns out that the difference (discrepancy) between these two processes is bounded. As soon as the expected total number of visits of parallel random walks to x up to t has exceeded the maximum possible discrepancy with respect to the rotor-router, we can be sure that node x has been visited by the rotor-router at least once up to time t. This is captured by the following lemma.

Lemma 1. *Take any graph G. Let t^* be such a time moment that*

$$\forall_{x \in V} \left(\sum_{\tau=0}^{t^*} \mathbf{M}^\tau \mathbf{n}_0 \right)(x) > \Psi_{t^*},$$

where

$$\Psi_t(G) = \max_{v \in V} \sum_{\tau=0}^{t} \sum_{(u_1, u_2) \in \vec{E}} |P_\tau(u_1, v) - P_\tau(u_2, v)|.$$

Then, the cover time of the k-agent rotor-router with arbitrary initialization on graph G satisfies $C_{rr}^k(G) \leq t^*$.

Before proceeding to prove the lemma, we remark that $\mathbf{M}^\tau \mathbf{n}_0$ is a vector describing the expected number of agents at nodes after τ steps of independent random walks on G, and that the expression $\left(\sum_{\tau=0}^{t^*} \mathbf{M}^\tau \mathbf{n}_0 \right)(x)$ on the left-hand side of the inequality is the before-mentioned expected total number of visits of random walks to x up to time t starting from initial agent placement. The expression Ψ_{t^*} is a generalization of the so-called 1-*discrepancy* Ψ of the graph, $\Psi = \lim_{t \to +\infty} \Psi_t$, introduced in [17]. The measure of 1-discrepancy is often applied when comparing a continuous and discrete process at a fixed moment of time t [5,12], whereas herein we compare the *total* distance of two processes over all steps up to time t.

Proof. Consider the total number of of visits $n_t(u)$ at vertex u until step t by the rotor-router. It may be expressed as the sum of the number of agents initially located in u and the number of agents that entered to u from its neighbors (cf. [19,14] for details of the argument):

$$n_t(u) = n_0(u) + \sum_{v \in \Gamma(u)} \left\lceil \frac{n_{t-1}(v) - port(v,u)}{\deg(v)} \right\rceil, \tag{1}$$

where $port(v,u) \in \{0, 1, \ldots \deg(v) - 1\}$ denotes the label of the port leading from v to u.

We can rewrite Equation (1) as follows

$$n_t(u) = \sum_{v \in \Gamma(u)} \frac{n_{t-1}(v)}{\deg(v)} + n_0(u) + \xi_t(u), \tag{2}$$

where ξ_t is an "error vector" defined as: $\xi_t(u) = \sum_{v \in \Gamma(u)} \alpha_t^{(v,u)}$, where $\alpha_t^{(v,u)} = \left(\left\lceil \frac{n_{t-1}(v) - port(v,u)}{\deg(v)} \right\rceil - \frac{n_{t-1}(v)}{\deg(v)} \right)$. Note that the values $\alpha_t^{(v,u)}$ are defined over directed arcs of the graph, $(v,u) \in \vec{E}$, satisfying $|\alpha_t^{(v,u)}| \leq 1$ and $\sum_{u \in \Gamma(v)} \alpha_t^{(v,u)} = 0$. Consequently, we have $\sum_{(v,u) \in \vec{E}} \alpha_t^{(v,u)} \mathbf{v} = \mathbf{0}$, and: $\xi_t = \sum_{(v,u) \in \vec{E}} \alpha_t^{(v,u)} \mathbf{u} = \sum_{(v,u) \in \vec{E}} \alpha_t^{(v,u)} \cdot (\mathbf{u} - \mathbf{v})$. Now, we rewrite (2) as follows:

$$\mathbf{n}_t = \mathbf{M} \mathbf{n}_{t-1} + (\mathbf{n}_0 + \xi_t), \tag{3}$$

where $\mathbf{M}$ is the transition matrix of the random walk on G. Expanding (3) we have:

$$\mathbf{n}_t = \sum_{\tau=0}^{t} \mathbf{M}^\tau \mathbf{n}_0 + \sum_{\tau=0}^{t} \mathbf{M}^\tau \xi_{t-\tau}. \tag{4}$$

We will now bound the absolute value of the maximum element of the vector $\sum_{\tau=0}^{t} \mathbf{M}^\tau \xi_{\tau-t}$.

We have

$$\left\| \sum_{\tau=0}^{t} \mathbf{M}^\tau \xi_{\tau-t} \right\|_\infty = \left\| \sum_{\tau=0}^{t} \left(\mathbf{M}^\tau \cdot \sum_{(v,u) \in \vec{E}} \alpha_{t-\tau}^{(v,u)} \cdot (\mathbf{u} - \mathbf{v}) \right) \right\|_\infty \leq$$

$$\leq \left\| \sum_{\tau=0}^{t} \sum_{(v,u) \in \vec{E}} \alpha_{t-\tau}^{(v,u)} \mathbf{M}^\tau \cdot (\mathbf{u} - \mathbf{v}) \right\|_\infty$$

Note that since $|\alpha_{t-\tau}^{(u,v)}| \leq 1$

$$\left\| \sum_{\tau=0}^{t} \mathbf{M}^\tau \xi_{\tau-t} \right\|_\infty \leq \left\| \sum_{\tau=0}^{t} \sum_{(v,u) \in \vec{E}} |\mathbf{M}^\tau \cdot (\mathbf{u} - \mathbf{v})| \right\|_\infty. \tag{5}$$

We rewrite the above in terms of probability distributions of of random walk on G after τ steps:

$$(\mathbf{M}^\tau \cdot (\mathbf{u} - \mathbf{v}))\,(w) = P_\tau(u, w) - P_\tau(v, w), \qquad\qquad (v, u) \in \vec{E}. \qquad (6)$$

In this way, we obtain for any $x \in V$:

$$\left|\left(\mathbf{n}_t - \sum_{\tau=0}^t \mathbf{M}^\tau \mathbf{n}_0\right)(x)\right| = \left|\left(\sum_{\tau=0}^t \mathbf{M}^\tau \xi_{\tau-t}\right)(x)\right| \le$$

$$\le \max_{w \in V} \sum_{\tau=0}^t \sum_{(v,u) \in \vec{E}} |P_\tau(u, w) - P_\tau(v, w)| = \Psi_t(G).$$

Thus, at time t any node the total number of visits in multi-agent rotor-router deviates from expected number of visits by multiple random walks by at most $\Psi_t(G)$. Since at time t^* at any node the expected number of visits by random walk is more than $\Psi_{t^*}(G)$ by assumption, all nodes have been visited at least once by the rotor-router.

4   Graphs with Small Mixing Time

Theorem 1. *The cover time $C_{rr}^k(G)$ of a k-agent rotor-router with arbitrary initialization on any graph G satisfies*

$$C_{rr}^k(G) \le t_{1/2}(G) + \frac{2\Delta}{\delta}\frac{n}{k}\Psi(G).$$

In order to apply Theorem 1 to special graph classes, we provide convenient bounds on the value of Ψ which hold for regular graphs. (All omitted proofs are deferred to the Appendix.)

Proposition 1. *For any d-regular graph G:*

(i) $\Psi(G) \le 4\sum_{t=0}^{\mathsf{MIX}_{1/4}(G)} \max_{v \in V} \sum_{\{u_1,u_2\} \in E} |P_t(u_1, v) - P_t(u_2, v)|$
(ii) $\Psi(G) = O(d\mathsf{MIX}_{1/4}(G))$.

By combining Theorem 1 and Proposition 1, we can obtain upper bounds on the cover time of the rotor-router in regular graphs. At this point we provide an auxiliary result, which allows us to extend all our considerations to almost-regular graphs, as well as to show that our bounds on cover time hold regardless of whether the considered graph has self-loops or not. The proof relies on a variant of the *delayed deployment* technique for the rotor-router, introduced in [14].

Proposition 2. *Consider a graph G' constructed from G by adding self-loops to vertices, so that in the port ordering at any vertex there are at most x consecutive self-loops. Then, $C_{rr}^k(G')/(x+1) \le C_{rr}^k(G) \le C_{rr}^k(G')$.*

Taking into account Theorem 1 and Propositions 1 and 2, we obtain an upper bound of $O(mD/k)$ on the cover time of the rotor-router in a wide class of almost regular graphs with small mixing time. The complementary lower bound $C_{rr}^k(G) = \Omega(mD/k)$ is due to [7]. These bounds hold for all k, until the trivial bound $C_{rr}^k(G) = \Omega(D)$ is reached.

Theorem 2. *For any graph G such that $t_{1/2}(G) = O(D)$, $\mathsf{MIX}_{1/4}(G) = O(D)$ and $\Delta/\delta = O(1)$ the cover time of the k-agent rotor-router in the worst-case initialization of the system is:*

$$C_{rr}^k(G) = \Theta\left(\max\left\{\frac{mD}{k}, D\right\}\right).$$

Theorem 2 immediately implies the results stated in Table 1 for the case of complete graphs, degree-constrained expanders, and Erdős-Renyi graphs with edge probability $p > (1 + \varepsilon)\frac{\log n}{n}$. For cliques it is easy to see that $t_{1/2}(G) = O(1)$. For degree-constrained expanders, and Erdős-Renyi graphs bound on value $t_{1/2}(G)$ can be found in [10] as $t_{1/2}(G)$ is upper bounded by the mixing time used there.

The classes of tori, cycles, and hypercubes require more careful analysis; we consider them in the following Sections.

5   The Torus

For the d-dimensional torus, Theorem 2 is not applicable, since the mixing time of the torus is $\mathsf{MIX}_{1/4}(G) = \Theta(n^{2/d})$ [15], for constant d, whereas its diameter is $D = \Theta(n^{1/d})$. In the range of $k \leq n^{1-1/d}$, we can apply Theorem 1, taking advantage of a known tight bound on $\Psi(G) = \Theta(n^{1/d})$. In this way, we obtain: $C_{rr}^k(G) = O\left(n^{2/d} + \frac{n^{1+1/d}}{k}\right) = O\left(\frac{mD}{k}\right)$. Moreover, the complementary lower bound $C_{rr}^k(G) = \Omega(mD/k)$ holds for all graphs by [7]. This resolves the case of $k \leq n^{1-1/d}$.

To bound the cover time for $k > n^{1-1/d}$, in view of Proposition 2, we can equivalently consider the torus with d self-loops added on each node. We will now rely on Lemma 1, taking into account tighter bounds on $\Psi_t(G)$ for small values of t. The following bound can be shown by a straightforward Markovian coupling argument.

Lemma 2. *If G' is a d-dimensional torus with d self-loops at each node, then $\Psi_t(G') \leq 24d\sqrt{t}$.*

Introducing the above bound into Lemma 1 and taking into account properties of the random walk in the torus, for $k = k'n^{1-1/d}$ ($k' > 1$), we eventually obtain a bound on cover time of the form $O\left(\frac{D^2}{\log k'}\right)$, Somewhat surprisingly, this bound is tight, and we propose an initialization of the rotor-router system which achieves this bound precisely. (The proof of tightness relies on a bound on cover time for the cycle, which is introduced in the following section.) In this way, we obtain a complete characterization of the speed-up of the rotor-router on the torus.

Theorem 3. *If G is a torus of constant dimension then cover time of k-agent rotor-router is*

(i) $C_{rr}^k(G) = \Theta\left(\frac{mD}{k}\right)$, *for $k \leq n^{1-1/d}$,*

(ii) $C_{rr}^k(G) = \Theta\left(\max\{\frac{D^2}{\log k'}, D\}\right)$, *for $k = k'n^{1-1/d}$, $k > n^{1-1/d}$.*

6   The Cycle

The general case result from [7] allows us to upper-bound the cover time of the k-rotor-router system on the cycle by $O\left(\max\{\frac{n^2}{\log k}, n\}\right)$, for any $k \geq 1$. On the other hand, the complementary lower bound of $\Omega\left(\frac{n^2}{\log k}\right)$ was only known to hold for $k < n^{1/11}$ [14]. In the following, we extend this lower bound to arbitrary values of k. The proof relies on a modification of the approach used in the proof of Lemma 1: whereas Lemma 1 can only be used to upper bound cover time, this time we perform a different transformation of (4) for a specific initialization of agents starting from a single node on the ring, for which we can show that the "error term" associated with vector $\xi_{t-\tau}$ is negative. Intuitively, this behaviour is due to an initialization of pointers which delays progress of the agents going along the path to the most distant node of the ring. We eventually obtain the following result.

Theorem 4. *If G is a cycle of size n then cover time of k-agent rotor-router is*

$$C_{rr}^k(G) = \Theta\left(\max\left\{\frac{n^2}{\log k}, n\right\}\right).$$

7   The Hypercube

For the hypercube with $n = 2^d$ vertices, the value of $\Psi(G)$ has been precisely derived in [5]. The corresponding asymptotic formula is $\Psi(G) = \Theta(\log^2 n)$. Using this result in combination with Theorem 1, we obtain the following corollary.

Corollary 1. *If G is a hypercube with n vertices then $C_{rr}^k(G) = \Theta\left(\frac{n\log^2 n}{k}\right) = \Theta\left(\frac{mD}{k}\right)$, for $k \leq n\frac{\log n}{\log\log n}$.*

The behavior of the rotor-router on the hypercube for $k > k_1 = n\frac{\log n}{\log\log n}$ is not completely understood. For $k = k_1$, the value of cover time is $O(\log n \log\log n)$. Interestingly, we can show that there exists a flat "plateau" region above k_1 in which the asymptotic cover time of the hypercube is precisely $\Theta(\log n \log\log n)$. The proof proceeds along slightly more complex lines than the proof of Theorem 4. We show that in the considered range of k, $\Theta(\log n \log\log n)$ time is required for k agents starting at one corner of the hypercube to reach the opposite corner, given an arrangement of ports at each node in which the pointer first traverses all ports leading the agent towards the starting vertex.

Theorem 5. *If G is a hypercube of size $n = 2^d$ then the cover time of k-agent rotor-router with $k \leq n \cdot 2^{\log^{1-\varepsilon} n}$ agents is $C_{rr}^k(G) > \frac{\varepsilon}{10}\log n \log\log n$, where $\varepsilon \in (0, 1)$ is an arbitrary fixed constant.*

We leave the question of the cover time of the rotor-router on the hypercube for $k > n \cdot 2^{\log^{1-\varepsilon} n}$ as open.

Acknowledgment. The authors thank Petra Berenbrink, Ralf Klasing, and Frederik Mallmann-Trenn for productive discussions on closely related topics.

References

1. Akbari, H., Berenbrink, P.: Parallel rotor walks on finite graphs and applications in discrete load balancing. In: SPAA, pp. 186–195 (2013)
2. Aldous, D., Fill, J.: Reversible markov chains and random walks on graphs (2001), http://stat-www.berkeley.edu/users/aldous/RWG/book.html
3. Alon, N., Avin, C., Koucký, M., Kozma, G., Lotker, Z., Tuttle, M.R.: Many random walks are faster than one. Combinatorics, Probability & Computing 20(4), 481–502 (2011)
4. Bampas, E., Gąsieniec, L., Hanusse, N., Ilcinkas, D., Klasing, R., Kosowski, A.: Euler tour lock-in problem in the rotor-router model. In: Keidar, I. (ed.) DISC 2009. LNCS, vol. 5805, pp. 423–435. Springer, Heidelberg (2009)
5. Berenbrink, P., Cooper, C., Friedetzky, T., Friedrich, T., Sauerwald, T.: Randomized diffusion for indivisible loads. In: Randall, D. (ed.) SODA, pp. 429–439. SIAM (2011)
6. Cooper, J.N., Doerr, B., Spencer, J.H., Tardos, G.: Deterministic random walks on the integers. Eur. J. Comb. 28(8), 2072–2090 (2007)
7. Dereniowski, D., Kosowski, A., Pajak, D., Uznanski, P.: Bounds on the cover time of parallel rotor walks. In: Mayr, E.W., Portier, N. (eds.) STACS. LIPIcs, vol. 25, pp. 263–275. Schloss Dagstuhl - Leibniz-Zentrum fuer Informatik (2014)
8. Doerr, B., Friedrich, T.: Deterministic random walks on the two-dimensional grid. Combinatorics, Probability, and Computing 18(1-2), 123–144 (2009)
9. Efremenko, K., Reingold, O.: How well do random walks parallelize? In: Dinur, I., Jansen, K., Naor, J., Rolim, J. (eds.) APPROX and RANDOM 2009. LNCS, vol. 5687, pp. 476–489. Springer, Heidelberg (2009)
10. Elsässer, R., Sauerwald, T.: Tight bounds for the cover time of multiple random walks. In: Albers, S., Marchetti-Spaccamela, A., Matias, Y., Nikoletseas, S., Thomas, W. (eds.) ICALP 2009, Part I. LNCS, vol. 5555, pp. 415–426. Springer, Heidelberg (2009)
11. Elsässer, R., Sauerwald, T.: Tight bounds for the cover time of multiple random walks. Theor. Comput. Sci. 412(24), 2623–2641 (2011)
12. Friedrich, T., Gairing, M., Sauerwald, T.: Quasirandom load balancing. SIAM J. Comput. 41(4), 747–771 (2012)
13. Kijima, S., Koga, K., Makino, K.: Deterministic random walks on finite graphs. In: Martinez, C., Hwang, H.-K. (eds.) ANALCO, pp. 18–27. SIAM (2012)
14. Klasing, R., Kosowski, A., Pajak, D., Sauerwald, T.: The multi-agent rotor-router on the ring: a deterministic alternative to parallel random walks. In: PODC, pp. 365–374 (2013)
15. Levin, D.A., Peres, Y., Wilmer, E.L.: Markov chains and mixing times. American Mathematical Society (2006)
16. Priezzhev, V., Dhar, D., Dhar, A., Krishnamurthy, S.: Eulerian walkers as a model of self-organized criticality. Phys. Rev. Lett. 77(25), 5079–5082 (1996)
17. Rabani, Y., Sinclair, A., Wanka, R.: Local divergence of markov chains and the analysis of iterative load balancing schemes. In: FOCS, pp. 694–705. IEEE Computer Society (1998)
18. Uznanski, P.: Personal communication (2014)
19. Yanovski, V., Wagner, I.A., Bruckstein, A.M.: A distributed ant algorithm for efficiently patrolling a network. Algorithmica 37(3), 165–186 (2003)

The Melbourne Shuffle:
Improving Oblivious Storage in the Cloud

Olga Ohrimenko[1], Michael T. Goodrich[2], Roberto Tamassia[3], and Eli Upfal[3]

[1] Microsoft Research Cambridge, UK
[2] University of California, Irvine, USA
[3] Brown University, USA

Abstract. We present a simple, efficient, and secure data-oblivious randomized shuffle algorithm. This is the first secure data-oblivious shuffle that is not based on sorting. Our method can be used to improve previous oblivious storage solutions for network-based outsourcing of data.

1 Introduction

One of the unmistakable recent trends in networked computation and distributed information management is that of *cloud storage* (e.g., see [14]), whereby users outsource data to external servers that manage and provide access to their data. Such services relieve users from the burden of backing up and having to maintain access to their data across multiple computing platforms. However, in return, such services also introduce privacy concerns since users no longer physically own the data. Thus, there is a need for algorithmic solutions that preserve the desirable properties of cloud storage while also providing privacy protection for user data.

Of course, users can encrypt data they outsource to the cloud, but this alone is not sufficient to achieve privacy protection, because the data access patterns that users exhibit can reveal information about the content of their data (e.g., see [13]). Therefore, there has been considerable amount of recent research on algorithms for *data-oblivious algorithms and storage*, which hide data access patterns for cloud-based network data management solutions (e.g., see [8, 9, 10, 11, 12, 15, 19, 20, 21, 22, 23]). Such solutions typically work by obfuscating a sequence of data accesses intended by a client by simulating it with the one that appears indistinguishable from a random sequence of data accesses. Often, such a simulation involves mixing the intended (real) accesses with a sequence of random "dummy" accesses. In addition, so as to never access the same address twice (which would reveal a correlation), such obscuring simulations also involve continually moving items around in the server's memory space. For this reason, the "inner-loop" computation required by such simulations is a *data-oblivious shuffling* operation, which moves a set of items to random locations in fashion that disallows the server to correlate the previous locations of items with their new locations. This inner-loop process requires putting items in new locations that are independent of their old locations while hiding the correlations between the two.

The most common way this inner-loop shuffling is implemented is, however, computationally expensive, since it involves assigning random (or pseudo-random) indices to items and then performing a data-oblivious sorting of these index-item pairs. Examples of such oblivious sorting algorithms include Batcher's sorting network [3], which

J. Esparza et al. (Eds.): ICALP 2014, Part II, LNCS 8573, pp. 556–567, 2014.
© Springer-Verlag Berlin Heidelberg 2014

requires $O(n(\log n)^2)$ I/Os to sort data of size n, or the AKS [1] or Zig-zag sorting [7] networks, which use $O(n \log n)$ I/Os, but with large constant factors that restrict their practicality. These algorithms are used in oblivious storage solutions by having a client use the server as an external memory, with the I/Os directing the client to issue commands to move items from the server to the client's private memory and from the client's private memory to the server. Though these solutions achieve a desired privacy level, they are expensive in their (amortized) access overhead time and also in their monetary cost when one considers a client outsourcing large volumes of data and accessing it from a cloud server that charges per every data request.

In this paper, therefore, we are interested in algorithmic improvements for oblivious storage solutions, in terms of their conceptual complexity, constant factors, and monetary costs. For instance, since cloud-storage servers typically charge users for each memory access but have fairly large bounds on the size of the messages for such I/Os, we allow for messages to have modest sizes, such as $O(\sqrt{n})$ for a storage of size n. This necessarily also implies that the client has an equally modest-sized private memory, in which to send and receive such messages (and also in which to perform internal swaps of data items away from the prying eyes of the server). Our goal in this research is to take advantage of such frameworks to replace data-oblivious sorting with simple oblivious data shuffling for the sake of providing simple, efficient, and cheap outsourced data management. Our framework, therefore, involves designing (or modifying) oblivious storage simulation algorithms where a client stores n items at the server and is allowed to issue a sequence of I/Os, each of which is a batch of reads and writes for the server's memory, for reasonable assumptions on message size and private memory size.

Related Work. A *shuffle* is an algorithm for rearranging an array to achieve a random permutation of its elements. Early shuffle methods were motivated by the problem of shuffling a deck of cards. Classic card shuffle methods (e.g., [2]) are not data-oblivious, however, as anyone observing card swaps or riffles (interleaving two subdecks) of such methods can learn the final output permutation. Goodrich and Mitzenmacher [9] showed that one can, in fact, shuffle a deck of n cards and guarantee that an observer cannot find a particular card in the output permutation with probability better than $O(1/n)$. However, this algorithm is not an effective shuffle for our purposes, since the output permutations produced by the algorithm are not all equally likely and there may be dependencies between large groups of cards that could be leaked. Most other existing efficient data-oblivious shuffling methods assign random values to the elements of the array and use a data-oblivious algorithm to sort the array according to these values.

Our Oblivious Shuffling Results. Our Melbourne shuffle[1] algorithm is instead the first data-oblivious shuffle method that is not based on a data-oblivious sorting algorithm. In Table 1, we compare the (optimized) Melbourne shuffle (Section 5), showing that it outperforms sorting-based shuffle methods. We refer the reader for detailed algorithms and the corresponding proofs of security to the full version of this paper in [17].

Improved Oblivious Storage. Oblivious storage and oblivious RAM (ORAM) simulation solutions aim at minimizing the *access overhead*, which is the amortized number of I/Os executed to perform a single storage access request while keeping reasonable

---

[1] The name of our algorithm is inspired by a "shuffle" dance technique.

Table 1. Comparison of data-oblivious sorting and shuffle algorithms over n items

	Randomized	Private Memory	Message Size	External Memory	I/Os
Batcher's network [3]		$O(1)$	$O(1)$	$O(n)$	$O(n(\log n)^2)$
Batcher's network I		$O(\sqrt{n})$	$O(\sqrt{n})$	$O(n)$	$O(\sqrt{n}(\log n)^2)$
Batcher's network II [8]		$O(\sqrt{n})$	$O(\sqrt[8]{n})$	$O(n)$	$O(n^{7/8})$
AKS [1], Zig-zag sort [7]		$O(1)$	$O(1)$	$O(n)$	$O(n \log n)$
Randomized shellsort [6]	✓	$O(\sqrt{n})$	$O(\sqrt{n})$	$O(n)$	$O(\sqrt{n} \log n)$
Melbourne shuffle	✓	$O(\sqrt{n})$	$O(\sqrt{n})$	$O(n)$	$O(\sqrt{n})$
Melbourne shuffle ($c \geq 3$)	✓	$O(\sqrt[c]{n})$	$O(\sqrt[c]{n})$	$O(n)$	$O(c\sqrt[c]{n^{c-1}})$

assumptions about the size of private memory and of messages exchanged between the client and the server. Goldreich and Ostrovsky [4, 5] give two oblivious storage solutions for a client with $O(1)$ private memory size: the square root method with $O(\sqrt{n})$ overhead and the hierarchical method with $O((\log n)^3)$ overhead. The hierarchical method was recently extended using techniques such as Bloom filters [22, 23] and cuckoo hash tables [8, 12, 15, 18]. E.g., the $\log n$-hierarchical solution of [12] uses $O(\sqrt[d]{n})$ temporary memory and achieves $O(\log n)$ access overhead, for $d \geq 2$. In [15] a similar method achieves $O((\log n)^2 / \log \log n)$ overhead with $O(1)$ private memory. All the above oblivious storage solutions rely on a periodic data-oblivious shuffle of the server storage, a task done using data-oblivious sorting. Thus, we can use our Melbourne shuffle to implement the shuffle steps of these algorithms.

Other oblivious storage solutions in [19, 20, 21] allow the user to have $o(n)$ private memory and then by applying the same solution recursively on this private memory bring it to $o(1)$ and adding a $\log n$ overhead. For example, Path ORAM [20] uses $O(\log n)$ (stateful) private memory and has $O((\log n)^2)$ access overhead.

As shown in Table 2, oblivious storage solutions based on our (optimized) Melbourne shuffle (Section 5) are efficient and practical.

Table 2. Comparison of oblivious storage solutions for n items

	Private Memory	Message Size	External Memory	Access Overhead
SquareRoot [5]	$O(1)$	$O(1)$	$O(n)$	$O(\sqrt{n})$
Path ORAM [20]	$O(\log n)$	$O(\log n)$	$O(n)$	$O((\log n)^2)$
Bucket Hash Hierarchical [5]	$O(1)$	$O(1)$	$O(n \log n)$	$O((\log n)^3)$
Cuckoo Hash Hierarchical [12] ($d \geq 2$)	$O(\sqrt[d]{n})$	$O(\sqrt[d]{n})$	$O(n)$	$O(\log n)$
SquareRoot with Melbourne shuffle	$O(\sqrt{n})$	$O(\sqrt{n})$	$O(n)$	$O(1)$
Hierarchical with Melbourne shuffle	$O(\sqrt[c]{n})$	$O(\sqrt[c]{n})$	$O(n)$	$O(c \log n)$
($c \geq 3$)	$O(\sqrt[c]{n} \log n)$	$O(\sqrt[c]{n} \log n)$	$O(n)$	$O(c)$

2   Preliminaries

We analyze the security of cryptographic primitives and our protocol in terms of the probability of success for an adversary in breaking them. Let k be a security parameter. We say that a scheme is secure if for every probabilistic polynomial time (in k) (PPT)

adversary $\mathcal{A}$, the probability of breaking the scheme is at most some *negligible function* negl(k), i.e., a function such that negl(k) < 1/|poly(k)| for every polynomial poly(k).

We use a *symmetric encryption scheme* ($\mathsf{Enc_{key}}, \mathsf{Dec_{key}}$) where key $\leftarrow \{0,1\}^k$. We require this scheme to be secure against the chosen-ciphertext attack (CPA) for multiple messages. Informally, in the *Enc-IND-CPA* security game, the adversary $\mathcal{A}$ picks two sequences of plaintexts and gives them to the challenger, who encrypts one of them and returns the sequence of ciphertexts, C, to the adversary. $\mathcal{A}$ is given full oracle access to $\mathsf{Enc_{key}}$ and limited oracle access to $\mathsf{Dec_{key}}$, where he cannot call $\mathsf{Dec_{key}}$ on ciphertexts in C. We say that ($\mathsf{Enc_{key}}, \mathsf{Dec_{key}}$) is secure if for every PPT adversary the probability of guessing which plaintext sequence was used to produce C is at most $1/2 + $ negl(k).

Consider an array of n elements that we wish to randomly rearrange and let $D = [1, n]$ be the set of indices of A. We use a family of secure and efficiently computable *pseudo-random permutations* (PRPs) $\Pi_{\mathsf{seed}} : D \rightarrow D$, keyed using a k-bit seed.

Given an array A of n (key,value) pairs (x, v) where $x \in [1, n]$, we denote the permutation π of A as $B = \pi(A)$, where $\pi = \Pi_{\mathsf{seed}}$ and $B[x] = A[\pi(x)], \forall x \in [1, n]$. We will use the same notation when A and B are encrypted. We refer to the original permutation of A, as permutation π_0. For A, sorted using x, π_0 is the identity.

We consider a cloud storage model where a client stores a dataset at a server while keeping a small amount of data in private memory. For simplicity, we assume that the dataset is an array of elements of equal size. The client encrypts each element and stores the elements at the server according to a PRP. The encryption key and the seed of the PRP are kept private by the client and are not revealed to the server.

The server supports a standard set of operations on an array S: getRange(S, loc, ℓ) returns the elements at locations in range loc, ..., loc+ℓ−1; putRange(S, loc, a) writes the elements of array a to locations loc, ..., loc + |a| − 1; putRangeDist(S, $\langle$loc$_1$, ..., loc$_c\rangle$, $\langle a_1, ..., a_c\rangle$) is a generalization of putRange that can write several arrays to non-sequential locations. We assume the server performs the above operations in time proportional to the number of elements read or written, but each operation takes one I/O.

3 Oblivious Shuffle Model

In this section, we introduce a formal model for the oblivious shuffle of an array.

Definition 1 (Shuffle). *A shuffle S is a pair of algorithms* (Setup, Shuffle), *as follows.*

- *$(s, S) \leftarrow$ Setup(1^k) Given security parameter k, run the key generation algorithm for a symmetric encryption scheme* (Enc, Dec) *and store the key in secret state s. Also, allocate an auxiliary datastore S.*
- *(Enc($\pi(A)$), α) $\leftarrow$ Shuffle(s, S, A, π) Given secret state s, auxiliary data store S, an array input A, and a permutation π, return (1) the encryption of the permutation of A according to π; (2) a transcript α of the operations that transform Enc(A) to Enc($\pi(A)$) using auxiliary space S.*

Transcript α is a sequence of l (request, response) pairs $\langle (r_1, g_1), ..., (r_l, g_l) \rangle$ that capture the evolution of the datastore via intermediate states $S_1, ..., S_{l+1}$. An invariant on each intermediate state is to store an encryption of some permutation of A along with any auxiliary data. For example S_1 contains Enc(A) and S_l contains Enc($\pi(A)$). Setting $S_1 \leftarrow \{$Enc(A), $S\}$, $s_0 \leftarrow s$, $g_0 \leftarrow \perp$ define the relationship between r_i and g_i:

$$\langle \, (s_i, r_i) \leftarrow \mathsf{GenRequest}(s_{i-1}, g_{i-1}), \quad (S_{i+1}, g_i) \leftarrow \mathsf{GenResponse}(S_i, r_i) \, \rangle.$$

Operations GenRequest *and* GenResponse *generate a request* r_i *and a corresponding response* g_i *and are defined as follows:*

- $(s_i, r_i) \leftarrow$ GenRequest(s_{i-1}, g_{i-1}) *Perform a computation based on a substructure of* $S_{i-1}, g_{i-1},$ *and generate next request to* $S_i, r_i.$
- $(S_{i+1}, g_i) \leftarrow$ GenResponse(S_i, r_i) *Generate the response to request* r_i *on* S_i: S_{i+1} *is the datastore* S_i *updated according to* r_i *and* g_i *is the response to* r_i *with respect to* $S_i.$ *For example, if* r_i *is a get request, then* $S_{i+1} = S_i$ *and* g_i *is the requested item. Also, if* r_i *is a put request, then* g_i *is empty.*

The private state s is updated if needed after every request.

In our cloud storage model, a shuffle $\mathcal{S}$ is a distributed computation executed by the user and the server. The user runs Setup to generate the encryption key and requests the server to allocate some space. He then runs Shuffle by accessing $\mathcal{S}$ through the server, that is, issuing requests to the server using GenRequest. The set of possible requests is defined by the storage model supported by the server. For every request r_i, the server executes GenResponse, locally updating S for put requests and returning to the user the queried items for get requests. (See Figure 1 for an illustration.)

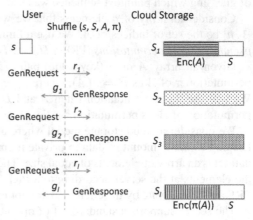

Fig. 1. Illustration of a shuffle $\mathcal{S}$ executed by the client and the cloud storage server (Section 3)

We capture the security of a shuffle $\mathcal{S}$ against a curious server in the cloud storage model as a game, *Shuffle-IND*, between $\mathcal{S}$ and a probabilistic polynomial-time bounded (PPT) adversary $\mathcal{A}$. In this game, the inputs and outputs of $\mathcal{S}$ that are revealed to the server in the cloud storage model are also revealed to $\mathcal{A}$. However, the secret state s kept by the client, any updates to it and computations inside of GenRequest are kept private, since in the cloud model they are also hidden and happen on the user side.

The game starts with $\mathcal{S}$ running Setup once, allocating at the server space to be used in subsequent computations. $\mathcal{A}$ then tries to "learn" how $\mathcal{S}$ performs the shuffle on a sequence of m_1 input arrays and permutations picked by $\mathcal{A}$. Based on what $\mathcal{A}$ learns, she picks two challenges (A_0, τ_0) and (A_1, τ_1) each consisting of a data array to be permuted using a corresponding permutation. $\mathcal{S}$ secretly picks one pair and performs the shuffle according to it. The adversary is then allowed to observe $\mathcal{S}$ shuffling another sequence of m_2 (input, permutation) pairs, also picked by $\mathcal{A}$. Finally, $\mathcal{A}$ has to guess which challenge pair (input, permutation) $\mathcal{S}$ picked to shuffle. Note that at any time, $\mathcal{A}$ can ask $\mathcal{S}$ to perform a shuffle on any combination of A_0 or A_1 and permutations τ_0 or τ_1.

Definition 2 (Oblivious Shuffle). *Let k be the security parameter and n be a polynomial in k. $\mathcal{S}$ is an oblivious shuffle over n items if for every PPT adversary, the probability of winning the Shuffle-IND game is at most $1/2 + \mathsf{negl}(k)$.*

4 The Melbourne Shuffle

In this section, we present a basic version of our Melbourne shuffle algorithm. An optimized version is given in the next section.

Overview. We assume that each element in the input array, A, is a key-value pair (x, v) for every $x \in D$. The algorithm has two phases: *distribution* and *clean-up*. For each phase, the data store S is split in several logical subparts: I, T and O. I is an array containing n encrypted items of the input A permuted according to some permutation π_0 (initially, π_0 is the identity). T is an encrypted temporary array used during the shuffle; after the shuffle is done, O contains the output, i.e., re-encrypted items of I permuted according to π. If the shuffle needs to be executed again, the user sets $I \leftarrow O$ and $\pi_0 \leftarrow \pi$. We further divide each subpart of S in buckets of equal size. The number of buckets and how it effects the runtime of the algorithm will be determined later.

During the distribution phase items of every bucket of I along with some dummy items are re-encrypted and distributed equally among buckets of T. Here, the distribution of item (x, v) is done according to its final location $\pi(x)$ in O. After the distribution phase the intermediate array T contains real and dummy items. Moreover, the items appear in correct buckets but not in correct positions within each bucket. The clean-up phase remedies this by reading one bucket at a time, removing dummy items, distributing the real items correctly within the bucket and writing the bucket to O.

The distribution phase alone cannot produce every possible permutation since the number of items sent from a bucket of I to a bucket of T is limited. E.g., the identity permutation cannot be achieved. To rectify this, we execute two shuffle passes. First, for a permutation π_1 picked uniformly at random and then for the desired permutation π. Although this framework still allows failures, our algorithm can produce every permutation, failing with very small probability independent of the desired permutation π.

Algorithm. The complete shuffle algorithm shuffle(I, π, O) is shown in Algorithm 1. The algorithm makes two calls to shuffle_pass first for a random permutation π_1 and then for the desired permutation π. We proceed with the description of shuffle_pass (I, T, ρ, O) where I and O are defined as in shuffle, T is a temporary array and ρ is the desired permutation. We use the convention of giving arrays I, T and O as inputs to the shuffle pass algorithm for the ease of explanation. In the cloud storage scenario that we consider here, one simply specifies the location where these arrays are stored remotely. Given an input array of size n, this method has messages and client's private memory of size $O(\sqrt{n} \log n)$ and server memory of size $O(n \log n)$. These user and server memory requirements are temporary and are reduced to $O(1)$ and n, respectively, when the shuffle is finished. As mentioned before, method shuffle_pass is split into a distribution phase and a clean-up phase.

Distribution Phase. The distribution phase of method shuffle_pass, shown in Figure 2, imitates throwing balls into bins by putting elements from every bucket of I to every bucket of T according to the permutation ρ. In particular, a batch of $p \log n$ encrypted elements from every bucket of I is put in every bucket of T (rev_bucket[id$_T$] in the pseudo-code). Here, p is a constant and is determined in the analysis.

Each batch contains real and dummy elements. The first batch is filled in with real elements (x, v) that would go to the first bucket in O according to ρ, i.e., the elements

---

Algorithm 1. The complete Melbourne shuffle algorithm, shuffle(I, π, O), where the user can read and store in private memory M up to $\sqrt{n} \times p \log n$ elements, $p \geq e$

I: array of n encrypted elements (x, v); π: permutation; O: permutation of I according to π, where every element is re-encrypted.

1. Let π_1 be a random permutation
2. Let T be an empty array of size $n \times p \log n$ stored remotely
3. shuffle_pass(I, T, π_1, O)
4. $I \leftarrow O$
5. shuffle_pass(I, T, π, O)

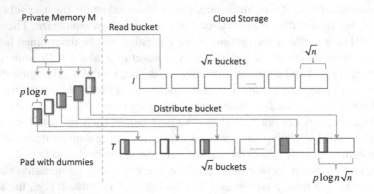

Fig. 2. Illustration of the distribution phase of shuffle_pass. Shadowed regions represent dummy values added to pad each batch to the size of $p \log n$. The batches are encrypted, hence, one cannot tell where and how many dummy values there are in each batch.

for which $\lfloor \rho(x)/\sqrt{n} \rfloor = 0$. Similarly for every other batch. Since a bucket of I contains only $\sqrt{n}$ elements and we put $\sqrt{n} \times p \log n$ elements in total in all buckets in T, most batches will have less than $p \log n$ elements. We pad such batches with dummy elements to hide where and how many elements of I's bucket are placed in T. Note that a batch is re-encrypted before it is written to T, completely hiding the content and making it impossible to recognize where dummy or real elements are. If according to ρ more than $p \log n$ elements are mapped from a bucket of I to a bucket of T, the algorithm fails. We later consider what happens in case of a failure. We note that all $\sqrt{n} \times p \log n$ elements from batches of a single bucket can be written to T using a single call to putRangeDist.

Clean-up Phase. The distribution phase leaves T with two problems: first, though the elements are in correct buckets according to ρ they are not in the correct locations inside the buckets, and second, T contains dummy elements. To remedy these problems, the clean-up phase, illustrated in Figure 3, proceeds by reading buckets of T of size $\sqrt{n} \times p \log n$ and writing in their place buckets of size $\sqrt{n}$.

When processing each bucket, the algorithm removes dummy elements, sorts the remaining content of every bucket according to their final location in O. It is important to note that each written bucket contains exactly $\sqrt{n}$ elements before it is being written back. This follows from the fact that elements were distributed to buckets according

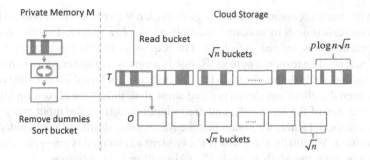

Fig. 3. An illustration of the clean-up phase of shuffle_pass. Shadowed regions represent dummy values that are removed during the clean-up phase.

to the permutation ρ and the algorithm failed in the distribution phase for those ρ that would have resulted in more than $\sqrt{n}$ elements in each bucket.

Performance. The performance of the Melbourne shuffle is summarized below.

Theorem 1. *Given an input array of size n, the Melbourne shuffle (Algorithm 1) executes $O(\sqrt{n})$ operations, each exchanging a message of size $O(\sqrt{n}\log n)$, between a user with private memory of size $O(\sqrt{n}\log n)$ and a server with storage of size $O(n\log n)$. Also, the user and server perform $O(n\log n)$ work.*

4.1 Security Analysis

In this section, we show that the Melbourne shuffle (Algorithm 1) is oblivious for every permutation π with high probability. For detailed proofs please refer to [17].

Definition 3. *Let A be an array of n elements such that every $x \in [1,n]$ is at location $\pi_0(x)$ in A. Let B be an array that stores a permutation π of elements in A, i.e., $B = \pi(A)$. Split A and B in $\sqrt{n}$ buckets of equal size and fix a constant $p \geq e$. Let π be a permutation on n elements where every bucket of B contains at most $p\log n$ elements of every bucket of A. We refer to the set of all such permutations as $P(\pi_0)$.*

Lemma 1. *The size of set $P(\pi_0)$ is $(1 - \mathsf{negl}(n)) \times n!$, for every permutation π_0.*

Lemma 2. *Let π_0 be the initial permutation of n elements in the input array I. Method shuffle_pass succeeds for all permutations $\rho \in P(\pi_0)$.*

Lemma 3. *Method shuffle(I, π, O) (Algorithm 1) is a randomized shuffle algorithm that succeeds with very high probability.*

We show that method shuffle_pass is oblivious by mapping it to the Oblivious Shuffle Model in Section 3, extracting the corresponding transcript and showing that the transcript reveals no information about the underlying permutation if the encryption scheme is CPA secure (see Section 2).

Method shuffle_pass corresponds to GenRequest. Calls to getRange and putRange trigger calls to GenResponse at the server. We do not describe GenResponse since it depends on the implementation of the storage provider. We are only interested in the fact that it uses server's state S to store and maintain arrays I, T and O. The transcript α of the shuffle execution is defined as follows. The request r_i is either getRange(S, i, l) or putRange(S, i, a). The response g_i to getRange is an array a. The response to putRange is empty. We first analyze the metadata (i.e., arguments not based on content) of every request between the client and the server, and show that, unless the algorithm fails, they depend on the size of the input only, and are independent from the input array and the desired permutation. Hence, we obtain that the Melbourne shuffle is a data independent shuffle algorithm. We finally show that if the content exchanged is encrypted, as it is in method shuffle_pass, the Melbourne shuffle (Algorithm 1) is oblivious.

Theorem 2. *The Melbourne Shuffle (Algorithm 1) is a randomized shuffle algorithm that succeeds with very high probability and is data-oblivious according to Definition 2.*

5   The Optimized Melbourne Shuffle

In this section we present an optimized version of the Melbourne shuffle that has smaller memory requirements and succeeds with higher probability than the basic version of the previous section. The main difference with the basic version lies in the shuffle pass.

Algorithm. As in the basic version, the shuffle pass splits the input array I and the output array O in consequent buckets of size $\sqrt{n}$. For auxiliary storage we use two temporary arrays T_1 and T_2 of size $p_1 n$ and $p_2 n$, respectively, where $p_1, p_2 > 1$ are constants to be determined in the analysis. We split T_1 and T_2 in buckets of size $p_1 \sqrt{n}$ and $p_2 \sqrt{n}$, respectively. The shuffle pass proceeds with two distribution phases, instead of one for the basic version, followed by a single clean-up phase in the end.

The first distribution phase moves elements from I to T_1, the second distribution phase moves elements from T_1 to T_2. We abstract the layout of elements in each array further by sequentially splitting buckets in chunks. The goal of the first distribution phase is to place elements in correct chunks and place them in correct buckets within the chunks in the second distribution phase. When a bucket is read, it is decrypted and any elements that are written back are re-encrypted. In the following, we denote with ρ the target distribution of the shuffle pass. For an array of size n, this algorithm assumes messages and client private memory of size $O(\sqrt{n})$ and server memory of size $O(n)$.

Distribution Phase I. We view a sequence of $\sqrt[4]{n}$ buckets in each array as a chunk. Hence, I, T_1, T_2 and O each have $\sqrt[4]{n}$ chunks. The goal of the first distribution phase is to place elements of I in T_1 in such a way that all elements that belong to the first chunk of O according to ρ can be found in the first chunk of T_1, similarly for the second chunk, and so on. For an illustration of this phase refer to Figure 4.

Distribution Phase II. Observe that elements in T_1 belong to the correct chunk but not the correct bucket within the chunk. The second distribution phase remedies this, such that by the end of this phase elements of chunks of T_1 are in their correct buckets in T_2.

Clean-up Phase. This phase is similar to the clean-up phase of the basic version. The elements are in the correct buckets but not in the correct spots. We remedy this by reading every bucket, removing dummy elements such that only $\sqrt{n}$ real elements are left, sorting it according to ρ, re-encrypting the elements and writing them back.

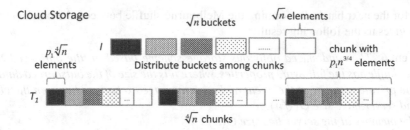

Fig. 4. Illustration of the arrangement of elements from the input I in the output T_1 after the first distribution phase of the Optimized Melbourne Shuffle (Section 5)

Performance. The performance and security properties of the optimized Melbourne shuffle are summarized in the following theorems. For detailed proofs please refer to [17].

Theorem 3. *Given an input array of size n, the optimized Melbourne shuffle executes $O(\sqrt{n})$ operations, each exchanging a message of size $O(\sqrt{n})$ between a user with private memory of size $O(\sqrt{n})$ and a server with storage of size $O(n)$. Also, the user and the server perform $O(n)$ work.*

Theorem 4. *The optimized Melbourne Shuffle is a randomized shuffle algorithm that succeeds with very high probability and is data-oblivious according to Definition 2.*

6 The Melbourne Shuffle with Small Messages

The Melbourne shuffle and its optimized version can be extended to work with messages and private memory of size $\sqrt[c]{n} \log n$ (or $\sqrt[c]{n}$ for the optimized version), for $c \geq 3$.

Theorem 5. *Given an integer constant $c \geq 3$ and an input array of size n, the optimized Melbourne shuffle executes $O(cn^{(c-1)/c})$ operations, each exchanging a message of size $O(\sqrt[c]{n})$ between a user with private memory of size $O(\sqrt[c]{n})$ and a server with storage of size $O(n)$. Also, the user and server perform $O(cn)$ work.*

7 Oblivious Storage

In this section, we give an overview of a secure and efficient oblivious storage scheme that uses the Melbourne shuffle. The oblivious storage (OS) we consider here follows the framework proposed in [5] and the follow-up work of [11]. The goal of the oblivious storage is to hide client's access pattern to his remotely stored data from anyone observing it, including the storage provider. Informally, OS transforms a virtual sequence of requests into a simulated one that appears to be data-independent. This is achieved by a mixture of accesses that are the same for every access sequence (e.g., the Melbourne shuffle) and of accesses that are randomized, and come from the same distribution, hence they appear to be independent of the access sequence.

Our OS scheme consists of *setup*, *access* and *rebuild* phases. The setup phase arranges, encrypts and outsources the data to the remote storage server. The access phase transforms a virtual request into a sequence of accesses to the remote storage. Once these accesses are performed, the requested element is returned. After a batch of requests, the data at the server is shuffled in order to be able to proceed with the access phase for the next batch. Performing the Melbourne shuffle between the batches of requests gives us the following result:

Theorem 6. *The randomized oblivious storage scheme based on the optimized Melbourne shuffle has the following properties, where n is the size of the outsourced dataset:*
- *The private memory at the client and each message exchanged between the client and server have size $O(\sqrt{n})$.*
- *The memory at the server has size $O(n)$.*
- *The access overhead to perform a storage request is $O(1)$.*

We apply recursion to the square root solution to support messages and private user memory of size $n^{1/c}$. This solution uses a cache of size $n^{1/c}$, which fits into private memory, and $c - 1$ levels of additional storage. Each level i is large enough to contain $n^{(i+1)/c}$ real elements and $n^{i/c}$ fake elements. The cache and levels 1 to $i - 1$ have a similar cache functionality for level i as the cache C in the square root solution, except they can store together $O(n^{i/c})$ previously accessed elements. Each level $i < c - 1$ contains $n^{(i+1)/c}$ buckets of size $O(\log n)$ that allows one to store $n^{(i+1)/c}$ elements in a hash table and avoid collisions with very high probability. Buckets with fewer than $\log n$ elements are filled in with dummies. The last level, level $c - 1$, has n elements, hence a permutation can be used to store and access the elements. After $n^{i/c}$ elements are accessed, level i is rebuilt in $O(in^{i/c} \log n)$ accesses using the optimized Melbourne Shuffle for small messages (see Section 6). This cost can be amortized, or deamortized using [10], to get $O(c \log n)$ access overhead per every element. We can increase message size from $O(\sqrt[c]{n})$ to $O(\sqrt[c]{n} \log n)$ to achieve a constant overhead from the rebuild.

Theorem 7. *The randomized oblivious storage scheme based on the Melbourne shuffle with small messages has the following properties, where n is the size of the outsourced dataset and c is a constant such that $c \geq 3$:*
- *The private memory at the client and each message exchanged between the client and server have size $O(\sqrt[c]{n})$ $(O(\sqrt[c]{n} \log n))$.*
- *The memory at the server has size $O(n)$.*
- *The access overhead to perform a storage request is $O(c \log n)$ $(O(c))$.*

Acknowledgements. This research was supported in part by the National Science Foundation under grants CNS–1011840, CNS–1012060, CNS–1228485, CNS–1228639, and IIS–124758, by the National Institutes of Health under grant R01-CA180776, and by the Office of Naval Research under grant N00014-08-1-1015. Olga Ohrimenko worked on this project in part while at Brown University.

References

[1] Ajtai, M., Komlós, J., Szemerédi, E.: An $O(n \log n)$ sorting network. In: ACM Symp on Theory of Computing (STOC), pp. 1–9 (1983)

[2] Aldous, D., Diaconis, P.: Shuffling cards and stopping times. The American Mathematical Monthly 93(5), 333–348 (1986)

[3] Batcher, K.E.: Sorting networks and their applications. In: Proc. 1968 Spring Joint Computer Conf., pp. 307–314. AFIPS Press (1968)

[4] Goldreich, O.: Towards a theory of software protection and simulation by oblivious RAMs. In: ACM Symp. on Theory of Computing. pp. 182–194 (1987)

[5] Goldreich, O., Ostrovsky, R.: Software protection and simulation on oblivious RAMs. J. ACM 43(3), 431–473 (1996)

[6] Goodrich, M.T.: Randomized Shellsort: A simple oblivious sorting algorithm. In: Proc. ACM-SIAM Sump. on Discrete Algorithms (SODA), pp. 1–16 (2010)

[7] Goodrich, M.T.: Zig-zag Sort: A Deterministic Data-Oblivious Sorting Algorithm Running in $O(n \log n)$ Time. In: ACM Symp. on Theory of Computing (2014)

[8] Goodrich, M.T., Mitzenmacher, M.: Privacy-Preserving Access of Outsourced Data via Oblivious RAM Simulation. In: Aceto, L., Henzinger, M., Sgall, J. (eds.) ICALP 2011, Part II. LNCS, vol. 6756, pp. 576–587. Springer, Heidelberg (2011)

[9] Goodrich, M.T., Mitzenmacher, M.: Anonymous card shuffling and its applications to parallel mixnets. In: Czumaj, A., Mehlhorn, K., Pitts, A., Wattenhofer, R. (eds.) ICALP 2012, Part II. LNCS, vol. 7392, pp. 549–560. Springer, Heidelberg (2012)

[10] Goodrich, M.T., Mitzenmacher, M., Ohrimenko, O., Tamassia, R.: Oblivious RAM simulation with efficient worst-case access overhead. In: Proc. ACM Workshop on Cloud Computing Security (CCSW), pp. 95–100 (2011)

[11] Goodrich, M.T., Mitzenmacher, M., Ohrimenko, O., Tamassia, R.: Practical oblivious storage. In: ACM Conf. on Data and Application Security and Privacy (CODASPY), pp. 13–24 (2012a)

[12] Goodrich, M.T., Mitzenmacher, M., Ohrimenko, O., Tamassia, R.: Privacy-preserving group data access via stateless oblivious RAM simulation. In: ACM-SIAM Symposium on Discrete Algorithms (SODA), pp. 157–167 (2012b)

[13] Islam, M.S., Kuzu, M., Kantarcioglu, M.: Access pattern disclosure on searchable encryption: Ramification, attack and mitigation. In: NDSS (2012)

[14] Kamara, S., Lauter, K.: Cryptographic cloud storage. In: Sion, R., Curtmola, R., Dietrich, S., Kiayias, A., Miret, J.M., Sako, K., Sebé, F. (eds.) RLCPS, WECSR, and WLC 2010. LNCS, vol. 6054, pp. 136–149. Springer, Heidelberg (2010)

[15] Kushilevitz, E., Lu, S., Ostrovsky, R.: On the (in)security of hash-based oblivious RAM and a new balancing scheme. In: ACM-SIAM Symposium on Discrete Algorithms (SODA), pp. 143–156 (2012)

[16] Mitzenmacher, M., Upfal, E.: Probability and Computing: Randomized Algorithms and Probabilistic Analysis. Cambridge University Press (2005)

[17] Ohrimenko, O., Goodrich, M.T., Tamassia, R., Upfal, E.: The Melbourne shuffle: Improving oblivious storage in the cloud. CoRR abs/1402.5524 (2014)

[18] Pinkas, B., Reinman, T.: Oblivious RAM revisited. In: Rabin, T. (ed.) CRYPTO 2010. LNCS, vol. 6223, pp. 502–519. Springer, Heidelberg (2010)

[19] Shi, E., Chan, T.-H.H., Stefanov, E., Li, M.: Oblivious RAM with $o((\log n)^3)$ worst-case cost. In: Lee, D.H., Wang, X. (eds.) ASIACRYPT 2011. LNCS, vol. 7073, pp. 197–214. Springer, Heidelberg (2011)

[20] Stefanov, E., van Dijk, M., Shi, E., Fletcher, C., Ren, L., Yu, X., Devadas, S.: Path ORAM: An Extremely Simple Oblivious RAM Protocol. In: ACM Conf. on Computer and Communications Security, CCS (2013)

[21] Stefanov, E., Shi, E., Song, D.: Towards Practical Oblivious RAM. In: Proc. Network and Distributed System Security Symposium (NDSS) (2012)

[22] Williams, P., Sion, R.: Single round access privacy on outsourced storage. In: ACM Conf. on Computer and Communications Security, pp. 293–304 (2012)

[23] Williams, P., Sion, R., Carbunar, B.: Building castles out of mud: practical access pattern privacy and correctness on untrusted storage. In: ACM Conference on Computer and Communications Security (CCS), pp. 139–148 (2008)

Sending Secrets Swiftly:
Approximation Algorithms
for Generalized Multicast Problems

Afshin Nikzad[1] and R. Ravi[2]

[1] MS&E Department, Stanford University, USA
nikzad@stanford.edu
[2] Tepper School of Business, Carnegie Mellon University, USA
ravi@cmu.edu

Abstract. We consider natural generalizations of the minimum broadcast time problem under the telephone model, where a rumor from a root node must be sent via phone calls to the whole graph in the minimum number of rounds; the telephone model implies that the set of edges involved in communicating in a round form a matching. The extensions we consider involve generalizing the number of calls that a vertex may participate in (the capacitated version), allowing conference calls (the hyperedge version) as well as a new multicommodity version we introduce where the rumors are no longer from a single node but from different sources and intended for specific destinations (the multicommodity version). Based on the ideas from [6,7], we present a very simple greedy algorithm for the basic multicast problem with logarithmic performance guarantee and adapt it to the extensions to design typically polylogarithmic approximation algorithms. For the multi-commodity version, we give the first approximation algorithm with performance ratio $2^{O\left(\log\log k \sqrt{\log k}\right)}$ for k source-sink pairs. We provide nearly matching lower bounds for the hypercasting problem. For the multicommodity multicasting problem, we present improved guarantees for other variants involving asymmetric capacities, small number of terminals and with larger additive guarantees.

Keywords: approximation algorithms, graph algorithms, b-matching, LP rounding.

1 Introduction and Motivation

Rumor spreading in networks has been an area of much study involving the gamut from finding the minimum possible number of messages to spread gossip around the network [23,2,12] to finding graphs with minimum number of edges that are able to spread rumors in the minimum possible time in the network [9]. An important NP-hard formulation asks to find a scheme that spreads a rumor from a single root node to all other nodes under the popular "telephone" model where every node can participate in a telephone call with at most one other neighbor in each round, and the goal is to minimize the number of rounds. This

J. Esparza et al. (Eds.): ICALP 2014, Part II, LNCS 8573, pp. 568–607, 2014.
© Springer-Verlag Berlin Heidelberg 2014

is the minimum broadcast time problem for which there has been active work in designing approximation algorithms [14,20,10,7]. We study generalizations of this problem that involve (i) sending the message to only a subset of receivers (multicasting), (ii) capacity constraints on the number of calls in which a node can participate (capacitated cases), (iii) allowing conference calls modeled by hyperedges (hypercasting problem), and (iv) multiple sources of rumors with different sets that are the targets for the different rumors (the multicommodity case). Our paper initiates work on the capacitated, hypercasting and multicommodity extensions of the "rapid rumor ramification" problem [20], bringing it to next range of generalizations of "sending secrets swiftly".

Problem Definition

Definition 1. *In the (minimum time) **Multicast** problem, we are given an undirected graph $G(V, E)$ which represents a telephone network on V, where two adjacent nodes can place a telephone call to each other. We are given a source vertex r and a set of terminals $R \subseteq V$. The source vertex has a message and it wants to inform all the terminals of the message. To do this, the vertices of the graph can communicate in rounds: In each round, we pick a matching of G and arrange a bidirected phone call between each vertex in the matching and its matched pair. If any of the two vertices knows the message before the phone call, the other one will also know it afterwards. The goal is to deliver the message to all the terminals in the minimum number of rounds.*

When $R = V$, the Multicast problem is known as the (minimum time) *broadcast* problem which is one of the most basic and well-studied problems in this setting. Applications of this problem arise commonly in multicasting in networks [21], keeping the information consistent across copies of replicated databases by broadcasting from the changed copy to the others [16,17], as well as in finding schemes that ensure that maximum information delay in problems modeled by vector clocks [15] is minimized.

In this paper, we investigate the Multicast Problem in the following three more general settings: i. Allowing multiple source vertices, i.e. the multi-commodity setting rather than the single-source setting. ii. Allowing conference calls (with possibly more than two participants) rather than just having phone calls. iii. Imposing capacities for the vertices, i.e. allowing a vertex to be in a number of phone calls in each round, which can not exceed its capacity. In all of these generalizations, the objective function remains minimizing the total number of rounds used in the solution.

The first generalization is having messages with arbitrary source and destination vertices, i.e. unlike the Multicast Problem, the messages do not need to share the same source vertex.

Definition 2. *In the **Multicommodity Multicast Problem** (MM), a graph $G(V, E)$ is given along with a set of pairs of nodes $P = \{(s_i, t_i) | 1 \leq i \leq k\}$, known as demand pairs. Each vertex s_i has a message m_i which needs be delivered to t_i. The vertices communicate similar to the Multicast problem, i.e. during a phone* *call, each vertex can pass (a copy of) all of the messages that it has to the other vertex.*

The second aspect in our model is having conference calls rather than having only phone calls, i.e. calls involving (possibly) more that two, rather than only two, persons. So, instead of a simple graph G, we are given a hypergraph the edges of which represent the potential conference calls. We call this problem the *Hypercast Problem*; we define this problem formally and study it extensively in Section D.

Finally, we bring in the notion of *capacity of a vertex* to our model for the Multicast Problem, and allow a vertex to be in possibly more than a single call in each round; the maximum number of phone calls that a vertex can have in each round is called the *capacity* of the vertex. The *Capacitated Multicast Problem* is formally defined and studied in Section C.

In this paper, we develop a unified solution framework that can incorporate each of the Hypercast, Multi-commodity Multicast, and Capacitated Multicast aspects, leading to the first approximation algorithms for any combination of these extensions such as the Capacitated Hypercast Problem, or Multi-commodity Hypercast Problem. In the rest of this paper, we let $n = |V|$, $k = |R|$ and OPT be the optimal number of rounds needed to solve the given Multicast instance, unless it is specified otherwise.

2   Related Work

Finding optimal broadcast schedules for trees was one of the first theoretical problems in this setting and was solved using Dynamic Programming [18]. For general graphs, Kortsarz et. al. developed an additive approximation algorithm which uses at most $c \cdot OPT + O(\sqrt{n})$ rounds for some constant c. Later, Ravi [20] provided a $O(\frac{\log^2 n}{\log \log n})$-approximation for the same problem using the result of Raghavan [19] for randomized rounding of an LP formulation for the concurrent multicommodity flow problem.

Guha et.al. [10] improved the approximation factor for Multicasting in general graphs to $O(\log k)$ where k is the number of terminals. To the best of our knowledge, the best approximation factor for the Multicast problem is $O(\frac{\log k}{\log \log k})$ [7]. Both of [7,10] present a recursive algorithm which reduces the total number of uninformed terminals in each step of the recursion, while using $O(OPT)$ number of rounds in that step. In [10], they reduce the number of uninformed terminals by a constant factor in each step and so they obtain a $O(\log k)$-approximation, but in [7], the number of uninformed terminals is reduced by a factor of OPT which gives a $O(\frac{\log k}{\log \log k})$-approximation due to the fact that $OPT = \Omega(\log k)$.

3   Our Results

Our main contribution is developing a framework to design approximation algorithms for various generalizations of the Multicast Problem. A summary of the key results in this paper is listed below. For the complete list of our results, see Tables 1 and 2 in Section 7.

1. In Section 5, we give a *very simple* $O(\log k)$-approximation for the Multicast Problem, which is based on the ideas of [6,7]. With a slight modification of this algorithm in Section C, we adapt it to solve the Capacitated Multicast Problem.
2. In Section 6, based on our simple algorithm for the Multicast Problem, we design a $2^{O(\log \log k \sqrt{\log k})}$-approximation for the Multi-commodity Multicast Problem (note that the approximation factor is, while being super-polylogarithmic, still smaller than any constant root of k).
3. In Section D, we develop our simple algorithm further and obtain a $O(\log k \cdot \log n \cdot D)$-approximation for the Hypercast Problem, where D is the maximum size of a hyperedge. Also, our hardness results for the Hypercast Problem show that the dependence on D is unavoidable (see Section I).
4. In Sections G and H, we explore the Multi-commodity Multicast problem further and design two polylogarithmic approximation algorithms which carry additional additive factors of $\sqrt{n}$ and $\Delta(G)$ (maximum degree).

While our algorithms for Multicast and Multi-commodity Multicast problems are purely combinatorial, the algorithm for Hypercast involves solving a Linear Program and randomized rounding of the LP solution.

Our results are not limited to this since our framework can handle more general cases of the Multicast Problem, e.g. Capacitated Multi-commodity Hypercast Problem; all of them are summarized in Section 7.

4 Preliminaries

4.1 The Multicast Problem

In the context of any of the problems discussed in Section 1, *sending vertex u to vertex v* means sending the information of u to v in a (potentially specified) number of rounds. Given that $P = \{(s_i, t_i) | 1 \leq i \leq k\}$ is the set of demand pairs in any single-source or multi-commodity instance, let $S = \{s_1, \ldots, s_k\}$ be the set of sources, $T = \{t_1, \ldots, t_k\}$ be the set of sinks, and $R = S \cup T$ be the set of *terminals*. Also, let $\mathcal{I}(G, P)$ denote the Multicast (Hypercast) instance defined by P on the graph (Hypergraph) G. We also use $\mathcal{I}$ when both G, P are clearly known from the context.

4.2 Schedules

A *schedule* is a sequence of rounds. The length of a schedule $\mathcal{S}$ is denoted by $|\mathcal{S}|$ and is the number of rounds it contains. A schedule for a (single-source or multi-commodity) Multicast instance is *non-lazy* if, in any round of it, any two idle and adjacent vertices have identical information in that round.

4.3 Graphs and Matchings

Suppose G is a simple graph and let $n = |V(G)|$. Let $N(v)$ be the set of the neighbors of a vertex v, and for any $S \subseteq V(G)$, let $N(S) = \cup_{v \in S} N(v)$. Denote the degree of the maximum-degree vertex in G by $\Delta(G)$. For any $X \subseteq V(G)$, let $G[X]$ be the induced subgraph of G on X. For any family of subgraphs of G, such as $\mathcal{F}$, let $V(\mathcal{F})$ denote the union of the vertices of the subgraphs in $\mathcal{F}$ and $E(\mathcal{F})$ denote the union of the edges of the subgraphs in $\mathcal{F}$.

The distance between two vertices of G, such as u and v, is represented by $d_G(u, v)$. If there is no path in G between u, v, then let $d_G(u, v) = \infty$. The diameter of G, denoted by $diam(G)$, is $\max\limits_{u, v \in V(G)} d_G(u, v)$. Also, define $diam_P(G) = \max\limits_{(s,t) \in P} d_G(s, t)$, where P is a set of pairs of vertices, e.g. the set of demand pairs.

Given a bipartite graph $H[X, Y]$ with partitions X and Y, a b-matching in H is a subset M of the edges of H such that each vertex of X is incident to exactly one edge of M and each vertex of Y is incident to at most b edges of M.

4.4   Spiders

Spiders are subgraphs that have been useful in designing algorithms for the Multicast and directed Broadcast problems in [10,5]. A *spider* S is a set of (almost) vertex-disjoint paths all starting at the same vertex, e.g. v, and sharing no vertex other than v. Define the *center of S* to be v. Also, let the *degree of S*, denoted by $deg(S)$, be the degree of v in S, and the *length of S*, denoted by $len(S)$, be $\max_{u \in V(S)} d_S(u, v)$.

It's very easy to verify the following lemma stated in [5].

Lemma 1. *Using a non-lazy schedule, the center of a spider S can send (broadcast) a message to the rest of its vertices in $deg(S) + len(S) - 1$ rounds.*

5   The Multicast Problem

In this section, we present a $O(\log k)$-approximation for the Multicast problem, which is obtained by simplifying the $O(\frac{\log k}{\log \log k})$-approximation given in [7]. Our simple algorithm plays an important role in our framework. Later, this algorithm will be developed further to design algorithms for the introduced generalizations of the Multicast problem. For instance, with a slight modification, it turns into a $O(\log k)$-approximation for the Capacitated Multicast problem (see Section C).

5.1   Outline of the Algorithm

Our algorithm accepts a parameter L as a part of the input, which is our guess for the optimal solution of the given multicast instance. Since n is an upper bound on the length of the optimal schedule, we can easily try all the possible values for L from 1 to n and run the algorithm once for each of these values. Our algorithm is guaranteed to return a schedule of length $O(L \cdot \log k)$ assuming that L is the length of the optimal schedule. So, from now on in this section, we think of L as the length of the optimal schedule W.L.O.G.

The algorithm is a recursive algorithm and has 4 phases. In Phase 1 of the algorithm, we reduce the given instance to a smaller instance. In Phase 2, we

solve the smaller instance recursively, and finally in Phase 3 and 4, we inform
the rest of the vertices which didn't receive the message in Phase 2. We explain
each of these phases briefly and after that, we'll see the full description of the
algorithm.

Phase 1. This phase starts with finding a family of *vertex disjoint* paths $\mathcal{P}$ each
of length at most $4L$ such that the endpoints of the paths belong to R. We find
these paths greedily, i.e. we start with $\mathcal{P} = \emptyset$ and using BFS, we search for a
new path of length at most $4L$ between the terminals. We continue until we can
add no more such paths to $\mathcal{P}$. Pick one endpoint from each path and designate
the picked set of vertices to be R'.

Phase 2. Solve the multicast problem for the new set of terminals R' recursively
and run the obtained schedule. (So, all the vertices in R' will receive the message
by the end of this phase.)

Phase 3. Inform all the vertices belonging to $V(\mathcal{P})$ in $4L$ rounds. This is possible
since in Phase 2, we have already informed at least one vertex of each path in $\mathcal{P}$.

Phase 4. For each of the uninformed terminals, namely $v \in R\backslash V(\mathcal{P})$, find a
path M_v which connects v to one of the informed vertices (note that the set of
the informed vertices is currently $V(\mathcal{P})$). These paths are guaranteed to satisfy
the following properties:

1. The length of each path is at most $2L$.
2. Only one vertex on each path belongs to $V(\mathcal{P})$, which is an endpoint of the
   path.
3. The paths won't share any vertices except possibly in the endpoints belonging to the set $V(\mathcal{P})$. Moreover the degree of any node in $V(\mathcal{P})$ due to these
   paths is at most L.

In other words, $M = \bigcup\limits_{v \in R\backslash V(\mathcal{P})} M_v$ is a union of vertex disjoint spiders of length
at most $2L$ and we will inform the vertices in $R\backslash V(\mathcal{P})$ using these spiders. (see
Figure 1)

Before presenting the algorithm formally, we describe Phase 4 in more details.

5.2   The Algorithm: Phase 4

Our goal in Phase 4, assuming that the vertices in $V(\mathcal{P})$ have received the message, is to inform the rest of the terminals in $O(L)$ rounds. The only assumptions
we need here are that the choice of $\mathcal{P}$ is maximal in Phase 1 and $V(\mathcal{P})$ is informed.

To inform the rest of the terminals, we find a family of vertex-disjoint spiders such that each of them has a length at most $2L$ and a center belonging to
$V(\mathcal{P})$. Moreover, we need the spiders to contain all the terminals in $R\backslash V(\mathcal{P})$.
To construct the family of spiders, we start with finding the paths M_v for
all $v \in R\backslash V(\mathcal{P})$. In order to find the paths, we construct a bipartite graph

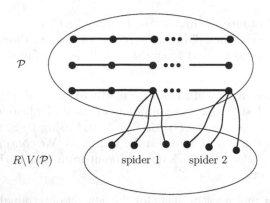

Fig. 1. Informing the vertices of $R\backslash V(\mathcal{P})$ using spiders

$H[R\backslash V(\mathcal{P}), V(\mathcal{P})]$ with edges defined as follows: There is an edge in H between two vertices $x \in R\backslash V(\mathcal{P})$ and $y \in V(\mathcal{P})$ if there is path of length at most $2L$ connecting x to y such that no other vertex of the path belongs to $V(\mathcal{P})$ except y. So, each edge in H is associated with a path in G (if there were many such paths for x, y, then choose one of them arbitrarily). Now, find a b-matching in H for the minimum possible integer b. Then, for all $v \in R\backslash V(\mathcal{P})$, define M_v to be a path in G which is associated with the edge incident to v in the b-matching. Let M be the subgraph of G defined as $M = \bigcup_{v \in R\backslash V(\mathcal{P})} M_v$, then, the following lemma holds for M.

Lemma 2. *M is a union of vertex disjoint spiders each with length at most $2L$ and degree at most L.*

Proof. First, we show that M is a union of vertex disjoint spiders. To prove this, note that for any two vertices $u, v \in R\backslash V(\mathcal{P})$, M_u and M_v can't share any vertex except possibly in the endpoints belonging to $V(\mathcal{P})$, since if they do, it contradicts the maximality of $\mathcal{P}$, i.e. there would have been a path of length at most $4L$ between u and v which could be added to $\mathcal{P}$.

So, M is a union of vertex-disjoint spiders, namely the family $\mathcal{M}$ of spiders, and since the length of each path M_v is at most $2L$, then $len(\mathcal{M}) \leq 2L$. It only remains to prove that $deg(\mathcal{M}) \leq L$. Consider the optimal multicast schedule which uses exactly L rounds. Let E' be the subset of the edges of G which are used in the optimal schedule and G' be the subgraph of G with E' as its edge set. It's easy to verify that G' is a tree of diameter at most $2L$ and maximum degree at most L [20]. Now, for each $v \in R\backslash V(\mathcal{P})$, define M'_v to be the unique path in G' from v to $f(v)$, where $f(v) = \arg\min_{u \in V(\mathcal{P})} d_{G'}(v, u)$. Note that $d_{G'}(v, f(v)) \leq 2L$ since the diameter of G' is at most $2L$, which implies $(v, f(v)) \in E(H)$ because of the existence of M_v.

Next, we prove that $\bigcup\limits_{v \in R \setminus V(\mathcal{P})} (v, f(v))$ is a L-matching in H, which shows the existence of a b-matching in H with $b \leq L$, and that's all we need to show that $deg(\mathcal{M}) \leq L$. To prove the claim, just note that the family of paths $\{M'_v : v \in R \setminus V(\mathcal{P})\}$ in G' are edge-disjoint, which means there can't be more than $\Delta(G')$ of these paths with the same endpoint, implying that $\bigcup\limits_{v \in R \setminus V(\mathcal{P})} (v, f(v))$ is a $\Delta(G')$-matching in H. The fact that $\Delta(G') \leq L$ finishes the proof.

From Lemmas 1 and 2 we conclude the following:

Lemma 3. *Assuming that the vertices in $V(\mathcal{P})$ have received the message, we can find a schedule in polynomial time which informs the rest of the terminals in $3L$ rounds.*

5.3 The Algorithm

The whole algorithm for the Multicast problem is presented more formally below.

Algorithm *Multicast*
Input: A graph G and a set of terminals R
1. $\mathcal{P}, R' \leftarrow \emptyset$
2. **for all** $(u, v) \in R \times R$ such that $u \neq v$ (∗ Phase 1 ∗)
3.   **do** Find the shortest path in $G[V(G) \setminus V(\mathcal{P})]$ from u to v, namely $Q_{u,v}$.
4.   **if** the length of $Q_{u,v}$ is not more than $4L$
5.   **then** $\mathcal{P} \leftarrow \mathcal{P} \cup Q_{u,v}$
6. $R' \leftarrow R' \cup \{u\}$
7. Inform R' recursively by calling *Multicast*(G, R'). (∗ Phase 2 ∗)
8. Inform $V(\mathcal{P})$ using the paths in $\mathcal{P}$ in at most $4L$ rounds. (∗ Phase 3 ∗)
9. Construct the bipartite graph $H[R \setminus V(\mathcal{P}), V(\mathcal{P})]$.(∗ Beginning of Phase 4 ∗)
10. Find the smallest integer b such that H has a b-matching.
11. Use the family of spiders $\mathcal{M}$ associated with the b-matching and inform $R \setminus V(\mathcal{P})$ in at most $deg(\mathcal{M}) + len(\mathcal{M}) - 1$ rounds.

Theorem 1. *Algorithm Multicast is a $7 \log k$-approximation for the Multicast problem.*

Proof. The proof of correctness is trivial. We just analyze the approximation factor of the algorithm here. Note that in each level of the recursion, the number of terminals are at least halved, which means there will be at most $\log k$ levels. Moreover, in each level we use at most $4L$ rounds in Line 9 and $deg(\mathcal{M}) + len(\mathcal{M}) - 1$ rounds in Line 12. By lemma 3 we have $deg(\mathcal{M}) + len(\mathcal{M}) - 1 \leq 3L$ implying that we use at most $7L$ rounds in each level of the recursion and $7L \cdot \log k$ rounds in total.

6 The Multicommodity Multicast Problem

In this section, we present a $2^{O\left(\log \log |R| \cdot \sqrt{\log |R|}\right)}$-approximation for the Multicommodity Multicast Problem; recall that R is the set of terminals.

6.1   Preleminaries: Multicast Schedules

Any *single-source* Multicast schedule can be represented by a directed tree such that there will be phone calls only on the tree edges. Given a multicast schedule, this tree is defined by choosing, for every vertex other than the source, the unique edge along which a message is conveyed to that vertex for the first time [20]. We state this fact in the proposition below.

Proposition 1. *Any single-source Multicast schedule can be represented by a subgraph of G which is a tree.*

Also, using Proposition 1, we can assume that the output of Algorithm *Multicast* is a tree:

Proposition 2. *W.L.O.G. the output of Algorithm Multicast can be assumed to be a tree, i.e. the set of edges used in the phone calls form a tree.*

6.2   Sparsification

Before describing our algorithm, we need a key lemma related to graph spanners, i.e. sparse subgraphs such that distances between adjacent nodes are preserved within a logarithmic factor in the subgraph. To the best of our knowledge, this result first appeared in [1] as Lemma 3.1. We only state and use a simple corollary of this lemma, for which an independent proof is given also in Section H of this paper.

Corollary 1. *Given a simple n-vertex graph G, we can find a subgraph $s(G)$ of G in polynomial time, such that $|E(s(G))| \leq 2n \log n$ and for each $(u,v) \in E(G)$, we have $d_{s(G)}(u,v) \leq 8 \log n \cdot d_G(u,v)$.*

Algorithm *Sparsify* follows as a consequence of Corollary 1. It takes a simple undirected graph $H(V, P)$ as its input and computes $s(H)$. We will use this subroutine later in our main algorithm.

Algorithm *Sparsify*
Input: A vertex set V, and a set of pairs of V, called P
Output: A subset of P
1.   Assuming that $H(V, P)$ is a simple undirected graph, use Corollary 1 to compute $s(H)$.
2.   Output $E(s(H))$.

6.3   The Algorithm

In the rest of this section, assume that we want to solve the instance $\mathcal{I}(G, P)$, and $\mathcal{L}$ is an optimal solution for this instance. Similar to Algorithm *Multicast*, our algorithm for the MM Problem accepts a parameter L in the input, which is our guess for the optimal solution.

The algorithm consists of 3 phases. In Phase 1, we (potentially) reduce the number of the demand pairs using Algorithm *Sparsify*, i.e. by calling *Sparsify*(R,P). Assuming that a subset of P, namely $\hat{P}$, is the output, we will see that repeating any feasible solution of $\mathcal{I}(G, \hat{P})$ for $O(\log n)$ times would give a feasible solution for $\mathcal{I}(G, P)$.

In Phase 2, we try to satisfy a large fraction of the demand pairs greedily. If it is done successfully, we repeat, otherwise, we go to Phase 3 and solve an instance with a fewer number of terminals recursively. A key new idea to handle the smaller instance is to use a fictitious multicast scheme to assign the terminal pairs not satisfied in this phase to one of the terminals that are.

Before seeing the formal description of the algorithm, we explain Phases 2 and 3 more precisely.

Phase 2. In this phase, we find a maximal family of vertex-disjoint paths, namely $\mathcal{P}$, where each path in $\mathcal{P}$ has a length at most L and connects s_i to t_i for some pair (s_i, t_i) in P. If $|\mathcal{P}|$ was large enough, then using these paths we satisfy a large fraction of the demand pairs in L rounds, and repeat Phase 2. Otherwise, we go to Phase 3.

More precisely, in the beginning of Phase 2, $\mathcal{P}$ is empty. We sort the pairs in P in some arbitrary order, say (s_i, t_i) for $1 \leq i \leq k$, and visit the pairs in this order. When visiting the i-th pair, we check if there exists a path of length at most L between s_i and t_i in $G[V(G) \backslash V(\mathcal{P})]$. If there was such a path, we add it to $\mathcal{P}$. After visiting all of the k pairs in P, assume $P' = \{(s'_1, t'_1), \ldots, (s'_{k'}, t'_{k'})\}$ is the subset of the pairs in P for which we were able to find the path of length at most L. Now, two possible cases can happen based on the size of $\mathcal{P}$. Define $\xi(x) = 2^{\log x - \sqrt{\log x}}$ for all positive x, then, if $|\mathcal{P}| > \xi(|R|)$ (sufficiently large, and hence sufficiently good progress), remove P' from P and repeat Phase 2. Otherwise, go to Phase 3.

Phase 3. In this Phase, we construct a smaller instance on the set of terminals $R' = \{s'_1, \ldots, s'_{k'}\}$, solve it recursively, and then provide a solution for the original problem using the solution of the smaller instance, as in our original framework for multicasting. Intuitively, we send each vertex in $R \backslash R'$ to a vertex in R' in a small number of rounds. By doing so, we create a new instance of the MM problem, i.e. the old instance induced on the set R'. We solve this new instance recursively, and convert its solution to a solution for the old instance. Note that since R' is "small", the recursive problem is much smaller than the original one and hence we have made progress, but at the expense of having to route all the remaining demands of the recursively picked demand terminals. Formally, we perform the following steps in Phase 3:

1. Construct a function $f : R \to R'$ and a schedule $\mathcal{S}$ such that $\mathcal{S}$ sends v to $f(v)$ for all $v \in R$, and also $|\mathcal{S}| \leq L$. Run the schedule $\mathcal{S}$ and send v to $f(v)$ for all $v \in R$.
2. Construct a new instance of the MM problem on G with the set of terminals R' and the set of demand pairs $P' = \{(f(u), f(v)) : \forall (u, v) \in P\}$. Solve this instance recursively and run the obtained schedule.

3. Run the schedule $\mathcal{S}$ in the reverse order to send $f(v)$ to v simultaneously for all $v \in R$.

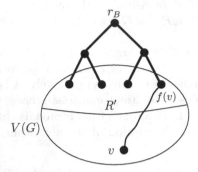

Fig. 2. The Multicast instance obtained from adding a dummy binary tree to the graph

Constructing $\mathcal{S}$ and f for Phase 3. To complete the above description for Phase 3, we need to find the schedule $\mathcal{S}$ and the function f. To do so, we construct and solve an *auxiliary Multicast instance*, the solution of which gives $\mathcal{S}$ and f.

In the beginning of Phase 3, construct an arbitrary binary tree B of height $O(\log |R'|)$ rooted at a dummy vertex r_B, such that the only common vertices between B and G are the leaves of B, which coincide with the set R' (see Figure 2).

Using Algorithm *Multicast*, solve the Multicast instance with the root r_B and the set of terminals $R \backslash R'$. Let the schedule $\mathcal{S}$ be the solution to this instance, which uses the tree $T_{\mathcal{S}}$ (recall that by Proposition 2, the solution provided by Algorithm *Multicast* is a tree). Then, define f as follows: If $v \in R'$, let $f(v) = v$, otherwise, consider the unique path in $T_{\mathcal{S}}$ from v to r_B. Define $f(v)$ to be the closest vertex to r_B which is on this path and belongs to $V(G)$ (see Figure 2).

Defining f and $\mathcal{S}$ completes the description of Phase 3. Now, we present our algorithm more formally below.

Algorithm MM
Input: A graph G and a set of pairs P
1.   **if** $P = \{\}$ **then return.**
2. $S \leftarrow \{s|\, \exists t : (s,t) \in P\}$
3. $T \leftarrow \{t|\, \exists s : (s,t) \in P\}$
4. $R \leftarrow S \cup T$
5.   **if** $|P| > 2|R|.\log|R|$ (∗ Phase 1 ∗)
6.   **then** $P = Sparsify(R, P)$
7.   **for** 1 **to** $8\log|R|$
8.   **do** $MM(G, P)$
9.   **return.**
10. $\mathcal{P}, X \leftarrow \emptyset$ (∗ Phase 2 ∗)
11. **for** $i \leftarrow 1$ **to** $|P|$
12.   **do** Find the shortest path in $G[V(G)\backslash V(\mathcal{P})]$ from s_i to t_i, namely Q_i.
13.   **if** the length of Q_i is not more than L

14.   **then** $\mathcal{P} \leftarrow \mathcal{P} \cup \{Q_i\}$
15. $X \leftarrow X \cup \{(s_i, t_i)\}$
16. **if** $|\mathcal{P}| \geq \xi(|R|)$
17.   **then** For all $(s_i, t_i) \in X$ simultaneously, send s_i to t_i through the path Q_i.
18. $P \leftarrow P \backslash X$
19.   Go to Line 10.
20.   **else** Construct and solve the auxiliary Multicast instance to obtain $\mathcal{S}$ and f. (∗ Phase 3 ∗)
21.   Run the schedule $\mathcal{S}$ in the reverse order to send v to $f(v)$ for all $v \in R$.
22. $Y \leftarrow \{(f(u), f(v)) | (u, v) \in P\}$
23. $MM(G, Y)$
24.   Run the schedule $\mathcal{S}$ to send $f(v)$ to v for all $v \in R$

Before analyzing the approximation ratio of the algorithm, we give the following lemma to bound the length of schedule $\mathcal{S}$ in terms of L and m, where $m = |R|$.

Lemma 4. $|\mathcal{S}| \leq 21 \log m \cdot L + 14 \log^2 m$.

Proof. Note that $\mathcal{S}$ is found by Algorithm *Multicast* as a solution to the auxiliary Multicast instance, and so by Theorem 1, $|\mathcal{S}| \leq 7 \log m \cdot L_{aux}$, where L_{aux} is the length of the optimal schedule for the auxiliary instance. Consequently, to prove the claimed bound, it's enough to prove that $L_{aux} \leq 3L + 2 \log m$. So, we show there exists a feasible schedule of length $3L + 2 \log m$ for the auxiliary instance. This schedule has 4 steps: 1. Use the dummy binary tree and send its root, i.e. r_B, to all of its leaves, i.e. the set R'. 2. Use the path Q_i and inform $V(Q_i)$ simultaneously for all $Q_i \in \mathcal{P}$. 3. Run the schedule $\mathcal{L}$. 4. Run the schedule $\mathcal{L}$ in the reverse order.

The suggested schedule has a length at most $3L + 2 \log m$ since Steps 1,2,3,4 have a length at most $2 \log m, L, L, L$ respectively. In the rest of the proof, we show the feasibility of this schedule. Note that after Step 1, the set of vertices R' is informed since they are the leaves of the binary tree. After Step 2, the set $V(\mathcal{P})$ is informed since each path $Q_i \in \mathcal{P}$ has an endpoint $s_i \in R'$. For the sake of contradiction, assume there is a terminal $r \in R \backslash R'$ which is not informed after Step 4. This terminal has to be in at least one demand pair, namely the pair (s, t). The schedule $\mathcal{L}$ sends s to t via a path in G, namely Q. The path Q must share at least a vertex with $\mathcal{P}$, due to the maximality of $\mathcal{P}$. This fact, and the fact that $V(\mathcal{P})$ is informed by the end of Step 2, imply that t should be informed after Step 3, and s should be informed after Step 4. Which is a contradiction with r not being informed by the end of Step 4.

Theorem 2. *Algorithm MM is a* $2^{O\left(\log \log |R| \cdot \sqrt{\log |R|}\right)}$*-approximation for the Multi-commodity Multicast Problem.*

Proof. Let $T(m)$ denote the approximation factor of our algorithm. By induction on m, we prove that $T(m) \leq 2^{\epsilon \log \log m \cdot \sqrt{\log m}}$, for any fixed $\epsilon > 6$. To do so, we provide an upperbound for $T(m)$ as follows:

$$T(m) \leq 8 \log m \cdot \left(\frac{2m \log m}{\xi(m)} \cdot L + 7 \log m \cdot (6L + 4 \log m) + T(\xi(m)) \cdot (43 \log m \cdot L + 28 \log^2 m) \right) \times \frac{1}{L}$$

Before analyzing this recurrence relation, we show that its right-hand side gives a valid upperbound on $T(m)$: The last coefficient in the right-hand side, i.e. $\frac{1}{L}$, is due to the definition of approximation ratio of an algorithm. The first coefficient, i.e. $8 \log m$, stands for Line 7 of the algorithm, as a result of the (potential) sparsification. It remains to analyze the middle coefficient.

The summand $\frac{2m \log m}{\xi(m)} \cdot L$ is an upperbound on the number of rounds used in Phase 2 of the algorithm. The two other summands bound the number of rounds used in Phase 3 of the algorithm. We justify the latter fact separately as follows.

The summand $7 \log m \cdot (6L + 4 \log m)$ is an upperbound on the number of rounds used in Lines 21 and 24 overall. Since only the schedule $\mathcal{S}$ is run in these lines, we equivalently show that $|\mathcal{S}| \leq 7 \log m \cdot (3L + 2 \log m)$, which is done in Lemma 4.

Finally, we verify that the summand $T(\xi(m)) \cdot (43 \log m \cdot L + 28 \log^2 m)$ is an upperbound on the number of rounds used in Line 23 of the algorithm: Let L_Y be the optimal number of rounds needed to solve $\mathcal{I}(G, Y)$. By the induction hypothesis, the number of rounds used in Line 23 is at most $T(\xi(m)) \cdot L_Y$. So, it's enough to show that $L_Y \leq 43 \log m \cdot L + 28 \log^2 m$. We do this by giving a feasible solution of length at most $43 \log m \cdot L + 28 \log^2 m$ for $\mathcal{I}(G, Y)$: Run the schedule $\mathcal{S}$, then run $\mathcal{L}$, and finally run the schedule $\mathcal{S}$ in the reverse order. The claimed upperbound for the length of this schedule simply follows from Lemma 4.

It only remains to analyze the recursive formula $T(m)$ and prove the claimed bound for it; we omit this part here, the complete proof appears in Section B.

7   Conclusion

Our main contribution is developing a unified recursive framework that we use to design approximation algorithms for various extensions of the Multicast Problem. In particular, we consider three generalizations: i. allowing conference calls involving possibly more than two participants; ii. allowing multiple source and destination vertices; and iii. allowing capacities for vertices .

A comprehensive summary of all our results can be seen in Tables 1 and 2. A descriptive summary of the tables also appears in Section A. For the cells in the tables which are marked with [∗], the proofs have been omitted in this paper, since they are very similar to the proofs that appear in the paper and can be derived from them with slight modifications.

Designing a poly-logarithmic approximation algorithm for the multicommodity multicast problem is the most important remaining open problem from our work.

Table 1. This table summarizes our approximation ratios and additive approximations for the Multicast Problem

Multicast	Single-source	Multi-commodity
Non-capacitated	$O(\log k/\log\log k)$ [7] $\Omega(3-\epsilon)$ [5]	$2^{O(\log\log k \cdot \sqrt{\log k})}$ $O\left(\log n \cdot OPT + \sqrt{n}\log^2 n\right)$ $O\left(\frac{\log^3 n}{\log\log n} \cdot (OPT + \Delta(G))\right)$
Capacitated	$O(\log k)$	$2^{O(\log\log k \cdot \sqrt{\log k})}$ [*]

Table 2. This table summarizes our hardness results and approximation ratios for the Hypercast Problem

Hypercast	Single-source	Multi-commodity
Non-capacitated	$O(\log k \cdot \log n \cdot D)$ $\Omega(D^{1/3})$	$2^{O(\sqrt{\log k}(\log\log n + \log D))}$ [*]
Capacitated	$O(\log k \cdot \log n \cdot D)$ [*]	$2^{O(\sqrt{\log k}(\log\log n + \log D))}$ [*]

Acknowledgments. We would like to thank Takuro Fukunaga for discussions in the early stages of this work. We also thank Subhash Khot for pointing out the hardness result of [11] on finding large independent sets and Guy Kortsarz for his comments about the choice of the presentation.

References

1. Awerbuch, B., Kutten, S., Peleg, D.: On buffer-economical store-and-forward deadlock prevention. IEEE Transactions on Communication 42, 2934–2937 (1994)
2. Baker, B., Shostak, R.: Gossips and telephones. Discrete Math 2, 191–193 (1972)
3. Censor-Hillel, K., Haeupler, B., Kelner, J., Maymounkov, P.: Global Computation in a Poorly Connected World: Fast Rumor Spreading No Dependence on Conductance. In: STOC: ACM Symposium on Theory of Computing (2012)
4. Dvork, T.: Chromatic Index of Hypergraphs and Shannons Theorem. European Journal of Combinatorics 21(5), 585–591 (2000)
5. Elkin, M., Kortsarz, M.G.: Combinatorial logarithmic approximation algorithm for directed telephone broadcast problem. In: Proceedings of the Thiry-fourth Annual ACM Symposium on Theory of Computing, STOC 2002 (2002)
6. Elkin, M., Kortsarz, G.: An approximation algorithm for the directed telephone multicast problem. Algorithmica 45(4), 569–583 (2006)
7. Elkin, M., Kortsarz, G.: Sublogarithmic approximation for telephone multicast. In: SODA 2003 Proceedings of the Fourteenth Annual ACM-SIAM Symposium on Discrete Algorithms, pp. 76–85 (2003)
8. Feige, U., Kilian, J.: Zero knowledge and the chromatic number. J. Comput. System Sci. 57, 187–199 (1998)
9. Grigni, M., Peleg, D.: Tight bounds on minimum broadcast networks. SIAM J. Discrete Math. 4, 207–222 (1991)

10. Guha, S., Bar-noy, A., Naor, J., Schieber, B.: Multicasting in heterogeneous networks. In: STOC 1998 Proceedings of the Thirtieth Annual ACM Symposium on Theory of Computing, pp. 448–453 (1998)
11. Guruswami, V., Sinop, A.K.: The complexity of finding independent sets in bounded degree (hyper)graphs of low chromatic number. In: SODA 2011 Proceedings of the 22nd Annual ACM-SIAM Symposium on Discrete Algorithms, pp. 1615–1626 (2011)
12. Hajnal, A., Milner, E.C., Szemeredi, E.: A cure for the telephone disease. Canad Math. Bull 15, 447–450 (1976)
13. Haeupler, B.: Simple, Fast, and Deterministic Gossip and Rumor Spreading. In: SODA: ACM-SIAM Symposium on Discrete Algorithms (2013)
14. Kortsarz, G., Peleg, D.: Approximation algorithms for minimum time broadcast. SIAM Journal on Discrete Methods 8, 401–427 (1995)
15. Kossinets, G., Kleinberg, J., Watts, D.: The structure of information pathways in a social communication network. In: Proceedings of the 14th SIGKDD International Conference on Knowledge Discovery and Data Mining, pp. 435–443 (2008)
16. Leighton, F.T., Lewin, D.M.: Global Hosting System, US Patent 6108703 (Issued August 22, 2000)
17. Onus, M., Richa, A.W.: Minimum maximum-degree publish-subscribe overlay network design. IEEE/ACM Transactions on Networking, TON (2011)
18. Proskurowski, A.: Minimum broadcast trees. IEEE Trans. Comput. C-30, 363 (1981)
19. Raghavan, P.: Probabilistic construction of deterministic algorithms: Approximating packing integer programs. In: 27th Annual Symposium on Foundations of Computer Science (FOCS 1986), pp. 10–18 (1986)
20. Ravi, R.: Rapid rumor ramification: approximating the minimum broadcast time. In: 35th Annual Symposium on Foundations of Computer Science, FOCS 1994 (1994)
21. Scheuermann, P., Wu, G.: Heuristic Algorithms for Broadcasting in Point-to-Point Computer Networks. IEEE Transactions on Computers 33(9), 804–811 (1984)
22. Schrijver, A.: Combinatorial optimization: Polyhedra and Eficiency, ch. 21. Springer (2003)
23. Tijdeman, R.: On a Telephone Problem. Nieuw Arch. Wisk. 19, 188–192 (1971)

A   Roadmap for the Appendix

The main results that appear in the Appendix are listed below.

1. In Section C, we present a very simple $O(\log k)$-approximation for the Capacitated Multicast Problem.
2. In Section D, we develop our simple algorithm further and obtain a $O(\log k \cdot \log n \cdot D)$-approximation for the Hypercast Problem, where D is the maximum size of a hyperedge. The dependence on D is natural due to our strong hardness results for the Hypercast Problem (see Section I): we prove that for any constant $\epsilon > 0$, the Hypercast Problem is $\Omega(n^{1-\epsilon})$-hard. But this hardness result, as our algorithm also suggests, only holds for large values of $D \geq \Omega(n^{1-\epsilon})$. The more natural and interesting case though, is when D is small, for which we provide a hardness ratio of $\Omega(D^{1/3})$ under Khot's 2-to-1 conjecture.

3. By modifying our algorithm for the Multi-commodity Multicast problem, we obtain two other algorithms for the same problem in Appendices G and H, which respectively produce solutions of length at most $O\left(\log n \cdot OPT + \sqrt{n} \log^2 n\right)$ and $O\left(\frac{\log^3 n}{\log \log n} \cdot (OPT + \Delta(G))\right)$.

4. In Section H we show an (approximate) equivalence between the Multicommodity Multicast problem and the following Minimum Poise Subgraph problem: find a subgraph H of G in which the *Poise* of H [20], namely, the maximum pairwise distance in H between all pairs (s_i, t_i) plus the maximum degree in H, is minimized. We prove that any α-approximation algorithm for either of these problems gives an $O\left(\alpha \cdot \frac{\log^3 n}{\log \log n}\right)$-approximation for the other one.

Below, we briefly discuss these results an their connection to each other.

The algorithm for the Capacitated Multicast Problem is presented in Section C. As it can be seen in the tables, the approximation ratio for the capacitated problems almost always matches the ratio for the associated non-capacitated version, which means our framework can handle capacities very well.

The algorithm for the Hypercast Problem appears in Section D. In the Hypercast Problem, note that despite the strong $\Omega(n^{1-\epsilon})$ hardness result, which holds only for the less interesting case of $D \geq \Omega(n^{1-\epsilon})$, we develop a $O(\log k \cdot \log n \cdot D)$-approximation. It is not hard to see that this result is tight up to logarithmic factors if $D \geq n^\epsilon$ for some constant $\epsilon > 0$ (the proof is very similar to the proof for the $\Omega(n^{1-\epsilon})$-hardness). For the general case when we have no restrictions on D, we can also prove an $\Omega(D^{1/3})$-hardness under Khot's 2-to-1 conjecture. All these hardness results appear in the Section I.

Although there is a significant gap between the existing hardness ratio and the approximation ratio provided for the Multicommodity Multicast Problem, we are able to tighten this gap when $OPT \geq \Omega(\sqrt{n})$. For this case, we have a $O(\log^2 n)$-approximation as a consequence of an alternate algorithm that we present in Section G, which uses at most $O(\log n \cdot OPT + \sqrt{n} \log^2 n)$ rounds.

On the way to obtaining this additive approximation, we prove results for the special case when the instance has a small number of terminals in Section F. For given demand pairs P, this variant finds a schedule of length $O(|P| + D)$ where D denotes $diam_P(G)$.

To derive the result for small number of terminals, we develop and employ an extension of our framework for asymmetric capacities in Section E. Note that our results on this model also apply for the GOSSIP models recently studied in the literature [3,13] to give *relative* approximation algorithms in contrast to the more absolute guarantees provided in these papers in terms of the size of the graph or its diameter.

We shed more light on the Multicommodity Multicast problem by showing that it is (approximately) equivalent to a Minimum Poise Subgraph Problem, i.e. the problem of finding a subgraph H of the graph G which minimizes $\Delta(H) + diam_P(H)$. More precisely, we show that any α-approximation algorithm for either of these problems gives an $O\left(\alpha \cdot \frac{\log^3 n}{\log \log n}\right)$-approximation for the other one. This equivalence is formally proved in Section H. As a consequence of this equivalence, we obtain an algorithm which guarantees to produce a schedule of length at most $O\left(\frac{\log^3 n}{\log\log n}\cdot(OPT+\Delta(G))\right)$. Note that this gives a poly-logarithmic approximation in the instances when $OPT=\Omega(\Delta(G))$.

B   Complete Proof of Theorem 2

Proof (of Theorem 2). Let $T(m)$ denote the approximation factor of our algorithm. By induction on m, we prove that $T(m)\leq 2^{\epsilon\log\log m\cdot\sqrt{\log m}}$, for any fixed $\epsilon>6$. To do so, we provide an upperbound for $T(m)$ as follows:

$$T(m)\leq 8\log m\cdot\left(\frac{2m\log m}{\xi(m)}\cdot L+7\log m\cdot(6L+4\log m)+T(\xi(m))\cdot(43\log m\cdot L+28\log^2 m)\right)\times\frac{1}{L}$$

Before analyzing this recurrence relation, we show that its right-hand side gives a valid upperbound on $T(m)$: The last coefficient in the right-hand side, i.e. $\frac{1}{L}$, is due to the definition of approximation ratio of an algorithm. The first coefficient, i.e. $8\log m$, stands for Line 7 of the algorithm, as a result of the (potential) sparsification. It remains to analyze the middle coefficient.

The summand $\frac{2m\log m}{\xi(m)}\cdot L$ is an upperbound on the number of rounds used in Phase 2 of the algorithm. The two other summands bound the number of rounds used in Phase 3 of the algorithm. We justify the latter fact separately as follows.

The summand $7\log m\cdot(6L+4\log m)$ is an upperbound on the number of rounds used in Lines 21 and 24 overall. Since only the schedule $\mathcal{S}$ is run in these lines, we equivalently show that $|\mathcal{S}|\leq 7\log m\cdot(3L+2\log m)$, which is done in Lemma 4.

Finally, we verify that the summand $T(\xi(m))\cdot(43\log m\cdot L+28\log^2 m)$ is an upperbound on the number of rounds used in Line 23 of the algorithm: Let L_Y be the optimal number of rounds needed to solve $\mathcal{I}(G,Y)$. By the induction hypothesis, the number of rounds used in Line 23 is at most $T(\xi(m))\cdot L_Y$. So, it's enough to show that $L_Y\leq 43\log m\cdot L+28\log^2 m$. We do this by giving a feasible solution of length at most $43\log m\cdot L+28\log^2 m$ for $\mathcal{I}(G,Y)$: Run the schedule $\mathcal{S}$, then run $\mathcal{L}$, and finally run the schedule $\mathcal{S}$ in the reverse order. The claimed upperbound for the length of this schedule simply follows from Lemma 4.

Now, we prove the claimed bound for $T(m)$. First, we simplify the above recurrence relation and write a slightly weaker version of it:

$$T(m)\leq 16\log^3 m\cdot\frac{m}{\xi(m)}+1128\log^3 m\cdot T(\xi(m)) \tag{1}$$

Recall that $\xi(m)=2^{\log m-\sqrt{\log m}}$. Use the induction hypothesis to bound the right-hand side of (1) by

$$\leq 16\log^3 m.(2^{\sqrt{\log m}}+2^{7+\epsilon\log\log\xi(m).\sqrt{\log\xi(m)}})$$

$$\leq 2^{12+3\log\log m+\epsilon\log\log\xi(m).\sqrt{\log\xi(m)}}$$

where in the last inequality, we have used the fact that $\sqrt{\log m} \leq 7 + \epsilon \log \log \xi(m) \cdot \sqrt{\log \xi(m)}$ for all $m \geq 1$. So, the proof is complete if we show that

$$12 + 3 \log \log m + \epsilon \log \log \xi(m) \cdot \sqrt{\log \xi(m)} \leq \epsilon \log \log m. \sqrt{\log m}$$

Observe that:

$$12 + 3 \log \log m + \epsilon \log \log \xi(m) \cdot \sqrt{\log \xi(m)} \leq \epsilon \log \log m \cdot \left(\frac{12}{\epsilon \log \log m} + \frac{3}{\epsilon} + \sqrt{\log \xi(m)} \right) \tag{2}$$

Since we have $\sqrt{\log m - \sqrt{\log m}} \leq \sqrt{\log m} - 0.5$ for all $m \geq 4$, then we can bound the right-hand side of (2) by

$$\leq \epsilon \log \log m \cdot \left(\frac{12}{\epsilon \log \log m} + \frac{3}{\epsilon} + \sqrt{\log m} - 0.5 \right) \tag{3}$$

And since for any fixed $\epsilon > 6$, there exists a fixed positive integer m_ϵ such that $\frac{12}{\epsilon \log \log m} + \frac{3}{\epsilon} < 0.5$ for all $m \geq m_\epsilon$, then we can bound (3) by $\epsilon \log \log m \cdot \sqrt{\log m}$, which finishes the proof.

C The Capacitated Multicast Problem

In this section, we bring in the notion of *capacity of a vertex* to our model for the Multicast Problem, and allow a vertex to be in possibly more than a single call in each round; the maximum number of phone calls that a vertex can have in each round is called the *capacity* of the vertex.

Definition 3. *In the Capacitated Multicast Problem (CM) we are given an instance of the Multicast problem along with an integer c_v for each vertex $v \in V$ as its capacity. The only difference with the Multicast Problem is that here, each vertex can be in up to c_v phone calls in a round, i.e. in each round, we can pick a subgraph H of G such that the degree of each vertex v in H is at most c_v, and arrange phone calls between the endpoints of all the edges in H. Note that we can assume w.l.o.g. that the capacities c_v are all at least one since we can delete nodes that have zero capacity from the problem.*

In this section, we present an $O(\log k)$-approximation for the Capacitated Multicast Problem.

C.1 Preliminaries

First, we need to define a new notion of b-matchings due to the presence of capacities. Given that c is a capacity vector for the vertices of Y, i.e. a $|Y|$-dimensional vector of positive integers such that c_y denotes the capacity of the vertex $y \in Y$, we define a c-matching in H to be a subset M of the edges of H such that each vertex of X is incident to exactly one edge of M and each vertex $y \in Y$ is incident to at most c_y edges of M.

We also need to refine the definition of the degree of a spider. When we have capacities c_v on the vertices, we can state a lemma similar to Lemma 1. Define the relative degree of a spider S, denoted by $rdeg(S)$, to be $\lceil \frac{deg(S)}{c_v} \rceil$ where v is the center of the spider. Then we have:

Lemma 5. *Using a non-lazy schedule, the center of a spider S can send (broadcast) a message to the rest of its vertices in $rdeg(S) + len(S) - 1$ rounds.*

C.2   The Algorithm

Our algorithm is an adaptation of Algorithm *Multicast* and the only difference between our algorithm and Algorithm *Multicast* is in Phase 4. Before proceeding to more details about Phase 4, first verify that Phases 1-3 are still valid and can be executed with the presence of capacities. Particularly in Phase 3, all we do is using matchings for sending the message through the paths in $\mathcal{P}$, and this is possible since all the capacities are at least 1.

Our goal in Phase 4, assuming that the vertices in $V(\mathcal{P})$ have received the message, is to inform the rest of the terminals in $O(L)$ rounds.

To do so, we find a family of vertex-disjoint spiders such that each of them has a length at most $2L$ and a center belonging to $V(\mathcal{P})$, moreover, we need the spiders to contain all the terminals in $R\backslash V(\mathcal{P})$. Assuming that S is such a family of spiders, by Lemma 5 we can inform the set $R\backslash V(\mathcal{P})$ in at most $rdeg(S) + len(S) - 1$ rounds, where $rdeg(S) = \max_{S \in \mathcal{S}} rdeg(S)$. We show we can find a family S with $rdeg(S) \leq L$ and $len(S) \leq 2L$, which implies the set $R\backslash V(\mathcal{P})$ can be informed in at most $3L - 1$ rounds in Phase 4. The other parts of the analysis will be identical to the analysis of Algorithm *Multicast*. So, in the rest of this section, we only show how to find the desired family of spiders.

Construct the bipartite graph $H[R\backslash V(\mathcal{P}), V(\mathcal{P})]$ similar as before, but instead of finding the smallest integer b such that H has a b-matching, find the smallest b such that H has a $(b \cdot c)$-matching, where c is the capacity vector for the vertices in $V(\mathcal{P})$. Then, using the $(b \cdot c)$-matching, construct the family of spiders $\mathcal{M}$ identical to the way we construct them in subsection 5.2. Following the proof of Lemma 2, it can be seen that $len(\mathcal{M}) \leq 2L$, and it only remains to show that $rdeg(\mathcal{M}) \leq L$.

Consider the optimal multicast schedule which uses exactly L rounds. Let E' be the subset of the edges of G which are used in the optimal schedule and G' be the subgraph of G with E' as its edge set. Note that G' is not necessarily a tree as in the previous section; however, the maximum degree of any node v in E' is $c_v \cdot L$ by the optimality of the schedule. Now, for each $v \in R\backslash V(\mathcal{P})$, define M'_v to be an arbitrary shortest path in G' from v to $f(v)$, where $f(v) = \arg\min_{u \in V(\mathcal{P})} d_{G'}(v, u)$.

We prove that $\bigcup_{v \in R\backslash V(\mathcal{P})} (v, f(v))$ is an $(L \cdot c)$-matching in H, which shows the existence of a $(b \cdot c)$-matching in H with $b \leq L$, which implies $rdeg(\mathcal{M}) \leq L$. To prove the claim, just note that the family of paths $\{M'_v : v \in R\backslash V(\mathcal{P})\}$ in G' are edge-disjoint (otherwise, it contradicts the maximality of $\mathcal{P}$). This means for any

$v \in V(\mathcal{P})$, there are no more than $c_v \cdot L$ of these paths with the same endpoint v, since otherwise, the degree of v in G' is more than $c_v \cdot L$, a contradiction. Consequently, $\bigcup_{v \in R \setminus V(\mathcal{P})} (v, f(v))$ is an $(L \cdot c)$-matching in H. This proves the claim.

D   The Hypercast Problem

In this section, we study the problem with conference calls, i.e. calls involving (possibly) more that two, rather than only two, persons. We formally define the Hypercast problem as follows.

Definition 4. *In the* **Hypercast** *problem, we are given a hypergraph $G(V, E)$ where there can be a conference call between two or more nodes if G has a hyperedge containing exactly these nodes. Similar to the Multicast problem, we are also given a source vertex r and a set of terminals R. Our goal is to deliver a message from the source vertex r to all the terminals in the minimum number of rounds. To do this, the vertices of the graph can communicate in rounds: In each round, we pick a matching of G, i.e. a set of vertex-disjoint edges, and for each edge of the matching, we arrange a conference call containing all the vertices in that edge. If any of the vertices in the conference call knows the message, the others will also know it after the call.*

In this section, we present an $O(\log k \cdot \log n \cdot D)$-approximation for the hypercast problem, where D denotes the maximum size of an edge in the hypergraph G. Unlike the algorithm for Multicast Problem which was purely combinatorial, this algorithm needs to set up a Linear Program for designing some parts of the multicast schedule.

The dependence on D in the approximation factor is natural due to our strong hardness results for the Hypercast Problem (see Section I): we prove that for any constant $\epsilon > 0$, the Hypercast Problem is $\Omega(n^{1-\epsilon})$-hard. But this hardness result, as our algorithm also suggests, only holds for large values of $D \geq \Omega(n^{1-\epsilon})$. The more natural and interesting case though, is when D is small, for which we provide a hardness ratio of $\Omega(D^{1/3})$ under Khot's 2-to-1 conjecture.

D.1   Preliminaries: Spiders and Hypergraphs

The set of vertices and (hyper)edges of a hypergraph G are respectively denoted by $V(G)$ and $E(G)$. We say an edge e intersects a subset of vertices $S \subseteq V(G)$ (or another edge e') if e contains at least one vertex of S (one vertex of e'). With abuse of notation, we denote this by $e \cap S \neq \emptyset$ (respectively $e \cap e' \neq \emptyset$). A subset $M \subseteq E(G)$ is called a matching if any two edges in M have an empty intersection.

A path P is a sequence of edges such as $e_0, \dots, e_m$ where $e_i \cap e_j \neq \emptyset$ iff $|i - j| \leq 1$. The length of P is denoted by $len(P)$ and is equal to m. Having the definition of the length of a path, the notions of connectivity, distance and diameter in a hypergraph are trivially adapted from simple graphs.

A spider S is a family of paths $P_1, \ldots, P_k$ such that the first edge of all of them is the same, and moreover, they will be a family of vertex-disjoint paths if their first edge is deleted. The set of vertices in the first edge (or simply, the first edge, when it's clear from the context) is called the center of S. By adapting the definition of length of a spider in simple graphs, we define $len(S) = \max_i len(P_i)$.

D.2   Outline of the Algorithm

Our algorithm is similar to Algorithm *Multicast* presented in Section 5; it is a recursive algorithm with 4 phases. In Phase 1 of the algorithm, we reduce the given instance to a smaller instance. In Phase 2, we solve the smaller instance recursively and as the result, inform a subset of the terminals. And finally in Phases 3 and 4, we inform the rest of the vertices that didn't receive the message in Phase 2. First we explain each of these phases briefly, and then present the full description of the algorithm and its analysis.

Phase 1. This phase starts with finding a family of vertex disjoint paths $\mathcal{P}$ each of length (number of the hyperedges) at most $4L$ such that the first and last edge of each path intersects R. Similar to Algorithm *Multicast*, we find these paths greedily, i.e. we start with $\mathcal{P} = \emptyset$ and using any shortest path algorithm for hypergraphs, we search for a new path in $G[V(G) \backslash V(\mathcal{P})]$ which connects two of the terminals. Moreover, length of the path must be at most $4L$. We continue until we can add no more such paths to $\mathcal{P}$. Then, pick an arbitrary terminal from the first edge of each path and let the obtained set of vertices be R'.

Phase 2. Solve the multicast problem for the new set of terminals R' recursively and run the obtained schedule. (So, all the vertices in R' will receive the message by the end of this phase.)

Phase 3. Inform all the vertices belonging to $V(\mathcal{P})$ in $4L$ rounds. This is possible since in Phase 2, we have already informed at least one vertex of each path in $\mathcal{P}$.

Phase 4. For each of the uninformed terminals, namely $v \in R \backslash V(\mathcal{P})$, find a path M_v which connects v to one of the informed vertices (note that the set of the informed vertices is currently $V(\mathcal{P})$). These paths will be guaranteed to satisfy the following properties:

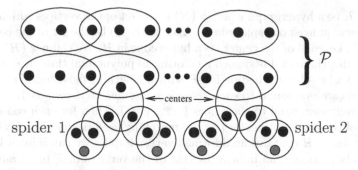

Fig. 3. The gray vertices are in $R\backslash V(\mathcal{P})$. Despite this figure, centers of the spiders can possibly intersect.

1. For each $v \in R\backslash V(\mathcal{P})$, the first edge of M_v contains v and the last edge intersects $V(\mathcal{P})$. Moreover, no other edge of M_v intersects $V(\mathcal{P})$.
2. The length of each path is at most $2L$.
3. The paths would be vertex disjoint if the last edge of each path is removed.

In other words, $M = \bigcup\limits_{v \in R\backslash V(\mathcal{P})} M_v$ is a union of spiders of length at most $2L$ and we will inform the vertices in $R\backslash V(\mathcal{P})$ using these spiders (see Figure 3). However, this is not as simple as it is in Section 5, since unlike there, the centers of spiders may intersect. So, bounding the degree and length of each of the spiders is not enough to bound the total number of rounds needed for informing $R\backslash V(\mathcal{P})$. We need the family of spiders to satisfy an additional constraint, which we call z-disjointness. We define this concept and then describe Phase 4 for the Algorithm. The other 3 Phases remain unchanged.

D.3   Disjoint and Fast Spiders

Definition 5. *A family $\mathcal{F}$ of spiders is z-disjoint if they satisfy the following conditions.*

- *If 2 edges which belong to 2 different spiders intersect, then both of them are the centers (also called first edges earlier) of spiders.*
- *No vertex is in more than z of the centers.*

The following lemma gives an upperbound on the broadcast time in a z-disjoint family of spiders. Recall that for a family $\mathcal{F}$ of spiders, $deg(\mathcal{F}) = \max_{S \in \mathcal{F}} deg(\mathcal{F})$ and $len(\mathcal{F}) = \max_{S \in \mathcal{F}} len(S)$.

Lemma 6. *We are given a z-disjoint family of spiders $\mathcal{F}$, such that for each spider, at least one vertex in its center is informed. Assuming that D is the maximum size of a hyperedge, we can find a broadcast schedule for $\mathcal{F}$ with a length at most $len(\mathcal{F}) + 2zD$ in polynomial time.*

Proof. Let $\mathcal{H}$ be a hypergraph where $V(\mathcal{H})$ is the set of the vertices which belong to the center of at least one spider in $\mathcal{F}$. Also, let $E(\mathcal{H})$ be the set of the centers of spiders in $\mathcal{F}$, i.e. each of the centers is a hyperedge in $\mathcal{H}$. We claim $\chi'(\mathcal{H}) \leq 2zD$. If we prove this claim and find such a coloring in polynomial time, we can easily convert it to a broadcast schedule for $\mathcal{F}$ with a length at most $len(\mathcal{F}) + 2zD$ as follows: Dedicate one round to each color, e.g. assume that rounds $1, \ldots, \chi'(G)$ are respectively dedicated to the colors $1, \ldots, \chi'(G)$. Then for each color i, pick the subset of the edges with color i as the matching used in round i. Having that each hyperedge in $\mathcal{H}$ contains an informed vertex implies we can inform $V(\mathcal{H})$ in $\chi'(\mathcal{H})$ rounds. Also, we can inform the rest of the vertices in $\mathcal{F}$ in an additional $len(\mathcal{F})$ rounds, due to the disjointness of the spiders legs. This means the length of the obtained broadcast schedule is at most $\chi'(\mathcal{H}) + len(\mathcal{F})$. To complete the proof, we show that $\chi'(\mathcal{H}) \leq 2zD$ and find a coloring with at most $2zD$ colors in polynomial time.

Let the strong degree of a vertex v, denoted by $\bar{d}(v)$, be the summation, over all $u \neq v$, of the number of hyperedges that contain both v and u. Also, let $\bar{\Delta}(\mathcal{H}) = \max_{v \in V(\mathcal{H})} \bar{d}(v)$. It's a well-known fact that $\chi'(\mathcal{H}) \leq 2\bar{\Delta}(\mathcal{H})$, and in fact, we can find such a coloring by a greedy algorithm in polynomial time [4].

Since $\mathcal{H}$ is loop-less, $\bar{d}(v)$ is at most the summation of the size of the hyperedges that contain v. Also, note that $\mathcal{F}$ is z-disjoint, which means v is in at most z hyperedges in $\mathcal{H}$. It implies $\bar{\Delta}(\mathcal{H}) \leq zD$ since the maximum size of a hyperedge is at most D. Consequently, $\chi'(\mathcal{H}) \leq 2zD$.

Recall that our goal in Phase 4 is informing the set of uninformed vertices, $R \backslash V(\mathcal{P})$, using a family $\mathcal{F}$ of spiders. If we can find a z-disjoint family of spiders $\mathcal{F}$ such that both z and $len(\mathcal{F})$ are small, then by Lemma 6 we get a short broadcast schedule for $\mathcal{F}$, and consequently, a short schedule for the original hypercast instance. The properties required for the family $\mathcal{F}$ are formally defined below:

Definition 6. *A family $\mathcal{F}$ of spiders is called* fast *if it satisfies the following properties:*

  i. $len(\mathcal{F}) \leq 2L$.
  ii. $\mathcal{F}$ is z-disjoint for $z = O(L \log n)$
  iii. The center of each spider in $\mathcal{F}$ has a non-empty intersection with $V(\mathcal{P})$.
  iv. Each vertex in $R \backslash V(\mathcal{P})$ belongs to exactly one of the spiders in $\mathcal{F}$.

Lemma 7. *If we can find a fast family of spiders for any given Hypercast instance, then we can design an $O(\log k \cdot \log n \cdot D)$-approximation for the Hypercast problem.*

Proof. Our recursive algorithm for the Hypercast problem reduces the number of terminals by at least a factor of 2 in each level of the recursion. This is easily observable by the definition of R' in Phase 2 of the Algorithm: from the two terminals connected by a path in $\mathcal{P}$, only one of them is in R'. So, the depth of the recursion is at most $\log k$.

The proof is complete if we show that each level of the recursion takes at most $O(LD \cdot \log n)$ number of rounds. First, see that the cost of the optimal solution does not increase when the number of terminals decreases in the next levels of recursion. So, if we find a fast family $\mathcal{F}$ at each level of the recursion, then by Lemma 6 we spend at most $O(LD \cdot \log n)$ rounds in that level. This means the length of the obtained schedule for the given Hypercast instance would be at most $O(LD \cdot \log k \cdot \log n)$.

So, by Lemma 7, all we need to do in Phase 4 is finding a fast family $\mathcal{F}$ and a broadcast schedule for it. In the rest of this section, we describe an algorithm for finding a fast family of spiders, given that $\mathcal{P}$ is maximal and $V(\mathcal{P})$ is informed.

D.4   Finding a Fast Family of Spiders

The idea for finding the family of spiders is similar to Section 5, except that instead of a bipartite graph, we construct a tripartite graph $H[A, B, C]$, where $A = R \backslash V(\mathcal{P})$, B has a vertex for every hyperedge which intersects both $V(\mathcal{P})$ and $V(G) \backslash V(\mathcal{P})$, and $C = V(G)$. Note that B is in fact the set of potential centers for the family of spiders $\mathcal{F}$.

The edges of H, which lie only between the partitions A, B and B, C, are defined as follows: There is an edge between the vertices $v \in A$ and $e \in B$ iff there is a path in $G[V(G) \backslash V(\mathcal{P})]$ of length at most $2L$, such that its first edge contains v and its last edge intersects with the hyperedge e. Also, there is an edge between the vertices $e \in B$ and $u \in C$ iff the hyperedge e contains the vertex u.

The induced bipartite graph between the partitions A, B, denoted by $H[A, B]$ is used in a way similar to Section 5.2, i.e. any $|A|$-matching in $H[A, B]$, e.g. M, represents a family of spiders, and more precisely, any connected component in M, represents a spider. Observe that any connected component of M forms a star. The vertex e in the center of the star, represents the center of the spider, and the edges of the star, represent the legs of the spider, i.e. the paths which are attached to the center of the spider. Recall that each edge of $H[A, B]$ is associated with a path in G. Any two legs, whether or not belonging to the same spider, are vertex-disjoint, due to the maximality of $\mathcal{P}$.

The family of spiders represented by M, namely $\mathcal{F}$, is z-disjoint for $z = |B|$. It is easy to verify that $\mathcal{F}$ is z-disjoint if each vertex in C is adjacent to at most z of the centers of the stars in M.

For $\mathcal{F}$ to be a fast family, we only need to choose M in such a way that $\mathcal{F}$ becomes z-disjoint for $z = O(L \log n)$, and this is where we use partition C of H. To be more formal, for any subset $S \subseteq B$, define the weight of S, denoted by $w(S)$, to be $\max_{v \in C} |S \cap N(v)|$. Now, finding a fast family $\mathcal{F}$ can be reformulated as finding a subset $S \subseteq B$, such that $N(S) = A$ and $w(S) \in O(L \log n)$. If we find such a subset S, then we can easily find a fast family by choosing any $|A|$-matching in $H[A, S]$. The family of spiders associated with the $|A|$-matching would be $w(S)$-disjoint, and consequently, fast.

All that remains to do is finding such a subset S. First, we show that there exists such S, and then we present an algorithm for finding it in Section D.5.

Lemma 8. *There exists a subset $S \subseteq B$ with $N(S) = A$ and $w(S) \leq L$.*

Proof. Let $\mathcal{L}$ be an optimal schedule for the given Hypercast instance. Note that $|\mathcal{L}| = L$. Also, let J be a hypergraph where $V(J) = V(G)$ and $E(J)$ is the set of hyperedges used in $\mathcal{L}$. Delete the isolated vertices of J. Then clearly, J is connected and $diam(J) \leq 2L$. So, for each vertex $v \in R \backslash V(\mathcal{P})$, J should contain a path of length at most $2L$ connecting v to a vertex in $V(\mathcal{P})$. If there are many such paths, then select one of them arbitrarily and define $g(v)$ to be the first edge on this path (closest to v) which intersects both $V(\mathcal{P})$ and $R \backslash V(\mathcal{P})$.

By the definition of partition B, it contains a vertex for each hyperedge $g(v)$, which we denote by $g'(v)$. We prove that $S = \bigcup\limits_{v \in R \backslash V(\mathcal{P})} g'(v)$ satisfies the requirements of the lemma: For all $v \in R \backslash V(\mathcal{P})$, since there is an edge between v and $g(v)$ in $H[A, B]$, then $N(S) = A$. Also, since each vertex $u \in V(G)$ is in at most L of the hyperedges of J, then $|N(u) \cap S| \leq L$ for all $u \in C$, which means $w(S) \leq L$.

D.5   An LP-Rounding Approach for Finding Fast Spiders

Given a tripartite graph $H[A, B, C]$, we want to find a subset $S \subseteq B$ which minimizes $w(S)$ and satisfies $N(S) = A$. Here we present a $8 \ln n$-approximation for this problem. This result, along with Lemma 8 implies that we can find a subset S such that $w(S) \leq 8L \ln n$ and $N(S) = A$.

To solve the above optimization problem, first we solve the fractional version of it by writing a linear program and then, using randomized rounding, we round the fractional solution of the linear program and obtain an integral one. To write this LP, a variable x_i is associated with each vertex $i \in B$.

$$\min z$$
$$\text{s.t.} \quad \sum_{i \in N(u)} x_i \geq 1, \; \forall u \in A \; (1)$$
$$z \geq \sum_{j \in N(v)} x_j, \; \forall v \in C \; (2)$$
$$0 \leq x_i \leq 1, \quad \forall i \in B \; (3)$$

Note that there is a one-to-one correspondence between the *integral* solutions of this LP and the subsets S such that $N(S) = A$. To see this, let x be an integral solution, then the set $S = \{i : x_i = 1\}$ would be a feasible solution for the original problem, because constraint (1) is enforcing $N(S) = A$. The reverse direction can also be verified easily: if $N(S) = A$ for a subset S, then $x = \mathbb{1}_S$, i.e. the characteristic vector of S, is a feasible integral solution for the LP. Also, note that constraint (2) is enforcing $z = w(S)$. However, we can't solve the integer program in polynomial time, and so, we relax the program by adding constraint (3) and use the optimal solution of the obtained LP as a lower bound on the optimal integral solution.

Let $\bar{x}$ be the optimal solution (assignment of values to the variables) for this LP. Also, let the optimal objective values for the integer program and the (fractional) LP be denoted by OPT, OPT_f respectively. We round $\bar{x}$ to obtain a feasible integral solution $\mathring{x}$ with objective value at most $8\ln n \cdot \max\{1, OPT_f\}$. This gives an $(8\ln n)$-approximation since $OPT_f \leq OPT$ and $1 \leq OPT$. The rounding procedure is described below.

The Rounding Procedure. For each $j \in B$, flip a coin $2\ln n$ times independently where the coin comes up heads with probability $\bar{x}_j$. If heads were observed at least once, then pick j to be in S. More formally, we create $2\ln n$ independent random variables $x_{j,i}$ such that $x_{j,i} = 1$ with probability $\bar{x}_j$ and $x_{j,i} = 0$ with probability $1 - \bar{x}_j$. We pick j to be in S iff $\sum_{i=1}^{2\ln n} x_{j,i} \geq 1$.

Lemma 9. *The rounding procedure outputs a set S such that with probability at least $1 - \frac{2}{n}$ both of the following conditions are satisfied: $N(S) = A$, and $w(S) \leq 8\ln n \cdot \max\{1, OPT_f\}$.*

Proof. First we prove that the $N(S) = A$ with high probability. For any $u \in A$, we compute the probability that u is not covered, i.e. none of the neighbors of u are in S:

$$\Pr\left[u \text{ is not covered}\right] = \prod_{j \in N(u)} (1 - \bar{x}_j)^{2\ln n} \leq \prod_{j \in N(u)} e^{-\bar{x}_j 2\ln n} \leq \frac{1}{n^2}$$

An application of the union bound over the nodes implies that the probability of one of the elements of A not being covered, i.e. $N(S) \neq A$, is at most $\frac{1}{n}$.

For the second condition, we claim that $w(S) \leq 8\ln n \cdot \max\{1, OPT_f\}$ with high probability. To prove this, we fix a vertex $v \in C$ and compute an upper bound on the probability of having $|N(v) \cap S| > 8\ln n \cdot \max\{1, OPT_f\}$. Then, we prove the main claim using a union bound over all v.

More formally, for any $v \in C$ we prove that

$$\Pr\left[\sum_{j \in N(v)} \mathring{x}_j > 2\alpha \ln n \cdot \max\{1, \bar{z}\}\right] \leq \frac{1}{n^2} \tag{4}$$

where $\alpha = 4$ is a fixed constant and $\bar{z} = OPT_f$. First, observe that

$$\Pr\left[\sum_{j \in N(v)} \mathring{x}_j > 2\alpha \ln n \cdot \bar{z}\right] \leq \Pr\left[\sum_{j \in N(v)} \mathring{x}_j > 2\alpha \ln n \cdot \sum_{j \in N(v)} \bar{x}_j\right]$$

$$\leq \Pr\left[\sum_{j \in N(v)} \sum_{i=1}^{2\ln n} x_{j,i} > \alpha \sum_{j \in N(v)} \sum_{i=1}^{2\ln n} \mathbb{E}\left[x_{j,i}\right]\right] \leq \left(\frac{e^{\alpha-1}}{\alpha^\alpha}\right)^\mu \tag{5}$$

where $\mu = \sum_{j \in N(v)} \sum_{i=1}^{2\ln n} \mathbb{E}\left[x_{j,i}\right]$. Note that (5) is a direct consequence of the Chernoff bound. Now we consider two cases: $\mu \geq 2\ln n$ and $\mu < 2\ln n$.

In the first case, having $\mu \geq 2\ln n$ implies $\left(\frac{e^{\alpha-1}}{\alpha^\alpha}\right)^\mu \leq \frac{1}{n^2}$, and this proves (4). On the other hand, if $\mu < 2\ln n$ then we have

$$\Pr\left[\sum_{j\in N(v)} \mathring{x}_j > 2\alpha\ln n\right] \leq \Pr\left[\sum_{j\in N(v)}\sum_{i=1}^{2\ln n} x_{j,i} > \frac{2\alpha\ln n}{\mu}\cdot\mu\right] \leq \left(\frac{e^{\beta-1}}{\beta^\beta}\right)^\mu \quad (6)$$

where $\beta = \frac{2\alpha\ln n}{\mu}$. Now, since $\mu < 2\ln n$ we have

$$\left(\frac{e^{\beta-1}}{\beta^\beta}\right)^\mu \leq \left(\frac{e}{\alpha}\right)^{2\alpha\ln n} \leq \frac{1}{n^2}$$

This fact, and (6) together imply the correctness of (4) in the second case as well. By a union bound over all $v \in C$ we have

$$\Pr\left[w(S) > 2\alpha\ln n \cdot \max\{1, \bar{z}\}\right] \leq \sum_{v\in C}\Pr\left[\sum_{j\in N(v)} \mathring{x}_j > 2\alpha\ln n \cdot \max\{1, \bar{z}\}\right] \leq \frac{1}{n}$$

which is due to (4). Consequently, the events $w(S) > 8\ln n \cdot \max\{1, OPT_f\}$ and $N(S) \neq A$ each happen with probability at most $\frac{1}{n}$. This proves the lemma.

E   Asymmetric Multicast Problem

In this section, we look at a variation of the Multicast Problem where the phone-calls are not bidirected. They are directed in the sense that only one vertex will be the sender, and the other one will be the receiver, i.e. a vertex u can call another vertex v and send (any number of) messages to v, but then v can not send any messages to u in the same phonecall. This variation is formally defined below:

Definition 7. *In the Asymmetric Multicast Problem, each vertex v has an out-capacity c_v^- as well as an in-capacity c_v^+, which are respectively the number of the vertices that v can send messages to, and receive messages from, in a single round. The objective is identical to the objective of the Multicast Problem.*

An asymmetric variation of the Multi-commodity Multicast Problem can also be defined in a natural way, which we call the *Asymmetric Multi-commodity Multicast* Problem (AMM). It is worth mentioning that a natural extension of Algorithm *MM* can solve AMM as long as all the capacities are non-zero. We do not state the algorithm, but the idea of the extension is very similar to the idea we used for designing the algorithm for the Capacitated Multicast Problem.

Below, we prove a lemma for comparing the lengths of the optimum schedules for an AMM instance and its corresponding MM instance. This Lemma will be used later in Section F.

Lemma 10. *Assume we are given an AMM instance $\mathcal{I}_A$ with the set of demand pairs P such that $c_v^- = c_v^+ = 1$ for all $v \in V(G)$. Define $\mathcal{I}(P,G)$ to be the corresponding MM instance. Also, let L_A, L respectively denote the length of optimal schedules for $\mathcal{I}_A, \mathcal{I}$. Then, we have $\frac{L}{3} \leq L_A \leq 2L$. Moreover, any schedule of length l for $\mathcal{I}$ can be converted to a schedule of length $2l$ for $\mathcal{I}_A$ in polynomial time, and also, any schedule of length l_A for $\mathcal{I}_A$ can be converted to a schedule of length $3l_A$ for $\mathcal{I}$ in polynomial time.*

Proof. First, observe that any schedule $\mathcal{S}$ for $\mathcal{I}$ can be easily turned into a schedule $\mathcal{S}_A$ for $\mathcal{I}_A$ as follows. For each round in $\mathcal{I}$, we have exactly two rounds in $\mathcal{I}_A$: In each of these two rounds, we use the same matching which is used in $\mathcal{I}$, e.g. the matching M, except that the edges of M will be used in different directions in each of these two rounds. Consequently, $L_A \leq 2L$.

To prove $\frac{L}{3} \leq L_A$, we show that given any feasible schedule $\mathcal{S}_A$ for the instance $\mathcal{I}_A$, we can construct a schedule $\mathcal{S}$ for the instance $\mathcal{I}$ such that $|\mathcal{S}| \leq 3|\mathcal{S}_A|$. This time, for each round in $\mathcal{S}_A$ we will have 3 rounds in $\mathcal{S}$.

To see this, first fix any arbitrary round in $\mathcal{S}_A$, and let M be the subset of the (directed) edges used in that round. Note that M is a union of paths and cycles, due to the fact that all the in-capacities and out-capacities are 1. Consequently, we can decompose the edges of M into 3 matchings, M_1, M_2, M_3, and use each of these matchings in one of the 3 rounds in $\mathcal{S}$. Clearly, $|\mathcal{S}| \leq 3|\mathcal{S}_A|$, and also, all of the demand pairs that were satisfied in $\mathcal{S}_A$ will also be satisfied in the obtained schedule $\mathcal{S}$.

F   An Algorithm for Small Numbers of Terminals

Given a Multi-commodity Multicast instance, we present an algorithm which finds a schedule of length $O(|P| + D)$ where D denotes $diam_P(G)$. Instead of solving this problem directly, we solve the Asymmetric MM instance where $c_v^+ = c_v^- = 1$ for all $v \in V(G)$, and then by Lemma 10, we can convert this schedule to a schedule for the original MM instance. So from now on, we think of the given instance as the Asymmetric MM instance described above.

Before presenting our algorithm, we need a few definitions. Let $m(u,v)$ denote the message that vertex u wants to send to v, given that $(u,v) \in P$. Also, let $M = \{m(u,v) | (u,v) \in P\}$ be the set of all of the messages.

F.1   The Algorithm

Our algorithm has two Phases. In Phase 1, we find a path for each message through which the message is sent. In Phase 2, a non-lazy schedule is used to send all the messages through the paths that were found in Phase 1. A non-lazy schedule in this context, is a schedule in which the messages are greedily sent to the next vertex on their paths, i.e. a message does not wait on a vertex if it can be sent to the next vertex on its path without violating the capacity constraints.

Phase 1. For each message $m(s,t)$ we find an s,t-shortest path. Moreover, these paths are found in such a way that the intersection of any two of them is a path itself (possibly empty). To find such a family of paths, we introduce the notion of *lightness* and choose the lightest path among all the s,t-shortest paths.

Definition 8. *Let $\prec$ be an arbitrary precedence relation defined on $E(G)$. For any path Q, let $\bar{Q}$ be the sequence of the edges of Q sorted in a decreasing order with respect to $\prec$. A path Q is lighter than Q' if $\bar{Q}$ is lexicographically (dictionary ordering) smaller than $\bar{Q}'$. We abuse the notation and denote this by $Q \prec Q'$. Also, we extend this definition in the natural way to compare any two subsets of the edges (not necessarily forming a path).*

Definition 9. *For any two vertices s,t, define the best s,t-path to be the lightest path among all the shortest paths between s and t.*

All we do in Phase 1, is finding the best s,t-path for each pair $(s,t) \in P$. We finish the description of this Phase by showing how to find these paths in polynomial time.

Lemma 11. *For any two vertices s,t, the best s,t-path can be found in polynomial time using Dijkstra's Algorithm.*

Proof. The proof has the same analysis as the analysis of Dijkstra's Algorithm. Define the weight of the i-th edge in the precedence relation to be $1 + 2^{-i-|E(G)|}$ and run Dijkstra's Algorithm to find the shortest (s,t)-path with respect to these weights, the obtained path would be the best (s,t)-path in G.

Phase 2. In Phase 2, we use an arbitrary non-lazy schedule which for all $(s,t) \in P$, sends $m(s,t)$ from s to t through the best s,t-path that was found in Phase 1.

The Algorithm is now completed by describing Phase 2. The schedule produced in Phase 2 is clearly feasible. In the analysis of the algorithm, we bound the length of this schedule by $2|P| + D$. But before moving to the analysis, we further exploit the structure of best paths.

F.2   On the Structure of Best Paths

Proposition 3. *Assume $A, B, C, D \subseteq E(G)$ such that $A \prec B$ and $C \prec D$. Moreover, we have $A \nsubseteq C, C \nsubseteq A, B \nsubseteq D$ and $D \nsubseteq B$. Then $(A \cup C) \prec (B \cup D)$.*

Lemma 12. *For any 4 (not necessarily disjoint) vertices s, s', t, t', the intersection of the best s,t-path and the best s',t'-path is a path itself (possibly empty).*

Proof. Let the best s,t-path and the best s',t'-path be respectively denoted by $P_{s,t}$ and $P_{s',t'}$. For the sake of contradiction, assume the intersection of $P_{s,t}$ and $P_{s',t'}$ is not a path, then, it is easy to verity that there exist two vertices a, b satisfying the following properties:

1. $a, b \in V(P_{s,t}) \cap V(P_{s',t'})$.
2. If we denote the path connecting a, b in $P_{s,t}$ by Q, and denote the path connecting a, b in $P_{s',t'}$ by Q', then Q and Q' are of the same length, $E(Q), E(Q')$ are non-empty, and $E(Q) \not\subseteq E(Q')$ and $E(Q') \not\subseteq E(Q)$.

Note that Q, Q' must be of the same length due to the fact that $P_{s,t}, P_{s',t'}$ are shortest paths. Let $\hat{P}_{s,t}$ be the path from s to t which is similar to $P_{s,t}$ except that it uses Q' instead of Q for connecting a and b. Similarly, Let $\hat{P}_{s',t'}$ be the path from s' to t' which is similar to $P_{s',t'}$ except that it uses Q instead of Q' for connecting a and b. Observe that $\hat{P}_{s,t}$ and $\hat{P}_{s',t'}$ can not have repeated vertices, again due to the fact that $P_{s,t}, P_{s',t'}$ are shortest paths.

We claim that $P_{s,t} \prec \hat{P}_{s,t}$ and $P_{s',t'} \prec \hat{P}_{s',t'}$. To verify this claim, first note that $E(P_{s,t}) \neq E(\hat{P}_{s,t})$ and $E(P_{s',t'}) \neq E(\hat{P}_{s',t'})$, due to the fact that $E(Q) \not\subseteq E(Q')$ and $E(Q') \not\subseteq E(Q)$. Also, see that $\hat{P}_{s,t}$ and $\hat{P}_{s',t'}$ are both shortest, but not the lightest shortest paths. This implies $E(P_{s,t}) \prec E(\hat{P}_{s,t})$ and $E(P_{s',t'}) \prec E(\hat{P}_{s',t'})$. Now, since the conditions stated in Proposition 3 apply, we can use this proposition and imply

$$(E(P_{s,t}) \cup E(P_{s',t'})) \prec (E(\hat{P}_{s,t}) \cup E(\hat{P}_{s',t'}))$$

But this is a contradiction, since we have

$$(E(P_{s,t}) \cup E(P_{s',t'})) = (E(\hat{P}_{s,t}) \cup E(\hat{P}_{s',t'}))$$

by the definition of $\hat{P}_{s,t}$ and $\hat{P}_{s',t'}$.

Analysis of the Algorithm. We say a message m is *out-waiting* for another message m' in round t, if the following conditions hold in that round:

i. m, m' have not reached their destinations and currently, they are on the same vertex u.
ii. The next vertices that m, m' should be sent to, are respectively v, v', and we have $v \neq v'$.
iii. In round t, we send m' to the next vertex by a phone-call from u to v'.

Similarly, we say a message m is *in-waiting* for another message m' in round t, if the following conditions hold in that round:

i. m, m' have not reached their destinations and currently, they are on two different vertices, u, u' respectively.
ii. The vertex v is the next vertex that m, m' should be sent to.
iii. In round t, we send m' to the vertex v by a phone-call from u' to v.

The *in-delay* of a message $m \in M$ in a given Multicast schedule Π is defined as the number of rounds in which m has been in-waiting for at least one other message, and it is denoted by $\xi_{\Pi}^{+}(m)$, or simply $\xi^{+}(m)$, whenever Π is clear from the context. Similarly, the *out-delay* of a message m in a given Multicast schedule Π is defined as the number of rounds in which m has been out-waiting for at least one other message, and is denoted by $\xi_\Pi^-(m)$. The *delay* of a message m, denoted by $\xi_\Pi(m)$, is simply equal to $\xi_\Pi^+(m) + \xi_\Pi^-(m)$.

The length of any non-lazy schedule Π is then clearly at most $\max_{(s,t) \in P} d_G$ $(s,t) + \xi(m(s,t))$, or simply, $D + \xi$ where ξ denotes $\max_{(s,t) \in P} \xi(s,t)$.

To bound the length of the non-lazy schedule obtained in Phase 2 by $2|P| + D$, its enough to show that $\xi \le 2|P|$.

Lemma 13. *For any message m we have $\xi(m) \le 2|P|$.*

Proof. We prove the claim by showing that $\xi^+(m), \xi^-(m) \le |P|$. To do so, we show that a message m would not be out-waiting for any other message m' in more than one round of the non-lazy schedule, which implies $\xi^-(m) \le |P|$. Similarly, we show that m would not be in-waiting for another message m' in more than one round of the schedule, and imply that $\xi^+(m) \le |P|$.

Assume m is out-waiting for m', and let Q, Q' respectively be the best paths associated with m, m'. Since m is out-waiting for m', then they are currently on the same vertex, e.g. u, but the vertices right after u on Q, Q' are two different vertices, e.g. respectively v, v'. Lemma 12 implies the uniqueness of u. In other words, it says that for any fixed m, m', there is no more than one 5-tuple (m, m', u, v, v') satisfying the properties above. Consequently, m would out-wait for m' in at most one round, which means $\xi^-(m) \le |P|$.

Now, assume m is in-waiting for m', and let Q, Q' respectively be the best paths associated with m, m'. Since m is in-waiting for m', then they are currently on different vertices, e.g. u, u', but the vertices right after u, u' on Q, Q' are the same, e.g. the vertex v. Again, Lemma 12 implies the uniqueness of the 5-tuple (m, m', u, u', v). Consequently, m would in-wait for m' in at most one round, which means $\xi^+(m) \le |P|$.

Theorem 3. *The length of the non-lazy schedule is at most $2|P| + D$.*

Proof. Recall that length of the schedule is bounded by $\xi + D$. By Lemma 13 we have $\xi \le 2|P|$, which implies length of the schedule is at most $2|P| + D$.

G   An Additive Approximation Algorithm

In this section, we provide an algorithm for the MM Problem which uses no more than $O(\log n \cdot OPT + \sqrt{n} \log^2 n)$ rounds. Before presenting the algorithm, we need the following definition.

Definition 10. *During the running time of a broadcast schedule, a pair of vertices (u, v) is called* active *if either (u, v) or (v, u) is an unsatisfied demand pair, i.e. the message from the source node is not yet received by the sink node. A subset of vertices X is called* active, *if there exist some active pair (u, v) such that $u \in X$ and $v \notin X$. In particular, a terminal u is called* active *if there is some vertex v such that (u, v) is active.*

Our algorithm has 4 phases. In Phase 1, we select a set C of at most $\sqrt{n}$ vertices from G, which we call the set of *centers*. In Phase 2 we find a schedule $\mathcal{S}$ with length $O(\sqrt{n})$ which sends each vertex $v \in R\backslash C$ to one of the centers, namely $f(v)$. Overall, our goal in Phases 1 and 2 is to send all the messages to the set of centers. In Phase 3, we solve another MM instance defined on G, in which the set of demand pairs is $P' = \{(f(u), f(v)) | (u, v) \in P\}$. In Phase 4, we simply run the schedule $\mathcal{S}$ in the reverse order to complete the information received by every $f(v)$ to the corresponding v. We describe each of these phases in more details, and then formally present the algorithm.

Phase 1. The goal in this phase is finding and running a schedule which we call $\mathcal{S}_0$. First, find two families of vertex-disjoint connected subgraphs of G, namely $\mathcal{T}$ and $\mathcal{T}'$. The subgraphs in $\mathcal{T}, \mathcal{T}'$ are called the *components* of $\mathcal{T}, \mathcal{T}'$, or simply components, whenever it's clear from the context. These components will be chosen in such a way that we can broadcast within each of them in at most $\sqrt{n}$ rounds, which also means this can be done simultaneously for all of them, since they're vertex disjoint.

To construct $\mathcal{T}$, repeatedly find vertex-disjoint trees of size $\sqrt{n}$: Order the vertices of G arbitrarily and visit the vertices in that order. When visiting a vertex v, run DFS to find a tree of size $\sqrt{n}$ rooted at v, which does share any vertices with the trees (of size $\sqrt{n}$) that are found previously. After visiting all the vertices of G, assume $\mathcal{T} = \{T_1, \ldots, T_k\}$ is the family of the trees we have found and $C = \{r_1, \ldots, r_k\}$ is the set of their roots where r_i is the root of T_i.

Find an arbitrary spanning tree in each component of $G[V(G)\backslash V(\mathcal{T})]$, and let the obtained set of trees be $\mathcal{T}' = \{T'_1, \ldots, T'_{k'}\}$. Clearly, any tree in $\mathcal{T}'$ has less than $\sqrt{n}$ vertices since otherwise, another tree could have been added to $\mathcal{T}$. So, there exist a broadcast schedule of length at most $\sqrt{n}$ to transmit messages from the root to all nodes in each component in $\mathcal{T}, \mathcal{T}'$. By running this schedule twice, first in the reverse order to the root and then regularly from the root, we can inform all the vertices in a component about the information of the rest of the vertices in that component. Moreover, this can be done simultaneously for all of the components since they are vertex disjoint. Let $\mathcal{S}_0$ be the final schedule accomplishing this. Run the schedule $\mathcal{S}_0$ in Phase 1.

Phase 2. This Phase is a preparation for Phase 3, where we solve a new MM instance defined on the set C as the set of new terminals. In Phase 2, we construct this instance by sending each active vertex $v \notin C$ to some vertex in C, namely $f(v)$. Note that this has already been done for all $v \in V(\mathcal{T})$ in Phase 1, i.e. we sent each vertex $v \in V(\mathcal{T})$ to one of the centers.

To accomplish this for all $v \notin V(\mathcal{T})$ as well, we find a schedule $\mathcal{S}$ which sends a vertex from each active components in $\mathcal{T}'$ to another arbitrary vertex in $V(\mathcal{T})$. To find $\mathcal{S}$, we find a b-matching in the bipartite graph $G[V(\mathcal{T}), V(\mathcal{T}')]$ in polynomial time, which saturates exactly one vertex from each active component T'_i, and minimizes b. Recall that b-matchings in bipartite graphs are like regular matchings, except that the degrees on one side of the partition, i.e. $V(\mathcal{T})$ in here, can be as large as b.

Before giving an algorithm for finding the b-matching, we show how to use it to construct $\mathcal{S}$. Assuming that a vertex $v \in V(\mathcal{T}')$ is matched to $m(v)$ in our b-matching, the schedule $\mathcal{S}$ sends a message from each vertex v to $m(v)$. This clearly can be done in b rounds by decomposing the edges of the b-matching into at most b disjoint matchings, so we have $|\mathcal{S}| \leq b$.

All we do in Phase 2, is run the schedule $\mathcal{S}$ and then $\mathcal{S}_0$. By doing so, we send each active vertex $v \in V(\mathcal{T}')$ to one of the centers, i.e. the vertex $f(v)$. To see why, note that in Phase 1, $\mathcal{S}_0$ already sent all the information from all the nodes in $V(\mathcal{T}')$ to all other nodes in the same component of $\mathcal{T}'$, and hence in particular to the node v in the component which has an edge of the b-matching incident on it. In this phase, $\mathcal{S}$ sends all the messages from v to its matched vertex $m(v)$ in $V(\mathcal{T})$, and $\mathcal{S}_0$ sends all the information from $m(v)$ in $V(\mathcal{T})$ to one of the centers.

Finally, it remains to find the optimal b-matching in polynomial time.

Lemma 14. *We can find the minimum integer b along with a b-matching in $G[V(\mathcal{T}), V(\mathcal{T}')]$ in polynomial time which saturates one vertex from each active component in $V(\mathcal{T}')$.*

Proof. Shrink each active component in $V(\mathcal{T}')$ and replace it with a single node. Also, remove all of the inactive components from the graph. Now, it's enough to find a minimum b-matching which saturates all of the shrunk nodes. This is doable by the standard algorithms for finding minimum b-matchings in bipartite graphs, e.g. see [22]. Note that since every shrunk node has at least one neighbor in $V(\mathcal{T})$ since the message from an active vertex in this node must transmit the message to its made outside the shrunk node, the existence of such a b-matching is guaranteed.

Phase 3. In this Phase, we reduce the original MM instance to a new instance with a smaller number of terminals. Naturally, since v is sent to $f(v)$ for every active terminal v, we want to reduce this instance to a new instance $\mathcal{I}'$ defined on the new set of terminals $R' = \{f(v)|v \in R\}$ along with the new set of demand pairs $P' = \{(f(u), f(v))|(u, v) \in P\}$.

To find a schedule for $\mathcal{I}'$, we can use Theorem 3, which guarantees a schedule of length $O(|P'| + diam_{P'}(G))$. But note that $|P'|$ can be as large as $\Omega(n)$, which makes this solution inefficient. To overcome this issue, we use an idea similar to Phase 1 of Algorithm MM. We can sparsify the demand pairs P' using Algorithm *Sparsify*, and obtain a subset $\hat{P}' \subseteq P'$, which satisfies the properties below:

1. $|\hat{P}'| \leq 2|R'| \log |R'|$
2. Assuming that $\hat{\mathcal{S}}'$ is a feasible schedule for $\mathcal{I}(G, \hat{P}')$, then repeating $\hat{\mathcal{S}}'$ for $8 \log |R'|$ times makes a feasible schedule for $\mathcal{I}(G, P')$.

The process of finding the sparsified subset of demand pairs, $\hat{P}'$, is identical to Algorithm MM. To summarize, all we need to do in Phase 3 is:

1. Construct the new set of demand pairs $P' = \{(f(u), f(v))|(u, v) \in P\}$.
2. Find the sparsified subset of demand pairs $\hat{P}'$.

3. By applying Theorem 3 on $\hat{P}'$, find a schedule $\hat{S}'$ for $\mathcal{I}(G, \hat{P}')$.
4. Run the schedule $\hat{S}'$ for $8 \log |R'|$ times to solve the instance $\mathcal{I}(G, P')$.

Phase 4. At this point, we know that for any $(u, v) \in P$, vertex $f(v)$ has received the message p_u, i.e. the message from vertex u. So, all we need to do to complete the solution is to inform v of what $f(v)$ knows. Note that by running Phase 1 and Phase 2, exactly the reverse happened, i.e. $f(v)$ was informed of what v knew. Consequently, all we need to do in Phase 4, is the same as we did in Phase 1 and Phase 2, but in the reverse order.

Theorem 4. *The given algorithm produces a schedule of length $O(\log n \cdot OPT + \sqrt{n} \log^2 n)$.*

Proof. Let S^* be the optimal schedule, so we have $|S^*| = OPT$. We provide an upper bound, in terms of OPT, on the number of rounds used in each of the four phases. In Phase 1, we only run the schedule S_0, for which we showed $|S_0| \leq 2\sqrt{n}$ in the description of the phase.

In Phase 2, we run the schedule S and then S_0, where we have $|S| \leq b$. Recall that b was the minimum integer for which there exists a b-matching saturating exactly one vertex from each active component T_i'. Here we prove $b \leq OPT$ which implies that $|S| \leq OPT$. To see this, we should consider the optimal multicast schedule, S^*.

Let $E^* \subseteq E(G)$ be the subset of the edges used in S^*. For any active component $T_i' \in \mathcal{T}'$, there must exist an edge $(x_i, y_i) \in E^*$ such that $x_i \in V(T_i')$ and $y_i \in V(\mathcal{T})$. This holds since otherwise, in the schedule S^*, the active component T_i' would be disconnected from the rest of the vertices, contradicting the feasibility of S^*. Now, let

$$M = \left\{ (x_i, y_i) \mid T_i' \in \mathcal{T}', \ T_i' \text{ is active} \right\}$$

Clearly, M defines a b-matching in $G[V(\mathcal{T}'), V(\mathcal{T})]$, but for what value of b? In other words, what is the maximum number of edges in M which share the same endpoint in $V(\mathcal{T})$? To answer this question, note that $M \subseteq E^*$, and no vertex in E^* has a degree more than OPT, due to the optimality of S^*. This implies $b \leq OPT$, and so, $|S| \leq OPT$. Recall that we run the schedule S and then S_0 in Phase 2, which means we use at most $OPT + 2\sqrt{n}$ rounds in this phase.

In Phase 3, we run the schedule $\hat{S}'$ for $8 \log |R'|$ times, so, the number of rounds used in this phase is at most $8 \log |R'| \cdot |\hat{S}'|$, which we are going to compute in terms of OPT. By applying Theorem 3 on $\hat{P}'$, we get

$$
\begin{aligned}
|\hat{S}'| &\leq O(|\hat{P}'| + diam_{\hat{P}'}(G)) \\
&\leq O(2|R'| \log |R'| + diam_{\hat{P}'}(G)) \quad &(7) \\
&\leq O(\sqrt{n} \log n + 4\sqrt{n} + OPT) \quad &(8)
\end{aligned}
$$

where (7) is due to the fact that $|\hat{P}'| \leq 2|R'| \log |R'|$ as a result of the sparsification, and (8) holds since $diam_{\hat{P}'}(G) \leq 4\sqrt{n} + OPT$. To see why, note that $d_G(v, f(v)) \leq 2\sqrt{n}$ for all v, which implies

$$
\begin{aligned}
d_G(f(u), f(v)) &\leq d_G(f(u), u) + d_G(u, v) + d_G(v, f(v)) \\
&\leq 4\sqrt{n} + d_G(u, v) \\
&\leq 4\sqrt{n} + OPT
\end{aligned}
$$

Recall that the number of rounds used in Phase 3 is at most $8 \log |R'| \cdot |\hat{S}'|$. If we plug in (8) we get

$$
8 \log |R'| \cdot |\hat{S}'| \leq O(\log n \cdot (\sqrt{n} \log n + OPT))
$$

Consequently, we use at most $O(\log n \cdot OPT + \sqrt{n} \log^2 n)$ rounds in Phase 3.

In Phase 4, we clearly use as many rounds as we use in Phase 1 and Phase 2 overall, i.e. $4\sqrt{n} + OPT$. By summing the upper bounds obtained on the length of each phase, it implies that the total length of the produced schedule is $O(\log n \cdot OPT + \sqrt{n} \log^2 n)$.

H   Equivalence with the Minimum Poise Subgraph Problem

The main result of this section is an (approximate) equivalence between the Multicommodity Multicast Problem and the *Minimum Poise Subgraph* Problem.

Definition 11. *In the Minimum Poise Subgraph Problem (MP), we are given a connected graph G along with a subset of pairs of its vertices, P. The goal is to find a subgraph H of G which minimizes $\Delta(H) + diam_P(H)$.*

We denote the quantity $\Delta(H) + diam_P(H)$ as the *Poise* of the subgraph H generalizing from the corresponding version for the broadcast problem in [20]. Recall that if two vertices u, v are not connected in H, then by convention, $d_H(u, v) = \infty$. So the subgraphs H which do not connect a pair $(u, v) \in P$ will be automatically excluded from the solution domain.

Assume we are given an MM instance, $\mathcal{I}(G, P)$, defined on the graph G with the set of demand pairs P. Also, let $\mathcal{J}$ denote the associated MP instance, i.e. the MP instance defined on the graph G with the subset of pairs P.

Theorem 5. *Given a feasible schedule for $\mathcal{I}$, namely $\mathcal{S}$, in polynomial time we can obtain a feasible solution for $\mathcal{J}$ with poise at most $O(|\mathcal{S}|)$. In the other direction, given a feasible solution for $\mathcal{J}$ with poise x, in polynomial time we can obtain a feasible schedule for $\mathcal{I}$ with length at most $O(x \cdot \frac{\log^3 n}{\log \log n})$.*

We need the following Lemma for the proof of Theorem 5.

Lemma 15. *There exist a collection of $\log |P|$ subgraphs of G, namely $\mathcal{F}$, satisfying the following properties.*

i. Each element of $\mathcal{F}$ is a forest,

ii. For each pair in P, e.g. (s_i, t_i), there exist a forest in $\mathcal{F}$, which connects both s_i and t_i, and

iii. For each $F \in \mathcal{F}$ we have $diam(F) \leq diam_P(G) \cdot 4 \log |P|$ and $\Delta(F) \leq OPT$ where OPT is the length of the optimum schedule for $\mathcal{I}(G, P)$.

Proof. Let $\mathcal{P} = \{P_1, \ldots, P_k\}$ be the set of paths in the optimal schedule for $\mathcal{I}(G, P)$, where P_i is the path through which the message from s_i is sent to t_i. For any $Q \in \mathcal{P}$, let $N(Q)$ denote the subset of the paths in $\mathcal{P}$ which have at least one vertex in common with Q. Also, for any subset $\mathcal{S}$ of $\mathcal{P}$, let $N(\mathcal{S}) = \{N(Q) | Q \in \mathcal{S}\} \backslash \mathcal{S}$.

Now, we use the following algorithm to find the first forest in $\mathcal{F}$. Later, we find the subsequent forests using the same algorithm.

Algorithm *Extract Forest*
Input: A graph G and a a family of paths $\mathcal{P}$
Output: A subgraph of G, which is a forest
1. $\mathcal{H}, \mathcal{X} \leftarrow \phi$ and $\mathcal{P}' \leftarrow \mathcal{P}$
2.   **repeat**
3.   Let $Q \in \mathcal{P}'$
4. $\mathcal{X} \leftarrow \mathcal{X} \cup \{Q\}$
5.   **while** $|N(\mathcal{X})| \geq |\mathcal{X}|$
6.   **do** $\mathcal{X} \leftarrow \mathcal{X} \cup N(\mathcal{X})$
7. $\mathcal{H} \leftarrow \mathcal{H} \cup \mathcal{X}$.
8. $\mathcal{P}' \leftarrow \mathcal{P}' \backslash (\mathcal{X} \cup N(\mathcal{X}))$
9.   **until** $\mathcal{P}' = \phi$
10.   Let H be a subgraph of G, with $V(H) = V(\mathcal{H})$ and $E(H) = E(\mathcal{H})$
11.   Using BFS, find a spanning tree with an arbitrary root in each connected component of H
12.   Output the collection of all trees found in the previous step

Let the forest F be the output of Algorithm *Extract Forest*. First, we bound the diameter of each connected component of F by $2 \log |\mathcal{P}| \cdot OPT$. See that each connected component of H, namely X, is made of a family of paths, namely $\mathcal{X}$. This family is formed by a consecutive execution of line 6 of the Algorithm *Extract Forest*. So, we can decompose $\mathcal{X}$ into a number of families $\mathcal{X}_1, \ldots, \mathcal{X}_j$, where for each i, $\mathcal{X}_i$ is the family of paths that was added to $\mathcal{X}$ in some iteration of line 6. W.L.O.G we can assume that $\mathcal{X}_a$ is added sooner than $\mathcal{X}_b$ iff $a < b$. Observe that after each iteration of line 6, the size of $\mathcal{X}$ is at least doubled, which implies $j \leq \log |\mathcal{P}|$.

Now, let u, v be two arbitrary vertices of X, such that $u \in V(\mathcal{X}_a)$ and $v \in V(\mathcal{X}_b)$ for some a, b. To prove our bound on the diameter, we show that $d_X(u, v) \leq 2 \log |\mathcal{P}| \cdot OPT$, as follows: Let w be an arbitrary vertex in $V(\mathcal{X}_1)$. Note that there is path in X between u and w with length at most $a \cdot OPT$. Similarly, there is a path between v and w with length at most $b \cdot OPT$. This implies $d_X(u, v) \leq (a + b) \cdot OPT$. Moreover, having $a, b \leq \log |\mathcal{P}|$ implies $d_X(u, v) \leq 2 \log |\mathcal{P}| \cdot OPT$. So, a BFS tree in X with an arbitrary root vertex, has a diameter at most $4 \log |\mathcal{P}| \cdot OPT$.

After we found the first forest, we update $\mathcal{P}$ by removing $\mathcal{H}$ from it. Then, by running the above algorithm on the updated $\mathcal{P}$, we find the second forest. We continue this process until $\mathcal{P}$ becomes empty. To bound the number of forests, we show that after extracting each forest, the size of $\mathcal{P}$ is reduced by at least a factor of 2. This will finish the proof, since it just means $\mathcal{P}$ becomes empty after finding at most $\log |\mathcal{P}|$ forests.

To prove the claim, simply see that in line 8 of the Algorithm *Extract Forest*, whenever we remove a subset of paths from $\mathcal{P}'$, we also add a subset of paths with at least the same size to $\mathcal{H}$, in line 7. It means $|\mathcal{H}| \geq \frac{|\mathcal{P}|}{2}$. Consequently, the size of the updated $\mathcal{P}$, i.e. $\mathcal{P} \backslash \mathcal{H}$, is at most $|\mathcal{P}|/2$. This finishes the proof.

Corollary 2. *Given a simple n-vertex graph G, we can find a subgraph $s(G)$ of G in polynomial time, such that $|E(s(G))| \leq 2n \log n$ and for each $(u, v) \in E(G)$, we have $d_{s(G)}(u, v) \leq 8 \log n$.*

Proof. We define an MM instance and then, we prove our claim by applying Lemma 15 on it. Let $\mathcal{I}(G, P)$ be an MM instance where $P = \{(u, v)|(u, v) \in E(G)\}$. By applying Lemma 15 on $\mathcal{I}$, we obtain a family of forests, namely $\mathcal{F}$, which is satisfying the properties (ii) and (iii) of Lemma 15. This means that for every $(u, v) \in P$, there exists a forest in $\mathcal{F}$, namely F, such that $d_F(u, v) \leq 4 \log |P|$. So, if we define $E(s(G))$ to be $E(\mathcal{F})$, then, $diam(s(G)) \leq 4 \log |P| \leq 8 \log n$, and moreover, $|E(s(G))| \leq (\log |P|).n \leq 2n \log n$, which finishes the proof.

Note that Algorithm *Sparsify* which was used in Section 6 is derived in the proof of the above corollary.

Proof (of Theorem 5). First we prove the easy direction. Given a schedule $\mathcal{S}$, we find a subgraph with poise at most $2|\mathcal{S}|$. Let $E^* \subseteq E(G)$ be the subset of edges used in $\mathcal{S}$. Let H be the subgraph of G with $E(H) = E^*$. Clearly, $\Delta(H) \leq |\mathcal{S}|$ and $diam_P(H) \leq |\mathcal{S}|$. This proves the claim.

Now, given a subgraph H with poise x, we construct a schedule $\mathcal{S}$ with $|\mathcal{S}| \leq O\left(x \cdot \frac{\log^3 n}{\log \log n}\right)$. Similar to Lemma 15, we show the existence a collection of $\log |P|$ subgraphs of H, namely $\mathcal{F}$, satisfying the following properties: *i*. Each element of $\mathcal{F}$ is a forest, *ii*. For each pair in P, e.g. (s_i, t_i), there exist a forest in $\mathcal{F}$, e.g. F, which connects s_i and t_i, and *iii*. For each $F \in \mathcal{F}$ we have $diam(F) \leq diam_P(H) \cdot 4 \log |P|$.

Proof of this claim is very similar to the proof in Lemma 15. We can obtain the family $\mathcal{F}$ by running Algorithm *Extract Forest* iteratively on the family of paths $\mathcal{P}^*$, which is defined as follows: The family $\mathcal{P}^*$ contains an arbitrary shortest path in H from u to v for all $(u, v) \in P$. It is easy to verify that the output will satisfy the required properties for $\mathcal{F}$, we do not repeat the proof here.

Given the family $\mathcal{F}$, we construct the schedule $\mathcal{S}$ as follows: Order the elements of $\mathcal{F}$ arbitrarily. Then, for each forest F in that order, broadcast within all components of F simultaneously. Since for each $(s_i, t_i) \in P$, there exists a forest in $\mathcal{F}$ which connects s_i and t_i, the schedule is clearly feasible. To provide an upper bound on $|\mathcal{S}|$, we show that we need at most $O(x \cdot \frac{\log^2 n}{\log \log n})$ rounds for each forest F. Then, given that $|\mathcal{F}| = O(\log n)$, we would have $|\mathcal{S}| \leq O(x \log^3 n)$.

To finish the proof, given a forest F, we prove that broadcasting within each component of F can be done in $O(x \cdot \frac{\log^2 n}{\log \log n})$ rounds. Note that this can be done simultaneously for all components since they are clearly vertex-disjoint. Ravi [20] has shown that the minimum broadcast time for a tree T is at most $\frac{\log n}{\log \log n} \cdot poise(T)$. Using dynamic programming, he also provides an Algorithm to find such a broadcast schedule in polynomial time. It just remains to see that each connected component in F is a tree, by the properties of $\mathcal{F}$, and has a poise at most $\Delta(H) + diam_P(G) \cdot 4 \log |P| = O(x \log n)$.

Corollary 3 (of Theorem 5). *Given an MM instance $\mathcal{I}(G, P)$, in polynomial time we can find a feasible schedule for it with length at most $O\left(\frac{\log^3 n}{\log \log n} \cdot (OPT + \Delta(G))\right)$.*

Proof. In polynomial time, we can find a subgraph H of G with poise at most $OPT + \Delta(G)$, i.e. a subgraph H satisfying $\Delta(H) + diam_P(H) \leq OPT + \Delta(G)$. To find such a subgraph, we just need to define H as the union of the (s, t)-shortest paths for all $(s, t) \in P$. Then, clearly we would have $\Delta(H) \leq \Delta(G)$ and $diam_P(H) \leq OPT$, which together imply the desired bound on the poise of H.

After we found H, all we need to do is applying Theorem 5 on H and obtain a schedule of length $O\left(\frac{\log^3 n}{\log \log n} \cdot (\Delta(H) + diam_P(H))\right)$, which is at most $O\left(\frac{\log^3 n}{\log \log n} \cdot (\Delta(G) + OPT)\right)$.

I Hardness

Definition 12. *In the Hyperedge-Coloring problem (HC), we are asked to color the edges of a given hypergraph H with the minimum number of colors such that no two intersecting edges have the same color.*

Lemma 16. *Assuming $ZPP \neq NP$, and for any positive $\epsilon < 1$, there are no $O(m^{1-\epsilon})$-approximations for the HC problem, where m is the number of the edges of the hypergraph.*

Proof. We reduce the problem to the vertex-coloring problem. It is known that the vertex-coloring problem is not approximable within a factor of $O(|V(G)|^{1-\epsilon})$ unless $ZPP = NP$ [8]. Given a graph G as the input for the vertex-coloring problem, we construct a hypergraph $\mathcal{H}$ such that $V(\mathcal{H}) = E(G)$, i.e. there is a vertex $\bar{e}$ in $\mathcal{H}$ for each edge $e \in E(G)$. Also, for each vertex $v \in V(G)$, there is a hyperedge $\bar{v}$ in $\mathcal{H}$, where $\bar{v}$ contains all the vertices $\bar{e}$ such that v is one of the endpoints of e in G, i.e. $\bar{v}$ contains the set of vertices $\{\bar{e}|\exists u : e = (u, v)\}$.

It's easy to verify that $\chi'(\mathcal{H}) = \chi(G)$: Observe that any valid vertex-coloring for G gives a valid edge-coloring for $\mathcal{H}$, and vice versa. This can be done by assigning the same colors to the vertex $v \in V(G)$ and the edge $\bar{v} \in E(\mathcal{H})$. Now, see that the existence of a $O(m^{1-\epsilon})$-approximation algorithm for the HC problem, implies we can approximate $\chi'(\mathcal{H})$ within a factor of $O(|E(\mathcal{H})|^{1-\epsilon})$, and this means we can approximate $\chi(G)$ within a factor of $O(|V(G)|^{1-\epsilon})$ due to the fact that $|E(\mathcal{H})| = |V(G)|$. Contradiction.

Theorem 6. *Assuming $P \neq NP$, and for any positive $\epsilon < 1$, there are no $O(\max\{n, m\}^{1-\epsilon})$-approximations for the Hypercast problem, where n, m are respectively the number of the vertices and edges of the hypergraph G given in the Hypecast instance.*

Proof. To prove the claim, given an instance of the HC problem with the hypergraph H, we reduce it to an instance of the Hypercast problem as follows. Let G be a hypergraph which contains a copy of all the edges and vertices of H. We will modify G step by step to obtain the desired Hypercast instance.

First, for each hyperedge $e \in E(H)$, insert two distinct new vertices e^- and e^+ in it. Then, add another new vertex r^- to G, and also, add a hyperedge which contains r^- and all the vertices e^-. Let r^- be the source node in our Hypercast instance and $R = \{e^+ | e \in E(H)\}$ be the set of terminals. We claim that any L-coloring of the hypergraph H corresponds to a solution of length $L+1$ for the Hypercast instance, and vice versa.

Assume we are given an L-coloring for H, then, we construct a schedule of length $L + 1$ as follows: In the first round of the schedule, use the only edge containing r^- to inform all the vertices e^-. Then, use one round for each color-class of the L-coloring, i.e. sort the colors in an arbitrary order and dedicate a round to each color with respect to that ordering. Use each edge of H in the round dedicated to its color. The obtained schedule has length $L + 1$ and is obviously feasible since each terminal $e^+ \in R$ is informed by e^- when the edge containing them is used in the round dedicated to its color.

To show the other direction, given a schedule of length $L+1$ for the Hypercast instance, we find an L-coloring for H. We can assume the first round is dedicated to the only edge containing r^- W.L.O.G. and that is because this edge intersects with all the other edges and also is the only edge containing the source node. Since each edge $e \in E(G)$ contains a distinct terminal, then e should be used in at least one of the rounds $2, \ldots, L + 1$. Pick one of the rounds in which e is used arbitrarily and let it be the color of e. Clearly, the number of used colors is L, and moreover, no two intersecting edges have the same color due to the feasibility of the given schedule.

The above argument implies that any α-approximation for the Hypercast problem gives an $O(\alpha)$-approximation for the HC problem. Since there are no $O(|V(H)|^{1-\epsilon})$-approximations for the HC problem, and since $n, m \in O(|V(H)|)$, then there are no $O(\max\{n, m\}^{1-\epsilon})$-approximations for the Hypercast problem.

Under Khot's 2-to-1 conjecture, we also provide a harness result in terms of D for the Hypercast Problem. To prove this result, we need the following Theorem of [11]:

Theorem 3.1. from [11] Assuming 2-to-1 conjecture, the following holds: Given any integer $k \geq 7$ and $\Delta \geq \Delta_0(k)$, it is NP-hard (under randomized reductions) to decide whether an unweighted graph G with maximum degree Δ is k-colorable or has largest independent set size at most $O(n/\Delta^{1-c/k-1})$

Proposition 4. *Under Khot's 2-to-1 conjecture, it is NP-hard to color a 7-colorable graph with $O(\Delta^{1/3})$ colors.*

Theorem 7. *Assuming $P \neq NP$, there are no $O(D^{1/3})$-approximations for the Hypercast problem, where D is the maximum size of a hyperedge in the hypergraph G.*

Proof. The proof is similar to the proof of Theorem 6. We use the exact same reduction to the vertex coloring problem, and then, instead of using the $\Omega(|V(G)|^{1-\epsilon})$ hardness ratio of [8], we use Proposition 4.

For contradiction, assume we are given an algorithm for the Hypercast problem, namely Algorithm $\mathcal{A}$, which has an approximation ratio of $O(D^{1/3})$. Using this algorithm, we will design an algorithm which, for any Δ, colors any 7-colorable graph with $O(\Delta^{1/3})$ colors. This will be a contradiction with Proposition 4.

Assuming that we are given a 7-colorable graph G as the input of the vertex coloring problem, we reduce this instance to an instance of the Hypercast Problem. This is done in exactly the similar way as it was done in Lemma 16 followed by Theorem 6. Note that the size of the hyperedges in obtained Hypercast instance would be bounded by Δ, i.e. $D = \Delta$.

By the properties of our reduction, since $\chi(G) \leq 7$, then the optimal schedule for the produced Hypercast instance has length $O(1)$, and so, Algorithm $\mathcal{A}$ generates a schedule of length at most $O(\Delta^{1/3})$. Due to our reduction, this schedule can be converted into a coloring of vertices for the graph G which uses no more than $O(\Delta^{1/3})$ colors. This is a contradiction with Proposition 4.

Bypassing Erdős' Girth Conjecture: Hybrid Stretch and Sourcewise Spanners*

Merav Parter

The Weizmann Institute of Science, Rehovot, Israel
merav.parter@weizmann.ac.il

Abstract. An (α, β)-spanner of an n-vertex graph $G = (V, E)$ is a subgraph H of G satisfying that $\text{dist}(u, v, H) \leq \alpha \cdot \text{dist}(u, v, G) + \beta$ for every pair $(u, v) \in V \times V$, where $\text{dist}(u, v, G')$ denotes the distance between u and v in $G' \subseteq G$. It is known that for every integer $k \geq 1$, every graph G has a polynomially constructible $(2k - 1, 0)$-spanner of size $O(n^{1+1/k})$. This size-stretch bound is essentially optimal by the girth conjecture. Yet, it is important to note that any argument based on the girth only applies to *adjacent vertices*. It is therefore intriguing to ask if one can "bypass" the conjecture by settling for a multiplicative stretch of $2k - 1$ only for *neighboring* vertex pairs, while maintaining a strictly *better* multiplicative stretch for the rest of the pairs. We answer this question in the affirmative and introduce the notion of k-*hybrid spanners*, in which non neighboring vertex pairs enjoy a *multiplicative k* stretch and the neighboring vertex pairs enjoy a *multiplicative $(2k - 1)$* stretch (hence, tight by the conjecture). We show that for every unweighted n-vertex graph G, there is a (polynomially constructible) k-hybrid spanner with $O(k^2 \cdot n^{1+1/k})$ edges. This should be compared against the current best (α, β) spanner construction of [5] that obtains $(k, k - 1)$ stretch with $O(k \cdot n^{1+1/k})$ edges. An alternative natural approach to bypass the girth conjecture is to allow ourself to take care only of a subset of pairs $S \times V$ for a given subset of vertices $S \subseteq V$ referred to here as *sources*. Spanners in which the distances in $S \times V$ are bounded are referred to as *sourcewise spanners*. Several constructions for this variant are provided (e.g., multiplicative sourcewise spanners, additive sourcewise spanners and more).

1 Introduction

1.1 Motivation

Graph spanners are sparse subgraphs that faithfully preserve the pairwise distances of a given graph and provide the underlying graph structure in communication networks, robotics, distributed systems and more [27]. The notion

---

* Recipient of the Google European Fellowship in distributed computing; research supported in part by this Fellowship. Supported in part by the Israel Science Foundation (grant 894/09), United States-Israel Binational Science Foundation (grant 2008348), Israel Ministry of Science and Technology (infrastructures grant), and Citi Foundation.

J. Esparza et al. (Eds.): ICALP 2014, Part II, LNCS 8573, pp. 608–619, 2014.

of graph spanners was introduced in [25,26] and have been studied extensively since. Spanners have a wide range of applications from distance oracles [31,8], labeling schemes [9] and routing [13] to solving linear systems [17] and spectral sparsification [19].

Given an undirected unweighted n-vertex graph $G = (V, E)$, a subgraph H of G is said to be a k-spanner if for every pair of vertices $(u, v) \in V \times V$ it holds that $\text{dist}(u, v, H) \le k \cdot \text{dist}(u, v, G)$. It is well known that one can efficiently construct a $(2k - 1)$-spanner with $O(n^{1+1/k})$ edges, even for weighted graphs [4,1]. This size-stretch ratio is conjectured to be tight based on the girth[1] conjecture of Erdős [18], which says that there exist graphs with $\Omega(n^{1+1/k})$ edges and girth $2k + 1$. If one removes an edge in such a graph, the distance between the edge endpoints increases from 1 to $2k$, implying that any α-spanner for $\alpha \le 2k - 1$ has $\Omega(n^{1+1/k})$ edges. This conjecture has been resolved for the special cases of $k = 1, 2, 3, 5$ [33].

Although the girth conjecture exactly characterizes the optimal tradeoff between sparseness and multiplicative stretch, it applies only to adjacent vertices (i.e., removing an edge (u, v) from a large cycle causes distortion to the edge endpoints). Indeed, Elkin and Peleg [15] showed that the girth bound (on multiplicative distortion) fails to hold even for vertices at distance 2. This limitation of the girth argument motivated distinguishing between *nearby* vertex pairs and "sufficiently distant" vertex pairs. This gave raise to the development of (α, β)-spanners which distort distances in G up to a multiplicative factor of α and an additive term β [15]. Formally, for an unweighted undirected graph $G = (V, E)$, a subgraph H of G is an (α, β)-spanner iff $\text{dist}(u, v, H) \le \alpha \cdot \text{dist}(u, v, G) + \beta$ for every $u, v \in V$. Note, that an (α, β)-spanner makes an implicit distinction between nearby vertex pairs and sufficiently distant vertex pairs. In particular, for "sufficiently distant" vertex pairs the (α, β)-spanner behaves similar to a pure multiplicative spanner, whereas for the remaining vertex pairs, the spanner behaves similar to an additive spanner [21]. The setting of (α, β)-spanners has been widely studied for various distortion-sparseness tradeoffs [16,32,15,5]. For example, [15] gave a construction for $(k - 1, 2k - O(1))$-spanners with size $O(k \cdot n^{1+1/k})$, with a number of refinements for short distances, and showed that for any $k \ge 2$ and $\epsilon > 0$, there exist $(1 + \epsilon, \beta)$-spanners with size $O(\beta \cdot n^{1+1/k})$, where β depends on ϵ and k but independent on n, implying that the size can be driven close to linear in n and the multiplicative stretch close to 1, at the cost of a large additive term in the stretch. Thorup and Zwick designed $(1 + \epsilon, \beta)$-spanners with $O(k \cdot n^{1+1/k})$ edges, with a multiplicative distortion that tends to 1 as the distance increases [32].

The best (α, β) spanner construction is due to [5] which achieves stretch of $(k, k - 1)$ with $O(k \cdot n^{1+1/k})$ edges, hence providing multiplicative stretch $2k - 1$ for neighboring vertices (which is the best possible by Erdős' conjecture) and a multiplicative stretch at most $3k/2$ for the remaining pairs.

Although (α, β)-spanners make an (implicit) distinction between "close" and "distant" vertex pairs, as the girth argument holds only for vertices at distance

---

[1] The girth is the smallest cycle length.

1, it seems that a tighter bound on the behavior of spanners may be obtained. In particular, it seems plausible that the multiplicative factor of k using $O(n^{1+1/k})$ edges, is not entirely unavoidable for non-neighboring vertex pairs, while providing multiplicative stretch of $2k - 1$ for the neighboring vertex pairs. The current paper confirms this intuition by introducing the notion of k-hybrid spanners, namely, subgraphs $H \subseteq G$ that obtain multiplicative stretch $2k-1$ for neighboring vertices, i.e., $\mathrm{dist}(u, v, H) \leq (2k - 1) \cdot \mathrm{dist}(u, v, G)$ for every $(u, v) \in E(G)$ and multiplicative stretch k for the remaining vertex pairs, i.e., $\mathrm{dist}(u, v, H) \leq k \cdot \mathrm{dist}(u, v, G)$ for every $(u, v) \notin E(G)$. Hence, hybrid spanners seem to pinpoint the minimum possible relaxation of the stretch requirement in spanners graphs so that the girth conjecture lower bound can be bypassed. The presented k-hybrid spanner with $O(k^2 \cdot n^{1+1/k})$ edges can be contrasted with several existing spanner constructions, e.g, k-spanners with $O(n^{1+2/(k+1)})$ edges (in which multiplicative stretch k is guaranteed also to neighboring pairs), the $\Omega(k^{-1} \cdot n^{1+1/k})$ lower-bound graph construction for $(2k - 1)$-additive spanners, and to the $(k, k - 1)$ spanner construction of [5] with $O(k \cdot n^{1+1/k})$ edges.

An alternative approach to bypass the conjecture is by focusing on a subset of pairs in $V \times V$. Following [10,28,12,20], we relax the requirement that small stretch in the subgraph must be guaranteed for *every* vertex pair from $V \times V$. Instead, we require it to hold only for pairs of vertices from a subset of $V \times V$. Specifically, given a subset of vertices $S \subseteq V$, referred to here as *sources*, our spanner H aims to bound only the distances between pairs of vertices from $S \times V$. For any other pair outside $S \times V$, the stretch in H can be arbitrary.

On the lower bound side, Woodruff [34] proved, independently of the Erdős' conjecture, the existence of graphs for which any spanner of size $\Omega(k^{-1}n^{1+1/k})$ has an additive stretch of at least $2k - 1$. Although sourcewise additive spanners have been studied by [28,12,20], currently there are no known lower bound constructions for this variant. We generalize Woodruff's construction to the sourcewise setting, providing a graph construction whose size has a smooth dependence with the number of sources.

1.2   Related Works

The notion of a sparse subgraph that preserves distances only for a subset of the $V \times V$ pairs has been initiated by Bollobás, Coopersmith and Elkin [9], who studied *pairwise preservers*, where the input is a graph $G = (V, E)$ along with a subset of vertex pairs $\mathcal{P} \subseteq V \times V$ and the problem is to construct a sparse subgraph H such that the $u - v$ distance for each $(u, v) \in \mathcal{P}$ is exactly preserved, i.e., $\mathrm{dist}(u, v, H) = \mathrm{dist}(u, v, G)$ for every $(u, v) \in \mathcal{P}$. They showed that one can construct a pairwise preserver with $O(\min\{|\mathcal{P}| \cdot \sqrt{n}, n \cdot \sqrt{|\mathcal{P}|}\})$ edges. At the end of their paper, they raised the question of constructing sparser subgraphs where distances between pairs in $\mathcal{P}$ are *approximately* preserved, or in other words, the problem of constructing sparse $\mathcal{P}$-spanners. Pettie [28] studied a certain type of $\mathcal{P}$-spanners, namely, additive sourcewise spanners. In this setting, one is given an unweighted graph $G = (V, E)$ and a subset of vertices $S \subseteq V$, termed as *sources*, whose size is conveniently parameterized to be $|S| = n^\varepsilon$, for $\varepsilon \in [0, 1]$, and the goal is to construct a sparse spanner H that maintains an additive approximation for the $S \times V$ distances. He showed a construction of $O(\log n)$-additive sourcewise spanners of size $O(n^{1+\varepsilon/2})$. Cygan et al. recently showed a stretch-size bound for $2k$-additive sourcewise spanners with $O(n^{1+(\varepsilon k+1)/(2k+1)})$ edges. The specific case of $k = 1$ has been studied recently by [20], providing a 2-additive sourcewise spanner with $\widetilde{O}(n^{5/4+\varepsilon/4})$ edges where $\varepsilon = \log |S| / \log n$.

Upper bounds for spanners with constant stretch are currently known for but a few stretch values. A $(1, 2)$ spanner with $O(n^{3/2})$ edges is presented in [2], a $(1, 6)$ spanner with $O(n^{4/3})$ edges is presented in [5], and a $(1, 4)$ spanner with $O(n^{7/5})$ edges is presented in [11]. The latter two constructions use the *path-buying* strategy, which is adopted in our additive sourcewise construction. Dor et al. [14] considered additive emulators, which may contain additional (possibly weighted) edges. They showed a construction of 4-additive emulator with $O(n^{4/3})$ edges. Finally, a well known application of α-spanners is *approximate distance oracles* [31,24,8,7,22]. The sourcewise variant, namely, *sourcewise approximate distance oracle* was devised by [29]. For a given input graph $G = (V, E)$ and a source set $S \subseteq V$, [29] provides a construction of a distance oracle of size $O(n^{1+\varepsilon/k})$ where $\varepsilon = \log |S| / \log n$ such that given a distance query $(s, v) \in S \times V$ returns in $O(k)$ time a $(2k - 1)$ approximation to dist(s, v, G).

1.3   Contributions

In this paper we initiate the study of k-hybrid spanners which seems to pinpoint the minimal condition for bypassing Erdős' Girth Conjecture. In addition, we also study the sourcewise variant of multiplicative spanners, additive spanners and additive emulators. The main results are summarized below.

Theorem 1 (Hybrid spanners). *For every integer $k \geq 2$ and unweighted undirected n-vertex graph $G = (V, E)$, there exists a (polynomially constructible) subgraph of size $O(k^2 \cdot n^{1+1/k})$ that provides multiplicative stretch $2k - 1$ for every pair of neighboring vertices u and v and a multiplicative stretch k for the rest of the pairs. (By Erdős' conjecture, providing a multiplicative stretch of k for all the pairs requires $\Omega(n^{1+2/(k+1)})$ edges.)*

Theorem 2 (Sourcewise spanners). *For every integer $k \geq 2$, and an unweighted undirected n-vertex graph $G = (V, E)$ and for every subset of sources $S \subseteq V$ of size $|S| = O(n^\varepsilon)$, there exists a (polynomially constructible) subgraph of size $O(k^2 \cdot n^{1+\varepsilon/k})$ that provides multiplicative stretch $2k - 1$ for every pair of neighboring vertices $(u, v) \in S \times V$ and a multiplicative stretch of $2k - 2$ for the rest of the pairs in $S \times V$. This subgraph is referred to here as sourcewise spanner.*

Theorem 3 (Lower bound for additive sourcewise spanners and emulators). *For every integer $k \in [2, O(\log n/ \log \log n)]$ and $\varepsilon \in [0, 1]$, there exists an n-vertex graph $G = (V, E)$ and a subset of sources $S \subseteq V$ of size $|S| = O(n^\varepsilon)$ such that any $(2k - 1)$-additive sourcewise spanner (i.e., subgraph that maintains a $(2k - 1)$-additive approximation for the $S \times V$ distances) has at least*

$\Omega(k^{-1} \cdot n^{1+\varepsilon/k})$ edges. The lower bound holds for additive emulators up to order $O(k)$. For 2-additive sourcewise emulators there is a matching upper bound.

Theorem 4 (Upper bound for additive sourcewise spanners). Let $k \geq 1$ be an integer. (1) For every unweighted undirected n-vertex graph $G = (V, E)$ and for every subset of sources $S \subseteq V$, $|S| = O(n^\varepsilon)$, there exists a (polynomially constructible) $2k$-additive sourcewise spanner with $\tilde{O}(k \cdot n^{1+(\varepsilon \cdot k+1)/(2k+2)})$ edges. (2) For $|S| = \Omega(n^{2/3})$, there exists a 4-additive sourcewise spanner with $O(n^{1+\varepsilon/2})$ edges (by the lower bound of Thm. 3, any 3-additive sourcewise spanner requires $\Omega(n^{1+\varepsilon/2})$ edges).

The time complexities of all our upper bound constructions are obviously polynomial; precise analysis is omitted from this extended abstract.

1.4   Preliminaries

We consider the following graph structures.

(α, β)-**spanners.** For a graph $G = (V, E)$, the subgraph $H \subseteq G$ is an (α, β)-spanner for G if for every $(u, v) \in V \times V$,

$$\text{dist}(u, v, H) \leq \alpha \cdot \text{dist}(u, v, G) + \beta . \tag{1}$$

$(\alpha, 0)$-spanners (resp., $(1, \beta)$-spanners) are referred to here as α-spanners (resp., β-additive spanners).

Hybrid Spanners. Given a graph $G = (V, E)$, a subgraph $H \subseteq G$ is a k-hybrid spanner iff for every $(u, v) \in V \times V$ it holds that

$$\text{dist}(u, v, H) \leq \begin{cases} (2k - 1) \cdot \text{dist}(u, v, G), & \text{if } (u, v) \in E(G); \\ k \cdot \text{dist}(u, v, G), & \text{otherwise.} \end{cases} \tag{2}$$

Sourcewise Spanners. Given an unweighted graph $G = (V, E)$ and a subset of vertices $S \subseteq V$, a subgraph $H \subseteq G$ is an (α, β, S)-*spanner* iff Eq. (1) is satisfied for every $(s, v) \in S \times V$. When $\beta = 0$ (resp., $\alpha = 1$), H is denoted by (α, S)-*sourcewise spanner* (resp., (β, S)-*additive sourcewise spanner*).

Emulators. Given an unweighted graph $G = (V, E)$, a weighted graph $H = (V, F)$ is an (α, β)-*emulator* of G iff $\text{dist}(u, v, G) \leq \text{dist}(u, v, H) \leq \alpha \cdot \text{dist}(u, v, G) + \beta$ for every $(u, v) \in V \times V$. $(1, \beta)$-emulators are referred to here as β-additive emulators. For a given subset of sources $S \subseteq V$, the graph $H = (V, F)$ is a (β, S)-*additive sourcewise emulator* if the $S \times V$ distances are bounded in H by an additive stretch of β.

1.5   Notation

For a subgraph $G' = (V', E') \subseteq G$ (where $V' \subseteq V$ and $E' \subseteq E$) and a pair of vertices $u, v \in V'$, let $\text{dist}(u, v, G')$ denote the shortest-path distance in edges between u and v in G'. Let $\Gamma(v, G) = \{u \mid (u, v) \in E(G)\}$ be the set of neighbors of v in G. For a subgraph $G' \subseteq G$, let $|G'| = |E(G')|$ denote the number of edges in G'. For a path $P = [v_1, \ldots, v_k]$, let $P[v_i, v_j]$ be the subpath of P from v_i to v_j. For paths P_1 and P_2, let $P_1 \circ P_2$ denote the path obtained by concatenating P_2 to P_1. Let $SP(s, v_i, G')$ be the set of $s - v_i$ shortest-paths in G'. When G' is the input graph G, let $\pi(x, y) \in SP(x, y, G)$ denote some arbitrary $x - y$ shortest path in G, hence $|\pi(x, y)| = \text{dist}(x, y, G)$. For a subset $V' \subseteq V$, let $\text{dist}(u, V', G) = \min_{u' \in V'} \text{dist}(u, u', G)$. Similarly, for subsets $V_1, V_2 \subseteq V$, $\text{dist}(V_1, V_2, G) = \min_{v_1 \in V_1, v_2 \in V_2} \text{dist}(v_1, v_2, G)$. When the graph G is clear from the context, we may omit it and simply write $\Gamma(u), \text{dist}(u, v), \text{dist}(u, V')$ and $\text{dist}(V_1, V_2)$.

A clustering $\mathcal{C} = \{C_1, \ldots, C_\ell\}$ is a collection of disjoint subsets of vertices, i.e., $C_i \subseteq V$ for every $C_i \in \mathcal{C}$ and $C_i \cap C_j = \emptyset$ for every $C_i, C_j \in \mathcal{C}$. Note that a clustering is not necessarily a partition of V, i.e., it is not required that $\bigcup_i C_i = V$. A cluster $C \in \mathcal{C}$ is said to be *connected* in G if the induced graph $G[C]$ is connected. For clusters C and C', let $E(C, C') = (C \times C') \cap E(G)$ be the set of edges between C and C' in G. For notational simplicity, let $E(v, C) = E(\{v\}, C)$. A vertex v is *incident* to a cluster C if $E(v, C) \neq \emptyset$. In a similar manner, two clusters C and C' are adjacent to each other if $E(C, C') \neq \emptyset$.

Organization. We start with upper bounds. Sec. 2 describes the construction of k-hybrid spanners. Sec. 3.1 presents the construction of (α, S) sourcewise spanners. Then, Sec. 3.2 presents a lower bound construction for (β, S) sourcewise additive spanners and emulators. Finally, Sec. 3.3 provides an upper bound for $(2k, S)$-additive sourcewise spanners for general values of k. In addition, it provides a tight construction for $(2, S)$-additive sourcewise emulators.

2   Hybrid Spanners

In this section, we establish Thm. 1. For clarity of presentation, we describe a randomized construction whose output spanner has $O(k^2 \cdot n^{1+1/k})$ edges in expectation. Using the techniques of [5], this construction can be derandomized with the same bound on the number of edges.

The algorithm. We begin by describing a basic procedure Cluster, slightly adapted from [5], that serves as a building block in our constructions. For an input unweighted graph $G = (V, E)$, a stretch parameter k and a density parameter μ, Algorithm Cluster iteratively constructs a sequence of $k+1$ clusterings $\mathcal{C}_0, \ldots, \mathcal{C}_k$ and a clustering graph $H_k \subseteq G$. Each clustering $\mathcal{C}_\tau$ consists of $m_\tau = n^{1-\tau \cdot \mu}$ disjoint subsets of vertices, $C_\tau = \{C_1^\tau, \ldots, C_{m_\tau}^\tau\}$. Each cluster $C_j^\tau \in \mathcal{C}_\tau$ is connected and has a *cluster center* z_j satisfying that $\text{dist}(u, z_j, G) \leq \tau$ for every $u \in C_j^\tau$. Denote the set of cluster centers of $\mathcal{C}_\tau$ by Z_τ. These cluster centers correspond to a sequence of samples taken from V with decreasing densities where $V = Z_0 \supseteq Z_1 \supseteq \ldots \supseteq Z_k$. On a high level, at each iteration τ, a clustering of radius-τ clusters is constructed and its shortest-path spanning forest (spanning all the vertices in the clusters), as well as an additional subset of edges Q_τ adjacent to unclustered vertices, are chosen to be added to the spanner H_τ. We now describe the algorithm $\mathsf{Cluster}(G, k, \mu)$ in detail. Assume some ordering on the vertices $V = \{v_1, \ldots, v_n\}$. Initially, the cluster centers are $Z_0 = V = \{v_1, \ldots, v_n\}$, where each vertex forms its own cluster of radius 0, hence $\mathcal{C}_0 = \{\{v\} \mid v \in V\}$ and the spanner is initiated to $H_0 = \emptyset$. At iteration $\tau \geq 1$, a clustering $\mathcal{C}_\tau$ is defined based on the cluster centers $Z_{\tau-1}$ of the previous iteration. Let $Z_\tau \subseteq Z_{\tau-1}$ be a sample of $m_\tau = O(n^{1-\tau \cdot \mu})$ vertices chosen uniformly at random from $Z_{\tau-1}$. The clustering $\mathcal{C}_\tau$ is obtained by assigning every vertex u that satisfies $\mathrm{dist}(u, Z_\tau, G) \leq \tau$ to its closest cluster center $z \in Z_\tau$, i.e., such that $\mathrm{dist}(u, z, G) = \mathrm{dist}(u, Z_\tau, G)$. If there are several cluster centers in Z_τ at distance $\mathrm{dist}(u, Z_\tau, G)$ from u, then the closest center with the minimal index is chosen.

Formally, for a vertex v and subset of vertices B, let $\mathtt{nearest}(v, B)$ be the closest vertex to v in B where ties are determined by the indices, i.e., letting $B' = \{v_1, \ldots, v_\ell\} \subseteq B$ be the set of closest vertices to v in B, namely, satisfying that $\mathrm{dist}(v, v_1) = \ldots = \mathrm{dist}(v, v_\ell) = \mathrm{dist}(v, B)$, then $\mathtt{nearest}(v, B) \in B'$ and has the minimal index in B'. Then v is assigned to the cluster of the center $\mathtt{nearest}(v, Z_\tau)$. Add to H_τ the forest F_τ consisting of the radius-τ spanning tree of each $C \in \mathcal{C}_\tau$. Note that the definition of the clusters immediately implies their connectivity. Next, an edge set Q_τ adjacent to unclustered vertices is added to H_τ as follows. Let Δ_τ denote the set of vertices that occur in each of the clusterings $\mathcal{C}_0, \ldots, \mathcal{C}_{\tau-1}$ but do not occur in $\mathcal{C}_\tau$. (Observe that such a vertex may re-appear again in some future clusterings.) Formally, let $\widehat{V}_\tau = \bigcup_{C \in \mathcal{C}_\tau} C$ be the set of vertices that occur in some cluster in the clustering $\mathcal{C}_\tau$. Then, $\Delta_\tau = \left(\bigcap_{j=0}^{\tau-1} \widehat{V}_j \right) \setminus \widehat{V}_\tau$. Note that by this definition, each vertex belongs to at most one set Δ_τ. For every vertex $v \in \Delta_\tau$ and every cluster $C \in \mathcal{C}_{\tau-1}$ that is adjacent to v, pick one vertex $u \in C$ adjacent to v and add the edge (u, v) to Q_τ. (In other words, an edge (u, v) is *not* added to Q_τ for $v \in \Delta_\tau$ if either $u \notin \widehat{V}_{\tau-1}$ or an edge (u', v) was added to Q_τ where u' and u are in the same cluster $C \in \mathcal{C}_{\tau-1}$.) Then add Q_τ to H_τ. This completes the description of Algorithm Cluster; a pseudocode is given below.

Algorithm $\mathsf{Cluster}(G, k, \mu)$.

(T1) Let $H_0 = \emptyset$ and $Z_0 = V$. Select a sample Z_τ uniformly at random from $Z_{\tau-1}$ with probability $n^{-\mu}$ for $\tau = 1$ to k (if $\mu = 1$ and $\tau = k$, set $Z_k = \emptyset$).

(T2) For $\tau = 1$ to k, define the clustering $\mathcal{C}_\tau$ by adding the τ-radius neighborhood for all cluster centers Z_τ, i.e., every $u \in V$ satisfying $\mathrm{dist}(u, Z_\tau) \leq \tau$ is connected to $\mathtt{nearest}(u, Z_\tau)$. Let F_τ denote the τ-radius neighborhood forest corresponding to $\mathcal{C}_\tau$.

(T3) For every vertex $v \in \Delta_\tau$ that was unclustered in the clustering $\mathcal{C}_\tau$ for the first time, let $e(v, C)$ be an arbitrary edge from $E(v, C)$ for every $C \in \mathcal{C}_{\tau-1}$.

(T4) $H_\tau = H_{\tau-1} \cup F_\tau \cup \{e(v, C) \mid v \in \Delta_\tau, C \in \mathcal{C}_{\tau-1}\}$.

The first step of Algorithm ConsHybrid applies Algorithm $\mathsf{Cluster}(G, k, \mu)$ for $\mu = 1/k$, resulting in the subgraph H_k. Note that by Thm. 3.1 of [5], H_k is a $(2k - 1)$ spanner. Hence, the stretch for neighboring vertices is $(2k - 1)$ as required. We now add two edge sets to H_k in order to provide a multiplicative stretch k for the remaining pairs. Let

$$t = \lfloor k/2 \rfloor \quad \text{and} \quad t' = k - 1 - t, \tag{3}$$

Note that $t' = t$ when k is odd and $t' = t - 1$ when k is even, so in general $t' \le t$.

The algorithm considers the collection of $Z_{t'} \times Z_t$ shortest paths $\mathcal{P} = \{\pi(z_i, z_j) \mid z_i \in Z_{t'} \text{ and } z_j \in Z_t\}$. Starting with $H = H_k$, for each path $\pi(z_i, z_j) \in \mathcal{P}$, it adds to H the ℓ_t last edges of $\pi(z_i, z_j)$ (closest to z_i), where

$$\ell_t = 7t + 8t^2 . \tag{4}$$

For every pair of clusters C_1, C_2, let $\pi(C_1, C_2)$ denote the shortest path in G between some closest vertices $u_1 \in C_1$ and $u_2 \in C_2$ (i.e., $\text{dist}(C_1, C_2, G) = \text{dist}(u_1, u_2, G)$). For every τ from 0 to $k - 1$, and for every pair $C_1 \in \mathcal{C}_\tau$ and $C_2 \in \mathcal{C}_{k-1-\tau}$, the algorithm adds to H, the ℓ last edges of $\pi(C_1, C_2)$, where $\ell = \ell_t$ for $\tau \in \{t', t\}$ and $\ell = 2k - 1$ otherwise. This completes the description of Algorithm ConsHybrid, whose summary is given below.

Algorithm ConsHybrid.

(S1) Let $H_k = \text{Cluster}(G, k, 1/k)$.

(S2) Let E_2 be the edge set containing the last ℓ_t edges of the path $\pi(z_i, z_j)$ for every $z_i \in Z_{t'}$ and $z_j \in Z_t$.

(S3) Let E_3 be the edges set containing, for every $\tau \in \{0, \ldots, k - 1\}$, and for every $C_1 \in \mathcal{C}_\tau$ and $C_2 \in \mathcal{C}_{k-1-\tau}$, the last ℓ edges of the path $\pi(C_1, C_2)$ where $\ell = \ell_t$ for $\tau \in \{t', t\}$ and $\ell = 2k - 1$ otherwise.

(S4) Let $H \leftarrow H_k \cup E_2 \cup E_3$.

In Section 2 of [23], we bound the size of H and the show correctness of Algorithm ConsHybrid. It is important to compare the $(k, k - 1)$ construction of [5] to the current construction. [5] constructs a $(k, k - 1)$ spanner with $O(k \cdot n^{1+1/k})$ edges. In contrast, Algorithm ConsHybrid provides a strictly better stretch for non-neighboring vertex pairs at the expense of having slightly more edges (e.g., $O(k^2 \cdot n^{1+1/k})$ vs. $O(k \cdot n^{1+1/k})$ edges). Indeed, Algorithm ConsHybrid bares some similarity to the $(k, k - 1)$ construction of [5] (e.g., similar cluster growing approach) but the analysis is different. The key difference between these two constructions is that in [5] only edges (i.e., shortest-path of length 1) are added between certain pairs of clusters. In contrast, in our construction, $O(k^2)$ edges are taken from each shortest-path connecting the close-most vertices coming from certain subset of clusters. This allows us to employ an inductive argument on the desired *purely* multiplicative stretch, without introducing an additional additive stretch term. Specifically, by adding paths of length ℓ_t between center pairs in $Z_{t'} \times Z_t$, a much better stretch guarantee can be provided for (non-neighboring) $Z_{t'} \times Z_t$ pairs: a multiplicative stretch k plus a *negative* additive term. This additive term is then increased but in a controlled manner (due to step (S3)), resulting in a *zero* additive term for *any* non-neighboring vertex pair in $V \times V$. Missing proofs for this section are deferred to the full version [23].

3   Sourcewise Spanners

In this section, we provide several constructions for sourcewise spanners and emulators.

3.1   Upper Bound for Multiplicative Stretch

In this section, we establish Thm. 2. For simplicity, we describe a randomized construction whose output spanner has $O(k^2 \cdot n^{1+\varepsilon/k})$ edges in expectation. Using [5], this construction can be derandomized with the same bound on the number of edges. We now show the construction of $(2k-1, S)$ sourcewise spanner which enjoys a "hybrid" stretch, though in a weaker sense than in Sec. 2. Specifically, we show that the neighbors of S enjoy a multiplicative stretch $2k-1$ and the remaining pairs enjoy a multiplicative stretch of $2k-2$.

The algorithm. The first phase of Algorithm ConsSWSpanner applies Algorithm Cluster(G, k, μ) for $\mu = \varepsilon/k$, resulting in a sequence of $k+1$ clusterings $\mathcal{C}_0, \ldots, \mathcal{C}_k$ and a cluster graph $H_k \subseteq G$. In the second phase of the algorithm, it considers the collection of $S \times Z_{k-1}$ shortest paths $\mathcal{P} = \{\pi(s_j, z_i) \mid s_j \in S \text{ and } z_i \in Z_{k-1}\}$. Starting with $H = H_k$, for each path $\pi(s_j, z_i) \in \mathcal{P}$, it adds to H the ℓ_k last edges of $\pi(s_j, z_i)$ (closest to z_i). Set

$$\ell_k = 2k^2 + 3k \quad \text{and} \quad \mu = \varepsilon/k . \tag{5}$$

In Section 3.1 of [23], we provide a complete analysis for the algorithm and establish Thm. 2.

3.2   Lower Bound for Additive Sourcewise Spanners and Emulators

We now turn to consider the lower bound side where we generalize the lower bound construction for additive spanners by Woodruff [34] to the sourcewise setting. In particular, we parameterize our bound for the $S \times V$ spanner in terms of the cardinality of the source set S. The basic idea underlying Woodruff's construction is to form a dense graph G by gluing (carefully) together many small complete bipartite graphs. For an additive stretch $2k-1 \geq 1$, the lower bound graph G consists of $k+1$ vertex levels, each with $O(n/k)$ vertices and $\Omega(n^{1+1/k})$ edges connecting the vertices of every two adjacent levels. In particular this is obtained by representing each vertex of level i as a coordinate in $\mathbb{Z}^{k+1}$, namely, $v = (a_1, \ldots, a_k, a_{k+1})$ and $a_j \in [1, O(n^{1/k})]$. Woodruff showed that if one omits in an additive spanner $H \subseteq G$, an $O(1/k)$ fraction of G edges, then there exists an $x - y$ path P in G of length k (i.e., x is on the first level and y is on the last level) whose all edges are omitted in H, and any alternative $x - y$ path in H is "much" longer than P. To adapt this construction to the sourcewise setting, some *asymmetry* in the structure of the $k+1$ levels should be introduced. In the following construction, the vertices of the first level correspond to the source set S, hence this level consists of $O(n^\varepsilon)$ vertices, while the remaining levels are of size

$O(n/k)$. This is achieved by breaking the symmetry between the first coordinate a_1 and the remaining $k - 1$ coordinates of each vertex $v = (a_1, \ldots, a_k, a_{k+1})$. Indeed, this careful minor adaptation in the graph definition is sufficient to generalize the bound, the analysis follows (almost) the exact same line as that of [34]. We show the following.

Theorem 5. *Let $1 \leq k \leq O(\ln r / \ln \ln r)$ for some integer $r \geq 1$. For every $\varepsilon \in [0, 1]$, there exists an unweighted undirected graph $G = (V, E)$ with $|V| = \Theta(r^\varepsilon + kr)$ vertices and a source set $S \subseteq V$ of size $\Theta(r^\varepsilon)$ such that any $(2k-1, S)$-additive sourcewise spanner $H \subseteq G$ has $\Omega(r^{1 + \frac{\varepsilon}{k}})$ edges. Similar bounds (up to factor $O(k)$) are achieved for $(2k - 1, S)$-additive sourcewise emulators.*

Note that Thm. 5 implies Thm. 3, since $n = \Theta(r^\varepsilon + kr)$ and hence $r^{1 + \frac{\varepsilon}{k}} = \Omega(k^{-1} \cdot n^{1 + \frac{\varepsilon}{k}})$. Note that by setting $\varepsilon = 1$, we get the exact same bounds as in Woodruff's construction.

The Construction. Let $N_1 = \lceil r^{\varepsilon/k} \rceil$ and $N_2 = \lceil (r/N_1^{k-1}) \rceil$. The graph G consists of vertices composed of $k + 1$ vertex-levels and connected through a series of k bipartite graphs. Each vertex $v = (a_1, a_2, \ldots, a_k, a_{k+1})$ represents a coordinate in $\mathbb{Z}^{k+1}$ where $a_{k+1} \in \{1, \ldots, k + 1\}$ is the *level* of v. The range of the other coordinates is as follows. For every $1 \leq j \leq k$, $a_j \in R_j$, where $R_1 = \{1, \ldots, N_1\}$ if $a_{k+1} = 1$ and $R_1 = \{1, \ldots, N_2\}$ otherwise. For $j \geq 2$, $R_j = \{1, \ldots, N_1\}$.
Edges in G join every level-i vertex $(a_1, \ldots, a_{i-1}, a_i, a_{i+1}, \ldots, a_k, i)$ to each of the level-$(i + 1)$ vertices of the form $(a_1, \ldots, a_{i-1}, c, a_{i+1}, \ldots, a_k, i + 1)$ for every $c \in \{1, \ldots, N_2\}$ if $i = 1$ and $c \in \{1, \ldots, N_1\}$ for $i \geq 2$. Let $L_i = \{(a_1, \ldots, a_k, i) \mid a_j \in R_j$ for $1 \leq j \leq k\}$ be the set of vertices on the ith level and let $n_i = |L_i|$ denote their cardinality. Then since $k = O(\ln r / \ln \ln r)$ it holds that $n_1 = N_1^k \leq (r^{\frac{\varepsilon}{k}} + 1)^k \leq e^{(k+1)/(r^{\varepsilon/k})} = \Theta(r^\varepsilon)$. and for every $i \in \{2, \ldots, k + 1\}$,

$$n_i = N_2 \cdot N_1^{k-1} \leq (r/N_1^k + 1)(r^{\frac{\varepsilon}{k}} + 1)^{k-1} \leq 2r^{1-\varepsilon/k} \cdot e^{k/(r^{\varepsilon/k})}$$
$$= r^{1-\varepsilon/k} \cdot \Theta(r^{\varepsilon/k}) = \Theta(r) ,$$

Overall, the total number of vertices is $|V(G)| = n_1 + k \cdot n_2 = \Theta(r^\varepsilon + k \cdot r)$.

Let g_i be the number of edges connecting the vertices of L_i to the vertices of L_{i+1}. Then $g_1 = N_2 \cdot n_1$ and $g_i = N_1 \cdot n_i$ for every $i \in \{2, \ldots, k\}$, thus $g_1 = g_2 = \ldots = g_k$. Hence $|E(G)| = \sum_{i=1}^{k+1} g_i = k \cdot N_1^k \cdot N_2 = \Theta(k \cdot r^{1+\varepsilon/k})$. Let the source set S be the vertex set of the first level, i.e., $S = L_1$, hence $|S| = n_1 = \Theta(r^\varepsilon)$. In Section 3.2 of [23], we analyze this graph construction and establish Thm. 3.

3.3 Upper Bound for Additive Sourcewise Spanners and Emulators

Additive sourcewise emulators. Recall that an emulator $H = (V, F)$ for graph G is a (possibly) weighted graph induced on the vertices of G, whose edges are not necessarily contained in G. In Thm. 5, we showed that every $(2, S)$-additive sourcewise emulator for a subset $S \subseteq V$ has $\Omega(n^{1+\varepsilon/2})$ edges, where

$\varepsilon = \log|S|/\log n$. In Section 3.3 of [23], we show that this is essentially tight (up to constants).

Theorem 6. *For every unweighted n-vertex graph $G = (V, E)$ and every subset $S \subseteq V$, there exists a (polynomially constructible) $(2, S)$-additive sourcewise emulator H of size $O(n^{1+\varepsilon/2})$ where $\varepsilon = \log|S|/\log n$.*

Additive sourcewise spanners. The construction of additive sourcewise spanners combines the path-buying technique of [5,12,20] and the 4-additive spanner techniques of [11].

Theorem 7. *Let $k \geq 1$ be an integer. For every unweighted n-vertex graph $G = (V, E)$ and every subset $S \subseteq V$, there exists a (polynomially constructible) $(2k, S)$-additive sourcewise spanner $H \subseteq G$ of size $\widetilde{O}(k \cdot n^{1+(k\varepsilon+1)/(2k+2)})$ where $\varepsilon = \log|S|/\log n$.*

Finally, we provide an "almost" tight construction for $(4, S)$-sourcewise additive spanners for a sufficiently large subset of sources S. We have the following.

Theorem 8. *For every unweighted n-vertex graph $G = (V, E)$ and a subset of sources $S \subseteq V$ such that $|S| = \Omega(n^{2/3})$, there exists a (polynomially constructible) $(4, S)$-additive sourcewise spanner $H \subseteq G$ with $O(n^{1+\varepsilon/2})$ edges.*

Acknowledgment. I am very grateful to my advisor, Prof. David Peleg, for many helpful discussions and for reviewing this paper. I would also like to thank Michael Dinitz and Eylon Yogev for useful comments and discussions.

References

1. Agarwal, R., Godfrey, P.B., Har-Peled, S.: Approximate distance queries and compact routing in sparse graphs. In: Proc. INFOCOM (2011)
2. Aingworth, D., Chekuri, C., Indyk, P., Motwani, R.: Fast estimation of diameter and shortest paths (without matrix multiplication). SIAM J. Comput. 28(4), 1167–1181 (1999)
3. Alon, N., Spencer, J.H.: The probabilistic method. Wiley, Chichester (1992)
4. Althöfer, I., Das, G., Dobkin, D., Joseph, D., Soares, J.: On sparse spanners of weighted graphs. Networks 9(1), 81–100 (1993)
5. Baswana, S., Kavitha, T., Mehlhorn, K., Pettie, S.: Additive spanners and (α, β)-spanners. ACM Trans. Algo. 7, A.5 (2010)
6. Baswana, S., Sen, S.: A simple Linear Time Randomized Algorithm for Computing Sparse Spanners in Weighted Graphs. Random Structures and Algorithms 30(4), 532–563 (2007)
7. Baswana, S., Kavitha, T.: Faster algorithms for approximate distance oracles and all-pairs small stretch paths. In: Proc. FOCS, pp. 591–602 (2006)
8. Baswana, S., Sen, S.: Approximate distance oracles for unweighted graphs in expected $O(n^2)$ time. ACM Transactions on Algorithms (TALG) 2(4), 557–577 (2006)
9. Bollobás, B., Coppersmith, D., Elkin, M.: Sparse distance preservers and additive spanners. SIAM Journal on Discrete Mathematics 19(4), 1029–1055 (2005)
10. Coppersmith, D., Elkin, M.: Sparse sourcewise and pairwise distance preservers. SIAM Journal on Discrete Mathematics 20(2), 463–501 (2006)

11. Chechik, S.: New Additive Spanners. In: Proc. SODA, vol. 29(5), pp. 498–512 (2013)
12. Cygan, M., Grandoni, F., Kavitha, T.: On Pairwise Spanners. In: Proc. STACS, pp. 209–220 (2013)
13. Gavoille, C., Peleg, D.: Compact and localized distributed data structures. Distributed Computing 16(2), 111–120 (2003)
14. Dor, D., Halperin, S., Zwick, U.: All-pairs almost shortest paths. SIAM on Computing 29(5), 1740–1759 (2000)
15. Elkin, M., Peleg, D.: $(1 + \varepsilon, \beta)$-Spanner Constructions for General Graphs. SIAM Journal on Computing 33(3), 608–631 (2004)
16. Elkin, M.: Computing almost shortest paths. ACM Transactions on Algorithms (TALG) 1(2), 283–323 (2005)
17. Elkin, M., Emek, Y., Spielman, D.A., Teng, S.H.: Lower stretch spanning trees. In: Proc. STOC, pp. 494–503 (2005)
18. Erdős, P.: Extremal problems in graph theory. In: Proc. Symp. Theory of Graphs and its Applications, pp. 29–36 (1963)
19. Kapralov, M., Panigrahy, R.: Spectral sparsification via random spanners. In: ITCS (2012)
20. Kavitha, T., Varma, N.M.: Small Stretch Pairwise Spanners. In: Fomin, F.V., Freivalds, R., Kwiatkowska, M., Peleg, D. (eds.) ICALP 2013, Part I. LNCS, vol. 7965, pp. 601–612. Springer, Heidelberg (2013)
21. Liestman, A.L., Shermer, T.C.: Additive graph spanners. Networks 23(4), 343–363 (1993)
22. Mendel, M., Naor, A.: Ramsey partitions and proximity data structures. In: FOCS, vol. 23(4), pp. 109–118 (2006)
23. Parter, M.: Bypassing Erdős' Girth Conjecture: Hybrid Stretch and Sourcewise Spanners (2014), http://arxiv.org/abs/1404.6835
24. Pătraşcu, M., Roditty, L.: Distance oracles beyond the Thorup-Zwick bound. In: FOCS, pp. 815–823 (2010)
25. Peleg, D., Schaffer, A.A.: Graph spanners. Journal of Graph Theory 12(1), 99–116 (1989)
26. Peleg, D., Ullman, J.D.: An optimal synchronizer for the hypercube. SIAM Journal on Computing 18(4), 740–747 (1989)
27. Peleg, D.: Distributed Computing: A Locality-Sensitive Approach. SIAM (2000)
28. Pettie, S.: Low distortion spanners. ACM Transactions on Algorithms (TALG) 6(1) (2009)
29. Roditty, L., Thorup, M., Zwick, U.: Deterministic constructions of approximate distance oracles and spanners. In: Caires, L., Italiano, G.F., Monteiro, L., Palamidessi, C., Yung, M. (eds.) ICALP 2005. LNCS, vol. 3580, pp. 261–272. Springer, Heidelberg (2005)
30. Thorup, M.: Undirected single-source shortest paths with positive integer weights in linear time. Journal of the ACM (JACM) 46(3), 362–394 (1999)
31. Thorup, M., Zwick, U.: Approximate distance oracles. Journal of the ACM (JACM) 52(1), 1–24 (2005)
32. Thorup, M., Zwick, U.: Spanners and emulators with sublinear distance errors. In: SODA, pp. 802–809 (2006)
33. Wenger, R.: Extremal graphs with no C4's, C6's, or C10's. Journal of Combinatorial Theory, 113–116 (1991)
34. Woodruff, D.P.: Lower bounds for additive spanners, emulators, and more. In: Proc. 47th Symp. on Foundations of Computer Science, pp. 389–398 (2006)

Author Index